SOIL AND ENVIRONMENTAL CHEMISTRY

SOIL AND ENVIRONMENTAL CHEMISTRY

SECOND EDITION

WILLIAM BLEAM
UW-Madison, Department of Soil Science
Madison, WI, United States

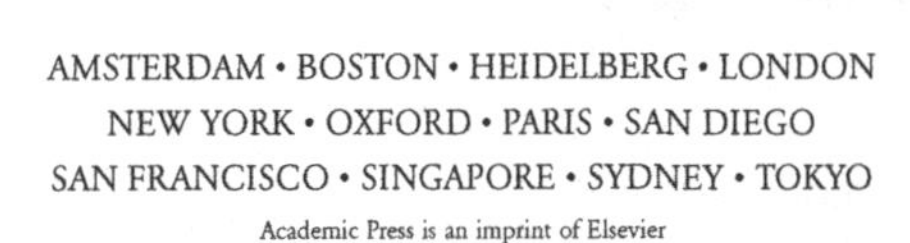
AMSTERDAM • BOSTON • HEIDELBERG • LONDON
NEW YORK • OXFORD • PARIS • SAN DIEGO
SAN FRANCISCO • SINGAPORE • SYDNEY • TOKYO
Academic Press is an imprint of Elsevier

Academic Press is an imprint of Elsevier
125 London Wall, London EC2Y 5AS, United Kingdom
525 B Street, Suite 1800, San Diego, CA 92101-4495, United States
50 Hampshire Street, 5th Floor, Cambridge, MA 02139, United States
The Boulevard, Langford Lane, Kidlington, Oxford OX5 1GB, United Kingdom

Notices

Knowledge and best practice in this field are constantly changing. As new research and experience broaden our understanding, changes in research methods, professional practices, or medical treatment may become necessary.

Practitioners and researchers must always rely on their own experience and knowledge in evaluating and using any information, methods, compounds, or experiments described herein. In using such information or methods they should be mindful of their own safety and the safety of others, including parties for whom they have a professional responsibility.

To the fullest extent of the law, neither the Publisher nor the authors, contributors, or editors, assume any liability for any injury and/or damage to persons or property as a matter of products liability, negligence or otherwise, or from any use or operation of any methods, products, instructions, or ideas contained in the material herein.

Library of Congress Cataloging-in-Publication Data
A catalog record for this book is available from the Library of Congress

British Library Cataloguing-in-Publication Data
A catalogue record for this book is available from the British Library

ISBN: 978-0-12-804178-9

For information on all Academic Press publications
visit our website at https://www.elsevier.com/

Publisher: Candice Janco
Acquisition Editor: Candice Janco
Editorial Project Manager: Emily Thomson
Production Project Manager: Paul Prasad Chandramohan
Cover Designer: Greg Harris

Typeset by SPi Global, India

Dedication

To Sharon, my wife

Contents

Second Edition Preface xiii

1. Element Abundance

1.1 Introduction 1
1.2 A Brief History of the Solar System and Planet Earth 3
1.3 Elemental Composition of Earth's Lithosphere and Soils 4
1.3.1 Relative Elemental Abundance 4
1.3.2 Elements and Isotopes 6
1.3.3 Nuclear Binding Energy 7
1.4 Enrichment and Depletion During Planetary Formation 12
1.4.1 Planetary Accretion 12
1.4.2 Planetary Stratification 13
1.5 The Rock Cycle 17
1.6 Soil Formation 19
1.7 Element Abundance Frequency Distributions 20
1.8 Estimating the Central Tendency of Logarithmic-Normal Distributions 25
1.9 Summary 32
Appendices 32
1.A Factors Governing Nuclear Stability and Nuclide Abundance 32
1.A.1 The Table of Nuclides 32
1.A.2 Nuclear Magic Numbers 34
1.B Random Sequential Dilution and the Law of Proportionate Effect 35
References 37

2. Chemical Hydrology

2.1 Introduction 40
2.2 Water Resources and the Hydrologic Cycle 40
2.2.1 Water Budgets 40
2.2.2 Residence Time 41
2.2.3 Porous Media: Volume and Mass Relations 44
2.2.4 Hydrologic Units 47
2.3 Saturated Zone Hydrology 48
2.3.1 Groundwater Flow Nets 51
2.4 Vadose Zone Hydrology 55
2.4.1 The Water Retention Curve 58
2.5 Soil Moisture Balance 60
2.5.1 Evapotranspiration Models 61
2.5.2 Soil Water Balance 63
2.6 Solute Transport 65
2.6.1 Retardation-Coefficient Transport Model 65
2.6.2 Solute Transport in the Saturated Zone 67
2.6.3 Solute Transport in the Vadose Zone 68
2.7 Summary 69
Appendices 70
2.A Hydraulic Heads and Gradients 70
2.B Capillary Forces 72
2.C Predicting Capillary Rise 74
2.D Empirical Water Retention and Unsaturated Hydraulic Conductivity Functions 76
2.D.1 Clapp-Hornberger Functions 76
2.D.2 van Genuchten Water Retention Function 78
2.E Day Length Calculation 79
2.F Solute Transport: Plate-Theory Transport Model 80
2.F.1 The Binomial Expansion 81
2.F.2 Plates and Transfers 83
2.F.3 Partitioning and the Movement of the Mobile Agent 84
2.F.4 Retardation-Coefficient and Plate-Theory Models Compared 85
References 85

3. Clay Mineralogy and Chemistry

3.1 Introduction 88
3.2 Mineral Weathering 90
3.2.1 Mineralogy 90
3.2.2 The Jackson Weathering Sequence 91

3.3 Silicate Minerals: Composition, Structure, and Chemistry 93
3.3.1 Coordination Polyhedra 94
3.3.2 Silicate Mineral Groups 97
3.3.3 Basicity and the Hydrolysis of Igneous Silicate Minerals 97
3.3.4 Oxidation of Iron(III) in Igneous Silicate Minerals 101
3.3.5 Solid-State Transformation of Pyribole and Feldspar Minerals Into Phyllosilicate Minerals 101
3.4 Clay Minerals 105
3.4.1 Composition and Structure of Phyllosilicate Layers 105
3.4.2 Structure of Neutral-Layer Minerals 106
3.4.3 Layer Structure of Mica Group Minerals 107
3.4.4 Layer Structure of Vermiculite and Smectite Group Minerals 110
3.4.5 Layer Structure of Chlorite Group Minerals 111
3.5 Clay Colloid Chemistry 115
3.5.1 Clay Mineral Plasticity 115
3.5.2 Hydration and Swelling of Smectite and Vermiculite Clay Minerals 116
3.5.3 Clay Plasticity and Soil Mechanical Properties 132
3.6 Summary 137
Appendices 138
3.A Bragg's Law and X-Ray Diffraction by Layer Silicates 138
3.B Determining Formal Oxidation States in Minerals 140
3.C Stoke's Law 142
3.D van't Hoff Factor i and Osmotic Coefficient ϕ 143
References 144

4. Ion Exchange

4.1 Introduction 148
4.2 The Discovery of Ion Exchange 148
4.3 The Ion Exchange Experiment 149
4.4 Selectivity Constants and the Ion Exchange Isotherm 152
4.4.1 The Vanselow Mole-Fraction Convention 153
4.4.2 The Gaines-Thomas Equivalent-Fraction Convention 154
4.4.3 The Gapon Convention 155
4.5 Interpreting the Ion Exchange Isotherm 155
4.5.1 The Exchange Isotherm for Symmetric Exchange 156
4.5.2 The Exchange Isotherm for Asymmetric Exchange 157
4.5.3 Effect of Electrolyte Concentration on the Ion Exchange Isotherm 159
4.5.4 Effect of Ion Selectivity on the Ion Exchange Isotherm 160
4.5.5 Physical Basis for Ion Selectivity 164
4.5.6 Other Influences on the Ion Exchange Isotherm 166
4.6 Summary 170
Appendices 170
4.A The Preparation and Characterization of Natural Ion Exchangers 170
4.A.1 Saturating an Exchanger With a Single Cation 171
4.A.2 Measuring Ion Exchange Capacity 171
4.B Non-Linear Least Square Fitting of Exchange Isotherms 172
4.B.1 Symmetric Exchange 172
4.B.2 Asymmetric Exchange 175
4.C Ideal and Real Mixtures: Significance for Ion Exchange Selectivity Constant Expressions 175
4.C.1 Ideal Mixtures 175
4.C.2 Real Mixtures 176
4.D Selectivity Constant Expressions: Ideal and Real Exchange-Complex Mixtures 179
4.E Equivalent Fraction-Dependent Selectivity Coefficients 181
4.F The Gapon Convention: Revisited 183
References 185

5. Water Chemistry

5.1 The Equilibrium Constant 190
5.1.1 Thermodynamic Functions for Chemical Reactions 190
5.1.2 Gibbs Energy of Reaction and the Equilibrium Constant 191
5.2 Activity and the Equilibrium Constant 193
5.2.1 Concentrations and Activity 193
5.2.2 Empirical Ion Activity Coefficient Expressions 194
5.3 Simple Equilibrium Systems 195
5.3.1 Setting Up Equilibrium Calculations to Replicate Solution Chemistry 195

5.3.2 Hydrolysis of a Weak Monoprotic Acid 197
5.3.3 Gas Solubility: Henry's Law 202
5.4 Complex Equilibrium Systems 204
5.4.1 Validating Water Chemistry Simulations 204
5.4.2 Gas Solubility: Fixed-Fugacity Method 208
5.4.3 Mineral Solubility 211
5.4.4 Interpreting Water Chemistry Simulations 225
5.4.5 Summary 232
Appendices 232
5.A Equilibrium Constants and Activity Coefficients: Notation and Units 232
5.B Fugacity: The Carbon Dioxide Case 234
5.C *ChemEQL* Input and Output File Formats 235
5.C.1 Understanding the *ChemEQL* Input Matrix: Components and Species 235
5.C.2 *ChemEQL* Output Data-File Format 238
5.D Solving Equilibrium Problems Using the *RICE* Table Method 240
5.D.1 Aqueous Solubility of an Ionic Compound 240
5.D.2 Aqueous Solubility of Gases 243
5.E Validating Water Chemistry Simulations 247
5.E.1 Charge Balance Validation 247
5.E.2 Mass Balance Validation 247
5.E.3 Ionic Strength Validation 248
5.E.4 Ion Activity Coefficient Validation 248
5.E.5 Activity Product Validation 248
5.E.6 Database Validation 250
References 250

6. Acid-Base Chemistry

6.1 Introduction 254
6.1.1 Exchangeable Acidity and Sodicity as Acid-Base Phenomena 254
6.1.2 Basicity and Alkalinity 255
6.2 Acid-Base Chemistry Fundamentals 256
6.2.1 Dissociation: The Arrhenius Acid-Base Model 258
6.2.2 Hydrogen Ion Transfer: The Brønsted-Lowry Acid-Base Model 258
6.2.3 Weak Acid and Base Conjugates 259
6.2.4 Water Reference Level: Acidity and Basicity 260
6.3 Natural and Anthropogenic Sources of Strong Acids and Bases 262
6.3.1 Chemical Rock Weathering 262
6.3.2 Atmospheric Deposition: Vulcanism and Combustion 272
6.4 Carbonate Chemistry 275
6.4.1 Carbonate Equilibrium Reactions 276
6.4.2 Aqueous Carbon Dioxide Reference Level 281
6.4.3 Alkalinity and Mineral Acidity 281
6.4.4 Geochemical Implications of Alkalinity 283
6.5 Alkali and Sodic Soils 289
6.5.1 Exchangeable Sodium and Clay Plasticity 290
6.5.2 Identifying Sodicity Risk 292
6.5.3 Predicting the Sodicity Risk of Irrigation Water 301
6.5.4 The Limiting Sodium Adsorption Ratio (*LSAR*) Sodicity Risk Parameter 304
6.6 Soil Acidity 309
6.6.1 Aluminum Chemistry 312
6.6.2 Exchangeable Aluminum 313
6.6.3 Nonexchangeable Aluminum 317
6.7 Summary 319
Appendices 319
6.A The pH Scale 319
6.B The Saturation Effect: Atmospheric Conversion of Sulfur Trioxide to Sulfuric Acid 320
6.C Acid-Base Implications of Nitrate and Sulfate Reduction 321
6.C.1 Nitrogen Cycling and Fertilization in Agricultural Landscapes 321
6.C.2 Marine Environment: Sulfate and Nitrate Reduction in the Deep Ocean 322
6.D Selectivity of (Na^+, Ca^{2+}) Exchange 323
6.E Fitting the Asymmetric (Al^{3+}, Ca^{2+}) Ion-Exchange Isotherm 324
References 327

7. Natural Organic Matter

7.1 Introduction 333
7.1.1 Organic Matter Extraction 334
7.1.2 Organic Matter Fractionation 335
7.2 Biological Attributes of Natural Organic Matter 337
7.2.1 Stoichiometric Composition of Natural Organic Matter 337

7.2.2 Organic Matter as Substrate for Microbial Growth 341
7.2.3 Exocellular Bioorganic Compounds 344
7.3 Organic Carbon Turnover and the Terrestrial Carbon Cycle 349
7.3.1 Carbon Fixation 350
7.3.2 Carbon Mineralization 351
7.3.3 Oxidation of Organic Compounds by Dioxygen 352
7.3.4 Organic-Matter Turnover Models 353
7.3.5 Soil Carbon Pools 362
7.4 Chemical Properties of Natural Organic Matter 363
7.4.1 Oxygen Functional Groups 363
7.4.2 Nitrogen Functional Groups 367
7.4.3 Phosphorus Functional Groups 370
7.4.4 Sulfur Functional Groups 372
7.4.5 Carbon Moieties and Functional Groups 375
7.4.6 Colloidal Properties 376
7.5 Summary 376
Appendices 377
7.A Limitations of Carbon K-Edge NEXAFS and STXM 377
7.B Spin Conservation and Spin Forbidden Reactions 378
References 380

8. Surface Chemistry and Adsorption

8.1 Introduction 385
8.2 Mineral and Organic Colloids as Environmental Adsorbents 386
8.3 Adsorption Isotherm Experiments 389
8.3.1 Area-Based Adsorption Isotherms 389
8.3.2 Mass-Based Adsorption Isotherms 392
8.3.3 Chemisorption and Physisorption 392
8.3.4 Langmuir Isotherm Model 394
8.3.5 Ion-Exchange Isotherm Model 399
8.3.6 Partitioning Isotherm Model 401
8.3.7 Freundlich Adsorption Isotherm 411
8.4 pH-Dependent Surface Charge 413
8.4.1 Weak Acid-Conjugate Base Proton Adsorption Model 414
8.4.2 Crystallographic Proton Adsorption Models 415
8.5 The Adsorption Envelope Experiment 423
8.5.1 Adsorption Edges 424
8.5.2 Interpreting Adsorption Envelopes 426
8.5.3 The Structure of Adsorption Complexes 429
8.5.4 Overview of Surface Complexation Models 430
8.6 Summary 432
Appendices 434
8.A Hydrophilic Colloids 434
8.B Bond-Valence Model 436
8.C Valence-Bond Proton Adsorption Model: Goethite (100) Surface 436
8.D Using the PBT Profiler 439
References 440

9. Reduction-Oxidation Chemistry

9.1 Introduction 445
9.2 Electrochemical Principles 446
9.2.1 Formal Oxidation States 446
9.2.2 Balancing Reduction Half-Reactions 450
9.2.3 Standard Electrochemical Potentials 453
9.2.4 The Nernst Equation 453
9.2.5 Standard Biological Potentials 455
9.3 Measurement and Interpretation of Electrochemical Potentials in Soils and Sediments 457
9.3.1 Electrochemical Stability Diagrams 457
9.3.2 Pourbaix Electrochemical Stability Diagrams 459
9.4 Microbial Respiration 470
9.4.1 Catabolism and Electron Transport Chains 473
9.4.2 Microbial Electron Transport Chains 475
9.4.3 Impact of Respiratory Efficiency on Carbon Use Efficiency 481
9.5 Methanogenesis 482
9.6 Summary 482
Appendices 483
9.A Reduction-Oxidation Reactions Without Electron Transfer 483
9.B Limitation of Platinum Oxidation-Reduction Electrodes 484
9.C Standard and Biological Electrochemical Potentials for Environmental and Biological Half-Reactions (Tables 9.C.1 and 9.C.2) 486
9.D Facultative and Obligate Anaerobes 486
9.E Fermentative Anaerobic Bacteria 487
References 488

10. Human Health and Ecological Risk Analysis

10.1 Introduction 492
10.2 The Federal Risk Assessment Paradigm 493
10.2.1 Risk Assessment 493
10.2.2 Risk Management and Mitigation 493
10.3 Dose-Response Assessment 493
10.3.1 Dose-Response Functions 494
10.3.2 Low-Dose Extrapolation 498
10.4 Exposure Pathway Assessment 502
10.4.1 Receptors 503
10.4.2 Exposure Routes 504
10.4.3 Exposure Points 505
10.4.4 Fate and Transport 506
10.4.5 Primary and Secondary Sources 507
10.4.6 Exposure Pathway Assessment 508
10.4.7 Exposure Factors 509
10.5 Intake Estimates 511
10.6 Risk Characterization 512
10.6.1 Target Cancer Risk 512
10.6.2 Cumulative Target Risk 513
10.6.3 Hazard Quotient 514
10.6.4 Cumulative Risk: Hazard Index 515
10.7 Exposure Mitigation 515
10.8 Risk-Based Screening Levels 516
10.9 Ecological Risk Assessment 519
10.9.1 Wildlife Risk Model 519
10.9.2 Ecological Soil Screening Levels 522
10.10 Summary 523
Appendices 524
10.A Factors Affecting Contaminant Transport by Surface Water 524
10.B Factors Affecting Contaminant Transport by Groundwater 526
10.C Factors Affecting Contaminant Transport by Soils or Sediments 528
10.D Water Ingestion Equation 529
10.E Soil Ingestion Equation 530
10.F Food Ingestion Equation 531
10.G Air Inhalation Equation 532
References 532

Soil and Environmental Chemistry: Exercises

1 Elemental Abundance 535
2 Chemical Hydrology 537
3 Clay Mineralogy and Chemistry 539
4 Ion Exchange 541
5 Water Chemistry 542
6 Acid-Base Chemistry 545
7 Natural Organic Matter 549
8 Surface Chemistry and Adsorption 551
9 Reduction-Oxidation Chemistry 555
10 Risk Analysis 558
References 561

Index 563

Second Edition Preface

I made numerous major changes in this the second edition of *Soil and Environmental Chemistry*. One that will be immediately apparent was a decision to adopt insofar as possible the notation and nomenclature recommended by the *International Union of Pure and Applied Chemistry* IUPAC. This was motivated, in part, to clarify the roots of soil chemistry, geochemistry, and environmental chemistry in pure chemistry and, in part, to make this particular field of applied chemistry more immediately accessible to students trained in pure chemistry.

There are a few exceptions, the first relating to partition constant notation. The environmental science community uses a special notation for the octanol-water partition constant that explicitly identifies octanol as the organic phase involved in physisorption by partition. Identifying the adsorbent is essential for the environmental science community because the same adsorptive will yield different partition constants for partitioning into octanol, the fatty tissue of living organisms (in the case of bioconcentration factors), adsorption by the solid phase soils, sediments or aquifers, and adsorption by the organic carbon component in the same. IUPAC does not identify the organic phase in their recommended partition constant $(K_D^{\phi})_A$ and, hence, the IUPAC recommended partition constant symbol is unable to represent legitimate distinctions recognized by the environmental science community.

The biochemistry community often uses a pH 7 standard state that differs from the pH 0 standard state used by the pure chemistry community. As such, IUPAC does not endorse a special notation for reduction half-reactions under biochemical standard state conditions. In the second edition I have adopted the notation used by Atkins and de Paula (2005) in *Physical Chemistry for the Life Sciences*.

Finally, the hydrology and soil physics notation used in Chapter 2 *(Chemical Hydrology)* and elsewhere follow Hillel (2004) in *Introduction to Environmental Soil Physics*.

Water residence time and water budgets have important implications for environmental chemistry. Chapter 2 (*Chemical Hydrology*) now includes a discussion of elementary potential evapotranspiration models that allow scientists to compute approximate water budgets and estimate soil water residence time using readily available temperature and precipitation data from thousands of weather stations worldwide. There are also significant revisions of the plate-theory transport model that will hopefully make the discussion of this model more transparent.

Reviewers requested further details on the mechanisms of silicate mineral chemical weathering. Chapter 3 (*Clay Mineralogy and Chemistry*) now includes a discussion of the solid-state transformation and surface-site limited reactions that many mineralogists believe explain the formation of low-temperature layer silicates through chemical weathering. This chapter also discusses the origins of silicate mineral alkalinity that is essential for the later chapter on acid-base chemistry. Finally, the discussion of clay swelling was completely rewritten to provide a more sound foundation to this very important phenomenon.

Chapter 4 (*Ion Exchange*) immediately follows the chapter on clay chemistry. This chapter now includes a thorough discussion of the Vanselow mole-fraction and the Gaines-Thomas equivalent-fraction conventions, demonstrating both the distinctions between the two equilibrium quotient expressions and their underlying unity. More advanced topics, such as the role of real and ideal mixtures in the ion exchange process and the fatal flaws of the Gapon convention, appear in appendixes.

Chapter 5 (*Water Chemistry*) places less emphasis on R.I.C.E. tables, scaling back the discussion and moving it to an appendix. This edition discusses the steps students can take to design input files that will faithfully simulate the chemistry of the component species and the importance of evaluating water analyses for reliability, a theme continued in Chapter 6 (*Acid-Base Chemistry*). Chapter 6 includes an expanded discussion of alkalinity and methods for incorporating alkalinity in water chemistry simulations.

Chapter 7 (*Natural Organic Matter*) is now positioned before the chapter on adsorption, following the logic of positioning clay chemistry before ion exchange. This chapter begins with a brief review of natural organic matter research for the past 100 years. The second edition expands the discussion of below ground carbon cycle models and now includes a discussion linking carbon-use efficiency to the mean carbon reduction in natural organic matter.

Some of the topics raised in Chapter 7 are expanded in Chapter 8 (*Surface Chemistry and Adsorption*) by drawing distinctions between physisorption and chemisorption, and importance of molecular association in organic matter extracts. The associative behavior of extracted organic matter is a key factor in the physisorption of organic substances by dissolved and adsorbed organic matter. Chapter 8 now includes a discussion of adsorption edge fitting and an expanded evaluation of surface-complexation models.

Reconciling the pure chemistry perspective of reduction-oxidation chemistry with the biological anaerobiosis driving reduction in soils and sediments is an important addition to Chapter 9 (*Reduction-Oxidation Chemistry*). This discussion provides a rational basis for the relative efficiency of ferric iron respiration and sulfate respiration and allows a deeper understanding of restrictions on carbon-use efficiency imposed by anaerobic respiration.

Chapter 10 (*Human Health and Ecological Risk Analysis*) now includes a more sound justification for the profoundly different treatment of carcinogens and noncarcinogens in risk analysis: the irreversible adverse effect of carcinogen intoxication and the presumed reversible effects of noncarcinogen intoxication. A brief discussion of ecotoxicity risk assessment reveals the parallels between human health and ecological risk assessment.

The number of chapter exercises is expanded by about half to more than one hundred exercises. None of the chapter exercises include solutions; exercise solutions appear in the *Instructor's Manual* (http://booksite.elsevier.com/9780128041789/) that accompanies the second edition. The second edition also attempts to strike a balance between topics appropriate for undergraduates with limited chemistry experience and more advanced topics intended for more advanced undergraduate students and graduate students who are not specializing in chemistry. Advanced topics generally appear in appendixes at the end of each chapter.

I wish to thank the following colleagues for their assistance and encouragement: Willim "Bill" Hickey, Robert W. Taylor, Beat Müller, Francisco Arriaga, Birl Lowery, and Alfred Hartimink.

William F. Bleam
Madison, Wisconsin

CHAPTER 1

Element Abundance

OUTLINE

1.1 Introduction 1
1.2 A Brief History of the Solar System and Planet Earth 3
1.3 Elemental Composition of Earth's Lithosphere and Soils 4
1.3.1 Relative Elemental Abundance 4
1.3.2 Elements and Isotopes 6
1.3.3 Nuclear Binding Energy 7
1.4 Enrichment and Depletion During Planetary Formation 12
1.4.1 Planetary Accretion 12
1.4.2 Planetary Stratification 13
1.5 The Rock Cycle 17
1.6 Soil Formation 19
1.7 Element Abundance Frequency Distributions 20
1.8 Estimating the Central Tendency of Logarithmic-Normal Distributions 25
1.9 Summary 32
Appendices 32
1.A Factors Governing Nuclear Stability and Nuclide Abundance 32
1.A.1 The Table of Nuclides 32
1.A.2 Nuclear Magic Numbers 34
1.B Random Sequential Dilution and the Law of Proportionate Effect 35
References 37

1.1 INTRODUCTION

Students encounter the Periodic Table of the Elements as early as high school and by college have come to understand the periodicity underlying its design, to associate letter symbols with atomic numbers, and how to find the atomic mass and electron configuration of each element. While it is likely students will be familiar with the isotope concept and understand

Soil and Environmental Chemistry
http://dx.doi.org/10.1016/B978-0-12-804178-9.00001-X

TABLE 1.1 Integer Atomic Masses for Elements That Have No Stable Nuclides

Atomic Number (Z)	Element Symbol	Element Name	Atomic Mass
43	Tc	Technetium	98
61	Pm	Promethium	145
84	Po	Polonium	209
85	At	Astatine	210
86	Rn	Radon	222
87	Fr	Francium	223
88	Ra	Radium	226
89	Ac	Actinium	227

what determines atomic mass of an element,[1] it is unlikely they are familiar with the Table of Nuclides.

The official atomic masses, appearing in a report released by *International Union of Pure and Applied Chemistry* (IUPAC) (Wieser, 2007), are based on isotopic abundances compiled by another *IUPAC* commission (Bohlke et al., 2005). Careful inspection of the atomic masses listed in the Periodic Table of the Elements reveals that, for some elements, the atomic mass is assigned integer atomic mass rather than a decimal atomic mass. These elements appear in Table 1.1. Several other actinide elements (thorium Th, protactinium Pa, and uranium U) are identified as having no stable isotopes yet *IUPAC* quotes decimal atomic masses. A complete understanding of atomic mass and isotope stability requires information compiled in the Table of Nuclides.

The Periodic Table of the Elements implies the existence of all naturally occurring elements from hydrogen to uranium and it is reasonable to assume these elements exist on planet Earth. In fact, every sample of water, rock, sediment, and soil contains every *stable* element and many of the *unstable* elements. The Periodic Table of the Elements, however, fails to provide sufficient information to understand natural abundance in the Universe, the Solar System, and planet Earth.

Natural elemental abundance has other implications besides cosmological and geological chemistry. The Environmental Working Group published an article in October 2008 entitled "Bottled water quality investigation: 10 major brands, 38 pollutants" (Naidenko et al., 2008). Among the pollutants found in bottled water sold in the United States was radioactive Sr-90 (0.02 Bq dm^{-3}),[2] radioactive radium[3] (0.02 Bq dm^{-3}), boron (60–90 mg dm^{-3}), and arsenic

[1] The atomic mass of an element, as it is most commonly reported, is a *weighted average*. It multiplies the mass of each stable isotope times its relative abundance (Bohlke et al., 2005).

[2] Strontium is the name given to the *element* with atomic number $Z = 38$. Strontium-90 ($^{90}_{38}Sr_{52}$) is the unstable (i.e., radioactive) *nuclide* with atomic number $Z = 38$, mass number $A = 90$, and neutron number $N = 52$. Strontium *isotopes* include all nuclides with atomic number $Z = 38$.

[3] Radioactive radium is reported as the combined activity of two unstable radium isotopes: $^{226}_{88}Ra_{138}$ and $^{228}_{88}Ra_{140}$.

(1 mg dm^{-3}). The reported concentrations, combined with a commentary listing the potential and actual toxicity of these substances, can be alarming. The important question is not whether drinking water, food, air, soil, or dust contains toxic elements; it most certainly does! The important questions are: Is the concentration of a toxic element in drinking water, food, or soils elevated relative to natural abundance or is it sufficient to produce an adverse health effect?

The US Environmental Protection Agency (*USEPA*) has established a maximum contaminant level (*MCL*) for beta emitters—such as strontium-90 or radium-228/radium-226—in public drinking water: 0.296 Bq dm^{-3}, which is tenfold higher than the *single* detection reported by Naidenko et al. (2008). The *USEPA* does not have a drinking water standard for boron, but the World Health Organization recommends boron levels in drinking water less than 500 mg dm^{-3}, and in 1998 the European Union adopted a drinking water standard of 100 mg dm^{-3}. Typical boron concentrations in US groundwater fall below 100 mg dm^{-3} and 90% below 40 mg dm^{-3}.

Of the contaminants featured by Naidenko et al. (2008), arsenic is the most troubling. Naidenko et al. (2008) reported a single case of 1 ppb arsenic[4] in bottled water; however, the mean arsenic concentration in US groundwater[5] is 13.9 μg dm^{-3} with a median arsenic concentration of 1.4 μg dm^{-3}. The original *USEPA* arsenic drinking water *MCL* of 50 μg dm^{-3} was lowered to the current *MCL* of 10 μg dm^{-3} in 2001.[6] Natural arsenic abundance in US groundwater means over 10% of the population is potentially exposed to arsenic at levels exceeding the current arsenic *MCL*.

Understanding the processes that determine relative elemental abundance and the statistical methods needed to estimate the elemental composition of rocks, sediments, and soils are the major themes of this chapter.

1.2 A BRIEF HISTORY OF THE SOLAR SYSTEM AND PLANET EARTH

Gravitational collapse of a primordial gas and dust cloud gave birth to the present-day Solar System. Conservation of angular momentum in the primordial cloud explains the rotation of the Sun and the primordial accretion disk that spawned the planets and other bodies that orbit the Sun.

Accretion and radioactive decay released sufficient heat to melt the primordial Earth, leading its stratification into a solid metallic core, a molten mantle, and a crystalline lithosphere. Planetary accretion and stratification altered the composition of Earth's lithosphere relative to the primordial cloud and imposed variability in the composition of each element as a direct consequence of each differentiation process.

Throughout its entire history, planet Earth has experienced continual transformation as plate tectonics generate new oceanic lithosphere and shift continental lithosphere plates around like jigsaw puzzle pieces. Plate tectonics and the hydrologic cycle drive a rock cycle that

[4] 1 ppb = 1 μg dm^{-3}.

[5] Geometric mean based on over 7199 samples (Newcomb and Rimstidt, 2002).

[6] http://water.epa.gov/lawsregs/rulesregs/sdwa/arsenic/regulations.cfm (September 17, 2015).

reworks portions of the oceanic and continental lithosphere through weathering, erosion, and sedimentation. The rock cycle imposes a new round of geochemical differentiation processes that alters the composition of the terrestrial land surface relative to the lithosphere from which it derives. The rock cycle and soil development, much like planetary formation and stratification early in Earth's history scale, impose additional variability in the abundance of each element. Transformations in the overall composition of planet Earth and the imposition of variability relative to the primordial gas cloud are among the important themes of this chapter.

1.3 ELEMENTAL COMPOSITION OF EARTH'S LITHOSPHERE AND SOILS

Most geology, soil science, and environmental science textbooks include a table listing the elemental composition of Earth's lithosphere or soil. There are several reasons for listing the elemental composition of Earth materials. Soil composition constrains the biological availability of each element essential for living organisms. Soils develop from the weathering of rocks and sediments and inherit much of their composition from the local geology. The most common elements—those accounting for 90–95% of the total composition—determine the dominant mineralogy of rocks and materials that form when rocks weather (residuum, sediments, and soils).

The soil at a particular location develop from parent material (residuum or sediments) under a variety of local influences—landform relief, climate, and biological community—over a period of centuries. Not surprisingly, soil composition is largely inherited from the parent material. The rocks or sediments in which a soil develops are usually not the basement rocks that compose the bulk of Earth's lithosphere. Regardless of geologic history, virtually every rock exposed at the Earth's terrestrial surface owes its composition to the crystalline igneous rocks of the lithosphere. The Earth's lithosphere inherits its composition from the overall composition of the planet and, by extension, the gas and dust cloud that gave rise to the Solar System as a whole. How much do the composition of soil, Earth's lithosphere, and the Solar System have in common? What can we learn about the processes of planetary formation, the rock cycle, and soil development from any differences in composition? What determines the relative abundance of elements in soil?

1.3.1 Relative Elemental Abundance

We are looking for patterns, and, unfortunately, data in tabular form usually do not reveal patterns in their most compelling form. Patterns in relative elemental abundance of the Solar System, Earth's lithosphere, and soils are best understood when abundance is plotted as a function of atomic number Z. Astronomers believe that the composition of the Sun's photosphere is a good representation of the primordial gas and dust cloud that gave rise to the Solar System because over 99% of the primordial cloud became the Sun. Solar System composition is usually recorded as the molar abundance of each element, normalized by the molar abundance of silicon $n(\mathrm{E})/n(\mathrm{Si})$ [mol mol^{-1}].

The composition of Earth's lithosphere and soil is usually recorded as the mass fraction of each element $w(\mathrm{E})$ [kg kg^{-1}]. If we are to compare the compositions of the Solar System, Earth's lithosphere, and soils, the abundance data must have the same units as most Solar System abundance datasets. The atom mole fraction relative to silicon fraction is found by dividing the mass fraction of each element—in lithosphere or soil—by its atomic mass (Eq. 1.1) then normalized by the atom mole fraction of silicon (Eq. 1.2).

$$m(\mathrm{E})\ [\mathrm{mol\ kg^{-1}}] = \frac{10^{+3}\ [\mathrm{g\ kg^{-1}}] \cdot w(\mathrm{E})\ [\mathrm{kg\ kg^{-1}}]}{\overline{m}_{\mathrm{a}}(\mathrm{E})\ [\mathrm{g\ mol^{-1}}]} \tag{1.1}$$

$$\frac{n(\mathrm{E})}{n(\mathrm{Si})} = \frac{m(\mathrm{E})\ [\mathrm{mol\ kg^{-1}}]}{m(\mathrm{Si})\ [\mathrm{mol\ kg^{-1}}]} \tag{1.2}$$

Plotting the *normalized* molar abundance $n(\mathrm{E})/n(\mathrm{Si})$ of the Solar System, Earth's lithosphere, and soil using a linear scale reveals little because 99.87% of the Solar System consists of hydrogen and helium. Plotting $\log_{10}(n(\mathrm{E})/n(\mathrm{Si}))$ (Figs. 1.1 and 1.2), however, reveals three important features. First, the abundance of the elements decreases exponentially with increasing atomic number Z (the decrease appears roughly linear when plotted using a logarithmic scale). Second, a zigzag pattern is superimposed on this general tread, known as the *even-odd effect*: elements with an even atomic number Z (or mass number A) are consistently more abundant than elements with an odd atomic number Z (or mass number A). Third, while the dataset for soil composition is missing many of the elements recorded for the Solar System

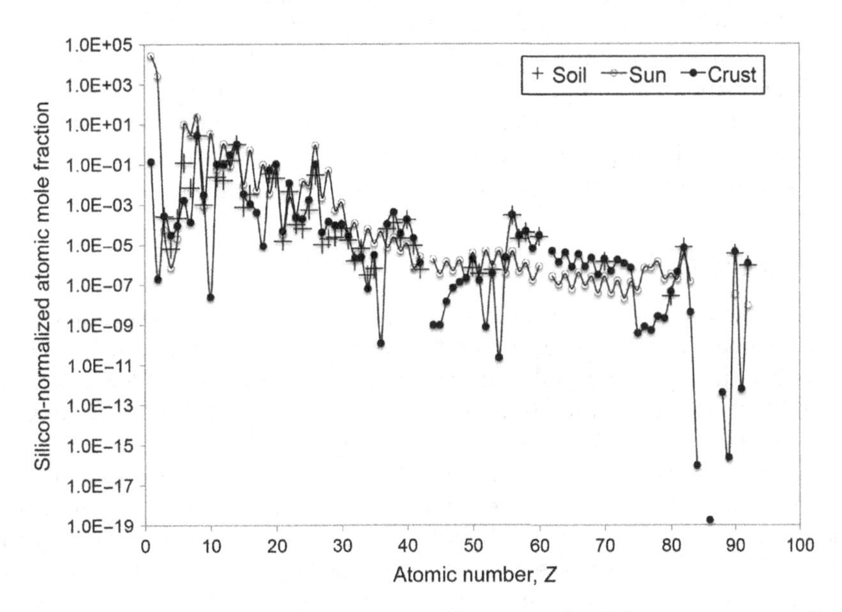

FIG. 1.1 Elemental abundance of the Solar System, Earth's crust, and soil decreases exponentially with atomic number Z (Shacklette and Boerngen, 1984; Lide, 2005).

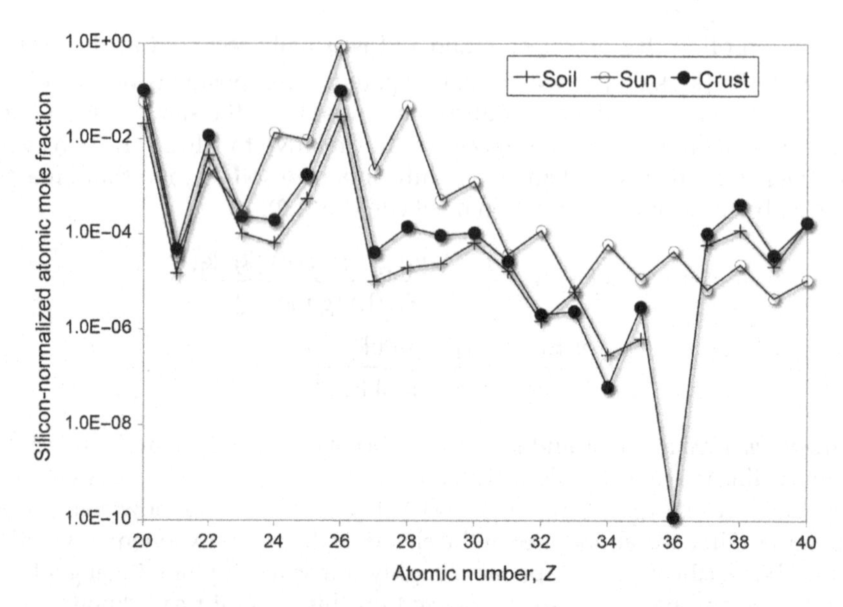

FIG. 1.2 Elemental abundance of the Solar System, Earth's crust, and soil from calcium (atomic number $Z = 20$) to zirconium (atomic number $Z = 40$), showing the even-odd effect (Shacklette and Boerngen, 1984; Lide, 2005).

and Earth's lithosphere, the exponential decrease in abundance with atomic number (Fig. 1.1) and the even-odd effect (Fig. 1.2) appear in all three datasets.

Cosmological processes, beginning with the so-called *Big Bang*, determined Solar System composition. The cosmological imprint—the exponential decrease in abundance with atomic number and the even-odd effect—is clearly discernible in the composition of the Earth's lithosphere and terrestrial soils despite the profound changes resulting from planetary formation, the rock cycle, and soil development.

1.3.2 Elements and Isotopes

The Periodic Table of the Elements organizes all known elements into groups and periods. Elements in the same chemical group have the same number of electrons in the outermost electronic shell but differ in the number of occupied electronic shells. Elements in the same chemical period have the same number of occupied electronic shells but differ in the number of electrons in the partially filled outermost electronic shell.

Period 1 contains elements H and He, the filling of atomic shell 1s. Atomic shell 2s is filled in Groups 1 and 2 of Period 2—elements Li and Be—while 2p is filled in Groups 13–18 of the same period. Elements of the same group exhibit similar chemical properties because their outermost or valence electronic shell has the same number of electrons. It is this characteristic—the valence electron configuration—that has the greatest influence on the chemistry of each element.

The Periodic Table of the Elements lists the symbol "E," atomic number Z and atomic mass $\overline{m}_{\mathrm{a}}(\mathrm{E})$ of each element. The atomic number Z is an integer equal to the number of positively charged protons in the nucleus. The atomic mass $\overline{m}_a(E)$, as noted earlier, is the mass of 1 mole of the pure element and accounts for the relative abundance of the stable and long-lived radioactive isotopes (see Appendix 1.A).

Figs. 1.1 and 1.2 plot the relative abundance of each element in the lithosphere and soil of planet Earth on a logarithmic vertical scale. We will have more to say about depletion and enrichment later, but for now we are most concerned with the processes that determine why some elements are more abundant than others. You will notice the abundance of adjacent elements fluctuate; Ca is more abundant than K or Sc, and Ti is more abundant than Sc or V. This is the even-odd effect mentioned earlier and results from the nuclear stability of each element, a topic discussed further in Appendix 1.A.

1.3.3 Nuclear Binding Energy

The even-odd effect, however, fails to explain the relative abundance of iron $_{26}Fe$. Fig. 1.3 is a plot of the nuclear binding energy per nucleon E_b/A for all known nuclides. It clearly shows that $^{56}_{26}Fe$ is the nuclide with the greatest binding energy per nucleon. The fusion of lighter nuclei to form heavier nuclei is exothermic when the product has a mass number $A < 56$ but is endothermic for all heavier nuclides. Notice that the energy release from fusion reactions diminishes rapidly as the mass number A increases to 20 and then gradually in the range from 20 to 56.

The mass of an electron (0.000549 *unified atomic mass units* [u]) is negligible compared to that of protons (1.007276 u) and neutrons (1.008665 u). The mass number A of each nuclide is an integral sum of protons Z and neutrons N in the nucleus, while the nuclide mass is the mass

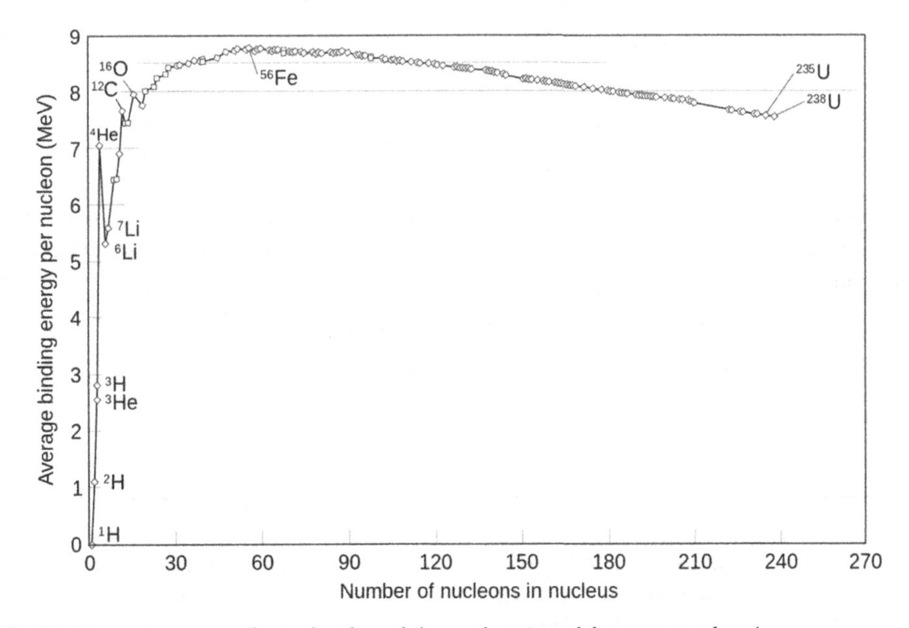

FIG. 1.3 Binding energy per nucleon of each nuclide as a function of the mass number A.

of 1 mole of pure nuclide. The mass of an nuclide with mass number A is always lower than the rest mass of Z protons plus the rest mass of N neutrons. Example 1.1 demonstrates how to determined the mass defect Δm and nuclear binding energy E_b of deuterium ^2_1H.

EXAMPLE 1.1

Example Permalink

http://soilenvirochem.net/4bXtPC

Calculate the mass defect Δm and binding energy per nucleon E_b/A of deuterium.

Deuterium ^2_1H is a stable hydrogen isotope whose nucleus consists of one proton ^1_1p and one neutron ^1_0n. The masses of the constituent particles in *unified atomic mass units*[7] [u] appear in Table 1.2.

TABLE 1.2 US National Institute of Standards and Technology (*NIST*) Values for Deuterium Mass ^2_1H, Proton Rest Mass m_p, and Neutron Rest Mass m_n

Particle	Particle Symbol	Rest Mass (u)
Deuterium	^2_1H	2.01410177812
Proton	^1_1p	1.00727646688
Neutron	^1_0n	1.00866491588

The mass of deuterium, however, is 2.0159413827 u; 0.0018396046 u *less* than the mass found by adding the rest mass of one proton and one neutron. This difference in known as the *mass defect* Δm.

$$m_p + m_n = 1.00727646688\ u + 1.00866491588\ u$$
$$m_p + m_n = 2.0159413827\ u$$
$$\Delta m = 2.0159413827\ u - 2.01410177812\ u$$
$$\Delta m = 1.8396046 \times 10^{-3}\ u$$

Multiplying the mass defect Δm by mass energy equivalent[8] 931.4941 MeV u^{-1} and dividing by 2 to account for the deuterium mass number A yields the nuclear binding energy per nucleon E_b/A for deuterium.

$$E_b = \left(1.8396046 \times 10^{-3}\ u\right)\left(931.4941\ MeV\ u^{-1}\right)$$
$$E_b/2 = (1.7135807\ MeV)/2 = 0.8567904\ MeV$$

[7] $1\ u = 1.660539040 \times 10^{-27}$ kg.
[8] $E_b = \Delta mc^2$.

Exothermic fusion reactions are the source of the energy output from stars, beginning with $^1_1H_0$ nuclei and other nuclei produced during the early stages of the universe following the *Big Bang*. Fig. 1.3 shows that exothermic fusion becomes increasingly inefficient as an energy source. Simply put: a star's nuclear fuel is depleted during its lifetime and eventually exhausted. Box 1.1 provides a few examples of the nuclear reactions that form nuclides found in our Solar System and plant Earth today. Some of those processes occurred during the early moments of the universe under conditions that no longer exist. Other processes continue to occur throughout the Universe because the necessary conditions exist in active stars or in the interstellar medium.

Nucleosynthesis is an accumulation process that gradually produces heavier nuclides from lighter nuclides, principally by exothermic fusion or neutron capture. It is for this reason lighter nuclides are usually more abundant than heavier nuclides. The extreme kinetic energy required for exothermic nuclear fusion and the complex interplay of neutron capture and radioactive decay, the principal nucleosynthesis processes, translate into an abundance profile (see Fig. 1.1) that embodies the relative stability of each nuclide.

BOX 1.1

NUCLEOSYNTHESIS

The particles that make up each element—protons $_{+1}^{1}p$, neutrons $_{0}^{1}n$, and electrons $_{-1}^{0}e$—condensed from more fundamental particles in the early instants following the *Big Bang*. In those early moments, none of the elements that now fill the Periodic Table of the Elements existed other than the particles just mentioned. All of the elements that now exist in the universe result from the nuclear reactions listed in Table 1.3.

The first two reactions—nuclear fusion and neutron-capture—should be considered synthesis reactions. Neutron capture proceeds at a very slow rate (s-process) throughout a star's lifetime. Fusion reactions produce proton-rich nuclides while neutron-capture reactions produce unstable neutron-rich nuclides which undergo beta-decay.

S-process neutron capture is slower than beta-decay at stellar temperatures generated by exothermic nuclear fusion reactions and is unable to bridge the gap between the long-lived bismuth isotopes ($_{82}^{208}Bi$, $_{82}^{209}Bi$, and $_{82}^{211}Bi$) to long-lived thorium isotopes ($_{90}^{230}Th$ and $_{90}^{232}Th$).[9] Most of the neutrons supporting s-process neutron capture are generated by the fusion of $_{6}^{13}C$ and $_{10}^{22}Ne$ with alpha particles $_{2}^{4}He$ (Table 1.3).

The following four reactions—alpha-decay, beta-decay, electron capture, and positron emission—are the spontaneous reactions of unstable (i.e., *radioactive*) nuclides formed by fusion and neutron capture. Electron capture[10]—the decay of unstable neutron-rich nuclides—converts a neutron into a proton by capturing an inner-shell $_{-1}^{0}e$. Positron emission[11] (beta-plus decay)—the decay of unstable proton-rich nuclides—converts a proton to a neutron in the nucleus coupled with the emission of a positron $_{+1}^{0}e$.

Exothermic fusion reactions generate stellar temperatures sufficient to destroy lithium, beryllium, and boron, hence their unusually low abundance relative to other low-mass elements (cf. Table 1.1). Lithium, beryllium, and boron formed during the initial few minutes following the *Big Bang* would have been destroyed in first-generation stars.

Continued

BOX 1.1 *(cont'd)*

TABLE 1.3 Nuclear Reactions Involved in Nucleosynthesis Yielding Elements and Their Isotopes

Nuclear Reaction	Example
Nuclear fusion	$^{1}_{1}H + ^{1}_{1}H \rightarrow ^{2}_{1}H + ^{0}_{+1}e + \nu_e$
	$^{1}_{1}H + ^{2}_{1}H \rightarrow ^{3}_{2}He + \gamma$
	$^{13}_{6}C + ^{4}_{2}He \rightarrow ^{16}_{8}O + ^{1}_{0}n$
	$^{22}_{10}Ne + ^{4}_{2}He \rightarrow ^{25}_{12}Mg + ^{1}_{0}n$
Neutron capture	$^{59}_{26}Fe + ^{1}_{0}n \rightarrow ^{60}_{26}Fe + \gamma$
	$^{60}_{27}Co + ^{1}_{0}n \rightarrow ^{61}_{27}Co + \gamma$
Alpha decay	$^{210}_{84}Po \rightarrow ^{206}_{82}Pb + ^{4}_{2}He$
Beta decay	$^{60}_{26}Fe \rightarrow ^{60}_{27}Co + ^{0}_{-1}e + \overline{\nu}_e$
	$^{61}_{27}Co \rightarrow ^{61}_{28}Ni + ^{0}_{-1}e + \overline{\nu}_e$
Electron capture	$^{7}_{4}Be + ^{0}_{-1}e \rightarrow ^{7}_{3}Li + \nu_e$
	$^{26}_{13}Al + ^{0}_{-1}e \rightarrow ^{26}_{12}Mg + \nu_e$
Positron emission	$^{13}_{7}N \rightarrow ^{13}_{6}C + ^{0}_{+1}e + \nu_e$
	$^{18}_{9}F \rightarrow ^{18}_{8}O + ^{0}_{+1}e + \nu_e$
Cosmic-ray spallation	$^{4}_{2}He + ^{12}_{6}C \rightarrow ^{7}_{3}Li + 2^{4}_{2}He + ^{1}_{1}H$
	$^{1}_{1}H + ^{14}_{7}N \rightarrow ^{9}_{4}Be + ^{4}_{2}He + 2^{1}_{1}H$
	$^{1}_{1}H + ^{12}_{6}C \rightarrow ^{11}_{5}B + 2^{1}_{1}H$
Thermal fission	$^{235}_{92}U + ^{1}_{0}n \rightarrow ^{90}_{38}Sr + ^{144}_{54}Xe + 2^{1}_{0}n + \gamma$
	$^{235}_{92}U + ^{1}_{0}n \rightarrow ^{96}_{39}Y + ^{137}_{55}Cs + 3^{1}_{0}n + 2^{0}_{-1}e + \gamma$

Cosmic rays—mostly protons and alpha particles—traveling at extremely high velocities[12] bath the interstellar medium seeded with gas and dust from earlier stellar generations. The kinetic energy of high-velocity cosmic rays colliding with heavy elements is sufficient to cause nuclear fragmentation or *spallation*. Cosmic-ray spallation occurs at low enough temperatures and densities to allow lithium, beryllium, and boron nuclides to survive once they have formed. Table 1.3 lists reactions representative of cosmic-ray spallation.

BOX 1.1 *(cont'd)*

Uranium-235 is the only naturally occurring fissile nuclide.[13] Fissile nuclides undergo thermal fission stimulated by neutron capture. Thermal fission (Table 1.3) generates two or more neutrons, sufficient to sustain the nuclear chain reaction harnessed by nuclear reactors and nuclear weapons. Unlike other decay reactions (cf. Table 1.3) fission produces large nuclear fragments. The fission yield of $^{233}_{92}U$ appears in Fig. 1.4.

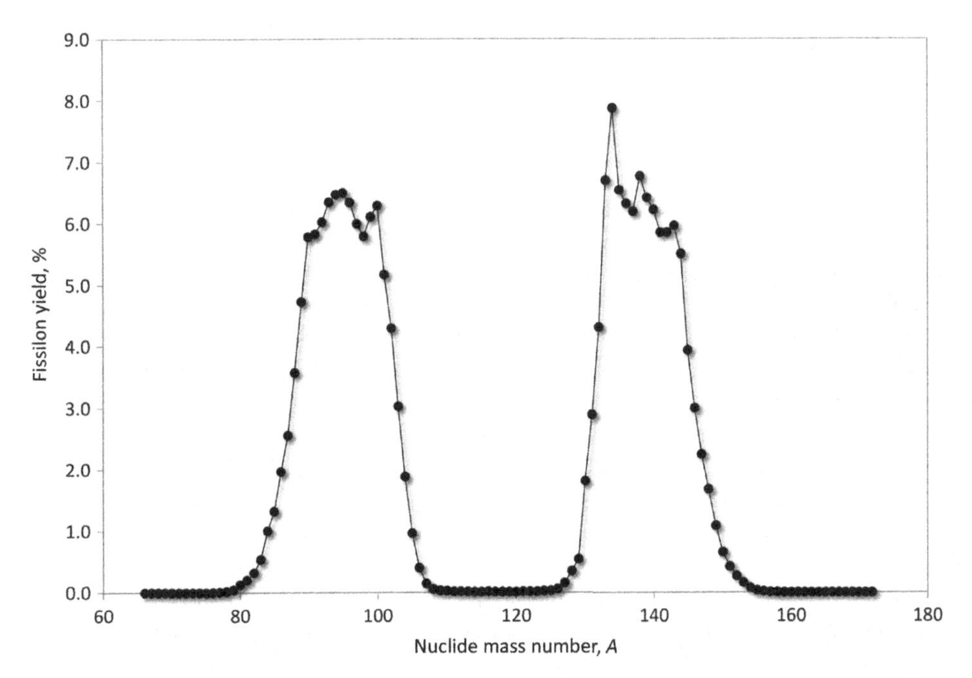

FIG. 1.4 Thermal-neutron fission yield of $^{235}_{92}U$ as a function of product nuclide mass number *A*.

[9]The Bi-to-Th gap is populated by extremely short-lived nuclides.
[10]Emission of an electron neutrino ν_e preserves nuclear spin.
[11]Emission of an electron neutrino ν_e to preserves nuclear spin.
[12]Cosmic-ray particles reach velocities on the order of 0.1 the speed of light.
[13]Other fissile nuclides (uranium isotope $^{233}_{92}U$ and plutonium isotopes $^{239}_{94}Pu$ and $^{241}_{94}Pu$) are bred by neutron capture in nuclear reactors.

1.4 ENRICHMENT AND DEPLETION DURING PLANETARY FORMATION

The random motion of gas molecules and dust particles in the primordial cloud ensured it remained well mixed. The cloud, being homogeneous, would yield normal concentration-frequency distribution for each element.

The various separation processes during planetary formation (accretion and stratification) and ongoing processes like the rock cycle and soil formation lead to the enrichment and depletion of individual elements, altering the abundance profile of Earth's lithosphere relative to the Sun and of the soil relative to the lithosphere (see Figs. 1.1 and 1.2). The weathered rock materials and soils that blanket the Earth's continental lithosphere bear the imprint of numerous separation processes and, as we will presently discover, yield abundance distributions that differ markedly from the normal distribution characteristic of the Sun.

1.4.1 Planetary Accretion

As noted earlier, the Solar System formed through the gravitational collapse of a primordial gas cloud. Preservation of angular momentum within the collapsing gas cloud concentrated dust particles into a dense rotating central body—the Sun—and a diffuse disc (Fig. 1.5, top) perpendicular to the axis of the Sun's rotation. Gravity led to the accretion of planetesimals within the disc that grew in size as they swept out orbits within the disc.

The Solar System formed about 4.5 billion years ago (Wollack, 2012) and contains 8 major planets and 153 confirmed moons, asteroids, meteors, comets, and uncounted planetesimals at its outer fringe. The innermost planets—Mercury, Venus, Earth, and Mars—are rocky ($3.94\ Mg\ m^{-3} < density < 5.43\ Mg\ m^{-3}$), while the outermost planets—Jupiter, Saturn, Uranus, and Neptune—are icy ($density < 1.64\ Mg\ m^{-3}$) (Fig. 1.5, bottom). Planet Earth is the largest of the rocky planets, with a radius of nearly 6370 km. Mars, mean radius 3390 km, has the most moons of the terrestrial planets: two. The Jovian planets are much larger—ranging from roughly 4 to 11 times the radius of Earth; are more massive—from 17 to 318 times the mass of Earth; and are surrounded by far more moons—ranging from 13 to 63.

Goldschmidt (1937) rationalized the enrichment and depletion of the Earth's lithosphere relative to Sun (cf. Fig. 1.6) by grouping element based on geochemical behavior: *lithophiles*, *chalcophiles*, *atmophiles*, and *siderophiles*.

The atmophiles are hydrogen, carbon, nitrogen, and the noble gases helium, neon, argon, krypton, and xenon. The noble gases do not combine with other elements and easily escape the relatively weak gravitational field of the rocky planets. Hydrogen, carbon, and nitrogen react with oxygen, but the products tend to be volatile gases. The gaseous planets, however, are cold enough and massive enough to capture atmophilic gases. Earth and the other rocky planets are depleted of the noble gas elements, hydrogen and other volatile elements relative to the Solar System, as shown in Fig. 1.6.

Depletion of chalcophiles, ferromagnetic metals, and noble metals from the Earth's lithosphere is discussed in the following section.

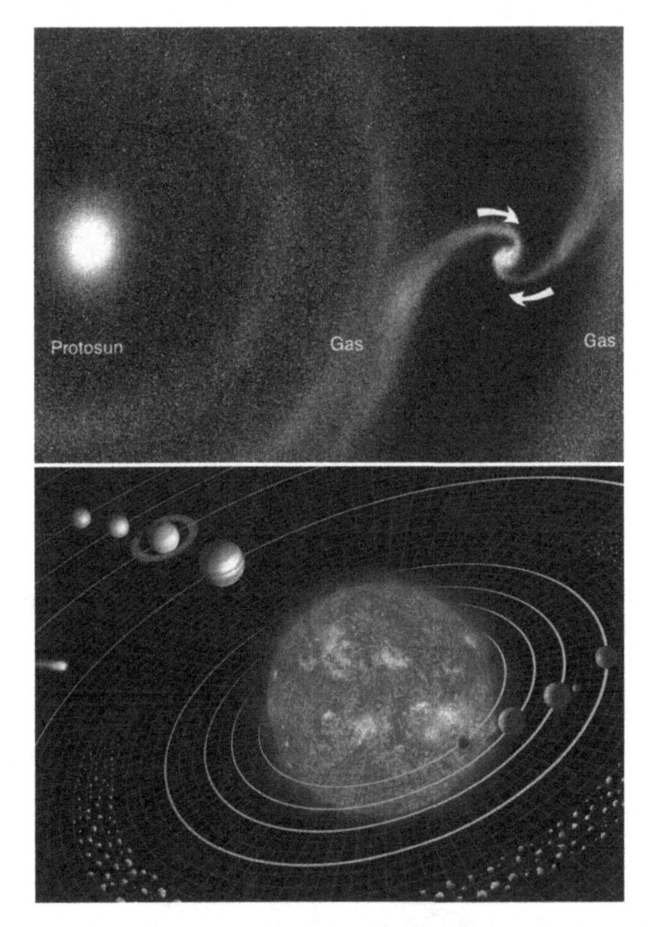

FIG. 1.5 Planetesimal formation in the accretion disc of a proto-sun (*top*) and the present Solar System (*bottom*).

1.4.2 Planetary Stratification

Planet Earth began stratifying into layers as the heat released by accretion and radioactive decay triggered melting. Modern Earth consists of a metallic core composed of ferromagnetic elements and noble metals and a two outer silicate-rich layers—sometimes called the basic silicate Earth—that further separated into a molten mantle and a crystalline lithosphere (Fig. 1.7).

The silicate-rich lithosphere is further differentiated into an oceanic lithosphere and a continental lithosphere. The oceanic lithosphere has a composition and density similar to the mantle, while the continental lithosphere is less dense, thicker, and far more rigid than the oceanic lithosphere. The composition of the continental lithosphere reflects the effect of planetary scale stratification but bears little imprint from the rock cycle because the mean age of the continental lithosphere is roughly the same as the age of planet Earth.

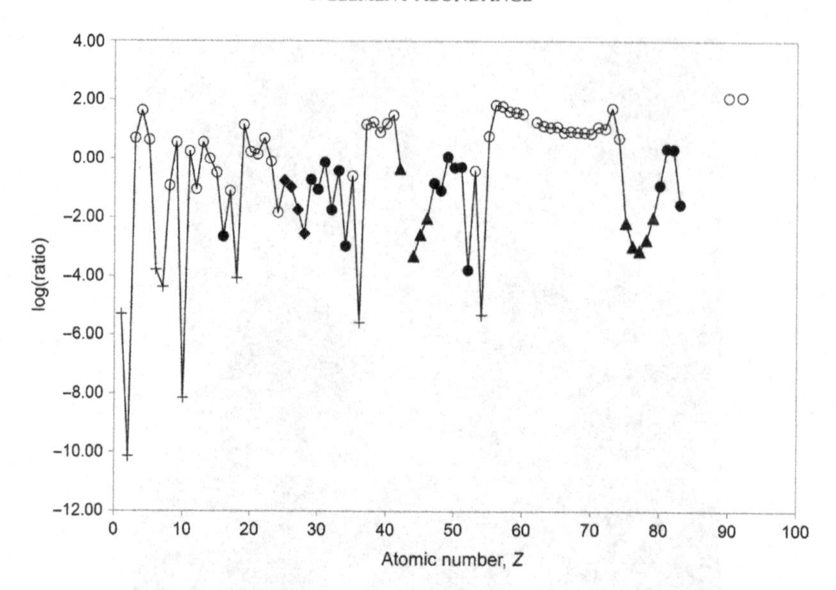

FIG. 1.6 Differentiation of the Earth's lithosphere relative to the Solar System: atmophilic elements (*crosses*), ferromagnetic metals (*diamonds*), noble metals (*triangles*), chalcophilic elements (*filled circles*), and lithophilic elements (*open circles*). Ferromagnetic transition metals and noble metals are grouped as siderophilic elements (Goldschmidt, 1937; Lide, 2005).

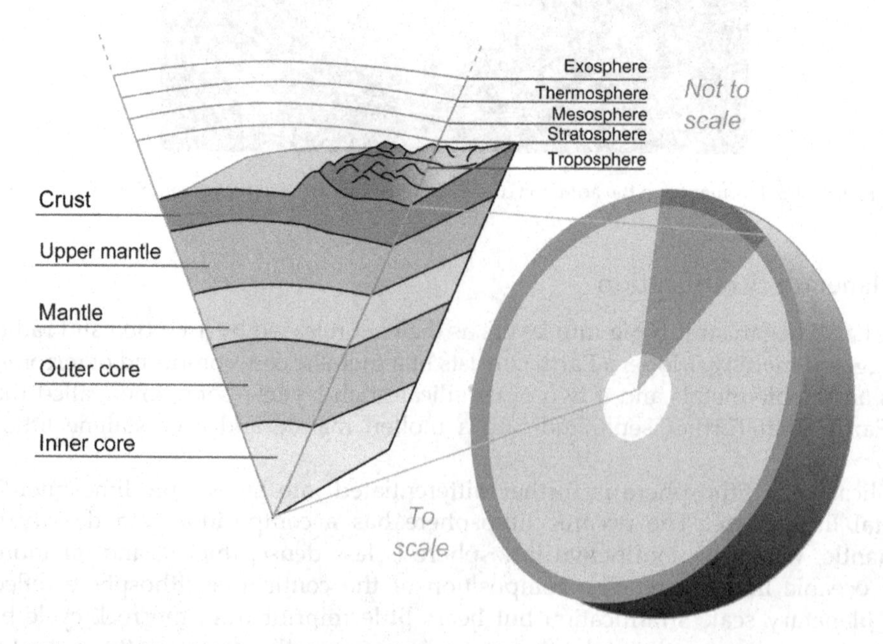

FIG. 1.7 Differentiated Earth: core, mantle, and lithosphere.

The ferromagnetic elements comprise four transition metals from the fourth period: manganese Mn, iron Fe, cobalt Co, and nickel Ni. These four metals are depleted in Earth's lithosphere (see Fig. 1.6), having become major components of the Earth's metallic core. The noble metals—ruthenium Ru, rhodium Rh, palladium Pd, silver Ag, rhenium Re, osmium Os, iridium Ir, platinum Pt, and gold Au—have little tendency to react with either oxygen or sulfur. These elements are also depleted from the basic silicate Earth, enriching the metallic core. Goldschmidt (1937) grouped the ferromagnetic transition metals and noble metals together as siderophilic metals.

The lithophilic elements (Goldschmidt, 1937) consist of those elements in the Periodic Table characterized by a strong tendency to react with oxygen. The chalcophilic elements—which include the elements from periods 4 (copper through selenium), 5 (silver through tellurium), and 6 (mercury through polonium)—combine strongly with sulfur. The basic silicate Earth became enriched in lithophilic and chalcophilic elements after formation of the metallic core.

Given the atomic mass of oxygen and sulfur and the relative atomic mass of lithophilic and chalcophilic elements in the same period; it should not be surprising that sulfide minerals are significantly denser than oxide minerals. Buoyancy in the Earth's gravitational field would cause lithophilic magma to migrate toward the Earth's surface, while chalcophilic magma would tend to sink toward the Earth's core, thereby depleting the lithosphere of chalcophilic elements (see Fig. 1.6).

EXAMPLE 1.2

Example Permalink

http://soilenvirochem.net/W587B4

Calculate the continental lithosphere charge balance based on its elemental composition.

Goldschmidt (1937) demonstrated the influence element chemistry exerts on lithosphere composition during the accretion and stratification of planet Earth by focusing on the affinity (or lack thereof) of different elements to combine as the lithosphere crystallized from the molten state.

This example will reveal chemical constraint on relative elemental abundance: the *charge balance condition*. Elements exchange electrons when they combine to form the ionic-oxide minerals that dominate the lithosphere (Pauling, 1929). We will test the following hypothesis: the charge neutrality condition of minerals constrains the relative elemental abundance of the lithosphere.

Step 1. Convert *mass-fraction* abundance to *molar* abundance (mass-based) listed in Table 1.4 column 2, dividing the mass fraction of each element $w(\text{E})$ [g kg^{-1}] by its atomic mass $\overline{m}_{\text{a}}(\text{E})$ [g mol^{-1}].

$$w(\text{Si}) = \frac{2.7 \times 10^{+5}\ \text{mg kg}^{-1}}{10^{+3}\ \text{mg g}^{-1}} = 270\ \text{g kg}^{-1}$$

$$m(\text{Si}) = \frac{w(\text{Si})}{\overline{m}_{\text{a}}(\text{Si})} = \frac{270\ \text{g kg}^{-1}}{28.0855\ \text{g mol}^{-1}} = 9.614\ \text{mol kg}^{-1}$$

Table 1.4 column 3 lists the molar abundance (mass-based) $m(\text{E})$ of the 15 most abundant lithosphere elements.

Step 2. Convert molar abundance $m(\mathrm{E})$ to *moles-of-charge* abundance $m_C(\mathrm{E})$ [$\mathrm{mol_c\ kg^{-1}}$] listed in Table 1.5 column 3, multiplying the molar abundance of each element by its typical ionic charge $q(\mathrm{E})$ [$\mathrm{mol_c\ mol^{-1}}$] in lithosphere minerals.

$$m_C(\mathrm{Si}) = q(\mathrm{Si}) \cdot m(\mathrm{Si})$$

$$m_C(\mathrm{Si}) = \left(9.614\ \mathrm{mol\ kg^{-1}}\right) \cdot \left(+4\ \mathrm{mol_c\ mol^{-1}}\right) = +38.454$$

TABLE 1.4 Elemental Abundance of the Earth's Lithosphere (Winter, 2007): Mass-Fraction $w(\mathrm{E})$ [$\mathrm{mg\ kg^{-1}}$] and Molar Concentration (Mass-Based) $m(\mathrm{E})$ [$\mathrm{mol\ kg^{-1}}$]

Element (E)	$w(\mathrm{E})$ ($\mathrm{mg\ kg^{-1}}$)	$m(\mathrm{E})$ ($\mathrm{mol\ kg^{-1}}$)
O	460,000	28.751
Si	270,000	9.614
Al	82,000	3.039
Fe	63,000	1.128
Ca	50,000	1.248
Mg	29,000	1.193
Na	23,000	1.000
K	15,000	0.384
Ti	6600	0.138
C	1800	0.150
H	1500	1.488
Mn	1100	0.020
P	1000	0.032
F	540	0.028
S	420	0.013

Iron in granite—the most abundant igneous rock of the lithosphere—is "mostly incorporated into magnetite" (Frost, 1991).[14] The average Fe valence in magnetite is: $q(\mathrm{Fe}) = +\frac{8}{3}\ \mathrm{mol_c\ mol^{-1}}$. Magnetite is one of the many spinel minerals[15] found in lithosphere rocks. Table 1.5 lists the typical manganese valence in spinel minerals: $q(\mathrm{Mn}) = +2\ \mathrm{mol_c \cdot mol^{-1}}$.

The charge balance (based on the 15 most abundant elements) yields $+2.126\ \mathrm{mol_c\ kg^{-1}}$, equivalent to an error of 1.8% when normalized by the total valence charge per kilogram of lithersphere: $117.24\ \mathrm{mol_c\ kg^{-1}}$.

Consider the magnitude of the charge balance error in light of the following: the mass-per-mass values in Table 1.5 are central-tendency values from the chemical analysis of many separate lithosphere specimens, subject to experimental error. Numerous sources have compiled lithosphere, soil and seawater composition central-tendency estimates and yet in every case the charge balance condition is satisfied to within a few percent.

TABLE 1.5 Elemental Abundance of the Earth's Lithosphere (Winter, 2007): Moles-of-Charge Units

Element (E)	q(E) ($mol_c\ mol^{-1}$)	m_c(E) ($mol_c\ kg^{-1}$)
O	−2	−57.502
Si	+4	+38.454
Al	+3	+9.117
Fe	+2.67	+3.008
Ca	+2	+2.495
Mg	+2	+2.386
Na	+1	+1.000
K	+1	+0.384
Ti	+4	+0.552
C	+4	+0.599
H	+1	+1.488
Mn	+2	+0.040
P	+5	+0.161
F	−1	−0.028
S	−2	−0.026

[14] Iron valence in granite is controlled by the fayalite-magentite-quartz redox buffer (Haggerty, 1976). Magnetite composition is: $FeO \cdot Fe_2O_3$ (s).
[15] The general composition of spinel minerals is: AB_2O_3 where cation A is divalent and cation B is trivalent.

1.5 THE ROCK CYCLE

The Earth's lithosphere is in a state of flux; convection currents in the mantle continuously forms new oceanic lithosphere along mid-oceanic ridges. The spreading of newly formed oceanic lithosphere drives the edges of the oceanic lithosphere underneath the more buoyant continental lithosphere along their boundaries, propelling slabs of continental lithosphere against each other. This movement—plate tectonics—drives the rock cycle (Fig. 1.8).

If we take a global view, the continental lithosphere is largely composed of granite, while the oceanic lithosphere is primarily basalt. Granite is a coarse-grained rock composed of aluminosilicate minerals that crystallizes slowly and at much higher temperatures than basalt. Basalt rock has a density comparable to the mantle; it solidifies along mid-oceanic ridges,

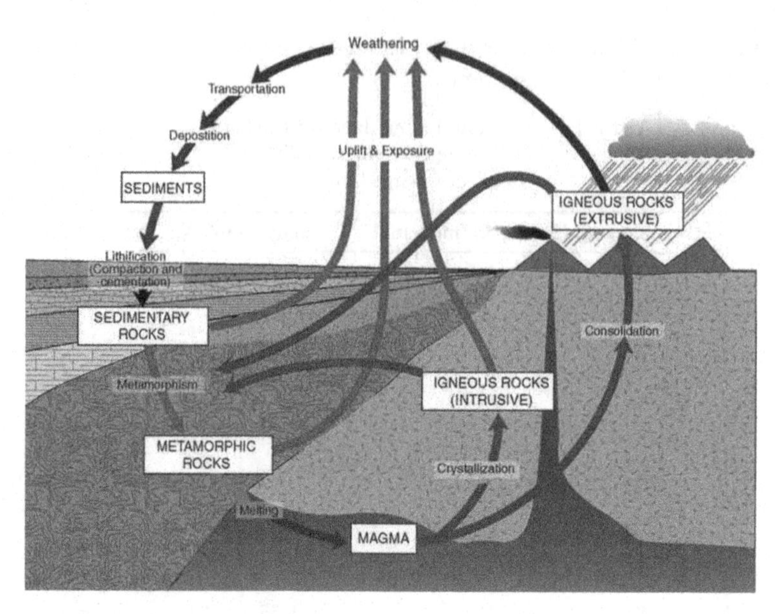

FIG. 1.8 The rock cycle (Monroe et al., 1992).

where convection currents in the mantle carry magma to the surface. Geologists classify granite, basalt, and other rocks that solidify from the molten state as igneous rocks.

A more detailed view of the lithosphere on the scale of kilometers to tens of kilometers reveals other rocks that owe their existence to volcanic activity, weathering, erosion, and sedimentation. Magma welling up from the mantle melts through the lithosphere. Rock formations along the margin of the melted zone are annealed; minerals in the rock recrystallize in a process geologists call metamorphism. Both the mineralogy and texture of metamorphic rocks reveal radical transformation under conditions just short of melting. Metamorphic rocks provide geologists with valuable clues about Earth history, but they are far less abundant than igneous rocks.

Inspection on the kilometer scale reveals a zone of rock weathering and sediment deposits covering both the continental and oceanic lithospheres. Stress fractures, caused by thermal expansion and contraction, create a pathway for liquid water to penetrate igneous and metamorphic rock formations. Freezing and thawing accelerate fracturing, exfoliating layers of rock and exposing it to erosion by flowing water, wind, gravity, and ice at the Earth surface. The water itself, along with oxygen, carbon dioxide, and other compounds dissolved in water, promote reactions that chemically degrade minerals formed at high temperatures in the absence of liquid water, precipitating new minerals that are stable in the presence of liquid water.

Burial, combined with chemical cementation, eventually transforms sedimentary deposits into sedimentary rocks. Primary minerals—minerals that crystallize at high temperatures where liquid water does not exist as a separate phase—comprise igneous and metamorphic rocks. Secondary minerals—those that form through the action of liquid water—form in surface deposits and sedimentary rocks.

The rock cycle (see Fig. 1.8) is completed as plate tectonics carry oceanic lithosphere and their sedimentary overburden downward into the mantle at the continental margins, or magma melting upward from the mantle returns rock into its original molten state.

1.6 SOIL FORMATION

The rock cycle leaves much of the continental lithosphere untouched. Soil development occurs as rock weathering alters rocks exposed on the terrestrial surface of the Earth's lithosphere. Weathered rock materials, sediments, and residuum, undergo further differentiation during the process of soil formation. Picture freshly deposited or exposed geologic formations (sediments deposited following a flood or landslide, terrain exposed by a retreating glacier, fresh volcanic ash, and lava deposits following a volcanic eruption). This fresh parent material transforms through the process of chemical weathering, whose intensity depends on climate (rainfall and temperature), vegetation characteristic of the climate zone, and topography (drainage and erosion).

The excavation of a pit into the soil reveals a sequence of layers or soil horizons with depth (Fig. 1.9). This sequence of soil horizons is known as the *soil profile* for that particular site. Soil horizons vary in depth, thickness, composition, physical properties, particle size distribution, color, and other properties. Soil formation takes hundreds to thousands of years, and, provided the site has remained undisturbed for that length of time, the soil profile reflects the natural history of the setting.

Plotting the abundance of each element in the Earth's lithosphere divided by its abundance in the Solar System reveals those elements enriched or depleted during formation of planet Earth (see Fig. 1.6). Plotting the abundance of each element soil divided by its abundance in the Earth's lithosphere reveals those elements enriched or depleted during the rock cycle and soil development. When both are displayed together on the same graph (Fig. 1.10), plotting the

FIG. 1.9 A northern temperate forest soil developed in glacial till (Northern Ireland).

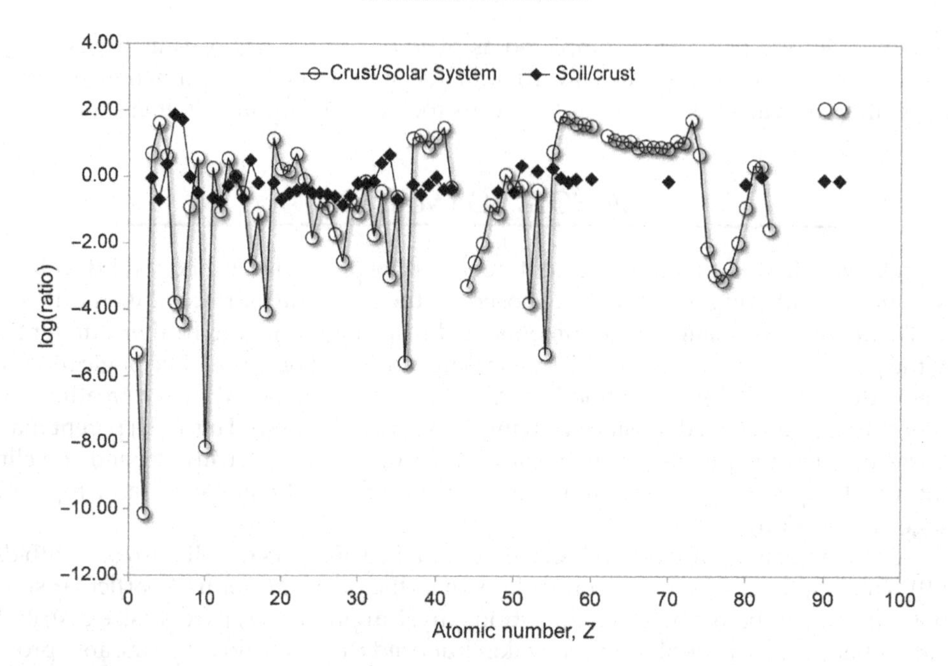

FIG. 1.10 Elemental abundance of the Solar System relative to the Earth's lithosphere (*open circles*) and soil relative to the Earth's lithosphere (*filled diamonds*) reveal the magnitude of planetary formation, the rock cycle, and soil formation (Shacklette and Boerngen, 1984; Lide, 2005).

logarithm of the abundance ratio of lithosphere to the Solar System and the soil-to-lithosphere abundance ratio, an important and not surprising result is immediately apparent: enrichment and depletion during planetary formation are orders of magnitude greater than those that take place during the rock cycle and soil formation. This explains the findings (Helmke, 2000) that a characteristic soil composition cannot be related to geographic region or soil taxonomic group.

Enrichment from biological activity is clearly apparent when plotting the logarithm of the soil-to-lithosphere abundance ratio (Fig. 1.11). Carbon and nitrogen, two elements strongly depleted in the Earth's lithosphere relative to the Solar System, are the most enriched in soil relative to the lithosphere. Both are major constituents of biomass. Selenium, sulfur, and boron—among the top six most enriched elements in soil—are essential for plant and animal growth. The relative enrichment of arsenic, which is not a nutrient, may have much to do with its chemical similarity with phosphorus—another essential element.

1.7 ELEMENT ABUNDANCE FREQUENCY DISTRIBUTIONS

Geochemical and biological processes enrich elements in certain zones while depleting it in others (cf. Figs. 1.9 and 1.11), ensuring no element is distributed uniformly in Earth materials. Geologists, soil scientists, environmental chemists, and others often seek to identify zones of enrichment or depletion. This requires collecting *aliquots* representative of

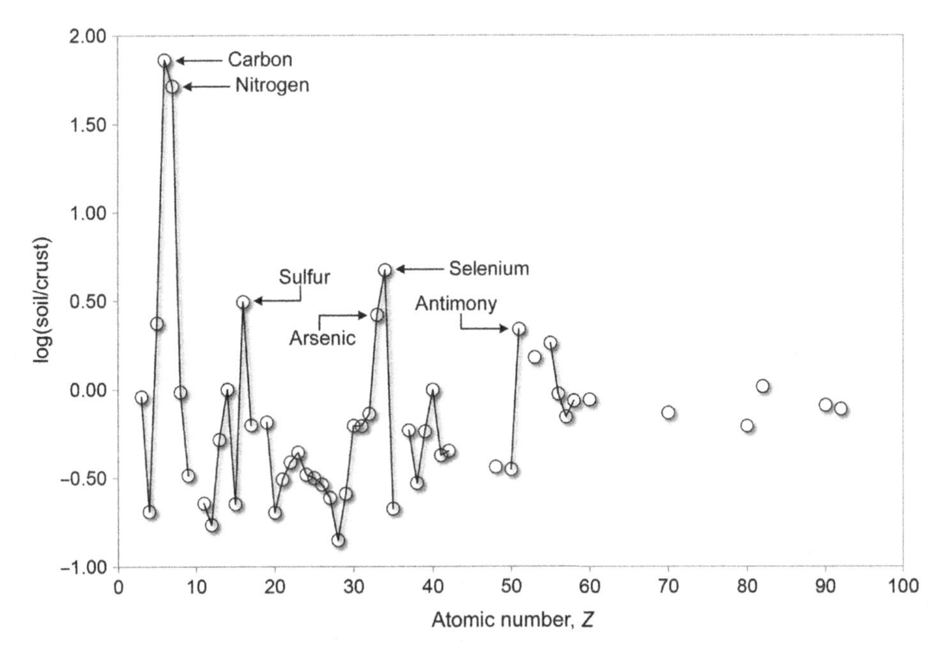

FIG. 1.11 Elemental abundance of soil relative to the Earth's lithosphere, showing enrichment of carbon, nitrogen, and sulfur caused by biological activity (Shacklette and Boerngen, 1984).

the site,[16] chemically analyzing each aliquot and performing a statistical analysis of the sample dataset to estimate the population central tendency[17] and variability.

Geologists delineate zones of enrichment (or depletion) using this approach when prospecting for economic mining deposits or reconstructing the geologic history of rock formations. Soil scientists rely on chemical analysis to classify certain soil horizons characterized by either the loss or accumulation (cf. Fig. 1.9). Environmental chemists map contamination and assess the effectiveness of remediation treatments by analyzing aliquots drawn from environmental media. In short, statistical analysis of a representative sample population will establish whether enrichment or depletion is significant.

Analyte concentration-frequency distributions reveal field site variability and enable the scientist to select the appropriate methods for estimating population parameters such as central tendency and standard deviation for each analyte in the sample.

Figs. 1.12, 1.13, and 1.14 (left) illustrate typical analyte concentration-frequency distributions for environmental media. The data in Fig. 1.12 represent media where geochemical processes dominate, while Figs. 1.13 and 1.14 (left) are media where biological processes exert a strong or dominant influence.

[16] Environmental media—rock formations, residuum zones, soil profiles, sediments, groundwater aquifers, and surface water bodies—are more or less continuous within a given depth range and spatial area. The entire body, however it is delineated, is the *population* and a representative *sample* is comprised of *aliquots* drawn from the body.

[17] The sample mean is one, but not the only, central tendency statistic. Median and mode are other statistics that serve as central tendency estimates depending on distribution skewness.

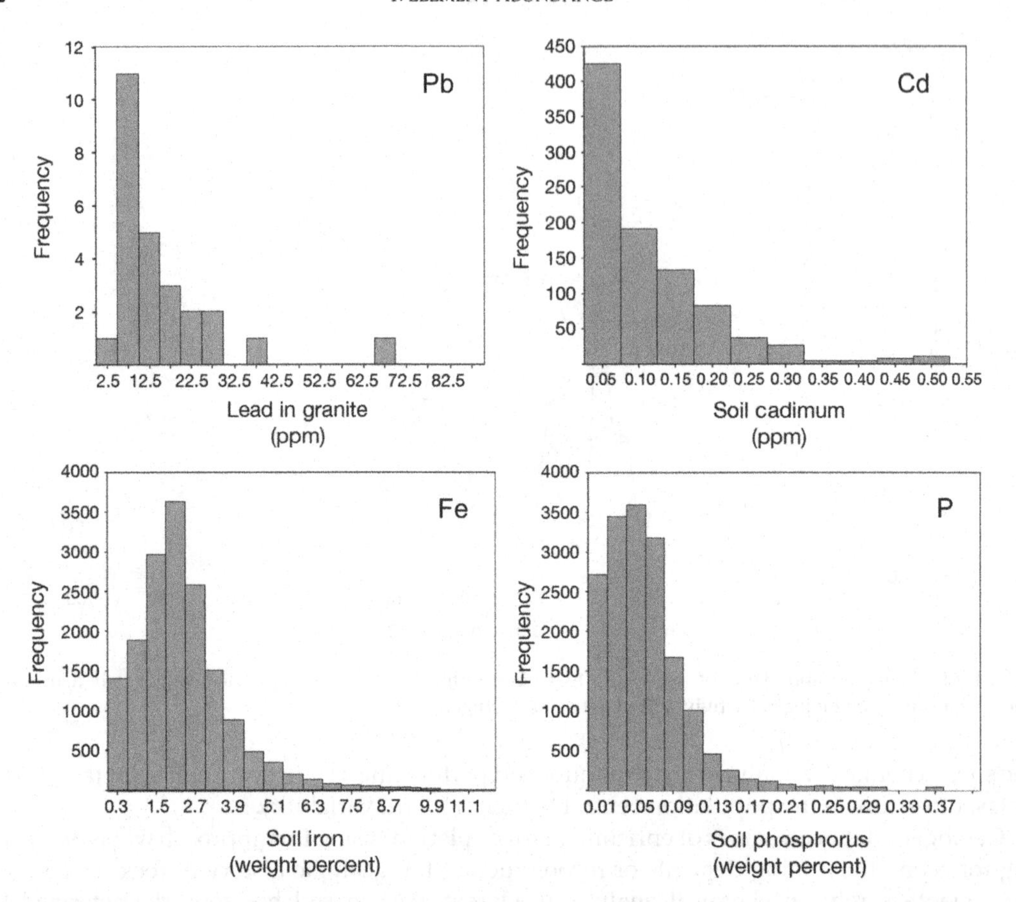

FIG. 1.12 Concentration-frequency distributions for the element Pb in Canadian granite specimens (Ahrens, 1954), the element Cd in soil specimens from Taiwan (Yang and Chang, 2005), and the elements Fe and P from the USGS National Geochemical Survey dataset of US soil and sediment specimens (Zhang et al., 2005). The number of samples (n) in each study are, respectively, 32 (Pb), 918 (Cd), and 16,511 (Fe and P).

Table 1.6 lists a number of studies reporting concentration-frequency distributions from the chemical analysis of rock, soil, water, atmosphere, and biota. The results in Figs. 1.12, 1.13, and 1.14 (left), along with the studies listed in Table 1.6, clearly show a characteristic property of concentration-frequency distributions for environmental media: a marked tendency to be skewed rather than symmetric (cf. Fig. 1.14, right).

Ahrens (1954) discovered the skewed concentration-frequency distributions usually encountered by geochemists when analyzing rock composition appear to follow a *logarithmic-normal* (or more commonly, *log-normal*) distribution. Biologists consider log-normal distributions typical of biological processes (Heath, 1967; Limpert et al., 2001; Limpert and Stahel, 2011).

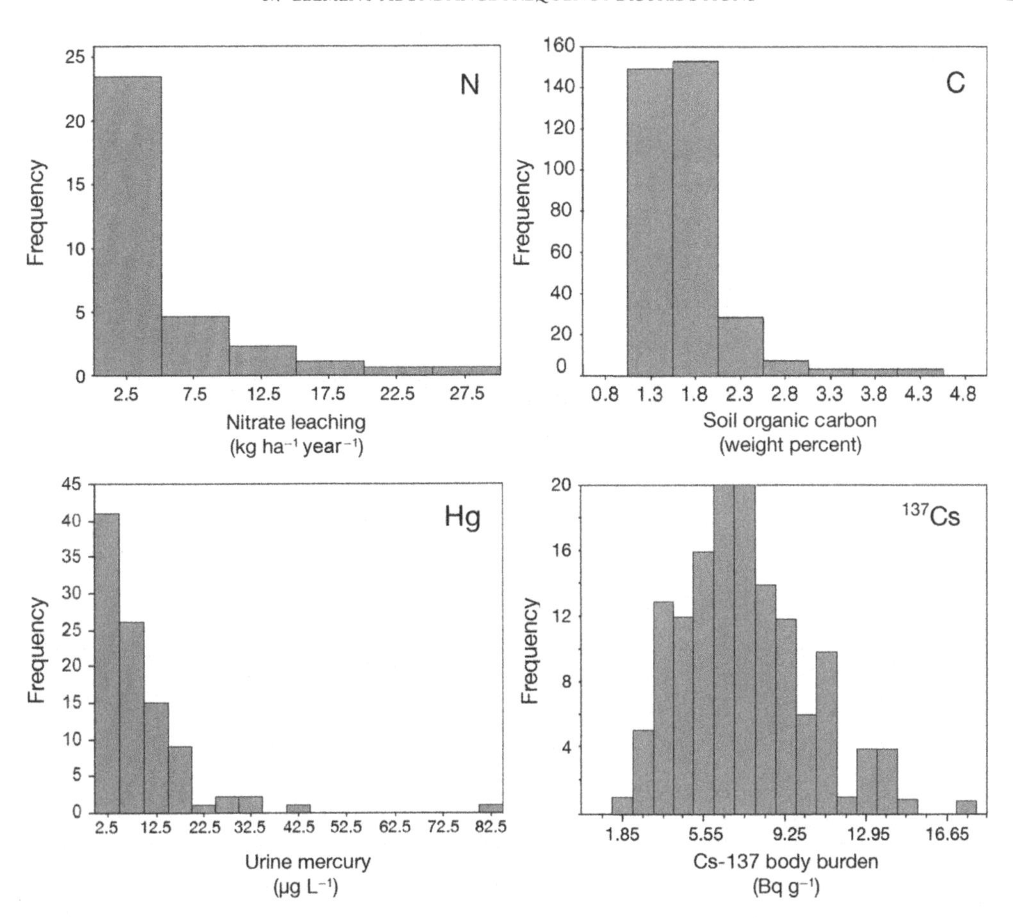

FIG. 1.13 Concentration-frequency distributions for nitrate nitrogen soil pore water (Borken and Matzner, 2004), soil organic carbon from China (Liu et al., 2006), urinary mercury levels in a human population in Rio-Rato, Brazil (Silva et al., 2004), and Cs-127 levels in a species of tree frogs (*Hyla cinerea*) near a nuclear facility in South Carolina (Dapson and Kaplan, 1975). The number of samples (n) in each study is, respectively, 57 (nitrate-nitrogen), 354 (soil organic carbon), 98 (Hg), and 141 (Cs-137).

The skewness of logarithmic-normal frequency distributions[18] is a distribution that under a logarithmic-transformation (i.e., a frequency distribution of logarithm-transformed values) becomes a symmetric or *normal* frequency distribution.

Some of the studies listed in Table 1.6 assume their dataset follows a lognormal distribution (Dudka et al., 1995; Chen et al., 1999; Budtz-Jorgensen et al., 2002), while others perform a statistical analysis to verify that the concentration distribution is best described by a log-normal distribution function (Hadley and Toumi, 2003). In some cases the concentration

[18] Logarithmic-normal frequency distributions are *right skewed*, the tail is toward higher values.

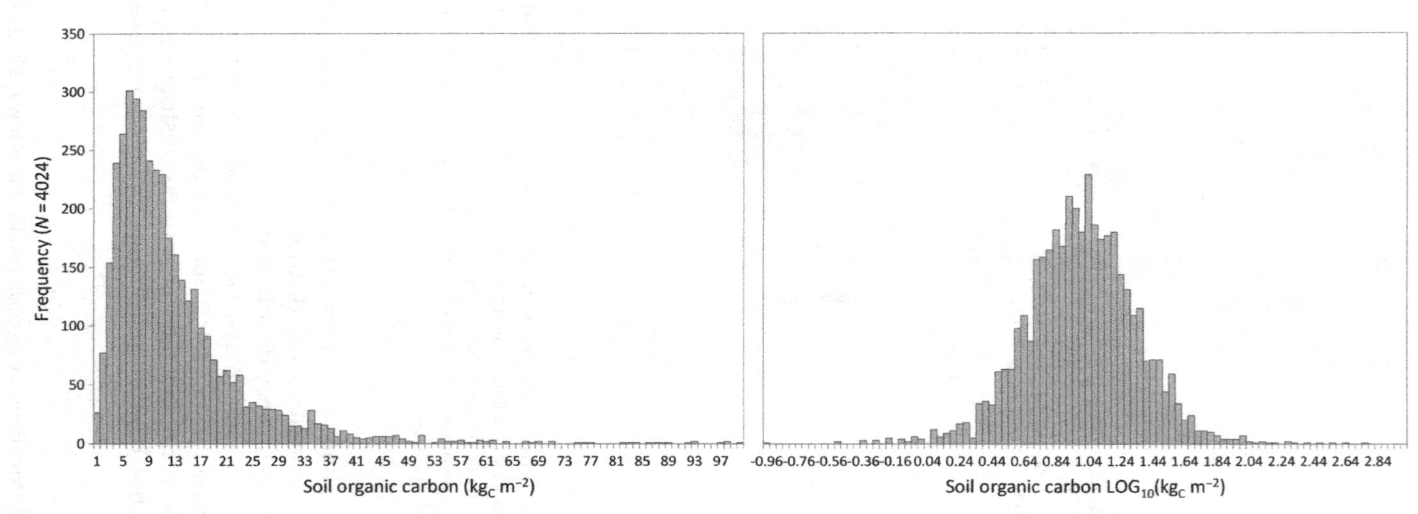

FIG. 1.14 Concentration-frequency distribution of organic carbon from soils worldwide (Zinke et al., 1984). The number of soil samples (n) is 4029. The frequency distribution on the *left* is plotted on a linear scale with units: kg$_C$ m^{-2}. The frequency distribution on the *right* is plotted on a base-10 logarithmic scale.

TABLE 1.6 Studies Reporting Concentration-Frequency Distributions for Environmental Media

Medium	Elements Analyzed	Sample Size	Reference
Granite	K, Rb, Sc, V, Co, Ga, Cr, Zr, La, Cs, F, Mo	158	Ahrens (1954)
Sediments, residuum, and soil	Al, Ca, Fe, K, Mg, Na, P, Ti, Ba, Ce, Co, Cr, Cu, Ga, La, Li, Mn, Nb, Nd, Ni, Pb, Sc, Sr, Th, V, Y, Zn	16,511	Zhang et al. (2005)
Soil	48 elements ($3 \leq Z \leq 92$)	297–1319	Shacklette and Boerngen (1984)
Soil	Cd, Co, Cu, Cr, Fe, Mn, Ni, S, Zn	73	Dudka et al. (1995)
Soil	Ag, As, Ba, Be, Cd, Cr, Cu, Hg, Mn, Mo, Ni, Pb, Sb, Se, Zn	448	Chen et al. (1999)
Soil	Cd	918	Yang and Chang (2005)
Soil	C	354	Liu et al. (2006)
		4029	Zinke et al. (1984)
Groundwater	Ag, As, Ba, Be, Cd, Co, Cr, Cu, Hg	104	Newcomb and Rimstidt (2002)
	Ni, Pb, Sb, Se, Tl, Sn, V, Zn	280	
Groundwater	NO_3^-	57	Borken and Matzner (2004)
Atmosphere	SO_2	365	Hadley and Toumi (2003)
Human tissue	Hg	140	Silva et al. (2004)
		1022	Budtz-Jorgensen et al. (2002)
Animal tissue	$^{137}_{55}Cs$	141	Dapson and Kaplan (1975)

distributions may not conform to a true log-normal distribution, yet they clearly deviate from a normal distribution (Yang and Chang, 2005; Zhang et al., 2005).

Appendix 1.B describes two models of environmental processes that result in log-normal concentration-frequency distributions. The simplest to understand is the *sequential random-dilution* model (Ott, 1990). The second model, known as the *Law of Proportionate Effect* (Kapteyn, 1903), is completely general.

1.8 ESTIMATING THE CENTRAL TENDENCY OF LOGARITHMIC-NORMAL DISTRIBUTIONS

Each type of frequency distribution results from processes with specific characteristics. For now we will rely on the same assumption as Ahrens (1954): a log-normal distribution is a reasonable representation of the concentration-frequency distribution of trace elements in environment media.

The best way to understand log-normal distributions is to consider real examples. Example 1.3 uses the dataset published by Ahrens (1954) consisting of Pb concentrations from a sample of Canadian granite specimens (Table 1.7; cf. Figs. 1.12 and 1.15).

EXAMPLE 1.3

Example Permalink

http://soilenvirochem.net/9fzK31

Estimate the arithmetic and geometric statistics for a sample of Canadian granite specimens.

Ahrens (1954) chemically analyzed 7 elements in 36 granite specimens collected in Canada.[19] Table 1.7 lists the Pb content of 32 granite specimens and Fig. 1.15 (left) plots the concentration-frequency distribution.

TABLE 1.7 The Pb Content in 32 Canadian Granite Specimens

w(Pb) (mg kg^{-1})	w(Pb) (mg kg^{-1})	w(Pb) (mg kg^{-1})
2.0	8.5	17.0
3.8	9.0	18.0
5.5	9.0	20.0
6.0	9.0	20.0
6.0	10.0	22.0
6.5	10.0	24.0
7.5	11.0	26.0
7.7	12.0	29.0
8.0	14.0	39.0
8.0	14.0	68.0
8.5	15.0	

From Ahrens, L.H., 1954. The lognormal distribution of the elements. Geochim. Cosmochim. Acta 5, 49–73.

Step 1. Compute the arithmetic mean and its standard error for the Canadian granite sample listed in Table 1.7.

Central Tendency Estimate: Arithmetic Mean

$$\overline{w}(\mathrm{Pb})_A = \tfrac{1}{32} \cdot \sum_{i=1}^{32} \left(w(\mathrm{Pb})_i \right) = 14.8$$

The histogram plot (Fig. 1.15, left) reveals 10 granite specimens have Pb contents in the range (8–12 mg kg^{-1}) while only 4 specimens have Pb contents in the range (12–16 mg kg^{-1}). Clearly the arithmetic mean ($\overline{w}(\mathrm{Pb})_A = 14.8$) is a poor estimate of central tendency for this concentration-frequency distribution.

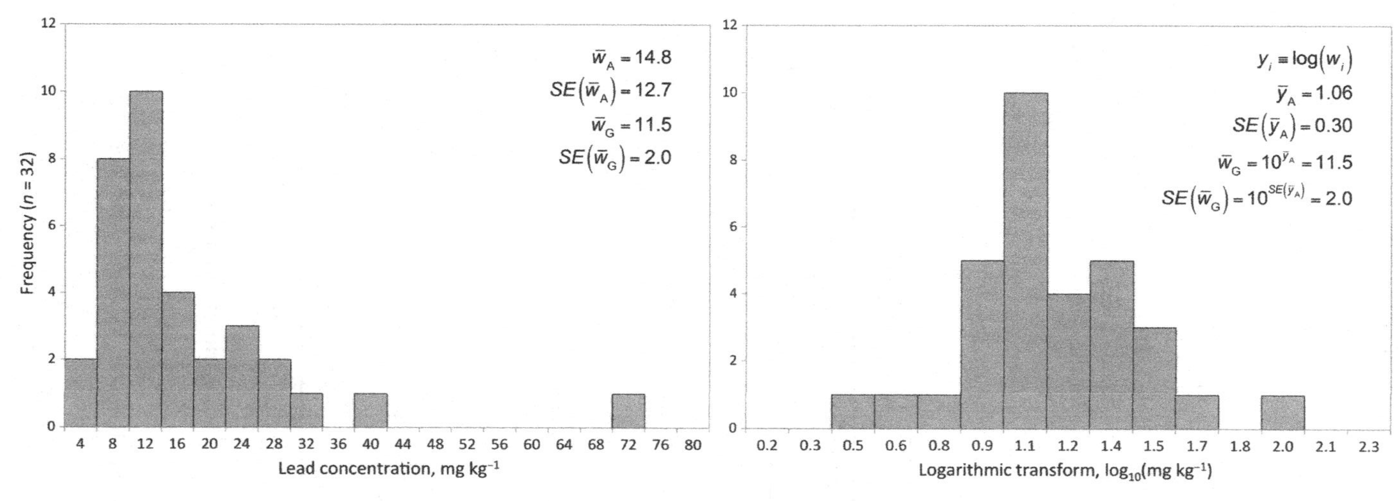

FIG. 1.15 Lead Pb concentration-frequency distributions in Canadian granite (Ahrens, 1954): w(Pb) distribution (*left*) and logarithmic-transformed distribution y(Pb) (*right*). The arithmetic mean and standard error for each distribution appear at the *upper right*. The geometric mean and the standard error of the geometric mean appear at *upper right* in both figures.

Standard Error

$$SE(\overline{w}(\mathrm{Pb})_A) = \sqrt{\frac{1}{32} \cdot \sum_{i=1}^{32} \left(\overline{w}(\mathrm{Pb})_A - w(\mathrm{Pb})_i\right)^2} = 12.7$$

The standard error of the arithmetic mean ($SE(\overline{w}(\mathrm{Pb})_A) = 12.7$) is almost as large as the arithmetic mean ($\overline{w}(\mathrm{Pb})_A = 14.8$) itself. Limpert and Stahel (2011) apply the "95% range check" to identify datasets that are not normally distributed. This is usually the case whenever the standard error of the arithmetic mean $SE(\overline{w}_A)$ is nearly equal to the arithmetic mean $\overline{w}_A$, as it is in this example. If a normal distribution were appropriate for the sample listed in Table 1.7, the probability of a *negative* Pb content would be 12%.

Step 2. Take the base-10 logarithm of each concentration $w(\mathrm{Pb})$ listed in Table 1.7 to generate a logarithmic-transformed dataset. Compute the arithmetic mean and standard error of the transformed dataset (Table 1.8).

TABLE 1.8 The Base-10 Logarithmic Transform of Pb Contents Listed in Table 1.7

$\log_{10}(\mathrm{mg\ kg^{-1}})$		
y(Pb)	**y(Pb)**	**y(Pb)**
0.30	0.93	1.23
0.58	0.95	1.26
0.74	0.95	1.30
0.78	0.95	1.30
0.78	1.00	1.34
0.81	1.00	1.38
0.88	1.04	1.41
0.89	1.08	1.46
0.90	1.15	1.59
0.90	1.15	1.83
0.93	1.18	

From Ahrens, L.H., 1954. The lognormal distribution of the elements. Geochim. Cosmochim. Acta 5, 49–73.

Logarithmic-Transform

$$y(\mathrm{Pb})_i \equiv \log_{10}(w(\mathrm{Pb})_i)$$

Central Tendency Estimate: Arithmetic Mean

$$\overline{y}(\mathrm{Pb})_A = \frac{1}{32} \cdot \sum_{i=1}^{32} \left(y(\mathrm{Pb})_i\right) = 1.06$$

Standard Error

$$SE(\overline{y}(\mathrm{Pb})_{\mathrm{A}}) = \sqrt{\frac{1}{32} \cdot \sum_{i=1}^{32} \left(\overline{y}(\mathrm{Pb})_{\mathrm{A}} - y(\mathrm{Pb})_i\right)^2} = 0.30$$

Step 3. Compute the geometric mean and its standard error for the Canadian granite sample listed in Table 1.7 using the logarithmic-transformed statistics from Step 2.

Central Tendency Estimate: Geometric Mean

$$\overline{w}(\mathrm{Pb})_{\mathrm{G}} = 10^{\overline{y}(\mathrm{Pb})_{\mathrm{A}}} = 10^{+1.06} = 11.5$$

Returning to the histogram plot in Fig. 1.15 (left), the geometric mean ($\overline{w}(\mathrm{Pb})_{\mathrm{G}} = 11.5$) is a superior central tendency estimate for this concentration-frequency distribution. The frequency of Pb contents in the range (8–12 mg kg^{-1}) exceeds by far any other interval plotted in the concentration-frequency distribution.

Standard Error

$$SE(\overline{w}(\mathrm{Pb})_{\mathrm{G}}) = 10^{SE(\overline{y}(\mathrm{Pb})_{\mathrm{A}})} = 10^{+0.30} = 2.0$$

Unlike the standard error of the arithmetic mean ($SE(\overline{w}(\mathrm{Pb})_{\mathrm{A}}) = 12.7$) the standard error of the geometric mean ($SE(\overline{w}(\mathrm{Pb})_{\mathrm{G}}) = 2.0$) is an order of magnitude smaller than the geometric mean $\overline{w}(\mathrm{Pb})_{\mathrm{G}} = 11.5$.

Technically, the logarithmic transformation of the entire dataset (Step 2) and the subsequent antilogarithmic transformation of the arithmetic statistics (Step 3) are unnecessary given the alternative expression for geometric mean.

Geometric Mean

$$\overline{w}_{\mathrm{G}} = \left(\prod_{i=1}^{n} w_i\right)^{1/n}$$

Calculation of the standard error of the geometric mean $SE(\overline{w}_{\mathrm{G}})$ is most efficiently computed, however, using Steps 2 and 3.

[19] The actual number of observations ranges from 23 to 32 because a complete analysis of all 7 elements was not reported for all 36 specimens.

Normal distributions are symmetric about the arithmetic mean, therefore the percentile the range is also symmetric. For example, a range extending from 1 standard deviation below the mean to 1 standard deviation above ($\overline{w}_{\mathrm{A}} - \sigma(\overline{w}_{\mathrm{A}}), \overline{w}_{\mathrm{A}} + \sigma(\overline{w}_{\mathrm{A}})$) covers 68.2% of the population. The range centered on the mean and spanning 4 standard deviations ($\overline{w}_{\mathrm{A}} - 2 \cdot \sigma(\overline{w}_{\mathrm{A}}), \overline{w}_{\mathrm{A}} +$

$2 \cdot \sigma(\overline{w}_A))$ is also symmetric and covers 95.4%. Example 1.3 used this normal-distribution property to estimate the probability of *negative* values.

Example 1.4 computes a central-tendency estimate for the chromium content in Canadian granite specimens (Ahrens, 1954) whose concentration-frequency distribution is a right-skew distribution. The central tendency estimate in this case is the geometric-mean chromium content of a sample based on chemical analysis of 27 granite specimens.

Example 1.4 also computes the 95% confidence interval for the central-tendency estimate. The right-skew of the Cr concentration-frequency distribution clearly demand the corresponding right-skew confidence interval provided by log-normal distribution statistics, not a symmetric confidence interval provided by normal distribution statistics.

EXAMPLE 1.4

Example Permalink

http://soilenvirochem.net/fzSx9M

Estimate the central-tendency chromium content of granite and the confidence interval for the central-tendency estimate based on the Canadian granite sample of Ahrens (1954).

Table 1.9 lists the Cr content of 27 granite specimens (Ahrens, 1954).

TABLE 1.9 The Cr Content in 27 Canadian Granite Specimens

$w(Cr)$ (mg kg^{-1})	$w(Cr)$ (mg kg^{-1})	$w(Cr)$ (mg kg^{-1})
22	22	38
40	410	17
30	2	19
120	24	7
25	15	3
3.5	35	6
5.3	28	5.5
5.8	95	27
17	59	2

From Ahrens, L.H., 1954. The lognormal distribution of the elements. Geochim. Cosmochim. Acta 5, 49–73.

Step 1. Compute the arithmetic mean and its standard error for dataset generated by the base-10 logarithms of the Cr contents listed in Table 1.9. The logarithmic-transformed values are not listed.

Logarithmic-Transform

$$y(\mathrm{Cr})_i \equiv \log_{10}(w(\mathrm{Cr})_i)$$

Central Tendency Estimate: Arithmetic Mean

$$\overline{y}(\mathrm{Cr})_{\mathrm{A}} = \tfrac{1}{27} \cdot \sum_{i=1}^{27} (y(\mathrm{Cr})_i) = 1.236$$

Standard Error

$$SE(\overline{y}(\mathrm{Cr})_{\mathrm{A}}) = \sqrt{\tfrac{1}{27} \cdot \sum_{i=1}^{27} (\overline{y}(\mathrm{Cr})_{\mathrm{A}} - y(\mathrm{Cr})_i)^2} = 0.551$$

Step 2. Compute the 95% confidence interval for the arithmetic mean estimate of the logarithmic-transformed dataset from Step 1.

The 95% confidence interval defines the range within which there is a 95% probability the true population central tendency will fall. Stated differently, their is a 5% probability α the central tendency estimate is not within the confidence interval.

The confidence interval is computed using the *Student's t-distribution* which is designed for normal distributions. The parameters required are the probability α of a Type I error, the standard error of the arithmetic mean $\overline{y}(\mathrm{Cr})_{\mathrm{A}}$, and the sample degrees of freedom ν.

$$\alpha = (1.00 - 0.95)/2 = 0.025$$
$$\nu = (27 - 1) = 26$$

Using a standard Student's t-table, select the row corresponding to $\nu = 26$ and a cumulative probability of 0.975[20]: $t_{0.025,26} = 2.056$.

$$\frac{t_{0.025,26} \cdot SE(\overline{y}(\mathrm{Cr})_{\mathrm{A}})}{\sqrt{n}} = 0.218$$

Step 3. Compute the 95% confidence interval for the geometric mean estimate of the population central tendency using the base-10 antilogarithm of the statistics from Step 2.

The sample estimate of the central-tendency Cr content of granite is the geometric mean $\overline{w}(\mathrm{Cr})_{\mathrm{G}}$ of the values listed in Table 1.9.

$$\overline{w}(\mathrm{Cr})_{\mathrm{G}} = 10^{\overline{y}(\mathrm{Cr})_{\mathrm{A}}} = 10^{+1.236} = 17.2$$

The 95% confidence interval for true central-tendency Cr content of granite $\mu(\mathrm{Cr})$ is the base-10 antilogarithm of the limit computed in Step 2.

$$10^{1.236-0.218} \leq \mu(\mathrm{Cr}) \leq 10^{1.236+0.218}$$
$$10.4 \leq \mu(\mathrm{Cr}) \leq 28.5$$

[20] A 95% confidence interval will a cumulative probability of 2.5% at the lower limit and 97.5% at the upper limit. T-tables only list the upper have of the Student's t-distribution.

1.9 SUMMARY

The evolution of the universe and the Solar System ultimately determined the composition of planet Earth. Gravitational collapse of primordial gas clouds formed first-generation stars, triggering exothermic fusion reactions that generated sufficient heat to forestall complete collapse and producing a series of heavier elements culminating in the nuclide $^{56}_{26}Fe$. The sequential exothermic fusion of lighter elements produces heavier elements with less efficiency in each subsequent stage. Eventually the first-generation stars depleted their nuclear fuel, either collapsing into obscurity or exploding into supernovae that seeded the interstellar medium with heavy elements.

Later stellar generations repeat the process, producing still heavier elements by neutron capture and beta-decay either slowly throughout the lifetime of the star (s-process) or in the paroxysm of a supernova (r-process). R-process neutron capture generates a host of radioactive nuclides, some of which have survived the formation of the Solar System and are found on Earth today, and others that decayed to stable nuclides by the present time.

Solar System differentiation occurred during planetary accretion, resulting in the formation of the four inner terrestrial planets, including Earth. Further differentiation took place as the Earth stratified into layers: core, mantle, and lithosphere. Separation processes continue to the present day, the result of the rock cycle and soil formation. Differentiation transformed the original homogeneous element distribution inherited from the collapsing gas cloud into the heterogeneous distributions we find in the Earth's terrestrial environment.

Samples drawn from the environmental media—rocks, residuum, soils, sediments, and the hydrosphere—reflect natural variability. Efforts to detect enrichment or depletion require statistical analysis of aliquots drawn from rock, soil, aquifer, and water bodies of sufficient sample size to capture natural variability. Chemical analysis of environmental samples typically yields datasets with right-skewed frequency distributions. Estimates of the central-tendency concentration of an element and the variation of the sample dataset usually rely on logarithmic transformation of the dataset, a practice justified because the natural concentration-frequency distributions tend to be log-normal.

APPENDICES

1.A FACTORS GOVERNING NUCLEAR STABILITY AND NUCLIDE ABUNDANCE

1.A.1 The Table of Nuclides

The Table of Nuclides (Fig. 1.A.1) organizes all known nuclides and elements into columns and rows, listing important information about each isotope: atomic number Z, neutron number N, mass number A, isotopic rest mass, binding energy, stability, half-life, and abundance.

Stable nuclides (black squares in Figs. 1.A.1 and 1.A.2) occupy a narrow band passing through the center of the known nuclides. Bordering this narrow peninsula of stability is a zone of radioactive (i.e., unstable) nuclides that are increasingly less stable (i.e., shorter half-lives)

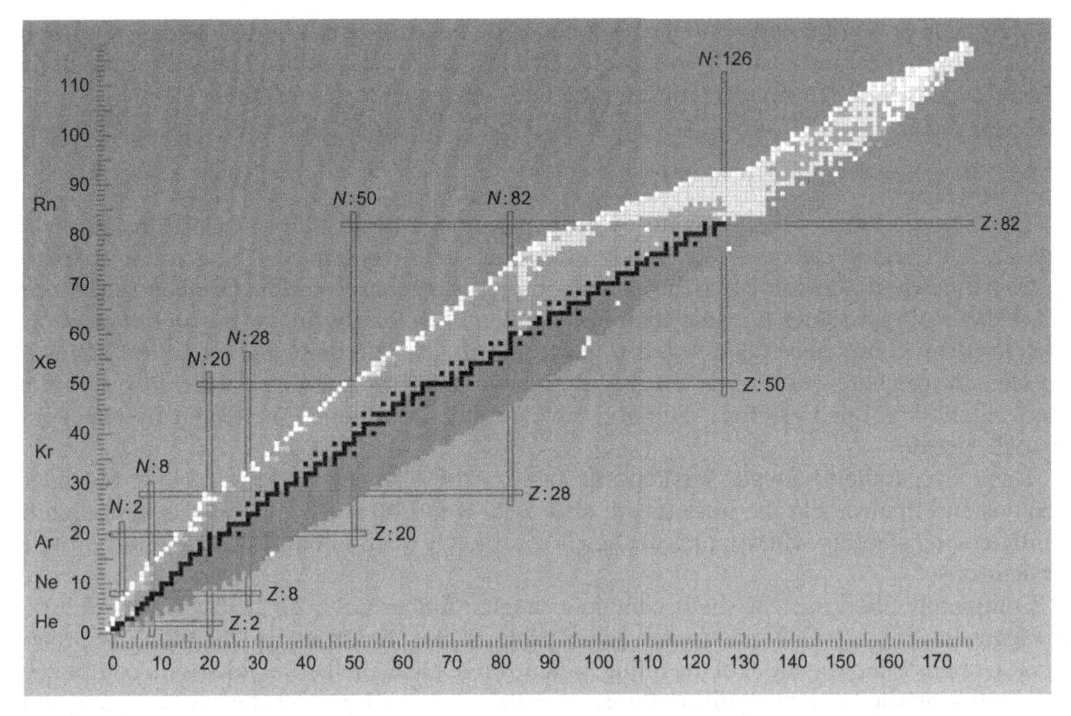

FIG. 1.A.1 The Table of Nuclides, with each *square* representing a known nuclide. *Black cells* indicate stable while *gray-shaded cells* indicate unstable nuclides. *Different shades of gray* further distinguish decay modes by emission type: beta-plus, beta-minus, and alpha.

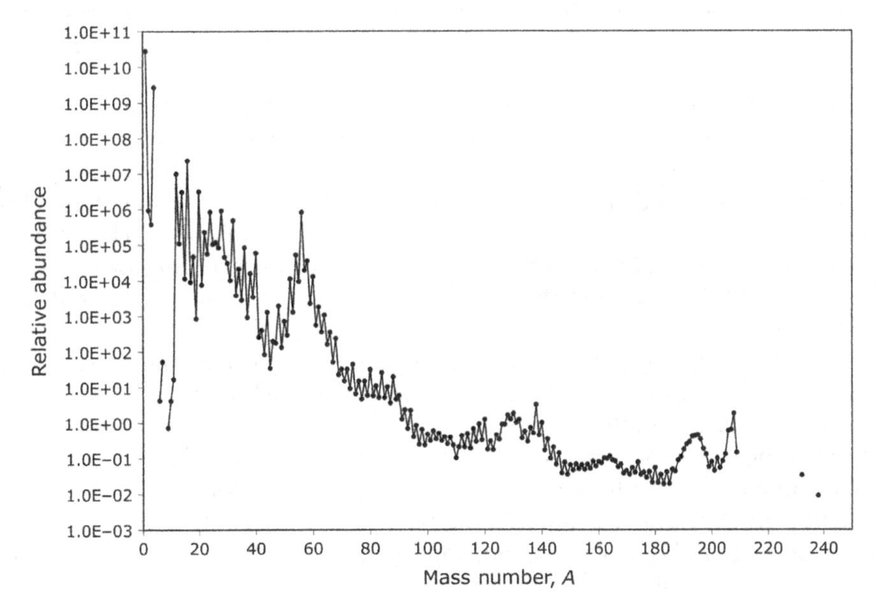

FIG. 1.A.2 Isotope abundance decreases with increasing mass number A. The abundance is the nuclide mole fraction normalized by 10^{+6} mol Si (Anders and Grevesse, 1989).

as the nuclides become increasingly proton-rich or neutron-rich. Lighter shades of gray in Fig. 1.A.2 represent increasingly less stable nuclides. Physicists and nuclear chemists call the several processes that form elements and nuclides nucleosynthesis (cf. Table 1.3).

1.A.2 Nuclear Magic Numbers

The Periodic Table of the Elements owes its periodicity to the filling of electronic shells. It takes 2, 8, 10, and 14 electrons to complete electronic shells designated s, p, d, and f. Chemists noticed that when elements react, they tend to lose, gain, or share a certain number of electrons. The Octet Rule associates unusual chemical stability with the sharing or transfer of electrons that result in completely filled s and p electronic shells. The preferred oxidation states of an element or charge of an ion can usually be explained by stability conferred by a filled electron shell. A filled electron shell and covalent chemical bonds increment by two spin-paired electrons.

There are parallels between the Periodic Table of the Elements and the Table of Nuclides. Protons and neutrons in the nucleus are also organized into shells; nuclides exhibit significantly greater stability when a nuclear shell is completely filled or half-filled by either protons or neutrons.

Completely filled nuclear shells contain a magic number (2, 8, 20, 28, 50, 82, and 126) of nucleons (cf. Fig. 1.A.1). Nucleons are either protons ${}^{1}_{1}p$ or neutrons ${}^{1}_{0}n$. The filling of proton nuclear shells is independent of the filling of neutron nuclear shells. Nuclides with completely filled nuclear shells have significantly greater stability—and greater abundance—than nuclides with partially filled nuclear shells. Nuclides with half-filled nuclear shells or shells with an even number of spin-paired nucleons also represent stable nuclear configurations.

The *even-odd* effect—which results from the stability of spin-paired nucleons—clearly apparent in the composition of the Solar System (Fig. 1.A.3) and, to a lesser degree in the Earth's lithosphere and soil (Fig. 1.A.3), and in the number of stable nuclides for each element.

To reiterate: atomic number Z determines the identity of each element, while neutron number N distinguishes each the isotopes of an element. A magic atomic number Z or a half-filled nuclear proton shell (an atomic number Z of 1, 4, 10, 14, etc.) or an even number of spin-paired protons confers added stability to the element that translates into greater abundance than elements lacking those characteristics. A magic neutron number N or a half-filled nuclear neutron shell (a neutron number N of 1, 4, 10, 14, etc.) or an even number of spin-paired neutrons confers added stability to the nuclide, which translates into greater stability than nuclides lacking those characteristics.

A portion of the Table of Isotopes appears in Fig. 1.A.3. The elements, represented by their atomic number Z, appear as columns in this version, while the nuclides, represented by the number of neutrons in the nucleus N, appear as rows. The columns and rows are not completely filled because only known nuclides are shown. Stable nuclides appear in black, while unstable nuclides are in lighter shade of gray and white, depending on the half-life of each radioactive nuclide. The half-life of a radioactive nuclide is the time it takes for half of the nuclide to undergo radioactive decay into another nuclide. Fig. 1.A.3 helps us understand how nuclide stability contributes to element abundance.

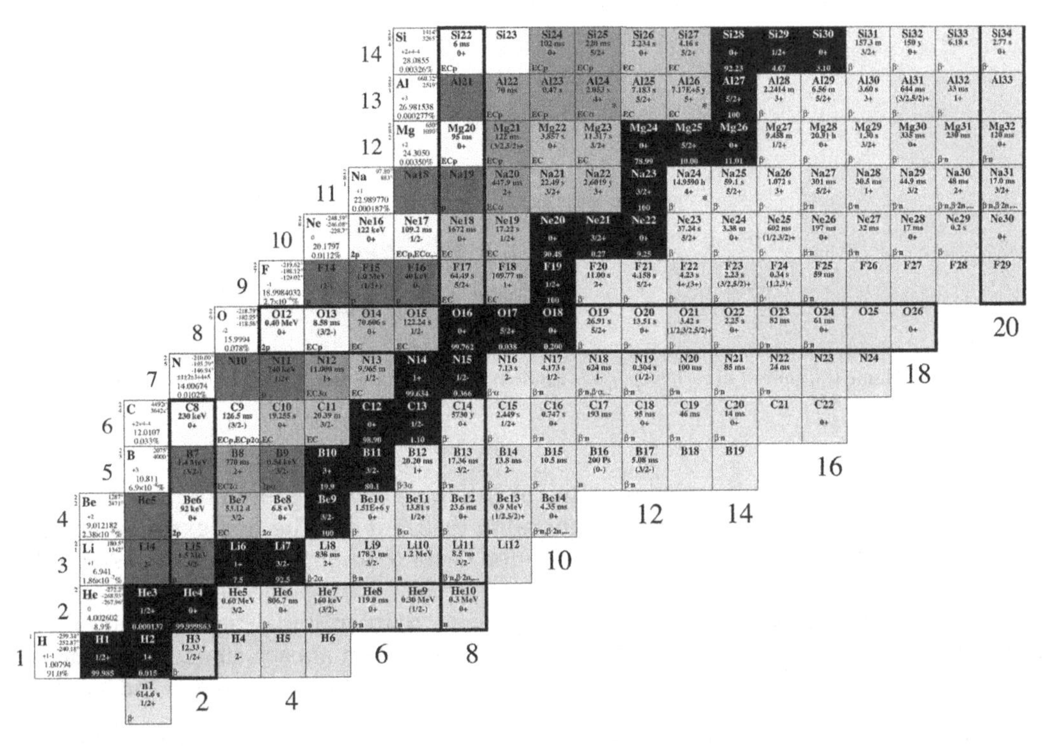

FIG. 1.A.3 A portion of the Table of Isotopes: the *darkest shade of gray* identifies stable isotopes, while the *lighter shades of gray* indicate radioactive nuclides with progressively shorter half-lives.

The longest-lived ($t_{1/2} = 12.73\ y$) radioactive hydrogen nuclide ${}_{1}^{3}H_{2}$ has a magic neutron number. All helium isotopes ${}_{2}^{A}He_{N}$ have a magic atomic number ($Z = 2$). Stable helium nuclides span a range from ${}_{2}^{3}He_{1}$ with a half-filled neutron shell to ${}_{2}^{10}He_{8}$ with a magic neutron number ($N = 8$) and the most abundant helium nuclide ${}_{2}^{4}He_{2}$ has a magic neutron number ($N = 2$).

All oxygen isotopes ${}_{8}^{A}O_{N}$ have a magic atomic number ($Z = 8$). The stable oxygen nuclides span a range from ${}_{8}^{16}O_{8}$ to ${}_{8}^{18}O_{10}$ The most abundant nuclide ${}_{8}^{16}O_{8}$ has a magic neutron number ($N = 8$) while ${}_{8}^{18}O_{10}$ has a half-filled neutron shell ($N = 20/2 = 10$).

1.B RANDOM SEQUENTIAL DILUTION AND THE LAW OF PROPORTIONATE EFFECT

Ott (1990) published a simple *random sequential-dilution* model to illustrate the type of environmental processes that would yield a log-normal concentration distribution. The random sequential-dilution model is actually a special case of a more general model known as the *Law of Proportionate Effect* first described by the Dutch astronomer and mathematician K.C. Kapteyn (Kapteyn, 1903).

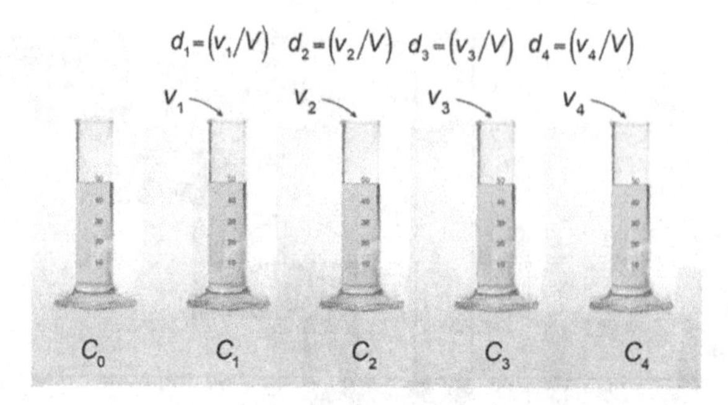

FIG. 1.A.4 A four-step random sequential dilution series begins with an initial concentration C_0 and uses a fixed volume V. The volume transferred in each step v_i is a random value, as is the resulting dilution factor d_i.

Ott described two simple examples that, in the limit of a large number of transformations, consistently yield log-normal concentration distribution. The first example imagines a sequential random dilution, in which the dilution factor ($0 < d_i < 0$) at each step is a random value.

The sequential random dilution (Fig. 1.A.4) begins with an initial concentration C_0 and transfers a random volume v_i in each step. The concentration at each stage C_i is a function of the transferred volume v_i, the concentration of the previous stage C_{i-1}, and the fixed total volume V.

$$C_i = \frac{v_i \cdot C_{i-1}}{V} = d_i \cdot C_{i-1} \tag{1.B.1}$$

The concentration C_n after n random volume transfers is a function of the initial concentration C_0 and the random transfer volumes v_i in all previous steps in the dilution sequence.

$$\begin{aligned} C_1 &= \frac{v_1 \cdot C_0}{V} = d_1 \cdot C_0 \\ C_2 &= d_2 \cdot C_1 = d_2 \cdot (d_1 \cdot C_0) \\ &\vdots \\ C_n &= d_n \cdot C_{n-1} = d_n \cdot (d_{n-1} \cdots (d_2 \cdot (d_1 \cdot C_0))) \end{aligned} \tag{1.B.2}$$

The logarithm of the product (Eq. 1.B.2) yields a sum of logarithmic terms (Eq. 1.B.3). Because each dilution factor d_i is a random number, the logarithm of the dilution factor $\ln(d_i)$ is itself a random number.

$$\ln(C_n) = \ln(C_0) + \sum_{i=1}^{n} \ln(d_i) \tag{1.B.3}$$

Boiroju and Reddy (2012) offer the following simple prescription for generating a standard normal distribution of random numbers.

The simplest way of generating normal variables is an application of the central limit theorem. The central limit theorem is a weak convergence result that expresses the fact that *any sum of many small independent random variables is approximately normally distributed.* Use of the central limit theorem on U(0,1)[21] random variables provide a simple method for closely approximating normal random variates.

Ott (1990) described a random sequential-dilution scenario that yields a log-normal distribution. Consider a fluid-filled chamber (gas or liquid) of volume V and a chemical substance in the chamber at an initial concentration C_0. The chamber has two openings: an entrance and an exit. Through the entrance a random volume of pure fluid v_1 enters, forcing the expulsion of both an equal fluid volume and a quantity of the chemical substance $v_1 \cdot C_0$ through the exit. A random sequential-dilution of the chemical substance from the initial concentration C_0 to some final concentration C_n results when n random pulses of fluid enter the chamber. The chamber could be either a pore in environmental media (e.g., soil, sediment, or aquifer) or an arbitrary volume of environmental media.

The sequential random dilution model (Ott, 1990) restricts the range of each random dilution factor to be less than 1: $0 < d_i < 1$. A more general formulation, known as the Law of Proportionate Effect (Kapteyn, 1903), imposes no such restriction on the random scaling factors: $0 < f_i < +\infty$.

$$\begin{aligned} C_1 &= f_1 \cdot C_0 \\ C_2 &= f_2 \cdot C_1 = f_2 \cdot (f_1 \cdot C_0) \\ &\vdots \\ C_n &= f_n \cdot C_{n-1} = f_n \cdot (f_{n-1} \cdots (f_2 \cdot (f_1 \cdot C_0))) \end{aligned} \tag{1.B.4}$$

As with the random sequential-dilution model, the logarithm of the product (Eq. 1.B.4) yields a sum of logarithmic terms (Eq. 1.B.5). Because each scaling factor f_i is a random number, the logarithm of the factor $\ln(f_i)$ is itself a random number.

$$\ln(C_n) = \ln(C_0) + \sum_{i=1}^{n} \ln(f_i) \tag{1.B.5}$$

Random concentration fluctuations in an arbitrary subvolume within a rock, sediment, soil, aquifer, or water body that is not uniformly mixed will generate a log-normal concentration-frequency distribution where the natural body is the population and the sample is a collection of n aliquots (i.e., arbitrary subvolumes) drawn from the body.

References

Ahrens, L.H., 1954. The lognormal distribution of the elements. Geochim. Cosmochim. Acta 5, 49–73.

Anders, E., Grevesse, N., 1989. Abundances of the elements: meteoritic and solar. Geochim. Cosmochim. Acta 53 (1), 197–214.

Bohlke, J.K., de Laeter, J.R., De Bievre, P., Hidaka, H., Peiser, H.S., Rosman, K.J.R., Taylor, P.D.P., 2005. Isotopic compositions of the elements, 2001. J. Phys. Chem. Ref. Data 34 (1), 57–67.

[21] A continuous uniform distribution U(a, b) is a distribution for which all intervals of the same length on the support are equally probable. The support is defined by the two parameters, a and b, which are its minimum and maximum values.

Boiroju, N.K., Reddy, M.K., 2012. Generation of standard normal random numbers. InterStat, Free eJournal, Date: May 2012, document #003, http://interstat.statjournals.net/YEAR/2012/articles/1205003.pdf.

Borken, W., Matzner, E., 2004. Nitrate leaching in forest soils: an analysis of long-term monitoring sites in Germany. J. Plant Nutr. Soil Sci. 167 (3), 277–283.

Budtz-Jorgensen, E., Keiding, N., Grandjean, P., Weihe, P., 2002. Estimation of health effects of prenatal methylmercury exposure using structural equation models. Environ. Health 1 (1), 2.

Chen, M., Ma, L.Q., Harris, W.G., 1999. Baseline concentrations of 15 trace elements in Florida surface soils. J. Environ. Qual. 28 (4), 1173–1181.

Dapson, R.W., Kaplan, L., 1975. Biological half-life and distribution of radiocesium in a contaminated population of green treefrogs *hyla cinerea*. Oikos 26 (1), 39–42.

Dudka, R., Ponce-Hernandez, S., Hutchinson, T.C., 1995. Current level of total element concentrations in the surface layer of Sudbury's soils. Sci. Total Environ. 162 (2–3), 161–171.

Frost, B.R., 1991. Introduction to oxygen fugacity and its petrologic importance. Rev. Mineral. 25, 1–9.

Goldschmidt, V.M., 1937. Principles of distribution of chemical elements in minerals and rocks. J. Chem. Soc. 655–673.

Hadley, A., Toumi, R., 2003. Assessing changes to the probability distribution of sulphur dioxide in the UK using a lognormal model. Atmos. Environ. 37 (11), 1461–1474.

Haggerty, S.E., 1976. Opaque mineral oxides in terrestrial igneous rocks. Rev. Mineral. 3, Hg101–Hg300.

Heath, D.F., 1967. Normal or log-normal: appropriate distributions. Nature 213, 1159–1160.

Helmke, P.A., 2000. The Chemical Composition of Soils. CRC Press, Boca Raton, FL, pp. B3–B24.

Kapteyn, J.C., 1903. Skew Frequency Curves in Biology and Statistics. Astronomical Laboratory at Groningen, Groningen, Netherlands.

Lide, D.R. (Ed.), 2005. Section 14: Geophysics, Astronomy, and Acoustics, Abundance of Elements in the Earth's Crust and in the Sea, CRC Handbook of Chemistry and Physics. CRC Press, Boca Raton, FL, p. 17.

Limpert, E., Stahel, W.A., 2011. Problems with using the normal distribution – and ways to improve quality and efficiency of data analysis. PLoS One 6 (7), e21403.

Limpert, E., Stahel, W.A., Abbt, M., 2001. Log-normal distributions across the sciences. Bioscience 51 (5), 341–352.

Liu, D., Wang, Z., Zhang, B., Song, K., Li, X., Li, J., Li, F., Duan, H., 2006. Spatial distribution of soil organic carbon and analysis of related factors in croplands of the black soil region, Northeast China. Agric. Ecosyst. Environ. 113 (1–4), 73–81.

Monroe, J., Wicander, R., Hazlett, R., 1992. Physical Geology: Exploring the Earth, first ed. West Publishing Co., St. Paul.

Naidenko, O., Leiba, N., Sharp, R., Houlihan, J., October 18, 2008. Bottled Water Quality Investigation: 10 Major Brands, 38 Pollutants. http://www.ewg.org/research/bottled-water-quality-investigation.

Newcomb, W.D., Rimstidt, J.D., 2002. Trace element distribution in US groundwaters: a probabilistic assessment using public domain data. Appl. Geochem. 17 (1), 49–57.

Ott, W.R., 1990. A physical explanation of the lognormality of pollutant concentrations. J. Air Waste Manage. Assoc. 40 (10), 1378–1383.

Pauling, L., 1929. The principles determining the structure of complex ionic crystals. J. Am. Chem. Soc. 51, 1010–1026.

Shacklette, H.T., Boerngen, J.G., 1984. Chemical Analyses of Soils and Other Surficial Materials of the Conterminous United States. U.S.G.S. Professional Paper No. 1270, 105 pp.

Silva, I.A., Nyland, J.F., Gorman, A., Perisse, A., Ventura, A.M., Santos, E.C.O., De Souza, J.M., Burek, C.L., Rose, N.R., Silbergeld, E.K., 2004. Mercury exposure, malaria, and serum antinuclear/antinucleolar antibodies in Amazon populations in Brazil: a cross-sectional study. Environ. Health 3 (1), 1.

Wieser, M.E., 2007. Atomic weights of the elements 2005. J. Phys. Chem. Ref. Data 36 (2), 485–496.

Winter, M., 2007. Abundance in Earth's Crust. http://www.webelements.com/periodicity/abundance_crust/.

Wollack, E.J., 2012. How Old is the Universe?. United States National Aeronautics and Space Administration, http://map.gsfc.nasa.gov/universe/uni_age.html.

Yang, S.-Y., Chang, W.-L., 2005. Use of finite mixture distribution theory to determine the criteria of cadmium concentrations in Taiwan farmland soils. Soil Sci. 170 (1), 55–62.

Zhang, C., Manheim, F.T., Hinde, J., Grossman, J.N., 2005. Statistical characterization of a large geochemical database and effect of sample size. Appl. Geochem. 20 (10), 1857–1874.

Zinke, P.J., Stangenberger, A.G., Post, W.M., Emanuel, W.R., Olson, J.S., 1984. Worldwide Organic Soil Carbon and Nitrogen Data ORNL/TM-8857; Order No. DE84012041. Oak Ridge Natl. Lab., Oak Ridge, TN

CHAPTER

2

Chemical Hydrology

OUTLINE

2.1 Introduction 40

2.2 Water Resources and the Hydrologic Cycle 40

2.2.1 Water Budgets 40

2.2.2 Residence Time 41

2.2.3 Porous Media: Volume and Mass Relations 44

2.2.4 Hydrologic Units 47

2.3 Saturated Zone Hydrology 48

2.3.1 Groundwater Flow Nets 51

2.4 Vadose Zone Hydrology 55

2.4.1 The Water Retention Curve 58

2.5 Soil Moisture Balance 60

2.5.1 Evapotranspiration Models 61

2.5.2 Soil Water Balance 63

2.6 Solute Transport 65

2.6.1 Retardation-Coefficient Transport Model 65

2.6.2 Solute Transport in the Saturated Zone 67

2.6.3 Solute Transport in the Vadose Zone 68

2.7 Summary 69

Appendices 70

2.A Hydraulic Heads and Gradients 70

2.B Capillary Forces 72

2.C Predicting Capillary Rise 74

2.D Empirical Water Retention and Unsaturated Hydraulic Conductivity Functions 76

2.D.1 Clapp-Hornberger Functions 76

2.D.2 van Genuchten Water Retention Function 78

2.E Day Length Calculation 79

2.F Solute Transport: Plate-Theory Transport Model 80

2.F.1 The Binomial Expansion 81

2.F.2 Plates and Transfers 83

2.F.3 Partitioning and the Movement of the Mobile Agent 84

2.F.4 Retardation-Coefficient and Plate-Theory Models Compared 85

References 85

Soil and Environmental Chemistry
http://dx.doi.org/10.1016/B978-0-12-804178-9.00002-1

2.1 INTRODUCTION

Water is the primary medium that transports nutrients and contaminants through terrestrial environments and a major environmental factor determining biological availability. This chapter examines the properties and behavior of water that is confined in the pores of soil and rocks. The selected topics focus on *chemical hydrology*.

Chemical hydrology focuses on the changes in water chemistry caused by exposure to soil and aquifers. *Residence time* influences pore water chemistry because mineral dissolution and precipitation reactions are often slow relative to the residence time. Water residence time estimates require an understanding of both the water volume in connected hydrologic environments and the water flux between those environments.

Groundwater flow paths control the migration pathway of dissolved and suspended substances. Flow paths above and at the water table are quite different from flow paths below the water table. Changes in aquifer texture alter water flow rates and, potentially, the direction of flow. Dissolved and suspended substances typically migrate more slowly than pore water itself because interactions with the aquifer retard migration.

2.2 WATER RESOURCES AND THE HYDROLOGIC CYCLE

Freshwater accounts for about 0.26% of the total water on Earth's surface (Table 2.1); and 70% of the freshwater is frozen in glaciers and polar icecaps. The remaining 30% of the fresh terrestrial water is either stored above-ground in lakes, rivers, and wetlands, or below-ground confined in the pores of soil and aquifers (water-bearing rock formations). The hydrologic cycle describes the exchanges of water between these reservoirs through precipitation, infiltration, evaporation, transpiration (water vapor loss from vegetation), surface runoff, and groundwater discharge.

2.2.1 Water Budgets

Hydrologists use water budgets to quantify water exchange in the hydrologic cycle. Conservation of mass ensures the total volume does not change at steady state. The example of a lake surrounded by a catchment (watershed) serves to illustrate how a water budget is used. At steady state the lake volume does not change, provided we neglect daily and seasonal fluctuations:

$$\frac{\Delta V}{\Delta t} = 0$$

Steady state does not mean that the lake and its catchment are static. Water enters the lake through mean annual precipitation $\overline{P}\,[\mathrm{m^3\,y^{-1}}]$, groundwater inflow $\overline{G}_{\mathrm{in}}\,[\mathrm{m^3\,y^{-1}}]$, and surface inflow (runoff) $\overline{R}_{\mathrm{in}}\,[\mathrm{m^3\,y^{-1}}]$ and is lost through evaporation and transpiration $\overline{ET}\,[\mathrm{m^3\,y^{-1}}]$, groundwater outflow $\overline{G}_{\mathrm{out}}\,[\mathrm{m^3\,y^{-1}}]$, and surface outflow $\overline{R}_{\mathrm{out}}\,[\mathrm{m^3\,y^{-1}}]$. The annual steady-state water budget for the lake is given by the expression (2.1).

$$\overline{P} - \overline{ET} + \left(\overline{G}_{\mathrm{in}} + \overline{R}_{\mathrm{in}}\right) - \left(\overline{G}_{\mathrm{out}} + \overline{R}_{\mathrm{out}}\right) = 0 \tag{2.1}$$

TABLE 2.1 World Water Resources

Water Source	Volume V (km^3)	Percent of Total (%)	Percent of Freshwater (%)
Terrestrial	47,971,610	0.346	
Fresh lakes	91,000	0.007	0.26
Saline lakes	85,400	0.006	
Rivers	2120	0.0002	0.006
Wetlands	11,470	0.0008	0.03
Soils	16,500	0.0001	0.05
Permafrost	300,000	0.022	0.86
Fresh groundwater	10,530,000	0.76	30.1
Saline groundwater	12,870,000	0.093	
Vegetation	1120	0.0001	0.003
Ice caps and glaciers	24,064,000	1.74	68.7
Ocean	1,338,000,000	96.54	
Atmosphere	12,900	0.001	0.04
Total	13,333,956,300	100.00	100.00

From Shiklomanov, I., 1993. World fresh water resources. In: Gleick, P.H. (Ed.), Water in Crisis: A Guide to the World's Fresh Water Resources. Oxford University Press, New York.

A simplified (expression 2.2) annual lake budget combines inflow[1] and outflow to yield net discharge $\overline{Q}\,[m^3\,y^{-1}]$.

$$\overline{P} - \overline{ET} - \overline{Q} = 0 \tag{2.2}$$

A catchment soil water balance (2.3) includes variations in soil water moisture content driven by precipitation and water loss by evapotranspiration, adding a term for soil water storage $\overline{\Delta S}$ and downward drainage from the soil $\overline{D}$. Later, we will learn how to determine soil water storage capacity from soil properties.

$$\overline{\Delta S} = \overline{P} - \overline{ET} - \overline{D} \tag{2.3}$$

2.2.2 Residence Time

Hydrologists use water budgets to estimate two important parameters of water systems: *discharge ratio* $\overline{Q}/\overline{P}$ and *residence time* t_r for water in a particular reservoir. These parameters can be defined on many scales, ranging from the smallest catchment to the global scale. Table 2.2 lists the runoff ratios for the continents.

[1] *Surface discharge* $\overline{R}$ combines surface runoff and stream discharge and excludes groundwater discharge.

TABLE 2.2 Annual Continental Water Budgets and Runoff Ratios

Continent	$\overline{P}$ (mm)	$\overline{R}$ (mm)	$\overline{ET}$ (mm)	$\overline{P}/\overline{R}$
Africa	690	140	550	0.20
Asia	720	290	430	0.40
Australia	740	230	510	0.31
Europe	730	320	410	0.44
North America	670	290	380	0.43
South America	1650	590	1060	0.36

From L'vovich, M.I., 1979. World Water Resources and Their Future. American Geophysical Union, Washington.

The average *global* surface-discharge ratio $\overline{R}/\overline{P}$ is 0.39, meaning evaporation-transpiration losses account for 61% of the global mean annual precipitation. Residence time t_r is the average time a particular hydrologic unit retains water (expression 2.4).

$$t_r = \frac{V}{Q} \tag{2.4}$$

Using the reservoir volumes listed in Table 2.1 and the flow rates in Table 2.3, we can estimate the mean global residence time (Table 2.4) for water in the atmosphere, surface waters, oceans, and subsurface aquifers. These residence times have considerable chemical significance.

For instance, the mean residence time for water in the troposphere t_r is 7.3 days. Chemical transformations in the troposphere are limited by the brief residence time and the lower intensity of ionizing radiation compared to the reactions possible in the stratosphere[2]. Similarly, the residence time in groundwater aquifers is sufficient to chemically weather aquifer minerals

TABLE 2.3 Global Hydrologic Cycle Flow Rates

Processes	Flow Rate Q [km^3 y^{-1}]
Terrestrial precipitation	119,000
Terrestrial evapotranspiration	72,600
Oceanic precipitation	458,200
Oceanic evaporation	504,600
Surface discharge	452,000
Groundwater discharge	1200

[2] The stratosphere contains 4–6 ppm(v) water vapor (Mote et al., 1996) and no liquid water aerosol. Stratosphere gas (e.g., nitrogen oxides NO_x (g), carbon dioxide CO_2 (g) and ozone O_3 (g)) residence time is on the order of 1–2 years.

TABLE 2.4 Mean Global Water Residence Times

Source	Residence Time t_r [y^{-1}]
Atmosphere	0.02
Oceans	2650
Lakes and rivers	4
Groundwater	20,000

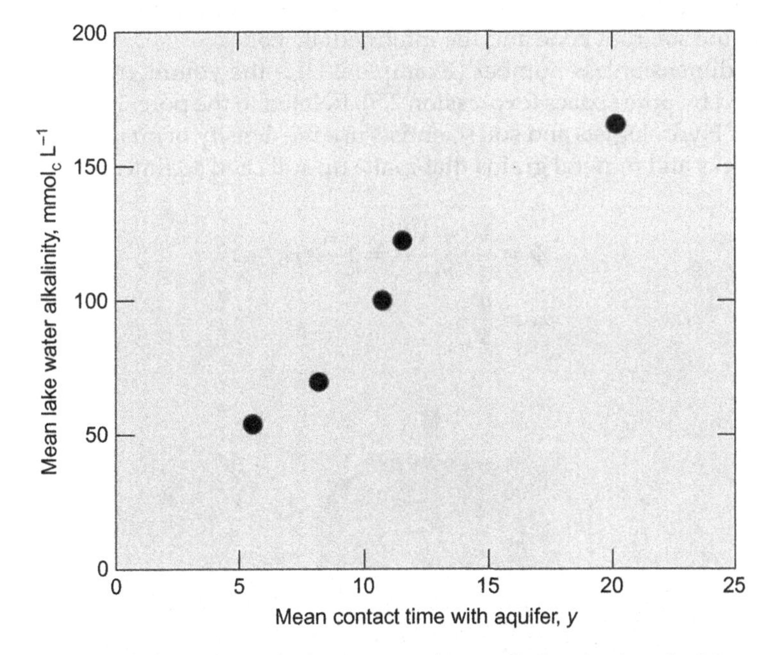

FIG. 2.1 Lake water alkalinity increases the longer groundwater discharging into the lake remains in contact with aquifer formations. *Wolock, D.M., Hornberger, G.M., Beven, K.J., Campbell, W.G., 1989. The relationship of catchment topography and soil hydraulic characteristics to lake alkalinity in the northeastern United States. Water Resour. Res. 25 (5), 829–837.*

to an extent that will not occur in soils or surface water bodies in humid climate zones. Evaporative water loss increases soil water residence time in arid climate zones, severely limiting solute leaching and increasing salinity.

Suppose we want to estimate the residence time for soil moisture, defined as subsurface water within the plant root zone. If we assume terrestrial evaporation-transpiration water flow comes primarily from soil (cf. Table 2.3) and use the global soil moisture volume listed in Table 2.1, the residence time of soil moisture is about 0.2 years, or 70 days.

Fig. 2.1 illustrates the effect that groundwater residence time has on lake alkalinity. A portion of the inflow that sustains lake volume comes from surface (i.e., stream) discharge, and a portion comes from groundwater discharge, the latter accounting for lake water alkalinity.

As the groundwater residence time increases, the alkalinity of the groundwater discharge and, consequently, lake water alkalinity increases. Alkalinity arises from the dissolution of silicate minerals into groundwater permeating the aquifer. The time scale on Fig. 2.1 illustrates the time scale required for mineral dissolution reactions. Notice that a 25-year residence time is still insufficient for alkalinity to reach a steady state.[3]

2.2.3 Porous Media: Volume and Mass Relations

Extending from the land surface downward to the water table is the *vadose zone* (Fig. 2.2). The water table marks the upper limit of the *saturated zone* (Fig. 2.2). The vadose zone is further subdivided into the soil root zone and the intermediate zone.

Porosity ϕ, a dimensionless number (Example 2.1), is the volumetric fraction of a porous medium occupied by pore spaces (expression 2.5). Related to the porosity ϕ is *dry* bulk density ρ_b (2.6) (Box 2.1). Hydrologists and soil scientists use the density of quartz $SiO_2(s)$ to represent the density of rocks and mineral grains that make up soils and sediments: $\rho_s = 2.65\,\mathrm{Mg\,m^{-3}}$.

$$\phi = \frac{V_a + V_w}{V_t} = 1 - (\rho_b/\rho_s) \tag{2.5}$$

$$\rho_b = \frac{m_s}{V_t} \tag{2.6}$$

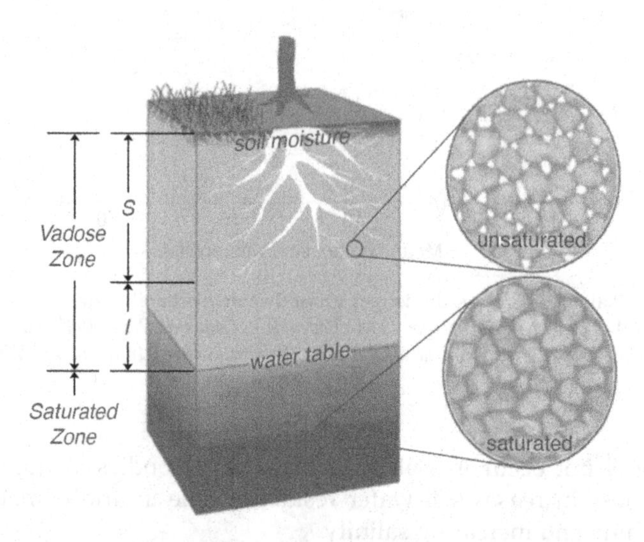

FIG. 2.2 The water table marks the boundary between the unsaturated or vadose zone and the saturated zone. The lower limit of the soil (*S*) zone is defined by the plant rooting depth. Transpiration, water loss by plant root uptake and evaporation from the plant canopy causes the soil water content to vary throughout the growing season. Extending below the rooting depth to the water table is the intermediate zone (*I*), the lower portion of the vadose zone.

[3] The residence time estimates of Wolock et al. (1989) are identified as "soil contact" time but the values plotted in Fig. 2.1 are more appropriate for shallow, unconfined aquifers.

EXAMPLE 2.1

Example Permalink

http://soilenvirochem.net/iqwFm5

Compute the soil porosity using wet bulk density ρ_t and dry bulk density ρ_b measurements.

Soil wet bulk density ρ_t and dry bulk density ρ_b reveal shrink-swell behavior. The simple expression (2.5) relating dry bulk density ρ_b to porosity ϕ cannot be used to calculate soil porosity in the field moist state.

The Ap-horizon of an Antigo soil from Polk County, Wisconsin[4] has the following properties: mass water content $w = 0.282\,\mathrm{Mg\,Mg^{-1}}$ and wet bulk density $\rho_t = 1.28\,\mathrm{Mg\,m^{-3}}$, both measured at $h_t = -33.3\,\mathrm{m}$ and dry bulk density $\rho_b = 1.37\,\mathrm{Mg\,m^{-3}}$. The density of the mineral fraction is $\rho_b = 2.65\,\mathrm{Mg\,m^{-3}}$.

Step 1. Compute the porosity of a dry soil using dry bulk density ρ_b using expression (2.5).

$$\phi_b = 1 - (\rho_b/\rho_s) = 1 - (1.37/2.56) = 0.483$$

Step 2. Compute the total mass of 1 $\mathrm{m^3}$ of moist soil.

$$m_s + m_w = \rho_t = 1.280\,\mathrm{Mg}$$

Step 3. Calculate solid mass for 1 $\mathrm{m^3}$ of moist soil.

$$m_s = \frac{\rho_t}{1+w} = \frac{1.280}{1.282} = 0.998$$

Step 4. Calculate solids volume for 1 $\mathrm{m^3}$ the moist soil.

$$V_s = \frac{m_s}{\rho_s} = \frac{0.998}{2.65} = 0.377\,\mathrm{m^3}$$

Step 5. Calculate the porosity (i.e., volume occupied by air and water) for the moist soil.

$$\phi = 1 - V_s = 1 - (0.377) = 0.623\,\mathrm{m^3\,m^{-3}}$$

Antigo soil at a tension head of $h_t = -33.3\,\mathrm{m}$ has 26% more porosity than when dried to the wilting point.

[4] NCSS Pedon ID 75WI095005, Pedon No. 40A1619, Layer 40A12780.

BOX 2.1

WET AND DRY BULK DENSITY

The National Soil Survey Center[5] measures and reports *wet bulk density* ρ_t (expression 2.7) and *dry bulk density* ρ_b values. The motivation for measuring dry bulk density ρ_b is found in expression (2.6); dry bulk density is the means for estimating soil porosity ϕ.

$$\rho_t = \frac{m_t}{V_t} \tag{2.7}$$

Geotechnical engineers express soil water content as the water-to-solids mass ratio w (expression 2.8), usually quoted as a percentage, where m_W and m_S are the water mass and dry soil mass, respectively (cf. Fig. 2.3). In general, soils shrink when dried. One of the Atterberg limits widely used by geotechnical engineers to measure soil shrinkage is the *swelling limit* w_{SL} (expression 2.9).

$$w = \frac{m_W}{m_S} \cdot 100 \tag{2.8}$$

$$w_{SL} = w - \frac{(V_W - V_d) \cdot \rho_W}{m_S} \cdot 100 \tag{2.9}$$

One parameter soil scientists use to quantify soil shrink-swell potential is the ratio of dry bulk density to moist bulk density (expression 2.10). Quantifying soil shrink-swell potential is, therefore, the motivation for measuring wet bulk density.[6]

$$\text{Shrink-Swell Potential} = \frac{\rho_b}{\rho_t} \tag{2.10}$$

[5] US Department of Agriculture.

[6] The National Soil Survey Handbook recommends quoting wet bulk density ρ_t at the −33 kPa (1/3-bar) moisture tension for clayey and loamy soil materials and −10 kPa (1/10-bar) for sandy materials. A tension of 1 bar is equivalent to 100 kPa.

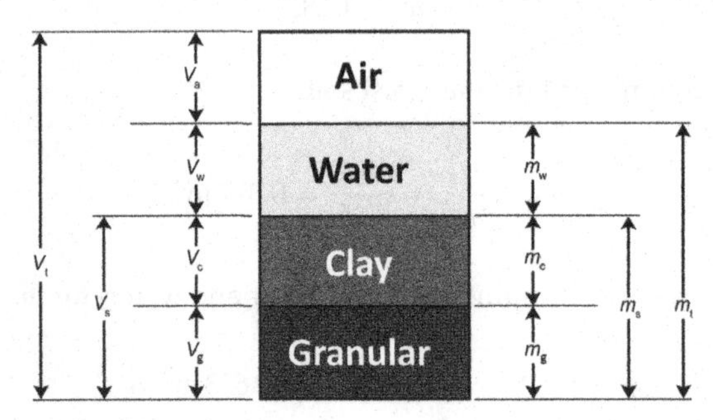

FIG. 2.3 This diagram displays the volume and mass of three phases: air (*a*), water (*w*), and mineral solids (*s*) in porous media (soils, sediments, aquifers, etc.). The mineral fraction is further subdivided into clay (*c*) and granular (*g*) fractions (silt plus sand).

The method for measuring bulk density suggests a simple method for measuring mass water content w (expression 2.11).[7]

$$w = \frac{m_w}{m_s} \tag{2.11}$$

In many cases, soil scientists and hydrologists need to know the water volume in a porous medium (2.12). Expression (2.13) is the relation between mass and volumetric water contents.

$$\theta = \frac{V_w}{V_t} \tag{2.12}$$

$$\theta = w \cdot \left(\frac{\rho_b}{\rho_w}\right) \tag{2.13}$$

Although it may not seem obvious at first, conversion from mass water content w to volumetric water content θ requires the dry bulk density ρ_b of the porous medium and the density of water ρ_w. Dividing a mass of a substance by its density yields the volume, as illustrated in expression (2.13).

Often soil scientists and hydrologists wish to determine water content relative to the water-holding capacity, a proportion called the *degree of saturation* s(expression 2.14).

$$s = \frac{V_w}{(V_w + V_a)} = \frac{V_w}{\phi} \tag{2.14}$$

2.2.4 Hydrologic Units

Hydrologists distinguish geologic formations in the saturated (or phreatic) zone, whose water-holding capacity and hydraulic conductivities are relevant to groundwater flow (Fig. 2.4). Formations that are effectively nonporous, holding negligible amounts of water and unable to conduct water, were formerly called *aquifuges*. Certain formations, usually containing significant amounts of clay, may hold substantial amounts of water, but their low permeability impedes water movement. These formations were formerly called *aquicludes*. Many hydrologists denote any formation with low permeability, regardless of porosity, as *confining bed*. An *aquifer* is any geologic formation, either consolidated rock or unconsolidated sediments, with substantial water-holding capacity and relatively high permeability. An unconfined, or water table, aquifer refers to any aquifer that is saturated in its lower depths but terminates at an unsaturated or vadose zone in its upper reaches, the two zones being separated by a water table. A confined, or artesian, aquifer is overlain by a confining bed. The hydraulic pressure in a confined aquifer can be substantial, causing water in wells drilled into the aquifer to rise above the top of the aquifer.

[7] Soil scientists and geotechnical engineers use w to indicate water content as a mass-fraction (expression 2.11) and as a percentage (expression 2.9).

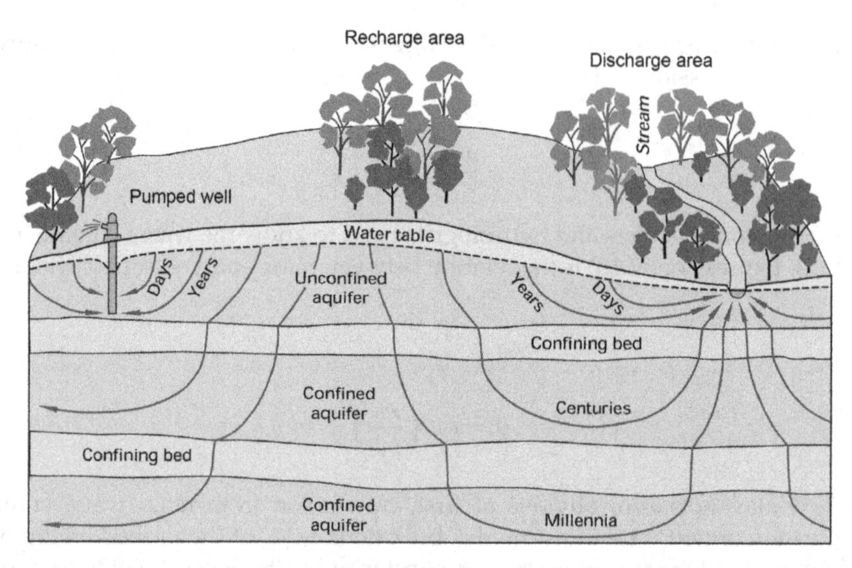

FIG. 2.4 Hydrologists recognize two distinct aquifer formations: unconfined aquifer and confined aquifer. Unconfined aquifers are underlain by a confining bed (formerly known variously as aquiclude, aquifuge, and aquitard). Confined aquifers are overlain and underlain by confining beds. In this chapter the only aquifer of direct interest is the unconfined or "water-table" aquifer. *Winter, T.C., Harvey, J.W., Franke, O.L., Alley, W.M., 1998. Ground Water and Surface Water: A Single Resource. United States Geological Survey Circular 1139. United States Government Printing Office, Denver, CO.*

2.3 SATURATED ZONE HYDROLOGY

For readers unfamiliar with the concepts of *total hydraulic* head h_T, *piezometric* (i.e., submergence) head h_p, and *elevation* head z, Appendix 2.A discusses the formalism used by hydrologists.

Darcy's Law (Box 2.2) can be applied directly to the example shown in Fig. 2.6. The only contribution to the hydraulic gradient Δh between points A and B in Fig. 2.6 is the elevation difference of the water surfaces at the two points. The submergence depth is zero at both points because A is located at the water table and B is located at the water surface in the valley bottom.

BOX 2.2

DARCY'S LAW

Fig. 2.5 shows the experimental apparatus used by Henry Darcy in 1856 to measure water discharge rates Q through unconsolidated porous Earth materials. Darcy discovered the water discharge rate (expression 2.15) is directly proportional to both the cross-sectional area A of the cylinder containing the porous material and the difference in the hydraulic head Δh between the two ends of the cylinder and inversely proportional to the length L of the cylinder.

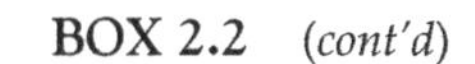

BOX 2.2 *(cont'd)*

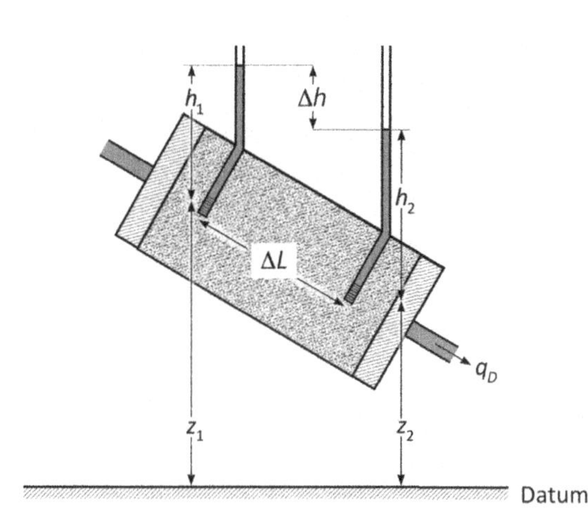

FIG. 2.5 The experimental apparatus Darcy used to measure water flow under a hydraulic gradient through porous material. *Reproduced with permission from Dingman, S.L., 1994. Physical Hydrology. Macmillan Pub. Co./Maxwell Macmillan Canada/Maxwell Macmillan International, New York/Toronto/New York.*

$$Q \propto \frac{A \cdot (h_1 - h_2)}{L} \tag{2.15}$$

$$Q_D\,[\mathrm{m^3\,s^{-1}}] = K_D\,[\mathrm{m\,s^{-1}}] \cdot \frac{A\,[\mathrm{m^2}] \cdot \Delta h\,[\mathrm{m}]}{L\,[\mathrm{m}]} \tag{2.16}$$

The proportionality coefficient K_D (expression 2.16) is called the *hydraulic conductivity* of the porous medium. Darcy could rotate the cylinder to vary the difference in the hydraulic head Δh, read by measuring the water level in two standpipes at either end of the cylinder relative to a datum (cf. Fig. 2.5).

The discharge Q between points A and B depends on the *flow path* distance L between the two points, the hydraulic conductivity of the formation K_D, and the hydraulic gradient Δh. The flow path distance L is not shown but is the distance between points A and B *traced along the water table surface* (*dashed line* in Fig. 2.6) and not the horizontal distance d_{AB}.

The effect of porosity requires the direct measurement of the hydraulic head using either stand tubes (cf. Fig. 2.5), which was how Darcy made the measurement, or piezometers (Fig. 2.7). Piezometers are basically stand tubes inserted into the saturated zone at or below

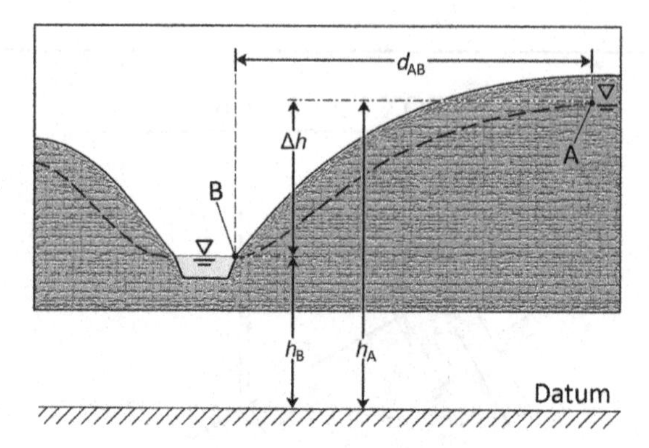

FIG. 2.6 The hydraulic gradient $\Delta h/L$ between two points at the water table of an unconfined aquifer is defined relative to an elevation datum. The elevation of water level at each point relative to a datum is the piezometric head (denoted h_A and h_B), while the difference in the elevation of two water levels is denoted Δh.

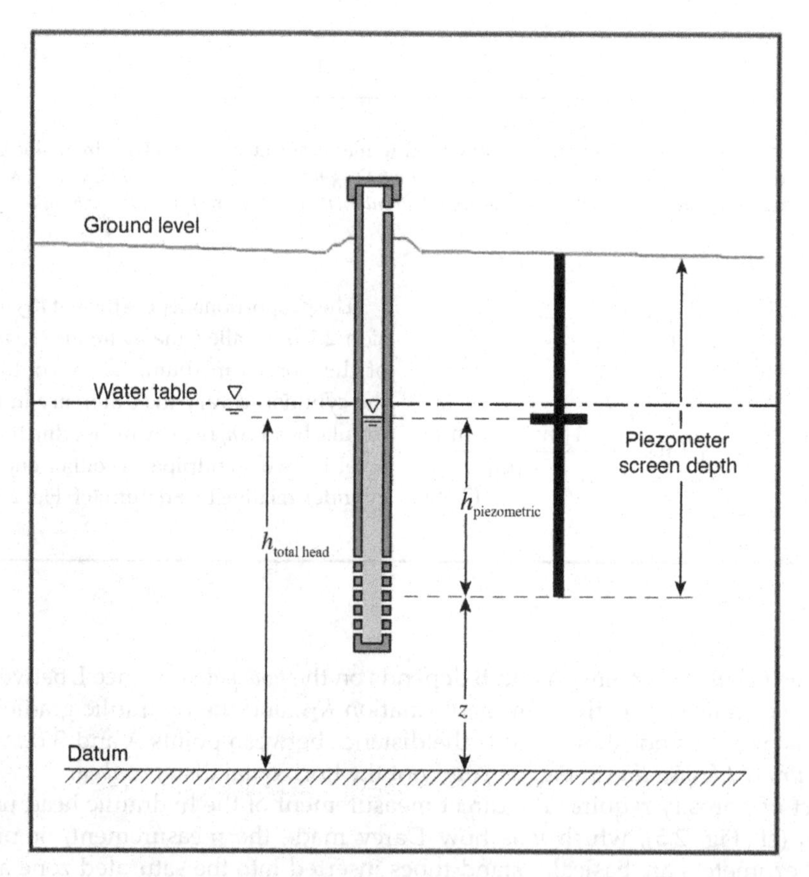

FIG. 2.7 The components and placement of a piezometer used to measure piezometric head h_P hydraulic gradient $\Delta h_P/L$. The screen depth and the water level in the piezometer tube are illustrated on the left and on the right the screen depth is denoted by the *heavy line* and the water level denoted by a *heavy crossbar*.

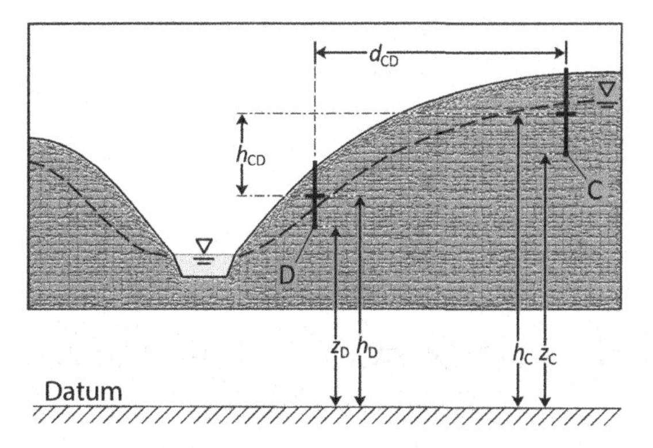

FIG. 2.8 The hydraulic gradient Δh between two points C and D below the water table in an unconfined aquifer requires piezometer measurements. The elevations of piezometer water levels relative to a datum are denoted h_C and h_C, while the difference in the elevation of two piezometer water levels is denoted h_{CD}. The elevations of the piezometer screens relative to a datum are denoted z_C and z_D.

the water table. A piezometer has a screened opening at its lower end that allows water to enter the tube. The water level in the piezometer—measured from the screen or *piezometric bottom*—is the submergence depth h_p at the screened opening.

Fig. 2.8 illustrates the measurement of the hydraulic gradient when both points are below the water table. The hydraulic gradient Δh simply reduces to the difference in the water levels of the two piezometers at points C and D. The relevance of the piezometric head h_p will become clearer when we use it to sketch hydraulic equipotential contours and flow nets in the saturated zone.

2.3.1 Groundwater Flow Nets

Water flow in aquifers is toward a lower hydraulic head h_T. By way of analogy, imagine two gas spheres connected through a tube fitted with a valve. Gas pressure in sphere A is higher than gas pressure in sphere B. Gas will flow from sphere A to sphere B when the valve is opened. This occurs regardless of whether the gas pressure in the spheres is greater than atmospheric pressure ($p_{gauge} > p_{atmospheric}$) or less than atmospheric pressure ($p_{gauge} < p_{atmospheric}$). The submergence pressure in the saturated zone is greater than zero ($p_{submergence} > p_{atmospheric}$) at depths below the water table and zero, by definition, at the water table. In the next section we will learn about water behavior in the unsaturated (vadose) zone, where the hydraulic pressure—and the hydraulic head—are negative. Regardless of whether the hydraulic head is positive or negative, water flows in the direction of lower hydraulic head.

Hydrologists can measure hydraulic head at any point at (cf. Box 2.3) or below the water table using piezometers (cf. Fig. 2.7). The piezometric head h_p is the water level measured to the screen at the bottom of the piezometer. Hydrologists typically insert nests of piezometers at selected locations below the land surface, each piezometer in the nest representing a different depth below the land surface. All measurements are relative to some geodetic datum, whether it is mean sea level (MSL) or some convenient local datum. The total hydraulic head h_T of the piezometer is the height of the water level in the piezometer tube relative to the datum.

BOX 2.3

DETERMINING GROUNDWATER FLOW USING WATER TABLE ELEVATIONS

Fig. 2.9 shows the elevation of the water table in meters (relative to MSL).

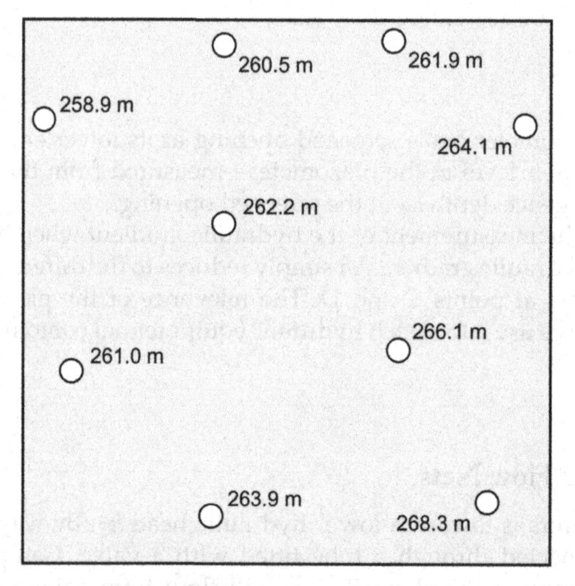

FIG. 2.9 Water table elevations in bore holes drilled into the saturated zone and screened over a wide depth.

The direction of groundwater flow at the water table relative to the water table elevation contours appear in Fig. 2.10. Water flow at the water table is "downhill" at a 90-degree angle to each water table contour.

Continued

BOX 2.3 *(cont'd)*

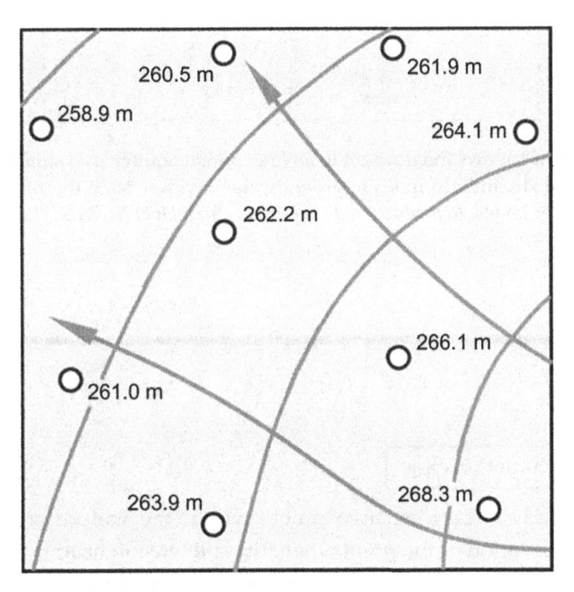

FIG. 2.10 Water table elevation contours and flow paths (with arrows). Each flow path crosses contours at a 90-degree angle to each elevation contour.

These components are also illustrated in Fig. 2.8. The geodetic datum is assigned a value of zero. MSL is a common geodetic datum because land surface elevations are usually reported relative to MSL. The hydraulic gradient Δh is not an absolute value and, as such, is independent of the datum.

Fig. 2.11 is an illustration adapted from Hubbert (1940) showing a land surface and the water table below it. A nest of three piezometers is located to the left of the center, and the water levels in the leftmost and center piezometers (indicated by the crossbar) are the same; both are lower than the rightmost piezometer. Water is flowing from the right piezometer toward the center and left piezometers because the hydraulic gradient Δh decreases in that direction (*cf.* Example 2.2). Although the water levels are the same in the center and left piezometers, the screens are not at the same depth. The *dashed line* running through the bottom of these two piezometers is an equipotential line connecting subsurface points with the same total hydraulic head h_T. The right piezometer indicates a higher hydraulic head and, therefore, lies on a different equipotential contour connecting points at a higher hydraulic head.

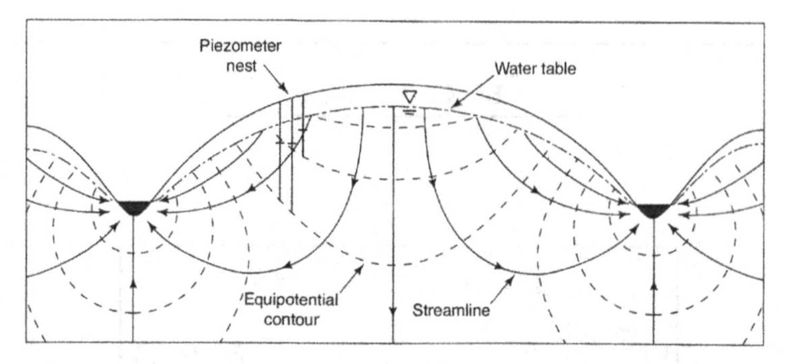

FIG. 2.11 Groundwater flow follows the flow net in an unconfined aquifer. The streamlines indicate the direction of groundwater flow. Vertical streamlines indicate groundwater divides. Note the relation between equipotential contours and piezometer water levels. *Reproduced with permission from Hubbert, M.K., 1940. The theory of ground-water motion. J. Geol. 48 (8), 785–944.*

EXAMPLE 2.2

Example Permalink

http://soilenvirochem.net/6A5pje

Determine hydraulic gradient between two points within the vadose zone using tension head measurements and the elevation of the point where the soil tension head is recorded.

Two tensiometers are inserted into the soil on the University of Wisconsin–Madison campus (elevation 281.9 m above MSL). The tensiometer whose porous cup was placed at a 15-cm depth below the surface recorded a tension head $h_{t(15)} = -435$ cm. The second tensiometer with its porous cup placed at a 30-cm depth below the surface recorded a tension head $h_{t(30)} = -350$ cm.

Step 1. Calculate the total hydraulic head h_T at the recording depth of each tensionmeter.

Rather than using MSL as the geodetic datum, the datum is zero at a 30-cm depth below the surface.

$$h_T = h_t + z$$
$$h_{T(15)} = -435\,\text{cm} + 15\,\text{cm} = -420\,\text{cm}$$
$$h_{T(30)} = -350\,\text{cm} + 0\,\text{cm} = -350\,\text{cm}$$

Step 2. Determine the hydraulic gradient between the two measurement points.

$$\frac{\Delta h_T}{\Delta L} = \frac{\left(h_{T(15)} - h_{T(30)}\right)}{\Delta L}$$
$$\frac{\Delta h_T}{\Delta L} = \frac{(-420 + 350)}{15} = -4.67\,\text{cm}\,\text{cm}^{-1}$$

The soil at a 15-cm depth has a more negative total hydraulic head than soil at a 30-cm depth. Water movement will be upward from the 30-cm depth to the 15-cm depth.

Several other dashed equipotential contours appear in Fig. 2.11 radiating outward from the position where the water table elevation is highest (generally in the uplands) and converging where the water table elevation is lowest (generally in the lowlands). The illustration also plots a set of streamlines that intersect successive equipotential contours at right angles. The system of hydraulic equipotential contours and streamlines constitutes the *groundwater flow net*.

Streamlines diverge from groundwater recharge areas and converge at discharge areas. A groundwater divide occurs where streamlines are vertical, centered on recharge and discharge areas. In reality, Fig. 2.11 illustrates two complete local flow cells, bordered by three groundwater divides, and portions of two more local flow cells on the left and right sides of the illustration. If the illustration is expanded vertically and horizontally to include regional trends in land surface elevation, then the features of regional groundwater flow would appear.

Local groundwater flow is dominated by water table topography and displays a complex pattern of local groundwater recharge and discharge areas. Regional groundwater flow is dominated by the orientation of geologic formations and large-scale changes in elevation. Regional groundwater flow is more strongly expressed in the lower depths of unconfined aquifers, while local groundwater flow is most evident in the vicinity of the water table.

The water level in a piezometer h_T is, in general, not equal to the water table elevation because the piezometer water level reports the hydraulic head h_T *at the piezometer screen opening*. The piezometer water level at point C in Fig. 2.8 is lower than the water table, indicating the hydraulic head h_T^C at the point C is lower than at the water table. This condition tells us water flow is downward at point C because the piezometer screen is located in a groundwater recharge area. The piezometer water level at point D in Fig. 2.8 is higher than the water table, indicating the hydraulic head h_T^D at point D is higher than at the water table. This condition tells us water flow is upward at the point D because the piezometer screen is located in a groundwater discharge area.

In summary, piezometer measurements allow hydrologists to determine the hydraulic head anywhere in an aquifer. Groundwater flow nets are mapped using piezometer nests. Flow nets reveal seepage forces and the direction of groundwater flow. Environmental scientists working with hydrologists can predict the direction and rate of contaminant migration using flow nets.

2.4 VADOSE ZONE HYDROLOGY

Fig. 2.2 illustrates the upper edge of an unconfined aquifer. An unsaturated or vadose zone lies between the land surface and the water table. Lying above the water table is a zone, called the capillary fringe zone, where the pores are mostly saturated but the water pressure is negative (i.e., under tension): $p_{\text{tension}} < p_{\text{atmospheric}}$. The capillary fringe is in direct contact with the water table where the hydraulic pressure equals atmospheric pressure: $p_{\text{hydraulic}} = p_{\text{atmospheric}}$.

The soil-water zone occupies the upper part of the vadose zone, representing that part inhabited by plant roots (cf. Fig. 2.2 soil (S) zone). The soil-water zone undergoes significant changes in water content as plants absorb water through their roots and transpire it into the atmosphere through their leaf canopy and as water infiltrates into the soil following precipitation. An intermediate zone (cf. Fig. 2.2 intermediate (I) zone) often lies between the saturated zone and the soil-water zone—below the plant root zone and above the water table—where little change in water content takes place.

Soil scientists recognize three water contents with special significance to water behavior in the vadose soil-water zone: saturation *s*, field capacity *fc*, and the permanent wilting point *wp*. Saturation is self-explanatory and usually does not persist long in upland sites but can persist at lowland sites where groundwater discharge occurs. Field capacity[8] is the water content of a soil after gravity has drained as much water from the soil as possible.

Soil water at field capacity is held by capillary forces against the force of gravity. A water tension of $p_{tension} = -10\,\text{kPa}$ is equivalent to a tension head of $h_t = -1.02\,\text{m}$ (expression 2.17).

$$h_{fc} = \frac{p_{fc}}{\rho_W \cdot g_0}$$

$$h_{fc} = \frac{(-10 \times 10^3\,\text{kg}\,\text{m}^{-1}\,\text{s}^{-1})}{(10^3\,\text{kg}\,\text{m}^{-3}) \cdot (9.807\,\text{m}\,\text{s}^{-2})} = 1.02\,\text{m} \tag{2.17}$$

The permanent wilting point is also self-explanatory to a degree, it being the water content at which plants can no longer extract water from the soil. Needless to say, the wilting point of desert plants is different from plant species adapted to wetter climates, so the permanent wilting point is defined as a water tension of $p_{tension} = -1500\,\text{kPa}$. A water tension at the permanent wilting point is equivalent to a tension head of $h_t = -153.0\,\text{m}$ (expression 2.18).

$$h_{wp} = \frac{(-1500 \times 10^3\,\text{kg}\,\text{m}^{-2}\,\text{s}^{-1})}{(10^3\,\text{kg}\,\text{m}^{-3}) \cdot (9.807\,\text{m}\,\text{s}^{-2})} = 153.0\,\text{m} \tag{2.18}$$

The *plant-available water content* (AWC) (Example 2.3) in a soil is the amount of water stored in a soil between field capacity and the permanent wilting point (2.19).

$$\theta_{AWC} = \theta_{fc} - \theta_{wp} \tag{2.19}$$

[8] The US Department of Agriculture Natural Resources Conservation Service quotes field capacity soil moisture content at a water tension of −33 or −10 kPa. The −33 kPa water tension value is best for fine (i.e., clayey and loamy) soil textures and −10 kPa is best for sandy materials.

EXAMPLE 2.3

Example Permalink

http://soilenvirochem.net/LgtUtu

Compute the available water-holding capacity AWC of the soil profile.

Expression (2.19) defines the volume available water-holding capacity θ_{AWC}. Field capacity θ_{fc} defines the upper soil moisture limit; gravity drainage prevents soil moisture from exceeding this limit. The wilting point θ_{wp} defines the lower soil moisture limit; plant roots are unable to draw down soil moisture below the permanent wilting point θ_{wp}.

Step 1. Convert mass water contents listed in Table 2.5 to volume water contents.

TABLE 2.5 Soil Dry Bulk Densities ρ_b and Mass Water Contents w at h_{fc} and h_{wp}, Listed by Horizon

Horizon	Depth Range (cm)	ρ_b Mg m^{-3}	w_{fc} Mg Mg^{-1}	w_{wp} Mg Mg^{-1}
A	0–30	1.28	0.28	0.08
B	30–70	1.40	0.30	0.15
C	70–120	1.95	0.20	0.05

The calculation is illustrated for horizon A only. Multiply each mass water content by the bulk-density to water-density ratio.

$$\theta = w \cdot \left(\frac{\rho_b}{\rho_w}\right)$$

$$w_{AWC} = (w_{fc} - w_{wp})$$

$$\theta^{A}_{AWC} = (0.28 - 0.08) \cdot \left(\frac{1.28}{1.00}\right) = 0.256\,\mathrm{m^3\,m^{-3}}$$

Precipitation is generally quoted in depth units rather than volume units. Therefore it is convenient to quote all volumetric moisture values (i.e., precipitation, evapotranspiration, soil moisture storage capacity, etc.) in depth units rather than volume units.

$$d_w\,[\mathrm{cm_{water}}] = \theta \cdot d_t\,[\mathrm{cm_{soil}}]$$

$$d_w/d_t = \theta\,[\mathrm{cm\,cm^{-1}}]$$

$$\theta^{A}_{AWC} = 0.256\,[\mathrm{cm\,cm^{-1}}]$$

Step 2. Convert volume water content θ_{AWC} to the *depth AWC* d_{AWC} for each horizon.

The available water content is not the same for each horizon and, furthermore, each horizon differs in thickness. The calculation is, once again, illustrated for horizon A only. Multiply the A-horizon volumetric AWC by horizon thickness.

$$d_{AWC} = \theta_{AWC} \cdot d_t$$

$$d^{A}_{AWC} = 0.256\,\mathrm{cm\,cm^{-1}} \cdot 30\,\mathrm{cm} = 7.7\,\mathrm{cm}$$

The depth AWC for the entire profile is the cumulative d_{AWC} for each horizon (cf. Table 2.6, column 4).

TABLE 2.6 Profile Available Water Content, Listed by Horizon

Horizon	Thickness cm	θ_{AWC} cm cm^{-1}	d_{AWC} cm
A	30	0.256	7.7
B	40	0.252	10.1
C	50	0.293	14.6

$$d_{AWC} = 7.7\,\text{cm} + 10.1\,\text{cm} + 14.6\,\text{cm} = 32.1\,\text{cm}$$

2.4.1 The Water Retention Curve

The soil tension head h_t quantifies water retention by capillary forces (cf. Appendix 2.C) in the vadose zone. Water rises higher in small-diameter capillaries, which shows that small-diameter pores hold water more strongly than large-diameter pores (cf. Box 2.4). As the water

BOX 2.4

MAXIMUM EFFECTIVE DIAMETER OF WATER-FILLED PORES AT FIELD CAPACITY

Appendix 2.C derives the expression (2.C.7) for predicting the height water will rise h_c in a glass capillary with radius r. The meniscus contact angle α for liquid water/silicate mineral interfaces in air are approximately zero (Guriyanova and Bonaccurso, 2008).

The tension head at field capacity (expression 2.17) is equivalent to the height of a liquid water column in a mineral capillary h_c. The following expression rearranges (2.C.7) so that the effective radius of the largest water-filled pores at field capacity $\bar{r}_{fc}$ is the dependent variable and the tension head at field capacity h_{fc} is the independent variable.

$$\bar{r}_{fc} = \frac{2 \cdot \gamma^{st}}{\rho_W \cdot h_{fc} \cdot g_0}$$

Table 2.7 lists values for the physical parameters required by expression (2.C.7).

$$\bar{r}_{fc} = \frac{2 \cdot 0.0728}{997.044 \cdot 1.02 \cdot 9.80665} = 1.49 \times 10^{-5}\,\text{m}$$

Soil scientists use $h_{fc} \approx -1$ m for coarse texture soils and $h_{fc} \approx -3.3$ m for fine-textured soils. Using these two tension heads at field capacity, the largest water-filled pores in coarse-textured soil is about 15 μm while the largest water-filled pores in fine-textured soil is about 4 μm. Tuller et al. (1999) provide a more detailed discussion of water film thickness in soils, examining geometries beyond the simple round capillary discussed here. Regardless, the results are essentially the same.

The maximum water film thickness ranges from 7.5 μm in coarse-textured soil to about

BOX 2.4 *(cont'd)*

2 μm in fine textured soil. The diameter of a bacterium is about 1 μm, meaning isolated soil bacteria are barely submerged in water at field capacity and bacterial colonies are only partially submerged.

TABLE 2.7 Physical Parameters for Computing h_{fc} of Water at 25 Degree in a Glass Capillary

Parameter	Symbol	Value	Units
Water density	ρ_W	997.044	$kg\,m^{-3}$
Contact angle	α_{water/SiO_2}	0.0	Radians
Water surface tension	γ^{st}	0.0728	$N\,m^{-1}$
Standard gravity	g_0	9.80665	$m\,m^{-2}$

content of a soil decreases, water drains from the largest pores first, followed by the pores with progressively smaller diameters. Consequently, the soil tension head $h_t(\theta)$ varies with the soil water content θ in a manner characteristic of the soil or porous medium (Example 2.4). The *water retention curve* reflects the unique pore size spectrum of each soil or surface material (cf. Appendix 2.D).

Likewise, the soil hydraulic conductivity $K_D(\theta)$ varies with the soil-water content θ because the mean pore velocity decreases with pore diameter, and the maximum pore diameter filled with water decreases with water content.

EXAMPLE 2.4

Example Permalink

http://soilenvirochem.net/3uIXO8

Calculate the water tension head h_t of a moist soil given the texture and volume water content θ.

The volume of water content of a fine-sand soil is $\theta = 0.20\,m^3\,m^{-3}$. Use the empirical Clapp-Hornberger water retention expression in Appendix 2.D to determine the water tension head h_t.

Step 1. Calculate the *degree of saturation* s of a fine-sand soil with a volume water content of $\theta = 0.20\,m^3\,m^{-3}$.

Table 2.D.1 lists the porosity of sand-textured soils: $\phi = 0.395\,m^3\,m^{-3}$.

$$s = \frac{\theta}{\phi} = \frac{0.200}{0.395} = 0.506$$

Step 2. Calculate the water tension of the fine-sand soil using expression (2.D.1) and parameters from Table 2.D.1.

The following two empirical parameters for sand-textured soils appear in Table 2.D.1: $|h_S| = 12.1$ and $b = 4.05$.

$$|h_t(s)| = h_S \cdot s^{-b}$$
$$|h_t(s)| = (12.1) \cdot (0.506)^{-4.05} = 190.5\,\text{cm}$$
$$h_t = -190.5\,\text{cm}$$

2.5 SOIL MOISTURE BALANCE

This section will discuss methods for assessing the soil moisture balance, focusing on the upper vadose zone occupied by plant roots. Water loss from rivers, lakes, reservoirs, and the like occurs exclusively by evaporation from an open water surface. Evaporation play a much smaller role in soil water loss. Plants withdraw water from root zone, transport it through the vascular (i.e., *zylem*) system to the leaf canopy where it evaporates from pores (i.e., *stomata*) on the leaf surface. Water loss through this process is called *evapotranspiration*. Soil water loss occurs, by far, through evapotranspiration.

The hydrologic water balance (expression 2.3) combines groundwater and surface discharge into a single term $\overline{D}$ with the understanding surface discharge (or runoff) is synonymous with stream discharge. This section will focus on soil storage $\overline{\Delta S}$.

Local weather data is readily available from the *National Oceanic and Atmospheric Administration* (NOAA)[9] and local surface discharge data is available from the *US Geologic Survey USGS*[10] and the *United Nations Global Environment Monitoring System Water Programme (GEMS)*[11]:

NOAA Climate Data Online

http://ncsslabdatamart.sc.egov.usda.gov/querypage.aspx

USGS National Water Information System

http://maps.waterdata.usgs.gov/mapper/index.html?

Most locations, often from multiple weather stations, provide monthly mean and daily precipitation—including snowfall, snow depth, and water equivalent of snow on the ground.

[9] US Department of Commerce.

[10] US Department of Interior.

[11] GENStat: http://gemstat.org/.

Local soils data is readily available from the *Natural Resource Conservation Service* (NRCS).[12] These soils data include soil water content at field capacity and wilting point (mass-basis w or volume basis θ) and bulk density ρ_b for all horizons of a particular soil series.

NRCS Web Soil Survey

http://websoilsurvey.sc.egov.usda.gov/App/WebSoilSurvey.aspx

A reasonably complete and accurate local soil water balance also requires a local monthly or daily estimate of evapotranspiration, the topic of the following section. Before attempting to simulate the local soil water balance using the cited databases let's pause to consider why a local soil water balance is important to soil chemists.

First, and foremost, a local soil water balance provides a reasonable estimate of soil-water residence time. Fig. 2.1 provides ample evidence that water residence time has a significant impact on groundwater and surface water chemistry. Estimates of soil water residence time are on the order of weeks to months in climate zones where annual precipitation exceeds evapotranspiration. Water residence time increases substantially in semi-arid and arid climate zones where annual precipitation is approximately equal to or less than evapotranspiration.

Second, local water balance is essential for assessing the likelihood that chemical leaching is confined to the soil zone or extends beyond the intermediate zone (which has no water storage capacity), reaching the water table. Soil moisture in many humid climate zones tends to fluctuate at or below field capacity with seasonal drainage from the soil through the intermediate zone to the water table. Arid and semi-arid soil experience a profoundly different moisture regime; soil moisture fluctuate at or above the wilting point. Drainage from the soil through the intermediate zone to the water table is limited to intense rain storms.

2.5.1 Evapotranspiration Models

Potential evapotranspiration PET models fall into two groups: temperature-based and radiation-based. NOAA weather data provides suitable data for temperature-based PET models. Radiation-based models require data typically not available from most NOAA weather stations. It is possible to estimate certain parameters required by radiation-based models using daily percent of possible sunshine and sky cover combined with day length computed form longitude and latitude, temperature-based PET models are sufficiently accurate to illustrate the modeling of local soil water balance.

Our discussion does not include local PET model calibration. This is accomplished by comparing annual or monthly mean precipitation and surface discharge from a local catchment (cf. USGS National Water Information System). A 10- or 20-year average provides accurate mean annual or monthly evapotranspiration ET that can calibrate either temperature-based or radiation-based PET models.

[12] US Department of Agriculture.

2.5.1.1 Thornthwaite Potential Evapotranspiration Model

Thornthwaite (1948) published the original PET equation, a temperature-based model (Tables 2.8 and 2.9). Thornthwaite assumed PET is zero if the temperature (monthly or daily mean) falls below 0°C.

Monthly Mean PET

$$\mathrm{PET_m}\ [\mathrm{cm\ month^{-1}}] = 1.6 \cdot \left(\frac{L_\mathrm{d}}{12}\right) \cdot \left(\frac{N}{30}\right) \cdot \left(\frac{10 \cdot \overline{T}_\mathrm{m}}{I}\right)^{a(I)} \tag{2.20}$$

Monthly Mean Heat Index

$$I = \sum_m \left(\frac{\overline{T}_\mathrm{m}}{5}\right)^{1.514} \tag{2.21}$$

Exponent a(I)

$$a(I) = a_0 \cdot I^0 + a_1 \cdot I^1 + a_2 \cdot I^2 + a_3 \cdot I^3 \tag{2.22}$$

TABLE 2.8 Parameters for Computing PET Using the Thornthwaite (1948) Equation

Parameter	Symbol	Units
Day length	L_d	$\mathrm{h\ day^{-1}}$
Days in the month	N	$\mathrm{Days\ month^{-1}}$
Mean temperature	$\overline{T}_\mathrm{m}$	°C
Heat index	I	Unitless
Exponent	$a(I)$	Unitless

The mean temperature T is the monthly mean if PET is simulated on a monthly basis or the daily mean if PET is simulated on a daily basis. The method for calculating day length L_d appears in Appendix 2.E.

TABLE 2.9 Coefficients for Exponent $a(I_\mathrm{m})$ Appearing in Expressions (2.20), (2.22)

Coefficient	Value
a_0	0.49239
a_1	1.7921×10^{-2}
a_2	7.711×10^{-5}
a_3	6.751×10^{-7}

2.5.2 Soil Water Balance

If precipitation exceeds evapotranspiration water loss the excess water will either run off or drain out the bottom of the soil profile. Since evapotranspiration cannot remove water from the intermediate zone any water entering the soil when the soil moisture content exceeds field capacity must drain all the way to the water table, regardless of the depth to water table. The distance traveled by the wetting front is crucial for computing chemical transport through the vadose zone.

Detailed soil water balance models attempt to predict the effect of land use and soil properties (i.e., soil texture) on surface run-off and infiltration. These models must take into account seasonal precipitation intensity. All PET models assume evapotranspiration water loss is zero when temperatures are at or below freezing. A corollary to this assumption would allow all precipitation to infiltrate when as long as the mean monthly air temperature is above freezing—effectively assuming the mean monthly air temperature is a reasonable estimate of soil temperature—and require all precipitation during the interval when the mean monthly air temperature is at or below freezing to runoff.

Soil moisture storage capacity will increase so long as water loss by evapotranspiration exceeds precipitation. A daily soil water balance would determine whether or not precipitation on a particular day will exceed the soil moisture storage capacity. If the soil water balance is updated on a monthly basis the simulation will identify those months where chemical leaching is most likely. If the soil water balance is updated on a daily basis the simulation will identify the risk of deep chemical leaching by linking daily precipitation to current water storage capacity.

Example 2.5 serves several purposes. Readily available weather records (temperature and precipitation, both monthly and daily means) along with essential soil properties (bulk density and soil moisture contents sufficient to define soil moisture capacity) make ET and soil water balance estimates possible most locations in the United States and probably many locations worldwide.

EXAMPLE 2.5

Example Permalink

http://soilenvirochem.net/5325Je

Compute the soil water balance and changes in soil moisture storage by taking into account precipitation and water-loss by evaporation during a specified time interval.

Table 2.10 lists all soil water balance parameters for the Oxford Experiment Station. Vegetation at the North Carolina Agricultural Experiment Station in Oxford used an average $ET_{\mathrm{d}} = 4.2\,\mathrm{mm\,d^{-1}}$ during May 2008.[13]

The soils at the Oxford Experiment Station have a depth of $d_{\mathrm{soil}} \approx 60\,\mathrm{cm}$ and a field-capacity soil moisture content $\theta_{\mathrm{fc}} = 0.437\,\mathrm{mm\,mm^{-1}}$. At the beginning of May 2008, the soil in the watershed held an average moisture content of $\theta_i = 0.412\,\mathrm{mm\,mm^{-1}}$.

TABLE 2.10 Soil Water Balance Parameters at the North Carolina Agricultural Experiment Station for May 1, 2008

Parameter	Symbol	Value	Units
Total monthly precipitation	P	66.9	mm
Total monthly evapotranspiration	ET	124.6	mm
Soil moisture at FC	θ_{fc}	0.437	mm mm^{-1}
Beginning soil moisture	θ_i	0.412	mm mm^{-1}

Step 1. Compute the soil moisture storage capacity ΔS_i on May 1, 2008.

The beginning soil moisture storage capacity is the difference between the moisture content at field capacity θ_{fc} and the soil moisture content measured on May 1, 2008 multiplied by the soil depth.

$$\Delta S_i = (\theta_{fc} - \theta_i) \cdot d_{soil}$$
$$\Delta S_i = (0.437 - 0.412) \cdot 600\,\text{mm} = 15.0\,\text{mm}$$

Step 2. Compute the soil moisture storage capacity ΔS_f on June 1, 2008.

Soil water loss by evapotranspiration $ET_m = 124.6\,\text{mm}$ exceeded precipitation at the Oxford Experiment Station during May 2008 (cf. Table 2.10). The soil moisture storage capacity increased to $\Delta S_f = 72.7\,\text{mm}$ by June 1, 2008.

$$\Delta S_f = (ET - P + \Delta S_i)\ \text{mm}$$
$$\Delta S_f = (124.6 - 66.9 + 15.0)\ \text{mm} = 72.7\,\text{mm}$$

Step 3. Estimate the mean soil moisture content θ_f on June 1, 2008.

The ending soil moisture content θ_f on June 1, 2008 is found by dividing the ending soil moisture storage S_f by the soil depth d_{soil}.

$$\theta_f = \theta_i + \left(\frac{ET - P}{d_{soil}}\right)$$
$$\theta_f = 0.412\,\text{mm mm}^{-1} + \left(\frac{-57.7\,\text{mm}}{600\,\text{mm}}\right) = 0.316\,\text{mm mm}^{-1}$$

The North Carolina Agricultural Station at Oxford reported an average soil moisture content of $\theta_f = 0.350\,\text{mm mm}^{-1}$ on May 30, 2008.

[13] North Carolina Climate Retrieval and Observation Network of the Southeast (CRONOS) Database.

The risk of chemical leaching during May 2008 at the Oxford Experiment Station (Example 2.5) is minimal however it is very likely that earlier in the growing season when soil moisture levels approach field capacity, as they often do during late winter and early spring, the risk is much higher because precipitation at that critical time could exceed soil moisture storage capacity allowing the wetting front to travel through the entire vadose zone to reach the water table.

As we will see in the following section, the rapid decline in soil organic matter content with soil depth (to say nothing of the very low levels typical of the intermediate zone) eliminate a key barrier limiting the transport of organic chemicals. The transport risk is further aggravated by the dramatic increase in the wetting front depth in soils with little or no moisture storage capacity.

2.6 SOLUTE TRANSPORT

2.6.1 Retardation-Coefficient Transport Model

Water percolating downward through the vadose zone or groundwater flow through the saturated zone carries dissolved chemical agents. Most agents do not travel at the same velocity as the water because adsorption to soil, residuum, sediments, or aquifer retard their movement. Bouwer (1991) derived a simple expression for the *retardation coefficient* R_f that first appeared as an empirical expression for solute transport through an adsorbent (Vermeulen and Hiester, 1952; Higgins, 1959). This section recapitulates the derivation by Bouwer and evaluates its limitations.

The retardation coefficient R_f expression derived by Bouwer is applicable to both saturated and unsaturated flow. Bouwer lists a single assumption: the timescale for water flow must be long relative to the timescale of the adsorption reaction that binds the agent to the solid phase of the soil or aquifer. In this assumption the soil-water partition constant $K^{\ominus}_{s/w}$ represents the adsorption reaction.[14]

The retardation coefficient R_f is defined as groundwater velocity divided by agent velocity or, alternatively, the distance groundwater travels L_W divided by the distance the dissolved chemical agent travels L_A in some time interval t (expression 2.23) and Fig. 2.12, right). This distance L_W can be written (expression 2.24) using Darcy's Law (expression 2.16).

$$R_f = \frac{L_W}{L_A} \tag{2.23}$$

$$L_W = \left(\frac{Q_D}{A}\right) \cdot t = q_D \cdot t \tag{2.24}$$

Consider a cylindrical volume of soil or aquifer centered on a streamline of water flow (Fig. 2.12). The base of the flow cylinder has an arbitrary area A. The *total* amount n_A [mol] of chemical agent A entering the flow-cylinder volume during a time interval t equals the product of the dissolved concentration c_A of agent A and the volume swept out by water

[14] The chemical basis for adsorption by partitioning and site-limited adsorption are covered in Chapter 8.

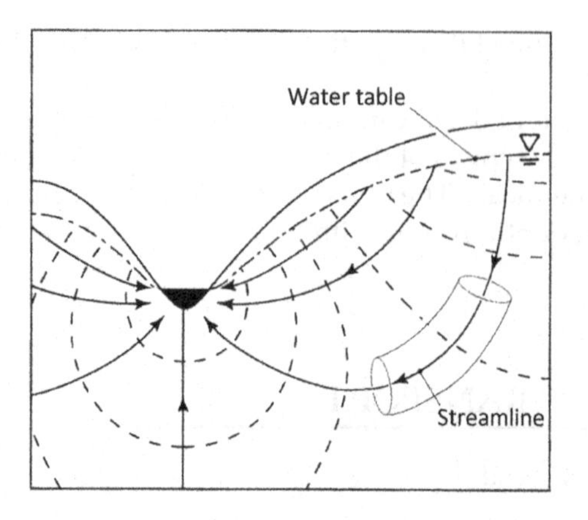

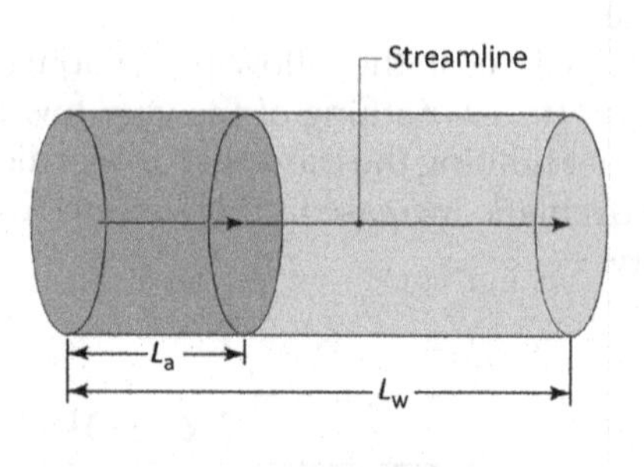

FIG. 2.12 Flow paths (i.e., *streamlines*) trace groundwater movement through an unconfined aquifer (left). Each streamline defines the axis of a cylindrical volume element. The *retardation-coefficient* model represents interactions between the a dissolved chemical agent and the aquifer retarding its movement, resulting in a shorter travel distance L_A than groundwater L_W during time interval t.

flowing along the streamline, proportional to L_W in Fig. 2.12. The volume of water content θ makes this expression valid for both saturated and unsaturated flow.

$$n_A = (L_W \cdot A) \cdot \theta \cdot c_A \tag{2.25}$$

The dissolved (i.e., *mobile*) agent in the flow cylinder is the product of the agent concentration c_A and the water volume V_W in the flow cylinder segment with length L_A in Fig. 2.12.

$$c_A \cdot V_W = c_A \cdot (L_A \cdot A) \cdot \theta \tag{2.26}$$

$$c_A \cdot V_W \leq n_A$$

The adsorbed (i.e., *stationary*) amount n_A^σ is simply the difference between the *total* amount in the flow cylinder n_A and the *mobile* amount in expression (2.26).[15]

$$n_A^\sigma = n_A - (c_A \cdot V_W)$$

$$n_A^\sigma = (L_W - L_A) \cdot A \cdot \theta \cdot c_A \tag{2.27}$$

The aquifer (adsorbent) mass m_s (cf. Fig. 2.3) is the product of the volume accessible to agent A during time interval t and the bulk density ρ_b of the aquifer.

$$m_s = L_W \cdot A \cdot \rho_b \tag{2.28}$$

[15] Designating the *dissolved* agent as *mobile* and the aquifer-bound agent as *stationary* allows us to use a common set of symbols for the retardation-coefficient and plate-theory models (Appendix 2.F) of solute transport.

Retardation of chemical agent A results from its adsorption from solution by the aquifer (cf. expressions 2.23[16] and 2.24[17]; Chapter 8).

$$\left(n_{A}^{\sigma}/m_{s}\right) = K_{s/w}^{\ominus} \cdot c_{A} \tag{2.29}$$

$$\left(n_{A}^{\sigma}/m_{s}\right) = f_{oc} \cdot K_{oc/w}^{\ominus} \cdot c_{A} \tag{2.30}$$

The total amount of agent A immobilized through adsorption by solids in the flow cylinder defined by L_{W} is given by expression (2.31). Substituting expression (2.29) into expression (2.31) and canceling duplicate terms yields expression (2.32).

$$\frac{n_{A}^{\sigma}}{m_{s}} = \frac{(L_{W} - L_{A}) \cdot A \cdot \theta \cdot c_{A}}{L_{W} \cdot A \cdot \rho_{b}} \tag{2.31}$$

$$\frac{n_{A}^{\sigma}/m_{s}}{c_{A}} = K_{s/w}^{\ominus} = \frac{(L_{W} - L_{A}) \cdot \theta}{L_{W} \cdot \rho_{b}} \tag{2.32}$$

Rearranging expression (2.32) and substituting expression (2.23) for the distance ratio L_{W}/L_{A} yields the retardation coefficient R_{f} expression (2.33).

$$R_{f} = \frac{L_{W}}{L_{A}} = \left(1 + \left(\frac{K_{s/w}^{\ominus} \cdot \rho_{b}}{\theta}\right)\right) = \left(1 + \left(\frac{f_{oc} \cdot K_{oc/w}^{\ominus} \cdot \rho_{b}}{\theta}\right)\right) \tag{2.33}$$

The retardation coefficient R_{f} applies to transport in the saturated zone and vadose zone. The saturated-zone water (i.e., mobile phase) content $\theta \equiv \phi$. The vadose-zone water content θ can vary from saturation ϕ to the wilting point θ_{wp}.

The *retardation-coefficient* model assumes the dissolved agent concentration c_{A} is uniform everywhere behind the solute front and zero in advance of the solute front (cf. Fig. 2.12). The *plate-theory* model (Appendix 2.F) uses multiple sequential partitioning to represent solute front dispersion.

2.6.2 Solute Transport in the Saturated Zone

Example 2.6 shows how to estimate the transport distance of 1,2-dibromoethane in a groundwater aquifer.

[16] Partition constant for adsorption by whole soil or aquifer.

[17] Partition constant for adsorption by organic carbon fraction in soil or aquifer where f_{oc} is the organic carbon mass fraction.

EXAMPLE 2.6

Example Permalink

http://soilenvirochem.net/gaM7u5

Estimate the transport distance of 1,2-dibromoethane (CAS Registry Number 106-93-4) in a groundwater aquifer.

The aquifer at Brookhaven National Laboratory (Upton, NY) is a compacted glacial till with a porosity of $\phi = 0.25$, a specific discharge of $q_D = 1.22\,\text{m}\,\text{day}^{-1}$, and an organic carbon content of $f_{oc} = 0.0041\,\text{Mg}_{oc}\,\text{Mg}_s^{-1}$.

Step 1. Using a organic-carbon partition constant $K^{\ominus}_{oc/w} = 21.90$ and expression (2.33), estimate the 1,2-dibromoethane retardation coefficient R_f for the Long Island glacial till aquifer.

$$\rho_b = \rho_s \cdot (1 - \phi)$$

$$\rho_b = \left(2.65\,\text{Mg}\,\text{m}^{-3}\right) \cdot (0.75) = 1.99\,\text{Mg}\,\text{m}^{-3}$$

$$R_f = 1 + \left(\frac{f_{oc} \cdot K^{\ominus}_{oc/w} \cdot \rho_b}{\phi}\right)$$

$$R_f = 1 + \left(\frac{0.0041 \cdot 21.90 \cdot 1.99}{0.25}\right) = 1.71$$

Step 2. Using the specific discharge of the aquifer $q_D = 1.22\,\text{m}\,\text{day}^{-1}$ and the retardation coefficient R_f estimated in Step 1, estimate the distance L_A 1,2-dibromoethane will have traveled in the 8 years between 1999 and 2007.

$$R_f = \frac{L_W}{L_A} = \frac{q_D \cdot t}{L_A}$$

$$L_A = \frac{q_D \cdot t}{R_f}$$

$$L_A = \frac{\left(1.22\,\text{m}\,\text{day}^{-1}\right) \cdot \left(365\,\text{day}\,\text{year}^{-1}\right) \cdot (8\,\text{year})}{1.71} = 2085\,\text{m}$$

2.6.3 Solute Transport in the Vadose Zone

Example 2.7 shows how to compute the leaching depth of a pesticide applied to the surface of a soil following a spring rainstorm.

EXAMPLE 2.7

Example Permalink

http://soilenvirochem.net/RW3SHs

Compute the leaching depth of a pesticide applied to the surface of a soil following a spring rainstorm.

The organophosphate insecticide *Phosmet* (CAS Registry Number 732-11-6) is applied to the surface of an Antigo soil (Langlade County, Wisconsin[18]). The *Phosmet* retardation factor is $R_f = 52.6$ for this Antigo soil. The Antigo soil, which has a field-capacity water content of $\theta_{fc} = 0.223$ in the surface horizon, has an initial water content of $\theta = 0.09$ when precipitation begins to fall. A total of 3.27 cm of precipitation falls during the storm.

Step 1. Estimate the wetting depth of the rainstorm assuming no runoff.

The precipitation wetting depth is determined by the soil moisture storage capacity. The Antigo surface horizon has a water storage capacity of $\Delta S = 0.223 - 0.090 = 0.133\,\text{cm cm}^{-1}$ at the start of the rainstorm.

$$L_W = \frac{P}{\Delta S} = \frac{P}{(\theta_{fc} - \theta)}$$

$$L_W = \frac{3.27\,\text{cm}}{0.133\,\text{cm cm}^{-1}} = 24.6\,\text{cm}$$

Step 2. Use the retardation coefficient for *Phosmet* in the surface horizon of the Antigo soil to estimate the *Phosmet* leaching depth.

The retardation coefficient for *Phosmet* in the surface horizon of the Antigo soil is $R_f = 52.5$.

$$R_f = \frac{L_W}{L_A}$$

$$L_A = \frac{L_W}{R_f} = \frac{24.6\,\text{cm}}{52.6} = 0.5\,\text{cm}$$

[18] National Cooperative Soil Survey Pedon ID S1981WI067340.

2.7 SUMMARY

Residence time, water-holding capacity, groundwater flow paths and rates, and transport models all contribute to our capacity to predict the impact of soil and aquifer properties on water chemistry and solute movement. This chapter introduced water budgets as a means to estimate groundwater residence time, a key factor in determining groundwater chemistry. Changes in hydraulic head within the saturated zone determine groundwater flow paths that vary at several spatial scales, although environmental chemists are most interested in local flow paths. Variation in water content in the vadose zone influences water retention by capillary forces (Appendix 2.B) and unsaturated hydraulic conductivity (Appendix 2.D). Darcy's Law (Box 2.2) provides a means to estimate pore water velocity that, in turn, influences solute transport rates. Finally, this chapter introduced two elementary transport models that simulate retarded solute movement along flow paths within the vadose and saturated zones caused by adsorption by the aquifer.

APPENDICES

2.A HYDRAULIC HEADS AND GRADIENTS

Hydraulic pressure is central to water flow dynamics in the saturated aquifers and open water. Hydraulic pressure p increases with depth below the water surface, the result of the mass of the water column. The downward-acting force per unit area in a water column is simply the water mass of column times the acceleration due to gravity g_0.

Hydrologists use the term *submergence* or *piezometric* head h_p when referring to the submergence depth below a free-water surface.[19] The mass of the water column at a depth h_p is simply water density ρ_w times the depth h_p. Expressions (2.A.1), (2.A.2) give the relation between the hydraulic pressure p and piezometric head h_p.

$$p = \frac{F\,[N]}{A\,[m^{-2}]}$$

$$p = \frac{m_w\,[Mg] \cdot g_0\,[m\,s^{-2}]}{A\,[m^{-2}]}$$

$$p = \left(\frac{m_w}{A \cdot h_p}\right) \cdot g_0 \cdot h_p$$

$$p = \rho_w \cdot g_0 \cdot h_p \tag{2.A.1}$$

$$h_p = \frac{p}{\rho_w \cdot g_0} \tag{2.A.2}$$

Expression (2.A.3) is the gravitational potential energy $E_{P(g)}$ $[kg\,m^2\,s^{-2}]$ of a mass of water m_w at an elevation z above a geodetic datum. Notice the units of $(E_{P(g)}/V_w)$ in expression (2.A.4) are identical to pressure units appearing in expression (2.A.1): $[N\,m^{-2}]$.

$$E_{P(g)} = (m_w\,[kg]) \cdot \left(g_0\,[m \cdot s^{-2}]\right) \cdot (z\,[m])$$

$$E_{P(g)} = (\rho_w \cdot V_w) \cdot g_0 \cdot z \tag{2.A.3}$$

$$\rho_w \cdot g_0 \cdot z = E_{P(g)}/V_w \tag{2.A.4}$$

$$z = \frac{E_{P(g)}/V_w}{\rho_w \cdot g_0} \tag{2.A.5}$$

Any change in elevation z represents a change in potential energy $E_{P(g)}$. The Bernoulli equation for fluid dynamics includes two primary contributions to the total hydraulic head

[19] Some hydrologists refer to this as the *pressure head*.

h_T in groundwater systems: the submergence depth h_p (expression 2.A.2) and the elevation head z that accounts for changes in the potential energy of water with elevation (expression 2.A.5).

Hydrologists and soil scientists measure hydraulic pressure relative to atmospheric pressure. The hydraulic head is zero at the surface of free water (Fig. 2.A.1, points A and B; Fig. 2.A.1, points R and S). The water table in an unconfined aquifer (Fig. 2.A.1, point R) is, by definition, the point where the piezometric head h_p is zero.

Fig. 2.A.1 also illustrates the difference between submergence depth h_p and elevation head z. A pressure measurement taken at the water surface above a dam (Fig. 2.A.1, point A) and at the water surface below a dam (Fig. 2.A.1, point B) both record atmospheric pressure, both of read zero on a pressure gauge. The water at point A above the dam is at a higher potential energy due to the difference in elevation z_{AB} between points A and B.

The total hydraulic head at point C includes contributions from both submergence depth h_{AC} and elevation head z_{BC}. As the submergence depth h_{AC} below the water surface increases the elevation z_{BC} decreases. Consequently, the total hydraulic head h^T remains unchanged relative to the total hydraulic head at point A.

Now, consider the saturated zone at and below the water table in Fig. 2.A.1 (points R and T). Pressure measurements taken at the water table (Fig. 2.A.1, point R) and at the water surface of

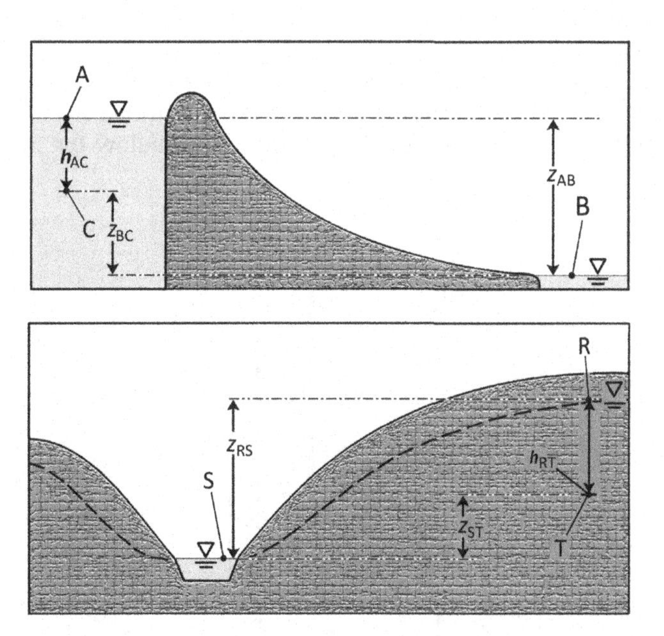

FIG. 2.A.1 Impounded water (upper) and an unconfined aquifer (lower) illustrating reference points and components of the total head (hydraulic and elevation). The water level elevation of impounded water at point A is denoted z_A. The water table elevation at point R is denoted z_R, while the elevation difference of two points A and B is denoted z_{AB}.

a stream in the valley bottom (Fig. 2.A.1, point S) both record atmospheric pressure. The water at point R at the water table up slope is at a higher potential energy due to the difference in elevation z_{RS} between point R and point S.

The submergence pressure at point T (below the water table in an unconfined aquifer) is greater than zero and equal weight of the overlying pore water. Unlike the case shown in Fig. 2.A.1 (point C), the piezometric head h_P at point T, depends on the porosity of the aquifer and, therefore, must be measured rather than inferred from the depth below the water table. The total hydraulic head h_T at point T includes contributions from both the piezometric head h_P and elevation head z.

2.B CAPILLARY FORCES

Capillary forces—interactions between thin films of water and mineral surfaces—dominate water behavior in the vadose zone. The basic principles behind capillary forces are illustrated by a simple experiment in which the tip of a small-diameter glass tube is inserted into a dish of water (Fig. 2.B.1).

Water is drawn into the tube as soon as the tip of the capillary tube touches the water surface, rising to a height h_c (point (a)) before it stops. The adhesion of water molecules to the glass surface and the cohesion binding water molecules to one another accounts for the capillary forces that draw water into small-diameter glass tubes. Soil pores behave like capillary tubes, drawing in water and, like the experiment illustrated in Fig. 2.B.1, holding the water in the pore against the force of gravity.

Our simplified description of capillary forces is cast in the same language used to describe forces in the saturated zone. The hydraulic pressure at the liquid surface in the dish is zero—relative to atmospheric pressure—and increases with depth below the liquid surface, as we

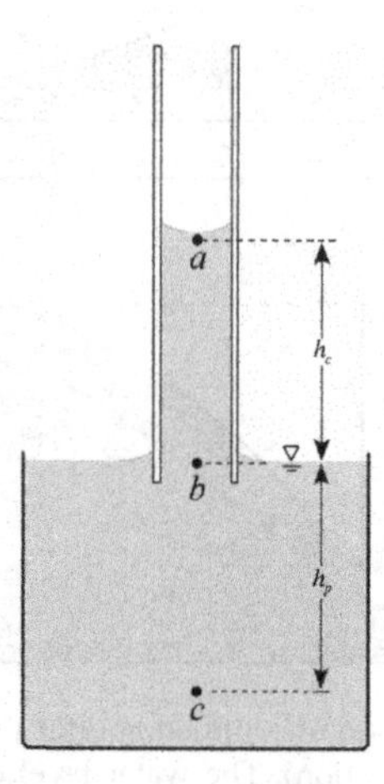

FIG. 2.B.1 The liquid pressure within and below a capillary tube illustrate capillary rise.

have already seen. The phenomenon we are interested in, however, is hydraulic pressure within the capillary tube (or by analogy, within a small pore in the vadose zone).

A clue emerges if we withdraw the capillary tube vertically from the water dish. The hydraulic pressure at the bottom tip of the capillary tube (point (b), Fig. 2.B.1) is zero—relative to atmospheric pressure—the same as the surface of the water outside the capillary tube. If the hydraulic pressure at a depth h_p below the water surface (point (c), Fig. 2.B.1) is positive because of the gravitational force of the water column, how can the hydraulic pressure at the bottom of the water column in the capillary tube (point (b), Fig. 2.B.1) be zero?

The gravitational force per unit area in unconfined water at a depth h_p below the water surface (point (c), Fig. 2.B.1) was previously identified as the *submergence pressure p* (2.B.1) resulting from the compressive force of the water column at the submerged depth h_p. We are selecting the submerged depth h_p equal to the height h_c at point (a) of water rise in the capillary tube.

$$p_p = \rho_w \cdot g_0 \cdot h_s \tag{2.B.1}$$

$$h_p = \frac{p}{\rho_w \cdot g_0} \tag{2.B.2}$$

Consider the change in hydraulic pressure along the vertical axis of the capillary from point (c), through point (b) at the level of the water surface outside the capillary, upward to point (a) inside the capillary at a height h_c. Since the total hydraulic pressure at any depth is uniform, and the hydraulic pressure at the unconfined water surface is atmospheric pressure, then the hydraulic pressure at point (b) must be zero, neglecting atmospheric pressure.

The hydraulic pressure at the bottom tip of the capillary tube at point (b) is zero regardless of whether the water is confined in the capillary tube or unconfined (outside the tube). The water in the capillary tube is, therefore, *suspended* in the tube by capillary forces, and those forces are balanced at the bottom tip.

The force suspending the column of water in the capillary tube against the force of gravity is a *tension* force. This means the hydraulic pressure at the meniscus height h_c in the capillary tube (point (a), 2.B.1) must be *equal in magnitude* to the hydraulic pressure at an equivalent distance h_p below the surface of the water dish (point (c), 2.B.1) but negative in sign (expression 2.B.3).

$$h_c \cdot (\rho_w \cdot g_0) = -h_p \cdot (\rho_w \cdot g_0) \tag{2.B.3}$$

A *negative hydraulic pressure* may, at first glance, seem counter intuitive, but the force acting on the water in the capillary is a tension force that, were it not for the cohesive forces that bind water molecules together, would tend to pull water molecules apart. Positive hydraulic pressure is a compressive force that tends to push water molecules closer together. The compressive hydraulic force will increase the density of water, while tensional force (caused by the pull of gravity) will decrease the density of water in the capillary tube. Appendix 2.C describes how to predict capillary rise based on the balance of forces, capillary diameter and the static surface tension γ^{st} of the air/fluid interface.

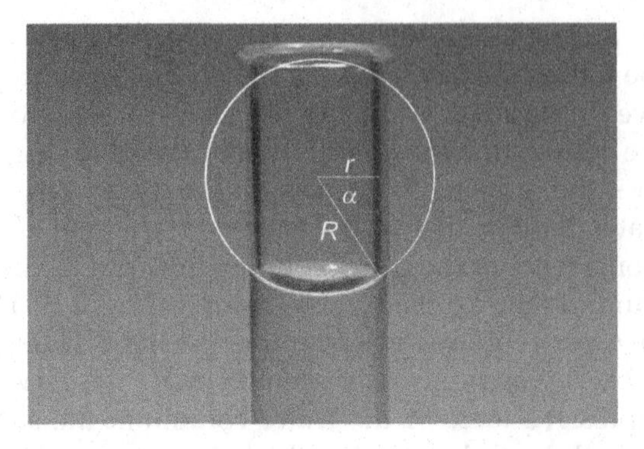

FIG. 2.C.1 The meniscus of a liquid in a capillary tube has a characteristic radius R and contact angle α between the liquid and the tube.

Henceforth we will refer to the *negative hydraulic head* in the vadose zone as $h_c < 0$ to emphasize role of *capillary forces* in the vadose zone.[20] We reserve piezometric head $h_p > 0$ for the saturated zone where pore water hydraulic pressure is positive relative to atmospheric pressure.

The tension force acting on water in a capillary tube depends on capillary meniscus height h_c, the meniscus radius R and capillary tube radius r (see Appendix 2.C). The hydraulic tension in an unsaturated porous medium is a complex function of the water content θ and texture of the medium as it influences porosity ϕ and the mean pore radius $\bar{r}_{pore}$.

2.C PREDICTING CAPILLARY RISE

Liquid in a capillary tube (Figs. 2.C.1 and 2.C.2) rises until the forces acting on the liquid in the tube are balanced at height h_c.

The downward *compression* force is simply the gravitational force $|\mathbf{F}_g|$ acting on the liquid column (2.C.1). The mass of the liquid column m_l is simply the liquid density ρ_l times the volume V_l of the liquid column, which is determined by the capillary radius r and the height h_c of the liquid column (expression 2.C.2).

$$|F_g| = m_l \cdot g_0 \tag{2.C.1}$$

$$|F_g| = \rho_l \cdot V_l \cdot g_0 = \rho_l \cdot \left(\pi \cdot r^2 \cdot h_c\right) \cdot g_0 \tag{2.C.2}$$

The upward *tension* force suspending the liquid column in the capillary results from the *static surface tension* γ^{st} of the gas/fluid interface. The surface tension γ^{st} is a force that acts at an angle relative to the tube axis known as the *contact angle* α.

[20] Soil physicists prefer the term *matric* for the negative tension of water held by capillary forces.

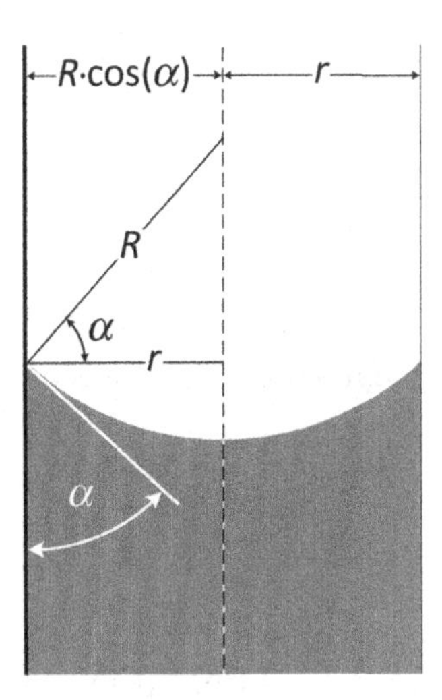

FIG. 2.C.2 The geometric relationship between the contact angle α of a liquid in a capillary tube, the meniscus radius R and capillary tube diameter r.

The surface tension force $|\mathbf{F}_{st}|$ is distributed along the circumference of the capillary tube $2\pi \cdot r$. The surface tension force vector $\vec{\mathbf{F}}_{st}$ is aligned parallel to the meniscus contact angle α and proportional to the static surface tension γ^{st} (expression 2.C.3).

$$|\mathbf{F}_{st}| = \gamma^{st} \cdot (2\pi \cdot r) \tag{2.C.3}$$

The static surface tension γ^{st} is a property of the gas/liquid interface, while the meniscus contact angle α is determined by the adhesion of the liquid to the capillary-tube material.

Expression (2.C.4) gives the vertical component of the surface tension force vector $\vec{\mathbf{F}}_{st}$ as determined by the geometry of the meniscus radius R and the capillary radius r. Fig. 2.C.2 illustrates the geometry of the meniscus contact angle α relative to the meniscus radius R and capillary radius r necessary to derive the vertical force vector.

$$|\mathbf{F}_{st}| = \gamma^{st} \cdot (2\pi \cdot r) \cdot \cos(\alpha) \tag{2.C.4}$$

At equilibrium, the vertical component of the two forces are equal and opposite (expression 2.C.5), enabling us to solve for the capillary rise h_c in a round tube as a function of the capillary radius r, liquid density ρ_l, the static surface tension of the air/liquid interface γ^{st}, and the meniscus contact angle α.

$$\left|\mathbf{F}_g\right| = |\mathbf{F}_{st}|$$

$$\rho_l \cdot \left(\pi \cdot r^2 \cdot h_c\right) \cdot g_0 = \gamma^{st} \cdot (2\pi \cdot r) \cdot \cos(\alpha) \tag{2.C.5}$$

$$h_c = \frac{\gamma^{st} \cdot 2\pi \cdot r \cdot \cos(\alpha)}{\rho_l \cdot \pi \cdot r^2 \cdot g_0}$$

$$h_c = \frac{\gamma^{st} \cdot 2 \cdot \cos(\alpha)}{\rho_l \cdot \cdot r \cdot g_0} \quad (2.C.6)$$

The meniscus contact angle α for the water/glass interface in air is approximately zero (i.e., $\cos(0) \approx 1$). Expression (2.C.7) is the relationship between capillary radius and the height h_c water will rise in a glass capillary.

$$h_c = \frac{2 \cdot \gamma^{st}}{\rho_w \cdot r \cdot g_0} \quad (2.C.7)$$

2.D EMPIRICAL WATER RETENTION AND UNSATURATED HYDRAULIC CONDUCTIVITY FUNCTIONS

2.D.1 Clapp-Hornberger Functions

Clapp and Hornberger (1978) published empirical functions that estimate hydraulic properties in the vadose zone. The two empirical functions estimate the absolute value of tension head $|h_t(s)|$ (expression 2.D.1) and the unsaturated hydraulic conductivity $K_D(s)$ (expression 2.D.2). Both functions use the degree of saturation s (expression 2.D.3) as the independent variable.

$$|h_t(s)| = h_s \cdot s^{-b} \quad (2.D.1)$$

$$K_D(s) = K_s \cdot s^{(2b+3)} \quad (2.D.2)$$

$$s = \frac{\theta}{\phi} \quad (2.D.3)$$

Substituting expression (2.D.3) into expression (2.D.1) and solving for the volume of water content θ yields the expression (2.D.4) typically used to plot water retention curves where $|h_t(s)|$ is the *absolute tension head.*

$$\theta(s) = \phi \cdot \left(\frac{h_s}{|h_t(s)|} \right)^{1/b} \quad (2.D.4)$$

Table 2.D.1 lists the empirical parameters $\{h_s, b\}$ for 11 soil textures. The Clapp-Hornberger expressions are used when the only available data on a particular soil or surficial specimen is its texture.

TABLE 2.D.1 Empirical Parameters for Clapp-Hornberger Expressions (2.D.1), (2.D.2), (2.D.4).

Texture	Clay Content c ($g \cdot g^{-1}$)	Porosity ϕ ($cm\ cm^{-1}$)	Power b	Tension h_S (cm)	Conductivity K_S ($cm\ s^{-1}$)
Sand	0.03	0.395	4.05	12.10	63.36
Loamy sand	0.06	0.410	4.38	9.00	56.28
Sandy loam	0.09	0.435	4.90	21.80	12.48
Silt loam	0.14	0.485	5.30	78.60	2.592
Loam	0.19	0.451	5.39	47.80	2.502
Sandy clay loam	0.28	0.420	7.12	29.90	2.268
Silty clay loam	0.34	0.477	7.75	35.60	0.612
Clay loam	0.34	0.476	8.52	63.00	0.882
Sandy clay	0.43	0.426	10.40	15.30	0.780
Silty clay	0.49	0.492	10.40	49.00	0.372
Clay	0.63	0.482	11.40	40.50	0.462

2.D.2 van Genuchten Water Retention Function

The National Cooperative Soil Survey NCSS[21] has compiled a vast soil data set that often, though not always, lists porosity ϕ, dry bulk density ρ_b, field-capacity water content θ_{fc} and permanent wilting-point water content θ_{wp}.
National Cooperative Soil Survey Soil Characterization Database

http://ncsslabdatamart.sc.egov.usda.gov/querypage.aspx

The NCSS uses the empirical 2-parameter $\{\alpha, n\}$ Van Genuchten (1980) function (expressions 2.D.5, 2.D.6) to generate an entire water retention curve based on two measured soil properties: porosity ϕ and wilting-point water content θ_{wp}.

$$m \equiv n + (1/n) \tag{2.D.5}$$

$$\theta(|h_t|) = \theta_{wp} + \frac{\phi - \theta_{wp}}{\left[1 + (\alpha \cdot |h_t|)^n\right]^m} \tag{2.D.6}$$

Expression (2.D.5) can be fitted to a 3-point soil data set $\{\phi, \theta_{fc}, \theta_{wp}\}$ (cf. Fig. 2.D.1) or a 2-point soil data set $\{\phi, \theta_{wp}\}$, depending available soil data.

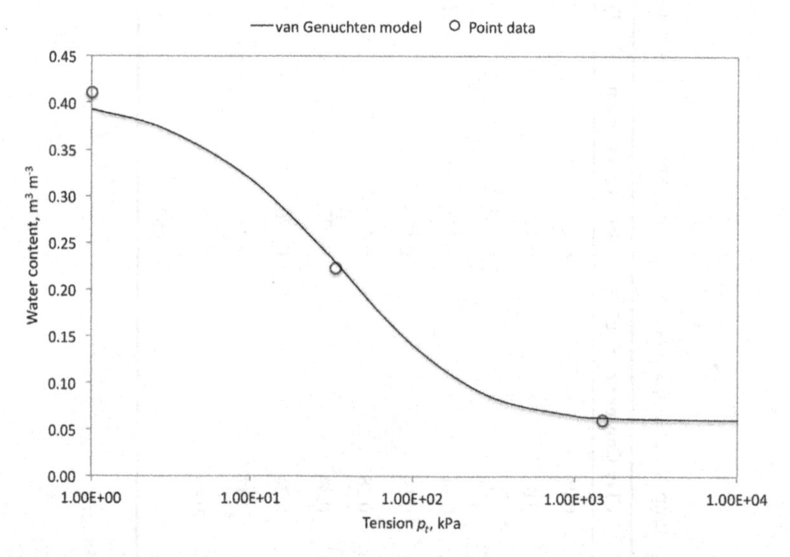

FIG. 2.D.1 The van Genuchten water retention curve plotted in this figure is a 3-point fit for an Antigo silt loam soil (NCSS Pedon ID: 81WI067340, Horizon: Ap). The three *open circles* represent three water contents: porosity ϕ, field capacity θ_{fc} and wilting point θ_{wp}.

[21] The US Department of Agriculture.

2.E DAY LENGTH CALCULATION

The Thornthwaite (1948) temperature-based potential evapotranspiration PET equation requires mean monthly temperature and day length to predict mean monthly PET. There are various other temperature- and radiation-based methods but the Thornthwaite equation will serve our purpose, which is to illustrate the role of empirical PET models in estimating the soil water balance.

Day length L_d at a particular location on a specific day is determined by the location's geographic coordinates (latitude and longitude) and solar declination δ on that day. Solar declination on a specific date δ depends on the number of days since the *vernal equinox* d_{vernal}. The location's latitude determines daily sunrise while the longitude determines the ordinal date of the vernal equinox (21 March in the northern hemisphere, 21 September in the southern hemisphere).

Vernal days are numbered relative the *ordinal day* of the hemispheric vernal equinox. The vernal equinox in the northern hemisphere is approximately ordinal day $d_{ordinal} = 80$ (Table 2.E.1).

Earth's equator is tilted 0.409 radians (Table 2.E.1) with respect to the *Equatorial Coordinate System.*[22] The vernal day d_{vernal} and latitude determine sunrise, computed at an angle referenced to the *Prime Meridian.*

TABLE 2.E.1 Parameters for Computing Day Length

Parameter	Value	Units
Northern vernal equinox	80	$d_{ordinal}$
Southern vernal equinox	264	$d_{ordinal}$
Vernal equinox	1	d_{vernal}
Solar declination	0.409	rad
Latitude		degrees

$$\delta = (0.409) \cdot \sin\left(2\pi \cdot (d_{Vernal}/365)\right) \tag{2.E.1}$$

$$\beta\,[\text{rad}] = \text{latitude}\,[^\circ] \cdot \frac{2\pi}{180} \tag{2.E.2}$$

$$\cos\tau = -\tan\beta \cdot \tan\delta \tag{2.E.3}$$

$$\tau = \arccos\tau \tag{2.E.4}$$

$$L_d\,[\text{h}] = \frac{24 \cdot \tau}{\pi} \tag{2.E.5}$$

[22] The Earth orbital plane around the Sun defines the astronomical *Equatorial Coordinate System.*

The statue of Abraham Lincoln on Bascom Hill (University of Wisconsin-Madison) in Madison, Wisconsin has geographic coordinates of (43.075326° N, 89.403697° W). May 6, the last day of classes in 2016, fell on ordinal day $d_{\text{ordinal}} = 126$ and vernal day $d_{\text{vernal}} = 47$. The approximate day length on May 6, 2016 was $L_d = 14.21\ \text{h}$.

2.F SOLUTE TRANSPORT: PLATE-THEORY TRANSPORT MODEL

Another solute transport model traces its origins to the chromatography field, where it is known as the *plate-theory* model (Martin and Synge, 1941). Plate theory represents the movement of a mobile chemical agent through a chromatographic column by subdividing the column into a series of plates. The agent partitions between the mobile phase and the stationary phase in each plate before the mobile phase advances, one plate at a time.

The migration of the mobile agent is determined by its relative affinity for the mobile and stationary phases. Plate theory is a discrete representation of the flow cylinder (Fig. 2.F.1)—dividing the cylinder into *theoretical* plates of equal length—with water being the mobile phase and soil or aquifer material being the stationary phase. Equilibrium partitioning of a mobile chemical agent A is established in each plate before the mobile phase carries the dissolved agent to the next plate.

The binomial expansion is a key feature of plate theory. A brief review of the binomial expansion will greatly simplify our discussion of the movement of a chemical agent through a sequence of discrete theoretical plates.

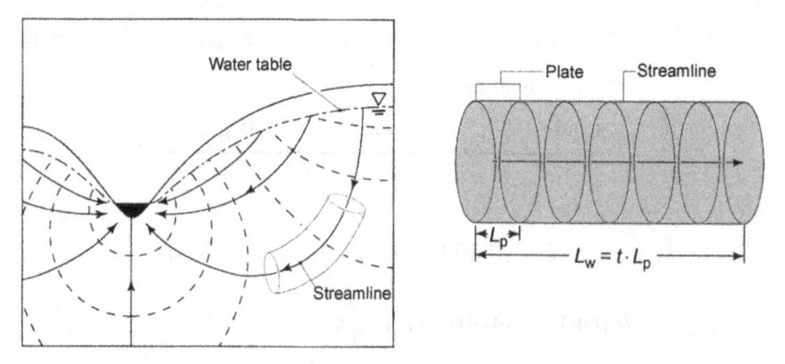

FIG. 2.F.1 Flow paths (i.e., *streamlines*) tracing groundwater movement through an unconfined aquifer (left). Each streamline forms the axis of cylindrical volume element. Plate theory divides the flow cylinder into a series of plates of thickness L_p. Groundwater travel distance L_W during time interval t is equivalent to $L_W = t \cdot L_p$. A dissolved chemical agent a undergoes partitioning between the mobile phase (i.e., flowing water) and the stationary phase (i.e., soil, residuum, or aquifer) in each *theoretical* plate before water movement carries a fraction of dissolved agent to the next plate.

2.F.1 The Binomial Expansion

A binomial is an algebraic expression for the sum (or difference) of two terms. Expression (2.F.1) is the plate-theory binomial consisting of a single independent variable α.

$$\alpha + (1 - \alpha) = 1 \tag{2.F.1}$$

According to the *binomial theorem*, expression $(\alpha + (1 - \alpha))^t$ can be expanded into a sum of terms: $a \cdot \alpha^b \cdot (1 - \alpha)^c$. Exponents b and c are *positive integers* with $t = (b + c)$. The coefficient a of each term in the expansion is a *specific positive* integer that depends on integers t and b.

$$(\alpha + (1 - \alpha))^t = 1 \tag{2.F.2}$$

For example, consider the expansion of binomial (2.F.2) for $t = 4$.

$$\begin{aligned}(\alpha + (1 - \alpha))^4 &= \alpha^4 + 4 \cdot \alpha^3 \cdot (1 - \alpha) + 6 \cdot \alpha^2 \cdot (1 - \alpha)^2 \\ &\quad + 4 \cdot \alpha \cdot (1 - \alpha)^3 + (1 - \alpha)^4 = 1\end{aligned} \tag{2.F.3}$$

Expression (2.F.4) is the general coefficient $f(n,k)$ for term $\alpha^{t-k} \cdot (1 - \alpha)^k$ where $\{k \in \text{integer} \mid 0 \le k \le t\}$. Identity (2.F.5) is required when $k = 0$ or $k = t$.

$$f(t,k) = \frac{t! \cdot (t-k)!}{k!} \tag{2.F.4}$$

$$0! \equiv 1 \tag{2.F.5}$$

The coefficients $f(n,k)$ can be arranged to form the familiar *Pascal's triangle*.

BOX 2.5

MOBILE-STATIONARY PHASE PARTITIONING IN A THEORETICAL PLATE

An illustration of a porous theoretical plate appears in Fig. 2.F.2. Plate volume $V_P = A \cdot L_P$ where the cross sectional area of the plate is arbitrary.

The mobile phase (i.e., water) volume V_m depends on the volume fluid content θ. The mass of the stationary phase (i.e., the soil or aquifer) m_S depends on the dry bulk density of the porous medium.

$$V_m = \theta \cdot V_P \tag{2.F.6}$$

$$m_S = \rho_b \cdot V_P \tag{2.F.7}$$

The plate-theory partition parameter α is the mole fraction of chemical agent A in the mobile phase. The mole fraction of chemical agent A adsorbed to the stationary phase is $(1 - \alpha)$.

$$x_{m(A)} = \frac{n_{m(A)}}{\left(n_{m(A)} + n_{s(A)}\right)} = \alpha \tag{2.F.8}$$

$$x_{s(A)} = \frac{n_{s(A)}}{\left(n_{m(A)} + n_{s(A)}\right)} = (1 - \alpha) \tag{2.F.9}$$

Continued

BOX 2.5 *(cont'd)*

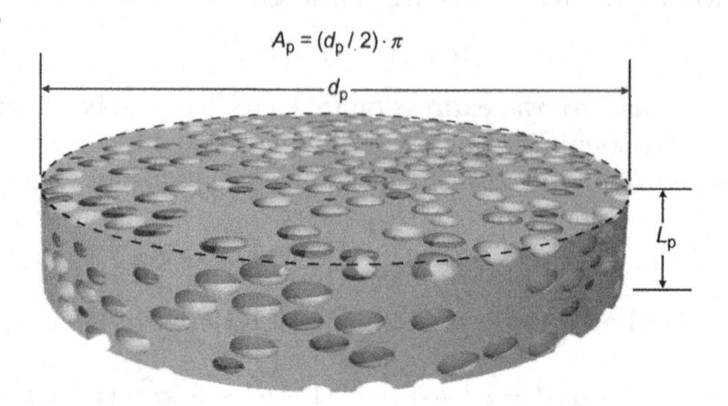

FIG. 2.F.2 Water-filled porous media.

The plate mass and volume relations are used to convert mole fractions to solution concentration c_A and the amount adsorbed to the stationary phase (i.e., specific surface excess concentration n_A^σ/m_s; cf. Chapter 8).

$$n_{m(A)} = c_A \cdot V_m \quad (2.F.10)$$

$$n_{s(A)} = \left(n_A^\sigma/m_s\right) \cdot m_s = n_A^\sigma \quad (2.F.11)$$

A simple expression,[23] reproduced as expression (2.F.12), give the relationship between solution concentration c_A and amount of agent A adsorbed to the stationary phase n_A^σ/m_s. Substituting expressions (2.F.10), (2.F.11) into expression (2.F.12) yields, after rearrangement, expression (2.F.13).

$$\frac{n_A^\sigma}{m_s} = K^{\ominus}_{s/w} \cdot c_A = f_{oc} \cdot K^{\ominus}_{oc/w} \cdot c_A \quad (2.F.12)$$

$$n_{s(A)} = \left(\frac{K^{\ominus}_{s/w} \cdot m_s}{V_m}\right) \cdot = n_{m(A)}$$

$$\left(\frac{f_{oc} \cdot K^{\ominus}_{s/w} \cdot m_s}{V_m}\right) \cdot n_{m(A)} \quad (2.F.13)$$

Replace $n_{m(A)}$ in the expression (2.F.13) using the mole-fraction definition of plate-theory partition parameter α (expression 2.F.8).

$$n_{m(A)} = \alpha \cdot \left(n_{m(A)} + n_{s(A)}\right)$$

$$n_{m(A)} = \alpha \cdot \left(n_{m(A)} + \left(\frac{K^{\ominus}_{s/w} \cdot m_s}{V_m}\right) \cdot n_{m(A)}\right)$$

$$1 = \alpha \cdot \left(1 + \left(\frac{K^{\ominus}_{s/w} \cdot m_s}{V_m}\right)\right)$$

After rearrangement of (2.F.14) we see the plate-theory partition parameter α is the inverse of the retardation coefficient R_f^{-1}.

$$\alpha^{-1} = \left(1 + \left(\frac{K^{\ominus}_{s/w} \cdot m_s}{V_m}\right)\right)$$

$$\alpha^{-1} = \left(1 + \left(\frac{K^{\ominus}_{s/w} \cdot \rho_b \cdot V_p}{\theta \cdot V_p}\right)\right)$$

$$\alpha^{-1} = \left(1 + \left(\frac{K^{\ominus}_{s/w} \cdot \rho_b}{\theta}\right)\right) = R_f \quad (2.F.14)$$

[23] cf. Chapter 8.

2.F.2 Plates and Transfers

Table 2.F.1 lists the first four expansions of the plate-theory binomial $(\alpha + (1 - \alpha))$. The expansion with exponent $t = 0$ is *unity*, a single term: $(\alpha + (1 - \alpha))^0 \equiv 1$.

TABLE 2.F.1 Binomial Expansion of $(\alpha + (1 - \alpha))^t$ Listing Exponents $0 \leq t \leq 4$ and Number of Terms in Each Expansion

Exponent (Transfer Index)	Binomial Expansion	Terms (Plate Count)
0	1	1
1	$\alpha + (1 - \alpha)$	2
2	$\alpha^2 + 2\alpha(1 - \alpha) + (1 - \alpha)^2$	3
3	$\alpha^3 + 3\alpha^2(1 - \alpha) + 3\alpha(1 - \alpha)^2 + (1 - \alpha)^3$	4
4	Expression (2.F.3)	5

The exponent t (Table 2.F.1, column 1) is the *transfer index t*. The number of terms in the binomial expansion (Table 2.F.1, column 3) is the *plate count*.

Consider transfer 0 (Table 2.F.1, row 1); the chemical agent enters plate 1. The plate count $p = 1$ corresponds to transfer index $t = 0$, meaning the chemical agent is confined to plate 1.

Transfer index $t = 1$ appears in Table 2.F.1, row 2. A fraction of mobile-phase agent has advanced from plate 1 to plate 2 leaving the stationary-phase agent in plate 1. Since the chemical agent is distributed over 2 plates the plate count is $p = 2$. The fraction of chemical agent in plate 1 is $(1 - \alpha)$ and the fraction advancing to plate 2 is α, hence the significance of the two terms in column 2.

Transfer index $t = 2$ is found in Table 2.F.1, row 3. A fraction of mobile-phase agent has advanced from plate 1 to plate 2 and another fraction advanced from plate 2 to plate 3. Some stationary-phase agent remains in plates 1 and plate 2 as mobile agent advances to plate 3. Since the chemical agent is distributed over 3 plates the plate count is $p = 3$. The fraction of chemical agent in plate 1 is $(1 - \alpha)^2$, the fraction in plate 2 is $2\alpha(1 - \alpha)$ and the fraction advancing to plate 3 is α^2. Once again, the significance of the three terms in column 2, row 3 should be apparent.

By now you should be able to fill in the details for transfer $t = 3$ (Table 2.F.1, row 4).

Finally, transfer 4 (Table 2.F.1, row 5) distributes the chemical agent over 5 plates. The fraction remaining in plate 1 after transfer 5 is the fifth term in expression (2.F.3): $(1 - \alpha)^3$. The fraction advancing to plate 5 is the first term in expression (2.F.3): α^3.

At each stage, whether stage 1 ($t = 0$; Table 2.F.1, row 1) or stage 4 ($t = 4$; Table 2.F.1, row 5), each term in the expansion gives the fraction of the total agent[24] in each plate. A fixed fraction of the total mobile agent in each plate, regardless of its absolute value, partitions between the mobile phase and the stationary phase; this is the significance of variable α.

[24] In this example the total agent is unity (cf. Table 2.F.1, row 1, column 2 or expression 2.F.1).

2.F.3 Partitioning and the Movement of the Mobile Agent

Returning to transfer 0 (Table 2.F.1, row 1). All of the mobile chemical agent is in plate 1; the fraction $(1-\alpha)$ partitions to the stationary phase while the remainder α remains in the mobile phase.

Transfer 1 (Table 2.F.1, row 2) advances the mobile-phase fraction α to plate 2 while the stationary-phase fraction $(1-\alpha)$ remains in plate 1. Following transfer 1 the mobile and stationary phases equilibrate in each plate. The $(1-\alpha)$ in plate 1 partitions in to a mobile fraction $\alpha(1-\alpha)$ and a stationary fraction $(1-\alpha)\cdot(1-\alpha)$. Likewise, the α in plate 2 partitions in to a mobile fraction $\alpha\cdot\alpha$ and a stationary fraction $\alpha\cdot(1-\alpha)$.

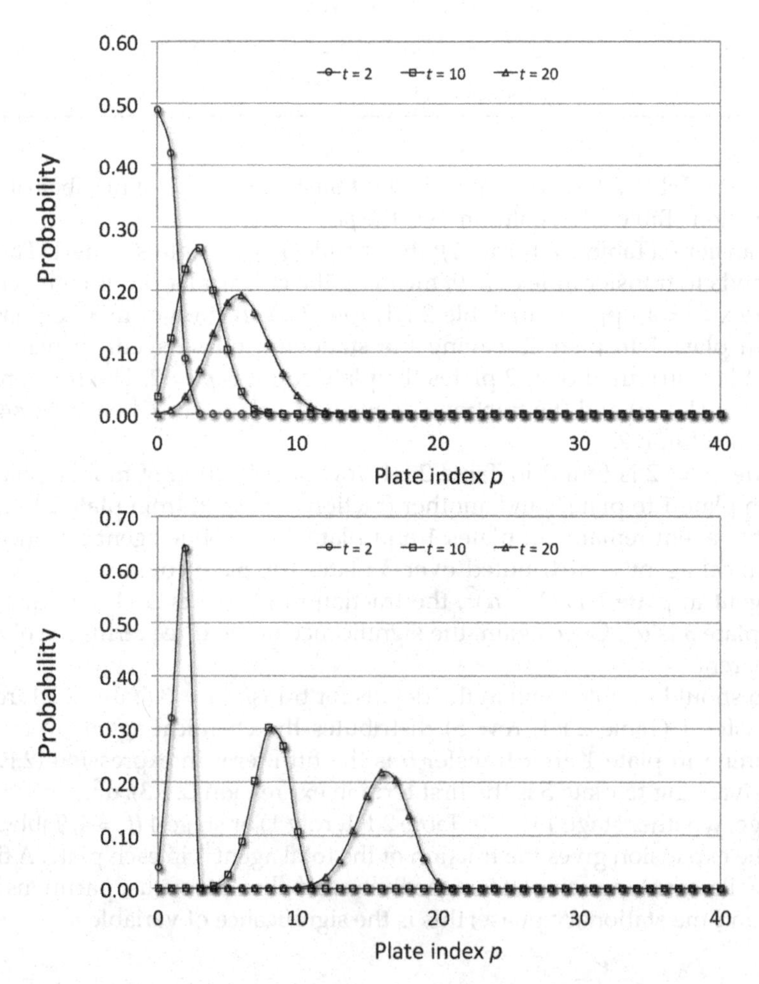

FIG. 2.F.3 Solute probability distributions for three different water advancement distances (2 transfers, 10 transfers, and 20 transfers) and two different partition coefficients (top : $\alpha = 0.333$ and $R_f = 3$; bottom: $\alpha = 0.8$ and $R_f = 1.35$).

Transfer 2 leaves the stationary fraction $(1-\alpha)^2$ in plate 1, and advances the mobile fraction α^2 in plate 3. Can you see the mobile fraction from plate 1 advancing to plate combined with the stationary fraction remaining in plate 2 combine to become $2\alpha(1-\alpha)$?

2.F.4 Retardation-Coefficient and Plate-Theory Models Compared

The binomial expansion is a discrete model of solute movement through a flow cylinder (cf. Fig. 2.F.1). Each theoretical plate represents discrete advances of water and mobile agent A, indicated by the transfer index t.

Fig. 2.F.3 plots the binomial probability distribution that tracks solute progress through the flow cylinder. The probability distribution of an agent with a lower the retardation coefficient (cf. Box 2.5 and Fig. 2.F.3, lower) advances further along the flow cylinder than an agent with a high-retardation coefficient (cf. Fig. 2.F.3, upper).

Fig. 2.F.3 can be deceiving. A portion of the mobile agent α^t reaches plate $(t+1)$ after t transfers, regardless of partition coefficient $\alpha = R_f^{-1}$. The plate index λ at the center of the binomial distribution (expression 2.F.15) is a function of the number of transfers t and the plate-theory partition coefficient α.

$$\lambda = 1 + \alpha \cdot t \tag{2.F.15}$$

$$L_\lambda = L_p \cdot (1 + \alpha \cdot t) \tag{2.F.16}$$

The transport distance L_A associated with retardation-coefficient model is most closely related to the center of the binomial distribution L_λ (expression 2.F.16).

References

Bouwer, H., 1991. Simple derivation of the retardation equation and application to preferential flow and macrodispersion. Ground Water 29 (1), 41–46.

Clapp, R.B., Hornberger, G.M., 1978. Empirical equations for some soil hydraulic properties. Water Resour. Res. 14 (4), 601–604.

Guriyanova, S., Bonaccurso, E., 2008. Influence of wettability and surface charge on the interaction between an aqueous electrolyte solution and a solid surface. Phys. Chem. Chem. Phys. 10 (32), 4871–4878.

Higgins, G.H., 1959. Evaluation of the Ground Water Contamination Hazard from Underground Nuclear Explosions, USAEC Report UCRL-5538, 23 pp.

Hubbert, M.K., 1940. The theory of ground-water motion. J. Geol. 48 (8), 785–944.

Martin, A.J.P., Synge, R.L.M., 1941. A new form of chromatogram employing two liquid phases: a theory of chromatography. 2. Application to the micro-determination of the higher monoamino-acids in proteins. Biochem. J. 35 (12), 1358.

Mote, P.W., Rosenlof, K.H., McIntyre, M.E., Carr, E.S., Gille, J.C., Holton, J.R., Kinnersley, J.S., Pumphrey, H.C., Russell, J.M., Waters, J.W., 1996. An atmospheric tape recorder: the imprint of tropical tropopause temperatures on stratospheric water vapor. J. Geophys. Res. Atmos. 101 (D2), 3989–4006. http://dx.doi.org/10.1029/95JD03422.

Thornthwaite, C.W., 1948. An approach toward a rational classification of climate. Geogr. Rev. 38 (1), 55–94.

Tuller, M., Or, D., Dudley, L.M., 1999. Adsorption and capillary condensation in porous media: liquid retention and interfacial configurations in angular pores. Water Resour. Res. 35 (7), 1949–1964.

Van Genuchten, M.Th., 1980. A closed-form equation for predicting the hydraulic conductivity of unsaturated soils. Soil Sci. Soc. Am. J. 44 (5), 892–898.

Vermeulen, T., Hiester, N.K., 1952. Ion-exchange chromatography of trace components. A design theory. Ind. Eng. Chem. 44, 636–651.

Wolock, D.M., Hornberger, G.M., Beven, K.J., Campbell, W.G., 1989. The relationship of catchment topography and soil hydraulic characteristics to lake alkalinity in the northeastern United States. Water Resour. Res. 25 (5), 829–837.

CHAPTER

3

Clay Mineralogy and Chemistry

OUTLINE

3.1 Introduction 88

3.2 Mineral Weathering 90
3.2.1 Mineralogy 90
3.2.2 The Jackson Weathering Sequence 91

3.3 Silicate Minerals: Composition, Structure, and Chemistry 93
3.3.1 Coordination Polyhedra 94
3.3.2 Silicate Mineral Groups 97
3.3.3 Basicity and the Hydrolysis of Igneous Silicate Minerals 97
3.3.4 Oxidation of Iron(III) in Igneous Silicate Minerals 101
3.3.5 Solid-State Transformation of Pyribole and Feldspar Minerals Into Phyllosilicate Minerals 101

3.4 Clay Minerals 105
3.4.1 Composition and Structure of Phyllosilicate Layers 105
3.4.2 Structure of Neutral-Layer Minerals 106
3.4.3 Layer Structure of Mica Group Minerals 107
3.4.4 Layer Structure of Vermiculite and Smectite Group Minerals 110
3.4.5 Layer Structure of Chlorite Group Minerals 111

3.5 Clay Colloid Chemistry 115
3.5.1 Clay Mineral Plasticity 115
3.5.2 Hydration and Swelling of Smectite and Vermiculite Clay Minerals 116
3.5.3 Clay Plasticity and Soil Mechanical Properties 132

3.6 Summary 137

Appendices 138

3.A Bragg's Law and X-Ray Diffraction by Layer Silicates 138

3.B Determining Formal Oxidation States in Minerals 140

3.C Stoke's Law 142

3.D van't Hoff Factor i and Osmotic Coefficient ϕ 143

References 144

Soil and Environmental Chemistry
http://dx.doi.org/10.1016/B978-0-12-804178-9.00003-3

3.1 INTRODUCTION

Humans in distant prehistory discovered earth collected from certain soils and sediments could be molded when moist and heated over hot coals to produce durable, heat-resistant vessels. The key component was a fine mineral fraction know as *clǽg* (Oxford English Dictionary, 1989).

Drawing on the review by Blott and Pye (2012), the soil physicist Whitney working at the Maryland Agricultural Experiment Station chose 5 μm as the grain size separating silt fraction from the clay fraction (Whitney, 1891). The Bureau of Soils adopted the Whitney particle-size classification when Whitney became its head. Swedish soil scientist Atterberg (1905), famous for his soil plasticity classification scheme (Atterberg, 1911), developed a particle-size classification independent of Whitney. Atterberg (1905) chose 2 μm as the grain size separating silt from clay. In 1938 the Bureau of Soils officially adopted 2 μm as the grain size separating silt from clay (Knight, 1938), making no mention of the Atterberg classification.[1] The following quote from Knight (1938) gives the rationale for changing the grain size separating silt from clay.

> "It is hoped that these changes will make the data from mechanical analysis more useful. The change to 2 microns for the upper limit for clay has the effect of bringing about a better correlation between field texture classification and classification from the data of mechanical analysis." ***Knight (1938)***

Mechanical analysis refers to a method that quantifies grain size by measuring sedimentation rate.[2] Field texture classification, on the other hand, relies on the *ribbon test* to distinguish silt from clay based on plasticity (cf. Box 3.1). The American Society for Testing and Materials ASTM method for measuring the *plastic limit* (ASTM, 2014) employs a similar technique.

BOX 3.1

THE RIBBON TEST

Soil mineral particles are typically separated into three particle-size fractions: sand (0.05–2.0 mm), silt (2–50 μm), and clay (<2 μm). Soil texture is usually a complex size distribution represented by the relative proportions of the three particle-size fractions (NCSS, 2014).

A highly plastic soil contains sufficient clay to be molded into a ribbon greater than 5 cm in length without cracking. A medium plastic soil forms a ribbon 2.5–5 cm in length without cracking. A slightly plastic soil forms a ribbon less than 2.5 cm without cracking (Fig. 3.1). Soil textural classes that are typically nonplastic include sand, loamy sand, and silt. These textural classes contain less than 7–15% clay, depending on silt content.

[1] The International Society of Soil Science adopted the Atterberg classification in 1928.

[2] cf. Appendix 3.C.

BOX 3.1 (*cont'd*)

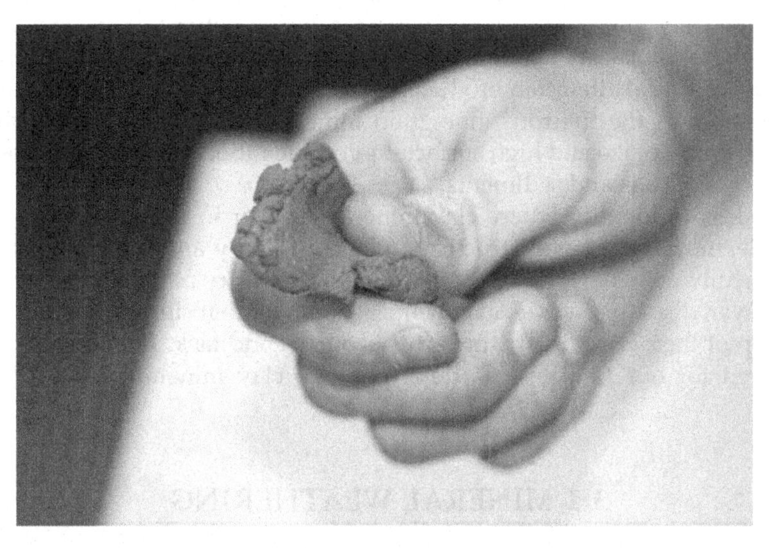

FIG. 3.1 Molding moist clay between forefinger and thumb produces a clay ribbon.

The 1938 revised standard for the upper limit in clay particle-size (Knight, 1938) is notable for two reasons. First, the modest shift from 5 to 2 μm recognized the importance of plasticity in defining clay particle size. Second, the 2-μm limit is identical to the limit chosen by Atterberg (1905) who clearly understood the importance of clay plasticity. A third major particle-size classification (Wentworth, 1922) explicitly recognized plasticity when naming the finest particle-size class.

> "After consideration of a number of alternative terms, the term *clay* has been selected as most likely to be acceptable to geologists for the finest clastic sediments. A few geologists objected to the term on the ground that it implied plasticity or that it referred to a definite chemical composition. It is the view of the writer and of many other geologists that nearly all clastic materials of this grade consist largely of the hydrous aluminum silicates which make up the clay of the chemist and also that the material is always more or less plastic. There is, therefore, in his opinion a common ground for the geologist and chemist without an insistence on the use of the term clay for the pure chemical compounds kaolin or other minerals of this group."
> ***Wentworth (1922)***

Besides emphasizing the importance of clay plasticity, Wentworth (1922) also refers to the emerging understanding of clay-fraction mineralogy. Petrographic microscopy could not resolve clay-size particles. Chemical analysis suggested the clay-size fraction had a chemical composition distinct from silt and sand mineral grains.[3] Early X-ray diffraction studies revealed the presence of crystalline minerals, resulting to the adoption of the term *clay minerals* to acknowledge the presence of yet to be identified minerals.

Today the *clay minerals* connotation encompasses several mineral groups within the *phyllosilicate* subclass. Clay minerals are phyllosilicates whose natural occurrence is confined to the clay particle-size class. Clay minerals differ from the phyllosilicates found in igneous rocks because they are chemical alteration products formed at Earth's surface, where the conditions favor the formation of the hydrous, fine-grain minerals. Clay minerals, by virtue of their extremely small particle size and high surface-to-volume ratio, are one of the most chemically active components of soils and sediments.

Our first task is to identify where clay minerals appear in the geochemical weathering sequence. Understanding mineral structures, while a worthy and interesting subject in and of itself, is generally not essential for environmental chemistry at the level of this book. Clay minerals, however, display physical and chemical behavior that simply cannot be appreciated without a grasp of their crystal structure. This is our second task. With these basics in hand, we are prepared for our final task: a description of clay mineral physical behavior and chemistry.

3.2 MINERAL WEATHERING

3.2.1 Mineralogy

Table 3.1 lists the major mineral classes. Silicate, phosphate, and sulfide minerals occur in all types of rock—igneous, sedimentary, and metamorphic—and weathering products. Oxyhydroxide, hydroxide, and certain hydrous silicate minerals are stable only at temperatures

TABLE 3.1 Major Mineral Classes

Mineral Class	Representative Mineral Formulas
Silicates	$MgSiO_4$, $CaMgSi_2O_6$
	$NaAlSi_3O_8$, $Al_2Si_2O_5(OH)_4$
Oxides	Al_2O_3, $Al(OH)_3$, MnOOH, TiO_2
Sulfides	FeS_2, PbS, HgS, Cu_2S
Carbonates	$CaCO_3$, $CaMg(CO_3)_2$, $FeCO_3$
Sulfates	$CaSO_4$, $KFe_3(SO_4)_2(OH)_6$
Phosphates	$Ca_5(PO_4)_3OH$, $AlPO_4 : 2H_2O$
Halides	NaCl, CaF_2
Elements	Cu, Ag, Au, S

[3] The mineral kaolinite was originally recognized solely on its chemical analysis (Johnson and Blake, 1867).

where liquid water can exist. Carbonate and sulfate minerals are found only in chemically weathered materials such as residuum, sediments, soils, and sedimentary rock formations.

3.2.2 The Jackson Weathering Sequence

Physical weathering and abrasion generate fine-grained sand and silt particles that retain the mineralogy of the rocks from which they form. Chemical weathering transforms *primary* (rock-forming) minerals into *secondary* minerals crystallize in the presence of liquid water under low-temperature conditions existing at Earth's surface. The clay particle-size class (diameter range: <2 μm) is, with few exceptions, comprised of secondary minerals that include the clay minerals. Mineralogy of the silt particle-size class (diameter range: 0.002–0.05 mm) often contains a mixture of primary and secondary minerals.

Jackson et al. (1948) outlined a geochemical weathering sequence based on the mineralogy of the fine (<5 μm) fraction in soils and sediments (Table 3.2). The minerals that are least resistant to chemical weathering (stages 1–7) are absent from the fine clay particle-size fraction

TABLE 3.2 Jackson Silt and Clay Weathering Stages

Weathering Stage	Clay Fraction Mineralogy
1	Gypsum, halite
2	Calcite, dolomite
3	Olivine, pyroxene, amphibole
4	Biotite (mica group), chlorite
5	Feldspars (plagioclase group, orthoclase, microcline)
6	Quartz
7	Muscovite (mica group), *illite*[a]
8	Vermiculite (clay mineral group)
9	Montmorillonite, bidellite (clay mineral group)
10	Kaolinite, halloysite (clay mineral group)
11	Diaspore, boehmite, gibbsite
12	Hematite, goethite, ferrihydrite[b]
13	Rutile, anatase, ilmenite

[a] *Illite is not an officially recognized mineral name because clay mineralogists cannot agree on an explanation for observed compositional differences. If, as many suspect, illite is an interstratified mica-clay mineral intergrade with mica layers alternating with vermiculite or montmorillonite layers (cf. Altaner et al., 1988) then a distinct mineral name would be unjustified.*

[b] *Jackson et al. (1948) listed* limonite *as a stage 12 mineral but this term has fallen out of use, replaced by* ferrihydrite. *The mineral name ferrihydrite is based solely on X-ray diffraction reflections and recognizes two variants: 2- and 6-line. The ferrihydrite crystal structure and composition are undefined.*

(<0.2 μm) and are confined to the coarse clay (0.2–2 μm) and fine silt (2–5 μm) size fractions. The minerals that are most resistant to chemical weathering (stages 8–13) occur predominantly in the clay fraction.

Minerals in stages 1–2 and 8–12 are exclusively secondary minerals, while the minerals in stages 3–7 and 13 are exclusively primary minerals. Making sense of the mineral weathering sequence requires some explanation because the stages represent chemical weathering as seen from a particular perspective.

Surficial deposits—soil, sediment, or residuum—whose fine-silt and coarse clay size fractions contain stage 1 or 2 minerals have undoubtedly undergone considerable chemical weathering from their original igneous precursors, but the presence of relatively soluble chloride, sulfate, and carbonate minerals indicates *current* chemical weathering conditions are not sufficient to dissolve these highly soluble minerals.

The stage 1 minerals (gypsum and halite) persist only in aridic climate zones where soil moisture content remains at the wilting point for a significant portion of the year. Climate zones whose annual evapotranspiration exceeds precipitation sharply reduces the leaching of chemical weathering products from the soil profile, slowing calcite and dolomite dissolution in the upper soil profile, leading to secondary calcite precipitation in the lower soil profile.[4] A line is drawn after stage 2 in Table 3.2, separating the first two stages from the subsequent stages as a reminder the Jackson stages reflect a pedogenic rather than geologic perspective.

Deposits whose fine-silt (2–5 μm) fraction contain stage 3 minerals represent materials that have undergone considerable physical weathering (possibly including erosion, transport and deposition) but little chemical alteration. Stage 3 minerals are common igneous rock minerals considered vulnerable to chemical weathering.

The absence of stage 3 minerals in the fine-silt-size fraction means that the recent chemical weathering history has progressed beyond stage 3. The chemical weathering of stage 3 minerals produces a variety of secondary minerals, but the key to advancing from stage to stage is the elimination of specific indicator minerals from the fine-silt (2–5 μm) or clay (<2 μm) particle-size fractions.

Stage 4–6 minerals are also igneous rock minerals but represent increasing resistance to chemical weathering. As chemical weathering dissolves and transforms minerals in the fine-silt to clay size-fractions, we witness the loss of biotite and chlorite (stage 5), feldspar minerals (stage 6), and, finally, quartz (stage 7). Regardless of whether these minerals occur in coarser (>5 μm) size fractions, the chemical weathering stage is dependent on their elimination from the fine (<5 μm) size fractions.

Muscovite, the last igneous aluminosilicate mineral to disappear from the fine-silt fraction, has been eliminated by stage 8, along with clay-grade, interstratified micas (cf. footnote 4). The indicator minerals of stages 8–10 occur largely in the clay-size fraction. Vermiculite, smectite group minerals,[5] and kaolinite group clay minerals,[6] commonly known as *clay minerals*, are the major topic of this chapter.

[4] Jackson et al. (1948) associates *caliche*, a pedogenic calcite, with stage 2 soils.

[5] Montmorillonite, bidellite, nontronite, etc.

[6] Kaolinite, halloysite, dickite, etc.

Chemical weathering of sufficient duration and intensity rarely results in fine-silt and clay fractions composed solely of insoluble oxide minerals: aluminum hydrous oxides (diaspore : α-AlOOH; boehmite : γ-AlOOH; gibbsite : $Al(OH)_3$), iron hydrous oxides (hematite : α-Fe_2O_3; goethite : α-FeOOH; ferrihydrite : $Fe(OH)_3$)), and titanium oxides (rutile-anatase : TiO_2, and ilmenite : $FeTiO_3$).

Indicator minerals from stages 1 to 5, inclusive, and stage 7 (mica group minerals) do not occur in the <2 μm size range. Quartz and illite particles do appear in the coarse clay fraction (0.2–2 μm) but are absent from the fine clay fraction (<0.2 μm). Although the remaining Jackson weathering stage indicator minerals, from 8 through 13 inclusive, are common clay fraction minerals only those of the clay mineral group (stages 8 through 10) exhibit plasticity.

In 1911 Atterberg published what eventually became a universal system for classifying soil plasticity. The Atterberg system defined six limits, three of which were later abandoned and the others renamed. Atterberg noticed changes in soil volume were related to plasticity and defined the *shrinkage limit* (SL) as the soil water content below which further water loss does not reduce soil volume. Water loss causes fine-grained nonplastic soils to shrink, but shrinkage will be less than in plastic soils. In short, plasticity is an indirect indication of the shrink-swell potential of a soil.

This chapter does not discuss the mineralogy and structure of the oxides, hydroxides, and hydrous oxides listed as indicator minerals for Jackson weathering stages 11 through 13 because these minerals do not influence soil plasticity or shrink-swell potential.

The crystal structure of silicate minerals associated with weathering stages 3 through 5, along with mica group minerals from stage 7, are discussed in the context of their transformation into the clay minerals listed in stages 8 through 10. Our interest in clay mineralogy and clay chemistry is motivated by the need to thoroughly understand plasticity, shrink-swell behavior and the ion exchange capacity of these particular mineral components of the clay fraction.

3.3 SILICATE MINERALS: COMPOSITION, STRUCTURE, AND CHEMISTRY

The three most abundant elements (oxygen, silicon, and aluminum) account for 83.3% of the total mass of Earth's outer crust (Taylor, 1964). Silicate and aluminosilicate minerals differ from other mineral classes because silicon and aluminum combine with oxygen to form tetrahedral units that polymerize into complex networks. This tendency to polymerize, more than elemental abundance, accounts for silicate mineral diversity—roughly 27% of the 4000 or so known minerals.

The structural diversity of silicate minerals parallels the diversity of carbon compounds and derives from similar chemical principles. Both carbon and silicon atoms have four valence electrons and tend to form four bonds with other atoms to satisfy the octet rule. Carbon and silicon also favor tetrahedral bond geometry, as rationalized by both the valence orbital hybridization and the valence shell electron-pair repulsion (VSPER; Gillespie and Nyholm, 1957) chemical bonding models. The tetrahedral bond geometry of carbon-based compounds rests on direct carbon-carbon bonds, while oxygen-rich silicate minerals base their tetrahedral geometry on the silicate tetrahedron SiO_4^{4-} and its tendency to polymerize.

General chemistry books discuss silicate minerals under a special category of network solids. Network solids favor crystal structures that are less dense than close-packed ionic solids because the dominant bonding in network solids has a highly directional, covalent nature. Silicate minerals are organized into classes, depending on the type of silicate network. As a general rule, the melting point, mineral hardness, and resistance to chemical weathering tends to increase as the dimensionality of the silicate network increases.

3.3.1 Coordination Polyhedra

Two leading scientists (Bragg, 1929; Pauling, 1929) simultaneously and independently published papers that would dramatically influence the application of X-ray crystallography to mineralogy. These papers described rules to guide crystallographers attempting to determine the atomic arrangement in minerals and other crystalline compounds using the, at that time, new method of X-ray diffraction.

Crystals *diffract* X-rays[7] into a pattern of bright spots that reveal the arrangement of all atoms lying within smallest repeating unit of the crystal, the *unit cell*. Diffraction is a special case of X-ray scattering which occurs when a collimated[8] X-ray beam scatter off the parallel atomic planes that define the composition, symmetry and periodicity of crystals. Bragg's Law defines the effect of X-ray wavelength, incident angle and the distance separating each atom plane on the constructive and destructive interference of the diffracted X-rays.

The bright spots recorded by X-ray diffractometers[9] reveal the symmetry, composition and the atomic arrangement within the unit cell. The crystal structure refinement process begins with a tentative structure. Bragg (1929), a physicist, suggested the oxygen ions in crystalline oxides are usually in a close-packed arrangement with metal and silicon ions occupying the tetrahedral and octahedral vacancies between close-packed spheres.

Pauling (1929) published five rules that went far beyond Bragg's close-packing concept. Pauling addressed the impact of cation radius on the preferred site within oxide minerals and offered three rules to guide cation placement in adjacent coordination sites. Burdett (Burdett and McLarnan, 1982, 1984; Burdett, 1988) reinterpreted Pauling's rule using molecular orbital theory, providing further insight into their chemical foundations.

It is impossible to overstate the influence of Pauling and Bragg on mineralogy. For example, chemists usually illustrate molecular structure using ball-and-stick models: atoms represented by balls and bonds represented by sticks drawn between bonded atoms. Pauling (1929) introduced an alternative polyhedral model widely used among mineralogists and solid-state chemists to this day. Oxygen ions occupy the vertices of coordination polyhedra and lines drawn between the vertices define the shape of the coordination polyhedron. Cations—usually not depicted—occupy the center of each coordination polyhedron (cf. Example 3.1).

[7] Modern diffraction crystallography also employs neutron and electron beams (cf. Appendix 3.A).

[8] Conventional X-ray tubes produce divergent X-rays which are collimated by capturing a beam after it passes through a pin-hole. Synchrotron X-ray sources naturally produce a brilliant (i.e., high intensity, low angular divergent) X-ray beam.

[9] Early scientists used X-ray film to quantify diffracted X-rays. Modern scientists use solid-state charge-coupled devices to record diffracted X-rays.

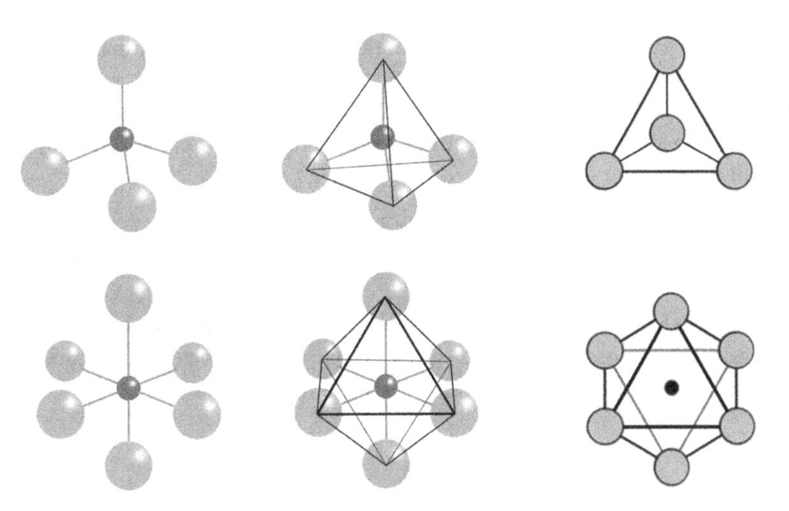

FIG. 3.2 Ball-and-stick and polyhedral models for fourfold and sixfold coordination.

Fig. 3.2 shows the correspondence between ball-and-stick and polyhedral models of the two most common coordination sites in silicate minerals. Ball-and-stick models of tetrahedral and octahedral coordination appear in the left column, and the same ball-and-stick models appear in the middle column, with lines defining the vertices of the two polyhedra. Idealized symbols of a tetrahedron and an octahedron appear in the right column, similar to the illustrations found in Pauling (1929) and Bragg (1929).

The idealized tetrahedron in the upper right of Fig. 3.2 is oriented with its threefold rotation axis perpendicular to the page. The idealized octahedron in the bottom right of Fig. 3.2 is oriented with its threefold rotation axis perpendicular to the page and its twofold rotation axes lie it the page, vertical and horizontal through the center of the octahedron.

EXAMPLE 3.1

Example Permalink

http://soilenvirochem.net/N4P5oB

Determine the cation-to-anion radius ratio for tetrahedral coordination.

Pauling (1929) proposed several rules governing mineral structures, the second being the *Radius Ratio* rule.

> "The Nature of the Coordinated Polyhedra. I. A polyhedron of anions surrounds each cation. The cation-anion distance is the sum of their respective radii and the coordination number of the cation is the determined by the radius ratio of cation to anion."

The Radius Ratio rule uses geometry to define the minimum radius ratio for stable coordination. Pauling (1929) based this rule on the notion that small cations fit into vacancies created by the packing together of large oxygen ions to form elementary polyhedra. Cations preferentially occupy polyhedra vacancies where the cation ionic radius just matches the space available or where it is somewhat too large, forcing the oxygen ions apart. *A cation rarely occupies vacancies where its ionic radius is smaller than the space available.*

Step 1. Determine the geometric relationships that define tetrahedral coordination.

Tetrahedral site geometry is best understood by visualizing a tetrahedron inside of a cube. The diagonal distance across the face of the cube is the length of the edge of the embedded tetrahedron. Since the coordinating spheres have unit radius, the cube face-diagonal is 2 and, by geometry, the cube edge e is the square root of 2.

$$\left(e^2 + e^2\right) = 2^2$$
$$e = \sqrt{2}$$

The center of the cube is also the center of the tetrahedron. Needless to say, the center is located at half the height of the enclosing cube $e/2$. The distance from any vertex of the cube to the center is sum of the radius of the anion and the radius of the cation (Fig. 3.3).

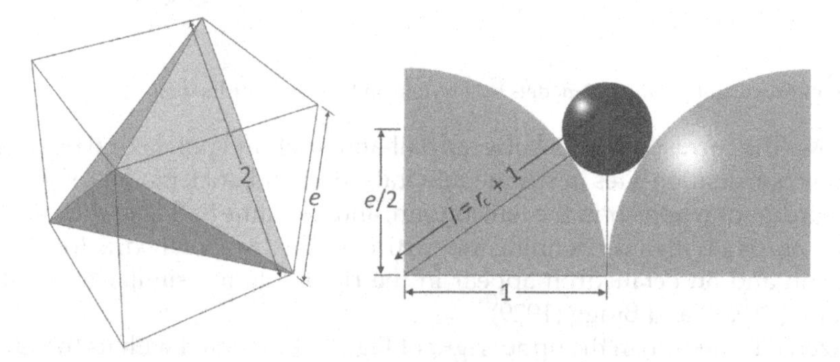

FIG. 3.3 Geometry for determining the radius ratio for tetrahedral coordination. A tetrahedron enclosed within a cube (left) illustrates the geometric relationship between the edge length of a tetrahedron and the edge length of the enclosing cube. To the right appears the geometric relationship between unit-radius spheres at the vertices of a tetrahedron and the vacancy at the center of the tetrahedron.

Step 2. Apply the geometric relationships that define tetrahedral coordination to compute the radius ratio for a sphere whose radius precisely fits the vacancy at the center of the coordination polyhedron.

The distance from any vertex of the cube to the center is the hypotenuse of a right triangle. The height of the right triangle is $e/2$ and the base is half the diagonal distance across the face of the cube, which is unity in this example.

$$l^2 = \left(\frac{e}{2}\right)^2 + 1^2 = \frac{6}{4}$$
$$l = \sqrt{\frac{6}{4}} = 1 + r_C$$
$$r_C = \left(\left(\frac{\sqrt{6}}{2}\right) + 1\right) = 0.2247\cdots$$

Pauling (1960) quotes an ionic radius of 140 pm for the divalent oxygen anion. The estimated Si–O bond length using the Radius Ratio method is 171 pm, which compares with 1.61 pm in the mineral quartz.

3.3.2 Silicate Mineral Groups

The composition and mineralogy of the continental lithosphere is very similar to granite and the oceanic lithosphere is very similar to basalt despite the diversity of igneous rock. The minerals commonly found in granite and basalt also occur in many metamorphic rocks. Igneous rocks crystallize from a molten state while the minerals in metamorphic rocks recrystallize under the combined influence of extreme pressure and temperature, approaching but never reaching the molten state. The conditions required for the formation of igneous and metamorphic rocks are notable because liquid water does not exist as a phase, therefore many of the minerals in igneous and metamorphic rocks contains minerals that cannot be crystallized in the presence of liquid water.

The following discussion focuses on the crystallographic similarities of silicate mineral groups and their chemical properties rather than emphasizing the differences that distinguish them. Pyroxenes, amphiboles and micas share common structural components that direct the solid-state transformation of these minerals into the clay minerals listed in Jackson weathering stages 7–10.

The silt-fraction indicator minerals for Jackson weathering stage 3 span three mineral groups: olivines, pyroxenes, and amphiboles (nesosilicates, single-chain inosilicates, and double-chain inosilicates, respectively). The sorosilicate and cyclosilicate groups are not featured in the list.

3.3.3 Basicity and the Hydrolysis of Igneous Silicate Minerals

Mineralogists classify silicate minerals based on the silicate polyanions that define their essential crystal structure (Table 3.3). Cation substitution of tetrahedral Si^{4+} by Al^{3+}, which leaves the essential aluminosilicate tetrahedral polyanion network intact, is rare or nonexistent in neso-, soro-, cylco-, and single-chain insosilicates.

The silicate anions listed in Table 3.3 do not occur in silicate minerals that crystallize from aqueous solution. They are very strong bases that react aggressively with liquid water. Igneous minerals crystallize directly from a high-temperature, complex *ionic liquid* known as *magma* below the Earth's surface and *lava* when it erupts at the Earth's surface. A liquid water phase cannot exist at the temperatures characteristic of magma and lava. Absent liquid water the silicate ions in magma and lava enter igneous minerals as the ionic liquid cools.

The formulas appearing in column two of Table 3.3 indicate the silicate anion charge as it exists in magma and a representative silicate mineral. The number of *terminal* oxygen atoms Si–O (cf. the nonbridging silicate tetrahedral vertices in Fig. 3.4) is equal to the silicate anion charge. The number of *bridging* oxygen atoms Si–O–Si linking silicate tetrahedra together is equal to the total oxygen atoms in the silicate anion minus the anion charge. For example, the insosilicate anion $Si_2O_6^{4-}$ contains four (4) terminal oxygen atoms and two (2) bridging oxygen atoms.

The basicity of a particular silicate mineral (igneous or otherwise) is determined by the silicate anion charge but is more easily quantified by counting the *moles of charge* summed over Groups 1 and 2 elements per formula unit.

TABLE 3.3 Silicate Mineral Groups

Silicate Group	Silicate Anion	Mineral Group
Nesosilicates	SiO_4^{4-}	Olivine
Sorosilicates	$Si_2O_7^{6-}$	Epidote
Cyclosilicates	$Si_3O_9^{6-}$	Cyclowollastonite
Cyclosilicates	$Si_6O_{18}^{12-}$	Beryl group
Insosilicate	$Si_2O_6^{4-}$	Pyroxene
Insosilicate	$Si_4O_{11}^{6-}$	Amphibole
Phyllosilicate	$AlSi_3O_{10}^{3-}$	Mica group
Phyllosilicate	$Si_4O_{10}^{4-}$	Kaolinite group
Tectosilicate	SiO_2^{0}	Quartz
Tectosilicate	$AlSi_3O_8^{1-}$	Alkali feldspars
Tectosilicate	$Al_2Si_2O_8^{2-}$	Plagioclase feldspar

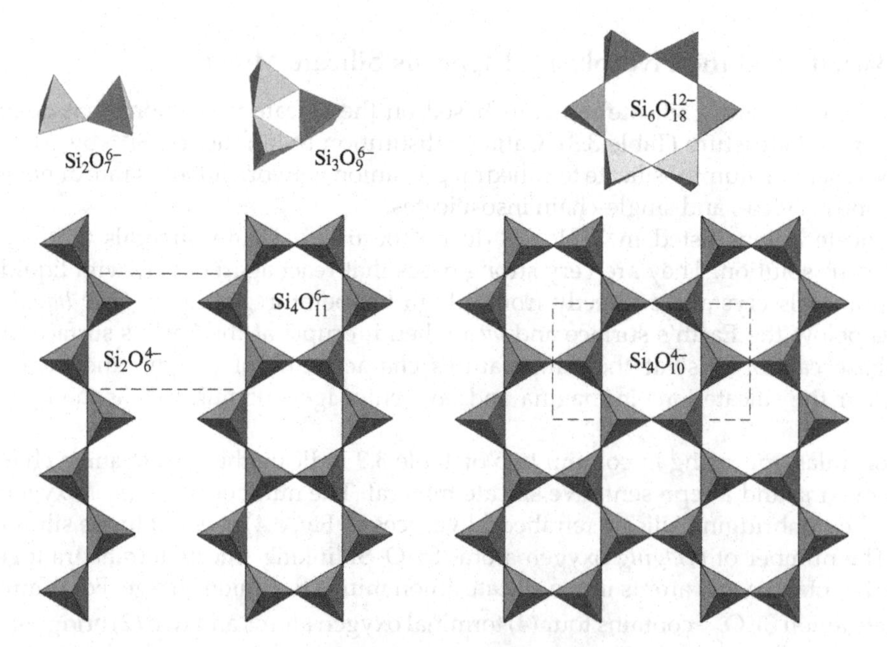

FIG. 3.4 Polymeric silicate anions of the six major silicate mineral groups.

The nesosilicate forsterite $Mg_2SiO_4(s)$ contains four (4) moles of base-charge per formula unit while the tectosilicate quartz $SiO_2(s)$ contains zero moles of base-charge per formula unit. To understand the chemical significance of *base-charge* in the silicate mineral context consider the net hydrolysis of forsterite where the chemical formula is written as a sum of oxides (cf. Example 6.1): $2\,MgO \cdot SiO_2(s)$.

$$\underset{\text{Forsterite}}{2\,MgO \cdot SiO_2(s)} + 4\,H_2O(l) \longrightarrow 2\,Mg^{2+}(aq) + 4\,OH^-(aq) + H_4SiO_4^0(aq) \quad (3.R1)$$

The orthosilicate anion SiO_4^{4-} in forsterite has four (4) terminal oxygen atoms and hydrolyzes in the presence of liquid water to produce four (4) moles of hydroxide OH^- per formula unit. There are two distinct hydrolysis reactions involving the silicate anions listed in Table 3.3 and water, reaction (3.R1) is an *acid-base* hydrolysis reaction involving terminal silicate oxygen atoms.

The inosilicate enstatite $Mg_2Si_2O_6(s)$ contains two (4) moles of base-charge per silicon, half as many as forsterite. To understand the chemical significance of bridging Si–O–Si oxygen atoms linking silicate tetrahedra consider the enstatite hydrolysis reaction (3.R2) for which, once again, the chemical formula is written as a sum of oxides: $2MgO \cdot 2SiO_2(s)$.

$$\underset{\text{Enstatite}}{2\,MgO \cdot 2\,SiO_2(s)} + 6\,H_2O(l) \longrightarrow 2\,Mg^{2+}(aq) + 4\,OH^-(aq) + 2\,H_4SiO_4^0(aq) \quad (3.R2)$$

Reaction (3.R2) contains the name number of terminal Si–O oxygen atoms as (3.R1) however, two additional moles of H_2O appear on the educt side of the reaction without adding to the hydroxide OH^- yield on the product side.

These additional H_2O molecules are required to cleave bridging Si–O–Si between silicate tetrahedra along the insosilicate chain. This second hydrolysis reaction *is not an acid-base hydrolysis reaction* such as the one that produces one mole of hydroxide OH^- for each mole of terminal Si–O oxygen in the formula unit.

The third hydrolysis reaction—an acid-base reaction related to the hydrolysis of terminal Si–O oxygen atoms (cf. 3.R1)—involves elements that precipitate in the presence of liquid water, forming insoluble oxides and hydroxides. To understand the difference between Groups 1 and 2 elements and the Group 13 element aluminum we will compare the hydrolysis of two very similar phyllosilicate minerals: talc and pyrophyllite.

The silicate anion $Si_4O_{10}^{4-}$ is identical in both minerals. The cations Mg^{2+} and Al^{3+} occupy identical octahedral coordination sites, the only difference is divalent Mg^{2+} occupies every octahedral site within the phyllosilicate layer while trivalent Al^{3+} occupies two-thirds of the sites. The complete hydrolysis of talc (reaction 3.R3) and pyprophyllite (reaction 3.R4) appear following, the formulas of both minerals written as sum-of-oxides.

$$\underset{\text{Talc}}{6\,MgO \cdot 8\,SiO_2 \cdot 2\,H_2O(s)} + 20\,H_2O(l) \longrightarrow$$

$$6\,Mg^{2+}(aq) + 12\,OH^-(aq) + 8\,H_4SiO_4^0(aq) \qquad (3.R3)$$

$$\underset{\text{Pyrophyllite}}{2\,Al_2O_3 \cdot 8\,SiO_2 \cdot 2\,H_2O(s)} + 20\,H_2O(l) \longrightarrow$$

$$\underset{\text{Gibbsite}}{4\,Al(OH)_3(s)} + 8\,H_4SiO_4^0(aq) \qquad (3.R4)$$

Talc hydrolysis, containing six Mg^{2+} cations per formula unit, yields twelve OH^- in reaction (3.R3). Talc $Mg_6Si_8O_{20}(OH)_4(s)$ contains the Group 2 cation Mg^{2+} and its terminal Si–O oxygen atoms by reaction (3.R1).

Pyrophyllite $Al_4Si_8O_{10}(OH)_4(s)$ contains the Group 13 cation Al^{3+}. The terminal Si–O oxygen atoms in pyrophyllite react with water through the third hydrolysis reaction because Al^{3+} forms insoluble oxides and hydroxides in the presence of liquid water. Pyrophyllite hydrolysis (reaction 3.R4) yields the same number of OH^- ions as talc hydrolysis, but Al^{3+} reacts with all of the OH^- ions produced to form the insoluble mineral gibbsite.

Reaction (3.R4) removes from aqueous solution three moles of silicate basicity for every Al^{3+} in the educt silicate, producing an insoluble conjugate base. To further illustrate the third hydrolysis reaction, consider the total hydrolysis of anorthite $CaAl_2Si_2O_8(s)$, a plagioclase feldspar, contains one Group 2 Ca^{2+} cation per formula unit and two Group 13 Al^{3+} cations. Unlike pyrophyllite all of the Al^{3+} in alkali and plagioclase feldspars replace tetrahedral Si^{4+}. The complete hydrolysis of anorthite appears in reaction (3.R5).

$$\underset{\text{Anorthite}}{CaO \cdot Al_2O_3 \cdot 2\,SiO_2(s)} + 8\,H_2O(l) \longleftrightarrow$$

$$Ca^{2+}(aq) + 2\,OH^-(aq) + \underset{\text{Gibbsite}}{2\,Al(OH)_3(s)} + 2\,H_4SiO_4^0(aq) \qquad (3.R5)$$

One mole of terminal Si–O oxygen atoms in silicate minerals yields one mole of OH^- through hydrolysis reactions (3.R1), (3.R3), (3.R4) and (3.R5), regardless. The difference between reaction (3.R4) and (3.R5), on the one hand, and reactions (3.R1) and (3.R3), on the other, is the fate of the OH^- produced. Aluminum and iron cations remove silicate basicity because trivalent cations of these two elements form insoluble oxides and hydroxides. The remaining basicity in silicate mineral, equal to the Groups 1 and 2 cation content when counted as *moles of charge*, remain in solution available to react with aqueous $CO_2(aq)$ for form alkalinity (cf. Chapter 6).

Basicity and the hydrolysis of silicate minerals has obvious implications for chemical weathering processes. To reiterate, every mole of terminal Si–O oxygen atoms in silicate anions (Table 3.3 and Fig. 3.3) yields one mole of OH^- through acid-base hydrolysis and the yield of soluble $OH^-(aq)$ depends on the chemical composition of the silicate mineral.

Silicate minerals resist chemical weathering to different degrees, but, in general, resistance is proportional to content of bridging Si–O–Si oxygen atoms in the silicate anion. Basalt contain high contents of easily weathered olivine, pyroxene, and amphibole group minerals. Granite contains a higher proportion of resistant minerals: micas, feldspar, and quartz. Ultimately, chemical weathering completely dissolves igneous silicate minerals by weathering stage 11 (cf. Table 3.2).

A later section will discuss in situ transformation of igneous pryribole (insosilicate) and feldspar (tectosilicate) minerals into clay minerals (clay-grade phyllosilicates). The justification for taking up clay mineral formation as a separate section is simply this: clay mineral crystals do not nucleate and grow from homogeneous from aqueous solution (i.e., from systems containing an aqueous solution phase and no mineral phases).

3.3.4 Oxidation of Iron(III) in Igneous Silicate Minerals

Shackleton (1978) reports the continental litherophere iron content as iron-oxide mass fractions: $w(\mathrm{FeO}) = 0.0178$ and $w(\mathrm{Fe_2O_3}) = 0.0154$. Iron valence in igneous granitic rocks is controlled by the *fayalite-magnetite-quartz* redox buffer (Haggerty, 1976). The iron(II) and iron(III) oxide compositions quoted by Shackleton (1978) can be formulated as magnetite and fayalite oxide mass fractions.

$$w(\mathrm{FeO \cdot Fe_2O_3}) = w(\mathrm{magnetite}) = 0.0154 \tag{3.1}$$

$$w(\mathrm{2\,FeO \cdot SiO_2}) = w(\mathrm{fayalite}) = \tfrac{1}{2} \cdot (0.0178 - 0.0154) = 0.0012 \tag{3.2}$$

The iron(II) mole-per-mass concentration $m(\mathrm{Fe(II)})$ as fayalite is 0.13, the remainder is 0.87 moles-per-mass magnetite iron, confirming the assessment of Frost (1991) "mostly incorporated into magnetite." Reaction (3.R6) is the total hydrolysis of fayalite $\mathrm{Fe_2SiO_4(s)}$, including the oxidation of iron(II) to iron(III) by $\mathrm{O_2(aq)}$.

$$\underset{\text{Fayalite}}{\mathrm{2\,FeO \cdot SiO_2(s)}} + \tfrac{1}{2}\mathrm{O_2(aq)} + 3\,\mathrm{H_2O(l)} \longrightarrow \underset{\text{Goethite}}{\mathrm{2\,FeOOH(s)}} + \mathrm{H_4SiO_4^0(aq)} \tag{3.R6}$$

Reaction (3.R6) has a similar impact as acid-base hydrolysis combined with precipitation of insoluble aluminum oxides and hydroxides (cf. reactions (3.R4) and (3.R5)) discussed in the previous section. The oxidation of iron(II) to iron(III) in the presence of water induces the precipitation of insoluble iron(III) oxides and hydroxides that draw iron out of silicate minerals. Much the same can be said of manganese, the 12th most abundant element in the lithosphere. We can anticipate a large fraction of the manganese(II) occurs in magnetite and other spinel minerals whose oxidation is of negligible consequence to silicate mineral weathering. The remaining manganese(II) occurs as a trace component of all silicate minerals, precipitating as insoluble manganese(III) and manganese(IV) oxides and oxyhydroxides upon hydrolysis and oxidation.

When viewed in perspective, the hydrolysis and oxidation of fayalite (reaction 3.R6) and other iron(II) bearing igneous silicate minerals is a relatively minor contribution to silicate rock weathering. Most of the iron in igneous rocks making up the lithosphere—between 80% and 90%—is magnetite whose oxidation is of no consequence to silicate mineral weathering.

3.3.5 Solid-State Transformation of Pyribole and Feldspar Minerals Into Phyllosilicate Minerals

This section summarizes the current understanding of how clay mineral form through the solid-state rearrangement of insosilicates and feldspar minerals into secondary clay minerals. Solid-state rearrangement, revealed by high-resolution, transmission electron microscopy

(TEM) of naturally weathered igneous silicates, resolves a critical dilemma in clay mineral formation: the lack of evidence for the nucleation and growth of clay minerals from homogeneous aqueous solution at low temperatures.

Laboratory studies led (Correns and von Engelhardt, 1938a,b; Correns, 1940) to suggest the formation of a depleted layer at silicate mineral surfaces during dissolution. Correns identified diffusion through the depleted layer as the rate limiting step in silicate mineral dissolution. The *diffusion-limited dissolution* mechanism—presumably involving cation exchange between $H^{+}(aq)$ diffusing into and through a depleted aluminosilicate surface layer and Groups 1 and 2 cations retained within the layer—was the accepted silicate mineral dissolution mechanism for 50 years (White and Brantley, 1995; Brantley and Stillings, 1996).

Scientists studying adsorption at oxide mineral surfaces developed an alternative, surface-site-limited model to describe oxide mineral dissolution. The advent of surface-site-limited adsorption models (cf. review by Wehrli et al., 1990) combined with the reinterpretation of laboratory dissolution studies (Lagache et al., 1961; Lagache, 1965; Holdren and Berner, 1979; Aagaard and Helgeson, 1982; Helgeson et al., 1984) led to growing interest in crystal defects exposed at mineral surfaces.

Attempts to gather experimental evidence of surface defects and their potential role in *surface-site-limited dissolution* (Helgeson et al., 1984) eventually led to high-resolution transmission electron microscopy (HRTEM) studies of naturally weathered igneous silicate minerals. An entirely new understanding of transformation of igneous silicates into clay minerals emerged from these studies.

Figs. 3.4 and 3.5 illustrate the idealized *I-beams* found in pyribole minerals. Scientists distinguish pyroxene and amphibole in HRTEM images using *I-beam* width. Phyllosilicate layers are structurally similar to pyribole *I-beams*; condensing amphibole *I-beams* to form sheets yields phyllosilicate layers with the 1.0-nm periodicity characteristic of many clay minerals.

Pyroxene and amphibole HRTEM images reveal a characteristic crystallographic defect—known variously as *Wadsley* defects (Wadsley and Andersson, 1970; Chisholm, 1973), *polysomatic* defects (Thompson, 1978), and zippers (Veblen and Buseck, 1980; Buseck and Veblen, 1981; Eggleton and Boland, 1982). These planar defects appear to propagate through pyroxene and amphibole structures (Fig. 3.6).

HRTEM images of naturally weathered pribole specimens also reveal direct transformation of pyribole *I-beams* into 1.0 nm phyllosilicate layers (Buseck and Veblen, 1981; Eggleton and Boland, 1982; Banfield et al., 1995). The transformation depicted in Fig. 3.7 (top) is a direct solid-state rearrangement of the common structural motif shared by pyribole and phyllosilicate minerals.

Fig. 3.7 also illustrates the structural similarity between pyribole *I-beams* (top) and the crystal structure of the clay mineral palygorskite (bottom). The double-chain *I-beams* in palygorskite are identical in structure to those in amphibole minerals, the difference being *I-beam* cross-linked through bridging Si–O–Si bonds in palygorskite. Sepiolite is a triple-chain *I-beam* variant similar to palygorskite.

Recently clay mineralogists realized a complete polysomatic series (i.e., crystalline intergrowth) linking palygorskite and sepiolite (Guggenheim and Krekeler, 2011; Suarez and Garcia-Romero, 2013). Krekeler et al. (2005) published HRTEM images demonstrating the epitaxial growth of the clay mineral smectite on palygorskite (cf. bottom, Fig. 3.7). The reader

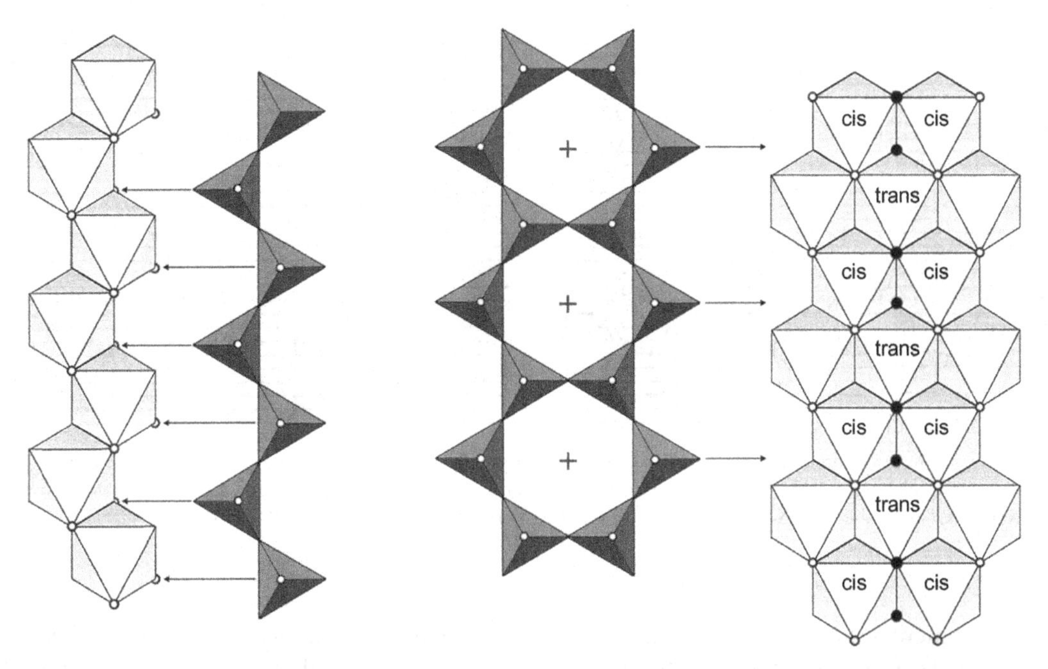

FIG. 3.5 Silicate anions and octahedral sites for pyroxene group (left) and amphibole group (right) minerals. The *open circles* in the pyroxene and amphibole octahedral chains match with the *open circles* of the respective silicate chains. The *filled circles* in the amphibole octahedral M-sites are OH^- ions that align with the crosses at the center of sixfold rings in the amphibole tetrahedral chain. A second tetrahedral chain covers octahedral M-sites from above for both mineral groups (cf. Fig. 3.6).

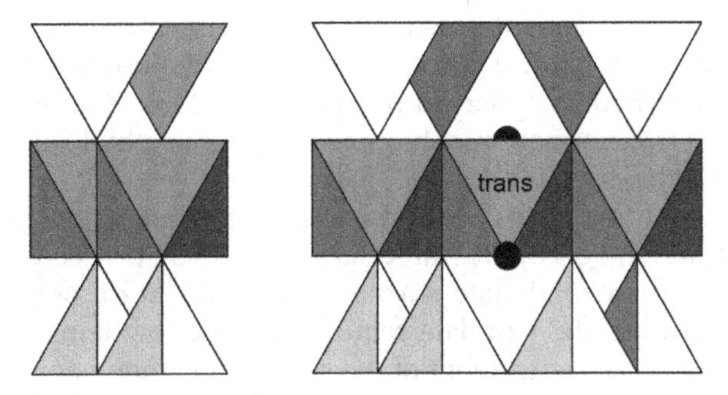

FIG. 3.6 *I-beams* formed by octahedral M-sites between paired silicate chains: pyroxene group (left) and amphibole group (right) minerals. The *I-beams* are viewed along the chain axes.

should recognize the role of palygorskite as an intermediate in the transformation of pyribole minerals into the clay minerals listed in Table 3.3.

Feldspar minerals, the indicator minerals of Jackson weathering stage 5 (Table 3.3), lack octahedral sites and, hence, the *I-beam* motif and Wadsley defects that prefigure the solid-state

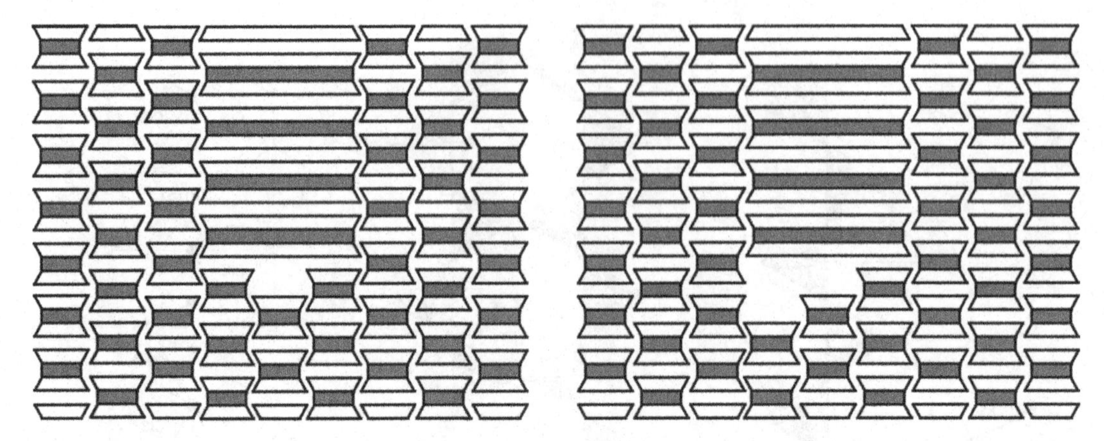

FIG. 3.7 Simple coherent zipper terminations for sextuple-chain zippers in anthophyllite. From left to right are experimental images TEM, models shown in the *I-beam* representation, and image calculations based on the models. *Veblen, D.R., Buseck, P.R., 1980. Microstructures and reaction mechanisms in biopyriboles. Am. Mineral. 65, 599–623.*

rearrangement into phyllosilicate layers observed in weathered pryribole minerals. Most geochemists agree feldspar weathering is a surface-site controlled process (Aagaard and Helgeson, 1982; Helgeson et al., 1984). The rate-limiting sites identified on the surface of naturally-weathered feldspar grains appear as etch pits whose origins are traced to crystal cleavages, dislocations, twin planes and assorted unidentified defects (Eggleton and Boland, 1982; Banfield and Eggleton, 1990).

HRTEM images reveal clay minerals—identified by their characteristic 0.7 and 1.0 nm spacings—at naturally-weathered feldspar surfaces (Eggleton and Boland, 1982; Tazaki, 1986; Banfield and Eggleton, 1990). The intermediates between the original feldspar and the final 1.0 nm clay mineral product include a "protocrystalline" aluminosilicate phase and a 0.7-nm-spacing component identified as spherical halloysite (cf. Table 3.4). Banfield and Eggleton (1990) also detected the transformation of biotite—an igneous phyllosilicate found in granite (cf. Table 3.4)—into illite (cf. footnote 4).

In summary, the low-temperature chemical weathering of all major igneous silicate minerals, regardless silicate mineral group (cf. Table 3.3 and Fig. 3.3), proceeds more or less directly to clay mineral products via solid-state rearrangement of silicate polymeric anions. Solid-state rearrangement supplies the template required for the low-temperature crystallization of clay minerals. Of course rearrangement of the silicate polymeric anion must involve acid-base hydrolysis of terminal Si–O oxygen atoms and the attendant expulsion of Groups 1 and 2 cations and the precipitation of insoluble iron and aluminum hydrous oxide minerals (cf. "cell-textured" material in Banfield and Eggleton (1990)).

All igneous silicate minerals have disappeared from the fine-silt and coarse-clay fraction by Jackson weathering stage 7, replaced by secondary clay minerals (Table 3.2). The stages where clay minerals appear as indicator minerals should not be misinterpreted as the stage where clay minerals form. The preceding discussion reveals the appearance of secondary clay minerals as early as stage 3, the earliest stage where igneous silicate minerals serve as indicator minerals. The gradual disappearance of stages 3 through 7 minerals clear the way for clay

TABLE 3.4 Representative Phyllosilicate Minerals (Fe Indicates Iron(II) Valence)

Mineral	Unit Cell Formula	Layer Type
Mica group		
Muscovite	$K_2 \cdot Al_4^{vi}(AlSi_3)_2^{iv}O_{20}(OH)_4(s)$	2:1
Biotite mica	$K_2 \cdot (Mg,Fe)_6^{vi}(AlSi_3)_2^{iv}O_{20}(OH)_4(s)$	2:1
Serpentine group		
Lizardite	$Mg_6^{vi}Si_2^{iv}O_{10}(OH)_8(s)$	1:1
Chlorite group		
Clinochlore	$(Mg_2Al)_2^{vi}(OH)_{12} \cdot (Mg,Fe)_6^{vi}(AlSi_3)_2^{iv}O_{20}(OH)_4(s)$	2:1:1
Chamosite	$(Fe_2Al)_2^{vi}(OH)_{12} \cdot (Mg,Fe)_6^{vi}(AlSi_3)_2^{iv}O_{20}(OH)_4(s)$	2:1:1
Clay minerals group		
Kaolinite	$Al_4^{vi}Si_4^{iv}O_{10}(OH)_8(s)$	1:1
Halloysite	$Al_4^{vi}Si_4^{iv}O_{10}(OH)_8 \cdot 4H_2O(s)$	1:1
Talc	$(Mg_6^{vi}Si_8^{iv}O_{20}(OH)_4(s)$	2:1
Pyrophyllite	$Al_4^{vi}Si_8^{iv}O_{20}(OH)_4(s)$	2:1
Vermiculite	$K_x \cdot Al_4^{vi}(Al_xSi_{8-x})^{iv}O_{20}(OH)_4(s)$	2:1
Montmorillonite	$K_x \cdot (Mg_xAl_{4-x})^{vi}Si_8^{iv}O_{20}(OH)_4(s)$	2:1

minerals to dominate the clay-size fraction in stages 8 through 10. The complete hydrolysis of clay minerals by stage 11 brings the Jackson weathering sequence to the final, most advanced chemical weathering stages.

3.4 CLAY MINERALS

3.4.1 Composition and Structure of Phyllosilicate Layers

Aluminosilicate tetrahedra polymerize to form a two-dimensional sheets that serves as a scaffold for crystal layers 0.70–1.4 nm thick separated by cleavage planes (Pauling, 1930a,b). A 0.7-nm layer-spacing is characteristic of minerals of the serpentine group and kaolinite clay mineral group, each layer composed of 1 tetrahedral sheet and 1 octahedral sheet (designated as the 1:1 layer type in Table 3.4). A 1.0-nm layer-spacing is characteristic of minerals of the mica group and clay mineral groups[10] with layers composed of 2 tetrahedral sheets and 1 octahedral sheet (designated as the 2:1 layer type in Table 3.4). A 1.4-nm layer-spacing is characteristic of chlorite group minerals with a single octahedral sheet alternating with 2:1 type layers (designated as the 2:1:1 layer type in Table 3.4).

[10] Clay minerals vermiculite and montmorillonite dehydrate when heated to 100°C, collapsing to 1.0 nm, otherwise these minerals swell to layer spacings exceeding 1.0 nm at 100% relative humidity.

Pyroxene, by virtue of its *I-beam* structure, contains only one type of octahedral site (cf. Figs. 3.4 and 3.5). The merging of two silicate chains to form the wider, double-chain amphibole *I-beam*, by contrast, contains three distinct octahedral sites: a central site with *trans* hydroxyl ions lying between two unlabeled sites and a pair of sites with *cis* hydroxyl ions (cf. Figs. 3.4 and 3.5). Condensing the amphibole double-silicate chains, forming tetrahedral sheets with the sixfold rings characteristic of phyllosilicates (cf. Fig. 3.3), reduces the number of distinct octahedral sites to two. Clay mineralogist designate the octahedral site with *trans* hydroxyl ions as site M1 and M2 as the octahedral site with *cis* hydroxyl ions.

If the phyllosilicate octahedral sheet is populated by divalent cations (e.g., Mg^{2+} or Fe^{2+}), then all octahedral sites must be occupied; clay mineralogists refer to these minerals as *trioctahedral*. Trioctahedral phyllosilicates listed in Table 3.4 include: biotite, lizardite, the chlorite group minerals, and talc. Clay mineralogists reserve the term *dioctahedral* for those phyllosilicates whose octahedral sheet is populated by trivalent cations (e.g., Al^{3+} or Fe^{3+}). Two-thirds of all octahedral sites are occupied in *dioctahedral* minerals: muscovite, kaolinite, pyrophyllite, vermiculite, and smectite group minerals (cf. Table 3.4). Example 3.2 demonstrates the procedure for calculating the unit cell formula for layer silicates (such as those appearing in Table 3.4) based on elemental composition.

3.4.2 Structure of Neutral-Layer Minerals

The indicator mineral for Jackson weathering stage 10 is the clay mineral kaolinite $Al_4Si_4O_{10}(OH)_8$, a clay mineral. Each kaolinite layer consists of one silicate tetrahedral sheet and one dioctahedral sheet (cf. Figs. 3.8 (top) and 3.9), with a layer spacing of 0.70 nm (Fig. 3.10).

Layer stacking periodicity distinguishes dickite, nacrite, and kaolnite. The kaolinite layer spacing is 0.71 nm because the orientation of all layers is identical. Dickite and nacrite have show two-layer periodicity (i.e., 1.43 and 1.44 nm, respectively) because the orientation of individual layers alternate.

The minerals kaolinite, dickite, and nacrite all have a planar crystal habit (Fig. 3.10) but halloysite adopts either a tubular or spherical crystal habit. The layer spacing in halloysite, significantly greater at 1.01 nm (Brindley and Goodyear, 1948) than the spacing in kaolinite, results from a water monolayer between the aluminosilicate layers of kaolinite. Dehydration by mild heating or low humidity results in a partial collapse to 0.75–0.79 nm.

Serpentine group minerals antigorite, chrysotile, and lizardite, with the ideal composition $Mg_6Si_4O_{10}(OH)_8$, represent the trioctahedral variation of the 1:1 layer structure found in kaolinite (Fig. 3.10). These minerals are very susceptible to chemical weathering, readily losing their highly soluble octahedral cations. Serpentine group minerals are largely limited to ultra mafic metamorphic rocks and, as such, are far less abundant in the continental lithosphere than mica group minerals found in granitic rocks.

The mineral talc $Mg_6(Si_4O_{10})_2(OH)_4$ is a familiar phyllosilicate typically found in ultra mafic rocks exposed to hydrothermic alteration as a serpentine alteration product. Each talc layer consists of two silicate tetrahedral sheets and one trioctahedral sheet (Fig. 3.11), with a layer spacing of 0.91–0.94 nm.

The mineral pyrophyllite $Al_4(Si_4O_{10})_2(OH)_4$, a dioctahedral mineral with a 2:1 layer structure identical to talc (Fig. 3.11), also belongs to the clay mineral group (cf. Table 3.4). Though a relatively rare mineral, it is most commonly found in metamorphic formations exposed to

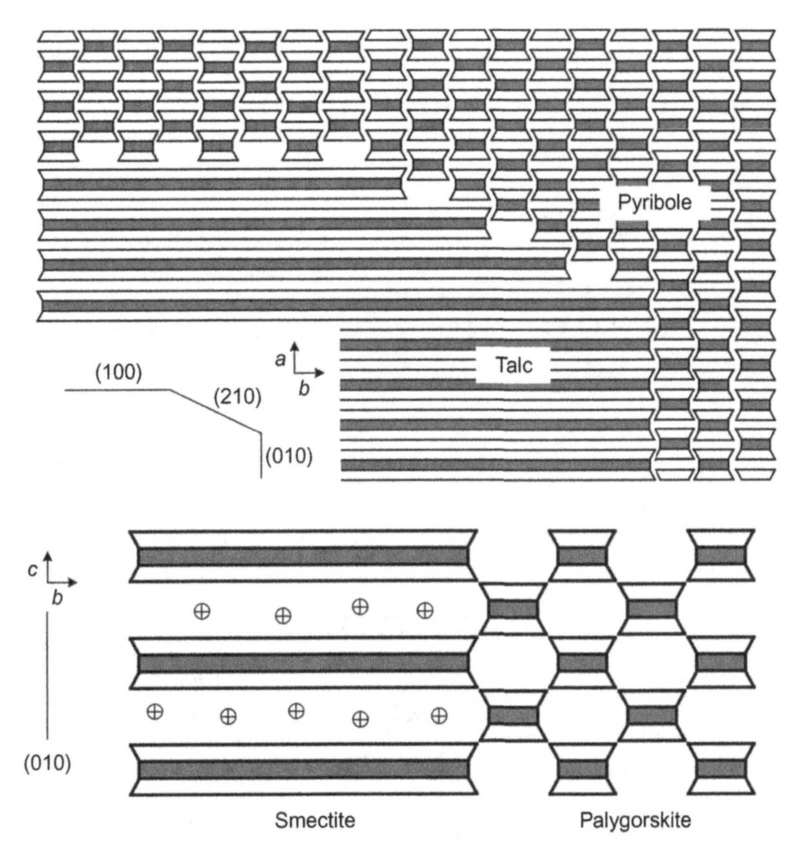

FIG. 3.8 Phyllosilicate minerals nucleate by expitaxy and grow by structural rearrangement. Top: *I-beam* model showing the three common types of coherent talc-pyribole boundaries (Fig. 17, Buseck and Veblen, 1981). Bottom: model showing growth of smectite perpendicular to the (010) crystal plane of palygorskite (Fig. 7, Krekeler et al., 2005). Cross-linking of amphibole double-chain *I-beams* generates the palygorskite structure.

low-temperature hydrothermal alteration which implies the presence of liquid water. Just as talc is often associated with serpentine group minerals, pyrophyllite is often found in formations that also contain kaolinite. When kaolinite and pyrophyllite occur together it appears the former is being transformed into the latter.

3.4.3 Layer Structure of Mica Group Minerals

Two types of mica are common in granite: the dioctahedral mineral muscovite and the trioctahedral mineral biotite. The layer structure (Fig. 3.12) of mica group minerals closely resembles pyrophyllite and talc, respectively (Fig. 3.11). There are, however, profound differences in composition (cf. Table 3.4).

Mica composition reveals Al^{3+} replaces about one of every four silicon ions Si^{4+} in the tetrahedral sheet and, furthermore, one K^{+} is added for each $Al^{3+} \longrightarrow Si^{4+}$ substitution (cf. Box 3.2). The charge neutrality condition for coupled cation substitution in mica constrains potassium stoichiometry: $\nu_{K} = 4 - \nu_{Si}$. Refinement of the mica crystal structure using

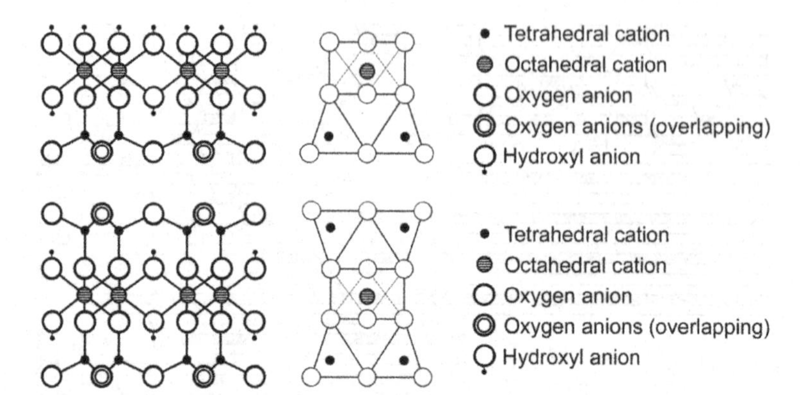

FIG. 3.9 Many phyllosilicate minerals are composed of layers containing both octaheral and tetrahedral sites configured much like those in pyribole minerals (cf. Figs. 3.4 and 3.5). Layers associated with 0.7-nm spacing (so-called 1:1 layers) appear at the top: ball-and-stick diagram (top left) and polyhedral diagram (top right). Layers associated with 1.0 nm spacing (so-called 2:1 layers) appear at the bottom: ball-and-stick diagram (bottom left) and polyhedral diagram (bottom right). The orientation of both layers emphasizes the vertical mirror plane oriented perpendicular to the page. *Adapted from Wells, A.F., 1976. Structural Inorganic Chemistry, fourth ed. Clarendon Press, Oxford.*

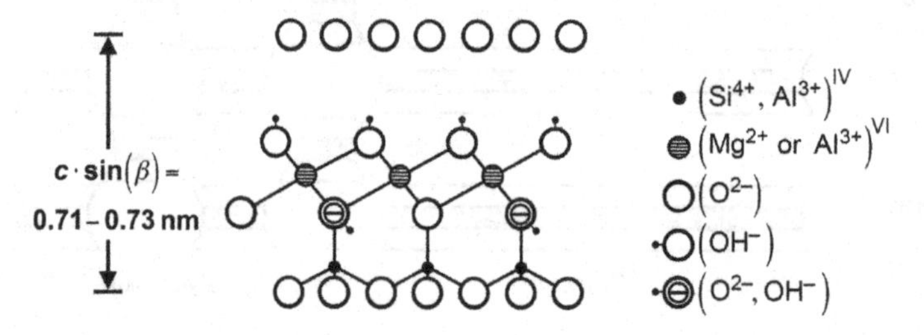

FIG. 3.10 Layer structure for minerals of the serpentine and kaolinite groups. The mirror plane in this layer orientation, unlike Fig. 3.9 (top), lies parallel to the page. *Adapted from Bailey, S.W., 1980. Structures of layer silicates. In: Brindley, G.W., Brown, G. (Eds.), Crystal Structures of Clay Minerals and their X-ray Identification, vol. 5. Mineralogical Society, London, pp. 1–123.*

single-crystal X-ray diffraction locates K^+ ions between mica layers. Although Al^{3+} cations replace Si^{4+} in the tetrahedral sheets crystal structure refinements reveal no apparent order to cation substitution in the tetrahedral sheet.

The coordination site occupied by K^+ in mica group minerals is similar to the large-cation site located between the base of amphibole *I-beams* (cf. the position marked by a cross in Fig. 3.4): each interlayer K^+ cation in mica group minerals is coordinated by tetrahedral-sheet sixfold rings (cf. Fig. 3.4) above and below.

Given the $Al^{3+} \longrightarrow Si^{4+}$ cation substitution rate (1-of-4), *every* sixfold ring in the tetrahedral sheets forming the interlayer gallery coordinate interlayer K^+ cations. Aluminum ions Al^{3+} cations replace 2-of-4 Si^{4+} cations in margarite,[11] a brittle mica. As a consequence one calcium ion Ca^{2+} occupies every sixfold ring of the tetrahedral sheet to balance layer charge.

[11] Margarite: $Ca \cdot Al_4^{vi}(Al_2Si_2)_2^{iv}O_{20}(OH)_4(s)$.

BOX 3.2

CATION SUBSTITUTION IN SILICATE MINERALS

Pauling's rules (Pauling, 1929) define broad restrictions governing cation substitution in minerals. For example, forsterite $Mg_2SiO_4(s)$ and fayalite $Fe(II)_2SiO_4(s)$ are olivine group minerals whose composition varies continuously.

forsterite-fayalite series: $Mg_{2-x}Fe(II)_xSiO_4(s)$

Solid-state chemists refer to the forsterite-fayalite series as a *complete binary solid solution* in which cation substitution allows composition to vary between the two end-members ($0 \leq x \leq 2$) while preserving the essential crystal structure. Likewise, pyroxene group minerals enstatite[12] and ferrosilite[13] are end-members of another complete binary solid solution. Cation substitution in solid-solution series, complete (cf. calcite-otavite and magnesite-siderite series) or incomplete (cf. magnesite-siderite series), involve cations with the same valence and similar ionic radii.

enstatite-ferrosilite series:

$Mg_{2-x}Fe(II)_xSi_2O_6(s)$

Tschermak substitution is a special type of *coupled* cation substitution that governs aluminum incorporation in amphiboles and, to a much lesser extent, pyroxenes.[14] The classic Tschermak substitution couples octahedral $Al^{3+} \longrightarrow Mg^{2+}$ substitution with tetrahedral $Al^{2+} \longrightarrow Si^{4+}$ minerals. The former generates excess positive charge within octahedra at the center of amphibole *I-beams* while the latter generates negative charge within the tetrahedral framework (cf. Figs. 3.4 and 3.5).

The charge neutrality condition demands a 1-to-1 coupling of each Tschermak substitution. As we will see below, Tschermak substitution couples octahedral $Al^{3+} \longrightarrow Mg^{2+}$ substitution in the interlayer "brucite" sheet of chlorite to tetrahedral $Al^{3+} \longrightarrow Si^{4+}$ substitution in the 2:1 layer (cf. Fig. 3.14).

Cations substitutions in mica group, vermiculite and smectite group minerals should be viewed as coupled substitutions similar in effect to Tschermak substitutions. Tschermak-like tetrahedral $Al^{3+} \longrightarrow Si^{4+}$ substitution deposits negative charge on the tetrahedral sheets of muscovite, biotite, vermiculite and the smectitie-group mineral bidellite. Likewise, Tschermak-like octahedral $Mg^{2+} \longrightarrow Al^{3+}$ substitution deposits negative charge on the octahedral sheet of montmorillonite, a smectite group mineral. Charge neutrality demands a 1-to-1 "substitution" of positive charge which, in the case of mica group minerals, vermiculite and smectite minerals, appears as an interlayer cation.

[12] $Mg_2Si_2O_6(s)$.
[13] $Fe(II)_2Si_2O_6(s)$.
[14] Tschermak substitution in pyroxene minerals $Mg_{2-x}Al_xSi_2O_6(s)$ is very limited: $x \ll 1$.

The combined effect of strong electrostatic forces binding negatively-charge mica layers together through interlayer K^+ cations, the low hydration enthalpy of K^+ cations, and the snug fit of bare K^+ cation into the sixfold rings of the tetrahedral sheets exclude water from the mica interlayer. In short, the layer spacing of mica group minerals remains unchanged when immersed in water. The fixed mica layer spacing—0.96–1.01 nm—is significantly greater than the layer spacing of talc and pyrophyllite because interlayer K^+ ions prop open the layers.

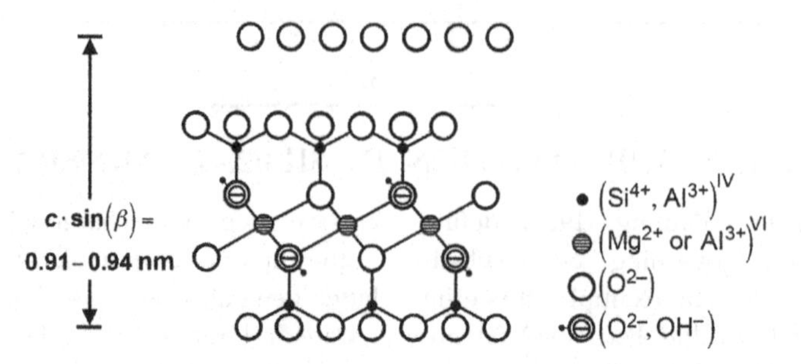

FIG. 3.11 Layer structure for the minerals talc and pyrophyllite. The mirror plane in this layer orientation, unlike Fig. 3.9 (top), lies parallel to the page. *Adapted from Bailey, S.W., 1980. Structures of layer silicates. In: Brindley, G.W., Brown, G. (Eds.), Crystal Structures of Clay Minerals and their X-ray Identification, vol. 5. Mineralogical Society, London, pp. 1–123.*

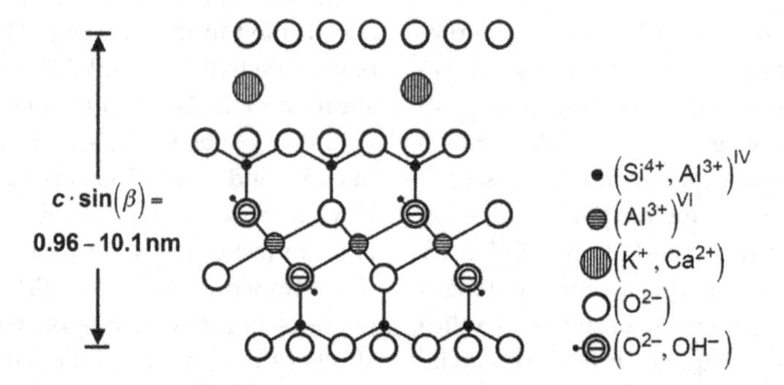

FIG. 3.12 Layer structure for mica group minerals. The interlayer cation is usually K^{+}. The mirror plane in this layer orientation, unlike Fig. 3.9 (top), lies parallel to the page. *Adapted from Bailey, S.W., 1980. Structures of layer silicates. In: Brindley, G.W., Brown, G. (Eds.), Crystal Structures of Clay Minerals and their X-ray Identification, vol. 5. Mineralogical Society, London, pp. 1–123.*

As noted at the beginning of this chapter (cf. footnote 4), some clay mineralogists use the term illite when referring to clay-fraction specimens whose composition and powder X-ray diffraction patterns resemble mica group minerals. Others believe specimens with these characteristics are not a distinct mineral but, rather, interstratified mixtures of mica and other 2:1 clay minerals (cf. Jackson weathering stages 8 and 9, Table 3.2).

3.4.4 Layer Structure of Vermiculite and Smectite Group Minerals

Cation substitution in vermiculite is largely $Al^{3+} \longrightarrow Si^{4+}$ replacement in the tetrahedral sheet, identical in location but significantly less than the 1-of-4 substitution rate in micas (Fig. 3.13). Smectite group minerals (Fig. 3.13), distinguished from vermiculite by lower layer charge, reveal cation substitution in both the octahedral sheet (e.g., $Mg^{2+} \longrightarrow Al^{3+}$ in montmorillonite) and $Al^{3+} \longrightarrow Si^{4+}$ substitution in the tetrahedral sheet (e.g., beidellite).

Table 3.5 lists the properties of vermiculite and some of the common smectite minerals found in nature. Layer structure and composition have much in common with mica group minerals (Fig. 3.13). The distinguishing properties of these clay minerals are: layer charge, accessible particle size (i.e., largely or exclusively confined to the clay-size fraction), and the interlayer water. The close association between layer charge magnitude and interlayer

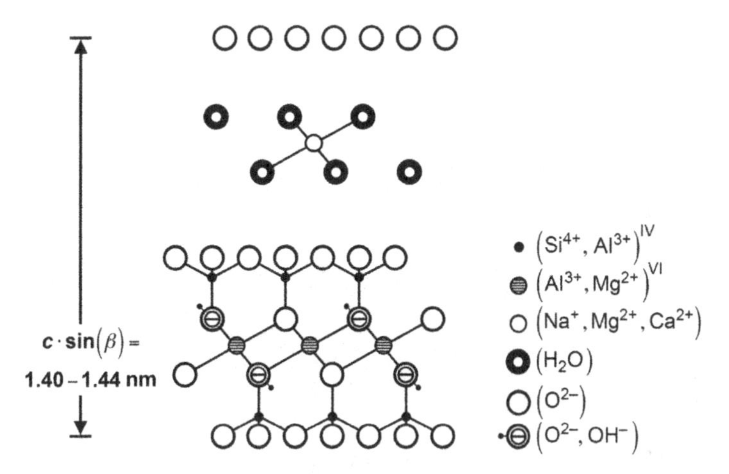

FIG. 3.13 Layer structure for vermiculite and smectite group minerals. The interlayer cation is exchangeable (i.e., interlayer cation composition is variable) and fully hydrated by water molecules. The mirror plane in this layer orientation, unlike Fig. 3.9 (top), lies parallel to the page. *Adapted from Bailey, S.W., 1980. Structures of layer silicates. In: Brindley, G.W., Brown, G. (Eds.), Crystal Structures of Clay Minerals and their X-ray Identification, vol. 5. Mineralogical Society, London, pp. 1–123.*

TABLE 3.5 Cation Substitution and Layer Charge Properties of Vermiculite and Smectite Group Minerals

Mineral	Substituting Cations	Stoichiometry [a]	CEC[b] ($cmol_c\ kg^{-1}$)
Montmorillonite	$(Mg^{2+} \longrightarrow Al^{3+})^{vi}$	0.50–1.2	60–150
Beidellite	$(Al^{3+} \longrightarrow Si^{4+})^{iv}$	0.50–1.2	60–150
Nontronite	$(Al^{3+} \longrightarrow Si^{4+})^{iv}$	0.50–1.2	60–150
Vermiculite	$(Al^{3+} \longrightarrow Si^{4+})^{iv}$	1.2–1.4	150–165

[a] *Cation substitution stoichiometry is based on the unit cell formula unit* $O_{20}(OH)_4$.
[b] *Cation exchange capacity.*

chemistry are taken up in the section entitled *Clay Chemistry*. Suffice to say: layer charge and interlayer chemistry are the signature properties of one of the most important mineral colloids in nature.

3.4.5 Layer Structure of Chlorite Group Minerals

X-ray diffraction identifies two clay minerals with a characteristic 1.4 nm layer spacing: the chlorite mineral group and a hydroxy-interlayered vermiculite or smectite. Hydroxy-interlayered smectite and vermiculite (cf. Chapter 6) are typically found in moderately weathered (i.e., Jackson weathering stages 8 and 9) acidic soils.

The crystal structure of clinoclore, the chlorite group mineral listed in Table 3.4, consists of two layer types as shown in Fig. 3.14. The 2:1 layer, structurally identical to those found in mica group minerals, alternates with a second layer *structurally* similar to the layers in brucite $Mg(OH)_2(s)$. Many clay mineralogists designate chlorite group minerals as 2:1:1 phylloslicate

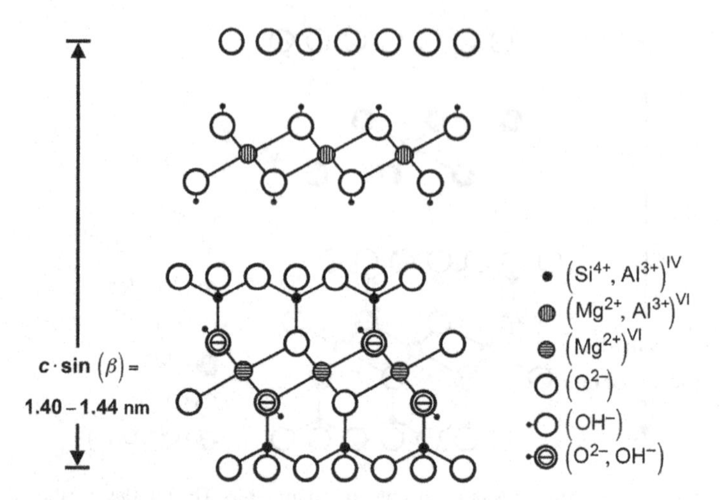

FIG. 3.14 Layer structure for chlorite group minerals. An octahedral layer equivalent to a single layer from the mineral brucite $Mg(OH)_3$ lies between 2:1 layers. Tschermak $Al^{3+} \longrightarrow Mg^{2+}$ substitutions in the brucite layer are coupled to $Al^{3+} \longrightarrow Si^{4+}$ substitutions in the tetrahedral sheets of the 2:1 layer. The mirror plane in this layer orientation, unlike Fig. 3.9 (top), lies parallel to the page. *Adapted from Bailey, S.W., 1980. Structures of layer silicates. In: Brindley, G.W., Brown, G. (Eds.), Crystal Structures of Clay Minerals and their X-ray Identification, vol. 5. Mineralogical Society, London, pp. 1–123.*

minerals (Table 3.4). Tschermak substitution in chlorite group minerals (cf. Box 3.2) couple tetrahedral $Al^{3+} \longrightarrow Si^{4+}$, substitutions in the 2:1 layer to octahedral $Al^{3+} \longrightarrow Mg^{2+}$, substitutions in the brucite-like layer.

EXAMPLE 3.2

Example Permalink

http://soilenvirochem.net/zvW0en

Given the chemical composition of a layer silicate specimen, compute the unit cell formula.

The unit cell formula cannot be computed without an essential parameter determined by X-ray diffraction: the number of oxygen atoms in the unit cell. The unit cell oxygen count is used to normalize the number of cation per unit cell. The unit cell formulas for many silicate minerals group cations according to coordination: tetrahedral, octahedral, and interlayer coordination for layer silicate minerals.

Step 1. Determine the number of each type of cation per unit cell.

This step begins with the chemical composition expressed as moles of oxide for each cation in the analysis. If your chemical analysis lists the mass fraction of each element $w(E)$ [$kg\,kg^{-1}$], you will need to convert that analysis to a mass fraction for each oxide (cf. Example 6.1).

Table 3.6 lists the chemical composition of Panther Creek montmorillonite (Aberdeen, MS) from the Panther Creek mine in Monroe County, Mississippi (Grim, 1928).[15]

TABLE 3.6 The Chemical Composition of the Panther Creek Montmorillonite Specimen (Foster, 1953), Expressed in Oxide Mass-Fraction Units

Oxide	M_f(oxide) g mol^{-1}	w(oxide) g g^{-1}	m(oxide) mol g^{-1}
$SiO_2(s)$	60.0844	0.5568	9.27×10^{-3}
$Al_2O_3(s)$	101.9614	0.1924	1.90×10^{-3}
$Fe_2O_3(s)$	159.6887	0.0651	4.08×10^{-4}
$TiO_2(s)$	79.8656	0.0080	1.00×10^{-4}
$MgO(s)$	40.3045	0.0230	5.71×10^{-4}
$Na_2O(s)$	61.9790	0.0248	4.00×10^{-4}

The *oxide* mole-per-mass concentration m(oxide) (column 4) is calculated by dividing the oxide mass fraction w(oxide) (column 3) by the oxide formula mass m_f(oxide) (column 2).

Although there are 24 oxygen atoms in the $O_{20}(OH)_4$ unit cell, the unit cell anion charge is equivalent to 22 divalent oxygen anions. Multiplying the m(oxide) values in column 4 of Table 3.6 by the oxygen stoichiometry of each oxide yields the values listed in column 2 of Table 3.7.

For example, the oxygen stoichiometry of SiO_2 requires multiplying 9.27×10^{-3} from Table 3.6 by 2 to yield 1.85×10^{-2} in Table 3.7. The sum over all m(O) values for each oxide yields $m(O) = 2.66 \times 10^{-2}$ mol g^{-1}. The normalizing factor f_N [cell g^{-1}] is the summed moles of oxygen per gram divided by the number of oxygen atoms in the unit cell.

$$f_N = \frac{2.66 \times 10^{-2}}{22} = 1.21 \times 10^{-3} \text{ cell g}^{-1}$$

TABLE 3.7 The Normalized Elemental Composition of the Panther Creek Montmorillonite Unit Cell

Oxide	m(O) mol$_O$ g^{-1}	m_N(O) mol$_O$ cell^{-1}	Cation	m_N(E) mol cell^{-1}
$SiO_2(s)$	1.85×10^{-2}	15.30	Si^{4+}	7.65
$Al_2O_3(s)$	5.71×10^{-3}	4.72	Al^{3+}	3.15
$Fe_2O_3(s)$	1.22×10^{-3}	1.01	Fe^{3+}	0.67
$TiO_2(s)$	2.00×10^{-4}	0.17	Ti^{4+}	0.08
$MgO(s)$	5.71×10^{-4}	0.47	Mg^{2+}	0.47
$Na_2O(s)$	4.00×10^{-4}	0.33	Na^{+}	0.66

The values listed in column 3 of Table 3.7 are computed by dividing the m(O) values in column 2 by the normalizing factor $f_N = 1.21 \times 10^{-3}$. A sum over the normalized values m_N(O) listed in column 3 should equal 22 if the normalization is correct. Finally, the normalized moles per unit cell for each cation m_N(E) listed in column 4 appear in column 5. These values are computed by multiplying the moles of oxygen per unit cell for each oxide by the moles of cation per oxide. For example, there is 1/2 moles of Si^{4+} in each mole of the oxide $SiO_2(s)$.

$$m_N(Si^{4+}) = \frac{1}{2} \cdot m(SiO_2) = \frac{1}{2} \cdot 15.30 = 7.65$$

Step 2. Assign the cations to each coordination site to determine the final unit cell formula.

In this case, as with all layer silicates, the rules for assigning cations is very simple. First, all Si^{4+} cations are assigned to unit cell tetrahedral coordination sites. Since there are a total of eight (8) tetrahedral sites in each unit cell, another cation must occupy the remaining 0.35 tetrahedral sites. Extensive research has revealed that of the remaining cations only Al^{3+} cations will occupy those remaining sits. The stoichiometry coefficient for tetrahedral Al^{3+} is 0.35.

All of the remaining cations except Na^+ can occupy octahedral sites; assign all 0.66 moles of Na^+ to the interlayer.

Each $O_{20}(OH)_4$ unit cell contains a total of 6 octahedral sites. Dioctahedral 2:1 layer silicates will have approximately 4 octahedral sites occupied by a combination of trivalent cations (e.g., Al^{3+} and Fe^{3+}) and divalent cations (e.g., Mg^{2+}). The total number of cations listed in column 4 of Table 3.7 that can occupy octahedral sites are 3.86.

$$m(Al^{3+}) + m(Fe^{3+}) + m(Ti^{4+}) + m(Mg^{2+}) = 4.03$$

Cations substitutions $(Al^{3+} \longrightarrow Si^{4+})^{iv}$ and $(Mg^{2+} \longrightarrow Al^{3+})^{vi}$ add negative layer charge while cation substitution $(Ti^{4+} \longrightarrow Al^{3+})^{vi}$ adds positive layer charge. Net negative layer charge equals 0.74 $mol_c\ cell^{-1}$, which is slightly larger than the 0.66 $mol_c\ cell^{-1}$ assigned to the interlayer based on the Na^+ analysis.

The unit cell formula based on the values in Table 3.7 and the assignments just discussed appears below.

$$[Al_{2.80}Fe_{0.67}Ti_{0.08}Mg_{0.47}][Al_{0.35}Si_{7.65}]O_{20}(OH)_4 \cdot Na_{0.66}$$

Foster (1953) ignores the titanium content when writing her unit cell formula for the Panther Creek montmorillonite.

$$[Al_{2.88}Fe_{0.68}Mg_{0.48}][Al_{0.28}Si_{7.72}]O_{20}(OH)_4 \cdot Na_{0.64}$$

[15] The Panther Creek mine is in the vicinity of the Itawamba mine, source of the Amory montmorillonite (API-22). Both samples came from the same geologic formation and have a similar compositions.

The structure and layer spacing of hydroxy-interlayered smectite and vermiculite resemble chlorite group minerals, but there are significant compositional differences. First, the extent and location of cation substitution in the 2:1 primary layers of these hydroxy-interlayered clay minerals are characteristic of vermiculite and smectite clay minerals. Second, the interlayer octahedral sheet is not continuous; the interlayer is populated by a combination of hydrated exchangeable Al^{3+} cations and *pillars* whose structure and composition resemble gibbsite $Al(OH)_3(s)$.

As we will see later, the exchangeable Al^{3+} cations and pillars of hydroxy-interlayered smectites results from the combined effects of Al^{3+} solubility at low pH, Al^{3+} cation-exchange selectivity and the hydrolysis of hydrated Al^{3+} cations within the interlayer of vermiculite and smectite minerals.

3.5 CLAY COLLOID CHEMISTRY

3.5.1 Clay Mineral Plasticity

Clay mineral plasticity (Box 3.3) is a direct result of crystal habit (Norton, 1948; Lubliner, 2008). Lubliner (2008) defines *plastic* materials as:

> "...materials, such as ductile metals, clay, or putty, which have the property that bodies made from them can have their shape easily changed by the application of appropriately directed forces, and retain their new shape upon removal of such forces."

Furthermore, phyllosilicate minerals posses an essential characteristic necessary for *plastic deformation* (Lubliner, 2008).

> "...plastic deformation is the result of relative motion, or slip, on specific crystallographic planes, in response to shear stress along these planes."

Plastic deformation under applied stress forces clay plates,[16] lubricated by adsorbed water, to slide past each other.

BOX 3.3

ATTERBERG PLASTICITY LIMITS.

In 1911 the Swedish chemist and agricultural scientist Atterberg devised scheme for quantifying and classifying soil plasticity that remains in use today (ASTM, 2014). Atterberg (1911) originally defined six limits in order of increasing water content, here we consider only two: the *rolling* or *plastic* limit and the *flow* or *liquid* limit. An Atterberg limit is a moisture content, expressed as water-to-soil mass ratio in percent.

$$m_W : \text{mass, water}$$

$$m_S : \text{mass, dry soil}$$

$$w \equiv \frac{m_W}{m_S} \cdot 100$$

Plasticity first appears at the plastic limit w_P, at lower water contents the soil or clay specimen is a moist semi-solid that cannot be formed into a ribbon (Box 3.1) or wire (ASTM, 2014).

The soil or clay remains malleable with increasing water content until it reaches the point where it begins to flow when unmolded; this is the liquid limit w_L.

Soil scientists and geotechnical engineers rely on three parameters to quantify the soil and clay specimen plasticity: plastic limit w_L, liquid limit w_L and *plasticity index PI* (expression 3.3).

$$PI = w_L - w_P \tag{3.3}$$

Soil scientists and civil engineers use *activity A* (expression 3.4) to reveal the influence of clay-fraction mineralogy on soil plasticity.

$$m_C : \text{mass, dry clay fraction}$$

$$c \equiv (m_C/m_S) \cdot 100$$

$$A = \frac{PI}{c} \tag{3.4}$$

[16] The crystallographic slip plane is parallel to clay layers but deformation can take place if plates comprised of many layers are being displaced.

Bain (1971) provides a concise survey of clay mineral plasticity. As noted earlier, clay-size aluminum, iron, and titanium oxides minerals defining Jackson weathering stages 11–13 do not exhibit plasticity.

Kaolinite clay specimens have w_P ranging from 30 to 40. Halloysite clay specimens have consistently higher w_P values, ranging from 60 to 70. This difference can be explained, in part, by the interlayer water (cf. Table 3.4 and Section 3.4.2) distinguishing halloysite from kaolinite (Bain, 1971). The PI for kaolinite and halloysite specimens both range from 10 to 40. The unique nanotube crystal habit apparently allows halloysite clays to absorb more water than kaolinite clays, accounting for higher w_P and PI values than could be assigned simply to interlayer water.[17]

Natural variation in kaolinite plasticity are linked to how the specimen formed. Hydrothermal kaolinite deposits, which yield particles that are rather large and well-crystallized, have very low plasticity PI. Kaolinite formed through chemical weathering, yielding smaller and more poorly-crystalline particles, tends to have higher plasticity PI. Complicating this assessment is the co-mingling of kaolinite and smectite in most chemical weathering environments. Distinguishing kaolinite particle-size or crystallinity effects from plasticity derived from trace amounts of smectite in the specimen is usually impossible.

A similar assessment applies to illite plasticity. Many clay mineralogists believe illite specimens are simply interstratified mica and smectite layers. Interstratified illite specimens would derive their plasticity from trace amounts of smectite in the sample rather than the distinct plasticity of an illite mineral. The plasticity of vermiculite and smectite group minerals is taken up below in a broader discussion of swelling behavior.

3.5.2 Hydration and Swelling of Smectite and Vermiculite Clay Minerals

3.5.2.1 Crystalline Swelling

A detailed picture of clay mineral swelling began to emerge in the early 1950s, as illustrated by the X-ray diffraction results in Fig. 3.15.[18] Increasing the water-to-clay ratio (Fig. 3.15A) causes the layer spacing —indicated by X-ray reflection {001}—to increase in a stepwise manner as water enters the interlayer gallery *one molecular layer at a time.*

Crystalline swelling is the term coined to describe these abrupt jumps in d_{001} since these increases in layer spacing equal the diameter of a water molecule: almost 0.25 nm. Crystalline swelling is a universal among vermiculite and smectite group clay minerals (disregarding the behavior of hydroxy-interlayered specimens).

The conditions prompting the insertion of a water monolayer and the number of monolayers taken up depend on the interlayer cation. For example, montmorillonite saturated by the weakly-hydrated cation K^+ add the first and second water monolayers at the same water content as Na^+-saturated montmorillonite (cf. Fig. 3.15A) but the former is unable reach the 4-layer hydrate accessible to the latter. Mooney et al. (1952) did not observe Cs^+-saturated

[17] Palygorskite and sepiolite lack the characteristic layer structure of other clay minerals (cf. Botton, Fig. 3.7). The cross-linked *I-beam* structure favors a fibrous crystal habit that retains water in a manner analogous to halloysite. Water retention by palygorskite and sepiolite fibers and by halloysite nanotubes, however, does not significantly enhance the plasticity.

[18] cf. Appendix 3.A for an explanation of how the distance between crystal planes is measured using X-ray diffraction.

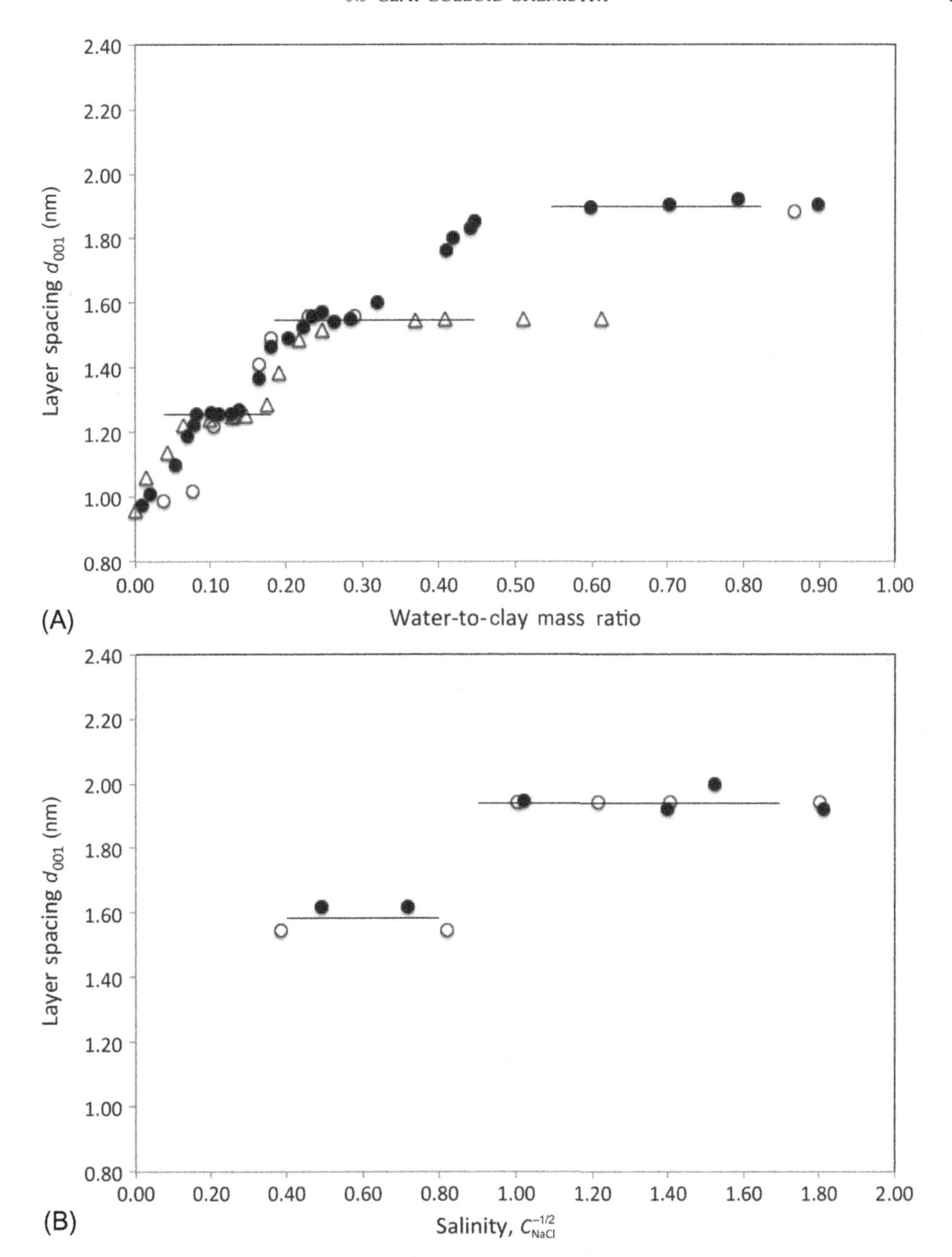

FIG. 3.15 Stepwise and crystalline swelling of sodium-saturated (*open and filled circles*) and potassium-saturated (*open triangles*) montmorillonite clay specimens as influenced by (A) the water-to-clay ratio and (B) water salinity. *Data from Norrish, K., 1954. The swelling of montmorillonite. Discuss. Faraday Soc. 18, 120–134 and Foster, W.R., Savins, J., Waite, J., 1954. Lattice expansion and rheological behavior relationships in water montmorillonite systems. Clays Clay Miner. 3, 296–316.*

montmorillonite swelling beyond the 1-layer hydrate (d_{001} ≈ 1.25nm. The crystalline swelling of vermiculite and smectite group minerals saturated by divalent interlayer cations such as Ca^{2+} and Mg^{2+} is usually restricted to 2-layer hydrates (Mooney et al., 1952; De la Calle et al., 1976).

The crystalline swelling results appearing in Fig. 3.15B reveal another notable swelling characteristic: vermiculite and smectite clay minerals immersed in saline water also swell as salinity decreases. Clay mineral swelling induced by decreased salinity is indistinguishable from swelling induced by adding water to salt-free specimens.

Interlayer swelling results from a fortuitous balance of forces. Electrostatic forces bind negatively-charged clay layers through positively-charged interlayer cations. Cation hydration increases the effective cation radius in liquid water; hydrating interlayer cations require increased layer spacing to accommodate the larger effective radius of hydrated interlayer cations (Fig. 3.16).

The layer charge in mica group minerals generates electrostatic forces binding interlayer cations to negatively-charge layers that are greater in magnitude than the forces that bind K^+ cations to water molecules; hence interlayer swelling cannot occur. Layer charge in vermiculite and smectite group minerals, however, is comparable in magnitude to the hydration energy of many cations. The hydration of interlayer cations forces clay layers apart (cf. Fig. 3.15 and Fig. 3.16), drawing water molecules into the interlayer gallery.

3.5.2.2 *Hydrated Interlayer of Swelling Clay Minerals*

Unfortunately, no currently-available experimental technique can reveal the detailed structure of hydrated smectite interlayers,[19] leading some clay chemists to rely on molecular dynamics simulations (Fig. 3.16).

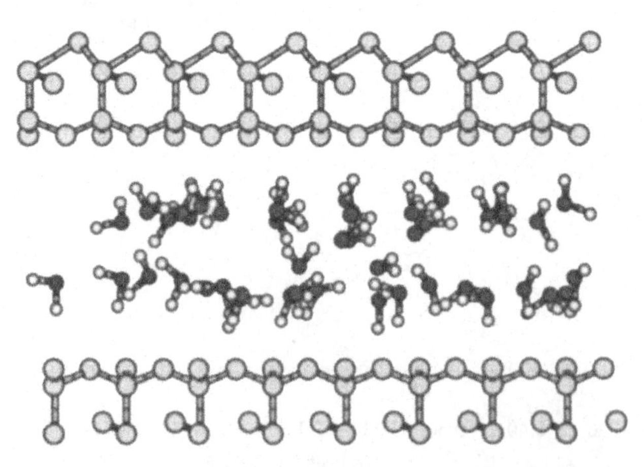

FIG. 3.16 A molecular dynamics simulation of a two-layer hydrate of Na^+ smectite showing a typical distribution of water molecules.*Chavez-Paez, M., Van Workum, K., de Pablo, L., de Pablo, J.J., 2001. Monte Carlo simulations of Wyoming sodium montmorillonite hydrates. J. Chem. Phys. 114 (3), 1405–1413.*

[19] Crystal structure refinements giving the position of interlayer cations and water molecules exist for vermiculite specimens (e.g., Mathieson and Walker, 1954; De la Calle et al., 1976).

Molecular dynamics simulations generate a host of predictions about the most probable structure of the interlayer: the detailed positions and orientations of water molecules, cation positions, interlayer spacing under specific conditions, and so on. To the extent these simulations match measurable parameters (e.g., interlayer spacing), we can cautiously accept interlayer structure predictions that are currently not verifiable by experiment.

Fig. 3.17 displays the probability distribution of hydrated interlayer K^+ for three crystalline swelling states. Interlayer K^+ ions (Fig. 3.17) remain near the clay surface as water enters the interlayer, regardless of the number of water layers being added. This behavior is consistent with the swelling behavior of K^+-saturated montmorillonite in Fig. 3.16A.

Fig. 3.18 displays probability distribution of interlayer Na^+ normal to the crystal layer for three crystalline swelling states. Unlike K^+, interlayer Na^+ ions move from the clay surface in the 1-layer hydrate to the center of the interlayer in the 2-layer hydrate (Fig. 3.18).

The electrostatic force between clay layer and interlayer cations not only depend on the hydrated radius of the cation, it also depends on the polyhedral sheet where cation substitution occurs. The strongest electrostatic force exists when cation substitution is in the tetrahedral sheet ($Al^{3+} \longrightarrow Si^{4+}$) and cation hydration energy is relatively weak. In this configuration the cation comes into close contact with the negatively-charge sites in the tetrahedral sheet. The weakest force exists when cation substitution is in the octahedral sheet ($Mg^{2+} \longrightarrow Al^{3+}$)

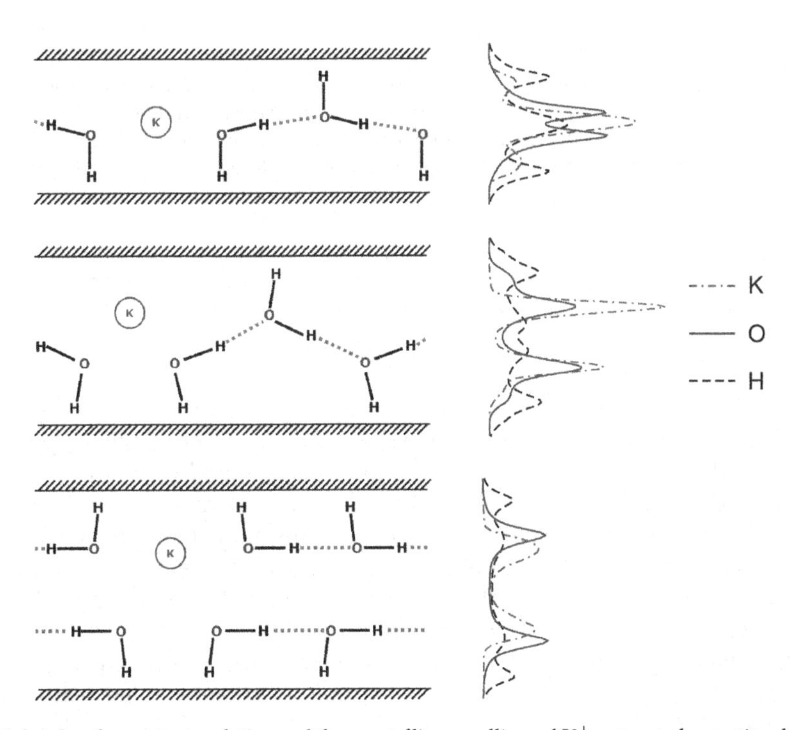

FIG. 3.17 Molecular dynamic simulations of the crystalline swelling of K^+-saturated smectite clay representing d_{001} = 1.25 nm, one-layer hydrate (top); d_{001} = 1.35 nm, intermediate hydrate (middle); and d_{001} = 1.45 nm, two-layer hydrate (bottom). The probability distributions normal to the clay layer appear on the right. *Tambach, T.J., Bolhuis, P.G., Hensen, E.J.M., Smit, B., 2006. Hysteresis in clay swelling induced by hydrogen bonding: accurate prediction of swelling states. Langmuir 22 (3), 1223–1234.*

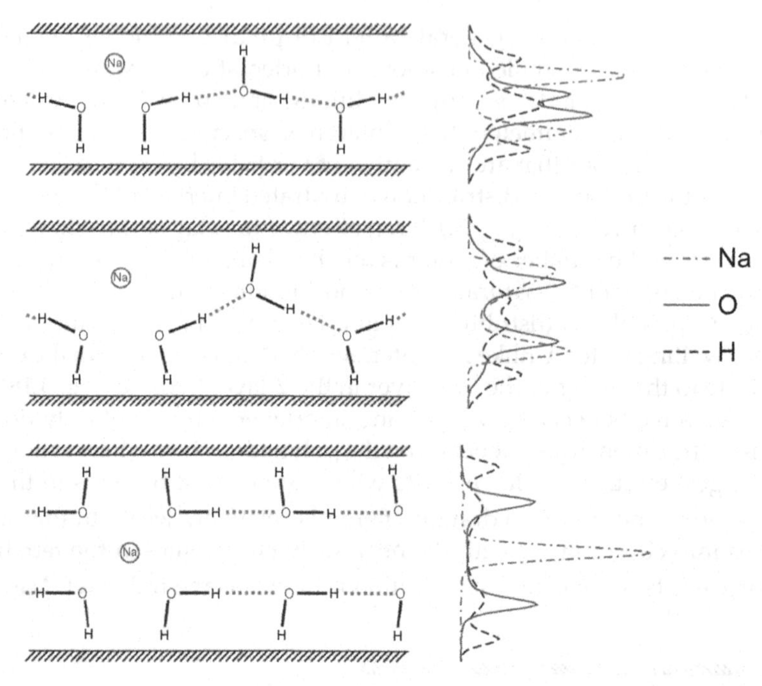

FIG. 3.18 Molecular dynamic simulations of the crystalline swelling of Na^+-saturated smectite clay representing $d_{001} = 1.25$ nm, one-layer hydrate (top); $d_{001} = 1.35$ nm, intermediate hydrate (middle); and $d_{001} = 1.45$ nm, two-layer hydrate (bottom). The probability distributions normal to the clay layer appear on the right. *Tambach, T.J., Bolhuis, P.G., Hensen, E.J.M., Smit, B., 2006. Hysteresis in clay swelling induced by hydrogen bonding: accurate prediction of swelling states. Langmuir 22 (3), 1223–1234.*

and cation hydration energy is strong. In this configuration the distance separating interlayer cations the negative layer charge is significantly greater.

3.5.2.3 *Free Swelling*

Free swelling is the continuous increase in layer spacing; stepwise increases in layer spacing as individual water monolayers enter the interlayer gallery are no longer apparent. The free swelling of Na^+-saturated montmorillonite specimens is plotted in Fig. 3.19.

Free swelling induced by increases in the water-to-clay mass ratio is indistinguishable from that induced by decreasing salinity because the same mechanism controls water uptake by interlayer cations in both cases. Free swelling vermiculite and smectite clay minerals diverge from crystalline swelling in this single characteristic: free swelling occurs only if the interlayer cation is a monovalent, strongly hydrated cation.

Vermiculite, which has a higher layer charge than smectite group clay minerals, will free swell if saturated by Li^+ but is restricted to crystalline swelling when saturated by Na^+ (Norrish and Rausell-Colom, 1963). Smectite group clay minerals free swell if saturated by Na^+ (cf. Fig. 3.19 and references cited in the caption) or Li^+ (Norrish, 1954; Foster et al., 1954) but are restricted to crystalline swelling when saturated by K^+ (Foster et al., 1954) or other larger radii Group 1 cations and all divalent cations (Examples 3.3 and 3.4).

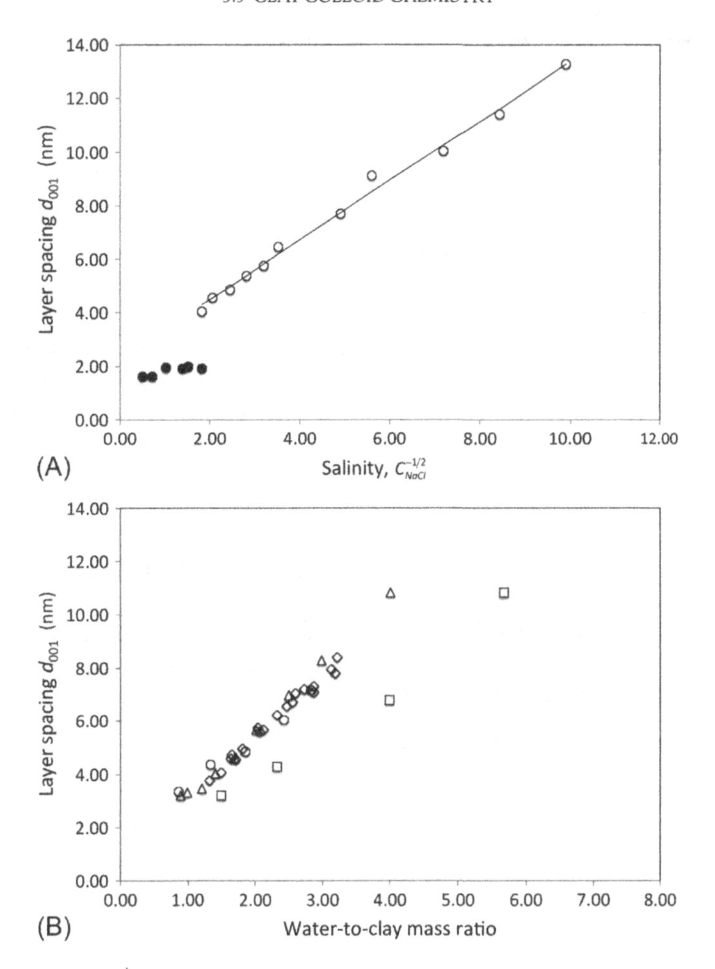

FIG. 3.19 Free swelling of Na^+-saturated montmorillonite clay specimens as influenced by NaCl(aq) concentration and the water-to-clay mass ratio. Upper graph (A) plots layer spacing d_{001} as a function of the inverse square root of the NaCl(aq) concentration: crystalline swelling (*filled circles*) and free swelling (*open circles*). Lower graph (B) plots layer spacing d_{001} as a function of water-to-clay mass ratio (Norrish, 1954; Foster et al., 1954; Hight et al., 1962). The open squares are from Fink et al. (1968)

3.5.2.4 Osmotic Equivalence of Electrolyte Solutions and Smectite Gels

The swelling of vermiculite and smectite group clay minerals results from interlayer cation hydration that draws water into the interlayer gallery, forcing the clay layers apart (Examples 3.3 and 3.4). The water-to-clay mass ratio w/c implies the tension head h_{tension} (cf. Chapter 2) will correlate with clay swelling.

This section introduces *osmotic head* h_{osmotic} to quantify the salinity effect on clay swelling. Fig. 3.22 plots data appearing in Fig. 3.15B and Fig. 3.19B with osmotic head h_{osmotic} replacing NaCl(aq) solution concentration.

EXAMPLE 3.3

Example Permalink

http://soilenvirochem.net/66XcLr

Use unit cell dimensions, layer spacing, and liquid water properties to estimate the water-to-clay mass ratio for sodium-saturated montmorillonite at the liquid limit.

Clay chemists studying smectite swelling find 90–95% of the water absorbed by smectite clay specimens is interlayer water, relatively little water is adsorbed to external surfaces (Fink et al., 1968). This example will use unit cell dimensions representative of montmorillonite clay specimens (Collis-George, 1955), the layer spacing of Na^{+}-saturated montorillonite swollen in pure water, and the molar volume of liquid water to estimate the water-to-clay mass ratio and compare this interlayer estimate with the empirical water-to-clay mass ratio of the liquid limit w_L.

Step 1. Determine the layer spacing d_{001} of a Na^{+}-saturated montorillonite swollen in pure water at the liquid limit w_L.

The liquid limit for Na^{+}-saturated montorillonite specimens collected from the Clay Spur bed in northeastern Wyoming range from $w/c = 4.99$ (Rioux, 1958; Bain, 1971; Magcogel bentonite) to $w/c = 6.25$ (White, 1949; Belle Fourche (API-27) montmorillonite). Norrish (1954), Foster (1954), Hight et al. (1962) and Fink et al. (1968) recorded the free swelling of Na^{+}-saturated montmorillonite specimens in pure water as function of water content, this example will use experimental results for the Belle Fourche montmorillonite (White, 1949; Foster et al., 1954).

$$\left[Al_{3.26}Fe_{0.34}Mg_{0.40}\right]\left[Al_{0.14}Si_{7.86}\right]O_{20}(OH)_4 \cdot Na_{0.56}$$

The area of a single unit cell is the product of unit-cell a- and b-dimensions in the layer plane: $a = 0.516\,\text{nm}$ and $b = 0.893\,\text{nm}$. Unit cell molar mass, however, is computed using atomic mass $M(\text{E})\ \left[\text{g mol}^{-1}\right]$ values, therefore the unit-cell area we require must be multiplied by the *Avogadro* constant $N_A = 6.022 \times 10^{+23}\ \text{mol}^{-1}$.

$$A_{\text{m, cell}} = (a \cdot b) \cdot N_A$$
$$A_{\text{m, cell}} = \left(0.516 \times 10^{-9}\ \text{m}\right) \cdot \left(0.893 \times 10^{-9}\ \text{m}\right) \cdot 6.022 \times 10^{+23}\ \text{mol}^{-1}$$
$$A_{\text{m, cell}} = 2.77 \times 10^{+5}\ \text{m}^2\ \text{mol}^{-1}$$

The unit cell volume $V_{\text{m, cell}}\ [\text{m}^3\ \text{mol}^{-1}]$ is computed using the layer spacing d_{001}. An estimate of the water-to-clay mass ratio w/c requires an estimate of the volume occupied by water. The interlayer gallery volume $V_{\text{m, IL}}\ [\text{m}^3\ \text{mol}^{-1}]$ found by subtracting layer thickness τ from the layer spacing d_{001}.

$$V_{\text{m, cell}} = A_{\text{m, cell}} \cdot d_{001}$$
$$V_{\text{m, IL}} = A_{\text{m, cell}} \cdot (d_{001} - \tau)$$

Step 2. Determine the layer spacing d_{001} of Na^{+}-saturated montmorillonite at the liquid limit w_L.

The source of our d_{001} estimate are data from of Foster et al. (1954). The Belle Fourche montmorillonite (API-27) is naturally Na^{+}-saturated (Kerr and Kulp, 1951; Kerr et al., 1951). While the Belle

Fourche liquid limit $w_L = 625$ (White, 1949) lies just beyond the highest w/c treatment measured by Foster et al. (1954), the fact that d_{001} and w_L were measured using the same specimen justifies extrapolation for our estimate.

The regression[20] plotted in Fig. 3.20 results in $d_{001} = 17.86$ nm at the liquid limit $w_L = 663$. Using a layer thickness $\tau = 0.985$ nm, we can compute the interlayer gallery volume at the Belle Fourche liquid limit w_L.

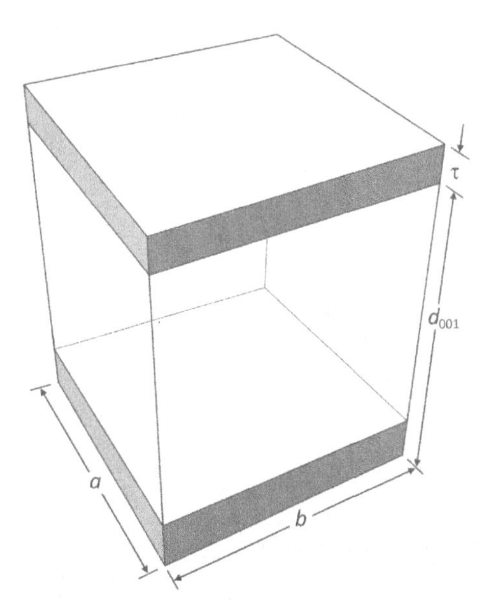

FIG. 3.20 The area of a single unit cell A_{cell} times the interlayer spacing d_{001} gives the unit cell volume V_{cell} for a swollen smectite or vermiculite. The interlayer gallery volume $V_{iterlayer}$ must subtract the thickness of a single 2:1 clay layer $\tau \approx 0.985$ nm.

$$V_{m,\ IL} = \left(2.77 \times 10^{+5}\ m^2\, mol^{-1}\right) \cdot (17.86 - 0.985) \cdot \left(10^{-9}\ m\right)$$
$$V_{m,\ IL} = 4.96 \times 10^{-3}\ m^3\, mol^{-1}$$

The Belle Fourche montmontmillonite unit cell molar mass $M_{f,\ cell} = 742.1\ g_c\ mol^{-1}$ is computed from the unit cell formula listed in Step 2. Dividing the interlayer gallery molar volume $V_{m,\ IL}$ by the unit cell molar mass $M_{f,\ cell}$ yields the interlayer gallery volume per clay mass (Fig. 3.21).

$$V_{IL} = \frac{V_{m,\ IL}}{M_{f,\ cell}}$$
$$V_{IL} = \frac{4.96 \times 10^{-3}\ m^3\, mol^{-1}}{742.1\ g_c \cdot mol^{-1}} = 6.31 \times 10^{-6}\ m^3 \cdot g_c^{-1}$$

Step 3. Determine the moles of water required to fill the interlayer gallery at the liquid limit using the interlayer gallery volume V_{IL} from Step 2 to estimate the water-to-clay mass ratio of Na^+-saturated Belle Fourche montmorillonite at its liquid limit w_L.

The molar volume of liquid water at 25°C is $V_{m,H_2O(l)} = 1.752 \times 10^{-5}\ m^3\ mol^{-1}$. Dividing the molar interlayer gallery volume V_{IL} by product of the molar volume $V_{m,H_2O(l)}$ and the molar mass M_{f,H_2O} yields the water-to-clay mass ratio w/c in the swollen state.

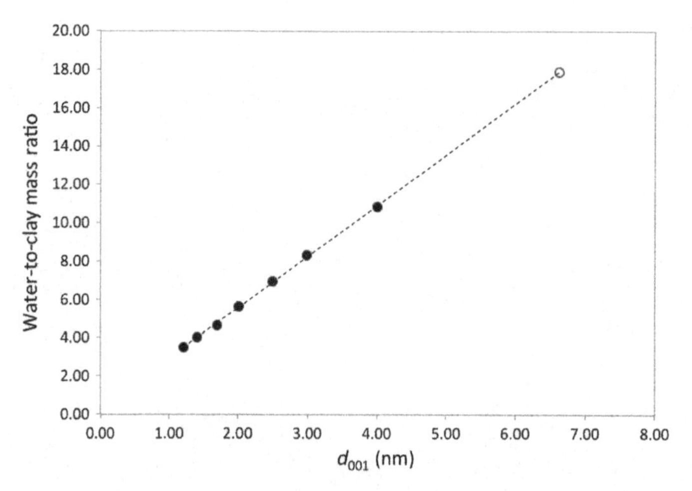

FIG. 3.21 Free swelling of Na^+-saturated Belle Fourche montmorillonite pure water (Foster et al., 1954). Experimental data are plotted as *filled circles*, the linear regression of the experimental data (the square of the correlation coefficient $r^2 = 0.9989$ is plotted as a *dashed line*, and the extrapolated layer spacing d_{001} at the w_L (White, 1949) is plotted as an open circle.

$$w/c = \frac{V_{IL}\left[m^3\, g_c^{-1}\right] \cdot \left(M_{f,H_2O}\left[g_W \cdot mol^{-1}\right]\right)}{\left(V_{m,H_2O(l)}\left[m^3\, mol^{-1}\right]\right)}$$

$$w/c = \frac{\left(6.31 \times 10^{-6}\ m^3\, g_c^{-1}\right) \cdot \left(18.02\ g_W\, mol^{-1}\right)}{\left(1.752 \times 10^{-5}\ m^3\, mol^{-1}\right)}$$

$$w/c = 6.49\ g_W\, g_c^{-1}$$

The water content based solely on interlayer volume is 5% lower than the empirical liquid limit $w_L = 663$. Stated differently, this example estimates the water adsorbed to the external surfaces of the Belle Fourche montmorillonite at the liquid limit w_L accounts for 5% of the total water-to-clay mass ratio w/c, consistent with estimates reported by Fink et al. (1968).

This example establishes a link between key Atterberg limits and montmorillonite layer spacing. The following section discusses how layer spacing responds to changes in salinity, which connects layer spacing to osmotic head. The osmotic head will allow us to infer the tension head for clays at the plastic limit w_P and the liquid w_L, Atterberg limits with physical significance for soil shear strength.

[20] Linear regression: d_{001} [nm] $= a \cdot \left(w/c \, [g_w \, g_c^{-1}]\right) + b$ where $a = 2.657$ nm $g_c \cdot g_w^{-1}$ and $b = 0.257$ nm.

EXAMPLE 3.4

Example Permalink

http://soilenvirochem.net/3t5XXc

Use unit cell parameters and layer spacing to compute the effect of swelling on montmorillonite clay density.

The soil science community uses the density of SiO_2(quartz) as the default density for the soil mineral fraction: $\rho_{quartz} = 2.65 \text{ Mg m}^{-3}$. Vermiculite and smectite group minerals swell when water enters the interlayer gallery, reducing the effective density of these clay minerals in the hydrated state. This example extends the calculations begun in Examples 3.2 and 3.3 to compute the density of a hydrated montmorillonite at the plastic limit w_P.

Step 1. Calculate the density of a dehydrated montmorillonite specimen using the formula and dimensions of the unit cell.

The unit cell formula for Na^+-saturated montmorillonite specimen identified as *Volclay*[21] appears below (Churchman, 1995). The unit cell molar mass corresponding to the formula below is: $M_{f,\text{ cell}} = 746.7 \, g_c \text{ mol}^{-1}$.

$$[Al_{3.10}Fe_{0.37}Ti_{0.02}Mg_{0.47}][Al_{0.13}Si_{7.87}]O_{20}(OH)_4 \cdot Na_{0.73}$$

From Example 3.4 (Step 1) we have the montmorillonite unit cell molar area $A_{m,\text{ cell}}$, which is simply the product of the unit cell a- and b-dimensions multiplied by the *Avogadro constant* N_A. Multiplying the unit cell molar volume $V_{m,\text{ cell}}$ by the layer spacing of dehydrated montmorillonite $d_{001} = 1.03 \times 10^{-9}$ m (Bailey, 1984) yields the unit cell molar volume $V_{m,\text{ cell}}$

$$A_{m,cell} = 2.77 \times 10^{+5} \text{ m}^2 \text{ mol}^{-1}$$

$$V_{m,dry} = \left(2.77 \times 10^{+5} \text{ m}^2 \text{ mol}^{-1}\right) \cdot \left(1.03 \times 10^{-9} \text{ m}\right) = 2.90 \times 10^{-4} \text{ m}^3 \text{ mol}^{-1}$$

The density for dehydrated *Volclay* montmorillonite is found by dividing the molar volume by the molar mass.

$$\rho_{dry} = \frac{V_{m,\text{ dry}}}{M_{f,\text{ cell}}}$$

$$\rho_{dry} = \frac{2.90 \times 10^{-4} \text{ m}^3 \text{ mol}^{-1}}{\left(746.7 \, g_c \cdot \text{mol}^{-1}\right) \times 10^{-6} \text{ Mg g}^{-1}} = 2.58 \text{ Mg}_c \text{ m}^{-3}$$

Step 2. Calculate the density of Ca^{2+}-saturated montmorillonite as a *2-layer hydrate* using results from Step 1.

Mooney et al. (1952) measured the crystalline swelling of a Ca^{2+}-saturated *Volclay*.[22] The layer spacing of the *2-layer hydrate* is $d_{001} = 1.54 \times 10^{-9}$ m.

$$V_{m,2\text{-layer}} = \left(2.77 \times 10^{+5}\ \mathrm{m^2\,mol^{-1}}\right) \cdot \left(1.54 \times 10^{-9}\ \mathrm{m}\right) = 4.26 \times 10^{-4}\ \mathrm{m^3\,mol^{-1}}$$

$$\rho_{2\text{-layer}} = \frac{4.26 \times 10^{-4}\ \mathrm{m^3\,mol^{-1}}}{\left(744.5\ g_C\,\mathrm{mol^{-1}}\right) \times 10^{-6}\ \mathrm{Mg\,g^{-1}}} = 2.07\ \mathrm{Mg}_C\,\mathrm{m^{-3}}$$

Calcium-saturated montmorillonite specimens from the Clay Spur bed have a plastic limit of $w_P \approx 30$ (equivalent to a water-to-clay mass ratio of 0.30) (Müller-Vonmoos and Løken, 1989). The effective density of Ca^{2+}-saturated montmorillonite at the plastic limit w_P is the density of the *2-layer hydrate* $\rho_{2-layer} = 2.07\ \mathrm{Mg}_C\,\mathrm{m^{-3}}$. Henceforth, we will use *plastic density* to when referring to hydrated montmorillonite at the plastic limit.[23]

$$\rho_{\text{plastic, } Ca^{2+}} \equiv 2.07\ \mathrm{Mg}_C\,\mathrm{m^{-3}}$$

[21] The source of *Volclay*, an American Colloid Company product, is the Clay Spur bed, the same geologic formation of montmorillonite specimens: Upton (API-25), Clay Spur (API-26), and Belle Fourche (API-27).

[22] The unit cell molar mass for Ca^{2+}-saturated *Volclay* montmorillonite is: $M_{f,\ cell} = 744.5\ g_C \cdot \mathrm{mol^{-1}}$. Replacing interlayer Na^{+} with Ca^{2+} must account for differences in atomic mass $M(E)$ and cation valance; the molar Ca^{2+} content will be half the molar Na^{+} content to balance layer charge.

[23] The plastic limit $w_P \approx 100$ for Na^{+}-saturated montmorillonite specimens, which occurs at a layer spacing of $d_{001} \approx 3.4$ nm. The *plastic density* of Na^{+}-saturated montmorillonite is significantly lower: $\rho_{\text{plastic, } Na^{+}} \approx 1.48\ \mathrm{Mg}_C\,\mathrm{m^{-3}}$.

To understand osmosis consider the two osmotic cells appearing in Fig. 3.23.

The initial state has both chambers of the osmotic cell filled to the same level, pure water in the right chamber and a solution in the left chamber (left osmotic cell, Fig. 3.23). The osmotic effect is illustrated by the equilibrium state (right osmotic cell, Fig. 3.23). The fluid level in the solution chamber is higher than the fluid level in the pure water chamber; the dissolved solutes have drawn water from the pure water chamber into the solution chamber.

The height of solution relative to the pure water at equilibrium is the osmotic head $h_{osmotic}$ and accounts for the elevated fluid pressure in the solution chamber commonly know as the osmotic pressure $p_{osmotic}$. van't Hoff (1887) derived the original osmotic pressure relation. Expression (3.5), which also appears in Fig. 3.23, includes a critical modification introduced by Morse et al. (1907).[24]

$$p_{osmotic} = i_{solute} \cdot b_{solute} \cdot R \cdot T \tag{3.5}$$

The terms appearing in the *van't Hoff-Morse* expression (3.5) are: the *van't Hoff factor*[25] i_{solute}, the solute concentration in *molal* units b_{solute}, the *ideal gas constant* R, and the absolute temperature K.

[24] Morse et al. (1907) replaced molar $[\mathrm{mol\,L^{-1}}]$ concentration units used by van't Hoff with molal $[\mathrm{mol\,kg^{-1}}]$ concentration units. Morse relied heavily on temperature-dependent osmotic pressure measurements; the use of molal concentration units eliminated solvent volume temperature-dependence from experimental results.

[25] cf. Appendix 3.D for an explanation of the *van't Hoff factor*.

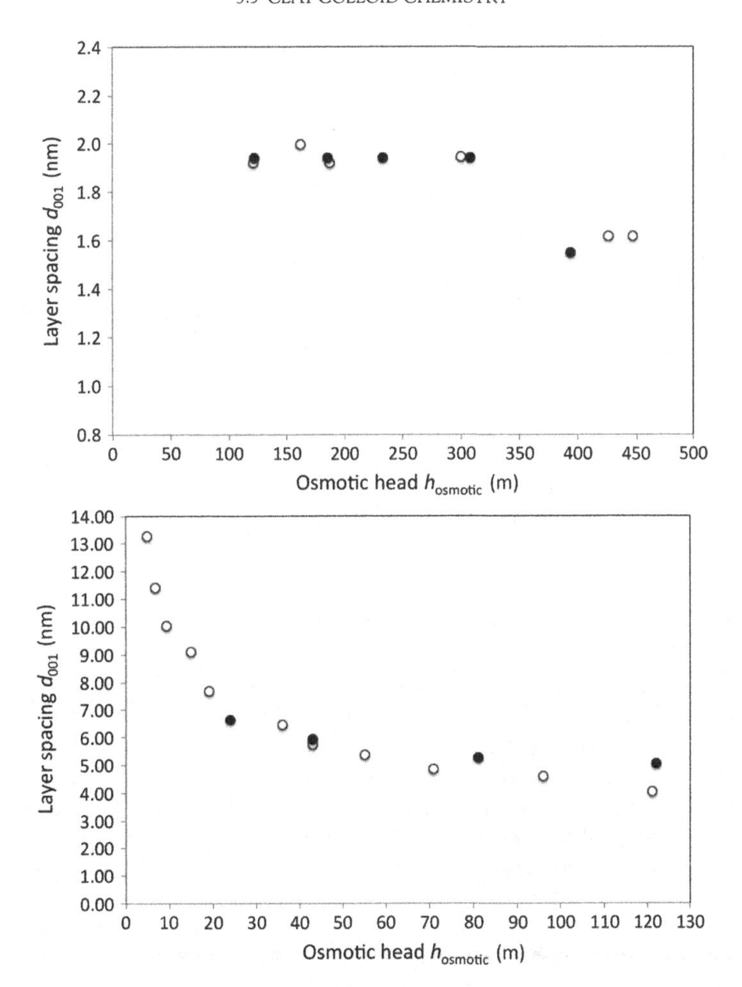

FIG. 3.22 The swelling of Na^+-saturated montmorillonite clay specimens NaCl(aq) solutions. Crystalline swelling (upper graph) and free swelling (lower graph) plots layer spacing d_{001} as a function of osmotic head $h_{osmotic}$. *Data from Norrish, K., 1954. The swelling of montmorillonite. Discuss. Faraday Soc. 18, 120–134, open circles; Foster, W.R., Savins, J., Waite, J., 1954. Lattice expansion and rheological behavior relationships in water montmorillonite systems. Clays Clay Miner. 3, 296–316, filled circles.*

To compute the osmotic head $h_{osmotic}$ of the NaCl(aq) solutions (Example 3.5) driving smectite clay swelling plotted in Figs. 3.16 and 3.19 you require a modified and rearranged *van't Hoff-Morse* equation (3.6).

$$h_{osmotic} \equiv \frac{i_{NaCl(aq)} \cdot \cdot b_{NaCl(aq)} \cdot R \cdot T}{\rho_{NaCl(aq)} \cdot g_0} \tag{3.6}$$

The osmotic pressure and head computed with expressions (3.5), (3.6), respectively, deviate from experimental measurement at high solute concentrations. Appendix 3.D discusses the adjustments require to accurately predict osmotic pressure and head. These adjustments are discussed in greater detail in Section 5.2 of Chapter 5.

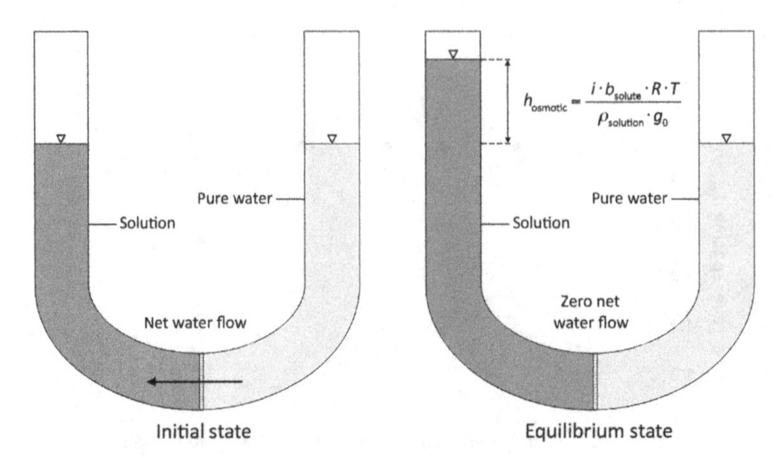

FIG. 3.23 A semipermeable membrane separates the two chambers of a U-shaped osmotic cell. The initial state—before water movement across the membrane–appears on the left; the liquid height in both chambers are equal. The final equilibrium state, the point where there is zero net water flow between chambers, appears on the right; the solution height (left chamber) is higher than the pure water (right chamber).

The osmotic head $h_{osmotic}$ appearing in Fig. 3.23 resembles the tension head $h_{tension}$ given by capillary rise (cf. Chapter 2). Although tension head $h_{tension}$ is explicitly *negative*, it would appear osmotic head $h_{osmotic}$ is inherently *positive* if for no other reason than the hydrostatic head in left-hand solution chamber in Fig. 3.23 (right). A more accurate analogy is this: capillary rise draws water to a height $h_{tension}$, likewise the osmotic effect draws water across a semipermeable membrane to a height $h_{osmotic}$. The two heads are equivalent despite the profound difference in the physical processes that drive the two processes (Examples 3.5 and 3.6).

EXAMPLE 3.5

Example Permalink

http://soilenvirochem.net/S85T2w

Compute the osmotic head $h_{osmotic}$ of a NaCl(aq) solution that is osmotically equivalent to a Na^+-saturated montmorillonite at its plastic limit w_P.

Step 1. Using swelling data for a Na^+-saturated montmorillonite, determine the layer spacing at the plastic limit w_P.

The crystalline-free swelling transition roughly corresponds to the plastic limit w_P of Na^+-saturated montmorillonite. This transition point occurs at $c_{NaCl} = 0.304 \text{ mol dm}^{-3}$, equivalent to $c_{NaCl}^{-1/2} = 1.81$ (cf. Fig. 3.19).

Step 2. Convert the molar concentration c_{NaCl} $[\text{mol dm}^{-3}]$ at the plastic limit w_P to the equivalent molal concentration b_{NaCl} $[\text{mol kg}^{-1}]$.

The molar-to-molal conversion requires the density of the $c_{NaCl} = 0.304\ \text{mol dm}^{-3}$ solution from Step 1. The NaCl(aq) solution density is computed using a linear regression of NaCl(aq) solution data from Vaslow (1966).

$$\rho_{NaCl(aq)} = \left(4.001 \times 10^{+1}\right) \cdot (0.304) + (997.41)$$

$$\rho_{NaCl(aq)} = 1009.57\ \text{g dm}^{-1}$$

The solution water mass $w(H_2O(l))/V$ [g dm^{-3}] is computed by subtracting the NaCl mass in the solution from the total solution mass.

$$w(H_2O(l))/V = \rho_{NaCl(aq)} - (c_{NaCl} \cdot M_f(NaCl))$$

$$w(H_2O(l))/V = \left(1009.57\ \text{g dm}^3\right) - \left(0.304\ \text{mol dm}^{-3} \cdot 55.84\ \text{g mol}^{-1}\right)$$

$$w(H_2O(l))/V = 991.81\ \text{g dm}^{-3}$$

The molal concentration b_{NaCl} [mol kg^{-1}] is simply the molar concentration c_{NaCl} [mol kg^{-1}] divided by the solution water mass.

$$b_{NaCl(aq)} = \frac{c_{NaCl}}{w(H_2O(l))\ [\text{kg dm}^3]}$$

$$b_{NaCl(aq)} = \frac{0.304\ \text{mol dm}^{-3}}{0.99181\ \text{kg dm}^{-3}} = 0.307\ \text{mol kg}^{-1}$$

Step 3. Use expression (3.6) to compute the osmotic head $h_{osmotic}$ of a $b_{NaCl(aq)} = 0.307\ \text{mol kg}^{-1}$ solution. Use the following parameters and constants: $i_{NaCl(aq)} = 2$, $R = 8.314 \times 10^{+3}\ \text{dm}^3\ \text{Pa K}^{-1}\ \text{mol}^{-1}$, $T = 298.15$ K, and $g_0 = 9.80655\ \text{m s}^{-2}$.

$$h_{osmotic} = \frac{i_{NaCl(aq)} \cdot b_{NaCl(aq)} \cdot R \cdot T}{\rho_{NaCl(aq)} \cdot g_0}$$

$$h_{osmotic} = \frac{2 \cdot (0.307) \cdot \left(8.314 \times 10^{+3}\right) \cdot 298.15}{(1009.57) \cdot (9.80655)}$$

$$h_{osmotic} = 152.2\ \text{m}$$

Interpretation. The layer spacing d_{001} of Na^+-saturated smectite at the plastic limit w_P roughly corresponds to the layer spacing d_{001} of Na^+-saturated smectite at the transition between crystalline and free swelling (cf. Fig. 3.19) when the water-to-clay mass ratio w/c is the sole determinant of clay hydration.

The osmotic head $h_{osmotic}$ for the same smectite at the crystalline-free swelling transition in an electrolyte solution is the osmotic head $h_{osmotic}$ at the plastic limit w_P when NaCl(aq) controls hydration. In this example the $h_{osmotic} = 152.2\ m$ at the plastic limit w_P.

Capillary forces and the osmotic effect are equivalent factors controlling clay hydration, therefore the tension head $h_{tension}$ of Na^+-saturated smectite at the plastic limit w_P is equivalent to the wilting point tension head $|h_{tension}| \approx 150$ m (Example 3.6). This equivalence is plotted in Fig. 3.20.

EXAMPLE 3.6

Example Permalink

http://soilenvirochem.net/yJwX7A

Given the tension head $h_{tension}$ of a Na^+-saturated montmorillonite clay gel compute the osmotic head $h_{osmotic}$ of an osmotically-equivalent NaCl(aq) solution.

Leonard and Low (1964) studied the swelling pressure of several Na^+-saturated montmorillonite specimens, measuring the liquid limit w_L and $h_{tension}$ at select clay contents in relatively dilute clay gels (Fig. 3.24). This example examines their results for the Cheto montmorillonite (CMS[26] SAz-1, Cheto, AZ).

Step 1. Determine the $h_{tension}$ at the Cheto montmorillonite liquid limit w_L.

Leonard and Low (1964) measured the liquid limit w_L of each Na^+-saturated montmorillonite specimen. The liquid limit $w_L = 550$ for specimen SAz-1, equivalent to a clay concentration of 15.38 wt% clay.

The tension head at $h_{tension}$ at the Cheto montmorillonite liquid limit w_L is estimated as a linear interpolation between $(\text{wt\% clay}, h_{tension}) = (13.5, 13.4)$ and $(\text{wt\% clay}, h_{tension}) = (15.6, 58.3)$: $h_{tension} \approx 53.5$ cm.

Step 2. Compute the tension $p_{tension}$ of a 15.4 wt% clay suspension of a Na^+-saturated Cheto montmorillonite.

Converting tension $p_{tension}$ to tension head $h_{tension}$ requires the density of the clay gel. The density of dry Cheto montmorillonite can be computed from the unit-cell formula (cf. Example 3.2) and typical unit cell dimensions for 2:1 layer silicates (cf. Example 3.3), but in this example we will use the density of quartz: $\rho_{quartz} = 2.65\ \text{g cm}^{-3}$.

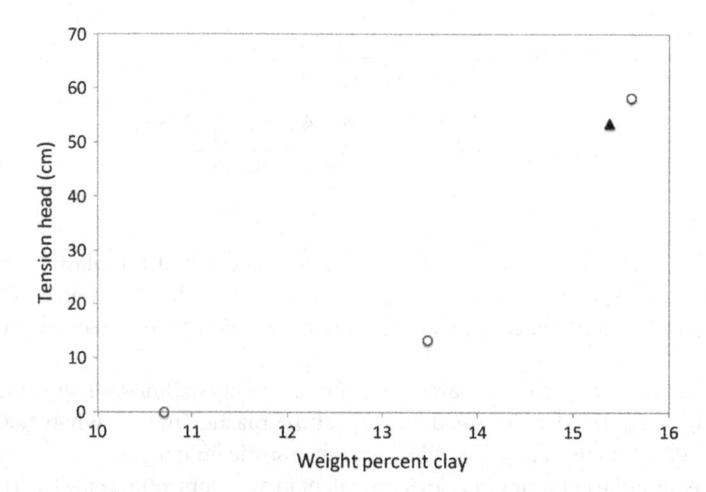

FIG. 3.24 Water tension head $h_{tension}$ [*cm*] plotted (*open circles*) as a function of the weight percent clay in three clay gels prepared from Na^+-saturated Cheto montmorillonite (CMS SAz-1). The liquid limit $w_L = 550$ for Na^+-saturated Cheto montmorillonite is equivalent to 15.38 weight-percent clay (filled triangle).

$$V_C = \frac{0.154\ g_C \cdot g_{gel}^{-1}}{2.65\ g_C \cdot cm^{-3}} = 5.81 \times 10^{-2}\ cm^3 \cdot g_{gel}^{-1}$$

$$V_W = \frac{(1.000 - 0.154)\ g_W \cdot g_{gel}^{-1}}{1.00\ g_W \cdot cm^{-3}} = 8.49 \times 10^{-1}\ cm^3 \cdot g_{gel}^{-1}$$

$$\rho_{gel} = \frac{1}{(5.81 \times 10^{-2}) + (8.49 \times 10^{-1})} = 1.103\ g_{gel} \cdot cm^3$$

The 15.4 wt% Cheto montmorillonite clay gel has a density that is about 10% greater than pure water. The clay gel tension $p_{tension}$ is computed using the usual expression.

$$p_{tension} = h_{tension} \cdot \rho_{gel} \cdot g_0$$

$$p_{tension} = (0.535\ m) \cdot \left(1.100 \times 10^3\ kg\,m^{-3}\right) \cdot \left(9.81\ m\,s^{-2}\right)$$

$$p_{tension} = 5.79 \times 10^{+3}\ Pa$$

Step 3. Compute the concentration and osmotic head $h_{osmotic}$ of a NaCl(aq) solution that is osmotically equivalent to Na^+-saturated Cheto montmorillonite at its liquid limit w_L.

The "*swelling head*" of Na^+-saturated Cheto montmorillonite at its liquid limit w_L was estimated[27] in Step 1: $h_{tension} = 0.535$ m. In Step 2 we computed the clay gel tension from the weight percent clay of the gel: $p_{tension} = 5.79$ kPa.

The osmotic pressure of a NaCl(aq) solution osmotically equivalent to the Cheto montmorillonite at its liquid limit w_L is identical to the gel tension: $p_{osmotic} = 5.79$ kPa. The osmotic head $h_{osmotic}$ of the NaCl(aq) solution, however, will not be equal the tension head of the gel because the two systems have different densities.

The NaCl(aq) solution concentration is computed by solving the *van't Hoff-Morse* equation (3.5) for $b_{NaCl(aq)}$.

$$b_{solute} = \frac{p_{osmotic}}{i_{solute} \cdot R \cdot T}$$

$$b_{solute} = \frac{5.79 \times 10^{+3}\ Pa}{2 \cdot \left(8.314 \times 10^3\ dm^3\ Pa\,K^{-1}\,mol^{-1}\right) \cdot 298.15\ K}$$

$$b_{solute} = 1.17 \times 10^{-3}\ mol\,kg^{-1}$$

The osmotic head of a $b_{NaCl(aq)} = 1.17 \times 10^{-3}\ mol\,kg^{-1}$ solution is computed using expression (3.6).

$$h_{osmotic} = \frac{2 \cdot \left(1.17 \times 10^{-3}\right) \cdot \left(8.314 \times 10^{+3}\right) \cdot 298.15}{(997.0\ kg\,m^3) \cdot (9.80655\ m\,s^{-2})}$$

$$h_{osmotic} = 0.591\ m$$

Illustrates the "*swelling head*" of Na^+-saturated Cheto montmorillonite at its liquid limit w_L and the osmotic head of a $b_{NaCl(aq)} = 1.17 \times 10^{-3}\ mol\,kg^{-1}$ solution that is osmotically equivalent to the clay gel. The osmotic head $h_{osmotic}$ of the NaCl(aq) is higher than the "*swelling head*" of Na^+-saturated Cheto montmorillonite because the former is less dense than the latter. The osmotic pressure is identical in both systems: $p_{osmotic} = 5.79$ kPa Fig. (3.25).

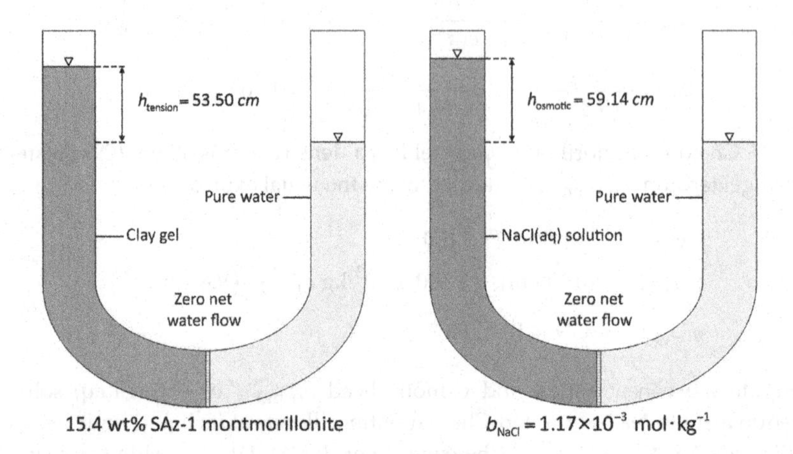

FIG. 3.25 A semipermeable membrane separates the two chambers of a U-shaped osmotic cell. An osmotic cell containing a Na^+-saturated montmorillonite (CMS SAz-1, Cheto, AZ) gel appears on the left. An osmotic cell containing a NaCl(aq) solution appears on the right. The osmotic head is $p_{osmotic}$ greater in the NaCl(aq) solution than in the montmorillonite gel.

[26] *Clay Minerals Society*, Source Clay Collection.
[27] By interpolation using data plotted in Fig. 3.24.

3.5.3 Clay Plasticity and Soil Mechanical Properties

Several soil mechanical properties—shear strength, hydraulic conductivity, and resistance to particle detachment during erosion—are influenced by the relative clay content. Clay-size mineral particles exhibit physical and chemical behavior unlike granular-size (i.e., silt and sand) particles. Although the specific particle diameter separating the clay- and granular-size fractions varies somewhat among earth science and engineering communities, the primary distinction appears when the surface-to-volume ratio or surface area per unit mass has a measurable impact on the adsorption of water and organic matter on mineral surfaces.

The soil mechanics field distinguishes between the granular-size fraction (silt plus sand: 0.002 mm $< d \leq 2$ mm) and the clay-size fraction ($d <$ 0.002 mm) because the granular-size fraction is never plastic, while the clay-size fraction *typically* exhibits plasticity over a range of water contents (cf. Box 3.3).

As we have seen, clay plasticity is not a universal property of clay-size mineral particles. The plasticity of clay minerals is unique among minerals of the clay-size fraction; hydrous aluminum and iron oxide minerals are not demonstrably plastic (cf. Section 3.5.1). In fact, kaolinite group minerals and vermiculite all have a relatively low plasticity index $PI \equiv w_L - w_P$. The plasticity of natural specimens of these minerals and interstratified mica-smectite (i.e., illite) is most likely due the smectite fraction found in nearly all natural clay deposits.[28] In short, the plasticity of the clay-size mineral fraction derives principally from the smectite component.

The US Department of Agriculture Natural Resource Conservation Service (NRCS) and the National Cooperative Soil Survey (NCSS) recognize 12 soil texture classes: sand, loamy sand, sandy loam, silt, silt loam, loam, sandy clay loam, silty clay loam, clay loam, sandy clay, silty, and clay (Fig. 3.26). Each texture class contains a combination of three particle classes: sand, silt, and clay. The important distinction in textural triangle is between the soil textures that exhibit negligible plasticity (sand, loamy sand, sandy loam, silt, silt loam, loam) and textures that exhibit plasticity (sandy clay loam, silty clay loam, clay loam, sandy clay, silty clay and clay). A heavy line traces the boundary separating these two groups of textural classes in Fig. 3.26. If the texture name contains *clay* as a modifier it can be molded into a wire or ribbon (cf. Box 3.3).

Soil textures that bordering the plasticity threshold—sandy clay loam, silty clay loam, clay loam, sandy clay, silty clay—derive their plasticity from clay coating the granular fraction. This section uses granular particle surface area and the density of a *2-layer* Ca^{2+}-saturated montmorillonite to estimate the clay film thickness coating granular particles. The heavy border delineated in Fig. 3.26 defines the threshold clay content required to confer plasticity to a soil and, hence, significant shear strength. Note, in particular, points *A* and *B* in Fig. 3.26 that lie on this boundary.

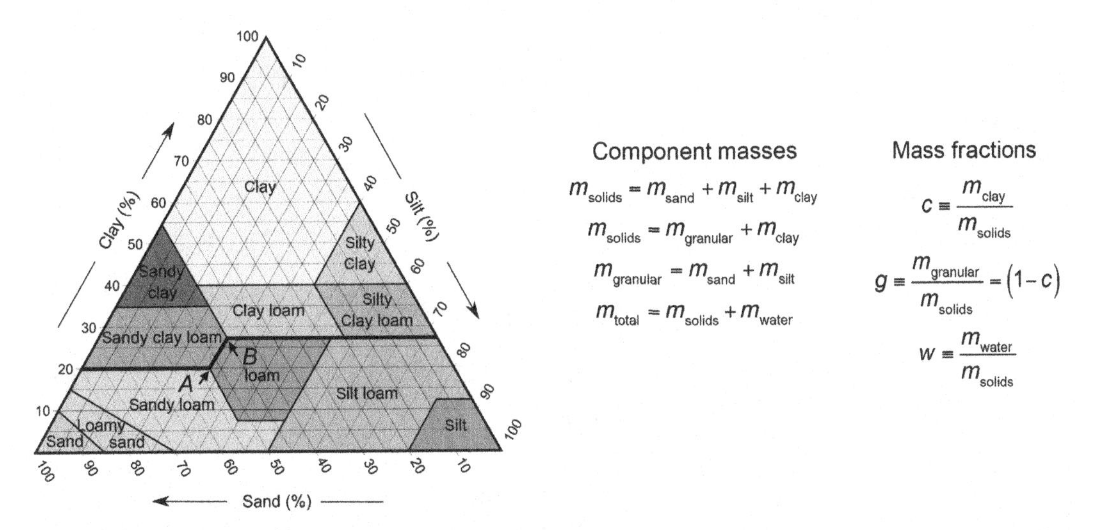

FIG. 3.26 Textural triangle (US Department of Agriculture) dislaying textural classes (right). Component fractions (granular, clay, solids, and water) as defined by the soil mechanics and geotechnical communities.

[28] cf. Section 3.3.5.

Point A lies at the intersection of three soil texture fields: *sandy loam, loam* and *sandy clay loam*. The clay content is 20 wt% at point A and along the boundary extending to the left. Point B lies at the intersection of three soil texture fields: *sandy clay loam, loam* and *clay loam*. The clay content is 27 wt% at point B and along the boundary extending to the right. Example 3.7 estimates the clay film thickness coating granular particles at point A.

EXAMPLE 3.7

Example Permalink

http://soilenvirochem.net/FxY1A4

Compute the clay film thickness coating granular (i.e., sand- and silt-size particles) given the plastic density of the clay fraction and soil texture at Point A in Fig. 3.26.

Unlike granular particles, the clay fraction exists as a plastic film. The clay film coating sand and silt particles at the intersection of sandy clay loam, sandy loam, and loam textures at Point A (Fig. 3.26) is the minimum clay film thickness for the soil texture at Point A (and along the heavy boundary delineated in Fig. 3.26) to manifest plasticity.

Step 1. Determine the specific surface area a_S [$m^2\,g^{-1}$] of sand (s) and silt (m) particles.

The diameter of medium sand, representative of the sand particle-size class, is: $d_S = 3.76\times10^{-4}$ m. The specific surface area a_S [$m^2\,g^{-1}$] is found by multiplying the area of an individual sphere times the number of spheres per unit mass. The grain count per unit mass n_{grain} [g^{-1}] depends on the volume of individual spheres and grain density: $\rho_{quartz} = 2.65 \times 10^{+6}\ g\,m^{-3}$.

$$V_S = \left(\frac{4\pi}{3}\right)\cdot\left(\frac{3.75\times10^{-4}\ m}{2}\right)^3 = 2.76\times10^{-11}\ m^3$$

$$n_{grain} = \frac{1}{V_{grain}\cdot\rho_{quartz}} \tag{3.7}$$

$$n_S = 1.37\times10^{+4}\ g^{-1}$$

$$A_S = 4\pi\cdot\left(\frac{3.75\times10^{-4}\ m}{2}\right)^2 = 4.42\times10^{-7}\ m^3$$

$$a_S = n_{grains}\ [g^{-1}]\cdot A_{grain}\ [m^3\,g^{-1}] \tag{3.8}$$

$$a_{S(s)} = 6.04\times10^{-3}\ m^2\,g^{-1}$$

Using a medium silt diameter of $d_m = 2.6\times10^{-5}$ m, the same method yields a specific surface area of $a_{S(m)} = 8.71\times10^{-2}\ m^2\,g^{-1}$. By way of comparison, a typical specimen of the clay mineral kaolinite will have a specific surface area of $a_S \approx 10\ m^2\,g^{-1}$.

Step 2. Using the specific surface area for sand and silt from Step 1 and the plastic density $\rho_{plastic,\,Ca^{2+}}$ from Example 3.4, compute the clay film thickness for soil with the texture at Point A in Fig. 3.26.

Although dehydrated clay minerals have a density comparable to the mineral density commonly used by soil scientist—ρ_{quartz}—smectite clay minerals absorb significant interlayer water at the plastic limit w_{P} lowering the effective density (Example 3.4).

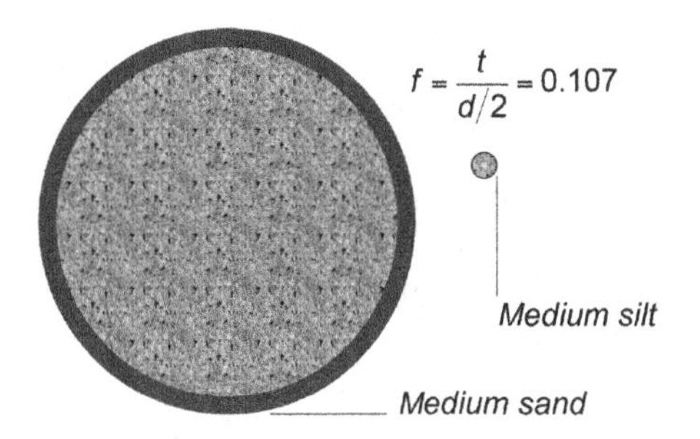

FIG. 3.27 Medium sand and medium silt particles, each with a *heavy line* drawn around the circumference to represent the clay film thickness. The relative diameter of the sand and silt particles and the clay-film thickness are all drawn to scale for 20 wt% clay (cf. Point *A*, Fig. 3.26). The ratio of clay film thickness *t* to particle radius *r* is a constant factor $f = 0.107$.

The clay volume at 20 wt% clay (cf. Point *A*, Fig. 3.26) is computed using the so-called *plastic density* from Example 3.4.

$$V_{\text{C}} = \frac{w(clay)\ [g_{C} \cdot g_{\text{soil}}^{-1}]}{\rho_{\text{plastic, Ca}^{2+}}\ [\text{g m}^{-3}]} \tag{3.9}$$

$$V_{\text{C}} = \frac{0.20\ g_{C} \cdot g_{\text{soil}}^{-1}}{2.07 \times 10^{+6}\ \text{g m}^{-3}} = 9.66 \times 10^{-8}\ \text{m}_{\text{C}}^{3} \cdot g_{\text{soil}}^{-1}$$

The clay volume V_{C} is equal to the specific surface area a_{S} multiplied by the clay film thickness t for sand and silt particles, respectively. Furthermore, assume the ratio of clay film thickness t to particle radius $r = d/2$ is a constant factor f.

$$V_{\text{C}} = \left((0.53) \cdot a_{s(\text{s})} \cdot t_{\text{S}}\right) + \left((0.27) \cdot a_{s(\text{m})} \cdot t_{\text{m}}\right)$$

$$f \equiv \frac{t_{\text{S}}}{d_{\text{S}}/2} = \frac{t_{\text{m}}}{d_{\text{m}}/2}$$

$$f = \frac{2 \cdot V_{\text{C}}}{\left(a_{s(\text{s})} \cdot d_{\text{S}}\right) + \left(a_{s(\text{m})} \cdot d_{\text{m}}\right)}$$

$$f = \frac{2 \cdot 9.66 \times 10^{-8}}{\left(3.20 \times 10^{-3} \cdot 3.75 \times 10^{-4}\right) + \left(2.35 \times 10^{-2} \cdot 2.6 \times 10^{-5}\right)} = 0.107$$

In summary, the clay film coating the granular fraction (Fig. 3.27) is about 10% of the particle radius and occupies a bulk volume that is about 30% greater than the volume occupied by the granular fraction if the smectite clay is Ca^{2+}-saturated.

Example 3.7 calculates the clay film thickness at the minimal clay content for a soil to manifest plasticity. Two assumptions underlie Example 3.7: the clay is entirely Ca^{2+}-saturated smectite and the clay is at its plastic limit w_P. Both assumptions are reasonable. Furthermore, the hydrated clay volume cannot increase further because Ca^{2+}-saturated clay has reached its swelling limit as the 2-layer hydrate.

The soil fabric derives its load-bearing and shear strength from the contact between mineral grains. The clay film coating granular particles at these contacts adds shear strength to the soil fabric in proportion to its plasticity and film thickness.

The soil pore structure results from particle displacement during wetting-drying cycles, freezing-thawing cycles, and bioturbation.[29] Clay film plasticity preserves an open soil fabric, promoting gas exchange between the soil and above-ground atmosphere, rapid water infiltration and improved hydraulic conductivity.

Calcium-saturated smectite clays are unable to reach their liquid limit w_L in soils with unrestricted drainage. As a consequence, soils containing Ca^{2+}-saturated smectite clays are unlikely to reach their liquid limit w_L, a point where the soil body deforms under the gravitational force acting on the soil mass.

Smectite clay swelling behavior leads us to two important conclusions. First, replacing interlayer Ca^{2+} with Na^+ increases the liquid limit w_L tenfold (Bain, 1971). Increasing the liquid limit w_L extends the water-content range over which the smectite is plastic. Second, the hydrated clay volume of Na^+-saturated smectite expands 100-fold relative to Ca^{2+}-saturated smectite at the liquid limit w_L. Hydrated Na^+-saturated smectite clay swells to occupy much of the soil void space with a gel that has little shear strength (Box 3.4).

The favorable soil physical properties we associate with clay plasticity should be qualified—the smectite interlayer is predominantly populated by Ca^{2+} cations and, as a consequence, clay hydration is limited to crystalline swelling states. If the smectite clay interlayer Na^+ population exceeds a certain threshold the clay will hydrate until it becomes a gel that occupies a much greater volume. Unlike soil containing Ca^{2+} smectite clays, soils containing Na^+ smectite clays may retain sufficient interlayer water to reach their liquid limit w_L even in soils without restricted drainage.

BOX 3.4

SMECTITE CLAY GELS

Soils with an appreciable clay fraction and virtually all clay minerals have a liquid limit w_L (cf. Box 3.3), a water content where the soil or clay mass flows under the influence of gravity (cf. ASTM method D4318-10e1). At and above the liquid limit w_L clay particles are effectively suspended in water, providing sufficient lubrication for the mass to flow.

Continued

[29] Bioturbation is the rearrangement of solid particles in soil and sediment caused by the activity of organisms.

BOX 3.4 *(cont'd)*

Free-swelling smectite group minerals[30] enter a *gel* state[31] above the w_L. A free-swelling Na^+-saturated smectite gel is a *solid colloidal system* composed of smectite clay particles dispersed in water. The smectite particles are sufficiently concentrated to come into physical contact to form a network spanning the entire dispersion. This clay-particle network gives the dispersion a small but nonzero shear strength; a smectite gel will temporarily retain its shape when unmolded as long as it remains undisturbed.

At a still higher water content particle-particle contacts can no longer maintain a coherent dispersion-spanning network. The water content where the clay-particle network breaks is the *sol-gel transition* point. Above the sol-gel transition point the colloidal clay-water system behaves like a true liquid with smectite clay particles dispersed throughout.

The gel state exists between the liquid limit w_L and the sol-gel transition. The liquid limit w_L for Na^+-saturated montmorillonite range from 500 to 700, equivalent to water-to-clay mass ratios ranging from 5 to 7. The sol-gel transition for free-swelling Na^+-saturated smectite occurs at about 2% clay, equivalent to a water-to-clay mass ratio of 50.

[30] Norrish and Rausell-Colom (1963) managed, after careful effort, to displace naturally-occurring interlayer cations with Li^+ in a few samples prepared from a Kenyan vermiculite specimen. Untreated samples swelled to no more than 1.5 nm when immersed in water, consistent with crystalline swelling. The Li^+-saturated vermiculite samples, however, underwent free swelling to 23.0 nm.

[31] Clay gels are widely used in oil-drilling as *drilling mud*. One common type of drilling mud is composed of Na^+-saturated montmorillonite combined with select additives. Barium is a common drilling mud component, but it is added as the highly insoluble $BaSO_4(s)$ compound barite. Drilling mud rheology derives from the Na^+-saturated montmorillonite gel. Soluble Ba^{2+}, Ca^{2+}, and Mg^{2+} are contaminants that can displace interlayer Na^+ and break the gel.

3.6 SUMMARY

Clay as a soil component is inextricably linked to plasticity, extending beyond the *ribbon test* as a field measurement of soil texture. The modern understanding of soil mechanical behavior is explicitly linked to plasticity through a reliance on Atterberg limits. Despite the unquestioned importance of the clay fraction to soil physical properties, it is essential to keep in mind clay plasticity derives solely from clay minerals group.

The clay minerals group are clay-grade phyllosilicates; phyllosilicate minerals whose natural occurrence is confined to the clay-size particle fraction. Clay minerals share another characteristic besides particle size: they form as low-temperature alteration products of igneous and high-temperature metamorphic minerals in the presence of liquid water. Mineralogists studying chemical weathering have come to realize clay minerals form in direct contact with igneous silicate minerals by solid-state rearrangement.

Clay minerals appear early in the chemical weathering sequence and disappear only in the later stages. The plasticity of smectite group minerals is sufficient to confer noticeable plasticity to soil even when these clay minerals are relatively minor components of the clay fraction. In fact, much of the plasticity associated with kaolinite, vermiculite, palygorskite, and illite clay specimens is believed to arise from trace amounts of smectite clays that co-occur with most clay mineral specimens. The co-occurrence of several clay mineral types is an expected consequence to the mechanism of clay mineral formation and transformation.

Smectite group minerals emerge as the key constituent soil scientists and geotechnical engineers associate with the clay fraction: plasticity, shrink-swell behavior, and shear strength. Smectite plasticity (i.e., quantitatively the *plasticity index PI* difference between the water content at the liquid limit w_L minus the water content at the plastic limit w_P) is very sensitive to the cation population occupying the hydrated interlayer. The role of interlayer cations comes down whether or not the clay will free swell when hydrated by low salinity water. Chapters 4 and 6 will continue to build on the clay chemistry foundation laid in this chapter.

APPENDICES

3.A BRAGG'S LAW AND X-RAY DIFFRACTION BY LAYER SILICATES

When highly collimated, monochromatic X-ray beam strike a single crystal the X-rays are diffracted by parallel atomic planes within the crystal. The intensity of the diffracted X-rays appear as a symmetric array of bright spots on an otherwise dark field. The bright spots represent constructively interfering reflected X-rays. Destructive interference extinguishes diffracted X-rays that would otherwise illuminate the dark field between the bright spots (Fig. 3.A.1, left).

If the collimated X-ray beam strikes a powder containing small, randomly-oriented crystals a series of bright concentric rings appears rather than an array of bright spots (Fig. 3.A.1, right).

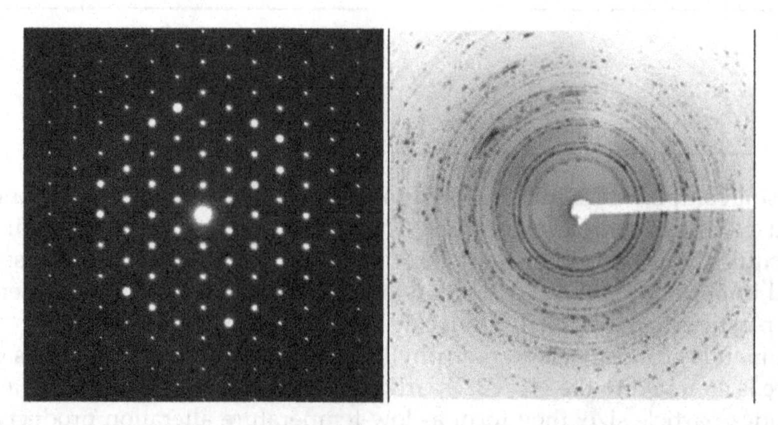

FIG. 3.A.1 X-ray diffraction patters from a single crystal of the mineral nepheline (left) and a powder of hydrocerussite.

Again, the bright rings represent constructively interfering reflected X-rays while destructive interference extinguishes diffracted X-rays.

Bragg's Law represents the interference of diffracted X-rays by a simple geometric relationship between X-ray wavelength λ and the spacing d_{hkl} between the parallel crystal planes that diffract the X-ray beam (Fig. 3.A.2).

Constructive interference occurs when the diffracted X-rays are in-phase (Fig. 3.A.2, left). Destructive interference results with the diffracted X-rays are out-of-phase (Fig. 3.A.2, right), extinguishing the diffracted X-rays.

Bragg quantified interference by the path length traveled by two diffracted X-rays (Fig. 3.A.3). Both incident X-rays strike the crystal planes at the same angle θ and the diffracted X-rays depart the same angle θ relative to the crystal plane.

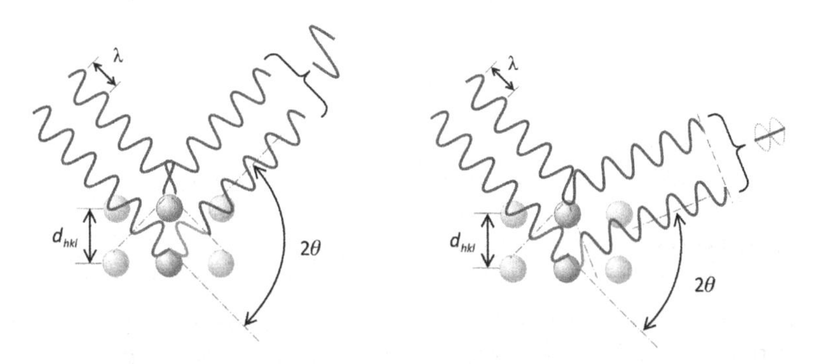

FIG. 3.A.2 Bragg's Law diffraction of monochromatic X-rays from crystal planes: constructive interference (left) and destructive interference (right). *Image modified from Chan, C.D.N., 2004. Bragg's Law. Wikimedia Commons. https://commons.wikimedia.org/wiki/File:Braggs_Law.svg.*

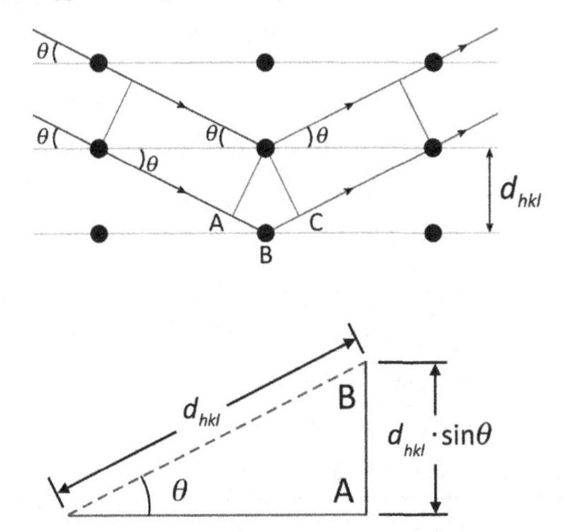

FIG. 3.A.3 The geometric relationship between the path-length difference Δl travelled by two X-rays diffracted by the symmetrically equivalent crystal planes $\{hkl\}$. *Image modified from Hadjiantonis, M., 2013. Bragg's Law Schematic. Wikimedia Commons. https://commons.wikimedia.org/wiki/File:Bragg's_law.svg.*

The path-length difference between the two diffracted beams is represented by the extra distance *ABC* traveled by the lower diffracted X-ray relative to the upper beam.

$$\Delta l = \overline{AB} + \overline{BC} \quad (3.A.1)$$

The two diffracted X-rays are in-phase when the when the path-length difference is an integer times the wavelength λ of the monochromatic X-rays.

$$\Delta l = n \cdot \lambda \quad (3.A.2)$$

The perpendicular distance separating the symmetrically equivalent crystal planes $\{hkl\}$ is d_{hkl}.

$$\overline{AB} = \overline{BC} = d_{hkl} \cdot \sin\theta \quad (3.A.3)$$

$$n \cdot \lambda = 2 \cdot d_{hkl} \cdot \sin\theta \quad (3.A.4)$$

Crystallographers identify X-ray reflections using $\{hkl\}$ to denote specific atomic planes within crystals. The {001} X-ray reflection in phyllosilicate minerals results from Bragg diffraction by atomic planes parallel to clay layers. The distance associated with each {001} X-ray reflection in phyllosilicates is the layer spacing, denoted d_{001}.

3.B DETERMINING FORMAL OXIDATION STATES IN MINERALS

Most general chemistry textbooks introduce a scheme for calculating the oxidation state of elements in compounds. Table 3.B.1 shows an abbreviated scheme adapted for mineralogy.

TABLE 3.B.1 Guidelines for Assigning Oxidation States in Minerals

1	The oxidation state of a free element is always 0
2	The oxidation state of hydrogen is usually +I
3	The oxidation state of oxygen is usually −II
4	The oxidation state of Group 1 elements (e.g., Na) is +I
5	The oxidation state of Group 2 elements (e.g., Mg) is +II
6	The oxidation state of Group 3 elements (e.g., Al) is +III
7	The oxidation state of Group 4 elements (e.g., Si) is +IV
8	The oxidation state of Group 7 elements (e.g., Cl) is −I
9	The sum of oxidation states for all the atoms in the formula unit of a mineral is 0
10	The sum of oxidation states in a polyatomic ion is equal to the charge of the ion

Olivine: $(Mg,Fe)_2SiO_4(s)$

The olivine formula places Mg and Fe in parentheses, equivalent to $Mg_{2-x}Fe_xSiO_4(s)$. This notation signifies Mg and Fe are present in varying rather than fixed proportions.

The oxidation state for oxygen is −II (Rule 3). The oxidation states of Mg (Rule 5) and Si (Rule 7) are +II and +IV, respectively. The charge neutrality condition (Rule 9) requires assignment of oxidation state +II to Fe (Table 3.B.2).

Hedenbergite: $FeCaSi_2O_6(s)$

The mineral hedenbergite belongs to the pyroxene mineral group (Table 3.B.3). The oxidation state for oxygen is -II (Rule 3). The oxidation states of Ca (Rule 5) and Si (Rule 7) are +II and +IV, respectively. The charge neutrality condition (Rule 9) requires, once again, assignment of oxidation state +II to Fe.

Goethite: $\alpha-FeOOH(s)$

The oxidation state for oxygen is −II and hydrogen is +I (Rules 2 and 3, respectively). The charge neutrality condition (Rule 9) requires, once again, assignment of oxidation state +III to Fe (Table 3.B.4).

TABLE 3.B.2 Oxidation States in the Mineral Olivine

Element	Stoichiometry	Oxidation State
O	4	−II
Mg	$(2-x)$	+II
Fe	x	+II
Si	1	+IV

TABLE 3.B.3 Oxidation States in the Mineral Hedenbergite

Element	Stoichiometry	Oxidation State
O	6	−II
Ca	1	+II
Fe	1	+II
Si	2	+IV

TABLE 3.B.4 Oxidation States in the Mineral Goethite

Element	Stoichiometry	Oxidation State
O	2	−II
H	1	+I
Fe	1	+III

3.C STOKE'S LAW

Stokes' Law predicts particle sedimentation rate in a fluid. Solid particles suspended in a liquid settle under the gravitational force. The sedimentation rate depends on particle buoyancy (the difference between the density of the particles and the density of the liquid), the viscosity of the liquid, and the size of the particles. The net force acting on a particle suspended in a liquid is the difference between the gravitational force F_g and the drag force F_d (3.C.1).

The gravitational force F_g (3.C.2) acting on a body moving at terminal velocity v_t equals the effective mass of the body times gravitational acceleration g [$\mathrm{m\,s^{-2}}$]. The effective mass of a particle suspended in a fluid is the particle volume times the difference between particle density ρ_{particle} and fluid density ρ_{fluid}.

$$F_{\text{net}} = F_g - F_d \tag{3.C.1}$$

$$F_g = \left(\frac{4 \cdot \pi \cdot r^3}{3}\right) \cdot \left(\rho_{\text{particle}} - \rho_{\text{fluid}}\right) \cdot g \tag{3.C.2}$$

Stokes (1851) used fluid dynamics to derive the drag force F_d acting on a spherical particle moving at a velocity v through a fluid. Assuming laminar flow, which is appropriate for small spherical particles, the drag force F_d (3.C.3) is proportional to the particle radius r, the dynamic viscosity μ of the fluid and the particle sedimentation velocity v.

$$F_d = (6 \cdot \pi \cdot r) \cdot \mu \cdot v \tag{3.C.3}$$

The particle will initially accelerate to a terminal velocity v_t at which point the drag force F_d, which opposes the settling of the suspended particle, equals the gravitational force.

$$F_d = F_g \tag{3.C.4}$$

$$(6 \cdot \pi \cdot r \cdot \mu) \cdot v_t = \left(\frac{4 \cdot \pi \cdot r^3}{3}\right) \cdot \left(\rho_{\text{particle}} - \rho_{\text{fluid}}\right) \cdot g \tag{3.C.5}$$

Solving expression (3.C.5) for the terminal or sedimentation velocity v_t we have expression (3.C.6), the *Stokes' Law* for the sedimentation velocity of small spherical particles in a fluid.

$$v_t\ [\mathrm{m\,s^{-1}}] = \left(\frac{2}{9}\right) \cdot \left(\frac{r^2 \cdot \Delta\rho \cdot g}{\mu}\right) \tag{3.C.6}$$

Soil scientists apply *Stokes' Law* to separate particles based on their radius. The separation of fine sand from silt particles is readily achieved by suspending particles in water ($\mu_{\text{water}} = 8.91 \times 10^{-4}\ \mathrm{kg\,m^{-1}\,s^{-1}}$) and relying on acceleration by standard gravity ($g_0 = 9.80665\ \mathrm{m \cdot s^{-2}}$). All particles whose diameter $r \geq 2.5 \times 10^{-5}$ m will settle a distance of 1 m in 396 s (i.e., 6.60 min).

$$t = \frac{9 \cdot \left(8.91 \times 10^{-4}\right)}{2 \cdot \left(2.5 \times 10^{-5}\right)^2 \cdot (2.65 - 1.00) \times 10^{-3} \cdot (9.80665)}$$

$$t = 396\ \mathrm{s}$$

Separating fine silt and clay particles using the same method would require $2.48 \times 10^{+5}$ s or 2.87 days! The settling rate is easily accelerated using a centrifuge. A centrifuge capable of $g = 20 \cdot g_0 \text{ m} \cdot \text{s}^{-2}$ will settle all silt particles through a distance of 5 cm in 903 s (i.e., 15.0 min).

3.D VAN'T HOFF FACTOR *i* AND OSMOTIC COEFFICIENT ϕ

The van't Hoff factor i was originally believed to be an integer equal to the number of ions formed when the electrolyte dissociates in solution. For example, the electrolyte $CaCl_2(aq)$ dissociates into three ions in aqueous solution: one $Ca^{2+}(aq)$ cation and two $Cl^-(aq)$ anions. The van't Hoff factor i for the solute $CaCl_2(aq)$ would be 3.

As chemists came to better understand electrolyte solutions they discovered the van't Hoff factor i was a real number (expression 3.D.1), the product of a true integer ν_{solute}—representing the original notion of solute stoichiometry—and a concentration-dependent term known as the *osmotic coefficient* ϕ (3.D.2).

$$i_{solute} \equiv \nu_{solute} \cdot \phi \tag{3.D.1}$$

$$\phi = 1 + \frac{\ln(\gamma_{z\pm})}{3} \tag{3.D.2}$$

The term $\gamma_{z\pm}$ appearing in expression (3.D.1) is the *ion activity coefficient*. The ion activity concept, ion activity coefficients, and ionic strength I are all discussed in Section 5.2 of Chapter 5.

Expression (3.D.3) is the original Debye-Hückel expression[32] for the *ion activity coefficient*. Expression (3.D.4) is *ionic strength*, the measure of electrolyte concentration in the ion activity context.

$$\ln(\gamma_{z\pm}) = -A_b^{DH} \cdot |z_+ \cdot z_-| \cdot \sqrt{I_b} \tag{3.D.3}$$

$$I_b = \tfrac{1}{2} \cdot \sum_j (z_j^2 \cdot b_j) \tag{3.D.4}$$

The van't Hoff-Morse equation expresses electrolyte concentrations using molal $[mol \cdot kg^{-1}]$ units to eliminate the effect of temperature on electrolyte concentration. The *molal* Debye-Hückel factor is: $A_b^{DH} \approx 0.5094$ at 25°C. Section 5.2 (Chapter 5) quotes the *molar* Debye-Hückel factor: $A_c^{DH} \approx 0.5102$ at 25°C.

The van't Hoff-Morse expression (3.D.5) makes the appropriate ion activity adjustments to the van't Hoff coefficient ϕ.

$$h_{osmotic} \equiv \frac{\nu_{NaCl(aq)} \cdot \phi \cdot b_{NaCl(aq)} \cdot R \cdot T}{\rho_{NaCl(aq)} \cdot g_0} \tag{3.D.5}$$

[32] cf. Section 5.2 for other ion activity coefficient expressions.

References

Aagaard, P., Helgeson, H.C., 1982. Thermodynamic and kinetic constraints on reaction rates among minerals and aqueous solutions. I. Theoretical considerations. Am. J. Sci. 282 (3), 237–285.

Altaner, S.P., Weiss, C.A., Kirkpatrick, R.J., 1988. Evidence from ^{29}Si NMR for the structure of mixed-layer illite/smectite clay minerals. Nature 331, 699–702.

Atterberg, A., 1905. Die rationelle Klassifikation der Sande und Kiese. Chem.-Ztg. 29, 195–198.

Atterberg, A., 1911. Über die physikalische Bodenuntersuchung und über die Plastizität der Tone. Int. Mitt. Bodenkd. 1, 10–43.

Bailey, S.W., 1984. Classification and structures of the micas. In: Bailey, S.W. (Ed.), Micas, vol. 13. Mineralogical Society of America, Washington, DC, pp. 1–12.

Bain, J.A., 1971. Plasticity chart as an aid to the identification and assessment of industrial clays. Clay Miner. 9 (1), 1–17.

Banfield, J.F., Eggleton, R.A., 1990. Analytical transmission electron microscope studies of plagioclase, muscovite, and potassium-feldspar weathering. Clays Clay Miner. 38 (1), 77–89.

Banfield, J.F., Ferruzzi, G.G., Casey, W.H., Westrich, H.R., 1995. Hrtem study comparing naturally and experimentally weathered pyroxenoids. Geochim. Cosmochim. Acta 59 (1), 19–31.

Blott, S.J., Pye, K., 2012. Particle size scales and classification of sediment types based on particle size distributions: review and recommended procedures. Sedimentology 59, 2071–2098.

Bragg, W.L., 1929. Atomic arrangement in the silicates. Trans. Faraday Soc. 25, 291–314.

Brantley, S.L., Stillings, L., 1996. Feldspar dissolution at 25°C and low pH. Am. J. Sci. 296 (2), 101–127.

Brindley, G.W., Goodyear, J., 1948. X-ray studies of halloysite and metahalloysite. part II. The transition of halloysite to metahalloysite in relation to relative humidity. Mineral. Mag. 28 (203), 407–422.

Burdett, J.K., 1988. Perspectives in structural chemistry. Chem. Rev. 88 (1), 3–30.

Burdett, J.K., McLarnan, T.J., 1982. An orbital explanation for Pauling's third rule. J. Am. Chem. Soc. 104 (19), 5229–5230.

Burdett, J.K., McLarnan, T.J., 1984. An orbital interpretation of Pauling's rules. Am. Mineral. 69 (7-8), 601–621.

Buseck, P.R., Veblen, D.R., 1981. Defects in minerals as observed with high-resolution transmission electron microscopy. Bull. Mineral. 104 (2–3), 249–60.

Chisholm, J.E., 1973. Planar defects in fibrous amphiboles. J. Mater. Sci. 8 (4), 475–483.

Dublin, C., Corrigan, P., Turney, T.W., 1995. Nature of the pillaring process during the formation of alumina-pillared clays. In: Churchman, G.J., Fitzpatrick, R.W., Eagleton, R.A. (Eds.), Clays: Controlling the Environment. C.S.I.R.O. Publishing, Adelaide, Australia, pp. 145–150.

Collis-George, N., 1955. The hydration and dehydration of sodium montmorillonite (Belle Fourche). J. Soil Sci. 6, 99–110.

Correns, C.W., 1940. Chemical weathering of silicates. Naturwissenschaften 28, 369–376.

Correns, C.W., von Engelhardt, W., 1938a. Weathering of potash feldspar. Chem. Erde 12, 1–22.

Correns, C.W., von Engelhardt, W., 1938b. Weathering of potassium feldspar. Naturwissenschaften 26, 137–138.

ASTM, 2014. Standard Test Methods for Liquid Limit, Plastic Limit, and Plasticity Index of Soils, D4318-10e1, vol. 04.08. American Society for Testing and Materials, p. 16.

De la Calle, C., Dubernat, J., Suquet, H., Pezerat, H., Gaultier, J., Mamy, J., 1976. Crystal structure of two-layer magnesium vermiculites and sodium, calcium vermiculites. In: Proceedings of the International Clay Conference. Appl. Publ. Ltd., pp. 201–209.

Eggleton, R.A., Boland, J.N., 1982. Weathering of enstatite to talc through a sequence of transitional phases. Clays Clay Miner. 30 (1), 11–20.

Fink, D.H., Rich, C.I., Thomas, G.W., 1968. Determination of internal surface area, external water, and amount of montmorillonite in clay-water systems. Soil Sci. 105 (2), 71–77.

Foster, M.D., 1953. Geochemical studies of clay minerals. 2. Relation between ionic substitution and swelling in montmorillonites. Am. Mineral. 38 (1-2), 994–1006.

Foster, M.D., 1954. The relation between composition and swelling in clays. Clays Clay Miner. 3, 205–220.

Foster, W.R., Savins, J.G., Waite, J.M., 1954. Lattice expansion and rheological behavior relationships in water-montmorillonite systems. Clays Clay Miner. 3, 296–316.

Frost, B.R., 1991. Introduction to oxygen fugacity and its petrologic importance. Rev. Mineral. 25, 1–9.

Gillespie, R.J., Nyholm, R.S., 1957. Inorganic stereochemistry. Quart. Rev. (London) 11, 339–380.

Grim, R.E., 1928. Preliminary Report on Bentonite in Mississippi, Bulletin 22. Mississippi State Geological Survey.
Guggenheim, S., Krekeler, M.P.S., 2011. The structures and microtextures of the palygorskite-sepiolite group minerals. Dev. Clay Sci. 3 (Developments in Palygorskite-Sepiolite Research), 3–32.
Haggerty, S.E., 1976. Opaque mineral oxides in terrestrial igneous rocks. Rev. Mineral. 3 (Oxide Minerals), Hg101–Hg300.
Helgeson, H.C., Murphy, W.M., Aagaard, P., 1984. Thermodynamic and kinetic constraints on reaction rates among minerals and aqueous solutions. II. Rate constants, effective surface area, and the hydrolysis of feldspar. Geochim. Cosmochim. Acta 48 (12), 2405–2432.
Hight Jr., R., Higdon, W.T., Darley, H.C.H., Schmidt, P.W., 1962. Small angle X-ray scattering from montmorillonite clay suspensions. II. J. Chem. Phys. 37 (3), 502–510.
Holdren Jr., G.R., Berner, R.A., 1979. Mechanism of feldspar weathering. I. Experimental studies. Geochim. Cosmochim. Acta 43 (8), 1161–1171.
Jackson, M.L., Tyler, S.A., Willis, A.L., Bourbeau, G.A., Pennington, R.P., 1948. Weathering sequence of clay-size minerals in soils and sediments. I. Fundamental generalizations. J. Phys. Colloid Chem. 52, 1237–1260.
Johnson, S.W., Blake, J.M., 1867. Contributions from the Sheffield Laboratory of Yale College. XIV. On kaolinite and pholerite. Am. J. Sci. 43 (129), 351–361.
Kerr, P.F., Kulp, J.L., 1951. Reference clay localities: United states. Reference Clay Minerals, American Petroleum Institute Project 49, Preliminary Report No. 2. Columbia University Press.
Kerr, P.F., Hamilton, P.K., Pill, R.J., Wheeler, G.V., Lewis, D.R., Burkhardt, W., Reno, D., Taylor, G.L., Mielenz, R.C., King, M.E., Schieltz, N.C., 1951. Analytical data on reference clay minerals. Reference Clay Minerals, American Petroleum Institute Project 49, Preliminary Report No. 7. Columbia University Press.
Knight, H.G., 1938. New size limits for silt and clay. Soil Sci. Soc. Am. Proc. 2, 2592.
Krekeler, M.P.S., Hammerly, E., Rakovan, J., Guggenheim, S., 2005. Microscopy studies of the palygorskite-to-smectite transformation. Clays Clay Miner. 53 (1), 92–99.
Lagache, M., 1965. Feldspar alteration in water between 100 and 200° under various CO_2 pressures and in the clay mineral syntheses. Bull. Soc. Fr. Mineral. Cristallogr. 88 (2), 223–253.
Lagache, M., Wyart, J., Sabatier, G., 1961. Dissolution des feldspaths alcalins dans l'eau pure ou chargeé de co_2 200°C (decomposition of alkali feldspars in pure or CO_2-containing water at 200°C). Compt. Rend. 253, 2019–2022.
Leonard, R.A., Low, P.F., 1964. Effect of gelation on the properties of water in clay systems. Clays Clay Miner. 12, 311–325.
Lubliner, J., 2008. Plasticity Theory. Dover Publications, Mineola, NY.
Mathieson, A.M., Walker, G.F., 1954. The crystal structure of magnesium vermiculite. Am. Mineral. 39, 231–255.
Mooney, R.W., Keenan, A.G., Wood, L.A., 1952. Adsorption of water vapor by montmorillonite. II. Effect of exchangeable ions and lattice swelling as measured by X-ray diffraction. J. Am. Chem. Soc. 74, 1371–1374.
Morse, H.N., Frazer, J.C.W., Dunbar, P.B., 1907. The osmotic pressure of cane-sugar solutions in the vicinity of 5°C. Am. Chem. J. 38, 175–226.
Müller-Vonmoos, M., Løken, T., 1989. The shearing behaviour of clays. Appl. Clay Sci. 4 (2), 125–141.
NCSS, 2014. Soil survey field and laboratory manual. In: Rebercca, B., Soil Survey Staff (Eds.), Soil Survey Investigations Report No. 51, Version 2, United States Environmental Protection Agency, Lincoln, Nebraska.
Norrish, K., 1954. The swelling of montmorillonite. Discuss. Faraday Soc. 18, 120–134.
Norrish, K., Rausell-Colom, J.A., 1963. Low-angle X-ray diffraction studies of the swelling of montmorillonite and vermiculite. Clays Clay Miner. 10, 123–149.
Norton, F.H., 1948. Fundamental study of clay. VIII. A new theory for the plasticity of clay-water masses. J. Am. Ceram. Soc. 31, 236–241.
Pauling, L., 1929. The principles determining the structure of complex ionic crystals. J. Am. Chem. Soc. 51, 1010–1026.
Pauling, L., 1930a. The structure of the chlorites. Proc. Natl. Acad. Sci. U. S. A. 16, 578–582.
Pauling, L., 1930b. Structure of the micas and related minerals. Proc. Natl. Acad. Sci. U. S. A. 16, 123–129.
Pauling, L., 1960. The Nature of the Chemical Bond, third ed. Cornell University Press, Ithaca, NY.
Rioux, R.L., 1958. Geology of the Spence-Kane Area, Big Horn County, Wyoming, Open-File Report 58-84. United States Geological Survey.
Shackleton, E.H., 1978. Lakeland Geology, fifth ed. Dalesman Publishing, Clapham, Lancashire.
Simpson, J.A. (Ed.), 1989. Oxford English Dictionary, second ed. Oxford University Press, Oxford/New York, NY.

Stokes, G.G., 1851. On the effect of the internal friction of fluids on the motion of pendulums. Trans. Camb. Philos. Soc. 9, 8–149.

Suarez, M., Garcia-Romero, E., 2013. Sepiolite-palygorskite: a continuous polysomatic series. Clays Clay Miner. 61 (5–6), 461–472.

Taylor, S.R., 1964. Abundance of chemical elements in the continental crust: a new table. Geochim. Cosmochim. Acta 28 (8), 1273–1285.

Tazaki, K., 1986. Observation of primitive clay precursors during microcline weathering. Contrib. Mineral. Petrol. 92 (1), 86–88.

Thompson Jr., J.B., 1978. Biopyriboles and polysomatic series. Am. Mineral. 63 (3-4), 239–49.

van't Hoff, J.H., 1887. Die rolle des osmotischen druckes in der analogie zwischen lösungen und gasen. Zeitschrift für physikalische Chemie 1, 481–508.

Vaslow, F., 1966. The apparent molal volumes of the alkali metal chlorides in aqueous solution and evidence for salt-induced structure transitions. J. Phys. Chem. 70 (7), 2286–2294.

Veblen, D.R., Buseck, P.R., 1980. Microstructures and reaction mechanisms in biopyriboles. Am. Mineral. 65, 599–623.

Wadsley, A.D., Andersson, S., 1970. Crystallographic shear, and the niobium oxide fluorides in the composition region MXx, $2.4 < x < 2.7$. Perspect. Struct. Chem. 3, 1–58.

Wehrli, B., Wieland, E., Furrer, G., 1990. Chemical mechanisms in the dissolution kinetics of minerals; the aspect of active sites. Aquat. Sci. 52 (1), 3–31.

Wentworth, C.K., 1922. A scale of grade and class terms for clastic sediments. J. Geol. 30, 377–392.

White, W.A., 1949. Atterberg plastic limits of clay minerals. Am. Mineral. 34, 508–512.

White, A.F., Brantley, S.L., 1995. Chemical weathering rates of silicate minerals: an overview. Rev. Mineral. 31, 1–22.

Whitney, M., 1891. Soil Investigations: VI. Mechanical Analysis of the Type Soils, Annual Report. Maryland Agricultural Experiment Station.

CHAPTER

4

Ion Exchange

OUTLINE

4.1 Introduction 148

4.2 The Discovery of Ion Exchange 148

4.3 The Ion Exchange Experiment 149

4.4 Selectivity Constants and the Ion Exchange Isotherm 152

4.4.1 The Vanselow Mole-Fraction Convention 153

4.4.2 The Gaines-Thomas Equivalent-Fraction Convention 154

4.4.3 The Gapon Convention 155

4.5 Interpreting the Ion Exchange Isotherm 155

4.5.1 The Exchange Isotherm for Symmetric Exchange 156

4.5.2 The Exchange Isotherm for Asymmetric Exchange 157

4.5.3 Effect of Electrolyte Concentration on the Ion Exchange Isotherm 159

4.5.4 Effect of Ion Selectivity on the Ion Exchange Isotherm 160

4.5.5 Physical Basis for Ion Selectivity 164

4.5.6 Other Influences on the Ion Exchange Isotherm 166

4.6 Summary 170

Appendices 170

4.A The Preparation and Characterization of Natural Ion Exchangers 170

4.A.1 Saturating an Exchanger With a Single Cation 171

4.A.2 Measuring Ion Exchange Capacity 171

4.B Non-Linear Least Square Fitting of Exchange Isotherms 172

4.B.1 Symmetric Exchange 172

4.B.2 Asymmetric Exchange 175

4.C Ideal and Real Mixtures: Significance for Ion Exchange Selectivity Constant Expressions 175

4.C.1 Ideal Mixtures 175

4.C.2 Real Mixtures 176

4.D Selectivity Constant Expressions: Ideal and Real Exchange-Complex Mixtures 179

4.E Equivalent Fraction-Dependent Selectivity Coefficients 181

4.F The Gapon Convention: Revisited 183

References 185

http://dx.doi.org/10.1016/B978-0-12-804178-9.00004-5

4.1 INTRODUCTION

While studying the chemistry of manure in soil, 19th-century agricultural chemists discovered ion exchange. This discovery was wholly unexpected and all the more puzzling because chemists at that time did not have an accurate understanding of electrolyte solutions. Regardless, a handful of scientists worked out the broad outlines of the ion exchange reaction within a remarkably short time. Ion exchange is now recognized as a very important chemical reaction with applications in nearly all chemistry fields. This chapter begins with a brief account of the discovery and the implications of ion exchange for environmental chemistry. The next topic is a complete description of an ion exchange experiment itself and the data collected from such an experiment. Scientists studying ion exchange reactions have developed several ways of writing the equilibrium quotient, and, not surprisingly, the different formulations are equivalent to each other. The final and most important topic is a discussion of the physical parameters that determine the ion exchange equilibrium.

4.2 THE DISCOVERY OF ION EXCHANGE

Soil fertility—the capacity of soil to supply essential nutrients to crops—has long been a major concern of agriculturalists. Nitrogen is one of the most important plant nutrients, and maintenance of soil nitrogen fertility is critical to crop yield. Manure is a rich and readily accessible nitrogen source, with the nitrogen in manure being released largely as ammonia through the microbial degradation process known generally as mineralization.

Much of the strong odor from animal manure comes from ammonia gas, and 19th-century agricultural chemists understood that ammonia loss from manure meant that there was less nitrogen for crop uptake. Thompson (1850) began studying methods to prevent ammonia loss from stored manure in 1845, including sulfuric acid additions that produced large amounts of ammonia sulfate. In a report to the Royal Agricultural Society, Thompson mentions an interesting property of soil: "*the power of retaining ammonia.*" Thompson designed experiments to measure the "*extent of this power and to ascertain whether it also extended to the* [sulfate]" by mixing ammonium sulfate with soil, filling a glass column with the soil, washing the soil column with water to simulate rainfall, and analyzing the drainage solution. Thompson was certainly aware of an observation made by Huxtable when liquid manure drained through a soil bed "*it went in manure and came out water*" (Way, 1850). Thompson found that the salt remaining after he evaporated the drainage solution "*proved to be chiefly gypsum.*"

> "This was a complete surprise. The large portion of gypsum…[in the drainage solution]…showed that a considerable portion of the sulphate of ammonia mixed with the soil had been decomposed and that this process was in some way connected with the presence of lime in the soil, as the sulphuric acid was washed out in combination with lime."

Thompson performed an experiment to observe the result when "*the absorptive powers of the soil were fully called into play by passing the filtered liquid repeatedly through the* [soil column]." He found that "*the whole of the ammonia was retained by the soil, whether applied in the form of sulphate or sesquicarbonate.*" Way (1850) published a report in the same publication of the Royal

Agricultural Society, discussing possible explanations for the "*absorptive power*" of soil and describing numerous experiments that extended those performed by Thompson. Way found that after adding ammonia sulfate solution to a soil column, the drainage solution was "*entirely free of the pungent smell of ammonia.*" Way also noticed that the soil had a fixed capacity to retain ammonia; after continued washing of the soil column, "*the ammonia would shortly have passed through, the soil being saturated with it.*" Way (1850, 1852) traced the "*absorbing power*" of soil to the clay fraction, but how clays retain cations would remain a mystery until Pauling (1930) solved the crystal structure of layer silicates. The most abundant material in nature with significant ion exchange capacity is smectite. A number of other minerals have measurable ion exchange capacity—for example, oxide clays, natural organic matter, zeolite minerals—but the overwhelming number of ion exchange studies involved smectite.

4.3 THE ION EXCHANGE EXPERIMENT

The simplest ion exchange experiment begins with a known volume of a well-mixed clay suspension containing negligible soluble electrolyte. The mass concentration of clay is known, and the clay is saturated with one or the other of the two cations undergoing exchange. In the example shown in Fig. 4.1, the clay is initially A^+-saturated. Solution aliquots from two stock solutions—A^+X^- and B^+X^-—are added to the clay suspension, yielding initial concentrations c'_{A^+} and c'_{B^+} (Fig. 4.1A). It is important that the electrolyte concentration of the initial solutions remains constant with only the concentration ratio of the two cations A^+ and B^+ being varied.

The exchange reaction is generally quite rapid at electrolyte concentrations greater than 10^{-2} mol dm^{-3}, but it can take weeks in more dilute solutions. It is best to continually agitate the clay suspension until equilibrium is reached (Fig. 4.1B). Analysis of the equilibrium solution to determine final concentrations c_{A^+} and c_{B^+} usually requires centrifugation to remove clay from suspension (Fig. 4.1C).

Analysis of the equilibrium solution becomes difficult under certain conditions: the clay particles are less than 2 μm, the exchanger is a smectite, and the saturating cation is either Li^+

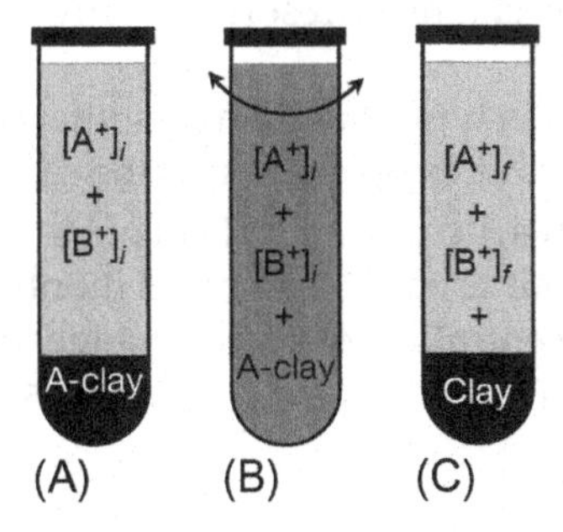

FIG. 4.1 Measuring the cation exchange isotherm involves three steps: (A) allowing a series of solutions containing varying concentrations of two salts A^+X^- and B^+X^- to react with A^+-saturated exchanger, (B) agitation to thoroughly blend the suspensions, and (C) separation of the exchanger from solution.

TABLE 4.1 Initial Conditions of a Symmetric (Na^+, K^+) Exchange Experiment

c'_{Na^+} (mmol$_c$ dm^{-3})	c'_{K^+} (mmol$_c$ dm^{-3})	$\left(n^{\sigma}_{Na^+}/V\right)'$ (mmol$_c$ dm^{-3})	$\left(n^{\sigma}_{K^+}/V\right)'$ (mmol$_c$ dm^{-3})
100.0	0.0	0.0	25.0
100.0	0.0	0.0	42.9
85.7	14.3	0.0	42.9
71.4	28.6	0.0	42.9
57.1	42.9	0.0	42.9
42.9	57.1	0.0	42.9
25.0	75.0	0.0	25.0
75.0	25.0	25.0	0.0
57.1	42.9	42.9	0.0
42.9	57.1	42.9	0.0
28.6	71.4	42.9	0.0
14.3	85.7	42.9	0.0
0.0	100.0	42.9	0.0
0.0	100.0	25.0	25.0

The cation exchanger is a montmorillonite clay mineral identified as bentonite-7.
Vanselow, A.P., 1932a. Equilibria of the base-exchange reactions of bentonites, permutites, soil colloids and zeolites. Soil Sci. 33, 95–113.

or Na^+. Chapter 3 discusses both crystalline and free swelling of clay minerals. *Vermiculite*[1] exhibits crystalline swelling, never free swelling, regardless of the saturating cation. Smectite exhibits both crystalline swelling and free swelling; the latter when the saturating cation is either Li^+ or Na^+, and the electrolyte concentration (or osmotic head) is low. A low osmotic-head suspension of either Li^+ or Na^+ saturated smectite typically forms colloidal dispersions that resist sedimentation in the operating range of conventional laboratory centrifuges. Under these circumstances, clay chemists resort to dialysis to isolate soluble cations for analysis.

Table 4.1 illustrates the design of a typical exchange isotherm experiment (Vanselow, 1932a). The exchanger Vanselow used for this experiment was the clay mineral montmorillonite[2] with a cation exchange capacity of n^{σ}_{+} 104.1 cmol$_c$ kg^{-1}. Half of the suspensions used Na^+-saturated montmorillonite, and the other half used K^+-saturated montmorillonite. Vanselow based the clay content on the volume concentration of the exchangeable cation—24.0 g dm^{-3} of K^+-saturated montmorillonite, yielding a K^+ concentration of 25.0 mmol$_c$ dm^{-3} as exchangeable

[1] *vermiculite*: $Na_x : Al_2\,(Si_{4-x}Al_x)O_{10}(OH)_2;\ 0.6 \le x \le 0.7$.

[2] *montmorillonite*: $Na_x : (Al_{(2-x)}Mg_x)\,Si_4O_{10}(OH)_2;\ 0.25 \le x \le 0.60$.

TABLE 4.2 Equilibrium Conditions of a Symmetric (Na^{+}, K^{+}) Exchange Experiment

c_{Na^+} (mmol$_c$ dm^{-3})	c_{K^+} (mmol$_c$ dm^{-3})	$\left(n^{\sigma}_{Na^+}/V\right)$ (mmol$_c$ dm^{-3})	$\left(n^{\sigma}_{K^+}/V\right)$ (mmol$_c$ dm^{-3})
86.3	14.9	13.7	10.1
81.1	20.6	18.9	22.3
71.9	29.3	13.8	27.9
61.8	39.5	9.6	32.0
50.5	50.6	6.6	35.2
38.7	62.2	4.2	37.8
23.7	76.9	1.3	23.1
85.3	15.6	14.7	9.4
79.8	22.3	20.2	20.6
70.4	31.5	15.4	25.6
60.0	42.1	11.5	29.3
48.4	53.1	8.8	32.6
37.0	64.7	5.9	35.3
22.7	78.3	2.3	46.7

The cation exchanger is a montmorillonite clay mineral identified as bentonite-7.
Vanselow, A.P., 1932a. Equilibria of the base-exchange reactions of bentonites, permutites, soil colloids and zeolites. Soil Sci. 33, 95–113.

K^+. For his experiment, he used different amounts of montmorillonite: 24.0 g dm^{-3} (equivalent to 25.0 mmol$_c$ dm^{-3}) and 41.2 g dm^{-3} (equivalent to 42.9 mmol$_c$ dm^{-3}).

$$c'_{Na^+} + \left(\frac{n^{\sigma}_{Na+}}{m}\right)' \cdot \left(\frac{m}{V}\right) = c_{Na^+} + \left(\frac{n^{\sigma}_{Na^+}}{m}\right) \cdot \left(\frac{m}{V}\right)$$

$$\left(\frac{n^{\sigma}_{Na^+}}{m}\right) \cdot \left(\frac{m}{V}\right) = c'_{Na^+} + \left(\frac{n^{\sigma}_{Na^+}}{m}\right)' \cdot \left(\frac{m}{V}\right) - c_{Na^+}$$

Table 4.2 lists the equilibrium composition of the clay mineral suspension, and Fig. 4.2 plots the exchange isotherms. Vanselow did not have to measure the amounts of Na^+ and K^+ bound to the clay mineral because he could determine those quantities by mass balance. The total Na^+ concentration is the sum of the soluble concentration c_{Na^+}[mmol$_c$ dm^{-3}] and the amount of clay-bound Na^+ expressed in volume concentration units $\left(n^{\sigma}_{Na^+}/m\right) \cdot (m/V)$ [mmol$_c$ dm^{-3}]. The initial and final total Na^+ concentration remains unchanged; only the relative soluble and clay-bound concentrations change.

A comparison of Tables 4.1 and 4.2 clearly shows that cation exchange has altered the solution concentration of both cations and the quantities of both cations bound to the clay. Row 5 in Table 4.2—where $c_{Na^+} \approx c_{K^+}$—also shows that the fraction of clay-bound Na^+ is significantly less than the Na^+ fraction in solution.

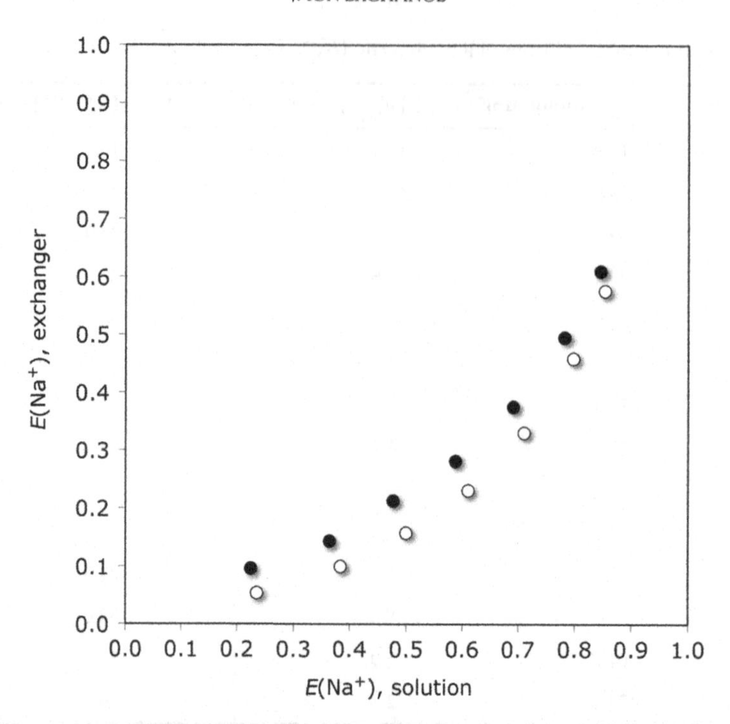

FIG. 4.2 The cation exchange isotherm for (Na^+, K^+) exchange on a montmorillonite clay specimen. *Vanselow, A.P., 1932a. Equilibria of the base-exchange reactions of bentonites, permutites, soil colloids and zeolites. Soil Sci. 33, 95–113.*

Each (Na^+,K^+) exchange isotherm (Fig. 4.2) plots the Na^+ exchanger-bound equivalent fraction $E_{\overline{Na^+}}$ as a function of the Na^+ solution equivalent fraction E_{Na^+}. This cation exchange experiment clearly demonstrates this montmorillonite specimen selectively accumulates K^+ from solution, yielding a clay-bound equivalent fraction that is higher than the equilibrium solution equivalent fraction: $E_{\overline{K^+}} > E_{K^+}$.

4.4 SELECTIVITY CONSTANTS AND THE ION EXCHANGE ISOTHERM

Electrolyte behavior in aqueous solution remained a mystery for nearly 40 years after the discovery of ion exchange until the late 1880s when three prominent chemists—Arrhenius, Ostwald, and van't Hoff—published a series of papers (Arrhenius, 1887; van't Hoff, 1887; Arrhenius, 1888; van't Hoff and Reicher, 1888; Ostwald, 1888) on the colligative properties of electrolyte solutions (freezing-point depression, vapor-pressure depression, and osmotic head). Those studies revealed many electrolytes fully dissociate to form solvated ions in aqueous solution.

Before chemists understood electrolyte dissociation they would have written ion exchange reactions listing dissolved electrolytes as NaCl(aq) and $CaCl_2$(aq), respectively, and denoted exchanger electrolytes as $\overline{KZ}$ and $\overline{CaZ_2}$, respectively.

Symmetric Exchange Reaction

$$\mathrm{NaCl(aq)} + \overline{\mathrm{KZ}} \longleftrightarrow \overline{\mathrm{NaZ}} + \mathrm{KCl(aq)} \tag{4.R1}$$

Asymmetric Exchange Reaction

$$\mathrm{2NaCl(aq)} + \overline{\mathrm{CaZ_2}} \longleftrightarrow 2\overline{\mathrm{NaZ}} + \mathrm{CaCl_2(aq)} \tag{4.R2}$$

After chemists understood electrolyte dissociation, they would write ion exchange reactions listing dissociated solution ions, and would consider it equally reasonable to list fully dissociated exchanger-bound ions: $\overline{Na^+}$ and $\overline{Ca^{2+}}$. If exchanger bound ions are fully dissociated the exchange-site co-ion would be written as: $\overline{Z^-}$.

$$\mathrm{Na^+(aq)} + \mathrm{Cl^-(aq)} + \overline{\mathrm{K^+Z^-}} \longleftrightarrow$$
$$\overline{\mathrm{Na^+Z^-}} + \mathrm{K^+(aq)} + \mathrm{Cl^-(aq)} \tag{4.R3}$$

$$\mathrm{2Na^+(aq)} + \mathrm{2Cl^-(aq)} + \overline{\mathrm{Ca^{2+}Z_2^-}} \longleftrightarrow$$
$$2\overline{\mathrm{Na^+Z^-}} + \mathrm{Ca^{2+}(aq)} + \mathrm{2Cl^-(aq)} \tag{4.R4}$$

Notice, co-ions Cl^-(aq) and $\overline{Z^-}$ appear on both sides of the cation exchange reaction written above. The ion exchange reaction is further simplified *without loss of generality* by deleting all chemical components that appear on both sides of the reaction equation. While this exercise may seem trivial it is necessary to identify the fatal flaw in the widely quoted Gapon convention (cf. Appendix 4.F).

$$\mathrm{Na^+(aq)} + \overline{\mathrm{K^+}} \longleftrightarrow \overline{\mathrm{Na^+}} + \mathrm{K^+(aq)} \tag{4.R5}$$

$$\mathrm{2Na^+(aq)} + \overline{\mathrm{Ca^{2+}}} \longleftrightarrow 2\overline{\mathrm{Na^+}} + \mathrm{Ca^{2+}(aq)} \tag{4.R6}$$

4.4.1 The Vanselow Mole-Fraction Convention

Vanselow (1932b) introduced the first accurate convention for writing ion exchange reaction *equilibrium quotient expressions*. The Vanselow ion exchange quotient expression assumes full dissociation of both dissolved and exchanger-bound electrolytes.

Dissolved ion concentrations (4.1) are assigned molar units c_{Na^+} mmol$_c$ dm^{-3}. The distinguishing feature of the *Vanselow mole-fraction convention* is the assignment of mole fraction units (4.2) to exchanger-bound ions: $x_{\overline{Na^{a+}}}$ mol mol^{-1}.

Vanselow mole-fraction convention

$$K^c_{Na/Ca} = \frac{x^2_{\overline{Na^+}} \cdot c_{Ca^{2+}}}{c^2_{Na^+} \cdot x_{\overline{Ca^{2+}}}} \tag{4.1}$$

$$x_{\overline{Na^+}} = \frac{n_{\overline{Na^+}}}{n_{\overline{Na^+}} + n_{\overline{Ca^{2+}}}} \tag{4.2}$$

$$x_{\overline{Na^+}} = 1 - x_{\overline{Ca^{2+}}}$$

Equilibrium constants for ion exchange reactions are typically called the *selectivity constants*. For the (Na^+,Ca^{2+}) ion exchange reaction (4.R6) with selectivity quotient expression (4.1) ion-exchange selectivity coefficient is denoted: $K^c_{Na/Ca}$. The notation adopted here[3] follows *IUPAC* guidelines (Cohen et al., 2007; Samuelson, 1972).

The following superscript c indicates $K^c_{Na/Ca}$ is a selectivity *coefficient*[4] because dissolved ions appear as molar concentrations[5] in reaction quotient expression (4.1). The following subscript Na/Ca indicates the cation exchange reaction is written with $Na^+(aq)$ and $\overline{Ca^{2+}}$ in the left-hand (i.e., *educt*) side of the reaction arrow.

The Vanselow mole-fraction convention is notable because it represents exchanger-bound ions as an *ideal* mixture.[6] In effect, the Vanselow ion exchange quotient expression (4.1) employs two very different conventions to represent reactions component. Following a practice that was already well established by 1900 for solubility and hydrolysis reactions, Vanselow represents *dissolved* ions using concentration notation. The Vanselow quotient expression then employs Raoult's mole-fraction notation to represent *exchanger-bound* ions.

4.4.2 The Gaines-Thomas Equivalent-Fraction Convention

The second most commonly encountered convention—the *Gaines-Thomas equivalent-fraction convention*—assigns equivalent fraction units $E_{\overline{Na^+}}$ $mol_c\, mol_c^{-1}$ to quantify exchanger-bound ions (4.3), otherwise it is very similar to the Vanselow mole-fraction convention.

$$E_{\overline{Na^+}} = \frac{n_{\overline{Na^+}}}{n_{\overline{Na^+}} + 2 \cdot n_{\overline{Ca^{2+}}}} \tag{4.3}$$

$$E_{\overline{Na^+}} = 1 - E_{\overline{Ca^{2+}}}$$

[3] The *Vanselow mole-fraction convention* is defined by the selectivity quotient *expression* not the selectivity constant (or coefficient) itself, therefore this book does not associate the selectivity constant (or coefficient) with a particular convention.

[4] cf. Appendix 4.D for an explanation of the difference between equilibrium constants and equilibrium coefficients or between selectivity constants and selectivity coefficients.

[5] cf. Appendix 4.D for a more advanced discussion of unit conventions and equilibrium constant expressions.

[6] cf. Appendix 4.C for a discussion of ideal and real mixtures.

The selectivity coefficient in the Gaines-Thomas selectivity quotient expression (4.4) is identical to the selectivity coefficient appearing in the Vanselow selectivity quotient expression (4.1).

Gaines-Thomas equivalent-fraction convention

$$K^{c}_{Na/Ca} = \frac{E^{2}_{\overline{Na^{+}}} \cdot c_{Ca^{2+}}}{c^{2}_{Na^{+}} \cdot E_{\overline{Ca^{2+}}}} \tag{4.4}$$

Ion exchange equilibrium quotient expressions written following the Vanselow convention are easily transformed to the Gaines-Thomas convention and *vice versa*. Relation (4.5) converts Vanselow convention mole-fraction units $x_{\overline{Na^{+}}}$ into Gaines-Thomas convention equivalent-fraction units $E_{\overline{Na^{+}}}$. Expression (4.5) is derived by combining expressions (4.2), (4.3).

$$E_{\overline{Na^{+}}} = \frac{1 \cdot x_{\overline{Na^{+}}}}{2 + (1 - 2) \cdot x_{\overline{Na^{+}}}} \tag{4.5}$$

Many chemists prefer the Vanselow convention because expressing exchanger-bound ions as mole-fractions retains the original ideal mixture concept and notation. Others, including this author, prefer the Gaines-Thomas convention because the equivalent fraction is the most natural independent variable for asymmetric ion exchange reactions (cf. Appendix 4.D for further details).

4.4.3 The Gapon Convention

The *Gapon convention* is another notable ion exchange convention mentioned in most soil chemistry textbooks and many research papers. Though discredited (Sposito, 1977; Evangelou and Phillips, 1987, 1988), soil chemists studying (Na^{+},Ca^{2+}) exchange in the context of saline and sodic soils (cf. Chapter 6) base the ubiquitous *sodium adsorption ratio* (SAR) on the Gapon convention (Sposito and Mattigod, 1977). As a consequence the Gapon convention has acquired an undeserved legitimacy. The flaws of the Gapon convention are discussed fully in Appendix 4.F.

4.5 INTERPRETING THE ION EXCHANGE ISOTHERM

The results in Table 4.2 and Fig. 4.2 are from a classic paper by Vanselow on ion exchange phenomena. The purpose of that and other studies was an attempt to understand the quantitative details of ion exchange reactions and the selective exchange of certain ions.

Much of the scientific research devoted to ion exchange seeks to properly formulate the equilibrium quotient for the ion exchange reaction. The historical developments in ion exchange research are beyond the scope of this book, but a summary of the most prominent ion exchange expressions is in order.

4.5.1 The Exchange Isotherm for Symmetric Exchange

Chemists commonly plot experimental data using an exchange isotherm rather than plotting the selectivity quotient. The exchange isotherm for symmetric (Na^+, K^+) exchange reaction (4.R7) is derived from the selectivity quotient (4.6) as follows:

$$Na^+(aq) + \overline{K^+} \longleftrightarrow \overline{Na^+} + K^+(aq) \quad (4.R7)$$

$$K^c_{Na/K} = \frac{E_{\overline{Na^+}} \cdot c_{K^+}}{c_{Na^+} \cdot E_{\overline{K^+}}} \quad (4.6)$$

$$\text{Define:} E_{Na^+} \equiv \frac{c_{Na^+}}{c_{Na^+} + c_{K^+}}$$

$$K^c_{Na/K} = \frac{E_{\overline{Na^+}} \cdot E_{K^+}}{E_{Na^+} \cdot E_{\overline{K^+}}} = \frac{E_{\overline{Na^+}} \cdot (1 - E_{Na^+})}{E_{Na^+} \cdot \left(1 - E_{\overline{Na^+}}\right)} \quad (4.7)$$

$$E_{\overline{Na^+}} = \frac{K^c_{Na/K} \cdot E_{Na^+}}{\left(1 - E_{Na^+} + \left(K^c_{Na/K} \cdot E_{Na^+}\right)\right)} \quad (4.8)$$

The symmetric exchange quotient expression (4.7)—which contains one adjustable parameter $K^c_{Na/K}$ and one independent variable E_{Na^+}—is readily rearranged to yield the symmetric exchange isotherm expression (4.8). Example 4.1 illustrates the single-point determination and interpretation of a symmetric selectivity coefficient.

EXAMPLE 4.1

Example Permalink

http://soilenvirochem.net/kf1Z0u

Estimate the selectivity coefficient $K^c_{Na/K}$ for the symmetric (Na^+, K^+) exchange results in Table 4.2.

Fig. 4.2 plots the exchange isotherm derived from data listed in Table 4.2. An estimate of the selectivity coefficient $K^c_{Na/K}$ can be made using any data point plotted in Fig. 4.2.

Taking the experimental results listed in row 5 of Table 4.2, first determine each term in the symmetric Gaines-Thomas selectivity quotient expression (4.6). The equivalent fractions of Na^+ and K^+ on the exchanger are determined as follows:

$$E_{\overline{Na^+}} = \frac{n_{\overline{Na^+}}}{n_{\overline{K^+}} + n_{\overline{Na^+}}}$$

$$E_{\overline{Na^+}} = \frac{6.60}{6.60 + 35.20} = 0.158$$

$$E_{\overline{K^+}} = 1 - 0.158 = 0.842$$

The corresponding equilibrium solution composition (Table 4.2, row 5) is $c_{Na^+} = 0.0505$ $mol\,dm^{-3}$ and $c'_{K^+} = 0.0506\ mol\,dm^{-3}$. Entering these values in expression (4.6) yields a single-point selectivity coefficient estimate for ion exchange reaction (Na^+, K^+).

$$K^c_{Na/K} \approx \frac{(0.158) \cdot \left(0.0506\ mol\,dm^{-3}\right)}{\left(0.0505\ mol\,dm^{-3}\right) \cdot (0.842)}$$

$$K^c_{Na/K} \approx \frac{0.158}{0.842} = 0.188$$

The selectivity coefficient $K^c_{Na/K} \approx 0.1880$ is for the ion exchange reaction as it is written above. Notice that the solution concentrations of the two cations at equilibrium are essentially equal for the selected data (Table 4.2, row 5). If the exchange were nonselective (i.e., $K^c_{Na/K} = 1$), then the exchanger equivalent fractions of the two cations would be equal. The exchange results appearing in Table 4.2 and Fig. 4.2, however, clearly indicate an enrichment of K^+ in the exchange complex ($E_{\overline{K^+}} = 0.842$) and a corresponding depletion of Na^+ ($E_{\overline{Na^+}} = 0.158$).

4.5.2 The Exchange Isotherm for Asymmetric Exchange

The exchange isotherm for asymmetric exchange reactions is more complex than the symmetric exchange isotherm because the stoichiometric coefficients (Example 4.2) in the numerator and denominator of the exchange selectivity quotient do not cancel (cf. asymmetric (Na^+, Ca^{2+}) exchange reaction (4.R6) and exchange selectivity quotient (4.4)). Example 4.2 illustrates the single-point determination of an asymmetric selectivity coefficient.

Replacing solution concentrations with solution equivalent fractions requires a new solution parameter N_0 that does not appear in the symmetric exchange isotherm. The new parameter N_0 equals the ion charge concentration summed over all ions involved in the exchange (4.9).

$$N_0 = \left(c_{Na^+}\right) + \left(2 \cdot c_{Ca^{2+}}\right) \tag{4.9}$$

Converting solution molar concentrations to equivalent fractions adds a factor of 2 to the Ca^{2+} solution equivalent fraction (4.10).

$$E_{Ca^{2+}} = \frac{2 \cdot c_{Ca^{2+}}}{N_0} \tag{4.10}$$

$$K^c_{Na/Ca} = \frac{E^2_{\overline{Na^+}} \cdot \left(\frac{N_0 \cdot E_{Ca^{2+}}}{2}\right)}{\left(N_0 \cdot E_{Na^+}\right)^2 \cdot E_{\overline{Ca^{2+}}}}$$

$$K^c_{Na/Ca} = \left(\frac{1}{2 \cdot N_0}\right) \cdot \frac{E^2_{\overline{Na^+}} \cdot \left(1 - E_{Na^+}\right)}{E^2_{Na^+} \cdot \left(1 - E_{\overline{Na^+}}\right)} \tag{4.11}$$

Rearranging expression (4.11) transforms the asymmetric (Na^+, Ca^{2+}) exchange quotient expression into a second-order polynomial (4.12) containing a second-order *exchange parameter* β_2 (4.13).[7] The equilibrium exchanger equivalent fraction $E_{\overline{Na^+}}$ is given by a quadratic equation (4.14).

$$2 \cdot N_0 \cdot K^c_{Na/Ca} = \frac{E^2_{\overline{Na^+}} \cdot (1 - E_{Na^+})}{E^2_{Na^+} \cdot \left(1 - E_{\overline{Na^+}}\right)}$$

$$E^2_{\overline{Na^+}} + \beta_2 \cdot E_{\overline{Na^+}} - \beta_2 = 0 \tag{4.12}$$

$$\beta_2 \equiv \left(\frac{2 \cdot N_0 \cdot K^c_{Na/Ca} \cdot E^2_{Na^+}}{1 - E_{Na^+}} \right) \tag{4.13}$$

$$E_{\overline{Na^+}} = \frac{-\beta_2 + \sqrt{\beta_2^2 + 4 \cdot \beta_2}}{2} \tag{4.14}$$

EXAMPLE 4.2

Example Permalink

http://soilenvirochem.net/Oy8thp

Estimate the selectivity coefficient for asymmetric (K^+,Ca^{2+}) exchange reaction on montmorillonite[8] clay specimen *American Petroleum Institute* API-23 (Chambers, Arizona).

Montmorillonite API-23 has a CEC = 126 $cmol_c\ kg^{-1}$. An asymmetric (K^+,Ca^{2+}) cation exchange experiment reports the following results: $c_{K^+} = 10.17$ mM, $c_{Ca^{2+}} = 0.565$ mM, exchangeable $c_{\overline{K^+}} =$ 63.0 $cmol_c\ kg^{-1}$, and exchangeable $c_{\overline{Ca^{2+}}} = 63.0\ cmol_c\ kg^{-1}$.

What is the value of the selectivity coefficient $K^c_{K/Ca}$ for (K^+,Ca^{2+}) exchange in this clay?

The asymmetric ion exchange reaction and corresponding selectivity quotient follow.

$$2K^+(aq) + \overline{Ca^{2+}} \longleftrightarrow \overline{2K^+} + Ca^{2+}(aq)$$

$$K^c_{K/Ca} = \frac{E^2_{\overline{K^+}} \cdot c_{Ca^{2+}}}{c^2_{K^+} \cdot E_{\overline{Ca^{2+}}}}$$

[7] The following subscript 2 distinguishes the *exchange parameter* appearing in a second-order polynomial from the exchange parameter β_3 appearing in a third-order polynomial associated with (Ca^{2+}, Al^{3+}) exchange.

As in Example 4.1, determine each term in the asymmetric quotient expression. The equivalent fraction of the univalent cation K^+ on the exchanger $E_{\overline{K^+}}$ is the simplest term because the denominator is the cation exchange capacity CEC.

$$E_{\overline{K^+}} = \frac{n_{\overline{K^+}}}{n_{\overline{K^+}} + 2 \cdot n_{\overline{Ca^{2+}}}} = \frac{n_{\overline{K^+}}}{CEC}$$

$$E_{\overline{K^+}} = \frac{63.0}{126.0} = 0.500$$

$$E_{\overline{Ca^{2+}}} = 1 - E_{\overline{K^+}} = 0.500$$

Entering these exchange equivalent fractions and cation concentrations in the (K^+,Ca^{2+}) exchange quotient expression yields a single-point selectivity coefficient estimate.

$$K^c_{K/Ca} \approx \frac{(0.500)^2 \cdot \left(5.65 \times 10^{-4}\ \text{mol dm}^{-3}\right)}{\left(1.017 \times 10^{-2}\ \text{mol dm}^{-3}\right)^2 \cdot (0.500)}$$

$$K^c_{K/Ca} \approx \frac{1.41 \times 10^{-4}}{5.17 \times 10^{-5}} = 2.731$$

[8] A layer silicate can be classified in the smectite mineral group based solely on X-ray diffraction analysis. Identifying a smectite clay specimen as the clay mineral montmorillonite requires chemical composition data.

4.5.3 Effect of Electrolyte Concentration on the Ion Exchange Isotherm

The symmetric exchange isotherm does not contain any terms explicitly influenced by electrolyte concentration. The selectivity coefficient $K^c_{A/B}$, however, includes an implicit electrolyte dependence.[9] Fig. 4.3 plots the experimental exchange isotherms for a symmetric (Na^+,K^+) cation exchange at two different electrolyte concentrations; the effect is clearly negligible relative to experimental error.

The asymmetric exchange isotherm expression (4.12) (and the associated *exchange parameter* β_2 (4.13)) contain a term explicitly influenced by electrolyte concentration: charge

[9] Electrolyte concentration can be expressed using different expressions, each suited to a different application. Ion charge-concentration N_0 (4.9) is the natural electrolyte concentration parameter for ion exchange isotherms. Chapter 5 offers another electrolyte concentration expression known as *ionic strength* I_c that is best suited for adjusting solution electrolyte concentrations for nonideal thermodynamic behavior. *Ionic strength* I_c, for instance, is used to correct the selectivity coefficients $K^c_{A/B}$ discussed in this chapter to yield true thermodynamic selectivity constants $K^a_{A/B}$ (where the trailing superscript *a* indicates solution activities appear in the equilibrium quotient). This correction will be explained fully in Chapter 5.

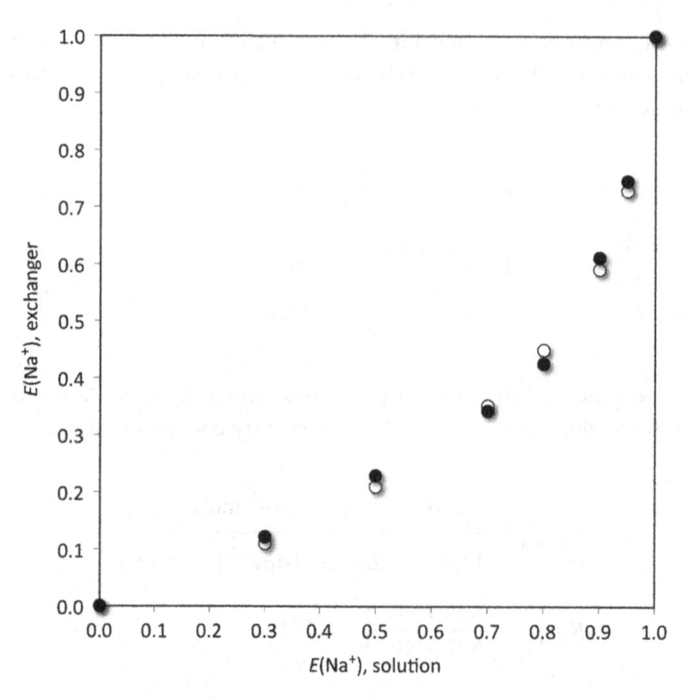

FIG. 4.3 The effect of electrolyte concentration on the symmetric (Na^+, K^+) exchange reaction (Jensen and Babcock, 1973). *Open circles* indicate $N_0 = 10^{-1}$ $mol_c\,dm^{-3}$ and *filled circles* indicate $N_0 = 10^{-2}$ $mol_c\,dm^{-3}$.

concentration parameter N_0 (4.9). Fig. 4.4 plots the experimental exchange isotherms for asymmetric (K^+,Ca^{2+}) cation exchange at two electrolyte concentrations: $N_0 = 10^{-2}$ $mol\,dm^3$ and $N_0 = 10^{-3}$ $mol\,dm^3$. Unlike the symmetric (Na^+, K^+) exchange isotherms in Fig. 4.3, the electrolyte concentration clearly affect the asymmetric (K^+, Ca^{2+}) exchange isotherms in Fig. 4.4.

Fig. 4.5 plots the general asymmetric (A^+, B^{2+}) exchange isotherms in solutions where the electrolyte concentrations are: 10^{-1}, 10^{-2}, and 10^{-3} $mol\,dm^{-3}$. The electrolyte concentration effect observed in Figs. 4.4 and 4.5 is sometimes referred to as the *Schofield Ratio Law* (Schofield, 1947).

4.5.4 Effect of Ion Selectivity on the Ion Exchange Isotherm

Figs. 4.2 and 4.3 illustrate the second major influence on the exchange isotherm: *ion selectivity*. In both of these experiments the exchanger (the clay mineral montmorillonite in Fig. 4.2, Yolo soil containing 18% clay dominated by montmorillonite and vermiculite in Fig. 4.3) exhibits exchange selectivity favoring K^+ over Na^+.

Quantifying *exchange selectivity* and interpreting the physical basis for exchange selectivity are the topics of this section. The approach presented here is based on a least sum-square regression of the exchange isotherm (cf. Appendix 4.B) on a symmetric exchange isotherm (4.8) with a single adjustable parameter: the exchange selectivity coefficient (Example 4.3).

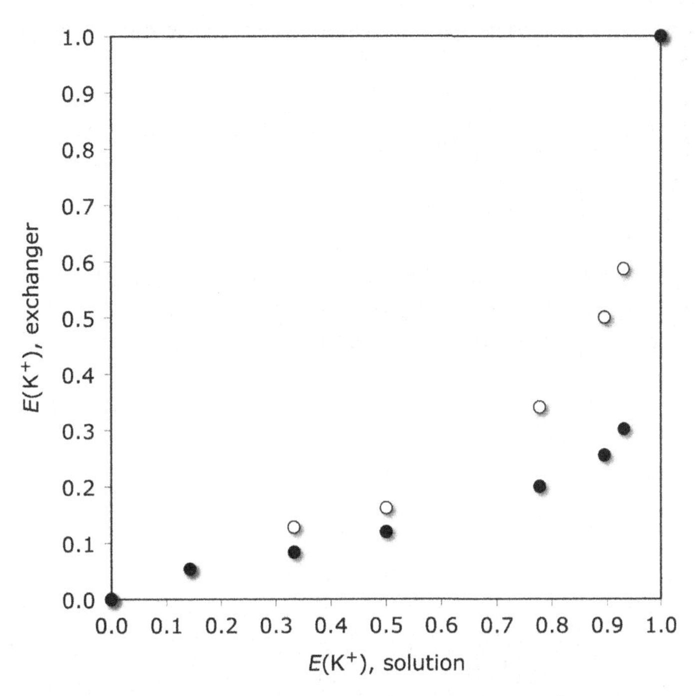

FIG. 4.4 The effect of electrolyte concentration on the asymmetric (K^+, Ca^{2+}) exchange reaction (Jensen and Babcock, 1973). *Open circles* indicate $N_0 = 10^{-1}$ $mol_c\ dm^{-3}$ and *filled circles* indicate $N_0 = 10^{-2}$ $mol_c\ dm^{-3}$.

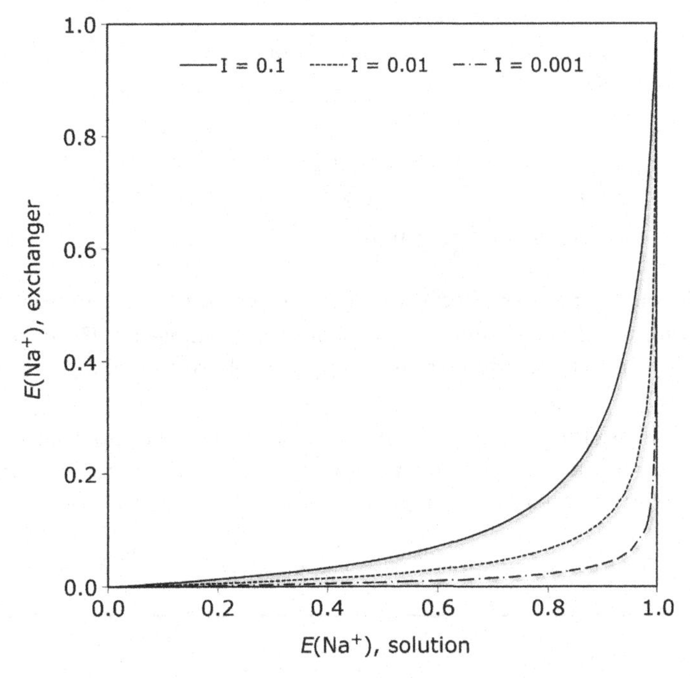

FIG. 4.5 Asymmetric (A^+, B^{2+}) exchange isotherms representing three electrolyte concentrations $N_0(10^{-1}, 10^{-2}$ and 10^{-3} molar). The isotherms are computed using a nonselective selectivity coefficient ($K^{\ominus}_{A/B} \equiv 1$).

EXAMPLE 4.3

Example Permalink

http://soilenvirochem.net/CA9TeL

Determine which cation—K^+ or Ca^{2+}—is selectively enriched during the ion exchange reaction described in Example 4.2.

The selectivity coefficient for the reaction in Example 4.2 is $K^c_{K/Ca} = 2.731$. Selectivity in asymmetric exchange reactions is most reliably found by comparing experimental equivalent fractions with the equivalent fraction of that same ion with the selectivity coefficient $K^c_{K/Ca}$ set equal to unity.[10]

$$2 \cdot K^+(aq) + \overline{Ca^{2+}} \longleftrightarrow 2 \cdot \overline{K^+} + Ca^{2+}(aq)$$

Define:

$$\frac{E^2_{\overline{K^+}} \cdot c_{Ca^{2+}}}{{c_{K^+}}^2 \cdot E_{\overline{Ca^{2+}}}} \equiv 1$$

$$\beta \equiv \left(\frac{2 \cdot N_0 \cdot 1 \cdot {E_{K^+}}^2}{1 - E_{K^+}} \right)$$

$$\beta = \left(\frac{2 \cdot (0.0113) \cdot 1 \cdot (0.500)^2}{(1 - 0.500)} \right) = 0.183$$

$$E^2_{\overline{K^+}} + (0.183) \cdot E_{\overline{K^+}} - (0.183) = 0$$

$$E_{\overline{K^+}} = \frac{-(0.183) + \sqrt{(0.183)^2 + 4 \cdot (0.183)}}{2} = 0.346$$

$$E_{\overline{Ca^{2+}}} = 1 - E_{\overline{K^+}} = 0.654$$

The exchanger-bound equivalent fraction for nonselective asymmetric exchange $E_{\overline{Ca^{2+}}}$ is predicted to be 0.654 for the solution composition specified in Example 4.2. The experimental equivalent fraction $E_{\overline{Ca^{2+}}}$ is 0.500, clearly demonstrating that selective exchange depletes Ca^{2+} (and enriches K^+).

This is an excellent example where basing the interpretation on the appearance of the exchange isotherm without reference to where the data plots relative to the nonselective isotherm (cf. Fig. 4.13) could easily lead to the wrong conclusion regarding exchange selectivity.

[10] The solution composition in both cases must be identical; only the computed equivalent fractions on the exchanger will differ.

Least sum-square regression of the symmetric exchange isotherm (4.8) to experimental (Mg^{2+},Ca^{2+}) exchange data from Jensen and Babcock (1973) appears in Fig. 4.6. The selectivity coefficient estimate $K^{c}_{Mg/Ca}$ for these data are: 0.616 ($N_0 = 10^{-3}$ mol dm^3) and 0.602 ($N_0 = 10^{-2}$ mol dm^3). The symmetric exchange isotherm (4.8), with $K^{c}_{Mg/Ca}$ as the sole fitting parameter, appears to be an acceptable model for the data in Fig. 4.6.

Fig. 4.7 plots the exchange isotherm a nonselective symmetric exchange reaction (i.e., $K^{c}_{A/B} \equiv$ unity). The nonselective isotherm extends from the lower left corner to the upper right corner. If data plot on the nonselective isotherm, the ion ratios in solution and in the exchange complex are identical for all compositions $\overline{E}_{A^{a+}} = E_{A^{a+}}$. Experimental symmetric exchange data rarely exhibit nonselective exchange; the results in Figs. 4.2, 4.3, and 4.6 are more typical.

Least sum-square fitting of experimental exchange data using the asymmetric exchange isotherm (cf. expression (4.12) with exchange parameter β_2 (4.13)) also permits estimates of $K^{c}_{K/Ca}$. A least sum-square regression of the asymmetric (K^+,Ca^{2+}) exchange from Jensen and

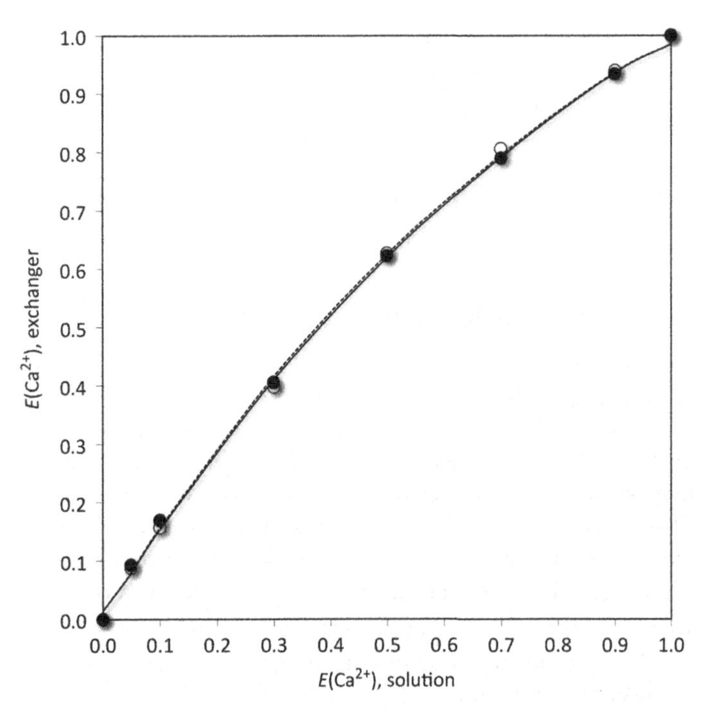

FIG. 4.6 Experimental exchange isotherm data for symmetric (Mg^{2+},Ca^{2+}) exchange (Jensen and Babcock, 1973) at two electrolyte concentrations. The $N_0 = 10^{-3}$ *molar* isotherm is plotted as *filled circles* (experimental points) and *solid line* (regression model 4.8). The $N_0 = 10^{-2}$ *molar* isotherm is plotted as *open circles* (experimental points) and *dashed line* (regression model 4.8).

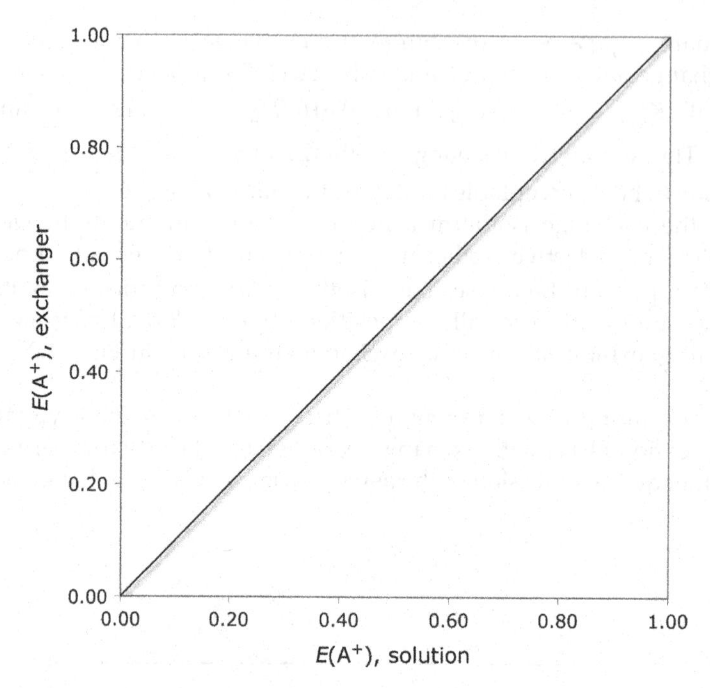

FIG. 4.7 The exchange isotherm for nonselective symmetric (A^{a+}, B^{b+}) exchange (i.e., $K^{\ominus}_{A/B} \equiv 1$) appears as a *straight line.*

Babcock appears in Fig. 4.8. The model estimates of selectivity coefficients $K^{c}_{K/Ca}$ for these data are: 6.77 ($N_0 = 10^{-3}$ mol dm^3) and 3.59 ($N_0 = 10^{-2}$ mol dm^3).

4.5.5 Physical Basis for Ion Selectivity

Ion selectivity appears to arise from differences in the Coulomb interaction between ions of the exchange complex and charged sites of the exchanger. The hydration of ions in the exchange complex limits how close they can approach surface-charge sites, which determines the energy that binds the ions in the exchange complex. Ions that can approach the surface more closely—ions surrounded by fewer water molecules—are selectively adsorbed relative to ions whose hydrated radius keeps them further from the surface-charge site.

Appendix 4.C discusses the case when the ion exchange complex behaves as *real* mixture. In this event the exchange selectivity coefficient ${}^{real}K^{a}_{A/B}$ varies as a function of the exchange complex composition and the true exchange selectivity constant $K^{\ominus}_{A/B}$ must be determined by integration, regardless of the chosen convention (Argersinger et al., 1950; Argersinger and Davidson, 1952; Gaines and Thomas, 1953).

Gast (1969) equated the standard Gibbs energy for symmetric ion exchange $\Delta_{ex}G^{\ominus}$ (4.15) with the change in electric potential energy given by expression (4.16) for an exchange reaction where solution ion A^{z+} (*aq*) displaces exchanger-bound cation $\overline{B^{z+}}$.

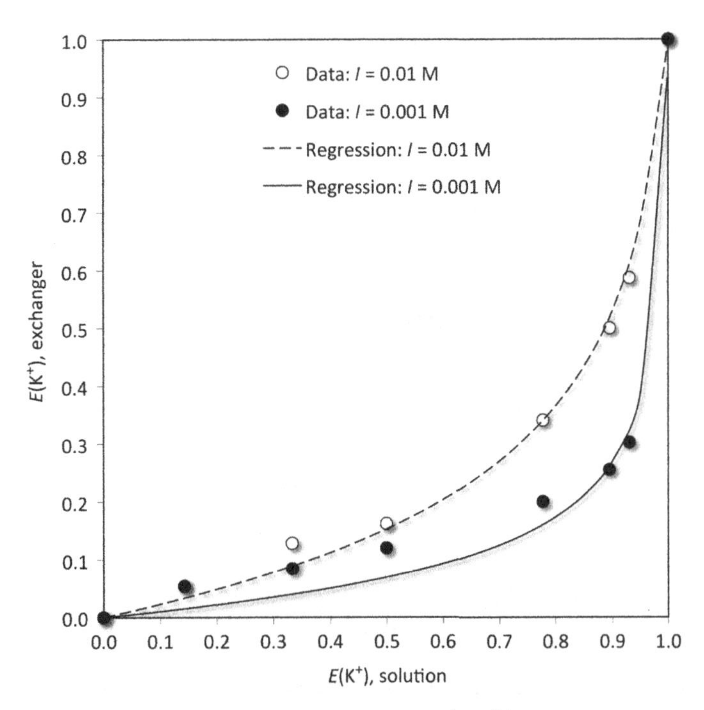

FIG. 4.8 Experimental exchange isotherm data for asymmetric (K^+,Ca^{2+}) exchange (Jensen and Babcock, 1973) at two electrolyte concentrations. The $N_0 = 10^{-3}$ *molar* isotherm is plotted as *filled circles* (experimental points) and *solid line* (regression model 4.12). The $N_0 = 10^{-2}$ *molar* isotherm is plotted as *open circles* (experimental points) and *dashed line* (regression model 4.12).

Radius a_i in expression (4.16) appears in an extension (Kielland, 1937) of the original Debye-Hückel activity coefficient equation (cf. Section 5.2.2). Brüll (1934) and Kielland (1937) defined a_i as "the effective diameter of the hydrated ion." Gast (1969) adopted a_i as the closest an ion could approach an exchange site. The remaining parameters in (4.16) are: elementary charge constant e, the electric constant (vacuum permittivity) ε_0, the relative permittivity of water ε_r.

$$\Delta_{ex}G^{\circ} = -R \cdot T \cdot \ln\left(K^{\circ}_{A/B}\right) \tag{4.15}$$

$$\Delta_{ex}G^{\circ} = \frac{(z \cdot e)^2}{4\pi\,\varepsilon_0\varepsilon_r} \cdot \left(\frac{1}{a_{B^{z+}}} - \frac{1}{a_{A^{z+}}}\right) \tag{4.16}$$

Fig. 4.9 illustrates the radius of closed approach a_{A^+} for exchangeable interlayer cations K^+ and Na^+ based on molecular dynamic simulations of interlayer cations and water in a smectite (Tambach et al., 2006). The disposition of K^+ cations place them closer to the clay layer (and the negative layer charge) than Na^+ at a the same interlayer spacing and water content. This difference in position of these two interlayer cations relative to the layer and its negative charge gives a distinct electrostatic potential energy advantage K^+ ions. The exchange of interlayer Na^+ by K^+ results in a negative $\Delta_{ex}G^{\circ}$.

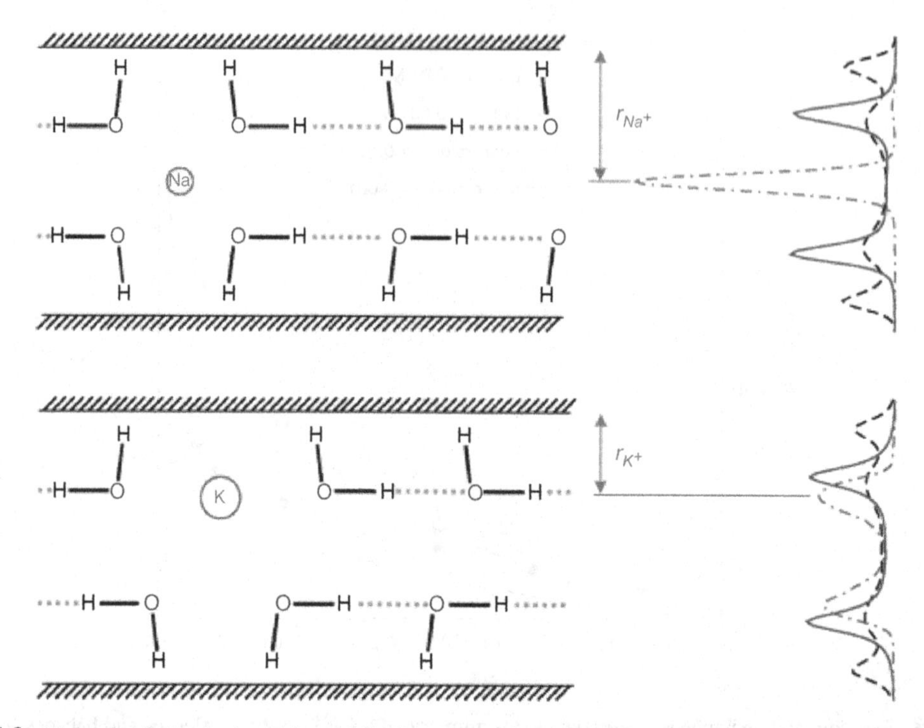

FIG. 4.9 Molecular dynamic simulations of the disposition of cations and water molecules in a smectite interlayer. The interlayer represents a two-layer hydrate saturated by Na^+ (top) and K^+ (bottom). The probability distributions normal to the clay layers appear on the right. *Reproduced with permission from Tambach, T.J., Hensen, E.J.M., Smit, B., 2004. Molecular simulations of swelling clay minerals. J. Phys. Chem. B 108, 7586–7596.*

The standard Gibbs energy $\Delta_{ex}G^{\circ}$ values for cation exchange listed in Table 4.3 are from Gast (1969) and Krishnamoorthy and Overstreet (1950). Recent research shows that the Debye-Hückel radii used by Gast lack a firm physical basis (Marcus, 1988; Ohtaki and Radnai, 1993; Abbas et al., 2002). A more reliable indication of ion hydration is the enthalpy of hydration (Marcus, 1987), also listed in Table 4.3.

Fig. 4.10 plots the Standard Gibbs energy $\Delta_{ex}G^{\circ}$ of symmetric exchange as a function of (a) the hydrated radius a_i and (b) the *absolute* standard enthalpy of hydration $\Delta_{hyd}H^{\circ}$ for Group 1A cations by a Wyoming beontonite. Fig. 4.11 plots the standard Gibbs energy $\Delta_{ex}G^{\circ}$ of symmetric exchanged exchange as a function of (a) the hydrated radius a_i and (b) the *absolute* standard enthalpy of hydration $\Delta_{hyd}H^{\circ}$ for selected divalent cations using selectivity coefficients from Krishnamoorthy and Overstreet (1950) and Wild and Keay (1964), the former using a bentonite from Utah and the latter using a vermiculite from the Transvaal, South Africa.

4.5.6 Other Influences on the Ion Exchange Isotherm

A exchange isotherm with a single-adjustable-parameter—the exchange selectivity coefficient (cf. expressions 4.8, 4.12)—often yields an excellent fit of experimental exchange isotherm

TABLE 4.3 Standard Gibbs Energy of Exchange $\Delta_{ex}G^{\circ}$ and *Absolute* Standard Enthalpy of Hydration

M^{+}	$\Delta_{ex}G^{\circ}$ kJ mol^{-1}	$\Delta_{hyd}H^{\circ}$ (kJ mol^{-1})	M^{2+}	$\Delta_{ex}G^{\circ}$ (kJ mol^{-1})	$\Delta_{hyd}H^{\circ}$ (kJ mol^{-1})
Cs^{+}	−4.52	−283	Ba^{2+}	−0.473	−1332
Rb^{+}	−2.65	−308	Sr^{2+}	−0.236	−1470
K^{+}	−1.28	−334	Ca^{2+}	+0.00	−1602
Na^{+}	+0.00	−416	Mg^{2+}	+0.207	−1948
Li^{+}	+0.20	−534	Cu^{2+}	+0.317	−2123

$\Delta_{hyd}H^{\circ}$ for Group 1A cations M^{+} on Wyoming bentonite (left three columns; Gast (1969)) and Group 2A cations (plus Cu^{2+}) M^{+} on Utah bentonite (right three columns; Krishnamoorthy and Overstreet (1950)).

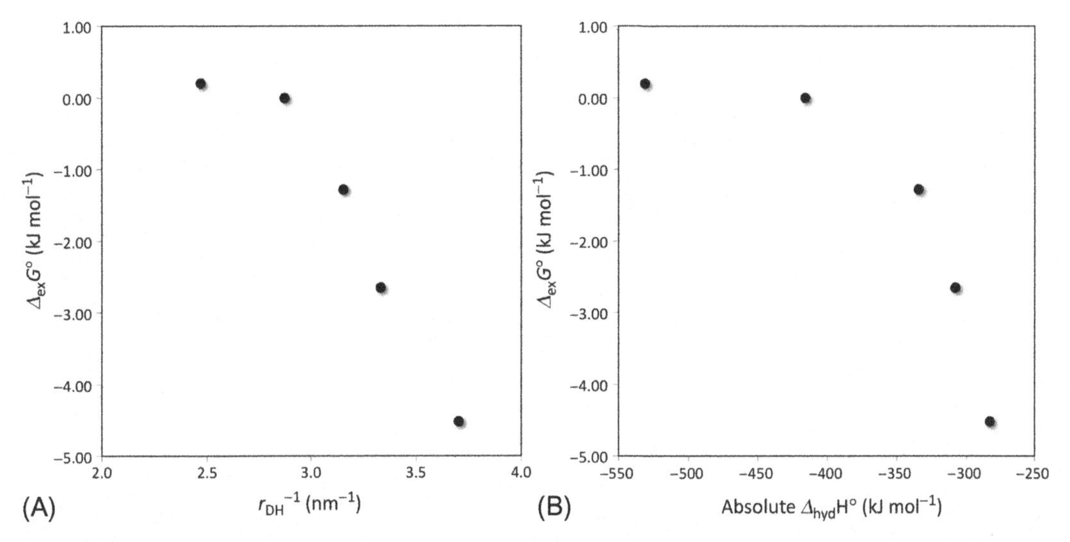

FIG. 4.10 The Gibbs energy of exchange $\Delta_{ex}G^{\circ}$ exchange—derived from the selectivity of Na^{+} exchange with Li^{+}, K^{+}, Rb^{+}, and Cs^{+} (Gast, 1969)—plotted as a function of: (A) the hydrated ionic radius $a_{M^{+}}$ and (B) the absolute standard enthalpy of hydration $\Delta_{hyd}H^{\circ}$ of each cation (Marcus, 1987).

data, but there are cases where experimental exchange isotherm data deviate from single-adjustable-parameter isotherm models.

One way to evaluate the relative importance of experimental deviations from single parameter isotherm models is to compare the deviation of experimental data from a sum-square regression using the simple isotherm models (4.8 or 4.12) to displacement of the exchange isotherm caused by ion selectivity or, in the asymmetric exchange case, electrolyte concentration effects.

Two high-quality symmetric exchange studies of report sufficient data to plot the exchange isotherms: (Na^{+}, K^{+}) exchange (Jensen and Babcock, 1973) and (Ca^{2+}, Fe^{2+}) exchange (Saeki et al., 2004). The exchange isotherms appear in Fig. 4.12A and B, respectively.

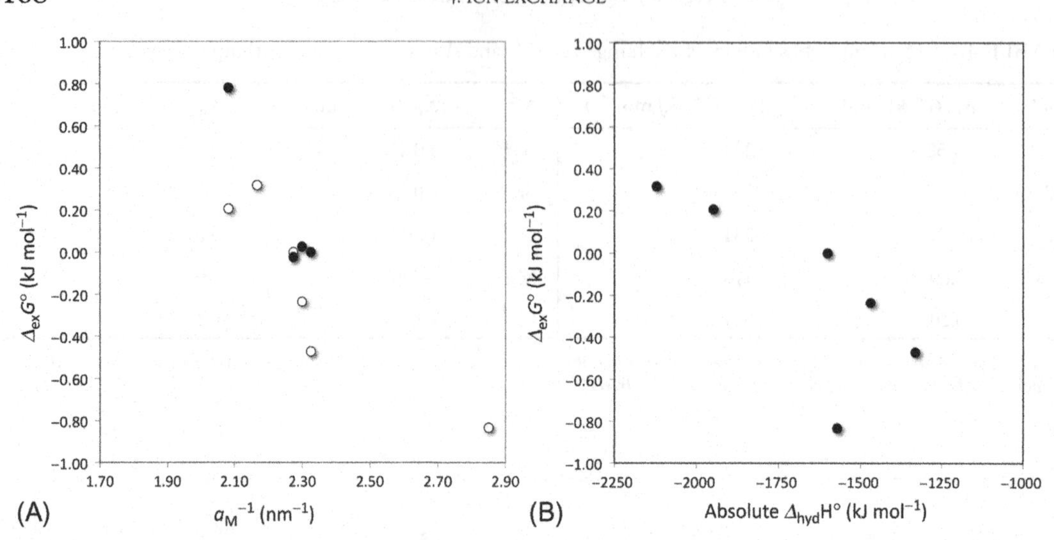

FIG. 4.11 The Gibbs energy of exchange $\Delta_{ex}G^{\circ}$ exchange—derived from the selectivity of Ca^{2+} exchange with Mg^{2+}, Sr^{2+}, Ba^{2+}, Pb^{2+} and Cu^{2+} (Krishnamoorthy and Overstreet, 1950; Wild and Keay, 1964)—plotted as a function of: (A) the hydrated ionic radius $a_{M^{2+}}$ and (B) the absolute standard enthalpy of hydration $\Delta_{hyd}H^{\circ}$ of each cation (Marcus, 1987).

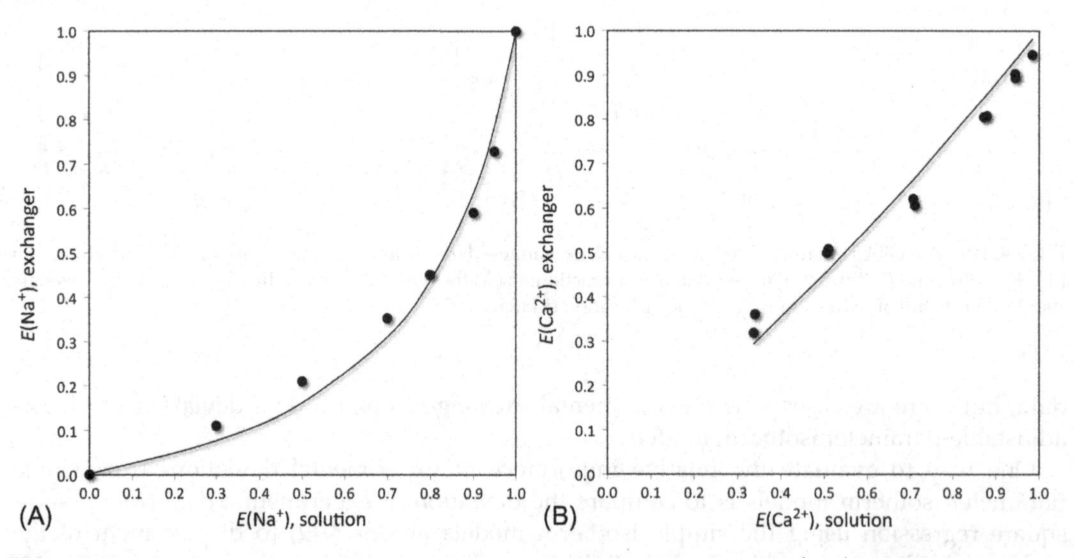

FIG. 4.12 Experimental cation exchange isotherm for symmetric exchange: (A) (Na^+, K^+) (Jensen and Babcock, 1973) and (B) (Ca^{2+}, Fe^{2+}) (Saeki et al., 2004). The experimental data are plotted as *filled circles* and the sum-square regression estimate of the selectivity coefficient using expression (4.8) is plotted as a *solid line* in both cases.

Displacement of the exchange isotherm in both cases indicates ion selectivity in the two exchange reactions (cf. nonselective symmetric exchange isotherm, Fig. 4.7). Notice that deviations of the data from the sum-square regression on symmetric exchange isotherm model (4.8) in the (Na^+,K^+) exchange example (Jensen and Babcock, 1973) are much smaller than the displacement caused by ion selectivity. Deviations in the (Ca^{2+},Fe^{2+}) exchange example (Saeki et al., 2004) are comparable to the displacement caused by ion selectivity because, in this case, the selectivity coefficient is rather small (≈ 1.2).

Two high-quality asymmetric exchange studies report sufficient data to plot the exchange isotherms at two different electrolyte concentrations (K^+,Ca^{2+}) exchange (Jensen and Babcock, 1973; Udo, 1978). The exchange isotherms appear in Fig. 4.13.

Displacement of the exchange isotherm in both cases indicates the effects of ion selectivity (*solid line*) on the exchange isotherm. The electrolyte concentration effect was illustrated earlier in Fig. 4.5. Experimental deviations from asymmetric exchange isotherm model (4.12) are much smaller than the displacement caused by both ion selectivity and electrolyte concentration.

Deviations of the exchange isotherm data in all of these cases from single-adjustable parameter exchange isotherm models (4.8) or (4.8) are statistically significant, but the magnitude of these deviations is small relative to the major influences we have identified: electrolyte concentration (asymmetric exchange only) and ion selectivity.

The deviations plotted in Figs. 4.12 and 4.13 appear to be systematic—the data points plot above the best-fit line as values of the independent variable approach zero and plot below

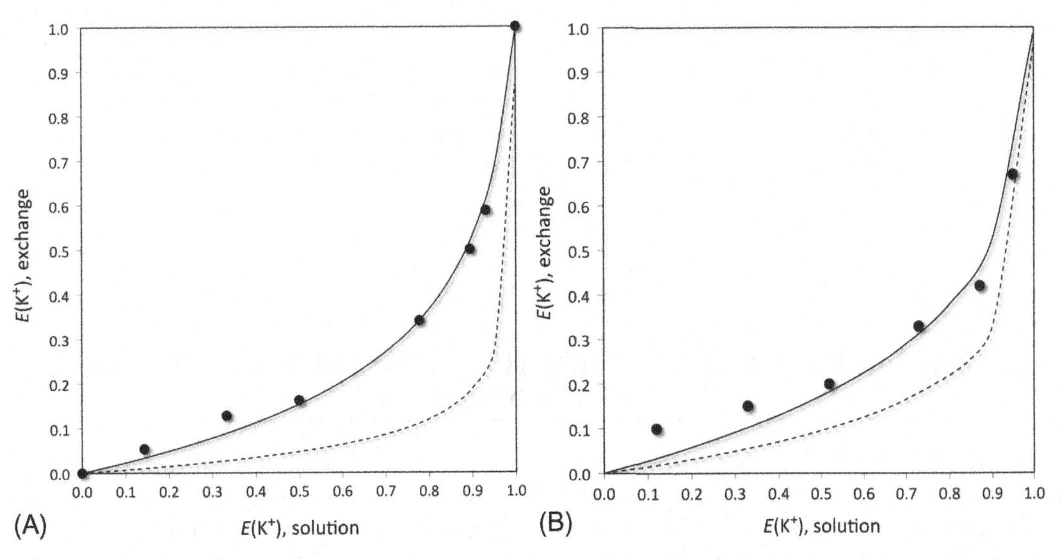

FIG. 4.13 Experimental cation exchange isotherms for asymmetric (K^+, Ca^{2+}) exchange: (A) Jensen and Babcock (1973) and (B) Udo (1978). The experimental data are plotted as *filled circles*, the sum-square regression estimate of the selectivity coefficient using expression (4.12) is plotted as a *solid line*, and the nonselective exchange isotherm is plotted as a *dashed line* in both cases.

the best-fit line as values of the independent variable approach unity—but do these small displacements have thermodynamic significance?

Systematic deviations of the type shown in Figs. 4.12 and 4.13 have little effect on the Gibbs energy of exchange. Ruvarac and Vesely (1970) found that the value of the selectivity coefficient determined when the equivalent fraction of either ion is 0.5 is a good estimate of the selectivity coefficient determined by the integral method of Argersinger et al. (1950).

A notable exception to the deviations from a single-adjustable-parameter symmetric exchange isotherm model (4.8) is the symmetric (Mg^{2+},Ca^{2+}) cation exchange on the Libby vermiculite (Peterson et al., 1965; Rhoades, 1967), which is discussed in Appendix 4.E.

4.6 SUMMARY

Ion exchange is an extremely important chemical reaction in soils, sediments, and aquifers, a process that has profound effects on the solution concentration of many ions. The ion exchange isotherm that describes the relation between solution composition and the ion exchange complex is influenced by two major factors: electrolyte concentration and ion selectivity.

Electrolyte concentration affects the exchange isotherm only for asymmetric exchange reactions, the effect being proportional to the ion charge concentration N_0. Ion selectivity influences symmetric and asymmetric exchange reactions and arises from Coulomb interactions between the ion and the exchange site (cf. Figs. 4.3 and 4.9).

The strength of the Coulomb interaction between ion and site is controlled by ion hydration; the hydration shell determines how closely each ion can approach the exchange site. Weakly hydrated ions that are able to shed their hydration shell can bind more strongly to the exchange site than strongly hydrated ions. As a general rule, single-adjustable-parameter exchange isotherms (4.8) or (4.8) provide an excellent fit of the experimental exchange isotherm. In certain cases (cf. Appendix 4.E) the selectivity coefficient adopts two limiting values representing preferred interlayer hydration states and undergoes an abrupt transition between the two limiting values when the exchange complex reaches a critical value.

APPENDICES

4.A THE PREPARATION AND CHARACTERIZATION OF NATURAL ION EXCHANGERS

The surface of all minerals has some ion exchange capacity, but unless the surface-to-mass ratio is extremely high, it is difficult to reliably measure the ion exchange capacity. If the ion exchange capacity of the material is very low, ion exchange reactions will have a negligible effect on the composition of the solution. The coarse clay particle-size fraction—2 μm spherical diameter—has a surface area of about 2 $m^2\ g^{-1}$. Particles much larger than this, with surface areas less than 1 $m^2\ g^{-1}$, have negligible ion exchange capacity per unit mass. For instance, the mineral kaolinite has a cation exchange capacity of 5 $cmol_c\ kg^{-1}$ and a surface area of about

$10\ m^2\ g^{-1}$, one of the lowest ion exchange capacities of naturally occurring ion exchangers and among the lowest surface areas of any clay mineral.

4.A.1 Saturating an Exchanger With a Single Cation

Naturally occurring ion exchanges found in the soils, sediments and aquifers are typically cation exchangers, with the notable exception of oxide clays under acidic conditions (cf. Chapter 8). Naturally occurring ion exchangers are prepared for ion exchange capacity measurement by first saturating the exchange complex with a single ion, preferably a cation that is easy to chemically analyze. Suspend the clay in a centrifuge bottle or tube containing a concentrated (e.g., $1\ mol\ dm^{-3}$) solution of a highly soluble salt (Fig. 4.A.1A). Agitate or shake for an hour or so to allow sufficient time for ion exchange (Fig. 4.A.1B). Centrifuge until the supernatant solution contains no suspended exchanger particles (Fig. 4.A.1C).

Discard the supernatant solution (Fig. 4.A.1D) and repeat (suspend, agitate, centrifuge, discard) several times. Finally, wash the clay repeatedly with pure water (suspend, agitate, centrifuge, discard). The clay is now saturated by a single cation and washed free of excess salts. All that remains is to suspend the clay in salt-free water and accurately determine the mass concentration of suspended clay. This is necessary because the ion exchange capacity is reported on a mass basis.

4.A.2 Measuring Ion Exchange Capacity

Measuring the ion exchange capacity of a natural exchanger requires advanced preparation that has saturated the exchanger with a single ion.

Transfer an aliquot of the exchanger to a centrifuge bottle or tube containing a concentrated solution of the new salt (Fig. 4.A.2A), the mass of the exchanger and final volume of the suspension must be precisely measured and recorded. Agitate or shake for an hour or so to allow sufficient time for the ion exchange to reach equilibrium (Fig. 4.A.2B).

Centrifuge until the supernatant solution contains no suspended exchanger particles (Fig. 4.A.2C). Reserve the supernatant solution by transferring it to another bottle (Fig. 4.A.2D) and repeat (suspend, agitate, centrifuge, reserve) three or four times. The reserved supernatant

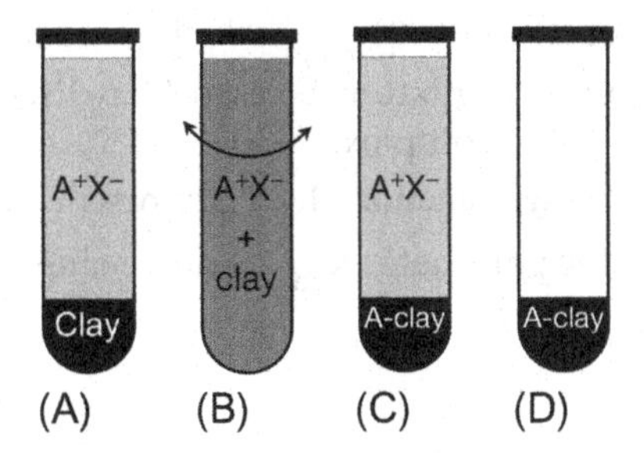

FIG. 4.A.1 Preparing clay saturated by cation A^+: (A) suspend clay in centrifuge tube containing a concentrated solution of a highly soluble salt A^+Z^-, (B) agitate to allow time for the ion exchange reaction, (C) centrifuge until the supernatant solution contains no clay particles, and (D) discard the supernatant solution and repeat.

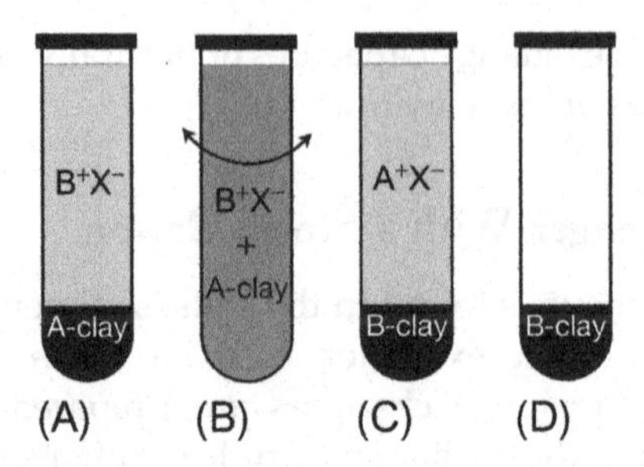

FIG. 4.A.2 Measuring *cation exchange capacity CEC* by quantitatively replacing all of the exchangeable A^+ from A^+-saturated clay with a new cation B^+: (A) Suspend A^+-saturated clay in centrifuge tube with a solution containing a new salt B^+Z^-, (B) agitate, (C) separate the supernatant solution from the clay, and (D) reserve the supernatant solution.

solutions are combined in a volumetric flask, diluted to the volume mark on the flask, and analyzed for the concentration of the *original* saturating ion.

The ion exchange capacity is the total moles of the original saturating ion in the reserved and combined supernatant solutions divided by the mass of exchanger in the suspension aliquot; the units are centimoles of charge per kilogram of clay: $\text{cmol}_c\ \text{kg}^{-1}$.

4.B NON-LINEAR LEAST SQUARE FITTING OF EXCHANGE ISOTHERMS

4.B.1 Symmetric Exchange

Consider the symmetric (Ca^{2+}, Mg^{2+}) exchange reaction.

$$Ca^{2+}(aq) + \overline{Mg^{2+}} \longleftrightarrow \overline{Ca^{2+}} + Mg^{2+}(aq)$$

The symmetric exchange isotherm (4.8) is the basis for the nonlinear least sum-square regression estimate of selectivity coefficient $K^c_{Ca/Mg}$. The simplest regression model uses a single value of the selectivity coefficient $\widehat{K}^c_{Ca/Mg}$, which is tantamount to assuming that the exchange complex behaves as an ideal mixture. Section 4.5.6 discusses the relative importance of nonideal behavior of the exchange complex.

The model for this cation exchange reaction (4.B.1) represents the dependent variable $\hat{E}_{\overline{Ca^{2+}}}$ as a function of a single adjustable parameter $\widehat{K}^c_{Ca/Mg}$ and a single independent variable[11]: the solution equivalent fraction $E_{Ca^{2+}}$.

[11] Fitting experimental data demands a distinction between experimentally measured variables $E_{\overline{Ca^{2+}}}$ and model estimates $\hat{E}_{\overline{Ca^{2+}}}$.

$$\hat{E}_{\overline{Ca^{2+}}} = \frac{\widehat{K}^{c}_{Ca/Mg} \cdot E_{Ca^{2+}}}{\left(1 - E_{Ca^{2+}} + \left(\widehat{K}^{c}_{Ca/Mg} \cdot E_{Ca^{2+}}\right)\right)} \tag{4.B.1}$$

The optimal selectivity coefficient parameter $\widehat{K}^{c}_{Ca/Mg}$ is found by minimizing the residual sum-square RSS[12] for all data points in the exchange isotherm (4.B.2). A least sum-square regression of symmetric (Mg^{2+}, Ca^{2+}) exchange isotherm using data from Jensen and Babcock (1973) appears in Fig. 4.6. Example 4.B.1 illustrates the multi-point least sum-square regression of a symmetric selectivity coefficient.

$$\mathrm{RSS} = \sum_{i} \left(\hat{E}_{\overline{Ca^{2+}}} - E_{\overline{Ca^{2+}}}\right)^2 \tag{4.B.2}$$

EXAMPLE 4.B.1

Example Permalink

http://soilenvirochem.net/DJ3fCU

Estimate the selectivity coefficient $K^{c}_{Na/K}$ for the symmetric (Na^+, K^+) exchange results in Table 4.2. In this example the selectivity coefficient $\widehat{K}^{c}_{Na/K}$ estimate is based on least sum-square fitting of all data points in Table 4.2 to the symmetric exchange isotherm model.

Step 1. Single-point selectivity coefficient (cf. Example 4.1).

The single-point selectivity coefficient estimate in Example 4.1 is based on the following ion exchange reaction.

$$Na^{+}(aq) + \overline{K^{+}} \longleftrightarrow \overline{Na^{+}} + K^{+}(aq)$$

Least sum-square fitting of experimental exchange isotherm data begins with an approximate single-point estimate: $\widehat{K}^{c}_{Na/K} = 0.1880$.

Step 2. Least sum-square fitting of symmetric exchange isotherm.

The model for the symmetric (Na^+, K^+) cation exchange reaction represents the dependent variable $\hat{E}_{\overline{Na^{+}}}$ as a function of a single adjustable parameter $\widehat{K}^{c}_{Na/K}$ and a single independent variable[13]: the solution equivalent fraction E_{Na^+}.

$$\hat{E}_{\overline{Na^{+}}} = \frac{\widehat{K}^{c}_{Na/K} \cdot E_{Na^{+}}}{\left(1 - E_{Na^{+}} + \left(\widehat{K}^{c}_{Na/K} \cdot E_{Na^{+}}\right)\right)}$$

The optimal least sum-square estimate of $K^{c}_{Na/K}$ is found by computing $\hat{E}_{\overline{Na^{+}}}$ and the squared difference $\left(\hat{E}_{\overline{Na^{+}}} - E_{\overline{Na^{+}}}\right)^2$, i.e., the square residual, for each experimental data point and

[12] The difference between an experimental value E' and its model estimate $\hat{E}$ is the *residual* while the square of the difference is the squared residual SR. The RSS is the sum over all SR for the entire dataset.

computing the residual sum-square RSS for each trial value of the fitting parameter $\widehat{K}^{c}_{Na/K}$, searching for the minimum RSS.

$$RSS = \sum_{i} \left(\hat{E}_{\overline{Na^{+}}} - E_{\overline{Na^{+}}} \right)^{2}$$

Trial-and-error will ultimately yield a reasonable estimate but is inefficient. A better procedure is to generate a handful of trial estimates centered on an initial estimate. For instance, the trial estimates in Table 4.B.1 cover a ±10% range centered on $\widehat{K}^{c}_{Na/K} = 0.1880$.

TABLE 4.B.1 Trial Selectivity Coefficients for Least Sum-Square Fitting of Experimental Exchange Isotherm (Example 4.B.1)

Weight w_r	$\widehat{K}^{c}_{Na/K}$	RSS
0.800	0.1504	0.02624
0.840	0.1579	0.01977
0.880	0.1654	0.01444
0.920	0.1730	0.01016
0.960	0.1805	0.00684
1.000	0.1880	0.00440
1.040	0.1955	0.00277
1.080	0.2030	0.00188
1.120	0.2106	0.00167
1.160	0.2181	0.00209
1.200	0.2256	0.00310

Column three lists residual sum-square (RSS) values for each trial estimate.

The initial estimate appears in column 2 in the same row as the *relative weight* equal to $w_r = 1.000$; the residual sum-square for this trial estimate is: RSS = 0.0044. The trial estimate $\widehat{K}^{c}_{Na/K} = 0.2106$ yields a lower residual sum-square: RSS = 0.0017.

The readjusting the center and range of the trial estimates yields an optimal estimate $\widehat{K}^{c}_{Na/K} = 0.2088$ represents the minimum sum-square error RSS. The least sum-square error estimate from fitting the exchange isotherm is always superior when compared to the a single-point estimate such as the one from Step 1: $\widehat{K}^{c}_{Na/K} \approx 0.188$.

[13] Fitting experimental data demands a distinction between experimentally measured variables $E'_{\overline{Na^{+}}}$ and model estimates $\hat{E}_{\overline{Na^{+}}}$.

4.B.2 Asymmetric Exchange

The selectivity coefficient $K^{c}_{A/B}$ estimate for asymmetric ion exchange requires a different least sum-square regression model designed to fit the experimental asymmetric exchange isotherm. The simplest model, once again, uses the asymmetric exchange isotherm (4.12) to represent the selectivity coefficient $K^{c}_{A/B}$. The asymmetric exchange isotherm model (4.12) requires an additional solution composition parameter—the electrolyte charge concentration parameter N_0 (4.9)—besides the independent solution equivalent fraction variable $E_{A^{a+}}$.

Consider the asymmetric (K^+, Ca^{2+}) exchange reaction.

$$2K^+(aq) + \overline{Ca^{2+}} \longleftrightarrow \overline{2K^+} + Ca^{2+}(aq)$$

The dependent variable in this example is the K^+ exchanger-bound equivalent $E_{\overline{K^+}}$ which varies as a function of the corresponding solution equivalent fraction E_{K^+}. The simple asymmetric exchange isotherm (4.12) is a second-order polynomial whose independent variable $\hat{E}_{\overline{K^+}}$ is found using the quadratic equation.

$$\hat{E}_{\overline{K^+}} = \frac{-\beta_2 + \sqrt{\beta_2^2 + 4 \cdot \beta_2}}{2} \tag{4.B.3}$$

$$\beta_2 \equiv \left(\frac{2 \cdot N_0 \cdot \widehat{K}^{c}_{K/Ca} \cdot E_{K^+}{}^2}{1 - E_{K^+}} \right)$$

The optimal selectivity coefficient $K^{c}_{K/Ca}$ is found by minimizing the residual sum-square RSS for all data points in the experimental exchange isotherm (4.B.4) using the quadratic expression (4.B.3) derived by expanding the simple asymmetric exchange isotherm (4.12).

$$\mathrm{RSS} = \sum_i \left(\hat{E}_{\overline{K^+}} - E_{\overline{K^+}} \right)^2 \tag{4.B.4}$$

A least sum-square regression of asymmetric (K^+, Ca^{2+}) exchange from the study by Jensen and Babcock (1973) appears in Fig. 4.13A.

4.C IDEAL AND REAL MIXTURES: SIGNIFICANCE FOR ION EXCHANGE SELECTIVITY CONSTANT EXPRESSIONS

4.C.1 Ideal Mixtures

The French chemist Raoult proposed an explanation for the vapor pressure of components in liquid mixtures that ultimately became known as Raoult's Law. Students generally encounter Raoult's Law in undergraduate general chemistry courses. Raoult's Law is important beyond the obvious success explaining component vapor pressures; it is a general model for mixtures both *ideal* and *real*.

The equilibrium vapor pressure of component A at a particular temperature $p_A(T)$ is the pressure of the vapor when the rate molecules escape the liquid equals the rate gaseous molecules condense to re-form liquid.

Experiments find that adding a solute B to pure liquid A lowers the vapor pressure of component A. Raoult reasoned the presence of solute B molecules dilute solvent A molecules at the liquid surface. Fewer solvent A molecules at the liquid surface means the solvent vapor pressure is lower than the vapor pressure of the pure liquid $p_A^*(T)$ at a given temperature.

Experiments reveal the vapor pressure of the solvent $p_A(T)$ is proportional to its mole-fraction x_A in the solution. Because $p_A(T) \propto x_A$ the equilibrium vapor pressure of the solvent over a solution is a product of a proportionality constant k and the mole fraction of the solvent in the mixture.

$$p_A(T) = k \cdot x_A \tag{4.C.1}$$

In the limit $x_A \rightarrow 1$ the solution becomes pure solvent whose vapor pressure is $p_A^*(T)$, revealing constant k to be the vapor pressure of the pure solvent.

$$\lim_{x_A \rightarrow 1} p_A(T) = p_A^*(T)$$

Expression (4.C.2) is known as *Raoult's Law*, named after Raoult (1830-1901). Ideal mixtures, by definition, obey Raoult's Law. Fig. 4.C.1 plots the component vapor pressures of a nearly ideal mixture of benzene and hexane at 25°C.

$$x_A = \frac{p_A(T)}{p_A^*(T)} \tag{4.C.2}$$

4.C.2 Real Mixtures

While a discussion of real liquid mixtures may seem out of place in the context of ion exchange reactions, the reader must fully understand the methods chemists apply to describe mixtures. These methods are applied in much the same manner when describing the exchange complex as a mixture.

Fig. 4.C.2 plots the component vapor pressures of a real mixture of acetone and dichloromethane at 25°C. Unlike the benzene-cyclohexane mixture (Fig. 4.C.1), the observed vapor pressures in the acetone-dichloromethane mixture (Fig. 4.C.2, *solid lines*) deviate significantly from Raoult's Law behavior (Fig. 4.C.2, *dashed lines*).

The relative vapor pressure of a *real* liquid-mixture has two limiting values (Fig. 4.C.3). At the limit as $x_A \rightarrow 1$ the ratio of the vapor pressure p_A to the vapor pressure of the pure liquid p_A^* converges on unity (expression 4.C.3 and Fig. 4.C.3). This is the *Raoult's Law limit* because the mixture converges on pure liquid A where $x_A = p_A(T) / p_A^*(T)$.

$$\lim_{x_A \rightarrow 1} \left(\frac{p_A}{x_A} \right) = 1 \tag{4.C.3}$$

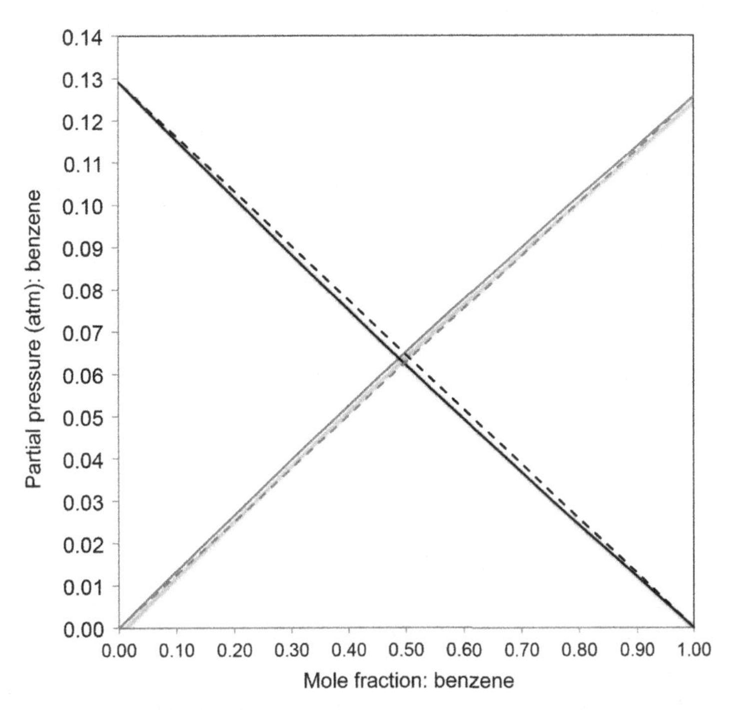

FIG. 4.C.1 A mixture of benzene and cyclohexane is very nearly an ideal mixture. The vapor pressure of each component is given by Raoult's Law (4.C.2).

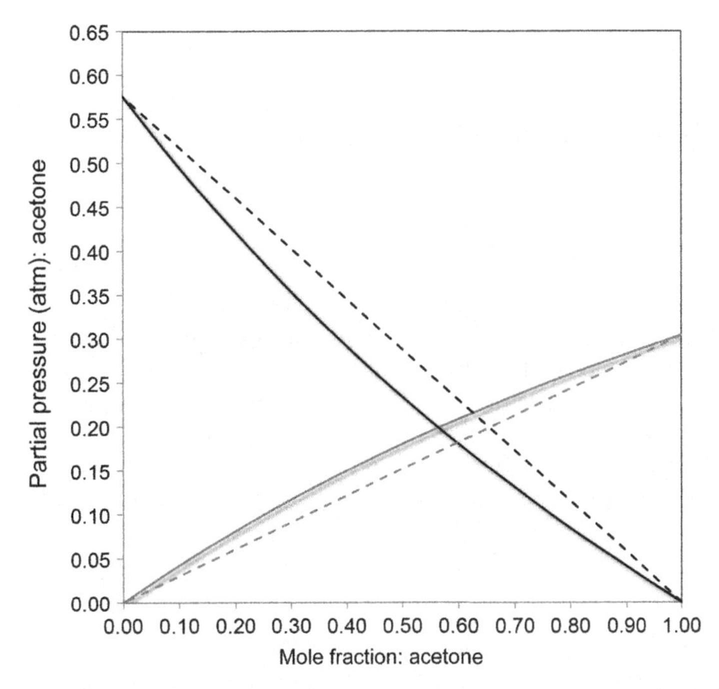

FIG. 4.C.2 A mixture of acetone and dichloromethane is a real mixture that deviates significantly from Raoult's Law (4.C.1). The vapor pressure of each component at the dilute limit is given by Henry's Law (4.C.3). The gray *solid line* plot the acetone partial pressure while the black *solid line* plots the dichloromethane partial pressure. The *dashed lines* indicate Raoult Law partial pressures.

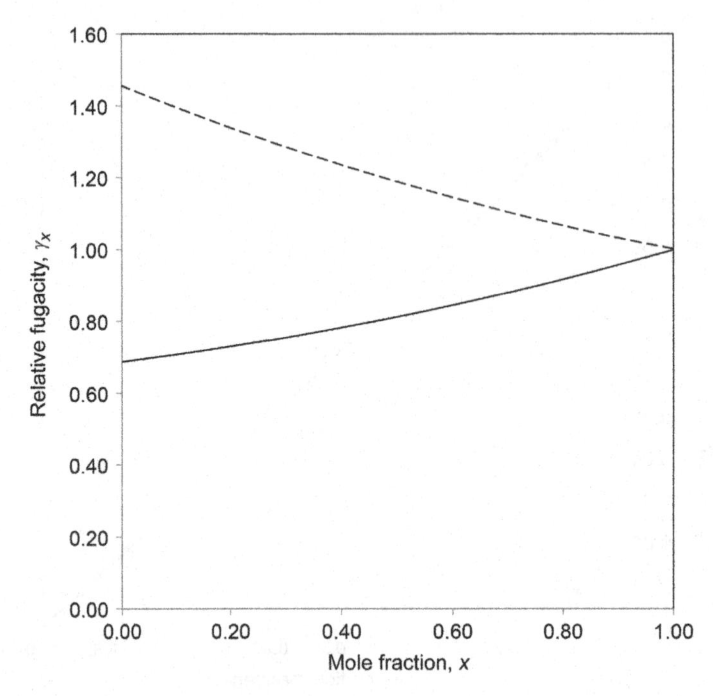

FIG. 4.C.3 The partial pressure for both components in a real mixture of acetone and dichloromethane approach the *Raoult Law Limit* as their mole fraction x_A converge on unity (4.C.3). The partial pressure for each component approach a different *Henry Law Limit* as their mole fraction x_A converges on zero: acetone (*dashed line*) and dichloromethane (*solid line*) (4.C.4).

At the limit as $x_A \rightarrow 0$ the ratio (p_A/x_A) converges on a constant (expression (4.C.4) and Fig. 4.C.3). This is the *Henry's Law limit* because the limiting constant is the *Henry's Law constant*[14]: $k_{H,A}^{px}$.

$$\lim_{x_A \rightarrow 0} \left(\frac{p_A}{x_A} \right) = k_{H,A}^{px} \tag{4.C.4}$$

The Henry's Law constant for gaseous acetone dissolved in dichloromethane (Fig. 4.C.3, *dashed line*) is 1.458 (mole-fraction basis) while the Henry's Law constant for gaseous dichloromethane dissolved in acetone (Fig. 4.C.3, *solid line*) is 0.688 (mole-fraction basis).

For an ideal mixture—be it a liquid mixture or an exchange complex—the activity of component *A* is equal to its mole fraction x_A (cf. Fig. 4.C.1). The deviation seen in Fig. 4.C.2 is typical of a real mixture where the activity of component *A* is the product of an activity

[14] The following superscript *px* indicates mole-fraction basis for liquid phase mixture.

coefficient $\gamma_{X, A}$ times the mole fraction x_A. Applying the activity coefficient correction would displace the *solid lines* in Fig. 4.C.2 onto the *dashed lines*, the latter representing an ideal mixture (cf. Appendix 4.D).

4.D SELECTIVITY CONSTANT EXPRESSIONS: IDEAL AND REAL EXCHANGE-COMPLEX MIXTURES

The equilibrium ion exchange constant (i.e., selectivity coefficient) expressions used throughout this chapter represent dissolved ions using molar concentration units and exchanger-bound ions using mole-fraction (or equivalent-fraction units. These units are *idealized*.

Concentration units $c_{A^{a+}}$ [mol dm^{-3}] fail to take into account the contribution of the electric potential of the dissolved electrolyte on the Gibbs energy of individual ions. The electric potential arising from electrolyte ions in solution makes a significant contribution to the Gibbs energy of solution ions under conditions commonly encountered in natural settings. Chapter 5 introduces the concept of ion activity coefficients as a means of adjusting the effect of electrolyte concentration in chemical equilibria involving ions dissolved in water. The reader will want to return to this appendix after studying the discussion of ion activity coefficients in Chapter 5.

Exchanger-bound ions are represented as a mixture, which can be either ideal or real (cf. Appendices 4.C and 4.E). Ideal mixtures are those mixtures where interactions between molecules in the mixture are identical to the molecular interactions of the components in their pure form. This is to say, the molecular interactions between benzene molecules and cyclohexane molecules in a benzene-cyclohexane liquid mixture are very similar to the molecular interactions in pure benzene and pure cyclohexane. As a result, the probability either molecule will escape the liquid phase and enter the gaseous phase are dependent solely on the relative abundance of the two components and unaffected by specific molecular interactions.

By contrast acetone is more likely to escape the liquid phase (and dichloro-methane is less likely) than the relative abundance of each component would suggest. The only explanation is that acetone-dichloromethane molecular interactions in the liquid phase increase the likelihood acetone will escape the liquid phase while those same interactions decrease the likelihood dichloromethane will vaporize.

The notion that interactions between component ions in the exchange complex influence the probability of exchange cannot be rejected out of hand. The following ion exchange equilibrium expressions apply the remedy used by chemists for liquid mixtures (cf. Appendix 4.C).

If the exchange complex (i.e., the exchanger-bound ions) behaves as a *real mixture* the exchange quotient expressions must include activity coefficients for the exchanger bound cations: Vanselow convention (4.D.1) and Gaines-Thomas convention (4.D.2). Furthermore, the exchange selectivity coefficient includes the leading superscript *real* to indicate the exchange complex is not an *ideal mixture*.

The dissolved ions are expressed using ion activities rather than concentrations in exchange quotient expressions (4.D.1), (4.D.2), therefore the exchange selectivity coefficients include the following superscript: *a*.

Vanselow mole-fraction convention

$$^{\text{real}}K^{a}_{\text{Na/K}} = \frac{\left(\gamma_{x,\overline{\text{Na}^+}} \cdot x_{\overline{\text{Na}^+}}\right) \cdot \left(\gamma_{1\pm} \cdot c_{\text{K}^+}\right)}{\left(\gamma_{1\pm} \cdot c_{\text{Na}^+}\right) \cdot \left(\gamma_{x,\overline{\text{K}^+}} \cdot x_{\overline{\text{K}^+}}\right)} \tag{4.D.1}$$

Gaines-Thomas equivalent-fraction convention

$$^{\text{real}}K^{a}_{\text{Na/K}} = \frac{\left(\gamma_{E,\overline{\text{Na}^+}} \cdot E_{\overline{\text{Na}^+}}\right) \cdot \left(\gamma_{1\pm} \cdot c_{\text{K}^+}\right)}{\left(\gamma_{1\pm} \cdot c_{\text{Na}^+}\right) \cdot \left(\gamma_{E,\overline{\text{K}^+}} \cdot E_{\overline{\text{K}^+}}\right)} \tag{4.D.2}$$

Expressions (4.D.1), (4.D.2) present a dilemma for novice soil and environmental chemists. The physics and chemistry of electric-potential interactions between exchanger-bound ions and proposed methods for computing exchanger-bound ion activity coefficients $\gamma_{x,\overline{A^{a+}}}$ and $\gamma_{E,\overline{A^{a+}}}$ requires an understanding of calculus and advanced chemical thermodynamics beyond the scope of this textbook.

Instead, we will adopt the following hypothesis: *the exchange complex is an ideal mixture* (the leading superscript *ideal* signifies: $\gamma_{x,\overline{A^{a+}}} = \gamma_{x,\overline{B^{b+}}} = 1$ and $\gamma_{E,\overline{A^{a+}}} = \gamma_{E,\overline{B^{b+}}} = 1$). This hypothesis is tested by employing the selectivity constant expression for *ideal exchange-complex mixtures* (4.D.3) (and the exchange isotherm derived from it) to fit experimental ion exchange results.

$$^{\text{ideal}}K^{c}_{\text{A/B}} = \frac{E^{b}_{\overline{A^{a+}}} \cdot c^{a}_{B^{b+}}}{c^{b}_{A^{a+}} \cdot E^{a}_{\overline{B^{b+}}}} = {}^{\text{ideal}}K^{a}_{\text{A/B}} \cdot \frac{\gamma^{b}_{A^{a+}}}{\gamma^{a}_{B^{b+}}} \tag{4.D.3}$$

Argersinger et al. (1950); Argersinger and Davidson (1952) and Gaines and Thomas (1953) showed the *thermodynamic selectivity constant* $K^{\circ}_{\text{A/B}}$ is defined by expression (4.D.4) and related to the selectivity coefficient $^{\text{real}}K^{a}_{\text{A/B}}$ through expression (4.D.5), regardless of unit convention.

$$\Delta_{\text{ex}}G^{\circ} \equiv -R \cdot T \cdot \ln\left(K^{\circ}_{\text{A/B}}\right) \tag{4.D.4}$$

$$\ln\left(K^{\circ}_{\text{A/B}}\right) = \int_0^1 \ln\left({}^{\text{real}}K^{a}_{\text{A/B}}\right) dE_{A^{a+}} \tag{4.D.5}$$

Of course, if the ideal mixture hypothesis is valid the selectivity constant $^{\text{ideal}}K^{\circ}_{\text{A/B}}$ is a constant—independent of $E_{A^{a+}}$—and (4.D.5) becomes (4.D.6).

$$\ln\left(K^{\circ}_{\text{A/B}}\right) = \ln\left({}^{\text{ideal}}K^{a}_{\text{A/B}}\right) \cdot \int_0^1 dE_{A^{a+}} \tag{4.D.6}$$

$$K^{\circ}_{\text{A/B}} \equiv {}^{\text{ideal}}K^{\circ}_{\text{A/B}} \tag{4.D.7}$$

To be specific, when the exchange complex is an *ideal* mixture—meaning a good fit of the experimental exchange isotherm is achieved with a constant fitting parameter $K^{a}_{A/B}$—then both (4.D.6) are (4.D.7) are valid. Furthermore, we can easily adjust selectivity coefficient $K^{c}_{A/B}$ using appropriate ion activity coefficients (cf. Section 5.2.2, *Water Chemistry*) to compute the thermodynamic selectivity constant $K^{a}_{A/B}$. This simple numerical adjustment has no impact on our working hypothesis.

Failure of our working hypothesis occurs when it is impossible to fit an experimental exchange isotherm with a constant selectivity coefficient (cf. Appendix 4.E). At this stage, it is more important to understand the *magnitude* of real-mixture effect relative to other factors influencing ion exchange than to explain in detail how ion-ion interactions in the exchange complex lead to real-mixture behavior.

4.E EQUIVALENT FRACTION-DEPENDENT SELECTIVITY COEFFICIENTS

Peterson et al. (1965) and Rhoades (1967) reported the symmetric (Mg^{2+}, Ca^{2+}) exchange reaction on a vermiculite specimen from Libby, Montana. Similar results were found for (Mg^{2+}, Ba^{2+}) exchange reaction involving another vermiculite specimen, identified as the *World* vermiculite, collected in Transvaal, South Africa (Wild and Keay, 1964).

The symmetric cation exchange reaction and selectivity quotient appear following.

$$Ca^{2+}(aq) + \overline{Mg^{2+}} \longleftrightarrow \overline{Ca^{2+}} + Mg^{2+}(aq)$$

$$K^{c}_{Ca/Mg} = \frac{E_{\overline{Ca^{2+}}} \cdot c_{Mg^{2+}}}{c_{Ca^{2+}} \cdot E_{\overline{Mg^{2+}}}} \tag{4.E.1}$$

The exchange isotherm Fig. 4.E.1 strongly deviates from the simple, single-adjustable-parameter symmetric exchange isotherm (4.8), jumping from $K^{c}_{Ca/Mg} \approx 0.6$ in the range $E_{\overline{Mg^{2+}}} < 0.40$ to $K^{c}_{Ca/Mg} \approx 10$ in the range $E_{\overline{Mg^{2+}}} > 0.70$.

The explanation for this dramatic change in cation exchange selectivity appears to be related to the effect of interlayer composition on the crystalline swelling state (cf. Chapter 3). Peterson et al. (1965) suggest the exceedingly structured (i.e., *crystal-like*) hydrated interlayer for Mg-saturated vermiculite apparently accounts for $K^{c}_{Ca/Mg} \approx 10$ when the interlayer composition is Mg^{2+} dominated, i.e., $E_{\overline{Mg^{2+}}} > 0.70$. In contrast, the interlayer for Ca-saturated vermiculite, i.e., $E_{\overline{Mg^{2+}}} < 0.40$ may be less structured (i.e., *fluid-like*). Differences in cation hydration and water structure influence interlayer cation positions and, therefore, exchange selectivity.

Picture the evolution of the hydrated interlayer of the Libby vermiculite as the interlayer equivalent fraction $E_{\overline{Mg^{2+}}}$ decreases from 1.0 (i.e., Mg^{2+}-saturated interlayer). Initially, the interlayer behaves as if it is a *sparsely*-substituted Mg^{2+}-dominated structure—the position of the interlayer cations and structured water both frozen—minimizing the effect of Ca^{2+} hydration on the remaining Mg^{2+} filling the interlayer. In this state, selectivity favors Mg^{2+} relative to Ca^{2+}.

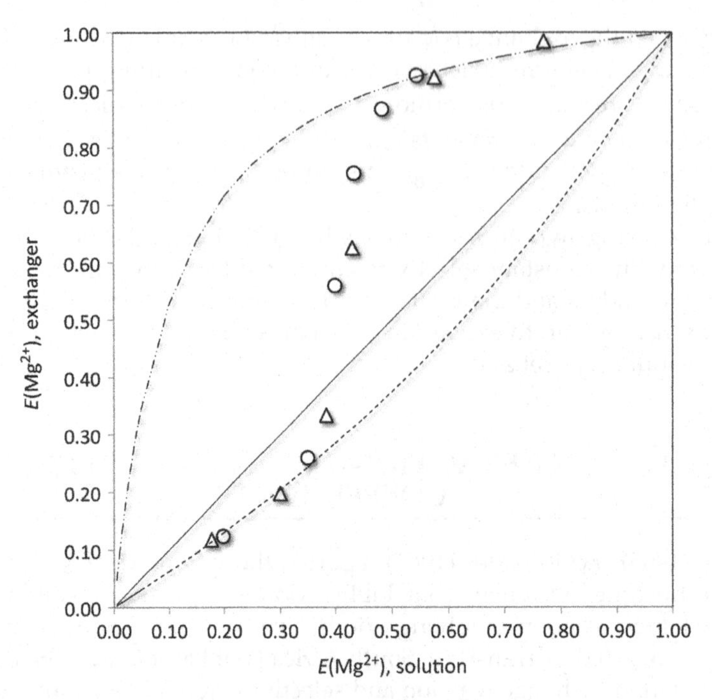

FIG. 4.E.1 The ion exchange isotherms for symmetric (Mg^{2+}, Ca^{2+}) exchange on a vermiculite specimen from Libby, Montana (Peterson et al., 1965), exhibit a dramatic change in selectivity.

When the Mg^{2+} interlayer equivalent fraction $E_{\overline{Mg^{2+}}}$ decreases below 0.7, the interlayer begins to melt, allowing interlayer cations to adopt positions relative to the clay layer that are consistent with the dimensions of their respective hydration shells. Interlayer melting is complete once the equivalent fraction $E_{\overline{Mg^{2+}}}$ decreases below 0.4. At this interlayer composition, the lower hydration energy of Ca^{2+} ions frees them to adopt positions closer to the clay layer, while the higher hydration energy of Mg^{2+} ions forces them to remain in positions further from the clay layer. As a consequence, exchange selectivity favors Ca^{2+} relative to Mg^{2+} for compositions where $E_{\overline{Mg^{2+}}} < 0.40$.

The extreme behavior seen in Fig. 4.E.1 appears to be unique to certain vermiculite specimens (cf. Chapter 3, vermiculite is a layer silicate whose layer charge permits crystalline swelling but prevents free swelling). Furthermore, the unique change in interlayer hydration requires cation exchange between a specific small, strongly-hydrated Group 2 cation Mg^{2+} and a much larger, weakly-hydrated cation (e.g., Ca^{2+} or Ba^{2+}). Symmetric and asymmetric exchange isotherms for reactions where Mg^{2+} is not involved appear to conform to the symmetric exchange isotherm (4.8) where selectivity is determined solely by the relative hydration enthalpy of the cations undergoing exchange.

4.F THE GAPON CONVENTION: REVISITED

Russian chemist Gapon published a series of five ion exchange papers in 1933 (Gapon, 1933a,b,c,d,e), introducing a convention for ion exchange equilibrium constant expressions unlike any other convention before or since. The scientific community studying role of (Na^+, Ca^{2+}) cation exchange in soil sodicity (cf. Chapter 6) quickly adopted the Gapon convention, modeling the ubiquitous *sodium adsorption ratio* (SAR) parameter after the Gapon convention. Despite being discredited (Sposito, 1977; Evangelou and Phillips, 1987, 1988) textbooks and research papers continue to promote the Gapon convention without reservation.

The scientific history of the Gapon convention is an intriguing and cautionary tale. Exposure to Gapon's Russian language papers outside of Russia can largely be traced to one extended abstract covering the first three papers (Gapon, 1933f) and two shorter abstracts covering the final two papers in the series (Gapon, 1933g,h). The error is subtle and difficult to identify but Sposito and Mattigod (1977) uncovered it decades after the Gapon convention and the SAR parameter had secured their hold on the scientific literature.

We begin with the cation exchange reaction as it appeared in the original paper (Gapon, 1933a).

$$\mathrm{NH_4{}^+(aq)} + \overline{\mathrm{Ca_{1/2}Z^-}} \longleftrightarrow \overline{\mathrm{NH_4Z^-}} + \tfrac{1}{2}\mathrm{Ca^{2+}(aq)} \qquad (4.F.R1)$$

Gapon wrote the equilibrium constant expression for the cation exchange reaction above where dissolved ions are listed as molar concentrations $c_{NH_4^+}$ [mol dm^{-3}] and exchanger electrolytes are listed as mole fractions $x_{\overline{NH_4^+}}$ [mol mol^{-1}].

$${}^{\mathrm{Gapon}}K^{c}_{\mathrm{NH_4/Ca}} = \frac{x_{\overline{\mathrm{NH_4{}^+Z}}} \cdot c^{1/2}_{\mathrm{Ca^{2+}}}}{c_{\mathrm{NH_4{}^+}} \cdot x_{\overline{\mathrm{Ca_{1/2}Z}}}} \qquad (4.F.1)$$

A more general cation exchange reaction appeared in the fourth paper (Gapon, 1933d) where the notation has been modified slightly for clarity. The most notable feature of these reactions is the cation stoichiometry that scales both cations to match the monovalent exchange site $\overline{Z^-}$.

$$\tfrac{1}{n}\,\mathrm{N}^{n+}(\mathrm{aq}) + \overline{\mathrm{M}_{1/m}\mathrm{Z}} \longleftrightarrow \overline{\mathrm{N}_{1/n}\mathrm{Z}} + \tfrac{1}{m}\,\mathrm{M}^{m+}(\mathrm{aq}) \qquad (4.F.R2)$$

Gapon (1933d) wrote the selectivity quotient expression for the general cation exchange reaction above using the parameters noted earlier. Expressions (4.F.1), (4.F.2) represent the *Gapon convention*.

$${}^{\mathrm{Gapon}}K^{c}_{\mathrm{N/M}} = \frac{x_{\overline{\mathrm{N}_{1/n}\mathrm{Z}}} \cdot c^{1/m}_{\mathrm{M}^{m+}(aq)}}{c^{1/n}_{\mathrm{N}^{n+}(aq)} \cdot x_{\overline{\mathrm{M}_{1/m}\mathrm{Z}}}} \qquad (4.F.2)$$

A key feature of the Gapon convention for writing the equilibrium quotient expression for ion exchange reactions is use of unit (i.e., 1) stoichiometry for the exchanger electrolyte (cf. expression (7a) in Gapon (1933a) and expression (1) in Gapon (1933d)).

Before we can evaluate this distinctive feature of the Gapon convention we must ensure every ion exchange reaction is fully balanced to reveal the *appropriate* stoichiometric coefficients for both reaction equations and ion exchange equilibrium quotient expressions. *The ion exchange reactions listed by* Gapon (1933a,d) are not balanced as written. Gapon omitted the dissolved co-ion but choose to include the exchange-site co-ion $\overline{Z^-}$. This is corrected in the following specific (reaction 4.F.R3) and general (reaction 4.F.R4) ion exchange reactions.

$$NH_4Cl(aq) + \overline{Ca^{2+}_{1/2}Z^-} \longleftrightarrow \overline{NH_4{}^+Z^-} + \tfrac{1}{2}\, CaCl_2(aq) \quad (4.F.R3)$$

$$\tfrac{1}{n}\, NCl_n(aq) + \overline{M^{m+}_{1/m}Z^-} \longleftrightarrow \overline{N^{n+}_{1/n}Z^-} + \tfrac{1}{m}\, MCl_m(aq) \quad (4.F.R4)$$

The next step recognizes and applies the principle that both dissolved and exchanger electrolytes are fully dissociated.

$$NH_4{}^+(aq) + Cl^-(aq) + \tfrac{1}{2}\, \overline{Ca^{2+}} + \overline{Z^-} \longleftrightarrow \overline{NH_4{}^+} + \overline{Z^-} + \tfrac{1}{2}\, Ca^{2+}(aq) + Cl^-(aq) \quad (4.F.R5)$$

$$\tfrac{1}{n}\, N^{n+}(aq) + Cl(aq) + \tfrac{1}{m}\, \overline{M^{m+}} + \overline{Z^-} \longleftrightarrow \tfrac{1}{n}\, \overline{N^{n+}} + \overline{Z^-} + \tfrac{1}{m}\, M^{m+}(aq) + Cl(aq) \quad (4.F.R6)$$

Having made a full accounting of all ions—dissolved and exchanger-bound—we can avoid errors when we simplify the cation exchange reactions by deleting co-ions appearing on both sides of the reaction arrow.

$$NH_4{}^+(aq) + \tfrac{1}{2}\, \overline{Ca^{2+}} \longleftrightarrow \overline{NH_4{}^+} + \tfrac{1}{2}\, Ca^{2+}(aq) \quad (4.F.R7)$$

$$\tfrac{1}{n}\, N^{n+}(aq) + \tfrac{1}{m}\, \overline{M^{m+}} \longleftrightarrow \tfrac{1}{n}\, \overline{N^{n+}} + \tfrac{1}{m}\, M^{m+}(aq) \quad (4.F.R8)$$

Deriving cation exchange reactions by following these steps and recognizing exchanger electrolytes are fully dissociated, reveals the error Gapon made. The inverse of their respective ion charges become the stoichiometric coefficients of the two exchanger-bound cations $\overline{N^{n+}}$

and $\overline{M^{m+}}$. Expressions (4.F.3), (4.F.4) are the corresponding equilibrium selectivity quotient expressions for the rewritten (and simplified) specific (4.F.R7) and general (4.F.R8) ion exchange reactions, respectively.

$$K^c_{NH_4/Ca} = \frac{x_{\overline{NH_4^+}} \cdot c^{1/2}_{Ca^{2+}}}{c_{NH_4^+} \cdot x^{1/2}_{\overline{Ca^{2+}}}} \tag{4.F.3}$$

$$K^c_{N/M} = \frac{x^{1/n}_{\overline{N^{n+}}} \cdot c^{1/m}_{M^{m+}(aq)}}{c^{1/n}_{N^{n+}(aq)} \cdot x^{1/m}_{\overline{M^{m+}}}} \tag{4.F.4}$$

Finally, multiplying both sides of the general ion exchange reaction (4.F.R8) by the product of the two ion valences ($n \times m$) transforms the ion exchange equilibrium quotient expression (4.F.4) into equilibrium quotient expression (4.F.5); which is Vanselow convention (cf. Section 4.1) for the general exchange reaction (4.F.R8). Notice, the stoichiometrically-correct exchange quotient expression (4.F.5) is *not* the same as the Gapon convention exchange quotient expression (4.F.2) for the general exchange reaction.

$$\frac{n \cdot m}{n} N^{n+}(aq) + \frac{n \cdot m}{m} \overline{M^{m+}} \longleftrightarrow \frac{n \cdot m}{n} \overline{N^{n+}} + \frac{n \cdot m}{m} M^{m+}(aq) \tag{4.F.R9}$$

$$\left(K^c_{N/M}\right)^{(n \cdot m)} = \left(\frac{x^{1/n}_{\overline{N^{n+}}} \cdot c^{1/m}_{M^{m+}(aq)}}{c^{1/n}_{N^{n+}(aq)} \cdot x^{1/m}_{\overline{M^{m+}}}}\right)^{(n \cdot m)}$$

$$\left(K^c_{N/M}\right)^{(n \cdot m)} = \frac{x^{m}_{\overline{N^{n+}}} \cdot c^{n}_{M^{m+}(aq)}}{c^{m}_{N^{n+}(aq)} \cdot x^{n}_{\overline{M^{m+}}}} \tag{4.F.5}$$

Sposito and Mattigod (1977) followed a slightly different line of reasoning to arrive at this conclusion: it is impossible to convert the Gapon convention (4.F.2) into either the Vanslow (4.F.5) or the Gaines-Thomas convention. The stoichiometry error is sufficient grounds to abandon the Gapon convention.

Evangelou and Phillips (1987, 1988) challenged the Gapon convention because it failed to adequately fit experimental data over the entire composition range. This challenge, however, carries less weight than the challenge advanced by Sposito and Mattigod (1977) because one could argue the appropriate choice of activity coefficients for exchanger bound cations (cf., Appendix 4.D, expression (4.D.1)) would compensate for unidentified deficiencies in the Gapon selectivity quotient expression when fitting experimental data.

References

Abbas, Z., Gunnarsson, M., Ahlberg, E., Nordholm, S., 2002. Corrected Debye-Huckel theory of salt solutions: size asymmetry and effective diameters. J. Phys. Chem. B 106 (6), 1403–1420.

Argersinger Jr., W.J., Davidson, A.W., Bonner, O.D., 1950. Thermodynamics and ion-exchange phenomena. Trans. Kans. Acad. Sci. 53, 404–410.

Argersinger Jr., W.J., Davidson, A.W., 1952. Experimental factors and activity coefficients in ion exchange equilibria. J. Phys. Chem. 56 (1), 92–96.

Arrhenius, S., 1887. Über die Dissociation der in Wasser gelösten Stoff. Z. Phys. Chem. 1, 631–648.

Arrhenius, S., 1888. Über den Gefrierpunkt verdünnter wässeriger Lösungen. Z. Phys. Chem. 2, 491–505.

Brüll, L., 1934. A consideration of the ionic radius in aqueous solutions of electrolytes. Gazz. Chim. Ital. 64, 624–634.

Cohen, E.R., Holmstrom, B., Mills, I., Stohner, J., Cvitas, T., Kuchitsu, K., Pavese, F., Strauss, H.L., Thor, A.J., Frey, J.G., Marquardt, R., Quack, M., Takami, M., et al., 2007. Quantities, Units and Symbols in Physical Chemistry, third ed. Royal Society of Chemistry, Cambridge, UK, 233 pp.

Evangelou, V.P., Phillips, R.E., 1987. Sensitivity analysis on the comparison between the Gapon and Vanselow exchange coefficients. Soil Sci. Soc. Am. J. 51 (6), 1473–1479.

Evangelou, V.P., Phillips, R.E., 1988. Comparison between the Gapon and Vanselow exchange selectivity coefficients. Soil Sci. Soc. Am. J. 52 (2), 379–382.

Gaines, George L., J., Thomas, H.C., 1953. Adsorption studies on clay minerals. II. A formulation of the thermodynamics of exchange adsorption. J. Chem. Phys. 21, 714–718.

Gapon, E.N., 1933a. K Tyeorii Obmyennoy Adsorbtsii V Pochvah. I. Zh. Obshch. Khim. 3, 144–152.

Gapon, E.N., 1933b. K Tyeorii Obmyennoy Adsorbtsii V Pochvah. II. Zh. Obshch. Khim. 3, 153–158.

Gapon, E.N., 1933c. K Tyeorii Obmyennoy Adsorbtsii V Pochvah. III. Zh. Obshch. Khim. 3, 159–163.

Gapon, E.N., 1933d. K Tyeorii Obmyennoy Adsorbtsii V Pochvah. IV. Zh. Obshch. Khim. 3, 660–666.

Gapon, E.N., 1933e. K Tyeorii Obmyennoy Adsorbtsii V Pochvah. V. Zh. Obshch. Khim. 3, 667–669.

Gapon, E.N., 1933f. Theory of exchange adsorption in soils. I. J. Gen. Chem. (U. S. S. R.) 3, 144–152,153–158,159–163.

Gapon, E.N., 1933g. Theory of exchange adsorption. IV. J. Gen. Chem. (U. S. S. R.) 3, 660–666.

Gapon, E.N., 1933h. Theory of exchange adsorption. V. J. Gen. Chem. (U. S. S. R.) 3, 667–669.

Gast, R.G., 1969. Standard free energies of exchange for alkali metal cations on Wyoming bentonite. Soil Sci. Soc. Am. Proc. 33 (1), 37–41.

Jensen, H.E., Babcock, K.L., 1973. Cation-exchange equilibria on a Yolo loam. Hilgardia 41 (16), 475–487.

Kielland, J., 1937. Individual activity coefficients of cations in aqueous solutions. J. Am. Chem. Soc. 59, 1675–1678.

Krishnamoorthy, C., Overstreet, R., 1950. An experimental evaluation of ion-exchange relationships. Soil Sci. 69, 41–55.

Marcus, Y., 1987. The thermodynamics of solvation of ions. Part 2. The enthalpy of hydration at 298.15 K. J. Chem. Soc., Faraday Trans. 1 83 (2), 339–349.

Marcus, Y., 1988. Ionic radii in aqueous solutions. Chem. Rev. 88 (8), 1475–1498.

Ohtaki, H., Radnai, T., 1993. Structure and dynamics of hydrated ions. Chem. Rev. 93 (3), 1157–1204.

Ostwald, W.F., 1888. Zur Theorie der Lösungen. Z. Phys. Chem. 2, 36–37.

Pauling, L., 1930. Structure of the micas and related minerals. Proc. Natl. Acad. Sci. U. S. A. 16, 123–129.

Peterson, F.F., Rhoades, J., Arca, M., Coleman, N.T., 1965. Selective adsorption of magnesium ions by vermiculite. Soil Sci. Soc. Am. Proc. 29 (3), 327–328.

Rhoades, J.D., 1967. Cation exchange reactions of soil and specimen vermiculites. Soil Sci. Soc. Am. Proc. 31 (3), 361–365.

Ruvarac, A., Vesely, V., 1970. Simple graphical determination of thermodynamic equilibrium constants of ion-exchange reactions. Z. Phys. Chem. (Frankfurt am Main) 73 (1–3), 1–6.

Saeki, K., Wada, S.-I., Shibata, M., 2004. Ca^{2+}-Fe^{2+} and Ca^{2+}-Mn^{2+} exchange selectivity of kaolinite, montmorillonite, and illite. Soil Sci. 169 (2), 125–132.

Samuelson, O., Bayer, E., Heliferich, F.G., 1972. Recommendations on ion-exchange nomenclature. Pure Appl. Chem. 29 (4), 619–624.

Schofield, R.K., 1947. A ratio law governing the equilibrium of cations in the soil solution. In: Proc. 11th Intern. Congr. Pure Applied Chem., London, vol. 3, pp. 257–261.

Sposito, G., 1977. The Gapon and the Vanselow selectivity coefficients. Soil Sci. Soc. Am. J. 41 (6), 1205–1209.

Sposito, G., Mattigod, S.V., 1977. On the chemical foundation of the sodium adsorption ratio. Soil Sci. Soc. Am. J. 41 (2), 323–329.

Tambach, T.J., Bolhuis, P.G., Hensen, E.J.M., Smit, B., 2006. Hysteresis in clay swelling induced by hydrogen bonding: accurate prediction of swelling states. Langmuir 22 (3), 1223–1234.

Thompson, H.S., 1850. On the adsorbent power of soils. J. R. Agric. Soc. Engl. 11, 68–71.

Udo, E.J., 1978. Thermodynamics of potassium-calcium and magnesium-calcium exchange reactions on a kaolinitic soil clay. Soil Sci. Soc. Am. J. 42 (4), 556–560.

Vanselow, A.P., 1932a. Equilibria of the base-exchange reactions of bentonites, permutites, soil colloids and zeolites. Soil Sci. 33, 95–113.
Vanselow, A.P., 1932b. The utilization of the base-exchange reaction for the determination of activity coefficients in mixed electrolytes. J. Am. Chem. Soc. 54, 1307–1311.
van't Hoff, J.H., 1887. Die Rolle osmotischen Druckes in der Analogie zwichen Lösungen und Gasen. Z. Phys. Chem. 1, 481–508.
van't Hoff, J.H., Reicher, L.Th., 1888. Über die Dissociationstheorie der Electrolyte. Z. Phys. Chem. 2, 77–781.
Way, J.T., 1850. On the power of soils to adsorb manure. J. R. Agric. Soc. Engl. 11, 313–379.
Way, J.T., 1852. On the power of soils to adsorb manure. J. R. Agric. Soc. Engl. 13, 123–143.
Wild, A., Keay, J., 1964. Cation-exchange equilibria with vermiculite. J. Soil Sci. 15 (2), 135–144.

CHAPTER

5

Water Chemistry

OUTLINE

5.1 **The Equilibrium Constant** 190
- 5.1.1 *Thermodynamic Functions for Chemical Reactions* 190
- 5.1.2 *Gibbs Energy of Reaction and the Equilibrium Constant* 191

5.2 **Activity and the Equilibrium Constant** 193
- 5.2.1 *Concentrations and Activity* 193
- 5.2.2 *Empirical Ion Activity Coefficient Expressions* 194

5.3 **Simple Equilibrium Systems** 195
- 5.3.1 *Setting Up Equilibrium Calculations to Replicate Solution Chemistry* 195
- 5.3.2 *Hydrolysis of a Weak Monoprotic Acid* 197
- 5.3.3 *Gas Solubility: Henry's Law* 202

5.4 **Complex Equilibrium Systems** 204
- 5.4.1 *Validating Water Chemistry Simulations* 204
- 5.4.2 *Gas Solubility: Fixed-Fugacity Method* 208
- 5.4.3 *Mineral Solubility* 211
- 5.4.4 *Interpreting Water Chemistry Simulations* 225
- 5.4.5 *Summary* 232

Appendices 232

5.A **Equilibrium Constants and Activity Coefficients: Notation and Units** 232

5.B **Fugacity: The Carbon Dioxide Case** 234

5.C ***ChemEQL* Input and Output File Formats** 235
- 5.C.1 *Understanding the ChemEQL Input Matrix: Components and Species* 235
- 5.C.2 *ChemEQL Output Data-File Format* 238

5.D **Solving Equilibrium Problems Using the RICE Table Method** 240
- 5.D.1 *Aqueous Solubility of an Ionic Compound* 240
- 5.D.2 *Aqueous Solubility of Gases* 243

5.E **Validating Water Chemistry Simulations** 247
- 5.E.1 *Charge Balance Validation* 247
- 5.E.2 *Mass Balance Validation* 247
- 5.E.3 *Ionic Strength Validation* 248
- 5.E.4 *Ion Activity Coefficient Validation* 248
- 5.E.5 *Activity Product Validation* 248
- 5.E.6 *Database Validation* 250

References 250

http://dx.doi.org/10.1016/B978-0-12-804178-9.00005-7

5.1 THE EQUILIBRIUM CONSTANT

5.1.1 Thermodynamic Functions for Chemical Reactions

Students encounter thermodynamics and the equilibrium concept in general chemistry. This initial encounter covers several important ideas while passing over others that, in the context of environmental chemistry, we must confront. Every chemical reaction involves the transformation of Gibbs energy G, redistributing chemical-bond energy—*enthaply H*—and kinetic energy (i.e., bond vibrations, molecular tumbling, and diffusion)—*entropy S*. The temperature in expression (5.1) is *absolute* temperature, reported in units of *Kelvin* [K] where $0°C \equiv 273.15\,K$.

$$G = H - T \cdot S \tag{5.1}$$

Scientist studying thermodynamics found that all *spontaneous* chemical reactions dissipate entropy S therefore Gibbs energy G decreases (5.2). Entropy S is dissipated to the surroundings through collisions involving bond vibrations, molecular tumbling, and diffusion.

$$\Delta_r G < 0 \tag{5.2}$$

Some general chemistry textbooks speak of *product-favored* reactions ($\Delta_r G < 0$) and *educt-favored* reactions ($\Delta_r G > 0$). In reality there is no such thing as a spontaneous edict-favored reaction because rewriting a "educt-favored" reaction so that educts[1] appear to the right side of the reaction arrow transforms it into a spontaneous "product-favored" reaction that satisfies the general requirement (5.2).

Though unable to measure absolute enthalpy H, absolute entropy S or absolute Gibbs energy G, chemists have methods to quantify changes in each of these thermodynamic variables. The systematic sharing of thermodynamic data led chemists to adopt a universal standard. The change in enthalpy $\Delta_f H$, entropy $\Delta_f S$ and Gibbs energy $\Delta_f G$ during the formation of any chemical compound is defined relative to a *standard state.*

The standard state convention specifies total pressure (viz. 1 *atmosphere* [atm]) but does not specify a temperature. This means the physical state (gas, solid, liquid, solute) of all reaction components—product compound and the elements from which it forms—should be the physical state consistent with the specified temperature T.

$$\underset{Educts}{H_2(g) + \tfrac{1}{2}\,O_2(g)} \longleftrightarrow \underset{Products}{H_2O(l)} \tag{5.R1}$$

For example, the physical state of water at 110°C (308.15 K) and 1 atm total pressure is $H_2O(g)$, therefore the standard state of water at 308.15 K is $H_2O(g)$. The physical state of water at 25°C (298.15 K) and 1 atm total pressure is $H_2O(l)$, therefore the standard state of water at 298.15 K is $H_2O(l)$. Finally, physical state of water at −10°C (263.15 K) is a specific crystalline form of ice ($H_2O(ice\ I_h)$), therefore the standard state of water at 263.15 K and 1 atm

[1] *Educts* are the precursors of chemical reaction products. Educts appear on the left side of the reaction arrow, products appear on the right side.

total pressure is $H_2O(ice\,I_h)$. The chemical reaction (5.R1) for the formation of water at 298.15 K and 1 atm total pressure lists all educts and the product in their respective standard states at 298.15 K.

$$\begin{aligned}
\Delta_f H^{\circ}\,(H_2O,\ l, 298.15\ K) &= -285.80\ \ kJ\,mol^{-1}\\
T \cdot \Delta_f S^{\circ}\,(H_2O,\ l,\ \ 298.15\ K) &= -48.60\ \ kJ\,mol^{-1}\\
\Delta_f G^{\circ}\,(H_2O,\ l,\ \ 298.15\ K) &= -237.34\ \ kJ\,mol^{-1}
\end{aligned}$$

The standard state convention also defines the enthalpy, entropy, and Gibbs energy of formation ($\Delta_f H^{\circ}$, $\Delta_f S^{\circ}$, and $\Delta_f G^{\circ}$, respectively) of all elements (regardless of their physical state) as zero. The enthalpy, entropy, and Gibbs energy of formation of $H_2(g)$ at 298.15 K are all zero.

$$\begin{aligned}
\Delta_f H^{\circ}\,(H_2,\ g,\ \ 298.15\ K) &= 0.00\ \ kJ\,mol^{-1}\\
T \cdot \Delta_f S^{\circ}\,(H_2,\ g,\ \ 298.15\ K) &= 0.00\ \ kJ\,mol^{-1}\\
\Delta_f G^{\circ}\,(H_2,\ g,\ \ 298.15\ K) &= 0.00\ kJ\,mol^{-1}
\end{aligned}$$

The Gibbs energy of reaction $\Delta_r G^{\circ}$ is computed by adding up the Gibbs energy of formation $\Delta_f G^{\circ}$ for all *products* (M, N, ...) then subtracting the sum over the Gibbs energy of formation $\Delta_f G^{\circ}$ for all *educts* (A, B, ...), where the symbol ν_i represents the stoichiometric coefficient for each compound involved in the chemical reaction.

$$\underbrace{\nu_A \cdot A + \nu_B \cdot B + \cdots}_{Educts} \longleftrightarrow \underbrace{\nu_M \cdot M + \nu_N \cdot N + \cdots}_{Products} \tag{5.R2}$$

$$\begin{aligned}
\Delta_r G^{\circ} = &\left(\nu_M \cdot \Delta_f G_M^{\circ} + \nu_N \cdot \Delta_f G_N^{\circ} + \cdots\right)\\
&- \left(\nu_A \cdot \Delta_f G_A^{\circ} + \nu_A \cdot \Delta_f G_B^{\circ} + \cdots\right)
\end{aligned} \tag{5.3}$$

5.1.2 Gibbs Energy of Reaction and the Equilibrium Constant

The standard state is not representative of prevailing conditions in the environment. Remarkably enough, a simple function (5.4) relates the Gibbs energy of reaction under prevailing conditions $\Delta_r G$ to the Gibbs energy of reaction under standard conditions $\Delta_r G^{\circ}$, where R is the *universal gas constant*.[2] and Q denotes the reaction quotient expression.[3]

[2] $R = 8.314\,472 \times 10^{-3}\ kJ\,K^{-1}\,mol^{-1}$.

[3] The reader is urged to read Appendix 5.A which explains the formal notation for writing reaction quotient expressions and equilibrium quotient expressions.

$$\Delta_r G = \Delta_r G^{\ominus}(T) + RT \cdot ln(Q) \tag{5.4}$$

$$Q^c = \frac{c_M^{\nu_M} \cdot c_N^{\nu_N} \cdots}{c_A^{\nu_A} \cdot c_B^{\nu} \cdots} \tag{5.5}$$

The general reaction (5.R2) is spontaneous *under ambient conditions* if: $\Delta_r G^{\ominus} < 0$. Initially there are no products; only educts are present. As the reaction proceeds educts combine to form products in the proportions determined by the stoichiometric coefficients, driven by a negative change in Gibbs energy.

At some point the reaction appears to stop; educt concentrations cease to decline and product concentrations cease to rise (Fig. 5.1). This state is *dynamic equilibrium*; the state where the rate educts combine to form products equals the rate products combine to re-form educts. The *equilibrium state* (5.6) is defined as the state where the Gibbs energy change is zero ($\Delta_r G = 0$) and Q equals a constant known as the *equilibrium constant*: $K^{\ominus}$ (cf. Appendix 5.A for an explanation of the following superscript $\ominus$).

$$0 = \Delta_r G^{\ominus}(T) + RT \cdot ln\left(K^{\ominus}\right) \quad \text{(equilibrium)} \tag{5.6}$$

$$\Delta_r G^{\ominus} \equiv -RT \cdot ln\left(K^{\ominus}\right) \tag{5.7}$$

Equilibrium constants $K^{\ominus}$ are rarely measured experimentally, instead chemists focus their efforts on accurately measuring the standard Gibbs energy of formation $\Delta_f G^{\ominus}$ for chemical

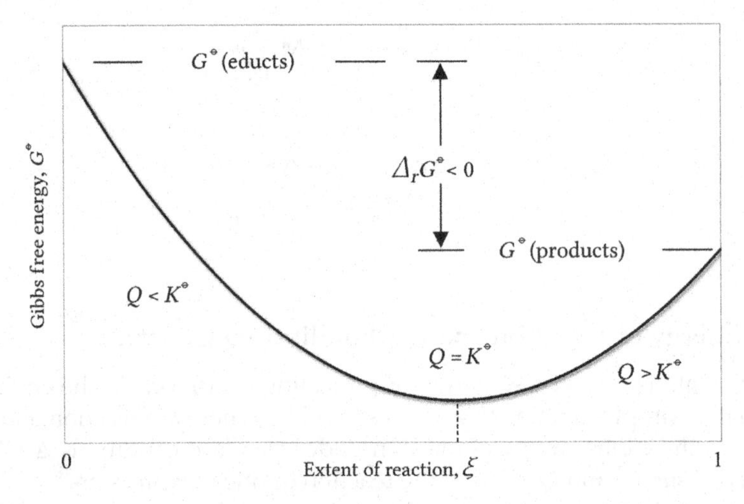

FIG. 5.1 The standard Gibbs energy *sum* of products $G^{\ominus}$ (*Products*) (plotted on the right axis) is less than the standard Gibbs energy *sum* of educts $G^{\ominus}$ (*Educts*) (plotted on the left axis). The reaction of educts to form products will occur spontaneously because $\Delta_r G^{\ominus} < 0$. The composition Q changes as the reaction proceeds to the right: $\Delta_r G$ steadily decreases toward a minimum value. When $\Delta_r G$ reaches its minimum the continued transformation of educts to products would cause $\Delta_r G$ to rise. Spontaneous reactions, however, can only decrease Gibbs energy; Gibbs energy never increases spontaneously. The composition of the system remains permanently at its equilibrium value $K^{\ominus}$.

species in solution and minerals that crystallize from solution. Drawing from databases that compile the standard Gibbs energy of formation for solutes and minerals (cf. Dick, 2008), chemists can write formation reactions for important solution and mineral species and calculate the equilibrium constants (5.8) for the species formation reactions.

$$K^{\circ} = e^{-\Delta_r G^{\circ}/RT} \tag{5.8}$$

5.2 ACTIVITY AND THE EQUILIBRIUM CONSTANT

5.2.1 Concentrations and Activity

The *equilibrium* quotient expression (5.9)[4] is often written using concentration terms (cf. Appendix 5.A for an explanation of the following superscript *c*).

$$K^c = \frac{c_M^{\nu_M} \cdot c_N^{\nu_N} \cdots}{c_A^{\nu_A} \cdot c_B^{\nu} \cdots} \tag{5.9}$$

By the end of the 19th century chemists realized that experimental K^c values for reactions involving electrolytes were actually *coefficients* that varied as a function of the electrolyte concentration. The dependence of K^c on electrolyte concentration occurred even if most of the ions dissolved in the solution were neither educts or products of the chemical reaction.

This prompted two chemists, Debye and Hückel (1923), to seek an explanation. Debye and Hückel recognized the electric potential ψ arising from electrolyte ions dissolved in solution generated a significant *electric potential energy* [5] U_E for all ions $A^{a\pm}(aq)$ in solution that was not included in the standard Gibbs energy $\Delta_f G^{\circ}\left(A^{a+}(aq),\ c^{\circ} = 1 \text{ molar},\ T\right)$, which is largely the Gibbs energy of ion hydration.

Debye and Hückel (1923) solved the *Boltzmann-Poisson* partial differential equation to derive an expression for electric potential ψ as a function of *ionic strength* I_c, the natural expression for electrolyte concentration to emerge from the derivation (Solomon, 2001).

$$I_c \equiv \tfrac{1}{2} \cdot \sum_i \left(z_i^2 \cdot c_i\right) \qquad \text{units:mol dm}^{-3} \tag{5.10}$$

Debye and Hückel (1923) include the electric potential energy contribution U_E to the standard Gibbs energy of formation $\Delta_f G^{\circ}$ of ion $A^{a\pm}(aq)$ by introducing the *electrolyte activity coefficient* $\gamma_{\pm}$ defined by the following expressions.

[4] The value of reaction quotient expression Q^c (5.5) is completely arbitrary but K^c (5.9) denotes the *equilibrium* quotient corresponding to the composition where $\Delta_r G$ is minimum (cf. Fig. 5.1).

[5] The electric potential energy U_E acting on solute ion $A^{z\pm}$ is equal to the product $k_e \cdot (\pm z) \cdot \psi$ where ψ is the electric potential of all ions dissolved in solution and $k_e = 1/4\pi\epsilon_r$.

$$\Delta_f G = \left(\Delta_f G^{\circ}\left(A^{a\pm}(aq), c^{\circ}, T\right) + U_E\right) + RT \cdot \ln\left(c_{A^{a\pm}}/c^{\circ}\right)$$

$$\text{Define:} U_E \equiv RT \cdot \ln(\gamma_\pm)$$

$$\Delta_f G = \Delta_f G^{\circ}\left(A^{a\pm}(aq), c^{\circ}, T\right) + RT \cdot \ln(\gamma_\pm) + RT \cdot \ln\left(c_{A^{a\pm}}/c^{\circ}\right)$$
$$\Delta_f G = \Delta_f G^{\circ}\left(A^{a\pm}(aq), c^{\circ}, T\right) + RT \cdot \ln\left(\gamma_\pm \cdot c_{A^{a\pm}}/c^{\circ}\right) \quad (5.11)$$

Debye and Hückel (1923) define the *activity* of ion $A^{a\pm}(aq)$ in expression (5.12). Notice activity—which appears in the second term on the right-hand side of expression (5.11)—is *unitless* by definition.

$$a_{A^{a\pm}} \equiv \gamma_{a\pm} \cdot \left(c_{A^{a\pm}}/c^{\circ}\right) \quad (5.12)$$

Appendix 5.A explains the significance of using unitless concentration ratios. Notice the *numerical* value of the unitless concentration ratio $\left(c_{A^{a\pm}}/c^{\circ}\right)$ is equal to the numerical value of the concentration $c_{A^{a\pm}}$. As a consequence it is acceptable to replace concentration ratios with concentrations in reaction quotient and equilibrium quotient expressions.

Applying the Debye-Hückel *ion activity coefficient* concept we are now prepared to write an equilibrium quotient expression (5.13) consistent with the requirement that K° is a true constant.[6]

$$K^{\circ} = \frac{a_M^{\nu_M} \cdot a_N^{\nu_N} \cdots}{a_A^{\nu_A} \cdot a_B^{\nu_B} \cdots} \quad (5.13)$$

5.2.2 Empirical Ion Activity Coefficient Expressions

All of the commonly used empirical ion activity coefficient expressions can be reduced to a simple function (5.14) of a temperature-dependent parameter[7] $A^{DH}(T)$ introduced by Debye and Hückel (1923) and ion charge $z\pm$. Distinctions between the different empirical expressions comes down to ionic strength I_c dependent functions[8] $f(I_c)$ in expression (5.14).

$$\log_{10}(\gamma_{a\pm}) = -A_c^{DH} \cdot (z_{a\pm})^2 \cdot f(I_c) \quad (5.14)$$

[6] The following superscript $\circ$ is reserved for equilibrium constants. Its use also indicates all solution and gaseous components are listed in the equilibrium quotient expression as activity ratios rather concentration ratios.

[7] $A_c^{DH} = 0.5102$ at 25°C.

[8] Solomon (2001) uses the *unitless* concentration ratio $\left(I_c/c^{\circ}\right)$ rather than I_c in activity coefficient expressions. Appendix 5.A explains why Solomon (2001) chose to normalize I_c with c°.

The simplest ion activity coefficient expression (5.15) is the Debye-Hückel limiting law, a single-parameter expression adequate for most freshwater environments where ionic strength I_c tends to be $\approx$ 0.01 molar. Guntelberg (1926) developed a single-parameter empirical expression (5.16) based on the Debye and Hückel limiting law, extending the ionic strength I_c range to $\approx$ 0.1 molar:

Debye and Hückel (1923)

$$f_{\mathrm{DH}}(I_c) = \sqrt{I_c} \tag{5.15}$$

Guntelberg (1926)

$$f_{\mathrm{G}}(I_c) = \frac{\sqrt{I_c}}{1+\sqrt{I_c}} \tag{5.16}$$

Davies (1938) developed an empirical expression (5.17) that extends the ionic strength range to $\approx$ 0.5 molar.

Davies (1938)

$$f_{\mathrm{D}}(I_c) = \frac{\sqrt{I_c}}{1+\sqrt{I_c}} + (0.2)\cdot(I_c) \tag{5.17}$$

None of these expressions listed above is adequate for seawater or brines found in extremely saline lake or groundwater. Beyond the *Davies limit* ($I_c \approx$ 0.5 molar) electric dipoles and quadruples from ion pairs and ion clusters become significant. Activity coefficients in seawater and brine solutions are sensitive to ion composition, not just concentration; no universal function appears to suffice.

How significant are ion activity coefficients under prevailing environmental conditions? Were Debye and Hückel, Güntelberg, and Davies engaged in minutiae, or would the use of concentration rather than activity yield significant error when applying equilibrium principles to environmental chemistry?

Figs. 5.2–5.4 illustrate the magnitude of the error. The ionic strength I_c of calcite-saturated water in contact with above-ground carbon dioxide partial pressures—which can be considered representative of electrolyte concentrations in natural waters—is $I_c \approx 10^{-3}$ molar.

The error is roughly 48% for trivalent ions and roughly 18% for divalent ions in $I_c \approx 10^{-3}$ molar solutions, significantly greater than experimental error that tends to be on the order of 1%. Based on a reasonable estimate of the ionic strength I_c, we are likely to find that the error introduced from reliance on concentrations rather than activities in most natural waters is much too great to neglect.

5.3 SIMPLE EQUILIBRIUM SYSTEMS

5.3.1 Setting Up Equilibrium Calculations to Replicate Solution Chemistry

General chemistry introduces students to a variety of simple equilibrium systems: the hydrolysis of weak monoprotic acids, the formation of solution ion pairs, or the dissolution

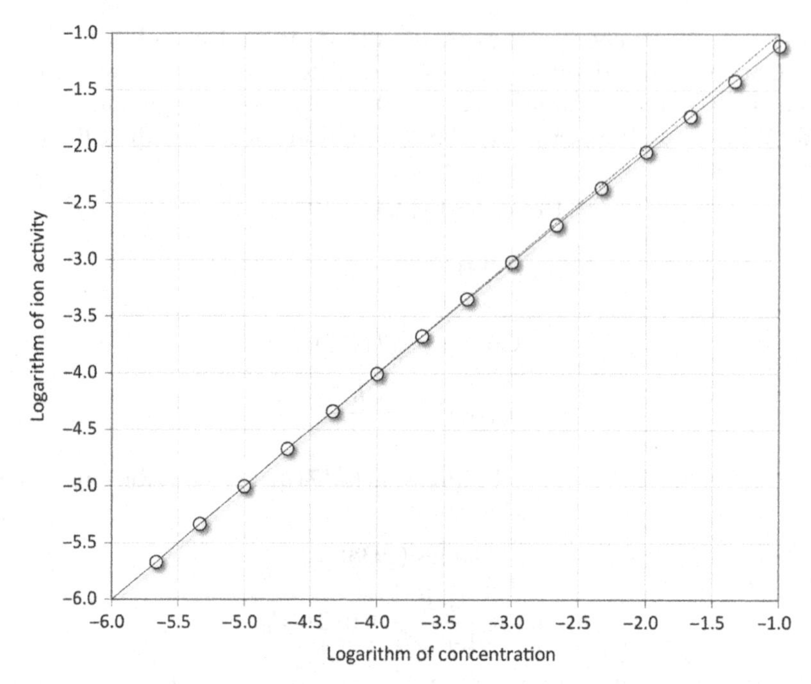

FIG. 5.2 The graph illustrates the case for monovalent ions $M^{\pm}$. The ion concentration $c_{M^{\pm}}$ (*curved solid line with open circles*) deviates from ion activity $a_{M^{\pm}}$ (*straight dashed line*) beginning at 10^{-2} molar.

of low solubility salts. Somewhat more complicated equilibrium systems—requiring either simplifying assumptions or complex algebraic manipulations—include the hydrolysis of weak diprotic acids, the effect of ion-pair formation on the solubility of a low solubility salt, or the simultaneous dissolution of two low solubility salts with a single common ion. None of these simple systems approach the complicated aqueous solutions commonly encountered by environmental chemists.

Perpetuating algebraic methods originally learned in general chemistry does little to prepare chemists to solve realistic water chemistry problems in an era where practitioners routinely use water chemistry simulation applications. What problem-solving skills must students develop to prepare them for a career that will include a heavy dose of water chemistry simulation?

As with any enterprise reliant on simulation models an accomplished practitioner must be prepared to validate simulation results. At one level validation exposes the simplifying assumptions built into the model. Two important benefits accrue from revealing the inner workings of the model: increases confidence in ways a particular model exchanges performance for accuracy and a clearer understanding of the performance limitations of the model. Validation also addresses more advanced issues: the particular choice of equilibrium constants in the model database or identifying questionable species and constants that should be eliminated from the database.

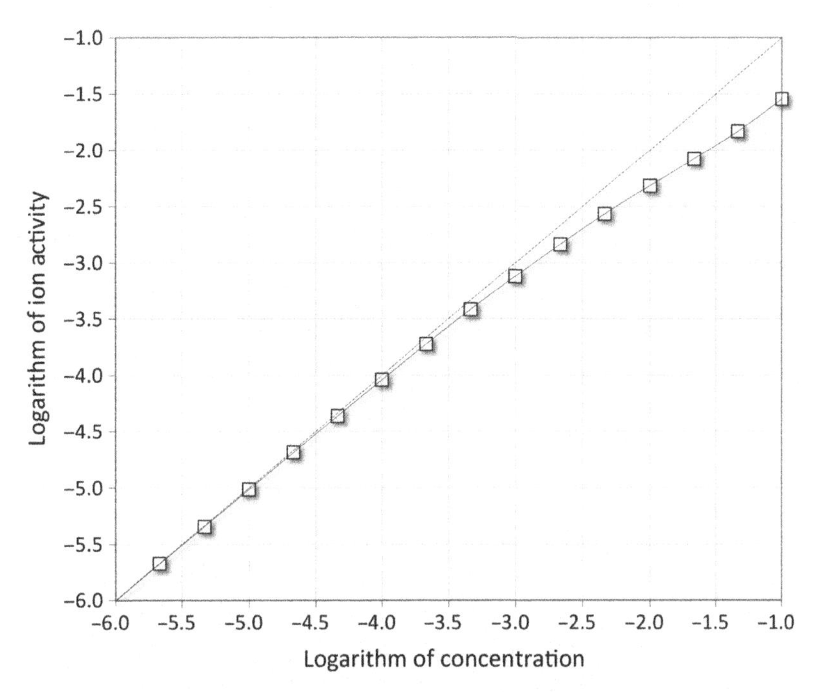

FIG. 5.3 The graph illustrates the case for divalent ions $D^{2\pm}$. The ion concentration $c_{D^{2\pm}}$ (*curved solid line with open circles*) deviates from ion activity $a_{D^{2\pm}}$ (*straight dashed line*) beginning at 10^{-4} molar.

Another issue that practitioners will eventually confront is water-analysis data quality. Chemists cannot expect reliable water chemistry simulations if the water analyses they are using is of low quality. Wary practitioners cannot assume that the analyses are error free. An experienced practitioner can use simulation results to assess water analyses for accuracy and to identify questionable data. Simulation validation is deferred until later in this chapter.

The following section addresses another challenge faced by chemists unfamiliar with water chemistry simulations: how to identify and avoid pitfalls when performing a simulation. The ultimate goal, after all, is to perform a simulation that faithfully reproduces the water chemistry as we understand it. *Water chemistry simulations may eliminate reliance on dated algebraic methods but the need to apply the fundamentals of equilibrium chemistry remains in full force.*

5.3.2 Hydrolysis of a Weak Monoprotic Acid

General chemistry textbooks examine two problem types involving the hydrolysis of weak monoprotic acids (or bases). One asks the reader to calculate the hydrolysis constant K_a° (or K_b°)

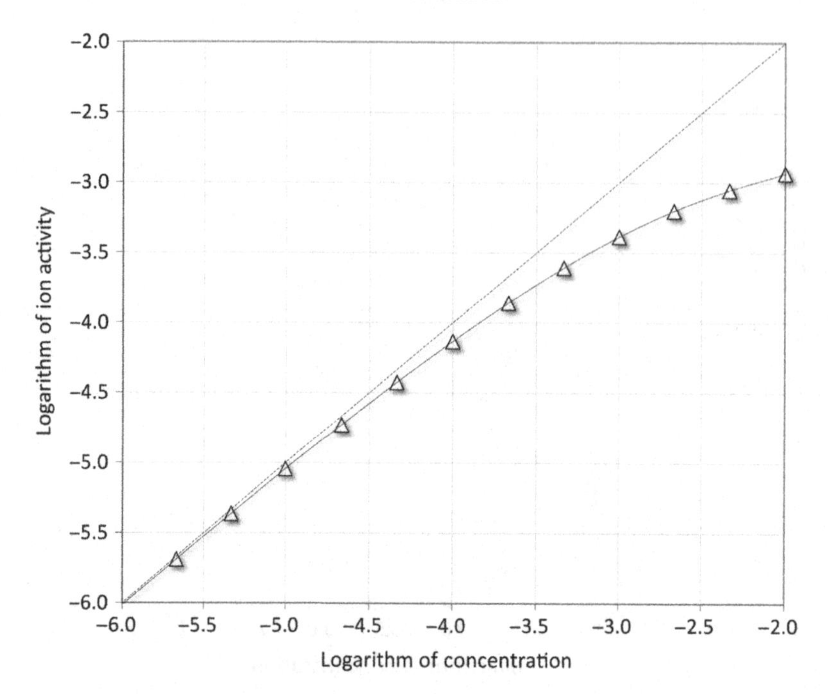

FIG. 5.4 The graph illustrates the case for trivalent ions $T^{3\pm}$. The ion concentration $c_{T^{3\pm}}$ (*curved solid line with open circles*) deviates from ion activity $a_{T^{3\pm}}$ (*straight dashed line*) beginning at $10^{-5.5}$ molar.

for a weak acid (or base) solution given the pH. The other asks the reader to calculate solution pH given the appropriate hydrolysis constant $K_a^{\ominus}$ (or $K_b^{\ominus}$).

The second of these two familiar problem types allows us to identify a common pitfall that can derail even a simple simulation and the steps you can take to avoid it. First, we will solve two problems using the *RICE table method* from general chemistry to emphasize the effect composition has on the result.

EXAMPLE 5.1

Example Permalink

http://soilenvirochem.net/3qY8ST

Calculate the pH of dilute acetic acid and acetate solutions given hydrolysis constants: $K_a^{\ominus} = 1.86 \times 10^{-5}$ and $K_b^{\ominus} = 5.38 \times 10^{-10}$.

The example calculates the pH of a 2×10^{-3} molar acetic acid solution and a 2×10^{-3} molar acetate solution using values listed in RICE Table 5.1.

The upper half of the RICE table lists the reaction and the terms for the equilibrium quotient expression for a 2×10^{-3} molar acetic acid solution. The lower half of table lists the terms for a 2×10^{-3} molar an acetate solution.

TABLE 5.1 RICE Tables for Acetic Acid Hydrolysis

Reaction →	Acetic Acid(aq) ($mol\ dm^{-3}$)		↔ Acetate$^-$(aq) ($mol\ dm^{-3}$)	H^+(aq) ($mol\ dm^{-3}$)
Initial	2×10^{-2}		0	0
Reaction	$-x$		$+x$	$+x$
Equilibrium	$2 \times 10^{-2} - x$		x	x
Reaction →	**Acetate$^-$(aq) ($mol\ dm^{-3}$)**	**H_2O(l)**	**↔ Acetic Acid(aq) ($mol\ dm^{-3}$)**	**OH^-(aq) ($mol\ dm^{-3}$)**
Initial	2×10^{-2}		0	0
Reaction	$-y$		$+y$	$+y$
Equilibrium	$2 \times 10^{-2} - y$		y	y

The equilibrium quotient expressions for the two reactions are very different but typically do not pose a potential pitfall; the educts and products for the two reactions are clearly different when the problem is solved by hand.

acetic acid(aq) : $C_{total} = 2 \times 10^{-3}\ mol\ dm^{-3}$

$$K_a^c = \frac{c_{acetate^-} \cdot c_{H^+}}{c_{acetic\ acid(aq)}} = \frac{x^2}{(C_{total} - x)}$$

$$x = \frac{-K_a^c \pm \sqrt{(K_a^c)^2 + (4 \cdot K_a^c \cdot C_{total})}}{2} = 1.84 \times 10^{-4}$$

$$pH = -\log_{10}\left(H^+(aq)\right) = -\log_{10}(x) = 3.74$$

acetate(aq) : $C_{total} = 2 \times 10^{-3}\ mol\ dm^{-3}$

$$K_b^c = \frac{c_{acetic\ acid(aq)} \cdot c_{OH^-}}{c_{acetate^-}} = \frac{y^2}{(C_{total} - y)}$$

$$y = \frac{-K_b^c \pm \sqrt{\left(K_b^c\right)^2 + \left(4 \cdot K_b^c \cdot C_{total}\right)}}{2} = 1.04 \times 10^{-6}$$

$$pH = 14 - \log_{10}\left(OH^-(aq)\right) = 14 - \log_{10}(y) = 8.02$$

The 2×10^{-3} molar *acetic acid* solution is pH = 3.74 while the *acetate* solution is pH = 8.02. The reason for this dramatic difference is the chemical properties of these two species; acetic acid is a weak acid and acetate is a weak base (cf. Chapter 6)

Novice users commonly make a simple error when solving the type of problem just illustrated using a numerical water chemistry application. The error occurs when the user is not aware of or does not take into account the way weak-acid (or weak-base) components are listed in the particular database they are using.

Numerical water chemistry applications utilize a common numerical kernel. Some models have more features, others have fewer features but the numerical kernel is essentially the same. The features a particular water chemistry application offers the user is not the only—nor is it the most important—distinction between models.

Each model (e.g., *MINTEQ, PHREEQC, EQ3/6, WATEQ4F*, etc.) requires a database that determines its scope. Each database includes a component list that determines the chemical compositions it can simulate. The database also lists solution species; defining the stoichiometric coefficients for the formation reaction for each specie (solute or mineral) using the database component list and equilibrium constants for each of the specie-formation reaction.

The same components, species, reactions, and equilibrium constants could be formatted for any numerical water chemistry application. Running the same simulation in any water chemistry application using the same database would yield similar if not identical simulation results.

Some databases (e.g., *EQ3/6*) define weak-acid components in their weak-acid form (e.g., acetic acid) while other databases (e.g., *MINTEQ* and *ChemEQL*) define weak-acid components in the conjugate-base form (e.g., acetate). *This distinction is significant because the way a component is defined in the model database determines its chemistry in the simulation.*

Example 5.2 illustrates how to simulate the two solutions listed in the RICE table when the input matrix (i.e., generated from the application's database) lists the weak acid in its conjugate base form: acetate.

EXAMPLE 5.2

Example Permalink

http://soilenvirochem.net/0pyz3i

Simulate the pH of dilute acetic acid and acetate solutions when the numerical water chemistry application lists the acetic-acid/acetate component in the conjugate base form.

Step 1. Simulate the solution pH of a 2×10^{-3} molar acetate solution.

The Step 1 *ChemEQL* components (i.e., input) table (Table 5.2, upper section) lists two chemical components in the first column: Acet- and H+.

TABLE 5.2 *ChemEQL* Components or Input Table (Top) and Species or Output Table (Bottom) for 2×10^{-3} molar Acetate Solution (Example 5.2, Step 1).

Component	Mode	Initial Concentration (mol dm^{-3})	
Acet-	total	2.00E−03	
H+	total	0.00E+00	
Species	**log(*K*)**	**Concentration (mol dm^{-3})**	**Activity (unitless)**
Acet-	0.00	2.00E−03	1.93E−03
HAcet(aq)	4.73	1.03E−06	1.03E−06
OH-	−14.00	1.04E−06	1.01E−06
H+	0.00	9.96E−09	9.96E−09

Component Acet- represents all species derived from acetic acid/acetate hydrolysis. The third column lists the initial concentration, which defines the total acetate (column 2) mass balance: 2×10^{-3} molar.

Component H+ represents all species forms from the hydrolysis of water $H_2O(l)$. The third column lists the initial concentration, which defines the total *strong acid* (column 2) mass balance: $C_{acid} =$ 0 molar.

Taken together, the input table represents a 2×10^{-3} molar acetate solution. Turning our attention to the bottom row of the species (output) table (Table 5.2, lower section), at equilibrium $a_{H^+} = 9.69 \times 10^{-9}$ and $pH = -\log(a_{H^+}) = 8.00$. The solution pH is consistent with the results from Example 5.1 for a 2×10^{-3} molar acetate solution.

Step 2. Simulate the solution pH of a 2×10^{-3} molar acetic acid solution.

TABLE 5.3 *ChemEQL* Components or Input Table (Top) and Species or Output Table (Bottom) for 2×10^{-3} molar Acetic acid Solution (Example 5.2, Step 2)

Component	Mode	Initial Concentration (mol dm^{-3})	
Acet-	total	2.00E−03	
H+	total	2.00E−03	
Species	**log(*K*)**	**Concentration (mol dm^{-3})**	**Activity (unitless)**
Acet-	0.00	1.85E−04	1.82E−04
HAcet(aq)	4.73	1.82E−03	1.82E−03
OH-	−14.00	5.48E−11	5.40E−11
H+	0.00	1.85E−04	1.85E−04

Step 2 components (i.e., input) table (Table 5.3, upper section) lists the same two chemical components in the first column as appear in Step 1 components (i.e., input) table (Table 5.2, upper section): Acet- and H+.

Table 5.3, (upper section), however, lists a different initial H+ concentration. The initial component H+ concentration, which defines the total *strong acid* (column 2) mass balance, is: $C_{acid} = 2 \times 10^{-3}$ molar.

To summarize our discussion of the input table, the component Acet- entry in Step 2 *ChemEQL* components table is identical to the entry listed in Step 1 components table, however sufficient *strong acid* must be added to neutralize the *conjugate-base* acetate, hence $C_{acid} = 2 \times 10^{-3}$ molar in row two of Components table.

The species (output) table (Table 5.3, lower section) lists the equilibrium $a_{H^+} = 1.85 \times 10^{-5}$, from which we find: $pH = -\log(a_{H^+}) = 3.73$. The solution pH is consistent with the results from Example 5.1 for a 2×10^{-3} molar acetic acid solution.

The practical significance of Example 5.2 is twofold. First, a novice model user must be aware of how weak-acid and weak-base components are listed—most notably carbonate and all organic acid components—and be prepared to make appropriate adjustments when designing solution pH simulations.

Second, users must anticipate *significant* weak-acid and weak-base (including the conjugate bases of weak acids) concentrations in natural waters. *Significant* in this context means sufficient to affect solution pH under field conditions. The confidence (or lack thereof) in water analysis data was raised in an earlier section, a thorough discussion of water analysis pH is deferred to Chapter 6.

5.3.3 Gas Solubility: Henry's Law

A system common in water chemistry is liquid water in contact with an atmosphere containing one or more gases. Expression (5.18) is the equilibrium quotient expression for gas solubility in a liquid (cf. Appendix 4.C). The unitless[9] equilibrium constant $k^{\ominus}_{H,B}$ is called the *Henry's Law constant*.

$$k^{\ominus}_{H,B} = \frac{f_B}{a_{c,B}} = \frac{\phi_B \cdot p_B}{\gamma_{c,B} \cdot c_B} \tag{5.18}$$

$$k^{pc}_{H,B} = \frac{p_B}{c_B} \tag{5.19}$$

Fugacity f_B, appearing in the numerator of expression (5.18) and representing the gas-phase equivalent of activity (i.e., *effective* pressure), is assigned pressure units. Pressure p_B, appearing in the numerator of expression (5.19), represents gas behavior under ambient conditions. *Real gas fugacity always registers as a lower pressure than ideal gas under identical conditions* (cf. Appendix 5.B).

Equilibrium between gas-phase components in the above ground atmosphere and solution-phase components in natural waters is most accurately represented at the *Henry's Law limit* (cf. Appendix 4.C). The ratio (f_B/x_B) converges on a constant known the *Henry's Law constant* at the dilute limit (5.20).

$$k^{px}_{H,B} = \lim_{x_B \to 0} \left(\frac{f_B}{x_B} \right) \tag{5.20}$$

Expression (5.20) is the mole-fraction based definition of the Henry's Law constant $k^{px}_{H,B}$, quoted in pressure units (SI pressure units are Pa. Here we use non-SI pressure units atm). Expression (5.19) is the concentration-based Henry's Law constant $k^{pc}_{H,B}$.

[9] cf. Appendix 5.A for a thorough discussion of the proper notation for equilibrium constants and component activities as unitless quantities. To avoid needless confusion, expression (5.18) adopts a more conventional notation.

TABLE 5.4 Henry's Law Gas-Solubility Constants H_B^{cp} (Expression (5.23)) for Select Inorganic Gases B: Solution-Phase Concentration Units are [mol dm^{-3}], Gas-Phase Pressure Units are [*atm*]

B	H_B^{cp}	B	H_B^{cp}
O_2	1.32×10^{-3}	NO	1.93×10^{-3}
N_2	6.48×10^{-4}	NO_2	1.42×10^{-2}
H_2	7.9×10^{-4}	SO_2	1.22×10^{0}
CH_4	1.42×10^{-3}	CO_2	3.34×10^{-2}
N_2O	2.43×10^{-2}	H_2S	9.22×10^{-2}

Expression (5.21) give the temperature-dependent[10] molar volume of water V_m, which is the factor used in expression (5.22) to convert $k_{H,B}^{px}$ into $k_{H,B}^{pc}$.

$$V_m(T) = \left(\rho_{H_2O}(T)/M_{H_2O}\right) \tag{5.21}$$

$$k_{H,B}^{pc} = V_m \cdot k_{H,B}^{px} \tag{5.22}$$

Expression (5.20) follows IUPAC recommended notation for Henry's Law constants $k_{H,B}^{\circ}$. Expression (5.20) is also known as *gas volatility* constant (Sander, 2015). Atmospheric chemists use an alternative notation (expressions 5.23, 5.24) to denote *gas solubility* constants.

$$H_B^{cp} = \frac{c_B}{p_B} = \left(k_{H,B}^{pc}\right)^{-1} \tag{5.23}$$

$$H_B^{xp} = \frac{x_B}{p_B} = \left(k_{H,B}^{px}\right)^{-1} \tag{5.24}$$

Henry's Law gas-solubility constants for several gases commonly dissolved in water in the environment appear in Table 5.4. Some of these gases (dioxygen O_2, dinitrogen N_2, methane CH_4, and nitric oxide NO) do not hydrolyze in water. Others (nitrogen dioxide NO_2, sulfur dioxide SO_2, and carbon dioxide CO_2, to name a few) combine with water to form weak and strong acids.

The Henry's Law constant H_X^{cp} is a good solubility predictor for a nonreactive gas B. The solubility of reactive gases are coupled to acid-base reactions occurring in solution in addition to the partial pressure of the gas in the gas phase (cf. Appendix 5.C).

[10] At 25°C the density of water is reported to be 997.0479 g dm^{-3} therefore molar volume of water is $V_m = 55.34455$ mol dm^{-3}

5.4 COMPLEX EQUILIBRIUM SYSTEMS

5.4.1 Validating Water Chemistry Simulations

Examples 5.1 and 5.2 and those in Appendix 5.C serve the following purposes. First, they review the equilibrium principles and problem-solving methods from general chemistry. Second, these examples demonstrate the challenges a water chemist faces when simple equilibrium systems become more complex—involving numerous minerals, soluble complexes, and hydrolysis reactions where two or more protons or hydroxyl ions dissociate from weak acids and bases. Complex water chemistry systems demand an entirely different numerical strategy to solve the numerous reactions in simultaneous equilibrium.

Environmental chemists can choose from several applications designed to simulate chemical equilibrium in water. This book uses *ChemEQL* (Müller, 2015) in all of the examples and problems. *ChemEQL* offers an opportunity to learn the basics of water chemistry simulation—setting up simulations and validating results (Table 5.4)—while illustrating important water chemistry principles.

Computer-based numerical models designed to solve simulate complex water chemistry systems relieve the practitioner of an enormous computational burden to focus their time and experience on assessing the reliability of the simulation. Before attempting to interpret simulation results, the practitioner must carefully consider whether the input data file will faithfully simulate water chemistry and assess the numerical and chemical validity of each simulation (Table 5.5).

Elementary validation (cf. Table 5.4, items 1–5) will expose simplifying assumptions built into numerical water chemistry applications; exchanging performance for accuracy inevitably creates performance limitations.

Another issue that practitioners will eventually confront is quality assessment, a simulation is no better than the water chemistry data the practitioner is using. Chemists should routinely assess the quality of water analyses; a wary practitioner cannot assume the chemical analyses are error free. Asessing the veracity of database constants (cf. Table 5.5, item 6) are advanced topics covered briefly at the close of this chapter.

The *ChemEQL* website offers an digital user's guide—also available for download using the permalink. The reader is encouraged to refer to it for detailed instructions.

ChemEQL Users Manual Permalink

http://soilenvirochem.net/U4gkv2

TABLE 5.5 Validation Checklist for Computer-Based Numerical Water Chemistry Applications

1	Validate charge balance
2	Validate mass balance for each constrained component
3	Validate the ionic strength estimate
4	Validate ion activity coefficients
5	Validate agreement between ion activity products from simulation results and equilibrium constants listed in the model database
6	Validate the thermodynamic equilibrium constants used in the model database, especially the most critical reactions in the simulation, against published values

5.4.1.1 *Charge Balance and Ionic Strength*

The first validation exercise—Example 5.3—revisits weak acid hydrolysis discussed in Example 5.2, calculating solution pH of solutions containing a weak acid and its conjugate base. Example 5.3 is very similar to the first example in p. 8 (*ChemEQL Manual*, version 3.0).

EXAMPLE 5.3

Example Permalink

http://soilenvirochem.net/l6WLkL

Validate the charge balance from numerical simulations of dilute acetic acid and acetate salt solutions.

Part 1. Simulation of 2×10^{-3} molar acetate solution.

TABLE 5.6 *ChemEQL* Components or Input Table (Upper) and Species or Output Table (Lower) for 2×10^{-3} molar Acetate Solution (Example 5.3, Part 1)

Component	Mode	Initial Concentration (mol dm^{-3})
Acet-	total	2.00E–03
H+	total	0.00E+00
Species	**Concentration (mol dm^{-3})**	**Concentration (mol$_c$ dm^{-3})**
Acet-	2.00E–03	−2.00E–03
HAcet(aq)	1.03E–06	0.00E–00
OH-	1.04E–06	−1.04E–06
H+	9.96E–09	+9.96E–09

The acetate solution charge balance is not satisfied in the simulation results (Table 5.6, lower section). The source of the charge-balance discrepancy is found in the input table (Table 5.6, upper section): the acetate anion component lacks an accompanying cation component.

Part 2. Simulation of 2×10^{-3} molar acetic acid solution.

The acetic acid solution charge balance is satisfied in the simulation results listed in Table 5.7 (lower section) because a cation component H+ is included in Table 5.7 (upper section). Without the H+ component the solution would be basic rather than acidic, failing to faithfully replicate the chemistry of an acetic acid solution. The H+ component also provides the missing cation charge required for satisfying the charge balance condition.

TABLE 5.7 *ChemEQL* Components or Input Table (Top) and Species or Output Table (Bottom) for 2×10^{-3} molar Acetic Acid Solution (Example 5.3, Part 2)

Component	Mode	Initial Concentration (mol dm^{-3})
Acet-	total	2.00E−03
H+	total	2.00E−03
Species	**Concentration (mol dm^{-3})**	**Concentration (mol$_c$ dm^{-3})**
Acet-	1.85E−04	−1.85E−04
HAcet(aq)	1.82E−03	0.00E−00
OH-	5.48E−11	−5.48E−11
H+	1.85E−04	+1.85E−04

Modifying the simulation in Example 5.3 (Part 1) by adding a spectator cation component such as Na+ or K+ would have no effect on solution pH but would supply the missing positive charge needed to satisfy the charge-balance condition. Failure to satisfy the charge-balance condition arises because *ChemEQL* does not enforce the charge-balance condition, allowing the user to omit nonacidic cations and nonbasic anions without significantly compromising solution pH simulation.

5.4.1.2 Ionic Strength and Ion Activity Coefficients

Allowing omission of spectator cation (and spectator anion) components routinely result in ionic strength I_c underestimates and inaccuracies in simulated ion activity coefficients $\gamma_{z\pm}$. The question facing the practitioner is whether component omissions allowed by the application have a significant impact on the overall simulation result. Example 5.4, using the same simulation output as Example 5.3, illustrates ionic strength I_c, ion activity $a_{acetate^-}$ and ion activity coefficient $\gamma_{z\pm}$ validation.

EXAMPLE 5.4

Example Permalink

http://soilenvirochem.net/tYVWqo

Validate the ionic strength I_c, ion activities and ion activity coefficients from numerical simulations of dilute acetic acid and acetate salt solutions.

Since the charge-balance condition is satisfied for acetic acid solutions simulated as in Example 5.3 this example will limit its analysis to the acetate solution, which registers a charge-balance error.

Part 1. Simulation of 2×10^{-3} molar acetate solution with two (2) components: Acet- and H+.

ChemEQL assigns all neutral species (e.g., HAcet(aq)) an activity coefficient equal to unity (Table 5.8, row 2, column 4). *ChemEQL* also assigns an activity coefficient equal to unity to the proton/hydronium ion H+ (Table 5.8, row 3, column 4). *ChemEQL* computes activity coefficients for all remaining ion species listed in the output table.

TABLE 5.8 *ChemEQL* Output Table for 2×10^{-3} molar Acetate Solution With Two (2) Components Acet- and H+ (Example 5.4, Part 1)

Species	Concentration (mol dm^{-3})	Activity (unitless)	Activity Coefficient (unitless)
Acet-	2.00E−03	1.93E−03	0.965
HAcet(aq)	1.03E−06	1.03E−06	1.000
OH-	1.04E−06	1.01E−06	0.965
H+	9.96E−09	9.96E−09	1.000

The values in column 4 are activity divided by concentration.

Part 2. Simulation of 2×10^{-3} molar sodium acetate solution with three (3) components: Acet-, nonacidic cation Na+ and H+.

Ion activities and activity coefficients from the three-component simulation (Table 5.9) differ from the two-component simulation (Table 5.8) because addition of 2.00×10^{-3} molar spectator cation Na+ doubles the ionic strength ($I_c = 2.00 \times 10^{-3}$ molar).

TABLE 5.9 *ChemEQL* Output Table for 2×10^{-3} molar Three Component Sodium Acetate Solution

Species	Concentration (mol dm^{-3})	Activity (unitless)	Activity Coefficient (unitless)
Na+	2.00E−03	1.90E−03	0.952
Na(OH)(aq)	1.19E−09	1.19E−09	1.000
Acet-	2.00E−03	1.90E−03	0.951
HAcet(aq)	1.03E−06	1.03E−06	1.000
OH-	1.04E−06	9.90E−07	0.951
H+	1.01E−08	1.01E−08	1.000

The values in column 4 are activity divided by concentration (Example 5.4, Part 2).

Part 3. Analysis and interpretation.

Activity coefficients in the two-component simulation ($\gamma_{1\pm} = 0.965$) are larger relative to those in the three-component system ($\gamma_{1\pm} = 0.951$). The difference—on the order of 1.5%—is negligible.

Allowing the omission of spectator cations (and spectator anions) results in a failure to satisfy the charge-balance condition and a systematic underestimate of ionic strength I_c in certain simulations. A thorough validation of simulation results, however, reveals such omissions have negligible effect on ion activity coefficients. Furthermore, output results demonstrate omission of spectator cations and (and spectator anions) has negligible effect on simulated solution pH.

5.4.1.3 *Ion Activity Products*

The logarithm of equilibrium acetic acid hydrolysis constant in the *ChemEQL* database is +4.73. The logarithm of the ion activity product IAP for acetic acid hydrolysis, computed using activities listed in Table 5.9, is +4.729. Validating agreement between the database constant and solute activities generated by the simulation.

ChemEQL Database Reaction

$$\text{acetate}^-(\text{aq}) + \text{H}^+(\text{aq}) \longleftrightarrow \text{acetic acid}(\text{aq}) \tag{5.R3}$$

$$\left(K_a^{\circ}\right)^{-1} = 10^{+4.73} = \frac{a_{\text{acetic acid(aq)}}}{a_{\text{H}^+} \cdot a_{\text{acetate}^-}}$$

$$IAP = 5.359 \times 10^{+4} = \frac{1.03 \times 10^{-6}}{9.96 \times 10^{-9} \cdot 1.93 \times 10^{-3}}$$

Most computer-based numerical water chemistry applications generate a file containing the results of each simulation. Transferring data from the results file to a spreadsheet application will greatly simplify the validation process. Appendix 5.B addresses file format issues specific to *ChemEQL*. Details on various validation tests and more examples appear in Appendix 5.D.

5.4.2 Gas Solubility: Fixed-Fugacity Method

Dissolved gases have an important impact on mineral solubility in most water chemistry systems. Carbonate chemistry, a major topic in aqueous geochemistry and many environmental problems, is deferred to Chapter 6.

Delany and Wolery (1984) introduced a scheme for modeling gas solubility in natural waters—the so-called *fixed-fugacity* scheme. This scheme represents the gas phase as limitless reservoir where gas partial pressures remain, by definition, constant (hence: *fixed fugacity*), allowing the gas phase to saturate the solution phase.

The *fixed-fugacity* scheme (Delany and Wolery, 1984) is implemented in virtually all currently available numerical water chemistry applications; it is not implemented in the *ChemEQL* model. The rationale behind the *fixed-fugacity* scheme (Delany and Wolery, 1984) and the approach adopted by *ChemEQL* is the topic of this section.

Here is the rationale for *fixed-fugacity* as given by Delany and Wolery (1984) who chose bicarbonate $HCO_3^-(aq)$ as the carbon dioxide component. Most, but not all, major water chemistry databases represent the carbon dioxide component as the ionic conjugate base (i.e., $HCO_3^-(aq)$) specie rather than the neutral weak acid (i.e., $CO_2(aq)$) specie.

Equilibrium between gas-phase carbon dioxide $CO_2(g)$ and aqueous-phase bicarbonate $HCO_3^-(aq)$ involves two steps: carbon dioxide dissolution into water (reaction 5.R4 and equilibrium reaction quotient 5.25) and aqueous carbon dioxide hydrolysis to form bicarbonate (reaction 5.R5 and equilibrium reaction quotient 5.27).

$$CO_2(g) \longleftrightarrow CO_2(aq) \tag{5.R4}$$

$$\left(k_{\mathrm{H,\,CO_2}}^{\ominus}\right)^{-1} = \frac{a_{\mathrm{CO_2}}}{f_{\mathrm{CO_2}}} = \frac{\left(\gamma_{c,\mathrm{CO_2}} \cdot c_{\mathrm{CO_2}}\right)}{\phi_{\mathrm{CO_2}} \cdot p_{\mathrm{CO_2}}} \tag{5.25}$$

The $CO_2(g)$ fugacity coefficient ϕ_{CO_2} (denominator of expression 5.25) in the above-ground atmosphere and soil pores is nearly unity (cf. Appendix 5.B) and therefore: $f_{CO_2} \approx p_{CO_2}$.

$$\mathrm{CO_2(aq) + H_2O(l) \longleftrightarrow H^+(aq) + HCO_3^-(aq)} \tag{5.R5}$$

$$K_{\mathrm{a1}}^{\ominus} = \frac{a_{\mathrm{H^+}} \cdot a_{\mathrm{HCO_3^-}}}{a_{\mathrm{CO_2}}} = \frac{a_{\mathrm{H^+}} \cdot \left(\gamma_{1\pm} \cdot c_{\mathrm{HCO_3^-}}\right)}{\left(\gamma_{c,\mathrm{CO_2}} \cdot a_{\mathrm{CO_2}}\right)} \tag{5.26}$$

The activity of solvent water $H_2O(l)$ is nearly unity for most natural waters, therefore: $a_{H_2O(l)} \approx 1$. Most numerical water chemistry applications do not compute activity coefficients for neutral molecules such as $CO_2(aq)$, therefore: $a_{CO_2} \approx 1 \cdot c_{CO_2}$.

Combining the dissolution reaction (5.R4) and hydrolysis reaction (5.R5) yields the following overall reaction (5.R6).

$$\mathrm{CO_2(g) + H_2O(l) \longleftrightarrow H^+(aq) + HCO_3^-(aq)} \tag{5.R6}$$

Applying the activity approximations discussed above and making the logarithm of the equilibrium quotient expression for the overall dissolution and hydrolysis of gas-phase carbon dioxide $CO_2(g)$ yields expression (5.27).

$$\log_{10}\left(a_{\mathrm{H^+}}\right) + \log_{10}\left(a_{\mathrm{HCO_3^-}}\right) - \log_{10}\left(p_{\mathrm{CO_2}}\right) = \log_{10}\left(K_{\mathrm{a1}}^{\ominus}/k_{\mathrm{H,\,CO_2}}^{\ominus}\right) \tag{5.27}$$

Delany and Wolery (1984) proposed reaction (5.R7) and equilibrium quotient expression (5.28) for the dissolution and hydrolysis of a *fictive solid*: $CO_2(s)$.

$$\mathrm{CO_2(s) + H_2O(l) \longleftrightarrow H^+(aq) + HCO_3^-(aq)} \tag{5.R7}$$

$$K_{\mathrm{a1}}^{\ddagger} = \frac{a_{\mathrm{H^+}} \cdot a_{\mathrm{HCO_3^-}}}{a_{\mathrm{CO_2(s)}} \cdot a_{\mathrm{H_2O(l)}}} \tag{5.28}$$

The activity of the *fictive solid* $CO_2(s)$ is unity, the convention for all pure phases: $a_{CO_2(s)} \equiv 1$. The activity of solvent water $H_2O(l)$ is also unity: $a_{H_2O(l)} \approx 1$. Applying the approximations then taking the logarithm of the equilibrium quotient expression for the hydrolysis of aqueous carbon dioxide $CO_2(aq)$ yields expression (5.29).

$$\log_{10}\left(a_{\mathrm{H^+}}\right) + \log_{10}\left(a_{\mathrm{HCO_3^-}}\right) - \log_{10}(1) = \log_{10}\left(K_{\mathrm{a1}}^{\ddagger}\right) \tag{5.29}$$

Subtracting the logarithmic expression listing gaseous carbon dioxide $CO_2(g)$ equilibrium hydrolysis quotient from the logarithmic expression listing fictive carbon dioxide solid $CO_2(s)$ yields expression (5.30) for the effective hydrolysis coefficient $K_{a1}^{\ddagger}$.

$$\log_{10}\left(K_{\mathrm{a1}}^{\ddagger}\right) = \log_{10}\left(K_{\mathrm{a1}}^{\ominus}/k_{\mathrm{H,\,CO_2}}^{\ominus}\right) + \log_{10}\left(p_{\mathrm{CO_2}}\right)$$

$$K_{\mathrm{a1}}^{\ddagger} = \frac{K_{\mathrm{a1}}^{\ominus} \cdot p_{\mathrm{CO_2}}}{k_{\mathrm{H,\,CO_2}}^{\ominus}} \tag{5.30}$$

Environmental chemists typically quote Henry's Law constants written as gas-solubility coefficients (5.23), neglecting the gas-phase fugacity coefficient ϕ_{CO_2} and solution-phase activity coefficient γ_{c,CO_2} in expression (5.25). The effective hydrolysis coefficient $K^{\ddagger}_{a1}$ using this alternative notation becomes expression (5.31).

$$K^{\ddagger}_{a1} = H^{cp}_{CO_2} \cdot K^{\circ}_{a1} \cdot p_{CO_2} \tag{5.31}$$

Numerical water chemistry applications that implement the *fixed-fugacity* scheme for dissolved gases automatically compute $K^{\ddagger}_{a1}$ coefficients from Henry's Law gas-solubility constants H^{cp}_{X} and hydrolysis constants K°_{a1} drawn from the database and user-supplied gas partial pressures when the user chooses this simulation option. The following example walks the reader through the *fixed-fugacity* method (Example 5.5).

EXAMPLE 5.5

Example Permalink

http://soilenvirochem.net/J2UmW7

The US *Occupational Safety and Health Administration* dihydrogen sulfide gas *Short-term Exposure Limit* (15-min exposure) is 15 ppmv (1.5×10^{-5} atm). Calculate dihydrogen sulfide $H_2S(aq)$ solubility and the corresponding *fixed-fugacity* coefficient $K^{\ddagger}_{a1}$ at the OSHA human exposure limit.

Part 1. Calculate dihydrogen sulfide $H_2S(aq)$ solubility at $p_{H_2S} = 1.5 \times 10^{-5}$ atm.

The dissolution of dihydrogen sulfide H_2S in water is given by reaction (5.R8). Expression (5.32) defines the Henry's Law $H_2S(g)$ solubility coefficient.

$$H_2S(g) \longleftrightarrow H_2S(aq) \tag{5.R8}$$

$$H^{cp}_{H_2S} = \frac{c_{H_2S}}{p_{H_2S}} = 9.22 \times 10^{-2} \tag{5.32}$$

The $H_2S(aq)$ concentration is found by using $H^{cp}_{H_2S} = 9.22 \times 10^{-2}$ entering $p_{H_2S} = 1.5 \times 10^{-5}$ atm.

$$c_{H_2S} = H^{cp}_{H_2S} \cdot p_{H_2S}$$

$$c_{H_2S} = \left(9.22 \times 10^{-2}\right) \cdot \left(1.5 \times 10^{-5}\right) = 1.38 \times 10^{-6} \text{ mol dm}^{-3}$$

Part 2. Calculate the *fixed-fugacity* dissolution-hydrolysis coefficient $K^{\ddagger}_{a1}$ for: $p_{H_2S} = 1.5 \times 10^{-5}$ atm.

The $H_2S(aq)$ first hydrolysis step (reaction 5.R9 and equilibrium quotient expression 5.33) produces the conjugate base hydrogen sulfide $HS^-(aq)$.

$$H_2S(aq) \longleftrightarrow H^+(aq) + HS^-(aq) \tag{5.R9}$$

$$K^{\circ}_{a1} = \frac{a_{H^+} \cdot a_{HS^-}}{a_{H_2S}} = 9.55 \times 10^{-8} \tag{5.33}$$

Combined dissolution and hydrolysis reactions represent $H_2S(g)$ solubility as a *fictive solid* (reaction 5.R10 and expression 5.34).

$$H_2S(s) \longleftrightarrow H^+(aq) + HS^-(aq) \tag{5.R10}$$

$$K_{a1}^{\ddagger} \equiv \left(H_{H_2S}^{cp} \cdot p_{H_2S}\right) \cdot K_{a1}^{\circ} \tag{5.34}$$

Entering the dihydrogen sulfide $H_2S(aq)$ concentration solubility from Part 1 in expression (5.33) gives the desired fixed-fugacity hydrolysis coefficient $K_{a1}^{\ddagger}$.

$$K_{a1}^{\ddagger} \equiv \left(1.38 \times 10^{-6}\right) \cdot \left(9.55 \times 10^{-8}\right) = 1.32 \times 10^{-13}$$

Example 5.D.5 demonstrates how to implement the *fixed-fugacity* method when using the *ChemEQL* water chemistry application (cf. Examples 5.D.3 and 5.D.4).

5.4.3 Mineral Solubility

A number of factors influence mineral solubility in natural waters. Here, we will examine three of the most important factors: solution pH, soluble complex-forming ligands, and simultaneous equilibrium involving two or more minerals that share a common component.

5.4.3.1 *Effect of pH on Mineral Solubility*

Example 5.D.2 calculates the solubility of GIBBSITE : $Al(OH)_3(s)$, a common soil mineral, using the RICE table method. The reader should also study Example 5.D.2—a *ChemEQL* simulation equivalent to Example 5.D.2—that provides a more complete picture of gibbsite solubility and aluminum chemistry in water.

While Examples 5.D.2 does not calculate pH-dependent gibbsite solubility, the equilibrium solubility expression below implies gibbsite solubility is highly pH-dependent.

$$\underset{\text{GIBBSITE}}{Al(OH)_3(s)} \longleftrightarrow Al^{3+}(aq) + 3OH^-(aq) \tag{5.R11}$$

$$K_{s0}^{\circ} = a_{Al^{3+}} \cdot a_{OH^-}^3 = 1.29 \times 10^{-34} = 10^{-33.89} \tag{5.35}$$

The water hydrolysis reaction (cf. Chapter 6) provides a means of transforming the gibbsite solubility quotient expression to more clearly reveal its pH-dependence.

$$H_2O(l) \longleftrightarrow H^+(aq) + OH^-(aq) \tag{5.R12}$$

$$K_w^{\circ} = a_{H^+} \cdot a_{OH^-} = 10^{-14.00} \tag{5.36}$$

Combining the gibbsite solubility and the water hydrolysis quotient expressions yields a transformed gibbsite solubility quotient expression that is explicitly dependent on proton activity.

$$\underset{\text{GIBBSITE}}{Al(OH)_3(s)} + 3H^+(aq) \longleftrightarrow Al^{3+}(aq) + 3H_2O(l) \quad (5.R13)$$

$$K^{\circ}_{net} = \frac{a_{Al^{3+}} \cdot a^3_{OH^-}}{\left(a_{H^+} \cdot a_{OH^-}\right)^3} = \frac{10^{-33.89}}{(10^{-14.00})^3} = 10^{+8.11} \quad (5.37)$$

$$K^{\circ}_{net} = a_{Al^{3+}} \cdot a^{-3}_{H^+} = \gamma_{3\pm} \cdot \gamma^{-3}_{1\pm} \cdot K^c_{net}$$

Taking the base-10 logarithm of the net equilibrium solubility yields a linear expression between the logarithm of $Al^{3+}(aq)$ activity and solution pH.

$$\log_{10}\left(K^{\circ}_{net}\right) = +8.11 = \log_{10}\left(a_{Al^{3+}}\right) - 3 \cdot \log_{10}\left(a_{H^+}\right)$$

$$\log_{10}\left(a_{Al^{3+}}\right) = 8.11 - 3 \cdot pH$$

Example 5.6 is a *ChemEQL* simulation designed to reveal pH-dependent mineral solubility.

EXAMPLE 5.6

Example Permalink

http://soilenvirochem.net/Tbk5lE

Use *ChemEQL* to simulate gibbsite solubility over the pH range: 4–9.

The input matrix is a very simple, containing only two components. Solution specie Al+++ is the component representing all dissolved aluminum species. Using the *ChemEQL* Access Library ... command (File menu), you can only create an input matrix from a list of solution species: choose Al+++, select mode = total and enter Concentration = 0.

The instructions asks for a simulation that constrains pH to cover a specified range. To impose this constraint *ChemEQL* requires the user to select mode = free for component H+. The user must enter a nonzero value in the Concentration field. Compile the input to generate the input matrix (Table 5.10, upper section).

Next, use *ChemEQL* Insert Solid ... command (Matrix menu). This displays the Insert Solid Phase window. Select Al+++ from the upper (replace) drop-down menu then select Al(OH)3 (Gibbsite) from the lower (with) drop-down menu and click the OK button. These actions result in the component list displayed in the lower section of Table 5.10.

The *ChemEQL Manual* (permalink: soilenvirochem.net/U4gkv2) provides detailed instructions for performing simulation where the user chooses to vary a component concentration over a specified range. The commands pH range ... and Component range ... appear under the Options menu. In this case, select pH range ... and enter the pH range specified above.

TABLE 5.10 *ChemEQL* Components Input Tables for pH-dependent Gibbsite Solubility

Component	Mode	Initial Concentration (mol dm^{-3})
Al+++	Total	0.00E+00
H+	Free	1.00E−07
Component	**Mode**	**Initial Concentration (mol dm^{-3})**
Al(OH)3 (Gibbsite)	Solid-phase	1
H+	Free	1.00E−07

Top: After compiling input parameters (cf. `Tbk51E(A).cql`).
Bottom: After using Insert Solid to replace solution specie Al+++ with solid-phase specie Al(OH)3(gibbsite) (cf. `Tbk51E(B).cql`).

The final step is set Activity parameters: approximation, ionic strength, Debye-Hückel factor and output values. In this simulation the parameters are: Davies approximation, $I_c = 1.0 \times 10^{-4}$ mol dm^{-3}, $A_c^{DH} = 0.510$, and output: activities.

Gibbsite solubility is more complicated than represented in Example 5.D.2 because the aluminum ion $Al^{3+}(aq)$ forms a complex with six (6) water molecules. This complex cation—the *hexaaquoaluminum*(III) complex cation $Al(H_2O)_6^{3+}(aq)$—behaves as a weak polyprotic acid that undergoes several hydrolysis steps.

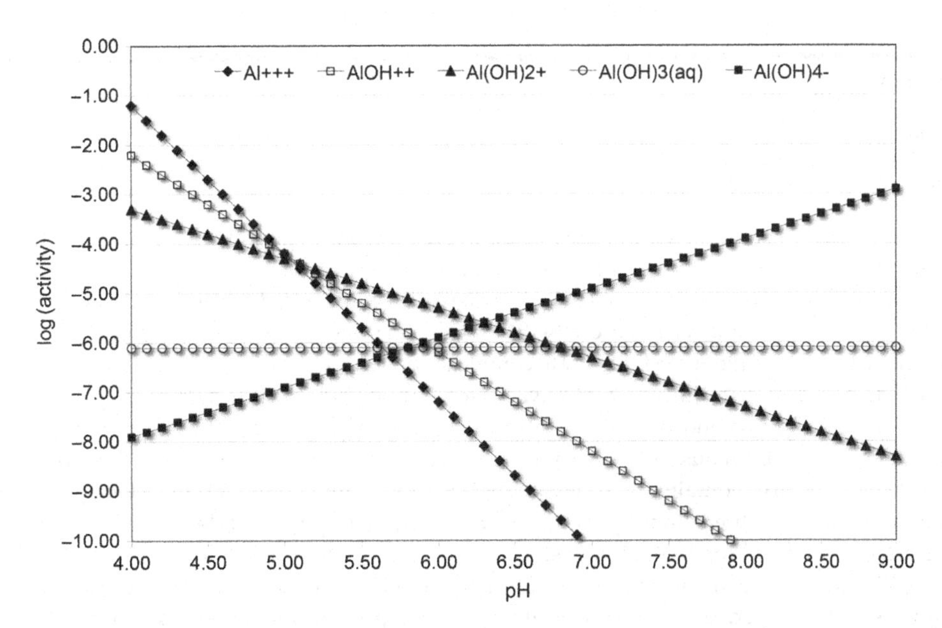

FIG. 5.5 The base-10 logarithm of ion activity for the five most abundant aluminum solution species are plotted as a function of pH. The Al^{3+} solubility is constrained by gibbsite solubility with H^+ is treated as an independent variable covering the range depicted in the graph.

We will return to the gibbsite solubility problem in Chapter 6, taking into full account the hydrolysis of the *hexaaquoaluminum*(III) complex cation and its conjugate bases.

Exercise 5.10 challenges the reader to plot the pH-dependent solubility of the calcium phospahte mineral HYDROXYAPATITE : $Ca_5(PO_4)OH(s)$. This exercise will familiarize the reader with the Insert Solid Phase ... command and the pH range ... option in *ChemEQL*.

5.4.3.2 *Effect of Ion Complexes on Mineral Solubility*

Bacteria, fungi, and plants secrete a host of organic acids and other compounds. Some of these compounds, e.g., organic acids, are respiration by-products or intermediates in a variety of biosynthetic pathways. Some are excreted specifically to increase the solubility of essential nutrients.

Lupine (*Lupinus* spp.), a genus of the legume family Fabaceae, is known for its ability to thrive in soils with low soluble phosphate, an essential plant nutrient. Phosphate-limiting conditions induce Lupines to excrete prodigious amounts of citrate into the rhizosphere—the soil in direct contact with plant roots. Example 5.7 examines the solubility of CALCITE : $CaCO_3(s)$ in a solution containing 10^{-3} mol dm^{-3} citric acid.

EXAMPLE 5.7

Example Permalink

http://soilenvirochem.net/C77JUe

Use *ChemEQL* to determine whether increased hydroxyapatite solubility in a solution containing 10^{-4} mol dm^{-3} citric acid is due to complexation or pH lowering.

Citric acid secretion by Lupine growing in calcareous soil is attributed to low phosphate biological availability (i.e., low solubility) in these soils. The citric acid secretion effect may increase phosphate solubility by forming soluble calcium citrate complexes or simply lowering solution pH.

Step 1. Use *ChemEQL* to simulate hydroxyapatite solubility absent citric acid.

The citric acid effect on phosphate solubility can only be quantified by first determining hydroxyapatite solubility in calcareous soils. Calcareous soil typically have a soil pore water pH around 8.4.

Step 1 input matrix contains three basis-species (i.e., components): Ca++, PO4---, and H+. Simulating hydroxyapatite solubility using *ChemEQL* requires setting mode = total and entering Concentration = 0 for all three components.

Compiling the input generates a properly formatted input matrix file listing solution species. The user must select Insert Solid ... from the Matrix menu to replace solution species with solid-phase species. The input window has drop-down menu listing basis-species available for replacement by solid phases. In this case, select PO4--- from the menu list and click the OK button (cf. *ChemEQL Manual* : permalink: soilenvirochem.net/U4gkv2.pdf).

The altered input matrix lists the original basis-species (i.e., components): Ca++, and H+ and the newly inserted solid-phase specie Ca5(PO4)3OH(hydroxyapatite). Select the ion activity coefficient approximation from the Options menu then select Go from the Run menu to execute the simulation.

Table 5.11 replicates the components list as written in *ChemEQL* output file for Step 1 simulation.

TABLE 5.11 *ChemEQL* Input Table for Hydroxyapatite Dissolution Into Pure Water (Example 5.7, Step 1)

Component	Mode	Initial Concentration (mol dm^{-3})	In or Out ...(mol dm^{-3})
Ca++	Total	0.00E+00	[-]
Ca5(PO4)3OH(s)	solid-Phase	Unlimited	2.86E−06
H+	Total	0.00E+00	[-]

The equilibrium pH is 9.05 ($a_{H^+} = 8.85 \times 10^{-10}$), slightly higher than the characteristic pH of calcareous soils. The simulated total phosphate concentration is $C_{phosphate} = 8.58 \times 10^{-6}$ mol dm^{-3}.

Step 2. Use *ChemEQL* to simulate hydroxyapatite solubility in a solution containing 10^{-4} mol dm^{-3} citric acid.

The Step 2 input matrix contains four basis-species (i.e., components): Ca++, PO4---, Citrate---, and H+. Simulating hydroxyapatite solubility in a citric acid solution requires setting mode = total for all basis-species. The Concentration settings for Ca++ and PO4--- are the same as in Step 1. The Concentration settings for Citrate--- and H+ are 1.00E-4 and 3.00E-4, respectively (cf. Example 5.2). Follow the steps in Step 1 to replace PO4--- with Ca5(PO4)3OH(s).

Table 5.12 replicates the components list as written in *ChemEQL* output file for Step 2 simulation.

TABLE 5.12 *ChemEQL* Input Table for Hydroxyapatite Dissolution Into a Solution Containing 10^{-4} mol dm^{-3} Citric Acid (Example 5.7, Step 2)

Component	Mode	Initial Concentration (mol dm^{-3})	In or Out ...(mol dm^{-3})
Ca++	Total	0.00E+00	[-]
Ca5(PO4)3OH(s)	Solid-phase	Unlimited	5.30E−05
Citrate---	Total	1.00E−04	[-]
H+	Total	3.00E−04	[-]

The equilibrium pH is 7.07 ($a_{H^+} = 8.57 \times 10^{-8}$), two pH units lower than the solution in Step 1. The simulated total phosphate concentration is: $C_{phosphate} = 1.59 \times 10^{-4}$ mol dm^{-3}, 18.5 time higher than Step 1. The only way to determine whether this increase is due to calcium citrate complexes or lowering pH is to proceed to Step 3.

Step 3. Use *ChemEQL* to simulate hydroxyapatite solubility in a solution at pH 7.07 but containing no citric acid.

The Step 3 input matrix is identical to Step 1 except for the mode setting and initial Concentration setting for basis-specie H+: mode = free and concentration = 8.57E−08. These basis-specie H+ settings constrain the pH to equal 7.07, thereby isolating the pH-lowering effect.

Table 5.13 replicates the components list as written in *ChemEQL* output file for Step 3 simulation.

TABLE 5.13 *ChemEQL* Input Table for Hydroxyapatite Dissolution Into a Solution Containing 10^{-4} mol dm^{-3} Citric Acid (Example 5.7, Step 3)

Component	Mode	Initial Concentration (mol dm^{-3})	In or Out ...(mol dm^{-3})
Ca++	Total	0.00E+00	[-]
Ca5(PO4)3OH(s)	Solid-phase	Unlimited	4.14E−05
H+	Free	8.57E−08	2.32E−04

The equilibrium pH is 7.07, the same as Step 2. The simulated total phosphate concentration is: $C_{phosphate} = 1.24 \times 10^{-4}$ mol dm^{-3}, 14.4 time higher than Step 1.

The drop in pH from 9.05 (Step 1) to 7.07 (Steps 2 and 3) accounts for 77.8% of the increased phosphate solubility, the remaining 22.2% is attributable to calcium citrate complexes. Clearly, citrate raises phosphate solubility in calcareous soils by lowering rhizosphere pH.

Example 5.8 is a more compelling example of mineral solubility influenced by a ligand capable of forming a much more stable complex than was the case of calcium citrate complexes in Example 5.7. Example 5.8 describes a very important soil chemistry reaction in which a siderophore (a bioorganic compound) forms a complex with iron(III) to increase its solubility. Bacteria, fungi, and grass plants excrete siderophores to meet their iron nutritional requirements.

EXAMPLE 5.8

Example Permalink

http://soilenvirochem.net/jbD24m

Use *ChemEQL* to determine the solubility of FERRIHYDRITE : $Fe(OH)_3(s)$ in soil pore water containing low levels of a bacterial siderophore.

Bacteria, fungi, and graminaceous plants excrete compounds known collectively as siderophores, water soluble organic compounds that form highly stable (and soluble) complexes with iron(III). The siderophore Ferrioxamine-B (CASRN 14836-73-8) is a hydroxamate siderophore secreted by

Streptomyces pilosus—a Gram-positive bacterium of the phylum and class actinobacteria. Hydroximate and other siderophores form soluble, extremely-stable complexes with iron(III) that promote uptake by complexing and transporting soluble ferri-siderphores (i.e., *iron(III)-containing* siderophores) to the organism.

Low iron solubility induces expression of bacterial genes required to synthesize desferri-siderophores (i.e., *iron-free* siderophores) and cell envelope siderophore-receptors that recognize, bind, and transport ferri-siderphore complexes into the cytoplasm. Soil chemists measure dissolved iron concentrations as high as 10^{-4} mol dm^{-3} in alkaline and circum-neutral soil pore water, far above the solubility limit of iron oxyhydroxide minerals in this pH range. Hence, a total siderophore concentration on the order of 10^{-5} mol dm^{-3} is entirely reasonable.

Step 1. Use *ChemEQL* to simulate the pH 7 solubility of ferric hydroxide (i.e., ferrihydrite) in a solution containing 10^{-5} mol dm^{-3} ferrioxamine-B.

The input matrix contains three basis-species (i.e., components): Fe+++, FerrioxamineB---, and H+. Simulating ferrihydrite solubility in a ferrioximine-B solution requires setting mode = total for the first two basis-species. The initial concentration for basis-specie Fe+++ and basis-specie FerrioxamineB--- are 0 and 1.0E−05, respectively. The pH 7 constraint requires setting mode(H+) = free and Concentration = 1.0E−07 for the remaining basis-specie H+.

Table 5.14 replicates the components list as written in *ChemEQL* output file for Step 4 simulation.

TABLE 5.14 *ChemEQL* Input Table for Ferrihydrite Dissolution Into a Solution Containing 10^{-5} mol dm^{-3} Ferrioxamine-B

Component	Mode	Initial Concentration (mol dm^{-3})	In or Out ...(mol dm^{-3})
Fe(OH)3(Ferrihydrite)	Total	0.00E+00	1.00E−05
FerrioxamineB---	Total	1.00E−05	[-]
H+	Free	1.00E−07	[-]

The maximum solubility of FERRIHYDRITE : $Fe(OH)_3(s)$ at pH 7 (absent siderophore ferrioxamine-B) is the concentration of the most abundant iron(III) specie: Fe(OH)3(aq) = 4.37E-10 molar.

The total soluble iron(III) at pH 7 in a solution containing 10^{-5} mol dm^{-3} ferrioxamine-B is—within rounding error—10^{-5} mol dm^{-3}. Furthermore, solution specie Fe(HFOB)+ accounts for 99.96% of the soluble iron(III) at pH 7.

Step 2. Use *ChemEQL* to simulate ferrihydrite solubility in solutions containing 10^{-5} mol dm^{-3} ferrioxamine-B over a pH range from 4 to 9 (cf. Appendix 5.D, Example 5.D.2 for a simulation that covers a defined pH range.).

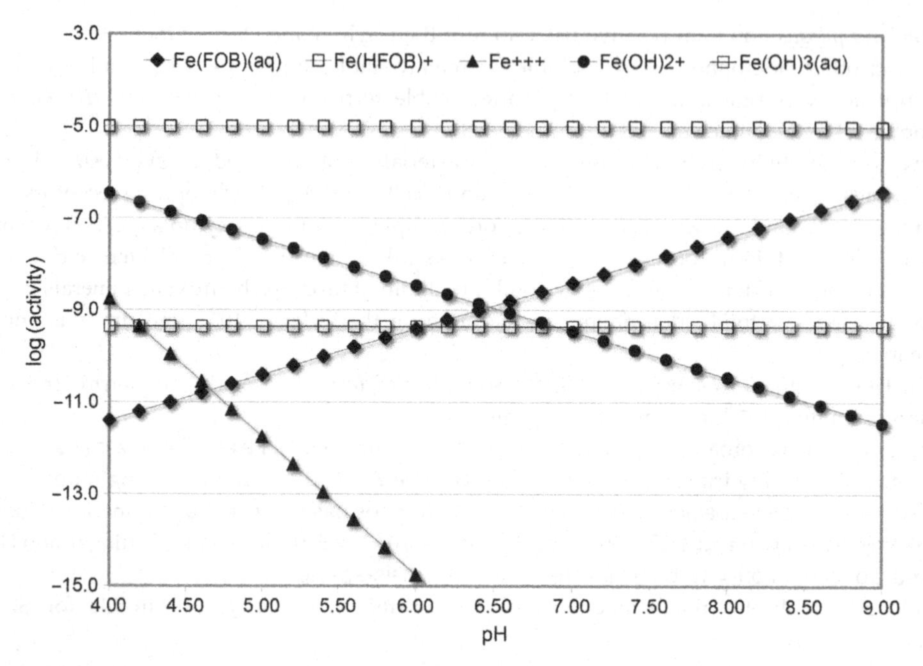

FIG. 5.6 The graph plots pH-dependent activities for the five most abundant iron(III) species in a solution containing 10^{-5} mol dm^{-3} ferrioxamine B and the mineral FERRIHYDRITE : $Fe(OH)_3$ (s). The concentration of solution specie Fe(HFOB)+ is virtually independent of pH, representing 99.96% of total dissolved iron(III) and ferrioxamine-B.

5.4.3.3 *The Gibbs Phase Rule*

Environmental chemists simulate the chemistry of water samples for many reasons, the most important being the identification of minerals that control the solute concentrations in water samples. The water chemistry simulations can predict minerals in solubility equilibrium with the solution. The control of solution chemistry implies a very specific and concrete concept: the activities of two or more solutes are constrained by equilibrium solubility and, therefore, are not independent of each other.

Some of the minerals in contact with water—minerals in the aquifer formation if the sample is groundwater, soil minerals if the sample is soil pore water, or minerals suspended in the water column or sediments if the sample is lake or river water—cannot attain solubility equilibrium. Many primary rock-forming minerals that crystallize from magma (igneous rock minerals) or sinter at temperatures approaching the mineral melting point (metamorphic rock minerals) are not stable in the presence of liquid water. Some low-temperature sedimentary minerals precipitate from water whose composition is very different from prevailing conditions (May et al., 1986). Solubility equilibrium can exist only if dissolution and precipitation reactions are *reversible*.

Another important instance when a mineral cannot control solution chemistry occurs when the residence time of the water is less than the time necessary for the mineral to dissolve sufficiently to saturate the solution. This is illustrated by the effect of groundwater discharge on lake water alkalinity (Fig. 2.2, Chapter 2). Lake water alkalinity increases proportionately with the mean residence time of groundwater discharged into the lake (Wolock et al., 1989).

Chemists can identify minerals controlling solubility but only for those minerals at chemical equilibrium with the solution. If water chemistry is controlled by the equilibrium solubility of a suite of minerals, it should be possible to identify the minerals involved by calculating the ion activity products (cf Appendix 5.C). Applying equilibrium methods requires water chemistry analyses that report reliable solution pH and concentrations. As we will see in Chapter 6, water analysis pH values are error prone and often unreliable.

Each element or compound in the analysis, including solution pH, is considered a component for simulation purposes. Chemical reactions distribute each solution component over a constellation of solution (and mineral) species through the species-formation reactions, many—but not all—may be included in the model database (cf Appendix 5.C). The number of minerals that control water chemistry is subject to the *Gibbs Phase Rule*, a thermodynamic rule defining the *degrees of freedom* F in a system containing C components and P phases. The Gibbs Phase Rule for a system at fixed pressure and temperature is a very simple expression.

$$F = C - P \geq 0 \tag{5.38}$$

A more general expression for the Gibbs Phase Rule adds an additional two degrees of freedom if temperature and pressure are allowed to vary, but most environmental chemistry models assume constant temperature and pressure. The phase rule, as we will see, provides much needed guidance when modeling equilibrium in natural water samples because the number of phases P must be less than or equal to the number of components C.

Suppose the system consists of two phases P—aqueous solution and a gaseous phase—and two components H_2O and CO_2. The degrees of freedom F for this system is: ($F = 2 - 2 = 0$). The $CO_2(g)$ partial pressure p_{CO_2} fixes both the activity of dissolved carbon dioxide $CO_2(aq)$ and solution pH.[11] In other words, solution species $CO_2(aq)$ and $H^+(aq)$ cannot vary independently of each other—there are no degrees of freedom F: solution pH is a direct consequence of p_{CO_2}.

Now, consider a system prepared from three components C: liquid water $H_2O(l)$, solute $Cd(OH)_2(aq)$, and dihydrogen sulfide $H_2S(aq)$. The number of phases P can range from 1 to 3, and, consequently, the degrees of freedom F can range from 2 to 0.

$$F = 3 - P \geq 0$$

A 1-phase, 3-component system ($F = 3 - 1 = 2$) would be $Cd(OH)_2(aq)$ and $H_2S(aq)$ dissolved in water absent contact with a gas phase. $Cd(OH)_2(aq)$ and $H_2S(aq)$ react each other, forming solute species through complexation reactions, and with water (the third component),

[11] Water chemistry applications typically denote $H_2O(l)$ using basis specie H+.

forming solute species through hydrolysis. Two degrees of freedom F allow two independent variables: $C_{cadmium}$ and $C_{sulfide}$ vary independently of each other, determining pH through hydrolysis reactions.

Two 2-phase, 3-component systems ($F = 3 - 2 = 1$) are possible: (1) a *gas-solution* system or (2) a *solution-mineral* system, the mineral being GREENOCKITE : CdS(s). The *gas-solution* 2-phase system has a single degree of freedom. Varying $C_{cadmium}$ determines both solution pH and hydrogen sulfide solubility $C_{sulfide}$.

The *solution-mineral* 2-phase system also has a single degree of freedom. Hydrogen sulfide solubility $C_{sulfide}$ and solution pH are both determined by $C_{cadmium}$ acting as the independent variable. Alternatively, $C_{cadmium}$ solubility $C_{sulfide}$ and solution pH will follow changes in total dissolved sulfide $C_{sulfide}$ acting as the independent variable.

A single 3-phase, 3-component systems ($F = 3 - 3 = 0$) is possible: a *gas-solution-mineral* system. Zero degree of freedom allows no independent variables; $C_{cadmium}$, $C_{sulfide}$, and pH are invariant when all three phases are present (Example 5.9).

EXAMPLE 5.9

Example Permalink

http://soilenvirochem.net/8ku6NB

This example applies the *Gibbs Phase Rule* to several 3-component systems, illustrating the degrees of freedom using *ChemEQL* input matrixes.

Part 1. A 3-component, 1-phase system consisting of a solution phase, has two (2) degrees of freedom.

The components are: liquid water $H_2O(l)$, solute $Cd(OH)_2(aq)$, and dihydrogen sulfide $H_2S(aq)$. The *ChemEQL* simulation of this system complies the input matrix using the following database components: Cd++, $H_2S(aq)$, H+. Once the input matrix is complied, use the Replace H+ by OH- command from the Matrix menu. The initial concentrations for components Cd++ and OH- in Table 5.15 are equivalent to a solution prepared from 2.00×10^{-8} mol dm^{-3} $Cd(OH)_2(aq)$.

TABLE 5.15 *ChemEQL* Input Table for a 3-Component, 1-Phase System

Component	Mode	Initial Concentration (mol dm^{-3})
Cd++	Total	2.00E−08
$H_2S(aq)$	Total	1.00E−07
OH-	Total	4.00E−08

A system consisting solely of a solution phase means the solution must be under-saturated with respect to the most insoluble cadmium sulfide mineral (i.e., GREENOCKITE : CdS(s)). Comparing the greenockite solubility constant ($K_s^\circ = 10^{-6.08}$) to the simulation ion activity product ($IAP = 10^{-6.287}$) confirms the solution is under-saturated and the system is 1-phase (solution).

$$IAP = \frac{a_{Cd^{2+}} \cdot a_{H_2S}}{a^2_{H^+}} = \frac{10^{-12.782} \times 10^{-7.353}}{10^{-13.874}} = 10^{-6.287}$$

Part 2. A 3-component, *gas-solution* system has one (1) degree of freedom.

The components are: liquid water $H_2O(l)$, solute $Cd(OH)_2(aq)$, and solute dihydrogen sulfide $H_2S(aq)$.[12] The *ChemEQL* simulation of this system uses an input matrix listing the following database components: Cd++, H2S(aq), OH- (cf. Part 1 for applying the Replace H+ by OH- command).

The initial concentrations for components Cd++ and OH- in Table 5.16 are equivalent to a solution prepared from 2.00×10^{-8} mol dm^{-3} $Cd(OH)_2(aq)$. The *fixed-fugacity* method is used to constrain H2S(aq): $p_{H_2S} = 100$ atm. The input matrix below treats gas-phase dihydrogen sulfide as a *fictive solid*.

TABLE 5.16 *ChemEQL* Input Table for a 3-Component, 2-Phase (Gas-Solution) System

Component	Mode	Initial Concentration (mol dm^{-3})
Cd++	Total	2.00E−08
H2S(aq)	Free	1.00E−07
OH-	Total	4.00E−08

[12] Gas-phase dihydrogen sulfide $H_2S(g)$ does not appear on the components list, its presence is implied by applying the *fixed fugacity* method.

A *gas-solution* 2-phase system must be under-saturated with respect to the most insoluble cadmium sulfide mineral (i.e., GREENOCKITE : CdS(s)). Comparing the greenockite solubility constant ($K_s^{\ominus} = 10^{-6.08}$) to the simulation ion activity product ($IAP = 10^{-6.292}$) confirms the solution is under-saturated and the system is 2-phase (gas-solution).

$$IAP = \frac{a_{Cd^{2+}} \cdot a_{H_2S}}{a^2_{H^+}} = \frac{10^{-13.000} \times 10^{-7.000}}{10^{-13.708}} = 10^{-6.292}$$

The *fixed fugacity* constraint on $H_2S(aq)$ results in a total sulfide concentration of 1.88×10^{-7} mol dm^{-3} at equilibrium between the solution and gas phase.

Part 3. A 3-component, *solution-mineral* system has one (1) degree of freedom.

The components are: liquid water $H_2O(l)$, solute $Cd(OH)_2(aq)$, and solid-phase GREENOCKITE : CdS(s)[13]. The *ChemEQL* simulation of this system complies the input matrix using the following database components: Cd++, H2S(aq), H+. Use the Insert Solid Phase ... to replace

[13] An equally acceptable component list would be: liquid water $H_2O(l)$, solute $H_2S(aq)$, and solid-phase GREENOCKITE : CdS(s)

Cd++ with CdS(greenockite)—do this before applying the Replace H+ by OH- command (cf. Step 4).

TABLE 5.17 *ChemEQL* Input Table for a 3-Component, 2-Phase (Solution-Mineral) System

Component	Mode	Initial Concentration (mol dm^{-3})
Cd++	Total	5.00E−08
CdS(s)	Solid-phase	Unlimited
OH-	Total	1.00E−07

The initial concentrations for components Cd++ and OH- in Table 5.17 are equivalent to a solution prepared from 5.00×10^{-8} mol dm^{-3} $Cd(OH)_2(aq)$.

A *mineral-solution* 2-phase system must be saturated with respect to the most insoluble cadmium sulfide mineral (i.e., GREENOCKITE : CdS(s)). The greenockite ion activity product IAP from the simulation is equal (within rounding error) to the solubility constant ($K_s^{\ominus} = 10^{-6.08}$), confirming the solution is a saturated and the system is 2-phase (solution-mineral).

Part 4. A 3-component, *gas-solution-mineral* system has zero (0) degrees of freedom.

The components are: liquid water $H_2O(l)$, solute $H_2S(aq)$,[14] and solid-phase GREENOCKITE : CdS(s). The *ChemEQL* simulation of this system complies the input matrix using the following database components: Cd++, H2S(aq), H+. Solution component Cd++ is replaced by solid-phase specie CdS(greenockite) by using the Insert Solid Phase ... command. The *fixed-fugacity* method is used to constrain H2S(aq): $p_{H_2S} = 100\ atm$. The input matrix below (Table 5.18) treats gas-phase dihydrogen sulfide as a *fictive solid*.

A *gas-mineral-solution* 3-phase system requires a solution saturated with respect to the most insoluble cadmium sulfide mineral (i.e., GREENOCKITE : CdS(s)). The greenockite ion activity product IAP form the simulation is equal (within rounding error) to the solubility constant ($K_s^{\ominus} = 10^{-6.08}$), confirming the solution is saturated and the system is 3-phase (gas-solution-mineral).

Applying the *Gibbs Phase Rule* in water chemistry simulations requires the user to pay careful attention to the replacement of solution components with solid-phase components. As

TABLE 5.18 *ChemEQL* Input Table for a 3-Component, 3-Phase (Gas-Solution-Mineral) System

Component	Mode	Initial Concentration (mol dm^{-3})
Cd++	Total	5.00E−08
H2S(aq)	Free	1.00E−7
H+	Total	0.00E+00

[14] Gas-phase dihydrogen sulfide $H_2S(g)$ does not appear on the components list; its presence is implied by applying the *fixed fugacity* method.

we will see in the following section, the Insert Solid Phase ... command in *ChemEQL* imposes certain restrictions that may seem arbitrary but which, in fact, are designed to avoid violating the *Gibbs Phase Rule*.

Numerical water chemistry applications are designed to simulate water chemistry by setting up and solving a system of linear equations. The numerical methods that solve the system of equations will generate output even if the input matrix violates the *Gibbs Phase Rule* or other important chemical equilibrium principles. Stated differently, if the user violates one or more chemistry principles the model will be none the wiser and the output will fail to correctly represent the water chemistry the user is attempting to simulate.

The following two sections describe methods for simulating water chemistry in systems with two or more mineral phases and methods for identifying the minerals that control solubility in natural waters.

5.4.3.4 Aqueous Solubility of Two or More Minerals

Example 5.10 illustrates equilibrium involving aqueous solution and two minerals sharing a common component, a 3-phase system. This example is representative of phosphate solubility in acidic soils, where the aluminum phosphate mineral variscite controls phosphate solubility and the mineral gibbsite controls aluminum solubility. *ChemEQL* requires the user to invoke the Insert Solid Phase ... command in a specific sequence when the system involves two or more minerals sharing a common component, a requirement imposed to avoid violating the *Gibbs Phase Rule*.

EXAMPLE 5.10

Example Permalink

http://soilenvirochem.net/JREH0s

Use *ChemEQL* to simulate the equilibrium chemistry of a solution in equilibrium with the minerals gibbsite and variscite.

Step 1. Invoke the File menu Access Library ... command to generate and compile the initial input matrix.

The simulation of a 3-phase system comprised of an aqueous solution and the minerals variscite and gibbsite require the following three components: Al+++, PO4--- and H+, where it is understood that *ChemEQL* employs component-specie H+ to represent all solution species formed from $H_2O(l)$.

Since there are three components *C* and three phases *P* and, furthermore, the simulation is at 25°C and 1 atm total pressure; the degrees of freedom *F* is zero. With zero degree of freedom *F*, the mode of all components must be set to total (Table 5.19).

TABLE 5.19 *ChemEQL* Input Table for a 3-Component, 3-Phase System for Simulating the Chemistry of a Solution in Equilibrium With Both Gibbsite and Variscite

Component	Mode	Initial Concentration (mol dm^{-3})
Al+++	Total	0.00E+00
PO4---	Total	0.00E+00
H+	Total	0.00E+00

Step 2. Invoke the Matrix menu Insert Solid Phase ... command to convert the complied input matrix into the simulation input matrix. Employing the Insert Solid Phase ... command, replace component-species Al+++ and PO4--- with solid-phase species Al(OH)3(gibbsite) and AlPO4:(H2O)2 (variscite).

Executing the Compile Matrix command generates a window that displays the input matrix contents. Next, select the Insert Solid Phase ... command from Matrix drop-down menu. The command window will display a Replace drop-down menu listing all three components in the input matrix and, below that, a with drop-down menu listing all Solid-Phase Species in the *ChemEQL* database. Table 5.20 lists the solid-phase species formation reactions for gibbsite and variscite.

TABLE 5.20 The Formation Reactions for Al(OH)3(Gibbsite) and AlPO4:(H2O)2 (Variscite) as They Appear in the Drop-Down Menu or as Listed When Invoking the Edit Solid Phases Lib. Species ... from the Libraries Menu

Solid-Phase Specie	Formation Reaction
Gibbsite	Al+++ ⟷ 3H+ + Al(OH)3 (gibbsite)
Variscite	Al+++ + PO4--- ⟷ AlPO4:(H2O)2 (variscite)

The user must decide which solid-phase specie will replace each component-specie *and* also decide the replacement sequence (i.e., which component-specie is replaced first, second, etc.)

The user could select either Al(OH)3(gibbsite) or AlPO4:(H2O)2 (variscite) to replace component-specie Al+++ because it appears as an educt in both formation reactions. The user, however, can only select AlPO4:(H2O)2 (variscite) to replace component-specie PO4---. The replacement choices are clear: Al(OH)3(gibbsite) replaces Al+++ and AlPO4:(H2O)2 (variscite) replaces component-specie PO4---.

The replacement sequence in this case comes down to choosing which component-specie to replace first. *ChemEQL* will allow a replacement only if the components listed in the formation reaction match the component-species remaining in the input matrix header. If the user's first replacement is Al+++ with Al(OH)3(gibbsite) then component-specie Al+++ no longer appears in the input matrix header and, therefore, the educts in the AlPO4:(H2O)2 (variscite) formation reaction do not match the active components in the input matrix header. An attempt to replace PO4--- with AlPO4:(H2O)2 (variscite) will result in an error message, forcing the user to start all over.

If the first replacement is PO4--- with AlPO4:(H2O)2 (variscite) then component-specie PO4--- no longer appears in the input matrix header. The active component-species listed in the input matrix header are Al+++ and H+, the educts in the Al(OH)3(gibbsite) formation reaction match the active components in the input matrix header. An attempt to replace Al+++ with Al(OH)3(gibbsite) is accepted. The resulting input matrix appears in Table 5.21.

The third component-specie H+ was assigned mode(H+) = total and a 0.00E+00 initial concentration in Step 6, consistent with a system with zero degree of freedom, *F*.

Step 3. Evaluate and interpret the simulated solubility results.

The simulation predicts this 3-phase system—with zero degree of freedom *F*—will be pH 5.33 at equilibrium. Any displacement of solution pH to higher or lower values requires an additional

TABLE 5.21 *ChemEQL* Input Table for a 3-Component, 3-Phase System for Simulating the Chemistry of a Solution in Equilibrium With Both Gibbsite and Variscite

Component	Mode	Initial Concentration (mol dm^{-3})
Al(OH)3(gibbsite)	Solid-phase	1.0
AlPO4:(H2O)2 (variscite)	Solid-phase	1.0
H+	Total	0.00E+00

component thereby increasing the degrees of freedom F to a value greater than zero. The phosphate concentration 4.83×10^{-6} mol dm^{-3} is more than tenfold higher than the aluminum concentration 1.39×10^{-7} mol dm^{-3}. This is a result of the "common ion" effect introduced in general chemistry.

5.4.4 Interpreting Water Chemistry Simulations

5.4.4.1 Reaction Quotients and Saturation Indices

The *ion activity product* (IAP) for gibbsite dissolution (5.R13) appears following. The IAP is computed from the activities of specific solutes in the reaction expression and can assume any positive numerical value for a system not at equilibrium.

$$\text{IAP} = a_{\text{Al}^{3+}} \cdot a_{\text{H}^+}^{-3} \neq \exp\left(-\Delta_r G^{\circ} / R \cdot T\right)$$

When gibbsite is in solubility equilibrium with aqueous solution, the IAP is numerically equal to the thermodynamic equilibrium constant K_s°.

$$K_s^{\circ} = a_{\text{Al}^{3+}} \cdot a_{\text{H}^+}^{-3} = \exp\left(-\Delta_r G^{\circ} / R \cdot T\right)$$

The solution may reach saturation as the IAP rises from an initial state of under-saturation[15]: $IAP < K_s^{\circ}$. A mineral may approach saturation by precipitating with IAP descending from an initial state of over-saturation: $IAP > K_s^{\circ}$.

The *saturation index* (SI) reveals whether solubility equilibrium exists between a mineral and the surrounding solution. The SI is defined as the ratio of the IAP to the equilibrium solubility constant K_s°.

$$\text{SI} \equiv \frac{\text{IAP}}{K_s^{\circ}} \tag{5.39}$$

[15] The IAP for a solubility reaction will increase in value as it advances from under-saturation to saturation if the reaction defining IAP lists the solid-phase as an educt.

If we recognize dissolution and precipitation rates diminish as saturation is approached, many chemists consider a solution *effectively saturated* if the SI is tenfold over-saturated or tenfold under-saturated.

Effective Saturation

$$10^{-1} \leq SI \leq 10^{+1}$$

Many chemists use the base-10 logarithm of the SI ratio: $\log_{10}(SI)$. In this form, a solution is *effectively saturated* if $\log_{10}(SI) = \pm 1$.

Effective Saturation

$$-1 \leq \log_{10}(SI) \leq +1$$

In some instances the IAP may have to exceed an over-saturation threshold (SI > 10) before a mineral will nucleate, the first step in crystallization. Furthermore, the solubility of extremely small, poorly-ordered crystals is often higher than large, well-crystallized particles. *Ostwald ripening* (Ostwald, 1896, 1897) occurs when small crystals dissolve at the expense of large crystals, gradually lowering solubility to a point independent of crystal size—the equilibrium saturated solution. Variations in chemical composition and crystallinity alter the apparent solubility of naturally occurring minerals (May et al., 1986), a further justification for applying the *effective saturation* definition.

Environmental chemists can identify minerals in solubility equilibrium with aqueous solution by performing a water chemistry simulation where precipitation is suppressed. This would mean not employing the Insert Solid Phase ... command under the *ChemEQL* Matrix menu after compiling the input matrix. If the output data file lists ion activities, saving the file an Excel document will allow the user to compute IAP and SI values for likely mineral candidates. The number of minerals potentially saturating the solution must obey the *Gibbs Phase Rule*.

Example 5.11 illustrates the use of saturation indexes SI to identify minerals controlling water chemistry. The water chemistry analyses come from a study of river and groundwater in the karst landscape of northern Florida (Crandall et al., 1999).

ChemEQL provides another method for assessing whether the saturation index SI of a particular mineral indicates under-saturation, saturation (solubility equilibrium), or over-saturation. This is described under the heading *Check for Precipitation* (cf. *ChemEQL Manual* permalink: soilenvirochem.net/U4gkv2.pdf). First, identify a candidate mineral and use the Insert Solid Phase... command to replace one of the solution components with the candidate mineral. This replaces the original component-specie with a solid-phase specie (i.e., a mineral) and sets solidPhase mode with a solid-phase activity of 1.0.

The user can change the mode from solidPhase to checkPrecip mode when the input matrix is displayed in the Matrix window, following instructions from the *ChemEQL Manual*. Finally, the user enters the water analysis concentration of the original component-specie below the checkPrecip mode listing. The resulting output data file will indicate whether the water analysis concentration of the solid-phase specie is over- or under-saturated by the results reported under In or Out of System.

EXAMPLE 5.11

Example Permalink

http://soilenvirochem.net/nNdZr2

This example simulates the chemistry of water samples collected in August and November 1995 when the Suwannee River was at low-flow stage and in February and April 1996 when the river was a high-flow stage.

Crandall et al. (1999) collected water samples from Wingate Sink, Suwannee County, Florida. Wingate Sink (LAT 30.03850° N, LON 82.99254° W) is visible in the US Geological Survey *O'Brien, FL* topographic map (*MRC: 30082A8*) just south of Florida County Road 349, 5.2 km west of O'Brien.

TABLE 5.22 Wingate Sink Water Samples: Sampling Dates and Water Analyses

Date	04-AUG-95	02-NOV-95	07-FEB-96	11-APR-96
Analyte	$mg\,kg^{-1}$	$mg\,kg^{-1}$	$mg\,kg^{-1}$	$mg\,kg^{-1}$
Calcium	60	65	59	40
Potassium	0.9	2.5	0.5	0.9
Magnesium	5.6	5.7	5.9	5.2
Sodium	2.7	2.7	2.6	3.3
Iron	8	< 3	< 3	150
Chloride	5.5	5	4.9	5.9
Bicarbonate	217	190	188	143
Sulfate	10	10	9.1	8.1
Silica	6.3	6.7	6.7	6.3
Nitrate-N	1.9	1.7	1.5	0.68
pH	7.31	7.55	7.33	7.39

Cf. Table 2 in Crandall et al. (1999).

Part 1. Convert the mass-concentration units listed in Table 5.22 to molar units compile *ChemEQL* input matrixes for each water sample date and run water chemistry simulation at the pH values reported by Crandall et al. (1999).

The following components are most likely controlled by mineral solubility: Ca++, Mg++, Fe+++, HCO3-, and SiO2(aq). A thorough analysis would calculate the saturation indexes SI for every mineral in the *ChemEQL* containing these components in their composition. Table 5.23 lists the saturation indexes SI for the most likely candidate minerals as reported by Crandall et al. (1999): calcite, DOLOMITE : $CaMg(CO_3)_2(s)$, QUARTZ : $SiO_2(s)$, and FERRIHYDRITE : $Fe(OH)_3(s)$.

TABLE 5.23 Saturation Indexes SI for Select Minerals Based on Wingate Sink Water Samples

Date	pH	Calcite	Dolomite	Ferrihydrite	Quartz
04-AUG-95	7.31	−0.05	−0.79	2.36	0.00
02-NOV-95	7.55	0.18	−0.35		0.03
7-FEB-96	7.33	−0.10	−0.84		0.03
11-APR-96	7.39	−0.29	−1.12	3.66	0.00

Cf. Table 2 in Crandall et al. (1999).

Part 2. Evaluate and interpret simulated water chemistry results using the pH values reported by Crandall et al. (1999).

Crandall et al. (1999)—using WATEQFP to simulate water chemistry for all samples in their study—found Wingate Sink water samples plot near calcite saturation regardless of flow stage.

"Saturation indices of water from the Suwannee River, Little River Springs, Wingate Sink …indicate under-saturation with respect to calcite and dolomite during high-flow conditions. Calcite SI values decreased from …from 0.17 to −0.30 in Wingate Sink …during high flow."

The saturation indexes SI for every water sample (Table 5.23) indicate calcite, dolomite, and quartz control calcium, magnesium, and silica solubility, confirming the simulation results reported by Crandall et al. (1999). Iron(III) solubility ranges from 200- to 4500-fold over-saturated, suggesting the groundwater flowing through Wingate Sink may contain significant amounts of iron(II).

A report by the Florida Center for Instructional Technology (FCIT, 2008) lists 55 sinkholes in Suwannee County, Florida[16] that appeared between 1969 to 2007, a formation rate of about one new sinkhole per year. Although Crandall et al. (1999) found groundwater calcite saturation varies over a narrow range throughout seasonal changes in Suwannee River stage and discharge, karst erosion is clearly ongoing through this landscape.

[16] Wingate Sink not included in this list.

5.4.4.2 *Logarithmic Activity Diagrams*

Figs. 5.5 through 5.10 plot solute activities using a logarithmic scale on both axes in what is commonly called an activity diagram. The use of activity diagrams is illustrated in Lindsay (1979) and Schwab (2005). The equilibrium solubility expression is converted into a linear expression by taking the logarithm of the equilibrium constant K_s° and the activity quotient in the matter demonstrated in Section 5.3.1. for the gibbsite solubility expression.

The log-log plot in Fig. 5.5 (cf. Example 5.6) shows the pH-dependent activity of several aluminum solution species, but the activity diagrams used to identify minerals controlling

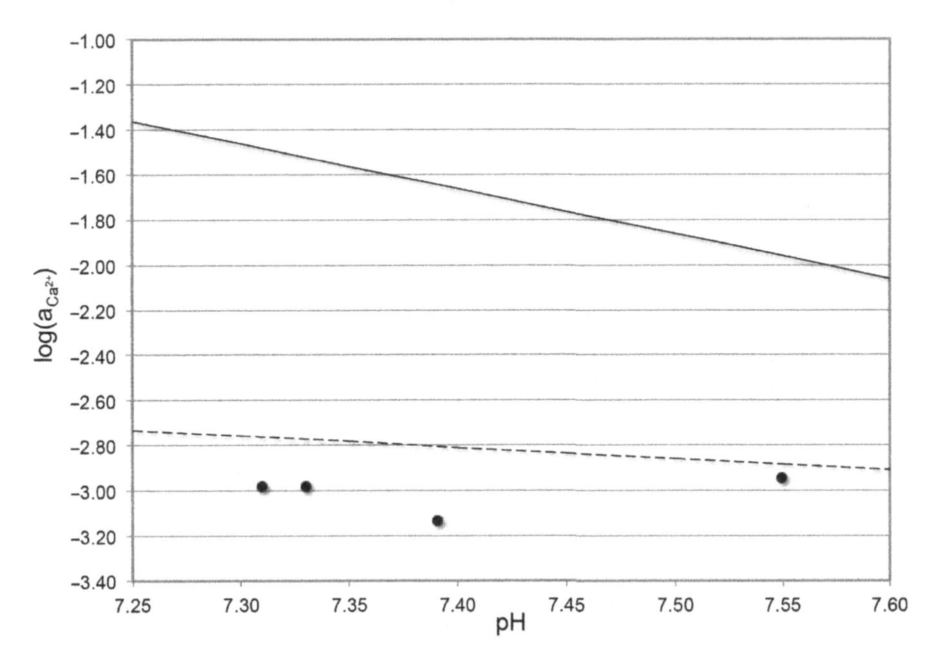

FIG. 5.7 This graph plots the logarithm of Ca^{2+} activity as a function of pH. The filled-circle symbols represent Wingate Sink water samples (Crandall et al., 1999) as simulated using *ChemEQL*. The *solid line* plots $\log_{10}(a_{Ca^{2+}})$ for a 3-phase system: solution, CALCITE and a gas phase containing $p_{CO_2} = 3.64 \times 10^{-4}$ *atm*. The *dashed line* plots $\log_{10}(a_{Ca^{2+}})$ for a 2-phase system: solution and CALCITE.

solute activities in environmental water samples typically plot one activity function; a single straight line for each plausible solubility-controlling mineral.

Numerical water chemistry applications play a critical role in graphical solubility assessment. After running solution species simulation the user plots select specie activities used as variables in the logarithmic activity diagram. If data points representing solute species in the water sample plot on or near the logarithmic activity expression representing a particular mineral, it is likely this mineral controls solute activities in the water sample. Figs. 5.7 through 5.10 illustrate the use of logarithmic activity diagrams to identify mineral phases the chemistry of natural water samples.

Fig. 5.7 plots the activity of cation specie $Ca^{2+}(aq)$ from the Wingate Sink simulations discussed in Example 5.11. The *dashed line* represents calcite solubility for a 2-phase (solution-calcite) system. Crandall et al. (1999) estimate 65% of the water flowing through Wingate Sink at peak flow is surface water from the nearby Suwannee River. The activity diagram (Fig. 5.7) confirms the calcite saturation index SI results in Example 5.11: the groundwater remains near calcite saturated throughout the years regardless of the seasonal mix of limestone aquifer and river water flowing through the caverns. Fig. 5.7, however, reveals a persistent undersaturation expected from an actively eroding karst aquifer.

Fig. 5.8 plots the activity of cation specie $Mg^{2+}(aq)$ from the Wingate water samples (cf. Example 5.11). The linear solubility line represents a three-phase (solution-dolomite-calcite) system, with calcite controlling $Ca^{2+}(aq)$ activity and dolomite controlling $Mg^{2+}(aq)$ activity.

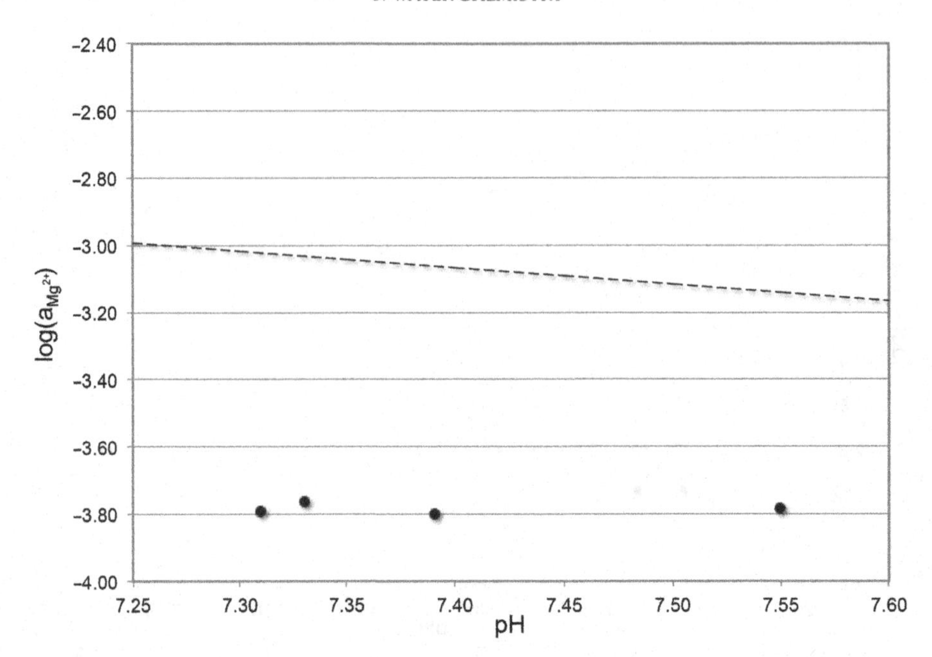

FIG. 5.8 This graph plots the logarithm of Mg^{2+} activity as a function of pH. The filled-circle symbols represent Wingate Sink water samples (Crandall et al., 1999) as simulated using *ChemEQL*. The *dashed line* plots $\log_{10}(a_{Mg^{2+}})$ for a 3-phase system: solution, DOLOMITE and CALCITE (the latter constraining Ca^{2+} activity) in this pH range.

Dolomite under-saturation in this water once again confirms the dolomite SI reported in Example 5.11. Clearly dolomite under-saturation (Fig. 5.8) is greater than calcite under-saturation.

The results in Figs. 5.7 and 5.8 demonstrate a significant point: saturation index SI values condense several variables into a single parameter. Graphical solubility diagrams provide a more comprehensive picture of water chemistry because two or three ion activities can be viewed simultaneously in two- and three-dimensional graphic plots.

Fig. 5.9 plots the activity of neutral silica specie $H_4SiO_4(aq)$ from the Wingate water samples (cf. Example 5.11). The linear logarithmic activity expression for quartz solubility is pH independent and passes through or very near simulated $H_4SiO_4(aq)$ activities from Wingate Sink. Quartz is the most insoluble of silicon dioxide minerals listed in the *ChemEQL* database.

Fig. 5.10 plots the activity of cation specie $Fe^{3+}(aq)$ from the Wingate water samples (cf. Example 5.11). The activity diagram plots two solubility lines for two candidate iron(III) hydrous oxide minerals representing the limits of anticipated iron(III) solubility: GOETHITE : $\alpha-FeOOH(s)$ and FERRIHYDRITE : $Fe(OH)_3(s)$. Water flowing through Wingate Sink is clearly over-saturated relative to ferrihydrite, the most soluble hydrous oxide iron(III) mineral. This result is intriguing but easy to explain.

Water draining through soils represents one recharge source for the karst aquifer in this landscape. The downward discharge of soil moisture combined with relatively high dissolved

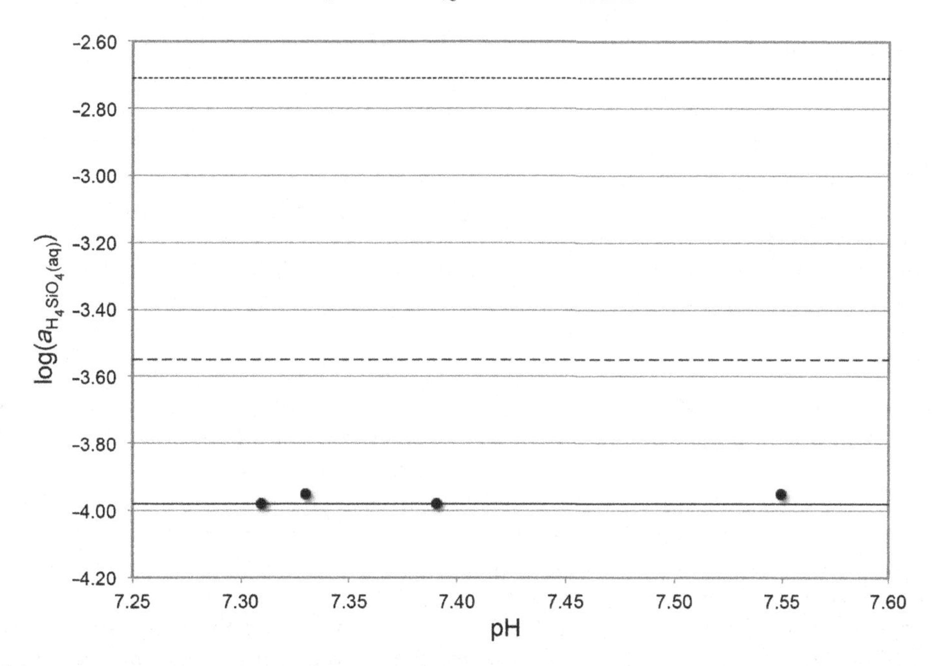

FIG. 5.9 This graph plots the logarithm of H_4SiO_4(aq) activity as a function of pH. The filled-circle symbols represent Wingate Sink water samples (Crandall et al., 1999) as simulated using *ChemEQL*. The *solid line* plots $\log_{10}(a_{H_4SiO_4(aq)})$ constrained by QUARTZ : SiO_2(s) solubility, the *dot-dashed line* indicates CHALCEDONY : SiO_2(s) solubility, and the *dashed line* indicates OPAL : SiO_2(s) solubility.

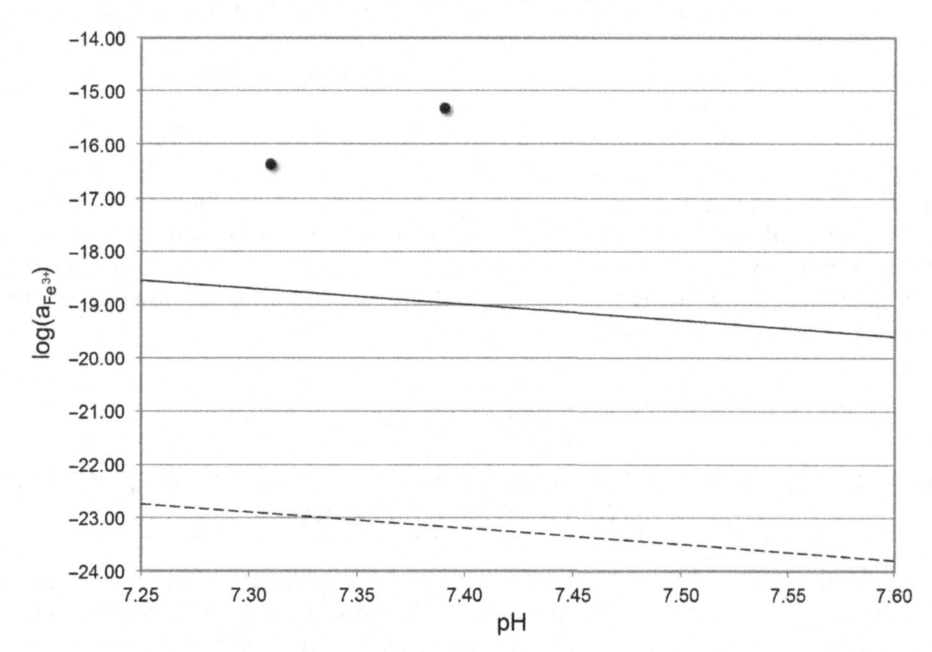

FIG. 5.10 This graph plots the logarithm of Fe^{3+}(aq) activity as a function of pH. The filled-circle symbols represent Wingate Sink water samples (Crandall et al., 1999) as simulated using *ChemEQL*. The *solid line* plots $\log_{10}(a_{Fe^{3+}})$ constrained by FERRIHYDRITE : $Fe(OH)_3$(s) solubility, and the *dashed line* indicates GOETHITE : α-FeOOH(s) solubility.

organic carbon contents in both the karst aquifer and the Suwannee River are the likely source of anaerobic conditions, characterized by high dissolved iron(II) concentrations.

Dioxygen $O_2(g)$ from the above-ground atmosphere dissolving in these waters would rapidly oxidize iron(II) to iron(III)—causing insoluble iron(III) hydrous oxide minerals to precipitate—were it not for what is apparently persistent anaerobic conditions. The water chemistry indicates precipitation of iron(III) hydrous oxide minerals is underway. A more careful chemical analysis of the water, capable of distinguishing dissolved iron(II) and iron(III), would probably result in a significant reduction in the over-saturation estimate.

5.4.5 Summary

This chapter discussed the thermodynamic basis for chemical equilibrium and the equilibrium constant K°. In most water chemistry problems we need to adjust concentrations using activity coefficients $\gamma_{z\pm}$, otherwise the *ion concentration product* will yield an equilibrium coefficient K^{c} whose numerical value depends on the concentration of dissolved ions in solution—the *ionic strength* I_c. Chemists have employed a handful of empirical expression to calculate ion activity coefficients (all of the water chemistry simulations discussed in this chapter are based on Davies (1938) equation).

General chemistry introduced several methods for solving simple equilibrium problems (e.g.,the RICE Table method) but few of these methods are practical when faced with actual water chemistry problems. The same numerical methods apply to complex systems with numerous mass balance expressions and a host of equilibrium species-formation expressions, but computer-based numerical methods are required to efficiently and reliably solve complex equilibrium systems. Your knowledge of chemistry and chemical equilibrium comes into play whenever you assess the validity of any water chemistry simulation. We covered the basic steps of a validity assessment, and you are strongly encouraged to develop the habit of validating any computer simulation.

The final topic we covered addresses a central issue in any simulation of natural water samples: the identification of minerals that control water chemistry. A simple rule—the Gibbs Phase Rule—places an upper limit on the number of mineral phases that can actively control water chemistry. With this rule in hand, you can use water chemistry simulations to compute the saturation index SI for candidate minerals or plot ion activities from your simulation in solubility diagrams. Either method is a suitable strategy for identifying minerals controlling the chemistry of the water samples.

APPENDICES

5.A EQUILIBRIUM CONSTANTS AND ACTIVITY COEFFICIENTS: NOTATION AND UNITS

This appendix summarizes the formal notation for equilibrium constants, equilibrium quotient expressions, and activity coefficients. The chemical notation adopted in this book follows guidelines recommended by the *International Union of Pure and Applied Chemistry* (IUPAC) (Cohen et al., 2007), supplemented by Solomon (2001) and other sources.

First, all equilibrium constants (e.g., $K^{\circ}, K^{\circ}_{A/B}, k^{\circ}_{H,X}$), equilibrium coefficients (e.g., $K^{c}, K^{c}_{A/B}, k^{pc}_{H,X}, H^{cp}_{X}$), reaction-quotient coefficients (e.g., Q^{c}) and activities (e.g., $a_{A^{a\pm}}, a_{x,B}, a_{c,B}$) are *unitless* by definition. This means all terms appearing in any equilibrium quotient or reaction-quotient expression must also be *unitless*.

Consider the following water chemistry reactions involving gaseous dihydrogen sulfide and the hydrolysis of aqueous dihydrogen sulfide to hydrogen sulfide anion.

$$H_2S(g) \longleftrightarrow H_2S(aq)$$

$$H_2S(aq) \longleftrightarrow H^+(aq) + HS^-(aq)$$

The equilibrium constant, equilibrium constant expression, and activity coefficients for these two reactions appear below; f_{H_2S} and ϕ_{H_2S} denotes the *fugacity* and *fugacity coefficient* of dihydrogen sulfide gas, respectively (Appendix 5.B).

$$k^{\circ}_{H,H_2S} = \frac{f_{H_2S}}{c_{H_2S}} = \frac{\phi_{H_2S} \cdot (p_{H_2S}/p^{\circ})}{\gamma_{c,H_2S} \cdot (c_{H_2S}/c^{\circ})}$$

$$K^{\circ}_{a1} = \frac{a_{H^+} \cdot a_{HS^-}}{a_{H_2S}} = \frac{\gamma_{1\pm} \cdot (c_{H^+}/c^{\circ}) \cdot \gamma_{1\pm} \cdot (c_{HS^-}/c^{\circ})}{\gamma_{H_2S} \cdot (c_{H_2S}/c^{\circ})}$$

The following three expressions define the standard states for molar and molal solution-phase concentration units and gas-phase pressure units.

$$\text{Molar standard state: } c^{\circ} = 1\,\text{mol dm}^{-3}$$

$$\text{Molal standard state: } b^{\circ} = 1\,\text{mol kg}^{-1}$$

$$\text{gas-phase standard state: } p^{\circ} = 1\,atm$$

Whenever equilibrium quotient expressions are written using molar concentration ratios, the equilibrium constants should include a following superscript *c* (if molal concentration units are used, the following superscript should be *b*).

$$H^{cp}_{H_2S} = \frac{(c_{H_2S}/c^{\circ})}{(p_{H_2S}/p^{\circ})} = \left(k^{pc}_{H,H_2S}\right)^{-1}$$

$$K^{c}_{a1} = \frac{(c_{H^+}/c^{\circ}) \cdot (c_{HS^-}/c^{\circ})}{(c_{H_2S}/c^{\circ})}$$

Second, and more significant, general chemistry textbooks typically list solute concentrations terms (e.g., c_{H_2S}) rather than *concentration ratios*: $c_{H_2S}/c^{\ominus}$. The standard state defined for solutes is: $c^{\ominus} \equiv 1$ mol dm^3. As a consequence the *unitless* concentration ratio $c_{H_2S}/c^{\ominus}$ is numerically equal to c_{H_2S}. This numerical equivalence is the reason most textbooks chose to list concentrations rather than concentration ratios in equilibrium quotient expressions.

Section 5.2.1 entitled *Concentrations and Activity* discusses the most common used empirical ion activity coefficient expressions. Solomon (2001) uses the unitless concentration ratio $I_c/c^{\ominus}$ when writing empirical ion activity coefficient expressions because ion activity coefficients $\gamma_{\pm}$ are *unitless*. The rationale for using $I_c/c^{\ominus}$ in ion activity coefficient expressions is the same as for using concentration ratios $c_{A^{a\pm}}/c^{\ominus}$ and pressure ratios $p_A/p^{\ominus}$ in reaction and equilibrium quotient expressions.

5.B FUGACITY: THE CARBON DIOXIDE CASE

Real gases differ from an ideal gas by registering a lower pressure under fixed conditions than the pressure calculated using the *Ideal Gas Law*. This appendix illustrates the fugacity concept by comparing the behavior of $CO_2(g)$ with that of an ideal gas under identical conditions (Example 5.B.1).

EXAMPLE 5.B.1

Example Permalink

http://soilenvirochem.net/HbCFaV

The first step is to calculate the volume occupied by 1 *mole n* of an ideal gas at 0°C and 1 atm. This calculation applies the *Ideal Gas Law*.

$$V = \frac{n \cdot R \cdot T}{p_{\text{ideal}}}$$

The standard temperature 25°C corresponds to an absolute temperature of 273.15 K. The standard pressure quoted as 1 atm is equal to:1.01325 × 10^5 *Pa* in SI units. Finally, the *ideal gas constant R* in SI units appropriate for calculation is: 8.3144621 Pa m^3 mol^{-1} K^{-1}.

$$V = \frac{1 \cdot (8.3144621) \cdot (273.15)}{1.01325 \times 10^5} = 2.241 \times 10^{-2}\ \text{m}^3$$

The final step calculates the effective pressure of CO_2 using the *van der Waals equation of state*. The van der Waals equation of state offers a simple method to simulate CO_2 fugacity under identical conditions quoted above.

$$f_{CO_2} = \left(\frac{R \cdot T}{(V/n) - b}\right) - \left(\frac{a \cdot n^2}{V^2}\right)$$

The fugacity is calculated using the volume occupied by 1 mole of ideal gas at 0°C (i.e., 298.15 K): $V = 2.241 \times 10^{-2}\ \mathrm{m}^3$.

The van der Waals coefficients (Weast, 1972) for $CO_2(g)$ are: $a = 0.3640\ \mathrm{m}^6 \cdot Pa \cdot \mathrm{mol}^{-2}$ and $b = 4.270 \times 10^5\ \mathrm{m}^3\ \mathrm{mol}^{-1}$.

$$f_{CO_2} = \left(\frac{(8.3144621)\cdot(273.15)}{(2.241 \times 10^{-2}) - (4.270 \times 10^5)}\right) - \left(\frac{0.3640}{(2.241 \times 10^{-2})^2}\right)$$

$$f_{CO_2} = 1.00794 \times 10^5$$

The $CO_2(g)$ fugacity coefficient ϕ_{CO_2} equals 0.995.

$$\phi_{CO_2} = \frac{f_{CO_2}}{p_{ideal}} = \frac{1.00794 \times 10^5\ Pa}{1.01325 \times 10^5\ Pa} = 0.995$$

Scientists have derived empirical expressions that yield more accurate estimates of $CO_2(g)$ fugacity but this simple exercise should be sufficient to explain the difference between pressure and fugacity and to reveal the nearly ideal behavior of real gases under ambient environmental conditions.

5.C *CHEMEQL* INPUT AND OUTPUT FILE FORMATS

5.C.1 Understanding the *ChemEQL* Input Matrix: Components and Species

The reader can find instructions in the *ChemEQL Manual* on how set up an input matrix (cf. Set up a matrix).

ChemEQL Users Manual Permalink

http://soilenvirochem.net/U4gkv2

The Select Components command window (cf. Fig. 5.C.1) appears when the user selects the Access Library ... command from the File drop-down menu. Listed under the heading ChemEQL Library on the left are all components in the database (Fig. 5.C.1). All solution and mineral (solid-phase) species are derived from this components list through formation reactions where each specie appears as the product and components appear as educts.

Every database used by the major numerical water chemistry applications employs a similar database structure: each specie (solutes are solution species, minerals are solid-phase species) is defined by a formation reaction along with a corresponding equilibrium constant. Despite notable differences, these databases consistently identify the *product* of these formation reactions as a specie.

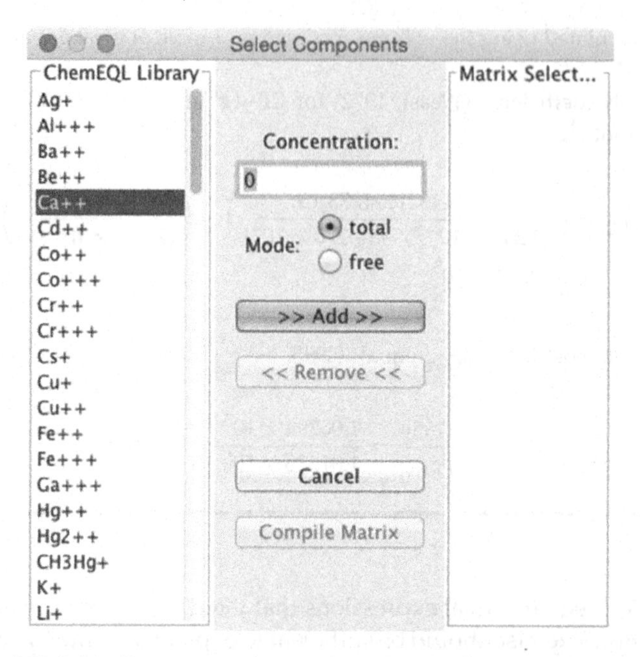

FIG. 5.C.1 The *ChemEQL* Select Components command window displays a scrolling component list on the left. The Concentration input field, Mode radio-buttons, and the Add button appear in the center. Once a component is selected, the concentration entered and model selected, clicking the Add button lists the component is transfer from the list on the left to the list on the right. Click the Compile Matrix button (center bottom) when all components have been entered.

Consider the two following specie formation reactions, the product of the first is a neutral ion-pair solute (i.e., a solution specie) while the product of the second is a mineral (i.e., a solid-phase specie) know as anhydrite.

$$Ca +++ SO4 -- <==> CaSO4(aq)$$
$$Ca +++ SO4 -- <==> CaSO4(Anhydrite)$$

The same two species—Ca++ and SO4 ---—appear as educts in both reactions. It should be obvious the only difference between educt species and product species is in where they appear in the formation reaction. *ChemEQL* uses the term component when referring to an educt specie.[17]

ChemEQL allows users to display (and edit[18]) the *ChemEQL* database components list by selecting either the Edit Regular Lib. Components ... command or the Edit Solid Phases Lib. Components ... command from the Libraries drop-down menu.

[17] *PHREEQC* (Parkhurst and Appelo, 2013) and *EQ3/6* (Wolery and Daveler, 1992) use term basis-specie while *MINTEQ* (Allison et al., 1991) and *WATEQ4F* (Ball et al., 1979) use term master-specie.

[18] *ChemEQL Manual* provides thorough instructions for adding components to and deleting components from the database using the commands listed under the Libraries menu.

Similarly, the user can display (and edit[19]) the *ChemEQL* database components list by selecting either the Edit Regular Lib. Species ... command or the Edit Solid Phases Lib. Species ... command from the Libraries drop-down menu. *Regular* Species are solution species while *Solid-Phase* Species are mineral species.

Returning to the Select Components input window shown above, a user can enter 0 in the Concentration field if the Mode = total radio button is selected but must enter a nonzero, positive number if the Mode = free radio button is selected. The Mode radio buttons determine how the mass balance constraint is imposed.

The easiest way to explain when to choose Mode = total or Mode = free is through an example (Example 5.C.1).

EXAMPLE 5.C.1

Example Permalink

http://soilenvirochem.net/RvjRNP

Simulate the effect of setting mode(H+) = total and mode(H+) = free when simulating dilute acetate/acetic acid solutions. The solution specie HAcet(aq) formation reaction appear below.

$$\text{Acetate}^{-} + \text{H}^{+} \longleftrightarrow \text{Acetic acid(aq)}$$

Part 1. Simulate the chemistry of a 1.0×10^{-5} mol dm^{-3} acetate/acetic acid solutions where: (1) initial component H+ concentration equals 1.0×10^{-5} mol dm^{-3} and (2) mode(H+) = total.

ChemEQL lists the conjugate base of acetic acid Acet- as a component (i.e., basis-specie or master-specie). The set-up described in the preceding paragraph represents a solution with equal concentrations of acetate and proton: an acetic acid solution.

Fig. 5.C.2 displays the output table for the Part 1 simulation. Note the mode designation and initial concentration for basis-specie H+ (i.e., component H+). A mass-balance summing the concentrations of specie H+ and specie HAcet(aq) equals the initial concentration assigned to basis-specie H+ (i.e., component H+): 1.0×10^{-5} mol dm^{-3}. The equilibrium pH is 5.14.

Part 2. Simulate the chemistry of a 1.0×10^{-5} mol dm^{-3} acetate/acetic acid solutions where: (1) initial component H+ concentration equals 1.0×10^{-5} mol dm^{-3} and (2) mode(H+) = free.

The set-up described in the preceding paragraph represents a solution which constrains the acetate mass balance to equal 1.0×10^{-5} mol dm^{-3} and the *equilibrium* concentration of basis-specie H+ (i.e., component H+) to be precisely: 1.0×10^{-5} mol dm^{-3}. A subtle but profoundly different constraint then the one applied in Part 2.

[19] *ChemEQL Manual* provides thorough instructions for adding solution (and solid-phase) species to and deleting species from the database using the commands listed under the Libraries menu.

Species	Stoich. Matrix	Log K	Conc. [mol/l]	Activity	Log conc.
Acet-	1 0	0.00	7.21E-06	7.19E-06	-5.14
HAcet(aq)	1 1	4.73	2.79E-06	2.79E-06	-5.56
OH-	0 -1	-14.00	1.39E-09	1.39E-09	-8.86
H+	0 1	0.00	7.22E-06	7.22E-06	-5.14
Components	Mode	Initial Conc.	In or out of system		
Acetate-	total	1.00E-05	----		
H+	**total**	1.00E-05	----		

FIG. 5.C.2 Output file for Part 1 as displayed by Excel.

Species	Stoich. Matrix	Log K	Conc. [mol/l]	Activity	Log conc.
Acet-	1 0	0.00	6.51E-06	6.49E-06	-5.19
HAcet(aq)	1 1	4.73	3.49E-06	3.49E-06	-5.46
OH-	0 -1	-14.00	1.00E-09	1.00E-09	-9.00
H+	0 1	0.00	1.00E-05	1.00E-05	-5.00
Components	Mode	Initial Conc.	In or out of system		
Acetate-	total	1.00E-05	----		
H+	**free**	1.00E-05	1.35E-05		

FIG. 5.C.3 Output file for Part 2 as displayed by Excel.

Fig. 5.C.3 displays the output table for the Part 2 simulation. Note the mode designation and initial concentration for basis-specie H+ (component H+). A mass-balance summing the concentrations of specie H+ and specie HAcet(aq) equals the H+ entering the system to meet constraint imposed by setting mode(H+) = free: 1.35×10^{-5} mol dm^{-3}. Notice the equilibrium concentration of specie H+ (cell with heavy outline) equals the initial concentration assigned basis-specie H+. The equilibrium pH is 5.00.

5.C.2 *ChemEQL* Output Data-File Format

An important attribute of *ChemEQL* is output data-file format. When the simulation is complete, a window appears displaying the results. Save the ChemEQL results data file by selecting Save Data ... from the File menu. You will be prompted (Fig. 5.C.4) to name the file (default: z.xls).

This naming convention allows you to open the file using the spreadsheet application Excel. The *ChemEQL* results data file default format (Fig. 5.C.5) is: txt.

Save the *ChemEQL* output data files after opening in Excel by selecting an appropriate workbook format (Fig. 5.C.6).

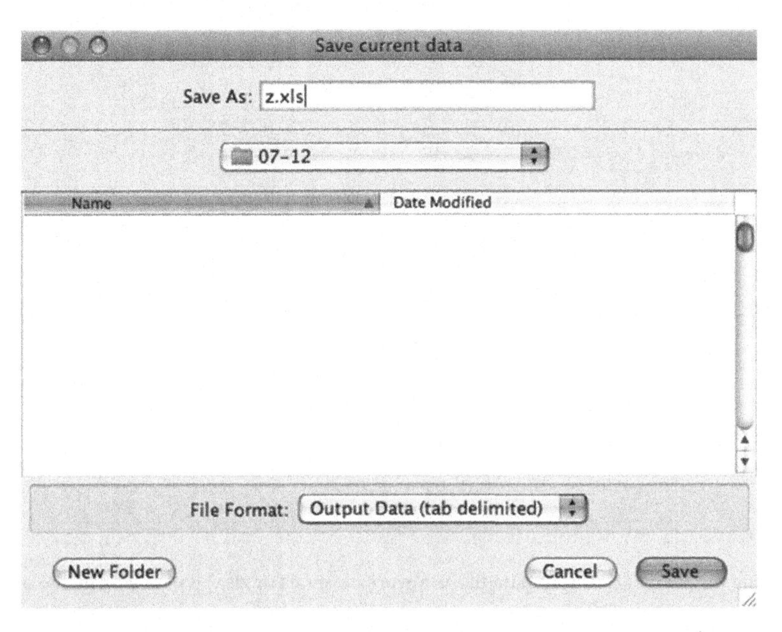

FIG. 5.C.4 Save the ChemEQL results data file by selecting Save Data ... from the File menu. You will be prompted to name the file (default: z.xls). This naming convention allows you to open the file using the spreadsheet application Excel.

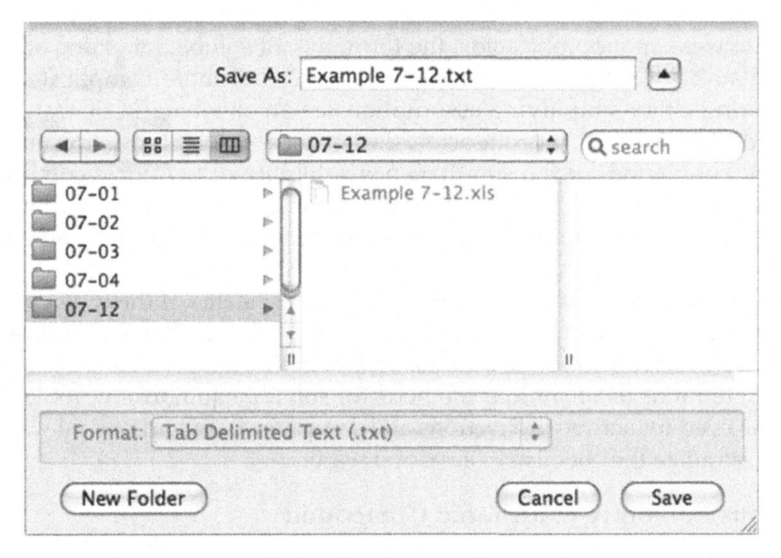

FIG. 5.C.5 Open the *ChemEQL* output file in Excel and select Save As ... from the Excel File menu. The dialog window reveals the actual *ChemEQL* results data file format: txt.

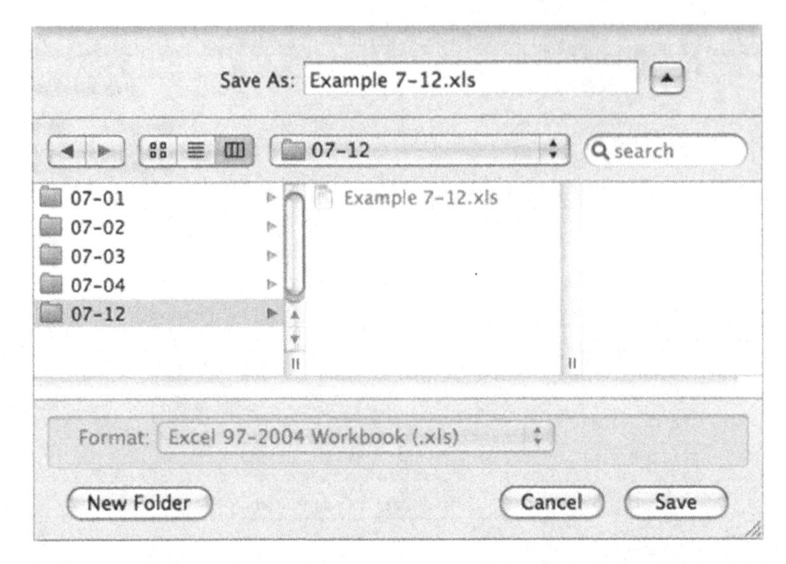

FIG. 5.C.6 Saving the *ChemEQL* output data file as a worksheet (xls or xlsx) will overwrite the original file.

5.D SOLVING EQUILIBRIUM PROBLEMS USING THE *RICE* TABLE METHOD

General chemistry textbooks discuss a variety of simple yet important equilibrium systems: the hydrolysis of weak monoprotic acids, the formation of soluble ion pairs, or the solubility of systems containing a single ionic compound. Somewhat more complicated equilibrium systems—requiring either simplifying assumptions or involved algebraic transformations—include: the hydrolysis of weak diprotic acids, the effect of ion-pair formation on the solubility of a single ionic compound, or the simultaneous solubility of two ionic compounds with a single common ion. The reader must understand: none of these simple equilibrium systems approach the complexity of natural waters commonly encountered by soil and environmental chemists.

In this appendix we review select simple equilibrium systems of the type you would have encountered in general chemistry, pausing to consider the simplifying assumptions used in general chemistry to solve what soon become very complicated problems. This review serves two purposes: it reminds us of familiar methods for solving equilibrium problems and it will demonstrate the need for activity corrections and computer-based numerical water chemistry applications to simulate the chemistry of natural waters.

5.D.1 Aqueous Solubility of an Ionic Compound

The solubility of ionic solids is a typical chemical equilibrium reaction discussed in general chemistry. Solubility is a key water chemistry process involving both the dissolution and precipitation of sparingly soluble minerals during the chemical weathering stage of the rock cycle and soil formation. The chemistry of many inorganic pollutants, as it pertains to both biological availability to living organisms and remediation treatment, involves solubility reactions (Example 5.D.1).

EXAMPLE 5.D.1

Example Permalink

http://soilenvirochem.net/qf0qNS

Calculate the solubility of the sparingly soluble mineral GYPSUM : $CaSO_4\ 2H_2O(s)$ using the RICE table method.

$$\underset{\text{GYPSUM}}{CaSO_4 2H_2O(s)} \longleftrightarrow Ca^{2+}(aq) + SO4^{2-}(l) \tag{5.D.R1}$$

$$K^c_{s0} = c_{Ca^{2+}} \cdot c_{SO_4{}^{2-}}$$

$$K^{\circ}_{s0} = a_{Ca^{2+}} \cdot a_{SO_4{}^{2-}} = \gamma^2_{2\pm} \cdot K^c_{s0}$$

Gypsum is common in arid and semiarid soils where chemical weathering is slowed by rainfall scarcity. Calculate the solubility of gypsum in aqueous solution given: $K^{\circ}_{s0} = 2.63 \times 10^{-5}$ and $\log_{10}\left(K^{\circ}_{s0}\right) = -4.58$. Use the RICE table (Table 5.D.1) to assign values to each component in the equilibrium solubility expression.

TABLE 5.D.1 RICE Table for Gypsum Dissolution

Reaction →	$CaSO_4\ 2H_2O(s)$	↔	Ca^{2+} mol dm^{-3}	$SO_4{}^{2-}$ mol dm^{-3}
Initial			0	0
Reaction			$+x$	$+x$
Equilibrium			x	x

$$K^{\circ}_{s0} = 2.63 \times 10^{-5} = x^2$$
$$a_{Ca^{2+}} = \sqrt{2.63 \times 10^{-5}} = 5.13 \times 10^{-3}$$

Gypsum solubility is sufficiently high to justify computing ion activity coefficients. Assuming the solubility can be used to compute ionic strength we find $I_c \approx 2.052 \times 10^{-2}$ and the ion activity coefficient for both $Ca^{2+}(aq)$ and $SO_4{}^{2-}(aq)$ is: $\gamma_{2\pm} = 0.746$.

$$I_c \approx \frac{\left((+2)^2 \cdot 5.13 \times 10^{-3} + (-2)^2 \cdot 5.13 \times 10^{-3}\right)}{2}$$
$$\gamma_{2\pm} = \exp\left(-0.510 \cdot 2^2 \cdot \sqrt{2.052 \times 10^{-2}}\right) = 0.746$$
$$c_{Ca^{2+}} = \frac{5.13 \times 10^{-3}}{0.746} = 6.87 \times 10^{-3}\ \text{mol dm}^{-3}$$

The following example calculates the solubility of a very insoluble aluminum hydroxide mineral gibbsite. The dissolution of one mole of gibbsite releases four moles of ions into solution: one mole of the ion Al^{3+} and three moles of the hydroxide ion OH^- ((Example 5.D.2)).

EXAMPLE 5.D.2

Example Permalink

http://soilenvirochem.net/wUStKl

Calculate gibbsite solubility in water using the RICE table method.

$$\underset{\text{GIBBSITE}}{Al(OH)_3(s)} \longleftrightarrow Al^{3+}(aq) + 3OH^- \tag{5.D.R2}$$

$$K_{s0}^{c} = c_{Al^{3+}} \cdot c_{OH^-}^{3}$$

$$K_{s0}^{\ominus} = a_{Al^{3+}} \cdot a_{OH^-}^{3} = \gamma_{3\pm} \cdot \gamma_{1\pm}^{3} \cdot K_{s0}^{c}$$

Gibbsite occurs in soils where chemical weathering has depleted most of the original rock-forming minerals. Calculate gibbsite solubility in aqueous solution given: $K_{s0}^{\ominus} = 1.29 \times 10^{-34}$ and $\log_{10}\left(K_{s0}^{\ominus}\right) = -33.89$. Use the RICE table (Table 5.D.2) to assign values to each component in the equilibrium solubility expression.

TABLE 5.D.2 RICE Table for Gibbsite Dissolution

Reaction →	$Al(OH)_3(s)$	↔	Al^{3+} mol dm^{-3}	3 $OH^-(aq)$ mol dm^{-3}
Initial			0	0
Reaction			$+x$	$+3 \cdot x$
Equilibrium			x	$3 \cdot x$

$$K_{s0}^{\ominus} = x \cdot (3 \cdot x)^3 = 27 \cdot x^4$$

$$a_{Al^{3+}} = \left(\frac{1.29 \times 10^{-34}}{27}\right)^{1/4} = 1.48 \times 10^{-9}$$

Gibbsite solubility is too low to justify computing ion activity coefficients.

Gibbsite solubility is much more complicated than represented in the preceding example. A *ChemEQL* simulation of this case reveals the aluminum ion Al^{3+} is actually a complex with six firmly bound water molecules. This complex *hexaquoaluminum(III)* cation $Al(H_2O)_6{}^{3+}$ has an unexpected but crucial characteristic; it is a polyprotic acid that undergoes several hydrolysis steps.

5.D.2 Aqueous Solubility of Gases

Natural water chemistry requires, in many instances, an account of dissolved gaseous. Surface waters and soil pore water are two systems where including dissolved gases in water chemistry simulations is absolutely essential.

The reader might assume groundwater simulations could disregard dissolved gases, yet even in where direct contact between water and a gas phase is not apparent dissolved gases cannot be neglected. For many important gases found in the above ground atmosphere the extremely long groundwater residence time is sufficient for gases to saturate the water. Dissolved dioxygen $O_2(aq)$ levels are often close to saturation, the exception being carbon-rich sediments and aquifers where bacteria consume virtually all dissolved oxygen.

This section covers two examples designed to illustrate how a user simulates natural water with dissolved gases. The first is a concrete example of the *fixed fugacity* method (Delany and Wolery, 1984). The second is the approach required by the *ChemEQL* model.

There is no question the *fixed fugacity* method implemented by most numerical water chemistry applications demands less from the user, the *ChemEQL* approach offers a decided advantage for novice users. *ChemEQL* requires the user to account for the detailed chemistry of gases that undergo hydrolysis when dissolve in water (e.g., carbon dioxide CO_2, dihydrogen sulfide H_2S and ammonium NH_3) (Example 5.D.3).

EXAMPLE 5.D.3

Example Permalink

http://soilenvirochem.net/F8o6pb

Calculate dihydrogen sulfide $H_2S(aq)$ solubility when the gas phase partial pressure is 4.34×10^{-3} *atm*. The Henry's Law gas-solubility constant $H^{cp}_{H_2S} = 9.22 \times 10^{-2}$ *atm* is listed in Table 5.4.

$$H_2S(g) \longleftrightarrow H_2S(aq) \qquad (5.D.R3)$$

$$\left(k^{\ominus}_{H,\,H_2S}\right)^{-1} = \frac{a_{H_2S}}{f_{H_2S}} = \frac{\gamma_{c,H_2S} \cdot c_{H_2S}}{\phi_{H_2S} \cdot p_{H_2S}} \approx H^{cp}_{H_2S}$$

The fugacity coefficient for $H_2S(g)$—calculated using the van der Waals equation of state (cf. Appendix 5.B)—is 0.993, hence we can assume: $\phi_{H_2S} \approx 1$. The dissolved gas $H_2S(aq)$ is a neutral molecule, therefore: $\gamma_{H_2S} \approx 1$.

$$c_{H_2S} = H^{cp}_{H_2S} \cdot p_{H_2S}$$

$$c_{H_2S} = 9.22 \times 10^{-2} \cdot 4.34 \times 10^{-3} = 4.00 \times 10^{-4}\ \text{mol dm}^{-3}$$

The *ChemEQL* model does not implement the *fixed fugacity* (Delany and Wolery, 1984) and requires another approach to simulating water in equilibrium with a limitless gas-phase reservoir. The following example will show aqueous dihydrogen sulfide $H_2S(aq)$ is independent of solution pH and depends solely on the dihydrogen sulfide $H_2S(g)$ partial pressure (Example 5.D.4).

EXAMPLE 5.D.4

Example Permalink

http://soilenvirochem.net/6SGCA0

Calculate the activities of all hydrogen sulfide species and total hydrogen sulfide solubility, over the pH range 4–9 for the dihydrogen sulfide partial pressure of 4.34×10^{-3} *atm*.

This example demonstrates $H_2S(aq)$ solubility depends solely on the dihydrogen sulfide gas partial pressure, entirely independent of solution pH. Solubility of the remaining two aqueous hydrogen sulfide species formed by $H_2S(aq)$ hydrolysis—$HS^-(aq)$, and $S^{2-}(aq)$—are pH dependent. As a consequence, total hydrogen sulfide solubility is pH dependent.

The two hydrolysis steps and their respective hydrolysis constants appear below.

$$S^{2-}(aq) + H_2O(l) \longleftrightarrow HS^-(aq) + OH^-(aq) \quad (5.D.R4)$$

$$K^{\circ}_{b1} = \frac{(\gamma_{1\pm} \cdot c_{HS^-}) \cdot (\gamma_{1\pm} \cdot c_{OH^-})}{\left(\gamma_{2\pm} \cdot c_{S^{2-}}\right)} = 7.94 \times 10^{-1}$$

$$HS^-(aq) + H_2O(l) \longleftrightarrow H_2S(aq) + OH^-(aq) \quad (5.D.R5)$$

$$K^{\circ}_{b2} = \frac{(\gamma_{c,H_2S} \cdot c_{H_2S}) \cdot (\gamma_{1\pm} \cdot c_{OH^-})}{(\gamma_{1\pm} \cdot c_{HS^-})} = 1.05 \times 10^{-7}$$

Rearranging the equilibrium quotient expression for the second base hydrolysis step yields the pH dependent expression for $HS^-(aq)$ solubility, where we apply: $\gamma_{H_2S(aq)} \approx 1$.

$$c_{HS^-} = \frac{(\gamma_{c,H_2S} \cdot c_{H_2S}) \cdot a_{OH^-}}{\gamma_{1\pm} \cdot K^{\ominus}_{b2}} \approx \frac{c_{H_2S} \cdot K^{\ominus}_{w}}{K^{\ominus}_{b2} \cdot \gamma_{1\pm} \cdot a_{H^+}}$$

Combining both hydrogen sulfide hydrolysis steps into a single reaction yields a composite hydrolysis constant $(K^{\ominus}_{b1} \cdot K^{\ominus}_{b2})$ for the formation of $S^{2-}(aq)$ from $H_2S(aq)$.

$$c_{S^{2-}} \approx \frac{c_{H_2S(aq)} \cdot a^2_{H^+}}{\left(K^{\ominus}_{a1} \cdot K^{\ominus}_{a2} \cdot K^{\ominus}_{w}\right) \cdot \gamma_{2\pm}}$$

Table 5.D.3 lists activities for each of the hydrogen sulfide species computed using the equations above.

TABLE 5.D.3 The Activities of Three Hydrogen Sulfide Species Covering the (Abridged) pH Range: 4–5

H2S(aq) mol dm^{-3}	HS- mol dm^{-3}	S- mol dm^{-3}	-LOG(H+) mol dm^{-3}
4.00E−04	3.81E−07	4.80E−17	4.00
4.00E−04	6.04E−07	1.21E−16	4.20
4.00E−04	9.57E−07	3.03E−16	4.40
4.00E−04	1.52E−06	7.61E−16	4.60
4.00E−04	2.40E−06	1.91E−15	4.80
4.00E−04	3.81E−06	4.80E−15	5.00

ChemEQL allows two methods for replicating the *fixed fugacity* method. The user—after calculating the partial pressure-dependent Henry's Law coefficient $k^{\ddagger}_{H,\,H_2S}$ as illustrated in the preceding example—can edit the database, replacing the equilibrium constant $k^{\ominus}_{H,\,H_2S}$ each time a simulation is performed. This method is not recommended for novice *ChemEQL* users.

Alternatively—using results from the preceding example—the user can calculate the pH-independent aqueous dihydrogen sulfide concentration c_{H_2S} for the chosen gaseous dihydrogen sulfide partial pressure p_{H_2S}. The user then prepares the input matrix (File : Access Library ...), selects $H_2S(aq)$ as the hydrogen sulfide component, enter the pH-independent aqueous dihydrogen sulfide concentration, and select free mode. This method—illustrated in the following example—requires defining duplicate components for each gas: H2S(aq) and HS-, CO2(aq) and HCO3-, NH3(aq) and NH4+, etc. (Example 5.D.5).

EXAMPLE 5.D.5

Example Permalink

http://soilenvirochem.net/84Cc6c

Simulate the activities of all hydrogen sulfide species and total hydrogen sulfide solubility, using the *ChemEQL* model, over the pH range 4–9 for the dihydrogen sulfide partial pressure of 4.34×10^{-3} *atm*.

The input file (84Cc6c.cql) sets the mode = free for both components (H2S(aq) and H+).

The mode of a component—as used in *ChemEQL*—requires explanation. A water chemistry simulation always has one or more components, in this example all hydrogen sulfide species are assigned to component H2S(aq) and water species are assigned to component H+, both of which appear in the database components list (cf. menu File : Access Library ...). Once the user selects a component, the input cell labeled Concentration and the Mode radio buttons are activated. This is the stage when the user sets the mass balance constraints on the simulation.

If the total radio button is selected; the simulation mass balance constrains the sum of all species formed from that component to equal the initial concentration entered in the field above the radio buttons. This is the simplest mass balance constraint.

The free mode constrains mass balance in a different way. If the user selects mode = free the initial concentration entered in the field above the radio buttons is assigned to the *specific specie* whose label is identical to the selected component. For example, if the selected component is H2S(aq) then choosing the free radio button constrains the final concentration of *specie* H2S(aq) to be identical to the initial concentration. Constraining specie H2S(aq) to remain constant, regardless of pH, is consistent with the intent of this simulation.

TABLE 5.D.4 *ChemEQL* Output Table (Abridged)

H2S(aq) **mol dm^{-3}**	HS- **mol dm^{-3}**	S– **mol dm^{-3}**	-LOG(H+) **mol dm^{-3}**
4.00E−04	3.82E−07	4.81E−17	4.00
4.00E−04	6.05E−07	1.21E−16	4.20
4.00E−04	9.60E−07	3.03E−16	4.40
4.00E−04	1.52E−06	7.62E−16	4.60
4.00E−04	2.41E−06	1.92E−15	4.80
4.00E−04	3.82E−06	4.81E−15	5.00

This simulation applies a second constraint: the pH is constrained to a series of values ranging from 4 to 9 in 0.2 log-unit steps (Table 5.D.4). In effect the output lists results from 26 separate simulations with each simulation running at a pre-selected pH value. Performing a simulation series such as this requires designated mode = free for component H+. *ChemEQL* also requires a nonzero concentration when a component is designated mode = free. In this case the initial H+ is a place-holder because the value appearing in the input matrix is overwritten when pH range ... is selected from the Options menu prior to running the simulation (cf. *ChemEQL Manual* : permalink: soilenvirochem.net/U4gkv2.pdf).

5.E VALIDATING WATER CHEMISTRY SIMULATIONS

Table 5.2 contains a checklist for validating computer-based water chemistry simulations. The list is not intended to be comprehensive. Consider it a framework that you can add to as your experience increases and you become more familiar with water chemistry simulations. Any practitioner should develop the habit of routinely validating the data used to simulate the process being studied and specific simulation results.

5.E.1 Charge Balance Validation

This assessment evaluates the solution charge-neutrality condition. It determines whether the charge sum over all cation species equals the charge sum over all anion species in solution. Computing this sum requires multiplying the concentration of each ion $c_{A^{a\pm}}$ [mol dm^{-3}] times its valence $\pm z_A$ to convert molar concentration to *moles of charge* concentration. The charge-balance condition is satisfied when a sum over all cation charges plus the sum over all anion charges is equal to zero to within rounding error.

$$\sum_i \underset{Cations}{(z_i \cdot c_i)} = \sum_i \underset{Anions}{(z_i \cdot c_i)}$$
$$\sum_i \underset{Cations}{(z_i \cdot c_i)} - \sum_i \underset{Anions}{(z_i \cdot c_i)} = 0$$

5.E.2 Mass Balance Validation

Consider the following hydrolysis and complexation reactions involving hydrogen sulfide and cadmium species.

$$\begin{aligned}
H_2S(aq) &\longleftrightarrow H^+(aq) + HS^-(aq)\\
HS^-(aq) &\longleftrightarrow H^+(aq) + S^{2-}(aq)\\
Cd^{2+}(aq) + H_2S(aq) &\longleftrightarrow 2H^+(aq) + CdS(aq)\\
Cd^{2+}(aq) + H_2S(aq) &\longleftrightarrow H^+(aq) + CdHS^+(aq)\\
Cd^{2+}(aq) + 2H_2S(aq) &\longleftrightarrow 2H^+(aq) + Cd(HS)_2(aq)
\end{aligned}$$

There is a separate mass-balance sum for each component—in this examples: proton, sulfide and cadmium—taking into account stoichiometry and composition of each component.

$$\begin{aligned}
C_{proton} &= 2 \cdot \left(c_{H_2S} + c_{Cd(HS)_2}\right) + c_{HS^-} + c_{CdHS^+} + c_{H^+}\\
C_S &= 2 \cdot \left(c_{Cd(HS)_2}\right) + c_{H_2S} + c_{HS^-} + c_{S^{2-}} + c_{CdS} + c_{CdHS^+}\\
C_{Cd} &= c_{Cd^{2+}} + c_{CdS} + c_{CdHS^+} + c_{Cd(HS)_2}
\end{aligned}$$

5.E.3 Ionic Strength Validation

The simulation may or may not record ionic strength I_c in the output file (*ChemEQL* does not), regardless the value of each ion activity coefficient is determined by I_c. If simulated species concentrations recorded in the output file are *comma separated values* that can be opened as spreadsheet document, the I_c sum is best performed using the following expression.

$$I_c = \sum_i \left(\frac{z_i^2 \cdot c_i}{2} \right)$$

5.E.4 Ion Activity Coefficient Validation

Numerical water chemistry applications may allow the user to select the approximation (e.g., Debye-Hückel, Güntelberg, Davies, etc.) and parameters (e.g., default *ChemEQL* value $A_c^{DH} = 0.5$ for 25°C but allows the user to enter an alternative value) required to calculate ion activity coefficients during simulation. *ChemEQL* offers users the option to record concentrations and activities in the output file. Divide the simulated ion activity by the simulated concentration to reveal the simulated ion activity coefficient.

Example 5.4 illustrates ionic strength validation for a solution prepared from acetate, the conjugate base of acetic acid. *ChemEQL* allows the user to omit spectator cations (and spectator anions) which can result in simulations that fail to satisfy the charge-balance condition (cf. Example 5.3) and underestimate I_c. Part 3 of Example 5.4 shows that underestimate I_c generally has negligible effect on simulated ion activity coefficients.

5.E.5 Activity Product Validation

The output from a water chemistry simulation lists the activities of the solution and mineral species formed from the components listed in the input file. Each specie in the model database is defined by its formation reaction and the equilibrium constant K° for the specie formation reaction. In effect there is, for each specie in the output file, an equilibrium quotient expression.

An *ion activity product* or IAP expression is none other than the reaction quotient expression introduced at the beginning of this chapter. The *ion activity product* or IAP is the *numerical value* calculated from a IAP expression (i.e., reaction quotient expression) populated by species activities from an equilibrium calculation—either based on a RICE calculation or a water chemistry simulation.

The following example illustrates activity product validation using a *ChemEQL* simulation of gibbsite solubility (Example 5.E.1).

EXAMPLE 5.E.1

Example Permalink

http://soilenvirochem.net/bU51VB

Simulate the solubility of the aluminum hydroxide mineral gibbsite.

Input matrix bU51VB(A).cql lists mode = free for both components (Al+++ and H+). After compiling the input matrix the menu command Matrix : Insert Solid Phase ... replaces solution component Al+++ with solid-phase component Al(OH)3 (Gibbsite) as appears in input matrix bU51VB(B).cql.

TABLE 5.E.1 *ChemEQL* Output Table (Abridged) for Simulated Gibbsite Solubility

Species	Concentration (mol dm^{-3})	Activity (unitless)	Activity Coefficient (unitless)
Al+++	2.02E−13	1.82E−13	0.901
AlOH++	1.70E−11	1.62E−11	0.955
Al(OH)2+	1.16E−09	1.15E−09	0.989
Al(OH)3(aq)	1.62E−09	1.62E−09	1.000
Al(OH)4-	2.32E−08	2.29E−08	0.988
OH-	9.02E−08	8.92E−08	0.988
H+	1.12E−07	1.12E−07	1.000

ChemEQL uses numbers to indicate stoichiometry and multiple plus or minus signs to indicate ion charge. For example Al+++ (first species) represents $Al^{3+}(aq)$, Al(OH)2+ (third species) represents $Al(OH)_2^+(aq)$, and Al(OH)4- (fifth species) represents $Al(OH)_4^-(aq)$.

Multiplying ion charge times concentration (second column, Table 5.E.1) and summing the product for all ions yields a charge balance of 2.65×10^{-11} $mol_c\ dm^{-3}$, which represents a negligible 0.11% relative error.

The ionic strength was constrained to the value: $I_c = 1 \times 10^{-3}$ mol dm^{-3}. The Davies ionic strength function appearing in the Davies ion activity coefficient expression equals: $f_D\left(I_c/c^{\circ}\right) = 9.93 \times 10^{-3}$. Table 5.E.2 lists validated ion activities (column 3) and ion activity coefficients (column 4) for the simulated species concentrations listed in column 2.

TABLE 5.E.2 Validated Ion Activities and Ion Activity Coefficients for the Major Solution Species (Abridged) From the *ChemEQL* Simulation of Gibbsite Solubility

Species	Concentration (mol · dm^{-3})	Activity unitless	Activity Coefficient unitless
Al+++	2.02E−13	1.82E−13	0.901
AlOH++	1.70E−11	1.62E−11	0.955
Al(OH)2+	1.16E−09	1.15E−09	0.988
Al(OH)4-	2.32E−08	2.29E−08	0.988
OH-	9.02E−08	8.91E−08	0.988
H+	1.12E−07	1.12E−07	1.000

The gibbsite solubility reaction below is as it appears in the *ChemEQL* database. The gibbsite ion activity product IAP computed using activities listed in the table above confirms the simulation is a faithful representation of gibbsite solubility in water.

$$Al^{3+}(aq) \longleftrightarrow Al(OH)_3(gibbsite) + 3 \cdot H^+(aq)$$

$$K^{\circ} = 10^{-8.11} = 7.76 \times 10^{-9}$$

$$IAP = \frac{a^3_{H^+}}{a_{Al^{3+}}} = \frac{\left(1.12 \times 10^{-7}\right)^3}{1.82 \times 10^{-13}} = 7.77 \times 10^{-9}$$

5.E.6 Database Validation

Every numerical water chemistry application uses an equilibrium reaction database listing the formation reaction for each solution and mineral species and the corresponding equilibrium constants. The database defines the scope of each water chemistry application. The user must keep in mind the numerical kernel that performs the simulation and the options offered by the model do not define the scope; this is defined by the database.

The formation reactions are written explicitly using only the components appearing on the master component list. The chemical identity of the components determine how the formation reactions are written and the numerical value of the equilibrium constants assigned to each formation reaction.

A careful practitioner will check the veracity of equilibrium constants appearing in the database, especially for those reactions most critical for the chemistry being simulated. There are cases where species listed in a database are not fully endorsed by the scientific community. In other cases the database lists empirical K^c coefficients measured at a particular ionic strength and have not been adjusted to the correct K° value (cf. desferrioxiamine-B hydrolysis and complexation reactions: Martell et al., 2004; Hernlem et al., 1996).

References

Allison, J.D., Brown, D.S., Novo-Gradac, K.J., 1991. MINTEQ2/PRODEFA2, A Geochemical Assessment Model for Environmental Systems: Version 3.0 User's Manual, EPA/600/3-91/021; Order No. PB91-182469. Environ. Res. Lab., 117 pp.

Ball, J.W., Jenne, E.A., Nordstrom, D.K., 1979. WATEQ2—a computerized chemical model for trace major element speciation and mineral equilibria of natural waters. ACS Symp. Ser. 93, 815–835.

Cohen, E.R., Holmstrom, B., Mills, I., Stohner, J., Cvitas, T., Kuchitsu, K., Pavese, F., Strauss, H.L., Thor, A.J., Frey, J.G., Marquardt, R., Quack, M., Takami, M., et al., 2007. Quantities, Units and Symbols in Physical Chemistry, third ed. Royal Society of Chemistry, Cambridge, UK, 233 pp.

Crandall, C.A., Katz, B.G., Hirten, J.J., 1999. Hydrochemical evidence for mixing of river water and groundwater during high-flow conditions, lower Suwannee River basin, Florida, USA. Hydrogeol. J. 7 (5), 454–467.

Davies, C.W., 1938. The extent of dissociation of salts in water. VIII. An equation for the mean ionic activity coefficient of an electrolyte in water, and a revision of the dissociation constants of some sulfates. J. Chem. Soc. 2093–2098.

Debye, P., Hückel, E., 1923. Zur Theorie der Elektrolyte. I. Gefrierpunktserniedrigung und verwandte Erscheinungen. Phys. Z. 24, 185–206.

Delany, J.M., Wolery, T.J., 1984. Fixed Fugacity Option for the EQ6 Geochemical Reaction Path Code, UCRL-53598; Order No. DE85016345. Lawrence Livermore Natl. Lab., 20 pp.

Dick, J.M., 2008. Calculation of the relative metastabilities of proteins using the CHNOSZ software package. Geochem. Trans. 9, 1–17.

FCIT, 2008. Sinkholes. University of South Florida, Tampa, FL.

Guntelberg, E., 1926. Untersuchungen über Ioneninteraktion. Z. Phys. Chem. 123, 199–247.

Hernlem, B.J., Vane, L.M., Sayles, G.D., 1996. Stability constants for complexes of the siderophore desferrioxamine B with selected heavy metal cations. Inorg. Chim. Acta 244 (2), 179–184.

Lindsay, W.L., 1979. Chemical Equilibria in Soils. Wiley, New York, 449 pp.

Martell, A.E., Smith, R.M., Motekaitis, R.J., 2004. NIST Critically Selected Stability Constants of Metal Complexes Database. Standard Reference Data Program, National Institute of Standards and Technology, U.S. Dept. of Commerce, Gaithersburg, MD.

May, H.M., Kinniburgh, D.G., Helmke, P.A., Jackson, M.L., 1986. Aqueous dissolution, solubilities and thermodynamic stabilities of common aluminosilicate clay minerals: kaolinite and smectites. Geochim. Cosmochim. Acta 50 (8), 1667–1677.

Müller, B., 2015. ChemEQL: A Software for the Calculation of Chemical Equilibria. http://www.eawag.ch/en/department/surf/projects/chemeql/.

Ostwald, W., 1896. Lehrbuch der Allgemeinen Chemie, vol. 2. W. Engelmann, Leipzig, Germany.

Ostwald, W., 1897. Studien über die Bildung und Umwandlung fester Körper. Z. Phys. Chem. 22, 289–330.

Parkhurst, D.L., Appelo, C.A.J., 2013. Description of Input and Examples for PHREEQC Version 3—A Computer Program for Speciation, Batch-Reaction, One-Dimensional Transport, and Inverse Geochemical Calculations, Book 6, Chapter A43. U.S. Geological Survey Techniques and Methods, 497 pp.

Sander, R., 2015. Compilation of Henry's law constants (version 4.0) for water as solvent. Atmos. Chem. Phys. 15 (8), 4399–4981.

Schwab, A.P., 2005. Chemical equilibria. In: Hillel, D. (Ed.), Encyclopedia of Soils in the Environment, vol. 1. Elsevier, Oxford, pp. 189–194.

Solomon, T., 2001. The definition and unit of ionic strength. J. Chem. Educ. 78 (12), 1691–1692.

Weast, R.C. (Ed.), 1972. Handbook of Chemistry and Physics, 53rd ed. Chemical Rubber Company, Cleveland, OH.

Wolery, T.J., Daveler, S.A., 1992. EQ6, A Computer Program for Reaction Path Modeling of Aqueous Geochemical Systems: Theoretical Manual, User's Guide, and Related Documentation (Version 7.0): Part 4, UCRL-MA-110662-Pt. 4; Order No. DE93007118. Lawrence Livermore Natl. Lab., 349 pp.

Wolock, D.M., Hornberger, G.M., Beven, K.J., Campbell, W.G., 1989. The relationship of catchment topography and soil hydraulic characteristics to lake alkalinity in the northeastern United States. Water Resour. Res. 25 (5), 829–837.

CHAPTER

6

Acid-Base Chemistry

OUTLINE

6.1 Introduction 254

6.1.1 Exchangeable Acidity and Sodicity as Acid-Base Phenomena 254

6.1.2 Basicity and Alkalinity 255

6.2 Acid-Base Chemistry Fundamentals 256

6.2.1 Dissociation: The Arrhenius Acid-Base Model 258

6.2.2 Hydrogen Ion Transfer: The Brønsted-Lowry Acid-Base Model 258

6.2.3 Weak Acid and Base Conjugates 259

6.2.4 Water Reference Level: Acidity and Basicity 260

6.3 Natural and Anthropogenic Sources of Strong Acids and Bases 262

6.3.1 Chemical Rock Weathering 262

6.3.2 Atmospheric Deposition: Vulcanism and Combustion 272

6.4 Carbonate Chemistry 275

6.4.1 Carbonate Equilibrium Reactions 276

6.4.2 Aqueous Carbon Dioxide Reference Level 281

6.4.3 Alkalinity and Mineral Acidity 281

6.4.4 Geochemical Implications of Alkalinity 283

6.5 Alkali and Sodic Soils 289

6.5.1 Exchangeable Sodium and Clay Plasticity 290

6.5.2 Identifying Sodicity Risk 292

6.5.3 Predicting the Sodicity Risk of Irrigation Water 301

6.5.4 The Limiting Sodium Adsorption Ratio (LSAR) Sodicity Risk Parameter 304

6.6 Soil Acidity 309

6.6.1 Aluminum Chemistry 312

6.6.2 Exchangeable Aluminum 313

6.6.3 Nonexchangeable Aluminum 317

6.7 Summary 319

Appendices 319

6.A The pH Scale 319

6.B The Saturation Effect: Atmospheric Conversion of Sulfur Trioxide to Sulfuric Acid 320

6.C Acid-Base Implications of Nitrate and Sulfate Reduction 321

6.C.1 Nitrogen Cycling and Fertilization in Agricultural Landscapes 321

6.C.2 Marine Environment: Sulfate and Nitrate Reduction in the Deep Ocean 322

6.D Selectivity of (Na^+, Ca^{2+}) Exchange 323

6.E Fitting the Asymmetric (Al^{3+}, Ca^{2+}) Ion-Exchange Isotherm 324

References 327

Soil and Environmental Chemistry
http://dx.doi.org/10.1016/B978-0-12-804178-9.00006-9

6.1 INTRODUCTION

This book extends its scope beyond the core topics essential for mastery of soil chemistry fundamentals and chemistry applications to soil science. Carbonate chemistry, in particular, receives expanded treatment in this edition because of its significance in allied fields of chemical hydrology, limnology, and environmental chemistry. The development of soil sodicity involves water chemistry changes linked to rock weathering, evaporation and the precipitation of calcite, processes long recognized as influencing surface water chemistry. The physical properties of sodic soils have much in common with quick clay deposits. Finally, recent developments in marine chemistry point to the importance of dissimilatory sulfate reduction as a basicity source that parallels the importance of nitrate reduction in cultivated soils.

6.1.1 Exchangeable Acidity and Sodicity as Acid-Base Phenomena

Some soil chemistry textbooks relegate exchangeable acidity and sodicity to separate chapters. This comes as no surprise because these two phenomena manifest at opposite extremes of the acid-base spectrum in very different climate zones and soil landscapes. Scientists studying these two processes often find themselves in separate scientific communities because the chemical processes appear to be profoundly different. One can ask, is there anything to be gained by conflating the treatment of these two phenomena into a single broad subject.

6.1.1.1 Exchangeable Acidity

Exchangeable acidity—a condition apparently unique to soil science—derives from exchangeable Al^{3+} cations. Exchangeable acidity accumulates as chemical weathering leads to ever decreasing soil pH and ever increasing aluminum solubility. Ultimately, Al^{3+} cations accumulate at the expense of other exchangeable cations, principally Ca^{2+}.

6.1.1.2 Sodicity

Soil sodicity, on the other hand, arises as Na^{+} cations accumulate at the expense of other exchangeable cations, principally Ca^{2+}. Unlike the weak-acid Al^{3+} cation, Na^{+} cations lack acid-base character but exchangeable Na^{+}, which accumulates as evaporation drives soil pH higher and calcite solubility lower, dramatically alters clay mineral plasticity and swelling.

Rock weathering simultaneously releases base equivalents and abundant (Groups 1 and 2) cations into natural waters. Evaporation in arid climate zones increases surface water and soil salinity, inducing steadily increasing solution pH which, as we will see, eventually triggers calcite precipitation. Evaporation-induced calcite precipitation can result in surface water composition classified as sodic. These processes (rock weathering and evaporation-induced increases in salinity and calcite precipitation) occur in soils, though with key differences compared to their operation in rivers and lakes.

6.1.1.3 *Impact of Rock Weathering and Mineral Solubility on Cation Exchange in Soils*

In the final analysis, the development of exchangeable acidity and sodicity both involve changes in water chemistry linked to primary mineral weathering, secondary mineral solubility, and the effect these two processes have on cation exchange. The differences distinguishing these two phenomena can be traced to the unique chemical properties of the exchangeable cations that accumulate in acid and alkaline soils.

6.1.2 Basicity and Alkalinity

Environmental chemistry and the earth sciences have a much different perspective on acid-base chemistry than the one you encountered in general chemistry. The most important difference is the reference point used to define acidity and basicity. General chemistry defines acidity and basicity relative to $a_{H^+} = 10^{-7}\ \mathrm{mol\,dm^{-3}}$ or pH 7; the pH of pure water.

Environmental chemists rely on the *alkalinity* concept to distinguish between solutions with an excess of strong bases (*alkaline* solutions) and solution with an excess of strong acids (*mineral acid* solutions) (Box 6.1). Unlike pure water where the reference point is pH 7, the reference point in natural waters is not a fixed pH value.

For instance, the mean atmospheric CO_2 (g) partial pressure for June 2015 at the Mauna Loa Observatory was $p_{CO_2} = 4.028 \times 10^{-4}$ atm. A two-phase system consisting of pure water in equilibrium with a gas phase containing CO_2 (g) at the mean partial pressure recorded for June 2015 at the Mauna Loa Observatory would be pH 5.61. Pure chemists consider water saturated by CO_2 (g) acidic. Environmental chemists consider pure water saturated by CO_2 (g) as water lacking excess strong acids and excess strong bases, choosing to ignore the ever present dissolved CO_2 (g) when tallying mineral acidity and alkalinity (Box 6.2).

The environmental interest in acid-base chemistry is focused on the capacity of natural waters and soils to resist pH changes resulting from human activity. Compounds dissolved in water and chemical reactions on the surfaces of mineral colloids buffer soil and surface water pH, yet this buffer capacity can be overwhelmed by acidity released by atmospheric deposition and soil fertility management. The magnitude and rate at which soil and surface water pH change in response to human activity influences ecosystem function and human exposure to inorganic toxicants.

BOX 6.1

ACID-BASE STRENGTH

Acid-base strength is quantified by the *negative base-10 logarithm* of the equilibrium constant $K_a^{\ominus}$ for acid dissociation[1]: $pK_a^{\ominus}$ (Table 6.1).

TABLE 6.1 Strong Acids, Weak Acids, Weak Bases, and Strong Bases in Aqueous Solutions

Acid or Base	$pK_a^{\ominus}$	Examples
Strong acids	$pK_a^{\ominus} < 1$	$HCl(aq)$, $HNO_3(aq)$, $H_2SO_4(aq)$
Weak acids	$1 \leq pK_a^{\ominus} < 7$	$CO_2(aq)$, $H_2S(aq)$, carboxyls, and phenols
Weak bases	$7 < pK_a^{\ominus} \leq 13$	$NH_3(aq)$ and amines
Strong bases	$13 < pK_a^{\ominus}$	Groups 1 and 2 (hydr)oxides[a]

[a] *The oxides and hydroxides Group 1 elements Na and K (e.g., $Na_2O(s)$ and $NaOH(s)$, respectively) are strong bases in water. The oxides and hydroxides of Group 2 elements Mg and Ca (e.g., CaO and $Ca(OH)_2$, respectively) are strong bases in water.*

$$HA(aq) + H_2O(aq) \longleftrightarrow H_3O^+(aq) + A^-(aq)p0100$$

$$K_a^{\ominus} = \frac{a_{H_3O^+} \cdot a_{A^-}}{a_{HA}}$$

$$M^+(aq) + 2H_2O(l) \longleftrightarrow H_3O^+(aq) + MOH(aq)$$

$$K_a^{\ominus} = \frac{a_{H_3O^+} \cdot a_{MOH}}{a_{M^+}}$$

[1] $pK_a^{\ominus} \equiv -\log_{10}(K_a^{\ominus})$.

6.2 ACID-BASE CHEMISTRY FUNDAMENTALS

Discussion of acid-base chemistry often intermingles terms and concepts from two very different models: the *dissociation* or *Arrhenius* model (Arrhenius, 1887) and the *proton-transfer* or *Brønsted-Lowry* model (Brønsted, 1923; Lowry, 1923). The Arrhenius acid-base model (Arrhenius, 1887), which appeared simultaneous with the discovery of electrolyte dissociation (Arrhenius, 1887, 1888; van't Hoff, 1887; van't Hoff and Reicher, 1888, 1889; Ostwald, 1888, 1889), shares terms and concepts with the emerging electrolyte dissociation model.

BOX 6.2

AQUEOUS CARBON DIOXIDE

The National Oceanic and Atmospheric Administration (NOAA) began recording atmospheric monthly mean carbon dioxide contents at the Manua Loa Observatory in March 1958 and maintains an RSS feed of monthly mean atmospheric concentrations (Tans and Keeling, 2015). Using the June 2015 mean monthly value of 402.80 ppm(v) CO_2 (g) we can simulated the pH of a two-phase gas-solution system[2] using *ChemEQL*.

The first step is apply Henry's Law to compute $c_{CO_2(aq)}$ for $p_{CO_2} = 4.0280 \times 10^{-4}$ atm.

$$H^{cp}_{CO_2} = \frac{a_{CO_2(aq)}}{f_{CO_2}} \approx \frac{c_{CO_2(aq)}}{p_{CO_2}}$$

$$c_{CO_2(aq)} = \left(3.40 \times 10^{-2}\right) \cdot \left(4.0280 \times 10^{-4}\right)$$

$$c_{CO_2(aq)} = 1.3695 \times 10^{-5} \text{ mol dm}^{-3}$$

Applying the fixed-fugacity method (Delany and Wolery, 1984) requires component-specie CO2(aq), mode(CO2(aq)) = free and an initial CO2(aq) concentration of: 1.37E−05 (Table 6.2).

The equilibrium pH of pure water in equilibrium with the above-ground atmosphere is 5.6.

TABLE 6.2 *ChemEQL* Species or Output Table (Bottom) for 1.3695×10^{-5} M Carbon Dioxide Solution

Species	log(*K*)	Concentration (mol dm^{-3})	Activity (Unitless)
CO2(aq)	0.00	1.37E−05	1.37E−05
H2CO3(aq)	−2.92	1.65E−08	1.65E−08
HCO3−	−6.37	2.38E−06	2.35E−06
CO3−	−16.70	4.64E−11	4.43E−11
OH−	−14.00	4.07E−09	4.03E−09
H+	0.00	2.48E−06	2.48E−06

[2] The solution phase pure water with aqueous carbon dioxide.

The reader should be warned, however, that some chemists oppose teaching the Arrhenius model to beginning chemistry students because it contains notable flaws and can lead to misunderstandings about acid-base chemistry.

6.2.1 Dissociation: The Arrhenius Acid-Base Model

The Arrhenius *dissociation* model defines an *acid* as any chemical substance that releases one or more hydrogen ions H^+ (aq) when it dissociates in solution (reaction 6.R1).

$$\underset{\text{Generic acid}}{HA(aq)} \longrightarrow H^+(aq) + A^-(aq) \quad (6.R1)$$

Arrhenis defined a *base* is any chemical substance that releases hydroxyl ions $OH^-(aq)$ when it dissociates in solution (reaction 6.R2).

$$\underset{\text{Generic base}}{MOH(aq)} \longrightarrow M^+(aq) + OH^-(aq) \quad (6.R2)$$

According to the Arrhenius model, when an acid HA(aq) reacts with a base MOH(aq) the hydrogen ion $H^+(aq)$ neutralizes the hydroxide ion $OH^-(aq)$, forming a water molecule, while the $M^+(aq)$ and $A^-(aq)$ remain as a dissociated neutral salt (reaction 6.R3).

$$HA(aq) + MOH(aq) \longrightarrow M^+(aq) + A^-(aq) + H_2O(l) \quad (6.R3)$$

Acidity is attributed entirely to the hydrogen ion $H^+(aq)$ and basicity to the hydroxide ion $OH^-(aq)$. A consequence (and notable flaw) of the Arrhenius model is its complete disregard of dissociated ions $M^+(aq)$ and $A^-(aq)$ as components with acid-base properties. Eliminating spectator ions $M^+(aq)$ and $A^-(aq)$ from the reaction above results in the *neutralization* reaction (reaction 6.R4) between the acid $H^+(aq)$ and base $OH^-(aq)$ yielding water as the product.

$$H^+(aq) + OH^-(aq) \longrightarrow H_2O(l) \quad (6.R4)$$

A second notable flaw of the Arrhenius model is its representation of aqueous ammonia as $NH_4OH(aq)$. This representation adheres to the Arrhenius model by listing the base anion $OH^-(aq)$; the ammonium cation $NH_4{}^+(aq)$ is considered a spectator ion. This representation is incorrect; the ammonium cation $NH_4{}^+(aq)$ is, in fact, a weak acid.

6.2.2 Hydrogen Ion Transfer: The Brønsted-Lowry Acid-Base Model

The Brønsted-Lowry model (Brønsted, 1923; Lowry, 1923) defines an acid as *any chemical substance containing a proton that can be removed* and defines a base as *any chemical substance that can remove protons from an acid.*

The Brønsted-Lowry model expands the definition of basicity to include chemical species other than hydroxyl ions $OH^-(aq)$. Acids react with bases to form conjugates through proton transfer reactions. Removing a proton from an acid (reaction 6.R5) transforms it into a *conjugate base* (e.g., acetate anion). The binding of a proton (reaction 6.R6) transforms a base into a *conjugate acid* (e.g., ammonium cation).

$$\underset{\text{Acetic acid}}{CH_3COOH(aq)} + H_2O(l) \longleftrightarrow \underset{\text{Conjugate base}}{CH_3COO^-(aq)} + H_3O^+(aq) \quad (6.R5)$$

$$\underset{\text{Ammonia}}{NH_3(aq)} + H_2O(l) \longleftrightarrow \underset{\text{Conjugate acid}}{NH_4{}^+(aq)} + OH^-(aq) \quad (6.R6)$$

One important consequence of the Brønsted-Lowry acid-base model, a feature lacking in the Arrhenius model, is a recognition that dissolved salts can exhibit acidity or basicity. Perhaps the most important contribution of the Brønsted-Lowry acid-base model is self-ionization of water (reaction 6.R7) that explains the electrical conductivity (EC) of pure water.[3]

$$2H_2O(l) \longleftrightarrow \underset{\text{Hydronium}}{H_3O^+(aq)} + OH^-(aq) \tag{6.R7}$$

$$K_w^\circ = a_{H_3O^+} \cdot a_{OH^-} \tag{6.1}$$

6.2.3 Weak Acid and Base Conjugates

Base hydrolysis reactions can be written listing the base specie (e.g., $NH_3(aq)$) as educt and $OH^-(aq)$ as product. For hydrolysis reactions written in this way the equilibrium constant K_b° has a following b-subscript to affirm the base specie appears as a product in the hydrolysis reaction (cf. reaction 6.R6).

$$K_b^\circ = \frac{a_{NH_4^+} \cdot a_{OH^-}}{a_{NH_3}} = 10^{-4.76} \tag{6.2}$$

Acid hydrolysis reactions, written with the conjugate-acid specie (e.g., $NH_4^+(aq)$) as educt and $H_3O^+(aq)$ as product (cf. reaction 6.R8), are denoted with a following a-subscript on the equilibrium constant K_a°, affirming the hydrolysis reaction is written with the acid specie appears as an educt.

$$\underset{\text{Conjugate acid}}{NH_4^+(aq)} + H_2O(l) \longleftrightarrow \underset{\text{Ammonia}}{NH_3(aq)} + H_3O^+(aq) \tag{6.R8}$$

$$K_a^\circ = \frac{a_{H_3O^+} \cdot a_{NH_3}}{a_{NH_4^+}} = 10^{-9.24} \tag{6.3}$$

Every acid-base hydrolysis reaction can be written in two forms, either listing hydronium ion $H_3O^+(aq)$ as product or hydroxide ion $OH^-(aq)$ as product. If $H_3O^+(aq)$ appears as product, the educt is the acid specie and the equilibrium constant is K_a° (cf. reaction 6.R5).

$$K_a^\circ = \frac{a_{H_3O^+} \cdot a_{\text{acetate}^-}}{a_{\text{acetic acid}}} = 10^{-4.73} \tag{6.4}$$

If $OH^-(aq)$ appears as product (reaction 6.R9), the educt is the conjugate-base specie and the equilibrium constant is K_b°.

$$\underset{\text{Conjugate base}}{CH_3COO^-(aq)} + H_2O(l) \longleftrightarrow \underset{\text{Acetic acid}}{CH_3COOH(aq)} + OH^-(aq) \tag{6.R9}$$

$$K_b^\circ = \frac{a_{\text{acetic acid}} \cdot a_{OH^-}}{a_{\text{acetate}^-}} = 10^{-9.27} \tag{6.5}$$

Each acid-base hydrolysis reaction quotient with hydronium in the numerator (i.e., the K_a° reaction quotient) is related to the alternate reaction quotient with hydroxyl in the numerator

[3] Equilibrium constant $K_w^\circ = 1.023 \times 10^{-14} \approx 10^{-14.00}$ at 25°C (Bandura and Lvov, 2006).

(i.e., the $K_b^{\ominus}$ reaction quotient) through the water self-ionization reaction (6.R7). The following equilibrium quotient expressions use expression (6.1) to convert an equilibrium quotient expression listing conjugate acid $NH_4^+(aq)$ as educt into an equilibrium quotient expression listing base $NH_3(aq)$ as educt (cf. expression 6.2).

$$\frac{K_a^{\ominus}}{K_w^{\ominus}} = \left(\frac{a_{H_3O^+} \cdot a_{NH_3}}{a_{NH_4^+}}\right) \cdot \left(\frac{1}{a_{H_3O^+} \cdot a_{OH^-}}\right) = \frac{10^{-9.24}}{10^{-14.00}}$$

$$10^{+4.76} = \left(\frac{a_{NH_3}}{a_{NH_4^+} \cdot a_{OH^-}}\right) = \left(K_b^{\ominus}\right)^{-1}$$

Expression (6.6) relates every acid hydrolysis constant $K_a^{\ominus}$ to its corresponding base hydrolysis constant $K_b^{\ominus}$.

$$K_w^{\ominus} = K_a^{\ominus} \cdot K_b^{\ominus} \tag{6.6}$$

6.2.4 Water Reference Level: Acidity and Basicity

Proton transfer to and from water molecules—forming $H_3O^+(aq)$ and $OH^-(aq)$, respectively—takes place in all aqueous acid-base reactions,[4] introducing a mass-conservation constraint known as the *proton balance*.[5]

BOX 6.3

WATER REFERENCE LEVEL

The *charge-balance* expression below is valid for solutions prepared from any combination of strong acids and strong bases—represented here by aqueous hydrogen chloride HCl(aq) and sodium hydroxide NaOH(aq), respectively—or the neutral salts that form when strong acids are neutralized by strong bases—represented here by sodium chloride NaCl(aq).

$$c_{Na^+} + c_{H_3O^+} = c_{Cl^-} + c_{OH^-} \tag{6.7}$$

Rearranging the charge-balance expression yields the following *proton-balance* expression.

$$c_{Na^+} - c_{Cl^-} = c_{OH^-} - c_{H_3O^+} \tag{6.8}$$

[4] The solvent does not participate in acid-base reactions in Arrhenius acid-base model. The neutralization reaction (6.R4) is nothing more than the annihilation of acid $H^+(aq)$ by base $OH^-(aq)$ and is not equivalent to the water self-ionization reaction (6.R7).

[5] In Chapter 5 we used mass-conservation constraints imposed by the input matrix and the charge-balance condition to validate water chemistry simulations.

BOX 6.3 *(cont'd)*

The ions on the left-hand side of the proton-balance expression are considered *spectator ions* because they do not participate in any aqueous proton-transfer reactions. Spectator cations such as Na^+ may originate from a strong base (e.g., sodium hydroxide NaOH(aq)) or a neutral salt (e.g., sodium chloride NaCl(aq)). Likewise, spectator anions such as Cl^-(aq) may originate from a strong acid (e.g., hydrogen chloride HCl(aq)) or a neutral salt (e.g., sodium chloride NaCl(aq)). No acid-base reactions couple spectator cations to spectator anions.

The ions on the right-hand side of the proton-balance expression H_3O^+(aq) and OH^-(aq) are the only acid-base species in solutions whose only other electrolytes are spectator ions. Unlike the spectator ions on the left-hand side of the expression, the H_3O^+ and OH^- ions are coupled to one another through an acid-base reaction identified previously as the water self-ionization reaction (6.R7). The relative activity of these two ions is constrained by the water self-ionization equilibrium quotient expression (6.1).

The *water reference level* is defined by relation (6.9), the state where the *spectator-ion charge balance*[6] equals zero.

$$c_{H_3O^+} = c_{OH^-} \tag{6.9}$$

If $c_{Na^+} > c_{Cl^-}$ the proton-balance expression tell us the excess positive charge in solution is balanced by an equal concentration of base anion OH^-(aq). Likewise, if $c_{Cl^-} > c_{Na^+}$ the proton-balance expression tell us the excess negative charge in solution is balanced by an equal concentration of acid cation H_3O^+(aq). At the water reference level the water self-ionization quotient (6.1) constrains the concentration of species H_3O^+(aq) and OH^-(aq).

The ion activity coefficients for H_3O^+ and OH^- are identical, therefore expression (6.9) can be written as follows.

$$\left(\gamma_{1\pm} \cdot c_{H_3O^+}\right) = \left(\gamma_{1\pm} \cdot c_{OH^-}\right)$$

$$c_{H_3O^+} = c_{OH^-}$$

Substituting the water reference level into the self-ionization equilibrium quotient (expression 6.1) yields the following expression.

$$a_{H_3O^+} = \left(K_W^{\circ}\right)^{1/2} = 10^{-7}$$

Chemists typically measure H_3O^+ activity directly using a potentiometric glass electrode (cf. Appendix 6.A), eliminating the need to apply an ion activity coefficient correction to the water reference level. If solution pH > 7 the solution is *basic* because $c_{OH^-} > c_{H_3O^+}$, consistent with an excess of spectator cations over spectator anions: $c_{Na^+} > c_{Cl^-}$.

If solution pH < 7 the solution is *acidic* because $c_{H_3O^+} > c_{OH^-}$. The concentration of spectator anions exceeds spectator cations in acidic solutions: $c_{Cl^-} > c_{Na^+}$.

We will revisit the proton-balance condition in Section 6.4. An extension of proton-balance condition to carbonate solutions provides the foundation for a very important natural water parameter known as *alkalinity*.

[6] The right-hand side of expression (6.8).

Pure chemistry defines acidity and basicity relative to the *water reference level*, a reference level determined by the proton balance of pure water (cf. Box 6.3).

6.3 NATURAL AND ANTHROPOGENIC SOURCES OF STRONG ACIDS AND BASES

Atmospheric deposition, biological activity, and chemical rock weathering control the acid-base chemistry of natural waters. The basicity in natural waters is overwhelmingly the product of rock weathering with negligible anthropogenic contributions. Acidity, however, has both natural and anthropogenic origins. In this section we examine each of the major acid-base source and the associated chemical processes.

6.3.1 Chemical Rock Weathering

Geologists classify rocks according to the geological process that led to their formation: igneous, sedimentary, and metamorphic. Major igneous rock types are further subdivided by texture and mineralogy, the latter determined by composition and prevailing conditions as magma crystallized. Sedimentary rocks are subdivided into two major groups: clastic and chemical. Clastic sedimentary rocks are composed of sediments that lithify through cementation and may be further transformed by elevated pressure and temperature during burial. Chemical sedimentary rocks are composed of minerals that have precipitated from low-temperature aqueous solution—either through evaporation or biomineralization—prior to lithifying. The mineralogy of evaporite sedimentary rocks is predominantly chloride and sulfate minerals. Precipitate sedimentary rocks consist of biogenic carbonate and silicate minerals—calcite, aragonite, dolomite, and diatomite—derived from the exoskeleton of aquatic and marine organisms.

In the acid-base chemistry context we can group igneous, metamorphic, and clastic sedimentary rocks into a single broad type: silicate rocks. Of the chemical sedimentary rocks, only carbonate rocks influence acid-base chemistry. As we will see below, the acid-base chemistry of natural waters is indifferent to evaporite and diatomite rock mineralogy.

6.3.1.1 Silicate Rocks

The chemical weathering of silicate rocks is most easily understood if we denote rock composition as a sum of oxides. This representation neglects mineralogical distinctions, favoring chemical composition when considering the global impact of chemical weathering on acid-base chemistry. Table 6.3 lists the oxide mole-per-mass fraction composition of Earth's lithosphere. A detailed discussion of how the composition of rocks or minerals is converted from a mass-fraction representation to a sum of oxides representation appears in Example 6.1.

TABLE 6.3 Oxide Mole-per-Mass m(Oxide) Composition of Earth's Lithosphere

Oxide	m(Oxide) (mol kg^{-1})
$SiO_2(s)$	9.68
$Al_2O_3(s)$	1.54
$FeO(s)$	0.37
$Fe_2O_3(s)$	0.37
$CaO(s)$	1.16
$Na_2O(s)$	0.49
$MgO(s)$	1.14
$K_2O(s)$	0.24

EXAMPLE 6.1

Example Permalink

http://soilenvirochem.net/pCtf10

Convert elemental composition of the continental lithosphere from mass-fraction units to mole-per-mass and oxide mass-fraction compositions.

Step 1. Convert the estimated continental lithosphere composition from commonly reported mass-fraction w(E) units, listed in Table 6.4, to more chemically relevant mole-per-mass concentration m(E) units.

TABLE 6.4 Estimated Elemental Abundance of the Continental Lithosphere of Planet Earth as Mass-Fraction w(E)

Element	$\overline{m}_a$(E) (g mol^{-1})	Valence (mol_c mol^{-1})	w(E) (g kg^{-1})
Oxygen	15.9994	−2	455.0
Silicon	28.0855	4	272.0
Aluminum	26.9815	3	83.0
Iron[a]	55.8452	2.7	62.0
Calcium	40.0784	2	46.6
Sodium	22.9898	1	22.7
Magnesium	24.3051	2	27.6
Potassium	39.0983	1	18.4

[a] *Iron valence in igneous granitic rocks is controlled by the fayalite-magnetite-quartz redox buffer (Haggerty, 1976) "mostly incorporated into magnetite" (Frost, 1991).* Magnetite: *$FeO \cdot Fe_2O_3(s)$ has a mean iron valence of 2.67.*

The unit transformation is very simple: divide the mass-fraction $w(E)$ in column 4 by the atomic mass $\overline{m}_a(E)$ in column 2. Table 6.5 lists the mole-per-mass concentration of each element $b(E)$ in column 2 and the moles-of-charge concentration $b_c(E)$ (the product of element valence and $b(E)$) in column 3.

TABLE 6.5 Estimated Elemental Abundance and Charge Balance of the Continental Lithosphere of Planet Earth Using Mole-per-Mass Units

Element	$b(E)$ ($mol_c\ kg^{-1}$)	$b_c(E)$ ($mol_c\ kg^{-1}$)
Oxygen	28.44	−56.877
Silicon	9.68	38.739
Aluminum	3.08	9.229
Iron(II)	0.37	2.961
Iron(III)	0.74	2.961
Calcium	1.16	2.325
Sodium	0.99	0.987
Magnesium	1.14	2.271
Potassium	0.47	0.471

The charge concentration of the continental lithosphere (found by summing terms in column 3) satisfies the charge-balance condition within 0.19% based solely on the nine (9) most abundant elements listed in this analysis.

Step 2. Convert the elemental moles-per-mass composition $b(E)$ of the continental lithosphere the oxide mass-fraction composition w(oxide).

The *elemental* mole-per-mass concentration $b(E)$ listed in Table 6.5 (column 2) is converted to oxide mole-per-mass content b(oxide) by counting the moles of the nonoxygen element. For example, the elemental aluminum content is 3.08 mol kg^{-1} therefore the Al_2O_3 oxide content will be half as much or 1.54 mol kg^{-1}.

TABLE 6.6 Estimated Elemental Abundance of the Continental Lithosphere of Planet Earth Using Oxide Mass-Fraction Units

Oxide	m_f(Oxide) ($g\ mol^{-1}$)	b(Oxide) ($mol\ kg^{-1}$)	w(Oxide) ($g\ kg^{-1}$)
$SiO_2(s)$	60.0844	9.68	581.9
$Al_2O_3(s)$	101.9614	1.54	156.8
$Na_2O(s)$	61.9790	0.49	30.6
$MgO(s)$	40.3045	1.14	45.8
$K_2O(s)$	94.1961	0.24	22.2
$CaO(s)$	56.0778	1.16	65.2
$FeO(s)$	71.8446	0.37	26.6
$Fe_2O_3(s)$	159.6887	0.37	59.1

The fayalite-magnetite-quartz redox buffer controls iron valence in granite. About one-third of the iron in MAGNETITE $FeO \cdot Fe_2O_3$ and, therefore, one-third of the iron in granite, is Fe(II). Assign the b(Fe(II)) content to FeO the same way the b(Ca) content is assigned to CaO. Likewise, assign the b(Fe(III)) content to Fe_2O_3 the same way the b(Al) content is assigned to Al_2O_3. The iron(III) oxide content $m(Fe_2O_3) = 0.37 \text{ mol kg}^{-1}$ is equal to the iron(II) oxide moles-per-mass content $b(FeO) = 0.37 \text{ mol kg}^{-1}$ at the fayalite-magnetite-quartz redox buffer.

The oxide mass-fraction w(oxide) (column 4) is calculated by multiplying the *oxide* mole-per-mass concentration b(oxide) (column 3) by the oxide formula mass m_f(oxide) (column 2). The oxide mass-fraction w(oxide) in Table 6.6 is very similar to the estimated worldwide composition of granite (Blatt et al., 2006), revealing the continental lithosphere of planet Earth is very similar to the composition of the igneous rock granite.

Chemical rock weathering occurs under oxidizing conditions and one of the most important chemical weathering reactions is the oxidation of iron(II) and manganese(II). Regardless of the mean iron or manganese valence in the continental lithosphere, the acid-base chemistry of natural water is indifferent to iron(II) and manganese(II) oxidation (reaction 6.R10) during chemical weathering.

$$2FeO(s) + \tfrac{1}{2}O_2(g) \longleftrightarrow Fe_2O_3(s) \qquad (6.R10)$$

Sum-of-oxide formulas for clay minerals listed in the Jackson clay weathering sequence (Jackson et al., 1948, 1952, cf. Chapter 3) reveal the progressive loss of Groups 1 and 2 oxide content in advanced stages of the Jackson sequence.

Clay fraction mineralogy is depleted of Groups 1 and 2 oxides (Fig. 6.1) in the final four stages: 10–13. Clay fraction indicator minerals for stage 11 through 13 are not only depleted of Groups 1 and 2 oxide content, these insoluble minerals—$Al_2O_3(s)$, $Fe_2O_3(s)$, and $TiO_2(s)$—are the end products of chemical weathering. The oxides $Al_2O_3(s)$, $Fe_2O_3(s)$, and $TiO_2(s)$ can be classified as *spectator oxides* in the lithosphere chemical weathering reaction, appearing on the educt side as oxide components of the lithosphere and the product side as end products of chemical weathering reaction.

Silicon dioxide $SiO_2(s)$ is also a spectator oxide, not because of insolubility, but because dissolution is essentially a hydration reaction. The substance geochemists call aqueous silica $SiO_2(aq)$ represents simple silicic acid $H_4SiO_4(aq)$ and a variety of poorly characterized polymeric silica compounds.

Silicate rock basicity is determined by its Groups 1 and 2 oxide content because these oxides are classified strong bases (cf. Chapter 3, Clay Mineralogy and Chemistry). The dissolution of Groups 1 and 2 oxides is a hydration reaction; every mole of Groups 1 and 2 oxide-oxygen combines with 1 mole of water to form 2 moles of $OH^-(aq)$. The following chemical weathering reaction is abridged[7] listing the stoichiometric results appearing in Example 6.1, shows the release of 6.05 moles $OH^-(aq)$ per kilogram of continental lithosphere.

[7] Spectator oxides $Al_2O_3(s)$ and $Fe_2O_3(s)$ do not appear.

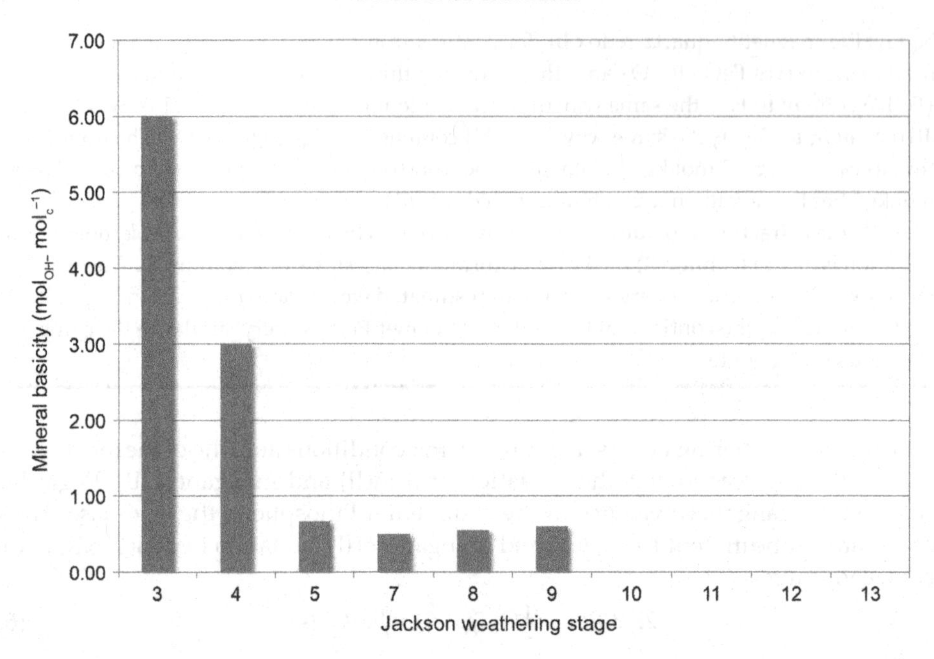

FIG. 6.1 Mineral basicity decreases from a high at weathering stage 3 (Jackson et al., 1948) to zero by weathering stage 10. Stages 1–2 are excluded to highlight the weathering of silicate rocks. Basicity is based on the formulas of indicator minerals.

$$9.68SiO_2(s) + 0.49Na_2O(s) + 0.24K_2O(s)$$

$$+1.14MgO(s) + 1.16CaO(s) + 3.03H_2O(l) \longleftrightarrow$$

$$9.68SiO_2(aq) + 0.99Na^+(aq) + 0.47K+(aq)$$

$$+1.14Mg^{2+}(aq) + 1.16Ca^{2+}(aq) + 6.05OH^-(aq) \qquad (6.R11)$$

Two processes control the carbonate dissolved in natural waters: carbon dioxide water solubility (cf. Henry's Law, Chapter 5, *Water Chemistry*) and acid-base neutralization reaction (6.R12) involving aqueous carbon dioxide and hydroxyl ions, the latter resulting from the chemical weathering of silicate minerals (cf. Example 6.1 and reaction 6.R11).

$$\underset{\text{Weak acid}}{CO_2(aq)} + OH^-(aq) \longleftrightarrow \underset{\text{Conjugate base}}{HCO_3{}^-(aq)} \qquad (6.R12)$$

Although direct contact ensures surface waters absorb carbon dioxide from the above-ground atmosphere, during the growing season biological activity throughout the surface water column and the soil profile generates carbon dioxide that super-saturates the surface and soil pore water above atmospheric levels. Garrels and Mackenzie (1971), in a discussion of the impact of rock weathering on groundwater chemistry, suggest dissolved bicarbonate concentrations "can be achieved if it is assumed that, on the average, silicate minerals produce one bicarbonate ion and two silica ions for each molecule of carbon dioxide." The Garrels-Mackenzie estimate of the bicarbonate-to-silica ratio derived from silicate rock weathering is close to the ratio based on the composition of Earth's continental lithosphere (cf. Example 6.1). Granite, the most abundant rock in the continental lithosphere, has a bicarbonate-to-silica ratio of about 0.3, based on the Blatt et al. (2006) estimate of most probable granite composition.

$$\text{Silicate rock weathering:} \frac{c_{HCO_3^-}}{c_{SiO_2(aq)}} \approx \frac{1}{2} \tag{6.10}$$

The Garrels-Mackenzie bicarbonate-to-silica ratio has implications beyond linking the composition of silicate rocks in the continental lithosphere to groundwater composition. The mean residence time of groundwater is 2×10^4 years, adequate time for groundwater to equilibrate with the above-ground atmosphere. Absent carbon dioxide, groundwater pH would range from 10 to 11 (cf. Example 6.2), much higher than found by groundwater analyses. The impact of dissolved carbon dioxide will be discussed thoroughly in Section 6.4; for now it is sufficient to acknowledge that carbon dioxide dissolution (Box 6.2) and hydrolysis reaction (6.R12) appear to reach equilibrium during the mean residence or, at least, closely approach equilibrium (Box 6.4).

EXAMPLE 6.2

Example Permalink

http://soilenvirochem.net/ofc13K

Calculate the pH a 1.0×10^{-3} M $Na_2CO_3(aq)$ solution using the *RICE* table method and the following base hydrolysis reaction and equilibrium reaction quotient.

$$CO_3^{2-}(aq) + H_2O(l) \longleftrightarrow HCO_3^-(aq) + OH^-(aq)$$

$$K_{b1}^{\ominus} = 2.14 \times 10^{-4} = \frac{a_{HCO_3^-} \cdot a_{OH^-}}{a_{CO_3^{2-}}}$$

Step 1. Prepare an *RICE* table (Table 6.7) representing $Na_2CO_3(aq)$ hydrolysis in solution.

Carbonate anions $CO_3^{2-}(aq)$ undergo two base hydrolysis steps. The second base hydrolysis constant $K_{b2}^{\ominus}$ is much smaller than the first base hydrolysis constant $K_{b1}^{\ominus}$, therefore the hydroxide ion $OH^-(aq)$ concentration results almost entirely from the first hydrolysis step.

TABLE 6.7 *RICE* Table for Carbonate Base Hydrolysis

Reaction →	$CO_3^{2-}(aq)$ (mol dm^{-3})	$H_2O(l)$	↔	$HCO_3^-(aq)$ (mol dm^{-3})	$OH^-(aq)$ (mol dm^{-3})
Initial	10^{-4}			0	0
Reaction	$-x$			$+x$	$+x$
Equilibrium	$10^{-4} - x$			x	x

Step 2. Using terms from the *RICE* table, write the equilibrium quotient expression for the first base hydrolysis step and solve the second-order polynomial using the quadratic equation.

$$x^2 = K_{b1}^{\ominus} \cdot \left(10^{-3} - x\right)$$

$$x^2 + K_{b1}^{\ominus} \cdot x - \left(K_{b1}^{\ominus} \times 10^{-3}\right) = 0$$

$$x_{+} = \frac{-K_{b1}^{\ominus} + \sqrt{\left(K_{b1}^{\ominus}\right)^2 + \left(4 \cdot K_{b1}^{\ominus} \times 10^{-3}\right)}}{2} = 3.68 \times 10^{-4}$$

$$\mathrm{pH} = 14 - \log_{10}\left(a_{OH^-}\right) = 14 + \log_{10}(x) = 10.57$$

Step 3. Simulate the pH of a 1.0×10^{-3} M $Na_2CO_3(aq)$ solution using bicarbonate HCO_3^- as the carbonate component.

A minimalist simulation, disregarding charge balance and spectator ions, contains two components: HCO3– and H+. Setting the initial HCO_3^- concentration equal to 1.0×10^{-3} M supplies the correct total carbonate but only half of the alkalinity of a 1.0×10^{-3} M $Na_2CO_3(aq)$ solution. Adding the additional alkalinity in a *ChemEQL* simulation requires replacing component H+ with OH–, setting mode(OH–) = total and assigning it an initial concentration of 1.0×10^{-3} M. The total alkalinity with these components and initial concentrations is the desired 1.0×10^{-3} M concentration.

The simulated pH value is 10.51, compared to the *RICE* table estimate of pH 10.57.

BOX 6.4

SIMPLE GIBBS MODEL

While chemical weathering (Section 6.3.1) is undoubtedly a major determinant of groundwater composition, surface waters are subjected to another significant process: evaporation. The combined impact of rock weathering and evaporation was dramatically illustrated in two "boomerang" graphs that first appeared in a paper by Gibbs (1970).

Gibbs (1970) used ppm concentration units ($mg\ kg^{-1}$) for all solutes and plotted total dissolved solids (TDS) (units: $mg\ kg^{-1}$) on the ordinate of both graphs.[8] Gibbs plotted the $f_{Na,Ca}$ mass-fraction ratio on the abscissa of the first graph and the f_{Cl,HCO_3} mass-fraction ratio on the abscissa of the second graph. The data plotted in these graphs were taken from a major compilation of surface water compositions (Livingstone, 1963) and study of 16 Amazon River tributaries (Gibbs, 1972).

$$f_{Na,Ca} \equiv \frac{w(\mathrm{Na})}{w(\mathrm{Na}) + w(\mathrm{Ca})}$$

$$f_{Cl,HCO_3} \equiv \frac{w(\mathrm{Cl})}{w(\mathrm{Cl}) + w(\mathrm{HCO_3})}$$

US Geological Survey (USGS) Professional Paper 440-G (Livingstone, 1963) lists 412 analyses suitable for plotting but Gibbs chose to plot the analyses for "135 major lakes and rivers around the world" (Gibbs, 1992) rather than results from all analyses. A plot of the entire Livingstone (1963) dataset does not reveal the "boomerang" trace so apparent in the Gibbs papers (Gibbs, 1970, 1971, 1992). Despite repeated defense of the model (Gibbs, 1971, 1992), a more complex picture emerges from an inclusive analysis of global surface water composition and valid objections to the model (Feth and Gibbs, 1971; Kilham, 1990; Eilers et al., 1992).

Continued

BOX 6.4 (*cont'd*)

Gibbs (1970) identified three end-points on both "boomerang" plots: meteoric-dominated waters, rock weathering-dominated waters, and evaporation/chemical precipitation-dominated waters.

The extremely low salinity end-point occurs where the $f_{Na,Ca}$ and f_{Cl,HCO_3} mass-fraction ratios approach unity. The surface waters at this end-point are dominated by meteoric water. The major ions in the extremely low salinity meteoric-dominated waters are chloride salts characteristic of marine aerosols.

The second end-point occurs at "moderate" salinity—100–300 ppm TDS—where the $f_{Na,Ca}$ and f_{Cl,HCO_3} mass-fraction ratios approach zero. Chemical rock weathering dominates surface water composition at this end-point. The major ions in the moderate salinity weathering-dominated waters are produced by weathering reactions (6.R11) and (6.R14).

The third end-point occurs at the extremely high salinity characteristic of major oceans and saline lakes where the $f_{Na,Ca}$ and f_{Cl,HCO_3} mass-fraction ratios veer back toward unit. The combined effects of evaporation and carbonate precipitation dominate surface water composition at this end-point. The major ions in the extremely high salinity evaporation/-chemical precipitation-dominated waters are the chloride salts characteristic of seawater.

A plot of all 412 analyses from Livingstone (1963) reveals a triangle with vertices at the three end-points identified by Gibbs (1970). The process missing from the simple Gibbs model was groundwater erosion of evaporite formations (Van Denburgh and Feth, 1965; Feth and Gibbs, 1971). Table 6.8 lists the EC at four *USGS* sites along the Pecos River in New Mexico. Pecos River salinity doubles from 1.86 $dS\ m^{-1}$ at Santa Rosa, New Mexico to 3.53 $dS\ m^{-1}$ at the Dark Canyon station near Carlsbad, New Mexico—a river distance of roughly 450 km. This salinity increase could be the result of evaporation, but the Gibbs model cannot explain the 2-fold salinity increase between Carlsbad and Malaga—a river distance of a mere 43 km—to reach 7.14 $dS\ m^{-1}$. Pecos River salinity increases 2-fold yet again to 15.04 $dS\ m^{-1}$ at Red Bluff, New Mexico—a river distance of 34 km.

TABLE 6.8 Mean Electrical Conductivity ($dS\ m^{-1}$) of Water Samples Collected at Four *USGS* Surface Water Sampling Sites Along the Pecos River, New Mexico, USA

Station	Station ID	River-Kilometer (km)	Electrical Conductivity ($dS\ m^{-1}$)
Santa Rose	08383000	1190	1.860
Carlsbad	08405200	739	3.525
Malaga	08406500	696	7.139
Red Bluff	08407500	662	15.043

BOX 6.4 *(cont'd)*

Rock weathering and evaporation-driven carbonate precipitation, as originally identified by Gibbs (1970), exert a strong influence on surface and soil pore water *acid-base chemistry*. The discharge of saline groundwater into the Pecos River (Van Denburgh and Feth, 1965; Feth and Gibbs, 1971) and other surface waters—to say nothing of groundwater seepage at the land surface—can have a profound effect on salinity in certain landscapes above and beyond salinity increases attributable to evaporation.

[8] Soil chemists often measure salinity using EC because it is a quick analytical method. EC becomes a more unreliable analytical method as salinity increases. Geochemists often encounter very saline water and rely on the more stable TDS method.

6.3.1.2 Sulfide Minerals

The sulfate dissolved in natural waters comes from two sources: the dissolution of GYPSUM: $CaSO_4 \cdot 2H_2O(s)$ in evaporite sedimentary rocks and sulfide mineral oxidation, the latter being a natural source of sulfuric acid. The atmospheric chemistry of sulfur oxides and the resulting atmospheric deposition of sulfuric acid are deferred to a later section; here, we focus on the reaction between dissolved molecular oxygen $O_2(aq)$ and sulfide ions.

The sulfur oxidation state $OS(S) = -1$ in the ferrous disulfide minerals PYRITE: $FeS_2(s)$ and MARCASITE: $FeS_2(s)$. The sulfide ion in these minerals is actually a disulfide: S_2^{2-}. The sulfide ion S^{2-} in the mineral TROILITE: $FeS(s)$ has a sulfur oxidation state $OS(S) = -2$.

Pyrite oxidation by molecular oxygen (reaction 6.R13), representative of oxidation reactions for other ferrous sulfide, yields a 1/2 mole of HEMATITE: $Fe_2O_3(s)$ and 2 moles of sulfuric acid H_2SO_4.

$$FeS_2(s) + \tfrac{15}{4}O_2(aq) + 2H_2O(l) \longleftrightarrow \tfrac{1}{2}\,Fe_2O_3(s) + 4 \cdot H^+(aq) + 2 \cdot {SO_4}^{2-}(aq) \qquad (6.R13)$$

The H^+ yield in reaction (6.R13) is determined by the *soluble* educt SO_4^{2-} yield. Ferrous disulfide $FeS_2(s)$ oxidation produces 2 moles SO_4^{2-} per formula unit and 4 moles of soluble H^+. Ferrous sulfide $FeS(s)$ oxidation produces 1 mole SO_4^{2-} per formula unit and 2 moles of soluble H^+.

The importance of the soluble oxidation products such as SO_4^{2-} is further illustrated by the oxidation of ferrous oxide $FeO(s)$, representative of the iron(II) oxide content of silicate minerals. Ferrous oxide $FeO(s)$ oxidation produces *insoluble* ferric (hydro)oxide products and, therefore, zero soluble acidity H^+.

A variety of other metallic and metalloid elements form sulfide and disulfide minerals, but the natural abundance of other sulfide-forming elements is much lower than iron, making them relatively insignificant contributors to the global acid-base chemistry of natural waters. Sulfur molar abundance (Lide, 2005) is a mere 0.3% of the combined abundance of Groups 1 and 2. In short, while sulfide mineral oxidation can be a significant strong acid source locally it is relatively insignificant globally.

6.3.1.3 Carbonate Rocks

Carbonate rock weathering of is major global factor in natural water acid-base chemistry. The chemical formula for CALCITE: $CaCO_3(s)$—the principle mineral in limestone—can be reformulated as a sum of oxides: $CaO \cdot CO_2(s)$.

Calcite dissolution in water containing $CO_2(aq)$ releases 2 moles of the strong base $OH^-(aq)$ which reacts with the weak acid $CO_2(aq)$ in solution (reaction 6.R12) to form 2 moles of the conjugate base bicarbonate. A thorough discussion of carbonate chemistry, including the dissolution and precipitation of carbonate minerals, is deferred to Section 6.4.

$$\underset{\text{Calcite}}{CaO \cdot CO_2(s)} + CO_2(aq) \longleftrightarrow Ca^{2+}(aq) + 2HCO_3^-(aq) \qquad (6.R14)$$

Garrels and Mackenzie (1971) used the following expression to estimate dissolved bicarbonate derived from carbonate rock dissolution.

$$\text{Carbonate rock weathering: } c_{HCO_3^-} \approx 2 \cdot \left(c_{Ca^{2+}} + c_{Mg^{2+}} - c_{SO_4^{2-}}\right) \qquad (6.11)$$

Garrels and Mackenzie (1971) combined the bicarbonate yield from silicate rock weathering (expression 6.10) with the bicarbonate yield from carbonate rock weathering (expression 6.11) to arrive at an estimate for the total dissolved bicarbonate (expression 6.12).

$$c_{HCO_3^-} \approx 2 \cdot \left(c_{Ca^{2+}} + c_{Mg^{2+}} - c_{SO_4^{2-}}\right) + \tfrac{1}{2} \cdot c_{SiO_2(aq)} \qquad (6.12)$$

The bicarbonate yield from silicate rock weathering (expression 6.10), which is determined by the composition of the continental lithosphere, is very different from the bicarbonate yield from carbonate rock weathering (expression 6.11). As we will discover in Section 6.4, expression (6.11) is related to the proton-balance expressions for natural waters (cf. Section 6.4.2). The absence of Cl^- in expression (6.11) merely reflects the relatively low chloride content of most carbonate rocks.

6.3.1.4 Evaporite Rocks

The highly soluble minerals typical of evaporite sedimentary rocks—HALITE: $NaCl(s)$, SYLVITE: $KCl(s)$, FLUORITE: $CaF_2(s)$, GYPSUM: $CaSO_4 \cdot 2H_2O(s)$ precipitate from concentrated salt solutions or brines. The congruent dissolution[9] of evaporite minerals generates neutral salt solutions because their ionic components are spectator ions.

We examined in this section the acid-base implications of chemical weathering of four major types of rocks: silicates, sulfides, carbonates, and evaporites. Chemical weathering of silicate rocks (6.R11) and carbonate rocks (6.R14) is the primary source of basicity entering terrestrial surface and ground waters. Chemical weathering involving disulfide (or sulfide) oxidation and ferric iron oxide precipitation produces soluble sulfate ions and acidity but this contribution to acid-base chemistry is local rather than global because of a low sulfur abundance in the continental lithosphere. Congruent dissolution of evaporite rock minerals is a source of dissolved chloride, sulfate, and fluoride ions but has no impact on natural water acid-base chemistry.

[9] Congruent dissolution reactions produce solutions where the mole ratio of the ions in solution is identical to their mole ratio in the solid.

6.3.2 Atmospheric Deposition: Vulcanism and Combustion

6.3.2.1 Halogens

Halmer et al. (2002) estimated volcanic HCl(g) emissions into the atmosphere varied between $1.2 \times 10^{+12}$ and $170 \times 10^{+12}$ g_{HCl} $year^{-1}$ over a 30-year period (1972–2000). Estimated HF(g) emissions into the atmosphere varied between $0.7 \times 10^{+12}$ and $8.6 \times 10^{+12}$ g_{HF} $year^{-1}$ over the same interval. Atmospheric halogen acid deposition is localized in the immediate vicinity because 99.99% of the halogen gases wash out of the volcanic plume before reaching the stratosphere.

6.3.2.2 Sulfur Oxides

Sulfur dioxide $SO_2(g)$ enters the atmosphere through both natural and anthropogenic processes. The estimated annual global sulfur emission from all sources (Bates et al., 1992; Halmer et al., 2002) is 10^{+14} g_S $year^{-1}$. Roughly 76% of the annual global atmospheric sulfur emission is anthropogenic carbon combustion (biomass and fossil fuels), 15% is emitted through the biological sulfur cycle, and about 9% is emitted by volcanic activity. Explosive volcanic eruptions cause significant annual fluctuations in atmospheric sulfur deposition (Halmer et al., 2002).

Halmer et al. (2002) estimate 25% of the dihydrogen sulfide and sulfur dioxide emitted by vulcanism washes out before reaching the stratosphere. Assuming uniform atmospheric deposition of stratospheric sulfur, annual $SO_2(g)$ should be on the order of 8.3 mol_{SO_2} ha^{-1}.

Volcanic $H_2S(g)$ emissions react rapidly with $SO_2(g)$ to form particulate sulfur by disproportionation (reaction 6.R15).

$$SO_2(g) + 2H_2S(g) \longleftrightarrow 3S(s) + 2H_2O(l) \qquad (6.R15)$$

Sulfide minerals and organosulfur compounds in coal undergo combustion reactions to produce $SO_2(g)$ in power plant emissions. Thiophene (structure **1**, Fig. 6.2), thioether (structure **2**, Fig. 6.2), and thiol (–SH) represent organosulfur species in biomass fuels, petroleum, and coal (George and Gorbaty, 1989; Gorbaty et al., 1989; Waldo et al., 1991, 1992).

Sulfur dioxide $SO_2(g)$ is very water-soluble (cf. Section 5.3.3). Were it not for enormous amounts of sulfur dioxide in explosive eruptions far more than 25% would washout before reaching the stratosphere. There is no evidence aqueous sulfur dioxide $SO_2(aq)$ reacts with a water to form the hydrate sulfurous acid $H_2SO_3(aq)$. The accepted hydrolysis reaction (6.R16) lists $SO_2(aq)$ as the weak acid and $HSO_3^-(aq)$ as the conjugate base.

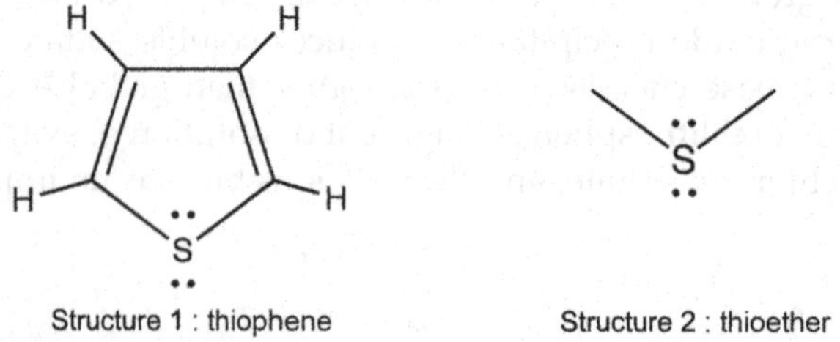

Structure 1 : thiophene

Structure 2 : thioether

FIG. 6.2 Molecular structures of thiophene (**1**) and thioether (**2**).

$$SO_2(aq) + H_2O(l) \longleftrightarrow H^+(aq) + HSO_3^-(aq) \tag{6.R16}$$

$$K_a^\ominus = 1.5 \times 10^{-2} = \frac{a_{H_3O^+} \cdot a_{HSO_3^-}}{a_{SO_2}} \tag{6.13}$$

The oxidation of $SO_2(g)$ to $SO_3(g)$ is slow reaction catalyzed by hydroxyl $HO^\bullet(g)$ and hydroperoxyl $HOO^\bullet(g)$ radicals in the stratosphere (Margitan, 1984).[10] The stratosphere is a distinctive environment with essentially no liquid water and intense ionizing radiation. This combination sustains appreciable levels of gaseous hydroxyl radicals. Margitan (1984) quotes a $SO_2(g)$ lifetime of 1–2 months for emissions during the March to April 1982 eruption of El Chichón (Mexico), a substantial time span considering the high potential for photolytic reaction in the stratosphere.

$$SO_2(g) + HO^\bullet(g) \longleftrightarrow \underset{\text{Hydroxysulfonyl}}{HSO_3^\bullet(g)} \tag{6.R17}$$

$$\underset{\text{Hydroxysulfonyl}}{HSO_3^\bullet(g)} + O_2(g) \longleftrightarrow SO_3(g) + \underset{\text{Hydroperoxyl}}{HOO^\bullet(g)} \tag{6.R18}$$

A major source of the hydroxyl radical $HO^\bullet(g)$ is the photolysis of ozone (Eq. 6.R19) to produce photo-excited atomic oxygen $O^*(g)$ which then reacts with water molecules (Eq. 6.R20) to produce hydroperoxyl radicals $HOO^\bullet(g)$.

$$O_3(g) \xrightarrow{\text{UV light}} O_2(g) + O^*(g) \tag{6.R19}$$

$$O^*(g) + H_2O(g) \longrightarrow 2\ \underset{\text{Hydroperoxyl}}{HOO^\bullet(g)} \tag{6.R20}$$

Another source of photo-excited atomic oxygen $O^*(g)$ is the photolysis (6.R21) of nitrogen dioxide $NO_2(g)$ (Li et al., 2008, 2009).

$$NO_2(g) \xrightarrow{\text{UV light}} NO(g) + O^*(g) \tag{6.R21}$$

Altshuller (1973) reports sulfur dioxide conversion to sulfuric acid in the troposphere for 18 urban centers and 12 rural sites. Tropospheric conversion—a solution-phase reaction occurring in water droplets—appears to be rate limited by ammonia and other less abundant weak bases (Hegg and Hobbs, 1978) dissolved in water droplets. Ammonia-dependent sulfur dioxide oxidation to sulfuric acid in the troposphere appears as a plateau in atmospheric sulfuric acid levels when sulfur dioxide levels exceed a certain threshold (cf. Appendix 6.A).

In summary, explosive volcanic eruption have a 25% sulfur dioxide washout rate, allowing a substantial fraction to reach the stratosphere. The sulfur dioxide washout on the order of 84–95% for biogenic and anthropogenic sources (Yoo et al., 2014). The oxidation of sulfur dioxide to sulfuric acid occurs primarily in water droplets which washout locally, a lesser fraction oxidizes photolytically in the stratosphere where the conversion rate is slower and global deposition is more uniform.

[10] The $SO_2(g)$ oxidation reactions (6.R17) and (6.R18) do not include the catalytic effect of nitrous oxide $NO(g)$.

Some of the strong base released by chemical weathering (Eq. 6.R11) is neutralized by atmospheric sulfuric acid deposition, and lesser amounts by sulfide mineral weathering (Eq. 6.R13). The product of this acid-base reaction (6.R22) is the following neutral salt solution that requires the sulfate correction in expression (expression 6.11) (Garrels and Mackenzie, 1971).

$$CaO(s) + 2H^{+}(aq) + SO_4^{2-}(aq) \longrightarrow Ca^{2+}(aq) + SO_4^{2-}(aq) + H_2O(l) \qquad (6.R22)$$

6.3.2.3 Nitrogen Oxides

Atmospheric chemists use the symbol NO_x to denote nitrogen monoxide NO(g) and nitrogen dioxide $NO_2(g)$ in Earth's atmosphere. Anthropogenic nitrogen oxide emissions from the combustion of petroleum and coal are predominantly NO_x compounds.

Biogenic nitrogen oxide emissions—dinitrogen oxide[11] $N_2O(g)$ and nitrogen monoxide NO(g)—are the product of *denitrification* by soil bacteria. Denitrifying bacteria are anaerobic bacteria that rely on nitrate respiration (cf. Chapter 9) when soils are waterlogged.

Nitrogen compounds in coal and petroleum undergo combustion reactions to produce the nitrogen oxides NO_x in power plant and auto emissions. Pyridine (structure **3**, Fig. 6.3) and pyrrole (structure **4**, Fig. 6.3) represent organonitrogen species in petroleum and coal (Snyder, 1970; George and Gorbaty, 1989; Gorbaty et al., 1989; Waldo et al., 1991, 1992).

Some of the atmospheric nitrogen dioxide $NO_2(g)$ is produced by nitrogen monoxide NO(g) oxidation (reaction 6.R23).

$$\underset{\text{OS(N)=+2}}{2\ NO(g)} + O_2(g) \longrightarrow \underset{\text{OS(N)=+4}}{2\ NO_2(g)} \qquad (6.R23)$$

Nitrogen dioxide $NO_2(aq)$ hydrolysis produces equal quantities of nitric $HNO_3(aq)$ and nitrous acid $HNO_2(aq)$ through disproportion (6.R24).

$$\underset{\text{OS(N)=+4}}{2\ NO_2(aq)} + H_2O(l) \longrightarrow \underset{\text{OS(N)=+3}}{HNO_2(aq)} + \underset{\text{OS(N)=+5}}{HNO_3(aq)} \qquad (6.R24)$$

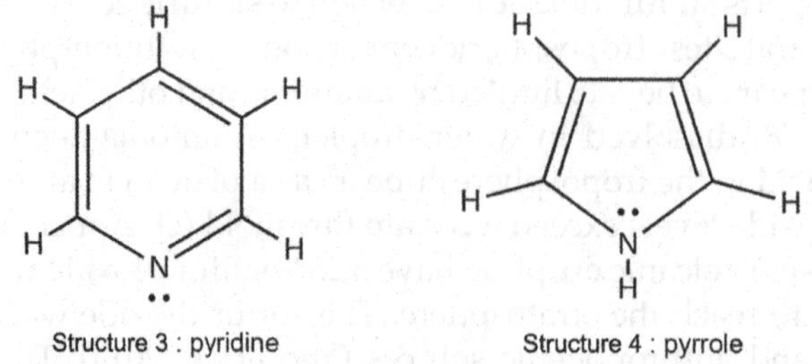

Structure 3 : pyridine

Structure 4 : pyrrole

FIG. 6.3 Molecular structures of pyridine (3) and pyrrole (4).

[11] Dinitrogen oxide $N_2O(g)$ is a relatively inert gas in the troposphere but reacts with ultraviolet radiation in the stratosphere to form nitrogen monoxide and, ultimately, nitric acid.

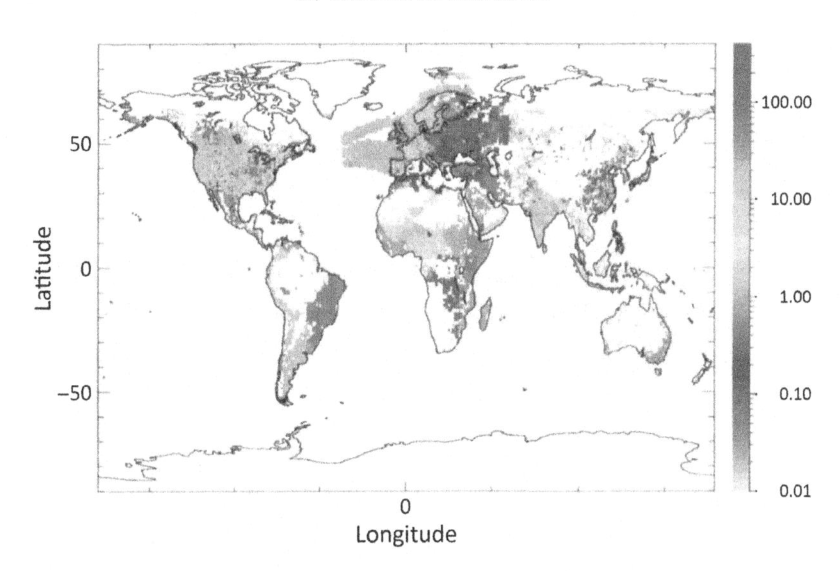

FIG. 6.4 Mole ratio of nitrogen to sulfur in atmospheric emissions reflects the relative importance of industrial and biogenic emissions. *(Reproduced with permission from Benkovitz, C.M., Scholtz, M.T., et al., 1996. Global gridded inventories of anthropogenic emissions of sulfur and nitrogen. J. Geophys. Res. 101(D22), 29239–29253.)*

(Benkovitz et al., 1996) found that atmospheric nitrogen emissions from anthropogenic and biogenic sources were roughly equal on the global scale, while atmospheric sulfur emissions are predominantly anthropogenic (cf. Section 3.2.2). The molar nitrogen-to-sulfur ratio for atmospheric emissions (Fig. 6.4) ranges from 10^{+4} to 10^{-4}.

Industrial and population centers with high sulfur oxide emissions are characterized by N/S ratios <0.01, while locations distant from industry and population centers with very low sulfur oxide emissions are characterized by N/S ratios >400. This variation in the nitrogen-to-sulfur ratio of atmospheric emissions reflects the importance of biogenic emissions.

Selecting data from Figs. 6.5 and 6.6 for sulfate and nitrate wet deposition and converting units from kg ha^{-1} to mol ha^{-1}, we can confirm the trend in the molar N/S depositional ratios reported by Benkovitz et al. (1996). The molar N/S depositional ratio for a site in western Kansas is 19 compared to a deposition ratio of 0.9 for a site in southeastern Ohio.

6.4 CARBONATE CHEMISTRY

Carbonate chemistry is an essential topic in any textbook or course on natural water chemistry—regardless of whether the emphasis is aquifer geochemistry, surface water limnology, marine chemistry, or soil chemistry. It is impossible to accurately simulate natural water chemistry or assess water analysis veracity absent a grasp of carbonate chemistry fundamentals. The following sections discuss the equilibrium reactions that control carbonate chemistry, alkalinity in relation to the aqueous carbon dioxide reference level, alkalinity measurement, and the geochemical implications of alkalinity.

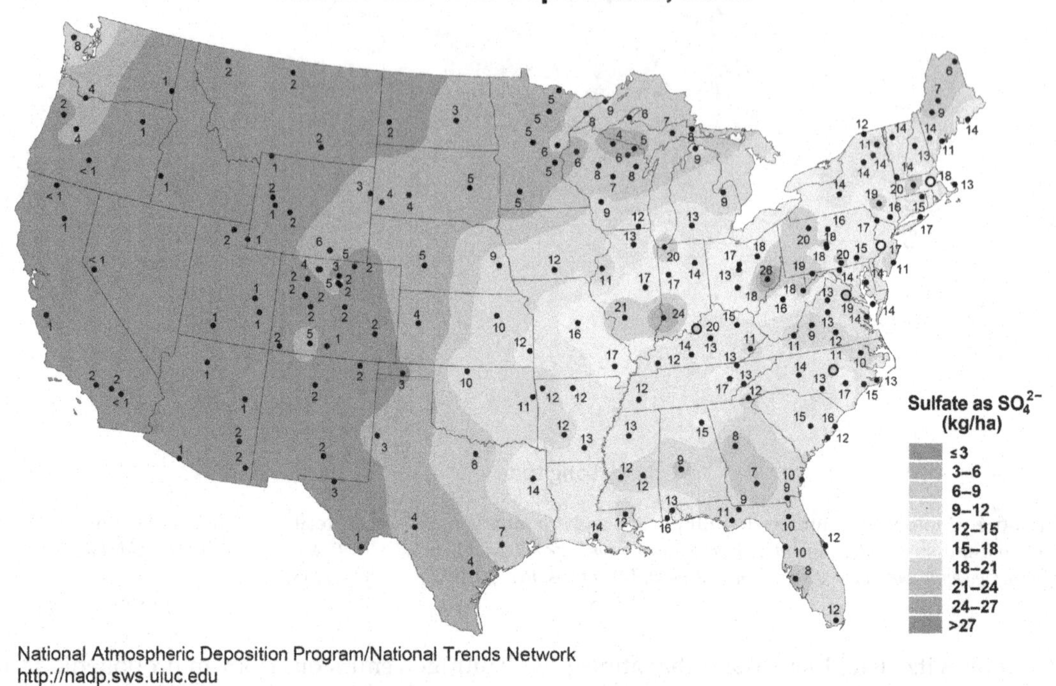

FIG. 6.5 Sulfate wet deposition in the contiguous United States during 2008 (National Atmospheric Deposition Program, National Trends Network).

6.4.1 Carbonate Equilibrium Reactions

Table 6.9 lists Earth's atmospheric composition. Of the 10 most abundant gases in Earth's atmosphere, carbon dioxide $CO_2(g)$ is the most water soluble. Regardless of its water solubility, atmospheric $CO_2(g)$ is not necessarily in solubility equilibrium with natural waters (Garrels and Mackenzie, 1971; Krešić, 2007).

Carbon dioxide dissolution in water (reaction 6.R25) is controlled by Henry's Law gas-solubility constant $H^{cp}_{CO_2} = 3.4 \times 10^{-2}$ (cf. Table 5.4; c_{CO_2} [mol dm^{-3}], p_{CO_2} [atm]).

$$CO_2(g) \longrightarrow CO_2(aq) \quad (6.R25)$$

Aqueous carbon dioxide $CO_2(aq)$ reacts water to form carbonic acid $H_2CO_3(aq)$ through reaction (6.R26). The equilibrium constant for hydration ($K^{\circ}_{\text{hydration}} = 1.18 \times 10^{-3}$) favors aqueous carbon dioxide $CO_2(aq)$ over $H_2CO_3(aq)$ by a ratio of 555-to-1 (Box 6.5).

$$CO_2(aq) + H_2O(l) \longrightarrow H_2CO_3(aq) \quad (6.R26)$$

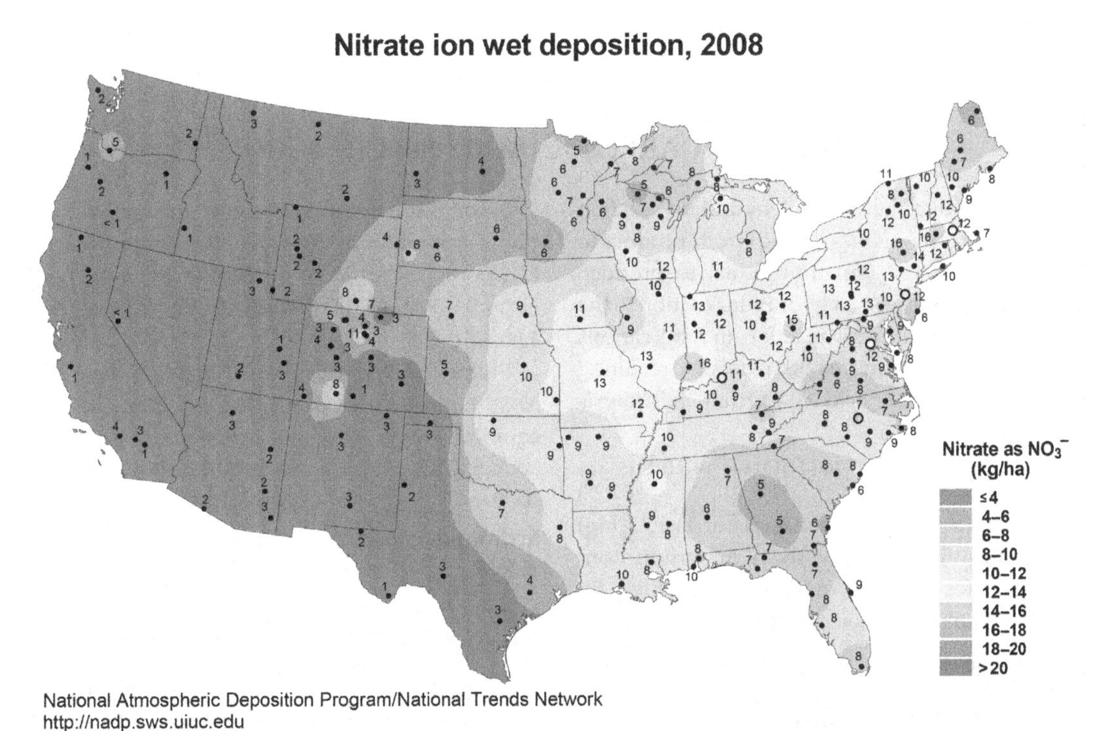

FIG. 6.6 Nitrate wet deposition in the contiguous United States during 2008 (National Atmospheric Deposition Program, National Trends Network).

TABLE 6.9 Volumetric Composition (ppm(v) Units: $cm^{-3}\ m^{-3}$) of the Most Abundant Gases in Earth's Atmosphere

Gas	Formula	Volume Fraction (ppm(v))
Dinitrogen	N_2	780,840
Dioxygen	O_2	209,460
Argon	Ar	9340.0
Carbon dioxide	CO_2	402.80
Neon	Ne	18.18
Helium	He	5.24
Methane	CH_4	1.79
Krypton	Kr	1.14
Dihydrogen	H_2	0.55
Nitrous oxide	N_2O	0.03

BOX 6.5

COMPOSITE CARBONIC ACID SPECIE $H_2CO_3^*$

Inverting the $CO_2(aq)$ hydration reaction (6.R26) converts $H_2CO_3(aq)$ from product to educt and $CO_2(aq)$ from educt to product. Adding the inverted hydration reaction to the $CO_2(aq)$ hydrolysis reaction (6.R28) cancels the $CO_2(aq)$ terms in both reactions, resulting in the carbonic acid $H_2CO_3(aq)$ hydrolysis reaction (6.R27).

$$H_2CO_3(aq) + H_2O(l) \longrightarrow H_3O^+(aq) + HCO_3^-(aq) \quad (6.R27)$$

The equilibrium constant for $H_2CO_3(aq)$ hydrolysis is simply the equilibrium constant for reaction (6.R28) $K^{\ominus}_{a1}$ divided by the equilibrium constant for reaction (6.R26) $K^{\ominus}_{hydration}$.

$$K^{\ominus}_{R\,29} = K^{\ominus}_{a1}/K^{\ominus}_{hydration} = 3.64 \times 10^{-4} \quad (6.14)$$

Some sources (Morel and Hering, 1993; Jaeschke, 2013) use $H_2CO_3^*$ to represent the combination of two weak acid species $CO_2(aq)$ and $H_2CO_3(aq)$ as a single weak carbonate acid. Other sources use $H_2O \cdot CO_2(aq)$ to denote the same combination. These notations are both misleading and confusing.

First, chemists have identified both $CO_2(aq)$ and $H_2CO_3(aq)$ species in solution. The same cannot be said about sulfur dioxide dissolved in water: the $SO_2(aq)$ specie does exist in aqueous solution but the $H_2SO_3(aq)$ has never been verified as a distinct aqueous specie.

Second, some sources assigned an equilibrium constant for hydrolysis of the composite $H_2CO_3^*$ species that is identical to the equilibrium constant of reaction (6.R28): $K^{\ominus}_{a1} = 4.30 \times 10^{-7}$.

Stumm and Morgan (2013) define a *composite* hydrolysis constant derived from the equilibrium constants for reactions (6.R27) and (6.R26). The *composite* equilibrium constant $K^{\ominus}_{composite}$ is slightly less than the hydrolysis constant $K^{\ominus}_{a1}$ listed for reaction (6.R28).

$$K^{\ominus}_{composite} = \frac{3.64 \times 10^{-4}}{1 + \left(1.18 \times 10^{-3}\right)^{-1}} = 4.29 \times 10^{-7} \quad (6.15)$$

Furthermore, Stumm and Morgan (2013) state

> *The true H_2CO_3 is a much stronger acid …than the composite $H_2CO_3^*$, because less than 0.3% of the CO_2 is hydrated at 25°C.*

To be completely clear, there is no need for the hydrolysis constant $K^{\ominus}_{composite}$. The four equilibrium constants listed in Table 6.10 completely represent the aqueous chemistry of the four species appearing in reactions (6.R25) through (6.R29). Hydrolysis reaction (6.R27) and its equilibrium constant $K^{\ominus}_{composite}$ (expression 6.14) are both derived from the equations mentioned earlier.

Continued

BOX 6.5 *(cont'd)*

TABLE 6.10 Equilibrium Constants for Key Dissolution, Hydration, and Hydrolysis Reactions Involving Carbonate Species

Reaction Number	Symbol	Equilibrium Constant
6.R25	$H^{cp}_{CO_2}$	3.40E−02
6.R26	$K^{\ominus}_{hydration}$	1.18E−03
6.R28	$K^{\ominus}_{a1}$	4.30E−07
6.R29	$K^{\ominus}_{a2}$	4.68E−11

In an earlier era where chemists used algebraic methods to solve water chemistry problems; defining composite species and equilibrium constants made sense because it simplified complex algebraic problems. The current era, however, relies on computed-based numerical methods where this approach yields no advantage and, more to the point, may confuse novice users.

The dominant weak carbonate acid is $CO_2(aq)$ whose hydrolysis is given by reaction (6.R28).

$$CO_2(aq) + 2 \cdot H_2O(l) \longrightarrow H_3O^+(aq) + HCO_3{}^-(aq) \quad (6.R28)$$

Reaction (6.R29) is the bicarbonate hydrolysis reaction yielding carbonate $CO_3{}^{2-}(aq)$, the dibasic conjugate of the weak acid $CO_2(aq)$.

$$HCO_3{}^-(aq) + H_2O(l) \longrightarrow H_3O^+(aq) + CO_3{}^{2-}(aq) \quad (6.R29)$$

Respiration by organisms—terrestrial and aquatic plant roots, and microbes living in soils and the water column—combined with sluggish gas exchange with the above-ground atmosphere result in $CO_2(aq)$ concentrations that exceed the passive $CO_2(g)$ dissolution from the above-ground atmosphere. Dissolved $CO_2(aq)$ concentrations from biological sources vary temporally, on both a daily and seasonal time scale, and spatially, depending on conditions influencing biological activity. For instance, soil $CO_2(aq)$ concentrations at the height of the growing season can be 10–100 times higher than predicted from the above-ground atmosphere $CO_2(g)$ partial pressure.

The variations in $CO_2(aq)$ concentration on carbonate chemistry are shown in Figs. 6.7 and 6.8.

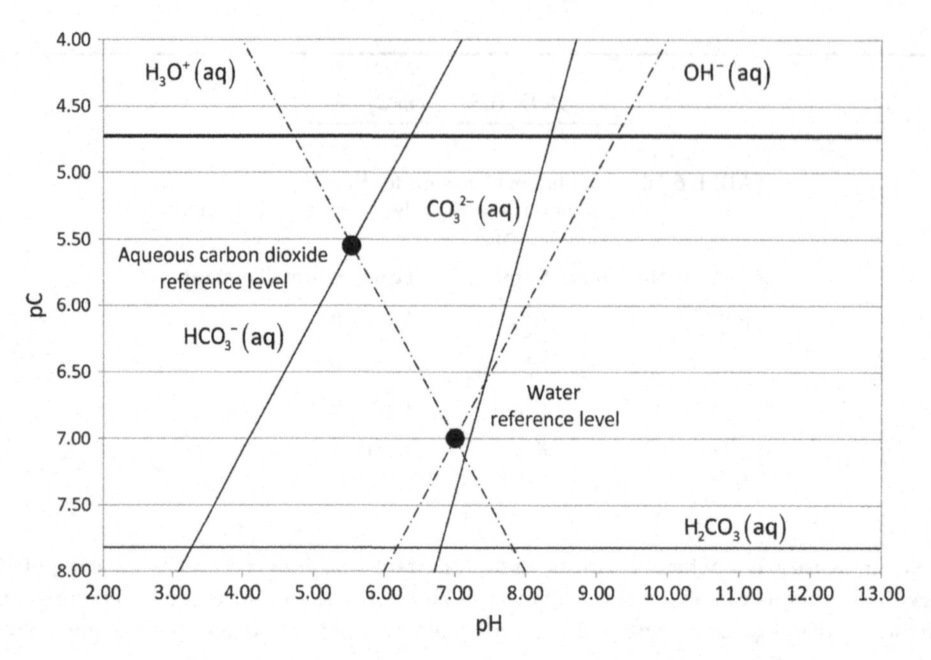

FIG. 6.7 The effect of pH on carbonate solution species, plotted as the negative logarithm of specie activity. Aqueous $CO_2(aq)$ activity is controlled by $CO_2(g)$ at a partial pressure representative of the above-ground atmosphere: $p_{CO_2} = 4.028 \times 10^{-4}$ atm.

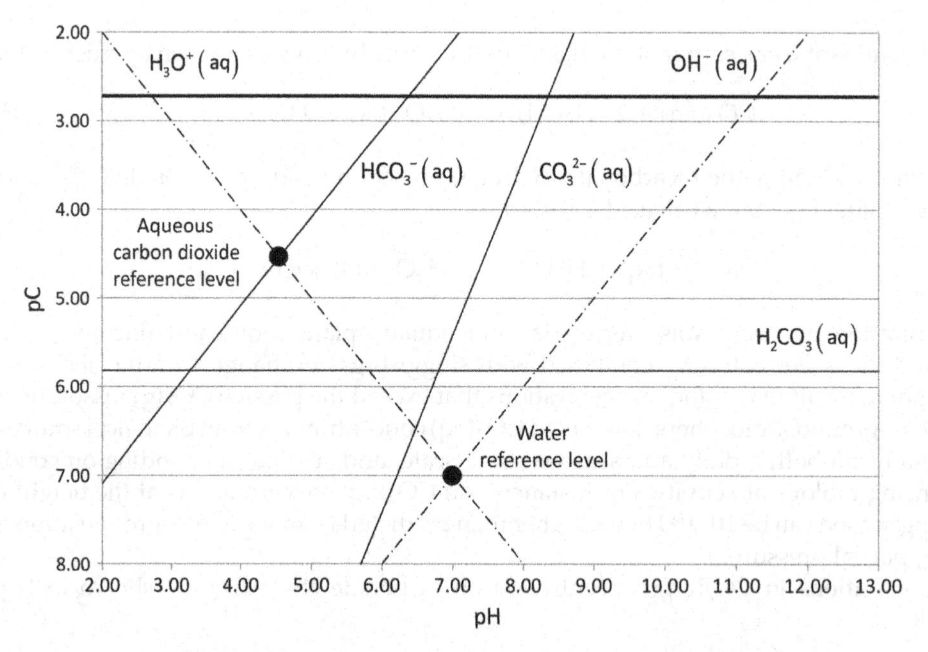

FIG. 6.8 The effect of pH on carbonate solution species, plotted as the negative logarithm of specie activity. Aqueous $CO_2(aq)$ activity is controlled by $CO_2(g)$ at a partial pressure representative of the soil atmosphere: $p_{CO_2} = 4 \times 10^{-2}$ atm.

6.4.2 Aqueous Carbon Dioxide Reference Level

Environmental chemists—recognizing the presence of aqueous carbon dioxide $CO_2(aq)$ in all natural waters and its effect on the pH—define *mineral acidity* and *alkalinity* relative to a reference point determined by the *proton balance* of water in equilibrium with carbon dioxide $CO_2(g)$. This reference point is the *aqueous carbon dioxide reference level*, indicated in both Figs. 6.7 and 6.8.

The *aqueous carbon dioxide reference level* is derived from the charge-balance and proton-balance expressions for a carbonate system containing spectator cations and anions.

The *charge-balance* expression is similar to expression (6.7) but contains the carbonate ion species defined by the reactions listed in Table 6.10.

$$c_{Na^+} + c_{H_3O^+} = c_{Cl^+} + c_{OH^-} + c_{HCO_3^-} + 2 \cdot c_{CO_3^{2-}} \tag{6.16}$$

Rearranging the charge-balance expression yields the following *proton-balance* expression. The proton-balance expression below is similar to expression (6.8) except it contains the additional carbonate ion species defined by reactions listed in Table 6.10.

$$c_{Na^+} - c_{Cl^-} = \left(c_{OH^-} + c_{HCO_3^-} + 2 \cdot c_{CO_3^{2-}}\right) - c_{H_3O^+} \tag{6.17}$$

The ions on the left-hand side of proton-balance expression (6.17) are spectator ions that do not participate in any aqueous proton-transfer reactions. The ions on the right-hand side of the proton-balance expression (6.17) are the only acid-base species in this solution. The relative activity of the ions $H_3O^+(aq)$ and $OH^-(aq)$ is constrained by the water self-ionization equilibrium quotient expression (6.1) while the equilibrium quotients for carbonate hydrolysis reactions (cf. reactions 6.R26–6.R29 and Table 6.10) constrain the relative concentration of the solution carbonate species.

6.4.3 Alkalinity and Mineral Acidity

Geochemists define *carbonate alkalinity* A_C relative to the carbon dioxide reference level. Alkalinity is by definition a positive value; alkalinity cannot be *negative*.

$$A_C \equiv \left(c_{OH^-} + c_{HCO_3^-} + 2 \cdot c_{CO_3^{2-}}\right) - c_{H_3O^+} > 0 \tag{6.18}$$

If the spectator ion charge-balance favors cations (i.e., $c_{Na^+} > c_{Cl^-}$ in expression 6.17) the proton-balance expression tell us the excess cation charge in solution is balanced by an equal negative charge supplied by the following three anions: base anion $OH^-(aq)$, conjugate base anion $HCO_3^-(aq)$, and conjugate base anion $CO_3^{2-}(aq)$. Water analysis typically quantifies alkalinity by titrating a water sample with a standard acid to the methyl-orange end-point (cf. Box 6.6)

BOX 6.6

QUANTIFYING ALKALINITY

Total alkalinity A_T is the equivalent concentration ($mol_c\ dm^{-3}$) of all weak-acid conjugate bases in solution. The carbonate conjugate bases bicarbonate HCO_3^- and carbonate CO_3^- typically dominate most natural waters but one cannot assume total alkalinity A_T and carbonate alkalinity A_C are identical.

Silicate conjugate bases become soluble only at high pH values that rarely occur in natural waters. Phosphate conjugate bases are typically insignificant compared to the carbonates and silicates in freshwater systems. Borate conjugate bases attain significant concentrations only in saline lakes and seawater.

The operational definition of total alkalinity is acid neutralizing capacity when the water sample is titrated to the methyl-orange (CAS 547-58-0) end-point by a strong acid.

$$K_a^{\circ} = 10^{-3.7} = \frac{a_{H_3O^+} \cdot a_{MethylOrange^-}}{a_{HMethylOrange}} \tag{6.19}$$

The color change associated with the methyl orange indicator begins at pH 4.4, which roughly 2 pH units below $pK_{a1}^{\circ} = 6.37$ (cf. Table 6.10, reaction 6.R28). At the methyl-orange end-point 99.9% of the neutralizing capacity of HCO_3^- has been neutralized.

Carbonate CO_3^{2-} is also quantified by acidometric titration to pH 8.3, the nominal phenolphthalein (CAS 77-09-8) end-point. At the phenolphthalein end-point 90% of the neutralizing capacity of CO_3^{2-} has been neutralized (cf. Table 6.10, reaction 6.R29).

$$K_a^{\circ} = 10^{-9.3} = \frac{a_{H_3O^+} \cdot a_{Phenolphthalein^-}}{a_{HPhenolphthalein}} \tag{6.20}$$

Water chemists frequently quote the results of a methyl-orange end-point titration as *acid neutralizing capacity* because it does not discriminate between carbonate alkalinity (6.18) and other conjugate bases in the water sample. A thorough carbonate alkalinity analysis would also measure dissolved inorganic carbon (DIC) and correct for dissolved carbon dioxide, besides measuring acid neutralizing capacity by acidometric titration.

Geochemists define *mineral acidity M* relative to the carbon dioxide reference level. Mineral acidity is by definition a positive value. If the total spectator anion charge exceeds the total spectator cation charge (i.e., $c_{Cl^-} > c_{Na^+}$) the proton-balance expression tell us the excess anion charge concentration in solution is balanced by an equal concentration of acid cation $H_3O^+(aq)$.

$$M \equiv c_{H_3O^+} - \left(c_{OH^-} + c_{HCO_3^-} + 2 \cdot c_{CO_3^{2-}}\right) > 0 \tag{6.21}$$

Mineral acidity *M* (e.g., acid rain or acidic mine drainage water) occurs in natural waters containing an excess of strong mineral acids.

6.4.4 Geochemical Implications of Alkalinity

The *carbon dioxide reference level* is defined at the pH where the spectator cation charge equals the spectator anion charge (expression 6.17). Figs. 6.7 and 6.8 reveal that while the *water reference level* is independent of $CO_2(aq)$ the *carbon dioxide reference level* shifts to lower pH values as $CO_2(aq)$ increases.

$$a_{H_3O^+} = a_{HCO_3^-} + 2 \cdot a_{CO_3^{2-}} + a_{OH^-} \tag{6.22}$$

Expression (6.22) has important consequences for water chemistry simulations based on water analysis data. Simply put, alkalinity by acidometric titration is a fairly reliable water parameter which can be validated by computing the spectator ion charge balance. By contrast, water analysis pH values should be considered suspect until validated by simulation (Example 6.3).

EXAMPLE 6.3

Example Permalink

http://soilenvirochem.net/nuT53r

Simulate solution alkalinity based on water analysis and evaluate the veracity of water analysis pH and alkalinity values.

The water sample was drawn from a well near Green Bay, Wisconsin in September 2003. Table 6.11 lists major ion concentrations from the water analysis in column 2. Additional water analysis values include: pH = 7.70 and the EC[12] $EC = 1.46\ \mathrm{dS\ m^{-1}}$.

TABLE 6.11 Water Analysis Results for Well Water Sample 213-1

Analyte	mg kg^{-1}	mol kg^{-1}
Ca	151	3.77E−03
K	14	3.58E−04
Mg	45	1.85E−03
Na	113	4.92E−03
Cl	8	2.26E−04
Bicarbonate	167.80	2.75E−03
Sulfate	333	3.47E−03
Phosphate	0.01	1.05E−07
Nitrate-N	0.44	3.14E−05

Part 1. Convert the mass-concentration units into molal units.

Water analysis major ion concentrations (units: mol kg^{-1}) appear in column 3 of Table 6.11.

Part 2. Prepare and compile the *ChemEQL* input matrix and run a water chemistry simulation at the water analysis pH value. This tests the veracity of the water analysis pH value.

Table 6.12 lists the $CO_2(aq)$ activity and the effective $CO_2(g)$ partial pressure that result from fixing the simulation pH at the water analysis value of pH 7.7. The $CO_2(g)$ partial pressure us almost 10-fold higher than the above-ground atmosphere. There is no indication biological activity or geochemical processes would account for elevated $CO_2(aq)$ for this particular groundwater source. Studies measuring indoor workstation $CO_2(g)$ partial pressure report values much higher atmospheric (Persily and Gorfain, 2008).

Part 3. Prepare and compile *ChemEQL* input matrix and run water chemistry simulation to estimate the solution pH.

Solution pH largely depends alkalinity and $CO_2(aq)$ levels. The water analysis reports $c_{HCO_3^-} = 2.75 \times 10^{-3}$ mol kg^{-1}, presumably by acidometric titration to the methyl-orange end-point. Deffeyes (1965) expression (6.23) yields an estimated alkalinity of $A_T = 9.23 \times 10^{-3}$ mol kg^{-1}, about threefold higher than the direct measurement.

TABLE 6.12 Simulated Aqueous Carbon Dioxide Activity and Effective Carbon Dioxide Partial Pressure Corresponding to the Water Analysis pH = 7.7

Specie	Intensive Variable	Value	Units
$CO_2(aq)$	a_{CO_2}	1.030E−04	mol dm^{-3}
$CO_2(g)$	p_{CO_2}	3.029E−03	atm
$CO_2(g)$	p_{CO_2}	3030	ppm(v)

One option would simulate aquifer pH for a one-phase (solution) system with total carbonate and alkalinity both set by water analysis HCO_3^- concentration. A second option would assume the aquifer $CO_2(aq)$ concentration is determined by the above-ground $CO_2(g)$ partial pressure, equivalent to assuming the aquifer residence time is sufficiently long for above-ground $CO_2(g)$ to saturate the aquifer and, furthermore, there are no extraneous biological or geological $CO_2(aq)$ sources. This is equivalent to the assumption behind expression (6.12) Garrels and Mackenzie (1971).

TABLE 6.13 Simulated Aqueous Carbon Dioxide Activity and Effective Carbon Dioxide Partial Pressure Corresponding to the Water Analysis $c_{HCO_3^-} = 2.75 \times 10^{-3}$ mol kg^{-1} and $CO_2(aq)$ Saturation by $p_{CO_2} = 373.2$ ppm(v)

Specie	Intensive Variable	Value	Units
$CO_2(aq)$	a_{CO_2}	1.269E−05	mol dm^{-3}
$CO_2(g)$	p_{CO_2}	3.732E−04	atm
$CO_2(g)$	p_{CO_2}	373.2	ppm(v)

Table 6.13 lists the $CO_2(aq)$ activity and the effective $CO_2(g)$ partial pressure that result from fixing the simulation alkalinity at the water analysis value of $A_C = 2.75 \times 10^{-3}$ mol kg^{-1} in

a one-phase (solution only) system. Since carbonate is represented by component CO2(aq) this simulation requires replacing H+ with OH−, setting mode(OH−) = total, and assigning OH− an initial concentration equal to the water analysis alkalinity.

Part 3 simulation yields pH = 8.57 and $A_C = 2.75 \times 10^{-3}$ mol kg^{-1}. Since the water analysis bicarbonate is measured by acidometric titration to the methyl-orange end-point, this validates the simulation.

[12] Soil scientists identify this measurement as *EC* while hydrologists refer to this analysis as *specific conductance*. The value is temperature corrected to 25°C.

Matthess (1982) makes the following statement regarding free dissolved (aggressive) CO_2 in groundwater.

> The usual range for free dissolved CO_2 in groundwater is 10–20 mg/L but more CO_2 can occur.

There are notable cases where dissolved $CO_2(aq)$ in groundwater is supersaturated relative to equilibrium with atmospheric $CO_2(g)$. Geologic processes that supersaturate groundwater include degassing from magma bodies (De Gregorio et al., 2011; Nordstrom et al., 2005) and neutralization reactions when acidic hydrothermal waters contact carbonate formations (Choi et al., 1998). Respiration by organisms living in soil, surface water sediments and the water column can elevate dissolved $CO_2(aq)$ 10- to 100-fold during the growing season. Anaerobic mineralization of buried organic matter in shallow sediments of the Bengal delta (Hasan et al., 2007; Sikdar and Chakraborty, 2008; Bhattacharya et al., 2009) result in free dissolved $CO_2(aq)$ well in excess of concentrations quoted by Matthess (1982).

The reader should expect in situ aqueous $CO_2(aq)$ concentrations in surface and soil water samples to differ considerably from values measured in the laboratory after transport and storage (Example 6.4). For instance, in situ biological respiration may support high $CO_2(aq)$ concentrations that drop as the sample degases during transport and storage. The surface water analyses assembled by Livingstone (1963) do not list pH analyses for this reason. The groundwater analyses assembled by (White et al., 1963) typically list pH analyses but the editors warn the reader.

> Determinations of pH and perhaps of bicarbonate must be made in the field when samples are collected, if they are to represent accurately the conditions in the aquifer. Practically no data of this type are available. Analyses [in this report] represent the usual laboratory determinations made after the samples had been stored for several days or weeks.

Crandall et al. (1999) simulated surface and groundwater chemistry using water analysis pH values that require adding substantial acidity to maintain pH (cf. simulation Example 5.11). This reveals a critical deficiency in the water analysis that could be assigned to one of the following: (1) analytic pH biased by atmospheric CO_2 dissolution during storage or laboratory analysis, (2) a 6–15% error in bicarbonate analysis, or (3) significant errors in major ion analysis.

Example 6.5 checks the relative veracity of water analysis pH and alkalinity for a groundwater sample drawn from a granitic aquifer. The water analysis reports pH 7.0 while two separate water chemistry pH simulations constrained by water analysis pH yield values a full pH unit higher than the water analysis pH. The impact of dissolved carbonates on groundwater chemistry is further illustrate by another pH simulation in Example 6.5 yielding an estimated pH 11 in the absence of dissolved carbonates.

Deffeyes (1965) regrouped the typical major ion charge balance, moving the spectator-ion charge balance on one side of the equation, results in an alternative alkalinity expression. Expression (6.23) allows the user to use major ion concentrations to independently validate water analysis alkalinity.

$$\left(c_{Na^+} + c_{K^+} + 2 \cdot c_{Ca^{2+}} + 2 \cdot c_{Mg^{2+}}\right) - \left(c_{Cl^-} + 2 \cdot c_{SO_4{}^{2-}}\right)$$
$$= \left(c_{HCO_3{}^-} + 2 \cdot c_{CO_3{}^{2-}} + c_{OH^-} - c_{H_3O^+}\right)$$
$$A_C = \left(c_{Na^+} + c_{K^+} + 2 \cdot c_{Ca^{2+}} + 2 \cdot c_{Mg^{2+}}\right) - \left(c_{Cl^-} + 2 \cdot c_{SO_4{}^{2-}}\right) \quad (6.23)$$

Table 6.11 lists the most abundant ions in seawater but does not report total alkalinity A_T. There are two ways, however, to estimate seawater alkalinity. First, we could assume the dissolved nitrogen is entirely nitrate $NO_3{}^-$. The spectator ion charge balance (expression 6.23) from this assumption yields a total alkalinity of $A_T = 2.299 \times 10^{-3}$ $mol_c\ dm^{-3}$.

Alternatively, we could assume dissolved carbon is entirely bicarbonate $HCO_3{}^-$. The spectator ion charge balance (expression 6.23) from this assumption yields a total alkalinity of $A_T = 2.331 \times 10^{-3}$ $mol_c\ dm^{-3}$. These two estimates differ by 1.4%. Keep in mind the values in 11 are not based on a single analysis but averaged over many analyses, yet as in the estimates of the continental lithosphere, deviation from the charge-balance condition is quite small (Examples 6.4 and 6.5).

EXAMPLE 6.4

Example Permalink

http://soilenvirochem.net/y1d803

Calculate the $CO_2(aq)$ concentration in a 1.0×10^{-3} M $Na_2CO_3(aq)$ solution (i.e., a one-phase, solution only, system) using the *RICE* table method and the following base hydrolysis reaction and equilibrium reaction quotient.

Example 6.2 estimated the pH of a 1.0×10^{-3} M $Na_2CO_3(aq)$ solution. This example estimates the $CO_2(aq)$ in the same solution. The $CO_2(aq)$ in a solution-only system is a product of the carbonate hydrolysis reactions listed in Table 6.10.

$$HCO_3{}^{2-}(aq) + H_2O(l) \longleftrightarrow CO_2(aq)(aq) + OH^-(aq)$$
$$K_{b2}^{\circ} = 2.33 \times 10^{-8} = \frac{a_{CO_2} \cdot a_{OH^-}}{a_{HCO_3{}^-}}$$

Step 1. Prepare an *RICE* table (Table 6.14) representing second-base hydrolysis step of $CO_3^-(aq)$.

The second-base hydrolysis constant $K^{\ominus}_{b2}$ is much smaller than the first-base hydrolysis constant $K^{\ominus}_{b1}$, therefore the hydroxide ion $OH^-(aq)$ concentration results almost entirely from the first hydrolysis step.

TABLE 6.14 *RICE* Table for Bicarbonate Base Hydrolysis

Reaction →	$HCO_3^-(aq)$ (mol dm^{-3})	$H_2O(l)$	↔	$CO_2(aq)$ (mol dm^{-3})	$OH^-(aq)$ (mol dm^{-3})
Initial	3.68×10^{-4}			0	3.68×10^{-4}
Reaction	$-y$			$+y$	$+y$
Equilibrium	$3.68 \times 10^{-4} - y$			y	$3.68 \times 10^{-4} + y$

Step 2. Using terms from the *RICE* table write the equilibrium quotient expression for the first-base hydrolysis step and solve the second-order polynomial using the quadratic equation.

$$y^2 + \left(3.68 \times 10^{-4}\right) \cdot y = K^{\ominus}_{b2} \cdot \left(3.68 \times 10^{-4} - y\right)$$

$$y^2 + \left(3.68 \times 10^{-4} + K^{\ominus}_{b2}\right) \cdot y - \left(3.68 \times 10^{-4} \cdot K^{\ominus}_{b2}\right) = 0$$

$$B = \left(3.68 \times 10^{-4} + K^{\ominus}_{b2}\right) \equiv 3.68 \times 10^{-4}$$

$$C = \left(3.68 \times 10^{-4} \cdot K^{\ominus}_{b2}\right) = -8.57 \times 10^{-12}$$

$$y_+ = \frac{-B + \sqrt{B^2 - (4 \cdot C)}}{2} = 3.68 \times 10^{-4}$$

$$CO_2(aq) = y = 3.68 \times 10^{-4}$$

This result confirms $HCO_3^-(aq)$ undergoes negligible base hydrolysis in a 1.0×10^{-3} M $Na_2CO_3(aq)$ solution-only system.

Step 3. Simulate the pH of a 1.0×10^{-3} M $Na_2CO_3(aq)$ solution using bicarbonate HCO_3^- as the carbonate component.

A minimalist simulation, disregarding charge balance and spectator ions, contains two components: HCO3− and H+. Setting the initial HCO_3^- concentration equal to 1.0×10^{-3} M supplies the correct total carbonate but only half of the alkalinity of a 1.0×10^{-3} M $Na_2CO_3(aq)$ solution. Adding the additional alkalinity in a *ChemEQL* simulation requires replacing component H+ with OH−, setting mode(OH−) = total and assigning it an initial concentration of 1.0×10^{-3} M. The total alkalinity with these components and initial concentrations is the desired 1.0×10^{-3} M concentration.

The simulated $CO_2(aq)$ concentration is 2.40×10^{-8} M, compared to the *RICE* table estimate of 3.68×10^{-8} M.

The pH and $CO_2(aq)$ concentrations in this example are not representative of natural groundwater systems. Groundwater residence time appear to be sufficient for above-ground $CO_2(g)$ to saturate groundwater to $CO_2(aq)$ concentrations on the order of 10^{-5} M, a 1000-fold higher than estimated for this one-phase system (Example 6.5).

EXAMPLE 6.5

Example Permalink

http://soilenvirochem.net/jHpi4Q

Simulate groundwater pH based solely on water analysis alkalinity.

The water sample was drawn from a well located in McCormick County, South Carolina in November 24, 1954 White et al. (1963). Table 6.15 lists major ion concentrations from the water analysis in column 2. Additional water analysis values include: pH = 7.00 and $EC = 0.076\ \mathrm{dS\ m^{-1}}$.

TABLE 6.15 Water Analysis Results for Well Water Sample Collected at John de la Howe School, McCormick, South Carolina

Analyte	$\mathrm{mg\ kg^{-1}}$	$\mathrm{mol\ kg^{-1}}$
Ca	13.0	3.244E−04
K	3.5	8.951E−05
Mg	4.3	1.769E−04
Na	8.4	3.654E−04
Cl	8.0	2.257E−04
F	0.2	1.053E−05
Bicarbonate	72.0	1.180E−03
Sulfate	6.90	7.183E−05
Phosphate	0.10	1.042E−06
Nitrate	0.40	6.452E−06
Silica	35.00	5.826E−04

Part 1. Convert the mass-concentration units into molal units.

Water analysis major ion concentrations (units: $\mathrm{mol\ kg^{-1}}$) appear in Table 6.15, column 3.

Part 2. Prepare and compile *ChemEQL* input matrix and run water chemistry simulation to estimate the solution pH.

Solution pH largely depends alkalinity and $CO_2(aq)$ levels. The water analysis reports $c_{HCO_3^-} = 1.18 \times 10^{-3}\ \mathrm{mol\ kg^{-1}}$, presumably by acidometric titration to the methylorange end-point. The Deffeyes (1965) expression (6.23) yields an estimated alkalinity of $A_T = 1.19 \times 10^{-3}\ \mathrm{mol\ kg^{-1}}$, validating the direct acidometric titration analysis.

One option would simulate aquifer pH as a one-phase (solution-only) system with alkalinity assigned to component HCO3− with mode(HCO3−) = total. The results of the first option are presented here because they are equivalent to results from the second option.

The second option assumes the aquifer $CO_2(aq)$ concentration is determined by the above-ground $CO_2(g)$ partial pressure, equivalent to assuming the aquifer residence time is sufficiently long for above-ground $CO_2(g)$ to saturate the aquifer and, furthermore, there are no extraneous biological or geological $CO_2(aq)$ sources. This is equivalent to the assumption behind expression (6.12) Garrels and Mackenzie (1971).

The two-phase (gas-solution) simulation designates CO2(aq) as the carbonate component, sets mode(CO2(aq)) = free, assigns the initial CO2(aq) concentration using the fixed-fugacity method for above-ground $CO_2(g)$ partial pressure. Alkalinity is simulated by replacing component H+ with component OH−, setting mode(OH−) = total, and assigning alkalinity as the initial OH− concentration.

TABLE 6.16 Simulated by Representing Carbonate Using Component HCO3− and Assigning It the Water Analysis $c_{HCO_3^-} = 1.18 \times 10^{-3} \text{mol kg}^{-1}$

Specie	Intensive Variable	Value	Units
$CO_2(aq)$	a_{CO_2}	1.75E−05	mol dm^{-3}
$CO_2(g)$	p_{CO_2}	5.14E−04	atm
$CO_2(g)$	p_{CO_2}	514	ppm(v)

Table 6.16 lists the $CO_2(aq)$ activity and the effective $CO_2(g)$ partial pressure that result from fixing the simulation alkalinity at the water analysis value of $A_C = 1.18 \times 10^{-3}$ mol kg^{-1}. The Part 2 simulation yields pH = 8.16, a full pH unit higher than the water analysis pH value.

Part 3. Prepare and compile *ChemEQL* input matrix and run water chemistry simulation to estimate the solution pH for a solution lacking dissolved carbonate.

The input matrix is identical to that prepared for Part 2 except for two changes: (1) the bicarbonate component HCO3− is omitted and (2) the water analysis alkalinity is assigned to component OH−. The second change requires replacement of component H+ by OH−, designating mode(OH−) = total and assigning component OH− an initial concentration equal to 1.18×10^{-3} mol kg^{-1}.

Absent dissolved carbonates, the simulated pH would be 11, much higher than observed in typical groundwater samples.

6.5 ALKALI AND SODIC SOILS

This section examines the impact of Na^+ ions on clay colloid chemistry. Given the relative abundance of the four most common Group 1 (Na^+ and K^+) and Group 2 cations (Ca^{2+} and Mg^{2+}) in Earth's lithosphere, soil pore water is typically dominated by Ca^{2+} ions unless saline groundwater discharge or mineral weathering in arid climates alters the pore water chemistry.

In Chapter 3 we learned that smectite clay minerals were capable of osmotic swelling if saturated by either Na^+ or Li^+ ions (the latter being inconsequential in most soil or groundwater settings). Smectite clays were limited to crystalline swelling if saturated by larger

Group 1 ions (e.g., K^+) and Group 2 ions. This section extends the clay chemical impact of Na^+ ions beyond swelling behavior and considers the role of soil alkalinity in altering pore water chemistry.

We begin with a discussion of the impact of clay content and mineralogy on soil mechanical and physical properties (soil strength, hydraulic conductivity, and erodibility) and continue with an examination of the effect of Na^+ ions on these soil mechanical properties. Finally, we discuss the pore water chemistry required to enrich the exchange complex in Na^+ ions by depleting it of Ca^{2+} ions.

6.5.1 Exchangeable Sodium and Clay Plasticity

Soil scientists apply several criteria to identify soils that either exhibit low hydraulic conductivity or are at risk for developing low hydraulic conductivity as a result of clay chemistry changes caused by Na^+ accumulating on the clay exchange complex. The loss of soil hydraulic conductivity is a particular concern in arid climate zones where precipitation is insufficient to leach soluble salts from the soil profile. Soil sodicity degrades soil physical properties through the effect of exchangeable Na^+ ions on clay behavior.

Sodicity is usually linked to either reduced hydraulic conductivity that complicates salinity leaching from the soil profile or surface sealing that prevents soil moisture infiltration. The increased potential for surface sealing adds yet another soil property degraded by sodicity: erodibility. Soil scientists have used clay activity A, however, along with the exchangeable sodium percentage (ESP) to identify highly erodible or dispersive soils[13] (Resendiz, 1977; Lebron et al., 1994). Erodibility criteria based solely on clay activity A fails to account for the impact of ESP on clay plasticity.

The empirical hydraulic conductivity parameters in Chapter 2 (cf. *Soil Moisture and Hydrology*, Appendix 2.F) represent the maximum hydraulic conductivity for each soil texture—that is, the clay fraction retains sufficient plasticity to maintain an open soil fabric with maximum pore size and, hence, hydraulic conductivity. Chemically altering soils by saturating the exchange complex with Na^+ ions increases the soil activity A and significantly lowering the air-water permeability ratio (Allison et al., 1954).

6.5.1.1 Soil Erodibility

Fig. 6.9 illustrates the effect of two soil treatments on the activity A of soil clay (Lebron et al., 1994). The natural soils (filled circles) are saline-sodic, with Na^+ ions occupying 90% or more of the exchange complex and a pore water EC in the range of 6.2–40.5 $dS\ m^{-1}$. Replacing Na^+ ions with Ca^{2+} ions by washing the soils with calcium chloride solutions (open triangles) increases the activity A of the clay fraction (cf. expression (3.4)). Washing the soils with pure water to lower salinity reduces the fraction of Na^+ ions occupying the exchange complex to 20–30% and significantly lowers clay activity A.

Geotechnical engineers and soil scientists (Resendiz, 1977; Lebron et al., 1994) associate soil erodibility with both clay content and activity A. Soil erosion begins with the detachment of soil particles by raindrop impact or the impact of fine windblown mineral particles bouncing

[13] Soil dispersion, the detachment of clay and fine silt particles, increases erodibility, surface sealing, and a loss of hydraulic conductivity.

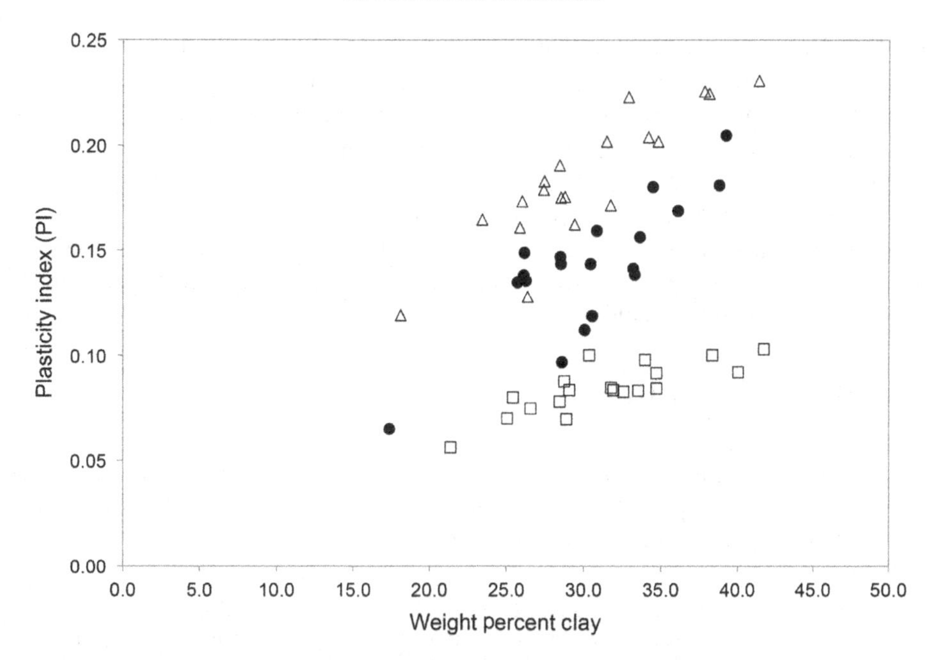

FIG. 6.9 The activity A of soil clay in 19 saline-sodic soils from Spain: (A) *filled circles* represent the untreated soil plasticity index (B) *open triangles* represent the plasticity index for soils washed with calcium chloride to replace exchangeable Na^+ ions with Ca^{2+} ions; and (C) *open squares* represent the plasticity index for soils washed with pure water to lower salinity. *(Modified from Lebron et al. (1994)). The slope of each treatment is the activity A (cf. expression (3.4)).*

across the soil surface. Soils with low clay contents or containing low-activity A clay are more susceptible to detachment and thus are more erodible. The accumulation of Na^+ ions in the exchange complex of soil clays is known to increase soil erodibility.

6.5.1.2 Shear Strength of Quick Clays

If we broaden out perspective, beyond soil science, to include other instances where exchangeable Na^+ alters clay behavior we would include *quick clays*. Quick clays are glaciomarine-clay formations notable for their instability under shear stress and high liquefaction risk. Quick-clay formations occur in coastal settings (e.g., coastal Scandinavia, St Lawrence, and Ottawa river valleys in the Ottawa-Montreal vicinity) submerged during the last ice age. Geotechnical engineers attribute quick clay instability under shear stress to the high exchangeable Na^+ contents typical of these marine deposits.

Bjerrum (1954) demonstrated the failure of conventional Atterberg limits (including both the plasticity index PI and the activity A) to adequately describe the shear strength of glaciomarine-clay formations. Skempton (1953) reports an activity A of 6.2 for the highly active clay mineral smectite, an A of 0.90 for the moderately active clay mineral illite and an A of 0.33 for the relatively inactive clay mineral kaolinite.

The A of quick clay specimens from the Oslo vicinity consistently fell in the inactive range despite the fact their mineralogy was predominantly illitic. Bjerrum (1954) notes that the unusually low A values were measured at in situ salinity levels, the salinity of some specimens were as high as 20×10^3 g dm^{-3} TDS (seawater salinity is 20×10^3 g dm^{-3}). The salinity of some quick-clay formations, however, was an order of magnitude lower (e.g., the salinity of the Toyen quick-clay specimen was 2.8×10^3 g dm^{-3} TDS, equivalent to a saturated-extract EC at the USDA salinity threshold of 4 dS m^{-1}). Although Bjerrum (1954) did not measure the major ion composition of the pore water, the dramatic loss of activity is the result two factors: the high exchangeable sodium content and low salinity. The low activity (i.e., low apparent plasticity) of quick-clay formations is a consequence of the high exchangeable sodium content of these glaciomarine-clay formations.

6.5.2 Identifying Sodicity Risk

The USDA Natural Resource Conservation Service rates sodicity risk using three criteria listed in Table 6.17 ESP, total dissolved salts as measured by saturated (soil) paste electrical conductivity EC_{se}, and soil pH. We will examine the implications of each criterion to better understanding of the chemical processes at work in sodic soils.

TABLE 6.17 USDA Natural Resource Conservation Service Salinity-Sodicity Classifications and Criteria (Allison et al., 1954): ESP, Saturated-Extract Electrical Conductivity EC_{se}, and Soil pH

Class	ESP (%)	EC_{se} (dS m^{-1})	Soil pH (Unitless)[a]
Nonsaline, nonsodic	<15	<4	<8.4
Saline	<15	>4	<8.4
Sodic	>15	<4	>8.4
Saline-sodic	>15	>4	<8.4

[a] *Three methods are acceptable for soil pH measurement in the salinity-sodicity context: (1) saturated soil paste, (2) 1-to-1 soil/water suspension, and (3) 1-to-2 soil/0.01 M $CaCl_2$ suspension (Allison et al., 1954; NCSS, 2014).*

6.5.2.1 Exchange Sodium Percentage

The ESP (expression 6.24) is an already familiar ion-exchange parameter: the exchangeable sodium equivalent fraction $E_{\overline{Na^+}}$ multiplied by 100. A common alternative parameter is the exchangeable sodium ratio (ESR) (expression 6.25).

$$ESP = E_{\overline{Na^+}} \times 100 \tag{6.24}$$

$$ESR \equiv \frac{E_{\overline{Na^+}}}{\left(1 - E_{\overline{Na^+}}\right)} = \frac{E_{\overline{Na^+}}}{\left(E_{\overline{Ca^{2+}}} + E_{\overline{Mg^{2+}}}\right)} \tag{6.25}$$

The ESP requires careful measurement of all exchangeable cations,[14] leading the USDA Salinity Laboratory Staff (Allison et al., 1954) to search for an empirical relation between

[14] Soil chemists at the publication of USDA *Agricultural Handbook 60* had yet to develop a reliable method for measuring exchangeable Ca^{2+} and Mg^{2+} in calcareous soils.

the solution composition and the composition of the exchange complex. They found a linear relation between the sodium adsorption ratio (SAR) (a solution parameter Box 6.7) defined by expression (6.26) and ESR (an exchange-complex parameter).

$$SAR \equiv \frac{c_{Na^+}}{\sqrt{c_{Ca^{2+}} + c_{Mg^{2+}}}} \tag{6.26}$$

BOX 6.7

SODIUM ADSORPTION RATIO (SAR)

The *SAR* expression first appeared in USDA *Agricultural Handbook 60* (Allison et al., 1954). The authors cite four earlier studies (Gapon, 1933a,b; Mattson and Wiklander, 1940; Davis, 1945; Schofield, 1947) prior to defining the *SAR* expression. A review of the mathematical foundations of *SAR* (Sposito and Mattigod, 1977) cites only the Gapon (1933a,b) study, ignoring the other three studies. You can find a detailed discussion of the flawed Gapon convention in Appendix 4.C.

Mattson and Wiklander (1940), Davis (1945), and Schofield (1947) describe asymmetric ion exchange either using expressions equivalent to (or describing exchange behavior[15]) consistent with the Vanselow convention discussed in Chapter 4. Davis (1945) also derived an expression identical to the Gapon convention but warns the derivation "indicates explicitly the probability that [the selectivity coefficient] will not be a constant in general over a wide range of variation of... added electrolyte. The simple assumptions made by Gapon do not provide for this possibility."

The *SAR* expression embodies the flawed Gapon convention, regardless of fact that most of the studies cited by USDA *Agricultural Handbook 60* (Mattson and Wiklander, 1940; Davis, 1945; Schofield, 1947) support what would later become known as the Vanselow convention.

[15]cf. *Schofield Ratio Law*, Section 4.5.3.

Fig. 6.10 plots (*SAR*, *ESR*) data from Allison et al. (1954) and the USDA Salinity Laboratory linear regression.[16] Most current references simplify the original empirical expression, dropping the intercept and shortening the proportionality factor to two significant digits. Relating *ESR* to *SAR* using expression (6.27) gives an alternative criterion for sodic soils: $SAR > 12$.

$$ESR = (-0.0126) + (0.01475) \cdot SAR \tag{6.27}$$

Appendix 6.D summarizes experimental results for (Na^+, Ca^{2+}) cation exchange for two closely related Wyoming smectite specimens: *American Petroleum Institute* specimen API-25 (Shainberg et al., 1980) and *Clay Minerals Society* specimen SWy-1 (Sposito et al., 1983a,b).

[16] $r^2 = 0.852$, cf. Allison et al. (1954, Fig. 9).

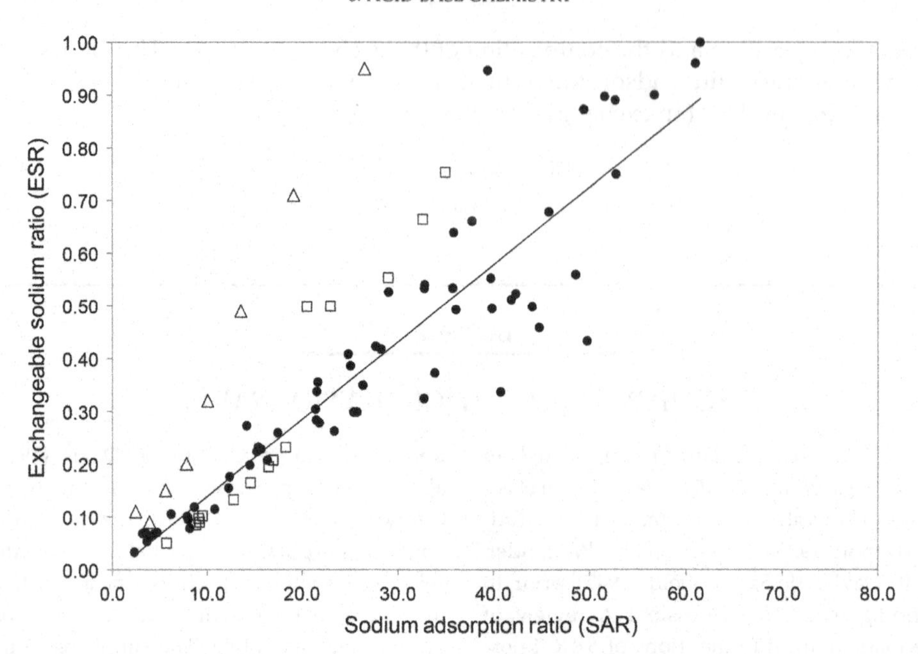

FIG. 6.10 *Filled circles* plot the exchangeable sodium ratio (ESR) as function of the sodium adsorption ratio (SAR) for arid US soils. The linear regression expression (6.27) of these data (Allison et al., 1954) appears as a *solid line*. The *open triangles* (Sposito et al., 1983a,b) and *open squares* (Shainberg et al., 1980) plot (Na^+, Ca^{2+}) exchange data for two smectite clay specimens (cf. Appendix 6.D).

Experimental results from Sposito et al. (1983a,b) provide a more complete coverage of the (Na^+, Ca^{2+}) exchange isotherm compared to the Shainberg et al. (1980) study. Despite the source proximity of the two smectite specimens (Upton, Wyoming is located 5 km south of the Crook-Weston County line in northeastern Wyoming) the (*SAR*, *ESR*) data from specimen SWy-1 (open triangle symbols, Fig. 6.10) has a significantly greater slope than specimen API-25 (open squares, Fig. 6.10).

Regardless of these differences, the reader should note the two variables plotted Fig. 6.10 provide a much different picture of exchange data than the equivalent fraction variables in an exchange isotherm plot. Equivalent fraction variables map the exchange reaction onto an interval lying between zero and unity. Fig. 6.10, by contrast, plots ratios that map the exchange reaction onto an interval lying between zero and infinity. Plotting ratios on both axes instead of fractions tends to linearize curvature in the data. A comparison of the conventional (Na^+, Ca^{2+}) exchange isotherm in Appendix 6.D with the same data in Fig. 6.10 should persuade the reader of this distortion.

More importantly, despite the difference between the (*SAR*, *ESR*) data from specimens SWy-1 and API-25, the slope of (*SAR*, *ESR*) expression (6.27) is determined by the selectivity constant $K^{\circ}_{Na/Ca}$ for (Na^+, Ca^{2+}) cation exchange. Furthermore, the effective soil cation-exchange selectivity coefficient $K^{c}_{Na/Ca}$ embodied in expression (6.27) is, in all probability, representative of smectite exchange selectivity in the "59 soil samples representing ...9 Western States" (Allison et al., 1954) (Example 6.6).

EXAMPLE 6.6

Example Permalink

http://soilenvirochem.net/3tQicu

Calculate the exchange selectivity coefficient $K^{c}_{Na/D}$ implied by expression (6.27) using soil analysis data.

These examples use analysis data for a soil sample collected from the Bssyz2 horizon of the DUKE soil series (fine, mixed, active, thermic Sodic Haplusterts). The site is located in Greer County, Oklahoma (34.738881° N, 99.518006° W).

Table 6.18 lists results from the soil analysis. Soil pH is 7.8, measured in a $CaCl_2(aq)$ suspension.[17] Major ion concentrations in the saturated-paste extract should be considered reliable; however, the Deffeyes' (1965) sum (expression 6.23) fails to validate the analysis bicarbonate value.

TABLE 6.18 Soil Analysis Data From Duke Soil (Horizon: Bssyz2; NSSC Pedon ID NSSC Pedon ID: s2002ok-055-004 Duke), Greer County, Oklahoma

Soil Saturated-Paste Extract	
Analyte	**$mmol_c\ dm^{-3}$**
Na^{+}	58.4
Mg^{2+}	29.8
K^{+}	0.20
Ca^{2+}	22.1
HCO_3^{-}	1.1
SO_4^{2-}	94.4
NO_3^{-}	1.2
Cl^{-}	20.8
F^{-}	0.3

Part 1. Simulate the soluble calcium concentration in the pore water of the DUKE soil assuming calcite and gypsum both saturate the soil solution.

The soil profile description and analysis both indicate the presence of both calcite and gypsum. A minimalist, four-component—Ca++, SO4−−, CO2(aq), H+—simulation employs a four-phase system: a gas phase containing $CO_2(g)$, a solution phase, and two mineral phases $CaO_3(s)$ and $CaSO_4 \cdot 2H_2O(s)$. Gypsum $CaSO_4 \cdot 2H_2O(s)$ controls SO_4^{2-} solubility, calcite $CaCO_3(s)$ controls Ca^{2+} and the $CO_2(g)$ partial pressure in the above-ground atmosphere controls $CO_2(aq)$ solubility. The Gibb's Phase Rule indicates the system has zero degrees of freedom.

TABLE 6.19 Results Summary From *ChemEQL* Simulation of Four-Phase System: Gas-Solution-Calcite-Gypsum

Specie or Component	Value	Units
Ca^{2+}	33.38	$mmol_c\ dm^{-3}$
HCO_3^-	0.34	$mmol_c\ dm^{-3}$
pH	7.69	Standard units

The results in Table 6.19 appear to validate the saturated-paste extract Ca concentration and the pH measured in the $CaCl_2$ suspension. Under the same conditions the three-phase system (gas-solution-calcite) would yield significantly higher values for solution pH 8.26, $A_C = 1.03 \times 10^{-3}$ mol dm^{-3} and a significantly lower dissolved calcium concentration. These changes in pH, alkalinity, and calcium solubility arise from the *common-ion effect* caused by the mineral gypsum.

Part 2. Determine the exchange selectivity coefficient $K^c_{Na/D}$—where D^{2+} represents Ca^{2+} and Mg^{2+} combined—implicit in expression (6.27) from USDA Handbook (Allison et al., 1954).

$$SAR = \frac{58.4}{\sqrt{(29.8 + 22.1)/2}} = 11.46$$

$$ESR = (-0.0126) + (0.01475) \cdot (11.46) = 0.16$$

$$E_{Na^+} = \frac{ESR}{1 + ESR} = \frac{0.16}{1.16} = 0.14$$

$$K^c_{Na/D} = \frac{(0.14)^2 \cdot \left(1.11 \times 10^{-2} + 1.49 \times 10^{-2}\right)}{\left(5.84 \times 10^{-2}\right)^2 \cdot (0.86)} = 0.16$$

Appendix 6.D fits (Na^+, Ca^{2+}) exchange isotherm data from Sposito et al. (1983a,b) to a selectivity coefficient $K^c_{Na/Ca} = 0.30$.

[17] The soil-to-solution ratio is 1-to-2, the solution is 10^{-2} mol dm^{-3} $CaCl_2(aq)$.

Up to this point all that we have shown is that the widely used solution parameter *SAR* traces its foundations to ion-exchange fundamentals. Furthermore, expression (6.27) linking solution parameter *SAR* to exchange parameter *ESR* is also based on ion-exchange fundamentals. What remains unexplained is the 15% *ESP* threshold separating nonsodic and sodic soils.[18]

The following is an extended quote from USDA Agricultural Handbook 60 (Allison et al., 1954).

> For porous media with fixed structure, such as sandstone or fired ceramic, measurements of intrinsic permeability with air, water, or organic liquids all give very nearly the same numerical value. Gravity, density, and the viscosity of the liquid are taken into account in the flow equation. However, if the intrinsic permeability for a soil as measured with air is markedly greater than the permeability of the same sample as subsequently measured using water, then it may be concluded that the action of water in the soil brings about a change in

[18] The 15% *ESP* threshold is equivalent to $ESR = 0.18$ threshold and $SAR = 13$.

structure indicated by the change in permeability. *The ratio of air to water permeability, therefore, is a measure of the structural stability of soils, a high ratio indicating low stability.* ...

The air-water permeability ratio increases greatly as the exchangeable sodium content of the soil increases, indicating that *exchangeable sodium decreased the water stability of the soil structure.* ...

It is to be expected that if the hydraulic conductivity of surface soil is as low as 0.1 cm/hr leaching and irrigation may present serious difficulties.

Experience has shown soils with $ESP > 15\%$ are at risk for decreased soil structure stability leading to an increase in the air-water permeability ratio and reduction of hydraulic conductivity below $100\ mm\ h^{-1}$. Evapotranspiration water loss increases soil salinity because all irrigation water delivers dissolved salts to soils. Irrigation management in climate zones where evapotranspiration water loss is high and available irrigation water contains appreciable dissolved salts requires water application exceeding crop evapotranspiration water losses to leach excess salts from the soil profile and prevent soil salination. Maintaining soil hydraulic conductivity (i.e., minimizing the air-water permeability ratio) is essential to soil salinity management. Soils with *natric* diagnostic horizons (Box 6.8) share similar characteristis as sodic soils (cf. Table 6.17).

BOX 6.8

NATRIC DIAGNOSTIC HORIZON

The USDA Soil Taxonomy employs diagnostic subsurface horizons to classify soils at the *Great Group* level. The *natric* horizon, one of the diagnostic subsurface horizons, is a variant of the *argillic* subsurface diagnostic horizon. The argillic horizon requirement is significant because ensures clay films coat granular fraction particles, soil pores, and aggregates. There reader should also note: the diagnostic criteria for the argillic horizon do not mention soil structure.

Natric horizons must also meet two other criteria: a morphological criterion based on soil structure and a chemical criterion.

The chemical criterion is an $ESP > 15$ (or an $SAR > 13$).[19] Entering $ESP = 15$ into expression (23) yields the $SAR = 13$ threshold. Both criteria were established by USDA Agricultural Handbook 60 (Allison et al., 1954).

The morphological criterion admits two possibilities:

"columns or prisms in some part (generally the upper part), which may break to blocks" or

"both blocky structure and eluvial materials, which contain uncoated silt or sand grains and extend more than 2.5 cm into the horizon."

Columnar or prismatic structure in a natric horizon represents end-stage sodicity, revealing significant soil structure degradation. Blocky structure with eluvial materials reveals incipient soil structure degradation.

Although elevated air-water permeability—which implies restricted water flow—is a common natric soil characteristic, the diagnostic criteria do not include either air-water permeability or hydraulic conductivity thresholds because both are difficult to measure in situ and laboratory measurements on soil cores do not accurately reflect internal drainage under field conditions.

BOX 6.8 *(cont'd)*

Although soil and geotechnical scientists distinguish between highly active clay minerals (i.e., smectite clays) and relatively inactive clay minerals (i.e., kaolinite clays) both clay minerals pass the ribbon test and both confer shear strength by welding granular particles together. For example, kaolinitic Natraqualfs occur throughout the Khorat and Sakon Nakhon basins in the Northeast Plateau of Thailand (Wongpokhom et al., 2008a,b; Kaewmano et al., 2009). The natric diagnostic horizon criteria make no mention of clay mineralogy.

[19] The *ESP* or *SAR* thresholds define the exchangeable sodium content at which we can anticipate a significant loss of clay plasticity as defined by the Atterberg plasticity limit (PL) and the activity *A*.

6.5.2.2 Salinity

Low annual rainfall and high evapotranspiration rates concentrate soluble salts in soil pore water, contributing to sodification through evaporation-driven calcite precipitation. This process, which leads to demonstrable changes in major ion composition of surface waters, occurs more intensely in soil because the surface-to-volume ratio is much greater than surface water bodies. The USDA Soil Taxonomy defines the aridic soil moisture regime as soils whose *moisture-control section* is as follows.

> Dry[20] in all parts for more than half of the cumulative days per year when the soil temperature at a depth of 50 cm from the soil surface is above 5°C and moist in some or all parts for less than 90 consecutive days when the soil temperature at a depth of 50 cm is above 8°C.

Empirical expression (6.28) is widely used by soil chemists to estimate ionic strength I_c (mol dm^{-3}) from electrical conductivity EC_{se} (dS m^{-1}) (Griffin and Jurinak, 1973). Previously (cf. Fig. 3.13 and Appendix 3.D), we discussed the effect of ionic strength I_c on the osmotic swelling of Na^+-saturated smectite clays which contributes to a reduction of clay plasticity correlated to high *ESP* and low salinity.

$$I_c \approx 0.013 \cdot EC_w \tag{6.28}$$

Soil scientists typically employ electrical conductivity EC_w to quantify salinity while geochemists and marine scientists rely on *TDS*. Electrical conductivity EC_w response is a function of ionic strength I_c (Fuoss and Onsager, 1957, 1958; Fuoss, 1959) and temperature while TDS is not.

The saturated-paste EC_w must fall below 4 dS m^{-1} to classify a soil as sodic. The significance of the salinity criterion is attributable to impact of exchangeable ions on clay plasticity.

[20] $p_{tension} \leq p_{wp} = -1500$ kPa or $h_{tension} \leq -153$ m.

Atterberg limits alone (Bjerrum, 1954) fail to account for the atypically low activity A of soils whose clay fraction has a high ESP. The activity A of high ESP soils decreases sharply at the low salinity limit (Bjerrum, 1954).

6.5.2.3 Extreme Alkalinity

The pH criterion used by the USDA to assess sodicity risk (Table 6.17), unlike the other two criteria, does not distinguish sodic from nonsodic behavior but serves as an ancillary sodicity risk indicator. As we will see in this section, there are circumstances where soil moisture or irrigation water can meet the first two criteria—ESP (or SAR) and EC_W criteria—and yet fail the pH $>$ 8.4 criterion.

The pH $>$ 8.4 criterion is valid if silicate and carbonate rock weathering control major ion chemistry. The Na/(Na + Ca) mass ratio of such waters will be rather low and SAR values will fall well below the sodicity threshold. Any SAR increase requires water loss by evaporation which drives the calcite saturation index $SI_{calcite}$ to over-saturation. The resulting calcite precipitation means dissolved calcium levels decrease and alkalinity increase at a slower rate than other major anions and saturated-extract pH will exceed the 8.4 threshold.

Evaporation-driven calcite precipitation, by itself, is not sufficient to exceed the sodicity SAR threshold. As noted in *Simple Gibbs Model* (Box 6.4), groundwater discharge into surface water bodies may contain salts eroded from subsurface evaporite formations. Two specific examples include the Pecos River at and below Malaga Bend, New Mexico, several rivers in the Missouri River basin in South Dakota (e.g., Grand and White rivers), and the Mun River (Khorat basin), Thailand. More importantly the occurrence of Natrustolls throughout central South Dakota and Natraqualfs along the Mun River in northeastern Thailand can be linked to the discharge of saline groundwater bearing a characteristic major ion signature.

Sodic water chemistry is best understood on a continuum between two end-points. The low-alkalinity end-point is saline groundwater. The salinity in most cases is eroded from subsurface evaporite formations, the exception being glaciomarine quick-clay formations (cf. Section 6.5.1.2). The Na/(Na + Ca) mass ratio or SAR of low-alkalinity end-point waters is derived from neutral salts: chlorides and sulfates.

The major cations in seawater (Table 6.20) originate from chemical weathering of Earth's lithosphere. The major anions result from acid deposition from volcanic activity—hydrochloric, sulfuric, and hydrofluoric acids—and atmospheric carbon dioxide. Acid deposition neutralizes most of the basicity resulting from chemical weathering of the silicate lithosphere.

Mean seawater composition does not quote values for either bicarbonate or nitrate yet both anions appear in Table 6.20. The nitrate concentration is estimated by assuming total dissolved nitrogen is dissolved inorganic nitrogen (DIN) and assigned entirely to NO_3^-. There is no way to verify the validity of this assumption.

On the other hand, we can verify whether assuming total dissolved carbon equals DIC, which is equivalent to assuming alkalinity is approximately equal to total dissolved carbon (i.e., HCO_3^- equals DIC). An independent alkalinity estimate is based on the spectator-ion charge balance: $A_{ion\text{-}sum} = 2.299 \times 10^{-3}$ mol kg^{-1}. The ion-sum value is 1.4% smaller than the DIC value listed in 10 above.

TABLE 6.20 Seawater Composition: Ionic Strength $I_c \approx 0.72$ $mol_c\ kg^{-1}$ and Total Alkalinity A_T = 2.4×10^{-3} $mol_c\ kg^{-1}$ (Lide, 2005)

Species	Concentration (mol kg^{-1})
Na^+	4.698E−01
Mg^{2+}	5.308E−02
Ca^{2+}	1.028E−02
K^+	1.021E−02
Sr^{2+}	9.017E−05
Cl^-	5.472E−01
SO_4^{2-}	2.822E−02
HCO_3^-	2.331E−03
$B(OH)_3^0$	4.107E−04
$H_4SiO_4^0$	7.833E−05
F^-	4.629E−05
NO_3^-	3.570E−05

Note: *Total dissolved carbon is assigned to HCO_3^- and total dissolved nitrogen to NO_3^-.*

Diluting seawater to $EC_W = 4$ dS m^{-1} achieves calcite saturation if alkalinity is allowed to increase by 8%. This results in pore water composition that satisfies the first 2 criteria in 10 but fails the pH criterion. The residual sodium carbonate (*RSC*) is negative, consistent with pH < 8.4.

The water chemistry in Table 6.21 is representative of the low-alkalinity end-point found in incompletely leached marine quick clay deposits and soils that develop sodicity from saline groundwater discharge or irrigation with saline groundwater.

The high-alkalinity end-point results solely from the chemical weathering of silicate rocks. Figs. 6.11 and 6.12 illustrate the dissolution of Earth's lithosphere by reaction (6.R11), where solution Ca^{2+}(aq) activity is constrained by calcite solubility. The process depicted in Fig. 6.11 is equivalent to the irreversible dissolution of ORTHOCLASE described by Helgeson et al. (1969) and Rhoades et al. (1968) measured significant silicate mineral weathering and increasing alkalinity when leaching arid-region soils with irrigation water.

Expression (6.23) (Deffeyes, 1965) defines alkalinity using the spectator ion charge balance—the dissolved positive charge from cations Ca^{2+}(aq), Mg^{2+}(aq), Na^+(aq), and K^+(aq) minus the dissolved negative charge from anions SO_4^{2-}(aq) and Cl^-(aq). Calcite solubility imposes no global constrain on alkalinity.

Eq. (6.23) suggests another basicity source besides silicate rock weathering: anaerobic dissimilatory sulfate reduction (i.e., biological sulfate reduction). Whittig and Janitzky (1963) identify anaerobic dissimilatory sulfate reduction as a plausible basicity source in soil landscapes where poor drainage, high levels of dissolved sulfate, and abundant soil organic matter are present.

Recent estimates suggest dissimilatory sulfate reduction in the deep ocean may account for 60% of global seawater alkalinity (Thomas et al., 2009). Geotechnical engineers studying glaciomarine quick-clay formations—which meet all the criteria listed by Whittig and Janitzky (1963)—have demonstrated dissimilatory sulfate reduction generates sufficient alkalinity to alter the pore water Na^+-to-Ca^{2+} ratio in glaciomarine quick-clay formations (Lessard, 1981).

6.5.3 Predicting the Sodicity Risk of Irrigation Water

Langelier (1936) developed a method for calculating the calcite saturation index $SI_{calcite}$. Given the water analysis total calcium concentration (expressed as the negative logarithm of the total soluble calcium concentration or $pCa \equiv -\log_{10}(c_{Ca})$) and alkalinity (expressed as the negative logarithm of the carbonate alkalinity or $pA_C \equiv -\log_{10}(A_C)$), Langelier derived an

TABLE 6.21 Seawater Diluted by a Factor of 0.076: Electrical Conductivity $EC_w \approx 4\ dS\ m^{-1}$, Ionic Strength $I_c = 5 \times 10^{-2}\ mol_c\ dm^{-3}$, Total Alkalinity $A_T = 1.47 \times 10^{-3}\ mol_c\ dm^{-3}$, and $c_{CO_2} = 1.37 \times 10^{-5}\ mol\ dm^{-3}$

Species	Concentration ($mol\ dm^{-3}$)
Na^+	3.41E−02
Mg^{2+}	3.85E−03
Ca^{2+}	7.46E−04
K^+	7.41E−04
Sr^{2+}	6.55E−06
Cl^-	3.97E−02
$SO_4{}^{2-}$	2.05E−03
$HCO_3{}^-$	1.47E−03
$B(OH)_3{}^0$	2.98E−05
$H_4SiO_4{}^0$	5.69E−06
F^-	3.36E−06
$NO_3{}^-$	2.59E−06

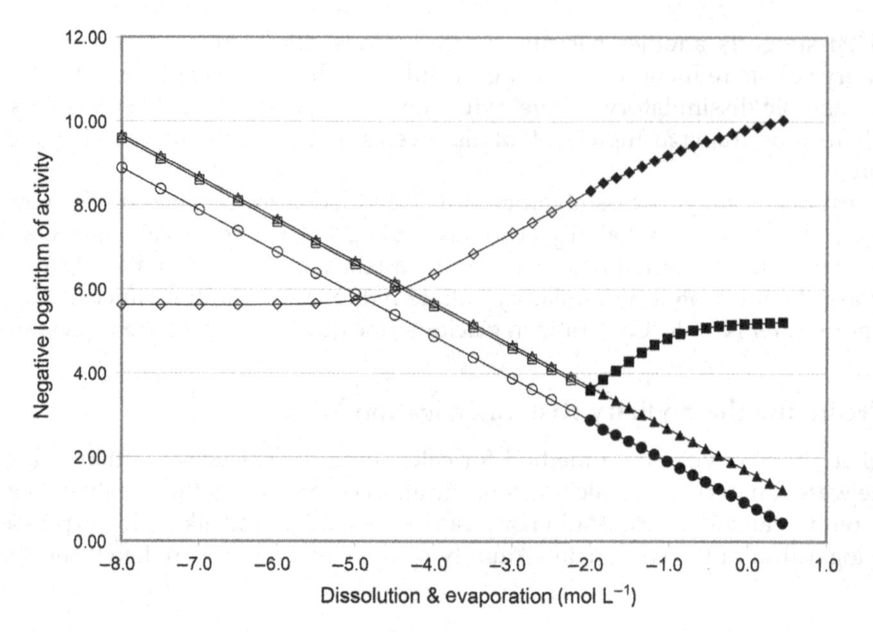

FIG. 6.11 The irreversible dissolution of Earth's lithosphere according to reaction (6.R23) neglects the solubility of $Al_2O_3(s)$, $Fe_2O_3(s)$, and $SiO_2(s)$. Calcium solubility is constrained by calcite solubility and $CO_2(aq)$ (1.37×10^{-5} mol dm^{-3}). Solution pH (*diamond symbol*) increases once 10^{-5} mol dm^{-3} of the lithosphere has dissolved. Maximum calcium solubility (*square symbol*) is attained at pH 8.4 when the solution becomes saturated by calcite (dissolution 10^{-2} mol dm^{-3}) then declines while pH and sodium solubility (*triangle symbols*) continue to increase.

expression for the pH at which the water is saturated by calcite: $\mathrm{pH_{cs}}$.

$$\mathrm{pH_{cs}} = \left(\mathrm{p}K^{c}_{\mathrm{a2}} - \mathrm{p}K^{c}_{\mathrm{s0}}\right) + \mathrm{pCa} + \mathrm{pA_C} \tag{6.29}$$

The $\mathrm{p}K^{c}_{\mathrm{a2}}$ term on the right-hand side of expression (6.29) is the negative base-10 logarithm of the hydrolysis coefficient (6.30) (cf. reaction 6.R29). The $\mathrm{p}K^{c}_{\mathrm{s0}}$ term is the negative base-10 logarithm of the calcite solubility coefficient (6.31).

$$K^{c}_{\mathrm{a2}} = \frac{K^{\ominus}_{\mathrm{a2}}}{\gamma_{2\pm}} = \left(\frac{c_{\mathrm{H_3O^+}} \cdot c_{\mathrm{CO_3{}^{2-}}}}{c_{\mathrm{HCO_3{}^-}}}\right) \tag{6.30}$$

$$K^{c}_{\mathrm{s0}} = \frac{K^{\ominus}_{\mathrm{s0}}}{\gamma^{2}_{2\pm}} = \left(c_{\mathrm{Ca^{2+}}} \cdot c_{\mathrm{CO_3{}^{2-}}}\right) \tag{6.31}$$

The empirical formula for calculating ion activity coefficients and the equilibrium constants for carbonate hydrolysis and calcite solubility have changed since Langelier published his paper.[21] More importantly, chemists studying calcite scale formation in water distribution

[21] Langelier: $\mathrm{p}K^{\ominus}_{\mathrm{s0}} = 10.26$ and $\mathrm{p}K^{\ominus}_{\mathrm{a1}} = 8.32$, ChemEQL: $\mathrm{p}K^{\ominus}_{\mathrm{s0}} = 10.33$ and $\mathrm{p}K^{\ominus}_{\mathrm{a1}} = 8.48$.

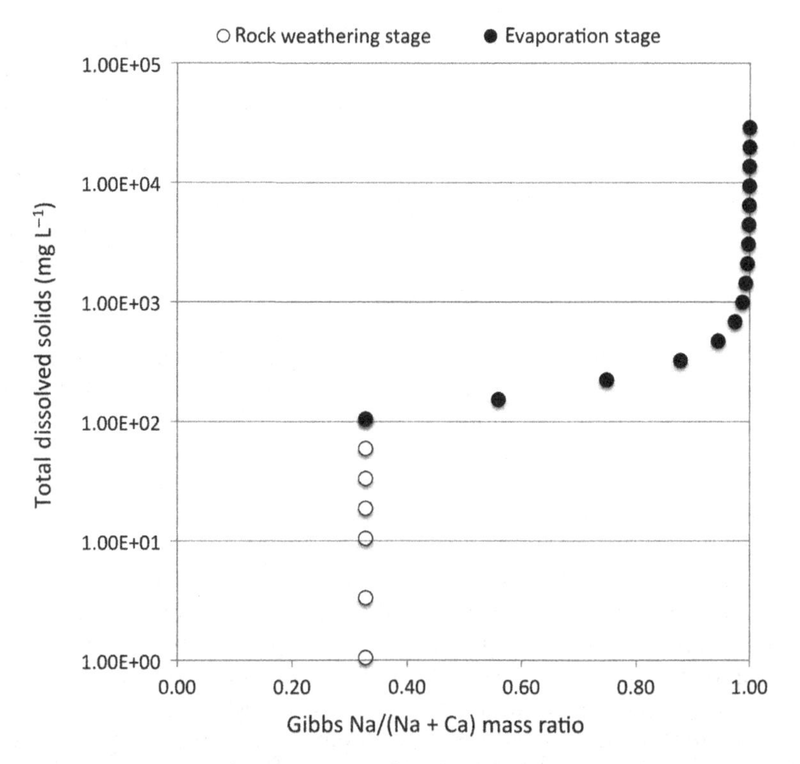

FIG. 6.12 Gibbs plot of the irreversible dissolution of Earth's lithosphere according to reaction (6.R23) neglects the solubility of $Al_2O_3(s)$, $Fe_2O_3(s)$, and $SiO_2(s)$. Calcium solubility is constrained by calcite solubility and $CO_2(aq)$ (1.37×10^{-5} mol dm^{-3}). Abscissa plots the Na/(Na + Ca) mass ratio while the ordinate plots the total dissolved solids (*TDS*).

systems recognized the *Langelier saturation index* (6.29) was fatally flawed (cf. Pisigan and Singley, 1985; Merrill et al., 1990).

Bower et al. (1965, 1968) used expression (6.32) to predict the effect of irrigation management on soil drainage water composition, the assumption being soil drainage water reflects soil pore water composition throughout the profile. First, Bower and co-workers assumed irrigation water SAR_{iw} eventually displaced native soil water *SAR*. Second, Bower and co-workers realized water loss by evapotranspiration concentrates salinity[22] by a factor F_c, ultimately leading to calcite over-saturation, and used the terms in parenthesis to predict whether calcite precipitation increases soil pore water *SAR* or calcite dissolution causes a decrease in soil pore water *SAR*.

$$SAR_{dw} = \sqrt{F_c} \cdot SAR_{iw} \cdot \left(1 + \left(8.4 - pH_{cs}\right)\right) \tag{6.32}$$

[22] Evaporative water loss concentrates aqueous salinity. Expressing water volumes in depth units, the concentration factor F_c is the water volume infiltrating the soil surface (irrigation water D_{iw} plus meteoric water D_{mw}) divided by the drainage water volume D_{dw}.

Bower and co-workers also assumed all irrigated soils are calcareous; soil pH buffer capacity immediately shifts irrigation water to pH 8.4, the pH typical of calcareous soils. The Bower equation (6.32) employs the Langelier equation (6.29) to compute parameter pH_{sc} from irrigation water composition.

Bower and co-workers address all of the processes that act in concert to alter irrigation water composition as it enters the soil: increased salinity caused by evaporative water loss and decreased dissolved calcite resulting from evaporation-induced calcite precipitation. Unfortunately, Bower et al. (1968) soon realized expression (6.32) grossly over-estimate SAR_{dw}. Despite its flaws, the global soil science community endorsed the Bower equation (Ayers and Westcot, 1977).

Suarez (1981) published the most recent attempt to predict the effect of evaporative water loss on drainage water SAR_{dw} and identify irrigation water sources with the potential for inducing sodicity. Suarez (1981) contribution attempts an algebraic solution of the coupled equations controlling calcite solubility equilibrium. Unfortunately, the derivation introduces a flaw,[23] heretofore overlooked, that renders the Suarez (1981) method invalid. Regardless, Ayers and Westcot (1985) withdrew their earlier endorsement of the Bower equation (Ayers and Westcot, 1977) and endorsed the Suarez method in its place.

It would seem there is no need for *SAR* indexes like the Bower equation (6.32) or the Suarez method. After all, water chemistry simulations allow soil scientists and irrigation practitioners to calculate the effect of evaporation-induced calcite precipitation on $c_{Ca^{2+}}$ and predict SAR_{dw} without resorting to tables such as those published by Bower and Suarez.

The need for a simple method to identify irrigation water sources that compromise salinity management by increasing soil sodicity exists, nonetheless. Technicians called upon to evaluate irrigation water quality rarely have the training and experience to perform the needed water chemistry simulations.

6.5.4 The Limiting Sodium Adsorption Ratio (*LSAR*) Sodicity Risk Parameter

Predicting the SAR of drainage water SAR_{dw} or soil saturated-extract SAR_e is an excellent example of a problem that water chemistry simulation cannot easily solve. Rhoades et al. (1968) found the silicate mineral weathering in arid soils[24] released 2.2×10^{-3} to 3.8×10^{-3} M $HCO_3^-(aq)$ over a 2-week interval. The alkalinity release by short-term mineral weathering can eliminate restrictions on evaporation-induced calcite precipitation from irrigation water with Ca-to-HCO_3^- ratios < 2.

[23] Suarez (1981) substitutes $HCO_3/Ca = 2$ into the calcite solubility quotient expression (Suarez 5) to derive equation (Suarez 13) then combines (Suarez 13) with the equilibrium quotient expressions for CO2(aq) solubility and the CO2(aq) hydrolysis to derive expression (Suarez 15a). Finally, Suarez (1981) states: "equation (15a) could be formulated for any other HCO_3/Ca ratio by substituting that ratio for the value (2) in the denominator." Table 1 Suarez (1981) lists solutions for HCO_3/Ca ratio ranging from 0.1 to 10. The correct solution limits the system of equations to the calcite solubility quotient (expression 6.31), the carbonate hydrolysis quotient expressions for reactions (6.R26) through (6.R29), and $CO_2(g)$ solubility quotient (reaction 6.R25); the system of equations *does not include the bicarbonate-to-calcium constraint.*

[24] San Joaquin Basin, California; series: Grangeville, Panoche, Panhill, Raynor, Hideaway.

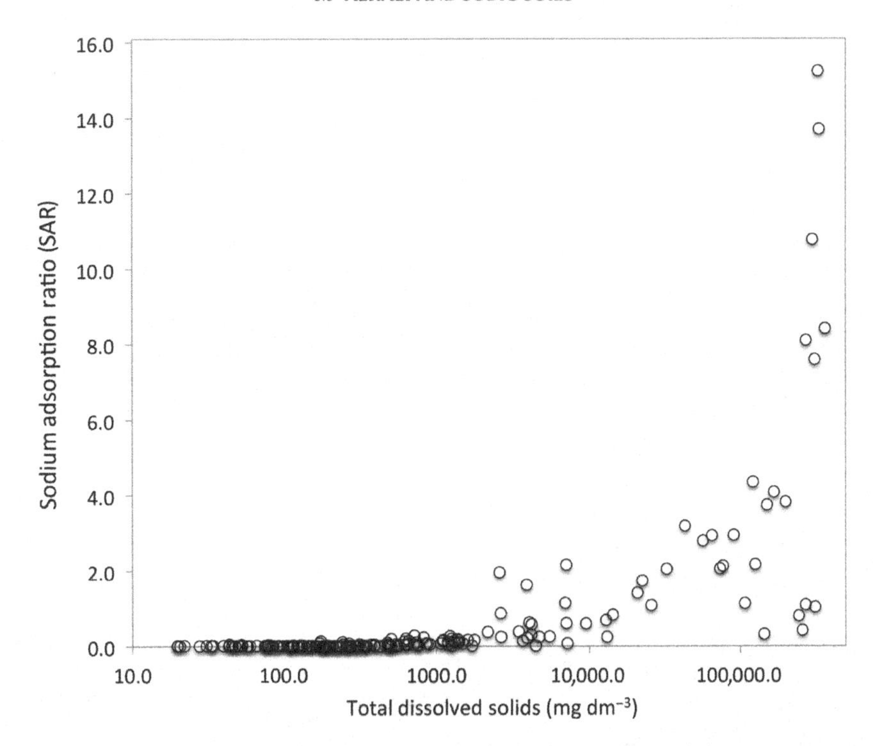

FIG. 6.13 Livingstone (1963) reported water analysis for 412 samples. The results plotted represent a subset of 270 analyses whose charge-balance error is less than 5%. Abscissa plots total dissolved solids (*TDS*) while the ordinate plots the sodium adsorption ratio (*SAR*).

Fig. 6.13 plots surface water data collected from rivers and lakes worldwide (Livingstone, 1963). As salinity increases (keeping in mind some of that increase results from evaporative water loss and some from saline groundwater discharge) the solution *SAR* generally increases; in some cases the increase is negligible but in other cases the increase is considerable. This figure illustrates the challenge chemists face as they attempt to distinguish irrigation water with a high sodicity risk from irrigation water with a low sodicity risk.

Fig. 6.14 plots soil data from a lysimeter study by Rhoades and Merrill (1976). Soil saturated-extracts collected periodically tracked increases in salinity (i.e., electrical conductivity EC_{se}) and pore water sodicity (i.e., sodium adsorption ratio SAR_{se}) with evapotranspiration water loss.

Briefly, soil water loss by evapotranspiration (D_{et}) increases irrigation water salinity once it enters the soil. The concentration of all soluble ions increases by a factor F_c (Eq. 6.33). Expressing all volumes in depth units, the water volume infiltrating the soil surface is irrigation water D_{iw} plus meteoric water D_{mw} while the water volume draining from the bottom of the soil profile is D_{dw}.

$$F_c = \frac{D_{mw} + D_{iw}}{D_{dw}} = \frac{D_{mw} + D_{iw}}{D_{mw} + D_{iw} - D_{et}} \tag{6.33}$$

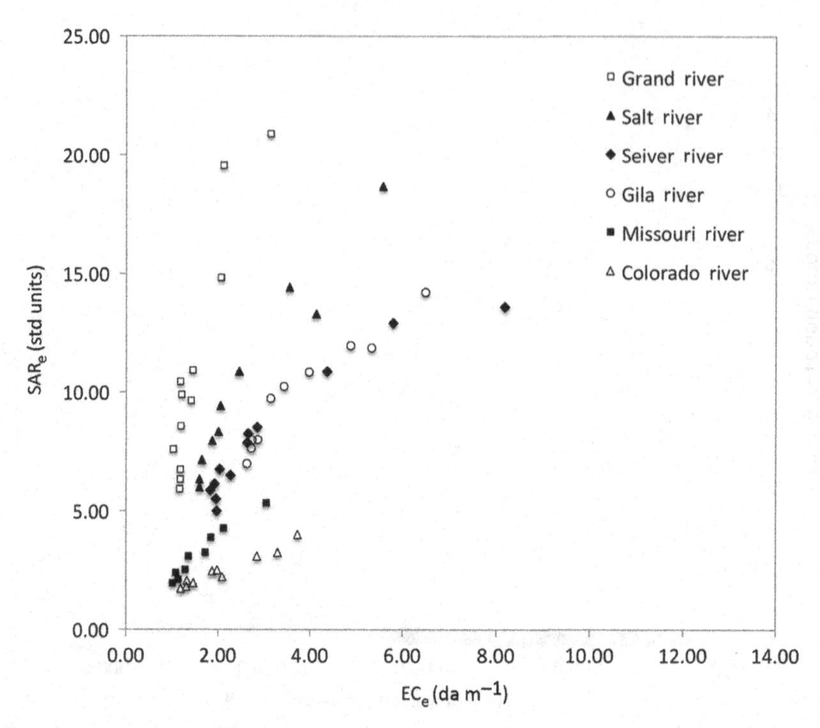

FIG. 6.14 Rhoades and Merrill (1976) reported soil analyses from an irrigation study. The lysimeters were filled with Pachappa soil series (Coarse-loamy, mixed, active, thermic Mollic Haploxeralfs) and irrigated with eight (8) types of river water, representing major rivers form the western United States. The results plotted represent a subset of six (6) irrigation waters. Abscissa plots saturated-paste electrical conductivity EC_{se} while the ordinate plots the sodium adsorption ratio SAR_{se} from saturated-paste extracts.

At some point the simultaneous increase in $c_{Ca^{2+}}$ and alkalinity reaches calcite saturation. While evaporative water loss drives salinity ever higher beyond this point, calcite precipitation leads to a sharp drop in soluble calcium (cf. Figs. 6.11 and 6.12). Since neither Na^+ or Mg^{2+} precipitate as evaporative water loss increases salinity, the *SAR* approaches a limiting value *LSAR* (Eq. 6.34).

$$\lim_{c_{Ca^{2+}} \to 0} \left(\frac{c_{Na^+}}{\sqrt{c_{Mg^{2+}} + c_{Ca^{2+}}}} \right) = \frac{c_{Na^+}}{\sqrt{c_{Mg^{2+}}}}$$

$$LSAR \equiv \sqrt{F_C} \cdot \left(\frac{c_{Na^+}}{\sqrt{c_{Mg^{2+}}}} \right) \tag{6.34}$$

Irrigation water representing the Grand River near Wakpala, South Dakota (open square symbols, Fig. 6.14, USGS station ID 06358000) and the Salt River below Stewart Mountain Dam, Arizona (filled triangle symbols, Fig. 6.14, USGS station ID 09502000) have a much higher *LSAR* than the Colorado River (at northerly international boundary above

Morelos Dam) near Andrade, California (open triangle symbols, Fig. 6.14, USGS station ID 09522000).

Evidence for alkalinity release by soil mineral weathering can be seen in the *LSAR* values for irrigation waters characterized by Ca-to-HCO_3^- ratios < 2: Gila River below Gillespie Dam, Arizona (USGS station ID 09519500), the Colorado River, and the Missouri River at Nebraska City, Nebraska (USGS station ID 06807000). Mineral weathering in the Pachappa soil releases sufficient alkalinity to deplete soil drainage water of soluble Ca^{2+} (Example 6.7).

EXAMPLE 6.7

Example Permalink

http://soilenvirochem.net/A2bn2e

Evaluate the sodicity risk of an irrigation water source.

The irrigation water source in this example is a Missouri River tributary. The White River, with its headwater in Souix County, Nebraska, flows eastward through southern South Dakota to join the Missouri River just south of Chamberlin, South Dakota. The water sample was collected in October 1949 at the USGS station (station ID 6447000) near Kadoka, South Dakota (43.7525° N, 101.524444 ° W).

TABLE 6.22 Water Analysis Results for Water Sample Collected October 1949 on the White River Near Kadoka, South Dakota

Analyte	mg kg^{-1}	mol kg^{-1}
Ca	18.0	4.49E−04
K	5.7	1.46E−04
Mg	1.7	6.99E−05
Na	108.0	4.70E−03
Cl	6.3	1.78E−04
Bicarbonate	249.0	4.08E−03
Carbonate	1.0	1.67E−05
Nitrate	1.8	3.14E−05
Sulfate	76.0	7.91E−04
$SiO_2(aq)$	41.0	6.82E−04

The EC of the White River near Kadoka in October 1949 was $EC_{iw} = 0.634$ dS m^{-1}. A threefold increase in salinity (cf. expression 6.33) caused by evapotranspiration water loss from irrigated soil raises salinity to $EC_{SW} = 2.0$ dS m^{-1}, well below the 4.0 dS m^{-1} salinity threshold listed in Table 6.17.

Part 1. Simulate the White River water chemistry after evaporation water loss has raised salinity to $EC_{SW} = 2.0$ dS m^{-1}.

The concentration factor $F_C = 3.15$ is the ratio of the soil salinity $EC_{SW} = 2.0\ \mathrm{dS\ m^{-1}}$ end-point to irrigation water salinity.

$$F_C = \frac{2.0\ \mathrm{dS\ m^{-1}}}{0.634\ \mathrm{dS\ m^{-1}}} = 3.15$$

TABLE 6.23 Water Composition Table 6.22 Concentrated by a Factor $F_C = 3.15$

Component	mg kg^{-1}	mol kg^{-1}
Ca	56.78	1.42E−03
K	17.98	4.60E−04
Mg	5.36	2.21E−04
Na	340.69	1.48E−02
Cl	19.87	5.61E−04
Bicarbonate	785.49	1.29E−02
Carbonate	3.15	5.26E−05
Nitrate	5.68	9.16E−05
Sulfate	239.75	2.50E−03
$SiO_2(aq)$	129.34	2.15E−03

The simulation uses major ion concentrations listed in Table 6.23, allowing for calcite precipitation. The soil pore water being simulated is a three-phase (gas-solution-calcite) system. The designated carbonate component for this simulation is CO2(aq), its initial concentration set by the fixed-fugacity method by assuming the $CO_2(g)$ partial pressure is 397 ppm(v).

$$H_{CO_2}^{cp} = 3.39 \times 10^{-2}$$

$$c_{CO_2(aq)} = H_{CO_2}^{cp} \cdot p_{CO_2}$$

$$c_{CO_2(aq)} = \left(3.39 \times 10^{-2}\right) \cdot \left(3.97 \times 10^{-4}\right) = 1.346 \times 10^{-5}\ \mathrm{mol\ dm^{-3}}$$

The carbonate alkalinity is assigned to component OH−, which replaces H+ for the three-phase (gas-solution-calcite) simulation.

$$A_C = \left(c_{HCO_3^-} + 2 \cdot c_{CO_3^{2-}}\right)$$

$$A_C = \left(1.29 \times 10^{-2}\right) + 2 \cdot \left(5.26 \times 10^{-5}\right) = 1.29 \times 10^{-2}\ \mathrm{mol\ dm^{-3}}$$

Results of the water chemistry simulation are discussed in Part 2.

Part 2. Compare the *SAR* computed from the water chemistry simulation (Part 1) and the *LSAR* estimate made without water chemistry simulation.

Table 6.24 summarizes the major ion concentrations from the simulation and the projected *SAR*. Calcite precipitation, which lowers carbonate alkalinity a mere 3.5%, removes 98.8% of the dissolved calcium. Although evaporation water loss increases salinity threefold, calcite precipitation results in a nearly fivefold increase in *SAR*.

TABLE 6.24 Simulated Major Ion Concentrations, SAR, and Alkalinity for White River Water Concentrated to a Salinity of 2 dS m^{-1}

Parameter	Value	Units
c_{Na^+}	14.53	mmol dm^{-3}
$c_{Ca^{2+}}$	0.01	mmol dm^{-3}
$c_{Mg^{2+}}$	0.14	mmol dm^{-3}
SAR	37.26	
LSAR	31.55	
A_C	1.25×10^{-2}	mol_c dm^{-3}

The *LSAR* calculated without water chemistry simulation using expression (6.34) is 15% smaller than predicted by water chemistry simulation. This example demonstrates the high sodicity risk of irrigating with water from the White River. Soils with natric subsurface diagnostic horizons are common in the vicinity of Kadoka, South Dakota.

6.6 SOIL ACIDITY

The earliest studies of soil acidity (Veitch, 1902; Hopkins et al., 1903) found the unidentified acids responsible for soil acidity could not be extracted by distilled water. Extracting soil acidity required adding some type of soluble salt to displace the insoluble acids. Veitch (1904) reported a comparison of his $Ca(OH)_2(aq)$ (limewater) soil acidity extract (Veitch, 1902) and the sodium-chloride extraction developed by Hopkins et al. (1903). Hopkins's sodium chloride extract released "considerable quantities of iron, alumina, and manganese" into solution that precipitated when the extract was titrated with sodium hydroxide to quantify acidity. Veitch (1904) found that extracting acid soils with a neutral sodium salt solution resulted in the "replacement of aluminum by sodium." He also found that potassium salts were more efficient at replacing aluminum than sodium.

Veitch coined the term *active acidity* to describe the acidity displaced by neutral salts such as NaCl(aq) and KCl(aq) in Hopkins's method and *total apparent acidity* to describe the much greater acidity displaced by $Ca(OH)_2(aq)$ in his limewater method.[25] The debate over soil acidity continued for another 60 years (Thomas, 1977), but no study during that span of years altered the conclusions one could easily draw—in hindsight—from the work of Veitch and Hopkins.

[25] The Veitch (1904) limewater method induces the compulsive exchange of acidic Al^{3+} cations by dissolved Ca^{2+} ions. The low alumina solubility in high-pH limewater solutions drives Al^{3+} from the exchange complex, allowing Ca^{2+} to take its place.

BOX 6.9

MEASURING EXCHANGEABLE ACIDITY AND CATION-EXCHANGE CAPACITY IN ACID SOILS

The National Soil Survey Center (NSSC) Laboratory Methods Manual (NCSS, 2014) lists several analytical methods for measuring ion-exchange capacity and exchangeable cations, some specifically designed for acid soils. Distinctions between each method and the reasons some are suited for acid soils while others are not require a understanding of exchangeable acidity, ion-exchange selectivity, and the chemistry of the reagents used by each method.

While anion-exchange capacity (AEC) can reach levels comparable to cation-exchange capacity (CEC) in certain highly weathered (Jackson-Sherman weathering stages greater than or equal to 11) acid soils dominated by oxide clays, soil AEC is generally much less than *CEC* and in most instances negligible. This discussion will not cover methods for measuring *AEC* and exchangeable anions in highly weathered, oxidic soils.

Table 6.25 lists several methods for measuring ion-exchange capacity and exchangeable ions, some are listed in the *NSSC* methods manual and others that are not. A complete picture of exchangeable and nonexchangeable acidity requires some familiarity with methods that measure *CEC* and exchangeable cations.

The *NSSC* routinely measures CEC7, a standard *CEC* analysis that serves as a reference soil parameter but which does not accurately reflect the actual *CEC* of acidic soils in the natural state. In particular, the reference CEC7 will be higher than *CEC* measured by methods designed to measure soil *CEC* in the natural state.

The CEC7 method extracts the soil with a 1 mol dm^{-3} ammonium acetate $NH_4OAc(aq)$ solution buffered at pH 7. The ammonium cation NH_4^+ is the *index ion* that displaces naturally occurring cations during the extraction stage and saturates the cation-exchange complex. Chemical analysis of the $NH_4OAc(aq)$ extract will quantify Groups 1 and 2 cations (Na^+, K^+, Mg^{2+}, Ca^{2+}). The method is not designed to quantify exchangeable or nonexchangeable acidity.

TABLE 6.25 Select Methods for Measuring CEC and Exchangeable Cations in Acidic Soils

Method	Reagent	Components Extracted
CEC7	$NH_4OAc(aq)$ + KCl	NH_4^+
CECb	$NH_4OAc(aq)$	$Na^+, K^+, Mg^{2+}, Ca^{2+}$
ECEC	$NH_4Cl(aq)$ + KCl	NH_4^+
ECEC	KCl(aq)	Al^{3+}, H^+
CEC8	$BaCl_2$(aq)-triethanolamine	Al^{3+}, H^+, Al-hydroxy polymers

Continued

BOX 6.9 *(cont'd)*

The initial $NH_4OAc(aq)$ extract is removed by vacuum filtration and the soil extracted a second time by 2 mol dm^{-3} KCl(aq) which quantitatively displaces the NH_4^+ index cation into solution for analysis. CEC7 is defined as the moles of NH_4^+ index cation adsorbed per soil mass. The pH 7 buffer is required for the compulsive exchange of exchangeable and nonexchangeable aluminum occupying the exchange complex with the NH_4^+ index cation. Aluminum hydroxide $Al(OH)_3(s)$ is extremely insoluble at pH 7 and its precipitation drives aluminum occupying the exchange complex (both exchangeable and nonexchangeable) into solution, allowing NH_4^+ to saturate the exchange complex.

Method CECb quantifies exchangeable Groups 1 and 2 cations (Na^+, K^+, Mg^{2+}, Ca^{2+}) using a variation of the CEC7 method. The CECb method extracts soil with 1 mol dm^{-3} $NH_4OAc(aq)$ buffered at pH 7. This extract displaces exchangeable Groups 1 and 2 cations for analysis in the soil extract. The soil science literature commonly (but erroneously) refers to Groups 1 and 2 cations as "base cations," hence CECb.

Methods for quantifying the *effective CEC* or ECEC of soil in its natural state employ unbuffered extracts. The two most common are 1 mol dm^{-3} ammonium chloride NH_4Cl or KCl. The absolute enthalpy of hydration of NH_4^+ and K^+ cations are very similar and have similar cation-exchange selectivity. The difference is that $NH_4Cl(aq)$ is not a neutral electrolyte—$NH_4^+(aq)$ is a weak conjugate acid—while KCl(aq) is a neutral electrolyte.

The *NSSC* Laboratory relies on two-step extraction method to measure ECEC by quantifying index cation NH_4^+. The first step extracts naturally occurring exchangeable cation using NH_4Cl to saturate the exchange complex with index cation NH_4^+. The second step extracts the soil with 2 mol dm^{-3} KCl(aq) to quantitatively displace the index NH_4^+ cation into solution for analysis, similar to the two-step CEC7 method above.

An alternative ECEC method extracts the soil with 1 mol dm^{-3} KCl to displace all exchangeable acid cations. The displaced acidity is quantified by acidometric titration of the soil extract to the phenolphthalein end-point. Acidometric titration of KCl soil extract cannot distinguish exchangeable Al^{3+} cations from exchangeable H_3O^+ cations. This requires reserving half of the extract for a separate chemical treatment and acidometric titration.

Half of the original KCl extract is reserved and treated by adding the neutral electrolyte NaF(aq) before a separate acidometric titration. The fluoride anion F^- forms very stable complexes with $Al^{3+}(aq)$ ions dissolved in the extract, displacing all water molecules bound to $Al^{3+}(aq)$ cations solution. The hexafluoride aluminum(III) $AlF_6^{3-}(aq)$ complex anion has no acid-base character and is stable over the pH range of the acidometric titration. The second acidometric titration of the NaF-treated extract measures H_3O^+ cations only, yielding exchangeable $Al^{3+}(aq)$ acidity as the difference between total exchangeable acidity (untreated KCl extract) and exchangeable H_3O^+ acidity (NaF-treated extract).

BOX 6.9 *(cont'd)*

The CEC7 will be greater than the ECEC for any acid soil. If one adds up the CECb (Groups 1 and 2 cations occupying the exchange complex) of an acid soil plus exchangeable acidity from ECEC analysis, the sum will still be smaller than CEC7. The discrepancy is that fraction of CEC occupied by *nonexchangeable aluminum hydroxide polymers*. These nonexchangeable polymeric cations cannot be displaced by neutral electrolyte extraction. Nonexchangeable aluminum-hydroxide polymers can be displaced by compulsive cation exchange using an extraction reagent buffered at high pH.

The final method we will discuss is *CEC* measured by extracting soil using a buffered barium chloride-triethanolamine $BaCl_2$—TEA reagent, reported as CEC8. The CEC8 method saturates the exchange complex with index cation Ba^{2+}, relying on the pH 8.2 buffer (triethanolamine TEA) for compulsive exchange of all aluminum (exchangeable and nonexchangeable) occupying the exchange complex in much the same way as the buffered NH_4OAc. The $BaCl_2$—TEA method employs acidometric titration of the soil extract to quantify exchangeable and nonexchangeable acidity. *NSSC* analyses characterize the result as "extractable" acidity (i.e., exchangeable and nonexchangeable acidity displaced by this particular extract).

It should be clear by now that no single analysis will adequately quantify the *effective CEC* and exchangeable cations of a acidic soil in its natural state. A combination of methods is required to separately quantify nonacidic exchangeable (i.e., Groups 1 and 2) cations, exchangeable acid cations (Al^{3+} and H_3O^+), and nonexchangeable (i.e., Al-hydroxide polymers) acidity and natural-state *CEC* in acid soils.

6.6.1 Aluminum Chemistry

Example 5.6 examines the solubility of a common hydrous aluminum oxide mineral: gibbsite. Here we use a simple model for exchangeable acidity based on gibbsite solubility (expression 5.37 and reaction 5.R13), but one could easily prepare similar models based on other naturally occurring aluminum oxide, hydroxide, or oxyhydroxide minerals.

Fig. 6.15 (cf. Fig. 5.3) plots the activity of the most abundant solution aluminum species in equilibrium with gibbsite. The trivalent $Al^{3+}(aq)$ species is the most abundant aluminum species below pH 5, where aluminum solubility exceeds 10^{-3} mol dm^{-3}. The univalent $Al(OH)_2^+(aq)$ species is the dominant species in a pH range—$5.10 < pH < 6.80$—where aluminum solubility remains below 3×10^{-5} mol dm^{-3}. The divalent $Al(OH)^{2+}(aq)$ species is never the most abundant species.

6.6.2 Exchangeable Aluminum

Soils containing calcite are buffered at roughly pH 8.4; they resist becoming more acidic as long as the soil moisture residence time sufficient for rock weathering rate to maintain nonzero alkalinity. Of course, once the soil has reaches a weathering stage where there minerals containing Groups 1 and 2 elements have been depleted (cf. Fig. 6.1), residence time becomes irrelevant.

Although chemical weathering in humid climates will eventually remove all traces of carbonate minerals and silicate minerals containing Groups 1 and 2 cations, the exchange complex of clay minerals and organic colloids will remain largely saturated by Ca^{2+} ions, a consequence of the natural abundance of this element. Absent calcite, there is little to prevent a drop in soil pH. Though continued chemical weathering and leaching drive the pH ever lower, soil colloids remain Ca^{2+}-saturated until pH 6.5, the pH of minimum aluminum solubility. As soil pH drifts below pH 6.5, aluminum solubility begins to gradually increase, displacing exchangeable Ca^{2+} ions. As we have already seen, asymmetric ion exchange tends to favor the ion with higher valence, shifting the advantage decidedly in the favor of Al^{3+} (Fig. 6.16).

6.6.2.1 *Asymmetric (Al^{3+}, Ca^{2+}) Exchange*

The relative abundance of $Al(OH)_2{}^+(aq)$, $Al(OH)^{2+}$, and Al^{3+} species as pH decreases and gibbsite solubility increases is very important because these solution species become increasingly competitive with Ca^{2+} for cation-exchange sites. If we assume a representative soil pore water ionic strength of 10^{-3} mol dm^{-3}, the univalent $Al(OH)_2{}^+(aq)$ species is never

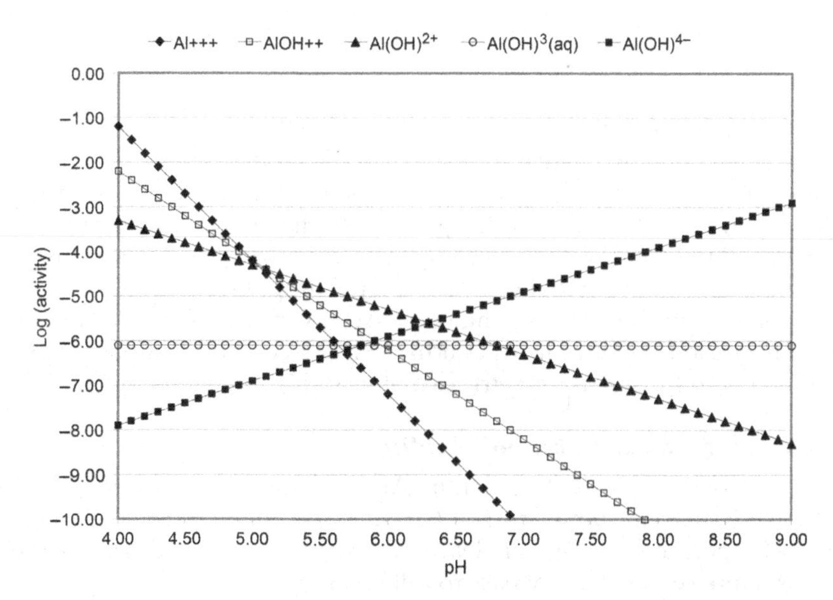

FIG. 6.15 This graph plots the logarithm of ion activity for the five most abundant aluminum solution species as a function of pH from simulation using *ChemEQL*. The solubility of $Al^{3+}(aq)$ is constrained by gibbsite solubility, as discussed in Example 5.6. Here $H^+(aq)$ is treated as an independent variable covering the range depicted in the graph.

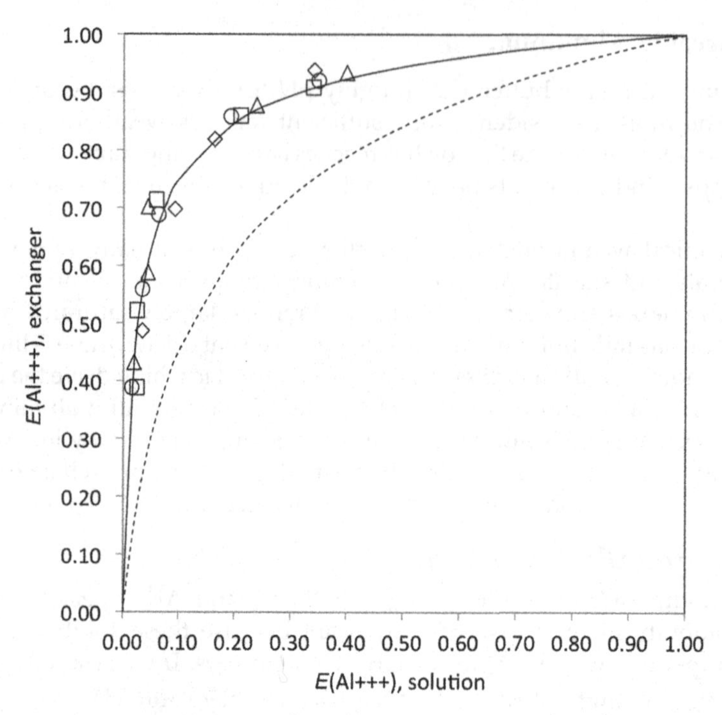

FIG. 6.16 This graph plots the asymmetric (Ca^{2+}, Al^{2+}) exchange isotherm from Coulter and Talibudeen (1968): $K^{c}_{Al/Ca} = 28.5$.

competitive for asymmetric cation exchange with Ca^{2+} in the pH range, where it is dominant because of its valence and low relative solubility. We can ignore symmetric ($Al(OH)^{2+}$, Ca^{2+}) cation exchange because $Al(OH)^{2+}(aq)$ is never the most abundant soluble aluminum species. By process of elimination, the cation-exchange reaction that is most significant to soil acidity is the asymmetric (Al^{3+}, Ca^{2+}) cation exchange where the advantage lies with Al^{3+} because of its valence and high relative solubility when soil pH drops below 5.

Numerical solution of the cubic asymmetric (Al^{3+}, Ca^{2+}) exchange isotherm using the Newton-Rapson method is found in Appendix 6.E. The combined effect of pH-dependent aluminum solubility (6.15) and asymmetric exchange (6.16) are plotted in Fig. 6.17.

6.6.2.2 Neutralizing Exchangeable Soil Acidity

Hydrolysis reactions (6.R30)–(6.R32) identify Al^{3+} as a weak acid with a high ion-exchange selectivity. Exchangeable calcium Ca^{2+} is a Group 2 spectator cation. The accumulation of exchangeable Al^{3+} portrayed in Fig. 6.17 leads to the accumulation of an exchangeable weak acid that requires three equivalents of base to fully neutralize.

$$Al(H_2O)_6{}^{3+}(aq) + H_2O(l) \longrightarrow Al(H_2O)_5OH^{2+}(aq) + H_3O^{+}(aq) \qquad (6.R30)$$

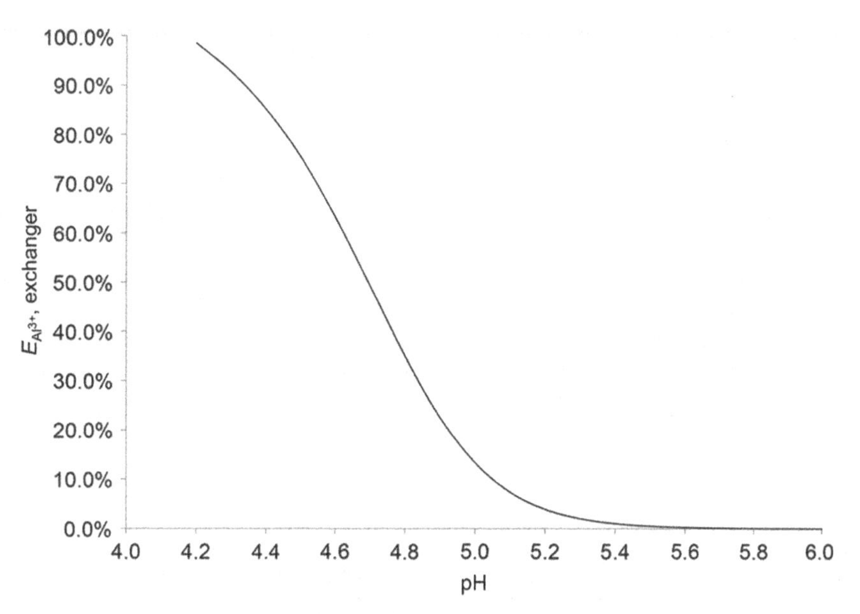

FIG. 6.17 This graph plots the exchangeable Al^{3+}(ex) equivalent fraction $E_{\overline{Al^{3+}}}$ for asymmetric (Al^{3+}, Ca^{2+}) exchange as a function of pH. Solution parameter N_0 is restricted to 1×10^{-3} $mol_c\,dm^{-3} < N_0 < 1 \times 10^{-3}$ $mol_c\,dm^{-3}$. The selectivity coefficient is $K^{\circ}_{Al/Ca} = 28.5$ (Coulter and Talibudeen, 1968). Gibbsite solubility is simulated using *ChemEQL* as discussed in Example 5.6.

$$Al(H_2O)_5OH^{2+}(aq) + H_2O(l) \longrightarrow Al(H_2O)_4(OH)_2{}^{+}(aq) + H_3O^{+}(aq) \quad (6.R31)$$

$$Al(H_2O)_4(OH)_2{}^{+}(aq) + H_2O(l) \longrightarrow Al(H_2O)_3(OH)_3{}^{0}(aq) + H_3O^{+}(aq) \quad (6.R32)$$

Combining reactions (6.R30) through (6.R32) with the conjugate base hydrolysis reaction (6.R14) yields the neutralization-exchange reaction (6.R33) between exchanger-bound Al^{3+} cations and the base $CaCO_3(s)$.

$$2\overline{Al^{3+}} + 3CaCO_3(s) + 3H_2O(l) \longrightarrow 2Al(OH)_3(s) + 3\overline{Ca^{2+}} + 3CO_2(aq) \quad (6.R33)$$

Calcite $CaCO_3(s)$ dissolution releases soluble $Ca^{2+}(aq)$ cations that undergoes asymmetric exchange with exchanger-bound Al^{3+} cations. The displacement of Al^{3+} by Ca^{2+} is precisely the reaction induced by the limewater extraction (Veitch, 1902) used to measure "total apparent acidity."

As Al^{3+} enter solution they reacts with the conjugate bases $HCO_3{}^{-}(aq)$ and $CO_3{}^{2-}(aq)$ resulting in $Al(OH)_3(s)$, precipitation, the very aluminum hydroxide precipitate described by Veitch (1902) (Example 6.8).

EXAMPLE 6.8

Example Permalink

http://soilenvirochem.net/OVqTO2

Calculate the grams $CaCO_3$(calcite) per kilogram soil needed to neutralize the exchangeable soil acidity.

The soil horizon in this example is from the TATUM soil series (Clayey, kaolinitic, thermic Typic Hapludults) in Richmond County, North Carolina (35.113608° N, 79.829000° W). A complete picture of exchangeable acidity in the TATUM soil emerges from measurements of buffered and unbuffered cation-exchange capacity—CEC7 and ECEC, respectively—and exchangeable ions (Table 6.26). Box 6.9 describes the analysis methods used to quantify exchangeable soil acidity. The stoichiometry of neutralization reaction between $CaCO_3$(calcite) and exchangeable Al^{3+} ions appears in reaction (6.R33).

TABLE 6.26 Results of Extracting the TATUM Soil With 1 mol dm^{-3} Ammonium Acetate and With 1 mol dm^{-3} Potassium Chloride

Ammonium Acetate Extraction	
Analyte	**$cmol_c\ kg^{-1}$**
Ca^{2+}	0.5
Mg^{2+}	0.9
Na^{+}	0.0
K^{+}	0.1
CECb	1.5
CEC7	7.1
Potassium Chloride Extraction	
Analyte	**$cmol_c\ kg^{-1}$**
Al^{3+}	1.9
Mn^{2+}	0.7
ECEC	3.4

The buffered NH_4OAc (i.e., CEC7) extract relies on compulsive cation exchange to displace exchangeable and nonexchangeable cations with index cation NH_4^+. The Groups 1 and 2 cations displaced into solution by the NH_4OAc are quantified by chemical analysis and summed to yield CECb. The complete CEC7 protocol extracts the NH_4^+-saturated soil a second time by 1 mol dm^{-3} potassium chloride to displace index cation NH_4^+. Cation-exchange capacity CEC7 based on the index cation NH_4^+ displaced by K^+ in the second extraction.

The unbuffered KCl (i.e., ECEC) extract relies solely on mass action from 1 mol dm^{-3} K^+ to displace exchangeable Al^{3+}, Mn^{2+}, along with all Groups 1 and 2 cations. Manganese(II) is measured because of its plant nutrition and physiological significance, not because it has acid-base properties.

$$n^{\sigma}_{Al^{3+}}/m = ECEC - CECb - n^{\sigma}_{Mn^{2+}}/m$$

$$n^{\sigma}_{Al^{3+}}/m = 3.4 - 1.5 - 0.7 = 1.2 \text{ cmol}_c \text{ kg}^{-1}$$

$$n^{\sigma}_{Al^{3+}}/m = 1.2 \times 10^{-2} \text{ mol}_c \text{ kg}^{-1}$$

The exchangeable acidity amounts to 1.2×10^{-2} mol$_c$ kg^{-1}. Calcite $CaCO_3(s)$ as a formula mass of $\overline{m}_f(CaCO_3) = 100.087$ g mol^{-1}. Since each $CO_3{}^{2-}$ anion is dibasic, then 50.043 g of $CaCO_3(s)$ will neutralize mole of acidity.

$$w(CaCO_3) = \left(1.2 \times 10^{-2} \text{ mol}_c \text{ kg}^{-1}\right) \cdot \left(50.43 \text{ g mol}^{-1}\right)$$

$$w(CaCO_3) = 0.60 \text{ g kg}^{-1}$$

The exchangeable soil acidity in the TATUM soil is neutralize by 1.2×10^{-2} moles of base per kilogram soil. Therefore 0.60 g of $CaCO_3(s)$ is required to neutralize the exchangeable acidity in each kilogram of TATUM soil. Stated differently, the $CaCO_3(s)$ requirement is 6.0×10^{-4} $Mg_{CaCO_3} \cdot Mg^{-1}_{soil}$.

The TATUM soil has an average bulk density of about 1.53 Mg m^{-3} in the upper 80 cm of soil depth, hence 1 ha of TATUM soil to a depth of 80 cm has a soil mass of about $1.22 \times 10^{+4}$ Mg. The $CaCO_3(s)$ required to neutralize the exchangeable acidity in the upper 80 cm of the TATUM soil is 7.4 Mg ha^{-1}.

The TATUM soil contains 3.7 cmol$_c$ kg^{-1} nonexchangeable aluminum. The nature of nonexchangeable aluminum will be discussed in the following section. With regard to exchangeable acidity, nonexchangeable aluminum-hydroxide polymers should be considered conjugate base of the weak acid $Al(H_2O)_6^{3+}$ and as such contribute little to exchangeable acidity.

$$CEC7 - ECEC = 7.1 - 3.4 = 3.7 \text{ cmol}_c \text{ kg}^{-1}$$

6.6.3 Nonexchangeable Aluminum

The accepted model for nonexchangeable acidity was proposed by Dixon and Jackson (1962) from a study of soil clay specimens described as "intergrades of chlorite-expansible 2:1 layer silicates." X-ray diffraction reveals the presence of a clay mineral component that would expand and collapse, depending on the chemical treatment, consistent with an expansible 2:1 mineral (i.e., vermiculite or smectite).

X-ray diffraction also revealed a second clay mineral component whose 1.4 nm layer spacing—typically associated with the 2:1:1 clay mineral chlorite—remained immune to chemical and heat treatments designed on induce swelling and collapse. Dixon and Jackson (1962) concluded the 1.4 nm component was not truly chlorite but a vermiculite or smectite clay mineral whose interlayer was partially occupied Al-hydroxy polymers. Complete filling of the interlayer gallery of an expansible 2:1 expansible clay would have dramatically reduced *CEC* but these clay mineral specimens retained *CEC* values about one-third of a typical

TABLE 6.27 Hypothetical Al-Hydroxy Polymeric Cations (Hsu and Bates, 1964)

Formula	Shared Edges	Al_6 Rings
$Al_2(OH)_2{}^{4+}$	1	0
$Al_3(OH)_4{}^{5+}$	2	0
$Al_4(OH)_6{}^{6+}$	3	0
$Al_5(OH)_8{}^{7+}$	4	0
$Al_6(OH)_{12}{}^{8+}$	6	1
$Al_{10}(OH)_{22}{}^{8+}$	11	2
$Al_{13}(OH)_{30}{}^{9+}$	15	3
$Al_{24}(OH)_{60}{}^{12+}$	30	7
$Al_{54}(OH)_{144}{}^{18+}$	72	19

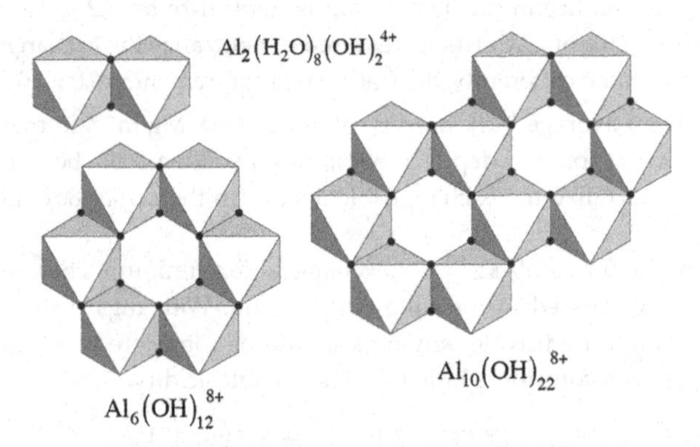

FIG. 6.18 Hypothetical Al-hydroxy polymeric cations representative of those listed in Table 6.27.

smectite. Furthermore, although NaOH(aq) extraction displaced Al^{3+} (and Fe^{3+}) into solution and doubling *CEC*, the treatment could not eliminate the 1.4 nm diffraction peak.

Soon after Hsu and Bates (1964) identified cationic Al-hydroxy polymers with varying Al-to-OH ratios, several examples are listed in Table 6.27. Fig. 6.18 illustrates three cationic polymers listed in Table 6.27: $Al_2(OH)_2{}^{4+}$ (two edge-sharing octaherdrons), $Al_6(OH)_{12}{}^{8+}$ (a ring of six edge-sharing octahedrons) and $Al_{10}(OH)_{22}{}^{8+}$ (a two rings of edge-sharing octahedrons).

The Al-hydroxy polymeric cations listed in Table 6.27 planar, edge-sharing $Al(H_2O)_{6-x}(OH)_x$ octahedrons with the same thickness as the interlayer brucite octahedral sheets found in the chlorite layer structure. The size and charge of these polymeric cations, combined with the added opportunity to form hydrogen bonds with the plane of oxygen atoms forming the base of each tetrahedral sheet, make these cations immune to cation exchange most Groups 1 and 2 cations. The only way these cations can be displaced from the interlayer is by compulsive cation exchange.

6.7 SUMMARY

Environmental acidity and basicity arise from numerous sources: wash-out of atmospheric gases in the troposphere and their subsequent wet deposition on the land surface, ammoniacal fertilizer use and biomass harvest in agriculture, dissimilatory nitrate and sulfate reduction, and chemical weathering of minerals in soils and aquifers. The criteria for measuring acidity or basicity in the environment must account for the dissolution of carbon dioxide in water and the production of bicarbonate ions—the conjugate base of the weak acid aqueous carbon dioxide. Bicarbonate ions are the basis of water alkalinity; the reference level used by environmental chemists shifts below pH 7 as hydroxyl ions are converted into bicarbonate ions by $CO_2(aq)$.

Climatic conditions and mineral weathering determine whether soils tend to be acidic or alkaline; the former are common in humid climates, while the latter are common in arid and semiarid climates. Land degradation can occur in arid and semiarid climates when the accumulation of sodium ions by clay minerals reduces the activity A of soil clays, reducing soil mechanical strength and—as a consequence—soil hydraulic conductivity. Sodification is a complex process that involves cation exchange, pH-dependent calcite solubility, and extreme alkalinity that increases the sodium-to-calcium ratio in solution by lowering calcium solubility. At the other extreme, land degradation in humid climates can result in soil acidification. This process also involves interactions between pH-dependent changes in aluminum hydrous oxide solubility and cation exchange, leading to the accumulation of exchangeable aluminum ions. Exchangeable aluminum ions behave as weak acids, and their accumulation amounts to a chemically active but insoluble form of acidity in humid region soils.

APPENDICES

6.A THE pH SCALE

Sørensen (1909) introduced the original pH scale and notation, two decades after Arrhenius (1887, 1888) proposed the first acid-base chemistry model but a decade before Brønsted (1923) and Lowry (1923) introduced the concept of water self-ionization and Debye and Hückel (1923) published the first paper on ion activity coefficients.

$$pH_{Sørensen} \equiv -\log_{10}\left(c_{H_3O^+}\right) \tag{6.A.1}$$

The chemical notation adopted in this book follows guidelines recommended by the *International Union of Pure and Applied Chemistry* (IUPAC) (Cohen, 2007). The recommended *IUPAC* definition for pH uses H_3O^+ activity rather than concentration.

$$pH \equiv -\log_{10}\left(a_{H_3O^+}\right) \tag{6.A.2}$$

The relation between the *IUPAC* pH scale and the Sørensen (1909) pH scale requires an ion activity coefficient correction.

$$pH = pH_{Sørensen} - \log_{10}(\gamma\pm)$$

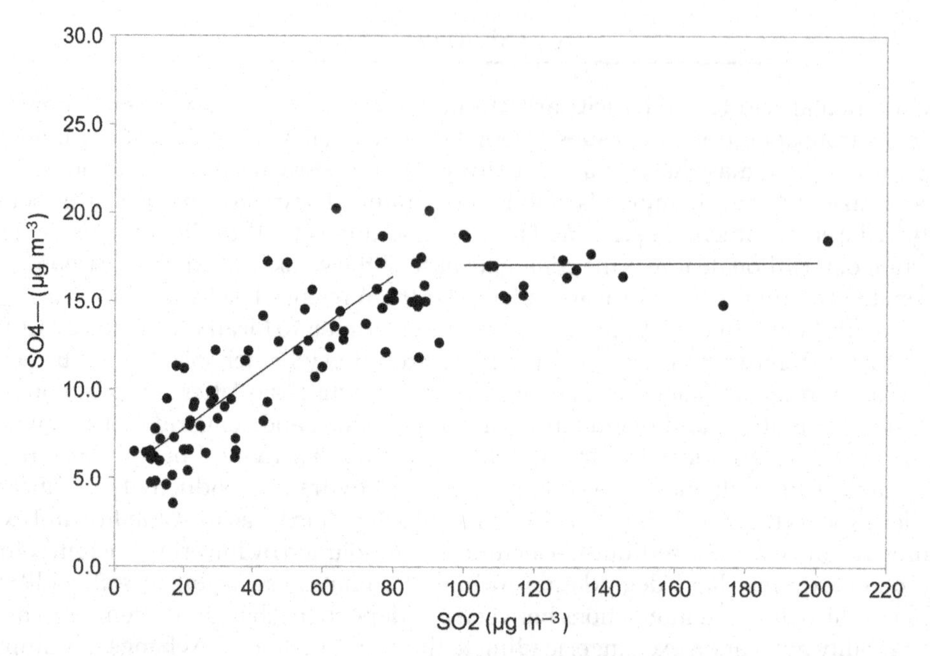

FIG. 6.B.1 The relationship between annual average troposphere sulfur dioxide concentration and troposphere sulfate concentration for 18 US cities. *(Reproduced with permission from Altshuller, A.P., 1973. Atmospheric sulfur dioxide and sulfate: distribution of concentration at urban and nonurban sites in United States. Environ. Sci. Technol. 7, 709–712.)*

The water reference level is identical on both the *IUPAC* pH scale and the Sørensen (1909) pH scale.

6.B THE SATURATION EFFECT: ATMOSPHERIC CONVERSION OF SULFUR TRIOXIDE TO SULFURIC ACID

Altshuller (1973) demonstrated the saturation effect that occurs when atmospheric sulfur dioxide concentrations exceed a certain threshold (Fig. 6.B.1. Below the threshold, the sulfur dioxide-to-sulfate ratio in the atmosphere of major cities increases linearly. Beyond the saturation threshold (80 mg m^{-3} in Fig. 6.B.1), atmospheric sulfuric acid levels cease to increase regardless of the sulfur dioxide concentration.

Ammonia and other alkaline substances in the atmosphere catalyze the oxidation of sulfur dioxide to sulfuric acid (Scott and Hobbs, 1967). Once sulfur dioxide levels reach a certain threshold, the acidity from the sulfuric acid exceeds the alkalinity, resulting in a drop in the pH of water droplets in the troposphere, effectively eliminating the further conversion of sulfur dioxide to sulfuric acid.

6.C ACID-BASE IMPLICATIONS OF NITRATE AND SULFATE REDUCTION

6.C.1 Nitrogen Cycling and Fertilization in Agricultural Landscapes

Soil acidification is a natural consequence of mineral weathering and leaching in humid climates. As a general rule, plant nutrient cycling is not a major contributor to soil acidification because components relevant to soil acid-base chemistry in standing biomass ultimately return to the soil when plant residue decomposes. Biomass harvesting by agriculture and forestry, however, does contribute to soil acidification because it exports plant ash from managed ecosystems. The import of ammonia-based fertilizers to replace soil nitrogen lost by harvest dramatically accelerates soil acidification.

The nitrogen cycle involves several reactions that transform nitrogen by electron transfer. Understanding the acid-base implications of the nitrogen cycle is, therefore, our point of departure. Nitrogen fixation (reaction 6.C.R1) consumes eight protons as it forms two $NH_4{}^+$ molecules for every N_2 molecule fixed, regardless of the specific fixation process—biological or abiotic atmospheric or industrial. Nitrogen fixation yields hydroxyl ions at the rate of 4 OH^- for each N atom fixed.

$$N_2(g) + 8\,H^+(aq) + 6\,e^- \xrightarrow{\text{Nitrogen fixation}} \overset{OS(N)=-3}{2\,NH_4^+(aq)} \tag{6.C.R1}$$

The *mineralization*[26] of biological nitrogen compounds, resulting in NH_3 assimilation, has no impact on environmental acid-base processes.

Nitrification (reaction 6.C.R2) is a biological oxidation process (see Chapter 8) that yields 10 protons for each $NH_4{}^+$ oxidized to $NO_3{}^-$.

$$\overset{OS(N)=-3}{2\,NH_4^+(aq)} + 6\,H_2O(aq) \xrightarrow{\text{Nitrification}} \overset{OS(N)=-5}{2\,NO_3^-(aq)} + 20\,H^+(aq) + 16\,e^- \tag{6.C.R2}$$

Plants, fungi, bacteria, and archaea must reduce $NO_3{}^-$ to $NH_4{}^+$ before then can assimilate $NO_3{}^-$ taken up from solution, effectively reversing the nitrification reaction (6.C.R3).

$$\overset{OS(N)=-5}{2\,NO_3^-(aq)} + 20\,H^+(aq) + 16\,e^- \xrightarrow{\text{Nitrate assimilation}} \overset{OS(N)=-3}{2\,NH_4^+(aq)} + 6\,H_2O(aq) \tag{6.C.R3}$$

Denitrification (reaction 6.C.R4) is a biological reduction process (see Chapter 8) that consumes 12 protons as it forms 1 N_2 molecule for every 2 $NO_3{}^-$ molecules reduced, completing the nitrogen cycle.

[26] Biologists refer to the (bio)chemical transformation of organic compounds into their inorganic components (e.g., the organic nitrogen and carbon in an amino acid to CO_2, NH_3, and H_2O) as *mineralization*. Earth scientists use the term *chemical weathering* when referring to mineral dissolution reactions.

$$\overset{OS(N)=-5}{2\,NO_3^-(aq)} + 12\,H^+(aq) + 10\,e^- \xrightarrow{\text{Denitrification}} \overset{OS(N)=0}{N_2(g)} + 6\,H_2O(aq) \qquad (6.C.R4)$$

The 8 protons consumed by nitrogen fixation (reaction 6.C.R1) and 16 protons consumed by denitrification (reaction 6.C.R4) balance the 20 protons produced during nitrification (reaction 6.C.R2). At steady state, nitrogen fixation (reaction 6.C.R1) replaces NO_3^- lost by leaching and denitrification. Nitrate leaching is effectively hydroxyl export from the soil profile because nitrate reduction by assimilation (reaction 6.C.R3) or denitrification (reaction 6.C.R4) occurs off-site.

Keeping in mind most soils have little anion-exchange capacity—humic colloids and clay minerals are cation exchangers—once NO_3^- has leached below the soil profile, there is little organic matter to sustain bacterial nitrate respiration (see Chapter 8). Remote nitrate reduction is most likely to occur in organic-rich sediments as groundwater discharges into a stream hyporheic (or lake hypolentic) zone.

The only significant contributions to soil acidification from nitrogen cycling in natural landscapes arise from nitrate leaching losses (i.e., remote nitrate reduction). Agricultural landscapes, by contrast, export biomass and rely on fertilization to replace nutrients lost by export. Nitrate fertilizer import constitutes hydroxyl import—released during assimilation (reaction 6.C.R3)—while the import of ammonia-based fertilizer (urea, ammonia, and ammonium salts) constitutes acid import—released during nitrification (reaction 6.C.R2).

Nitrate fertilization is equivalent to OH− import because nitrate must be reduced to OS(N) = −3 following uptake before assimilation can occur. The net OH− yield during assimilatory nitrate reduction is: 10 OH^-/N. Nitrate fertilizer lost by leaching has no on-site acid-base impact, although it does result in off-site OH^--deposition in hyporheic or hypolentic zones.

Nitrogen fertilization by ammoniacal compounds (ammonia, ammonium salts, urea, sewage sludge, livestock manure) is equivalent to H^+-import because nitrogen fertilizer use is less than 100% efficient. Ammoniacal nitrogen that is not take up is oxidized to OS(N) = +5 on-site by biological nitrification, yielding 10 H^+/N. Aerobic chemoautotrophic soil bacteria capture metabolic energy through biological nitrification, competing with the crop for ammoniacal nitrogen. Nitrate generated by on-site nitrification of ammoniacal nitrogen fertilizer does yield OH^- in the standing biomass during assimilatory nitrate reduction but biomass harvest exports these OH^- equivalents off-site. The net on-site effect of ammoniacal nitrogen fertilization is on-site H^+ deposition far exceeding atmospheric H^+ deposition from atmospheric sources described in Sections 6.3.2.2 and 6.3.2.3.

6.C.2 Marine Environment: Sulfate and Nitrate Reduction in the Deep Ocean

Chen and Wang (1999) and Thomas et al. (2009) identify anaerobic carbon mineralization in the deep ocean as a significant biological alkalinity source. Sulfate and nitrate enter the oceans through direct atmospheric deposition and surface runoff from the continents. These two compounds serve as electron acceptors for anaerobic respiration in the deep ocean.

The carbon oxidation state in dissolved organic carbon (DOC)—both terrestrial and marine—is OS(C) $\approx$ 0. Reaction (6.C.R5) represents DOC mineralization by deep ocean anaerobic nitrate respiration.

$$\underset{DOC}{5CH_2O} + 4NO_3^-(aq) \xrightarrow{\text{Nitrate respiration}} 5CO_2(g) + 2N_2(g) + 3H_2O(aq) + 4OH^- \qquad (6.C.R5)$$

The alkalinity yield is 1 mole HCO_3^- per mole of NO_3^- reduced to N_2, which escapes the ocean with scant probability it will be biologically fixed in the marine water column. Nitrate, however, is a negligible alkalinity source relative to sulfate primarily because marine nitrate concentrations are typically 100-fold less than sulfate.

Reaction (6.C.R6) represents DOC mineralization by anaerobic sulfate respiration in the deep ocean.

$$\underset{DOC}{2CH_2O} + SO_4^{2-}(aq) \xrightarrow{\text{Sulfate respiration}} 2CO_2(g) + H_2S(aq) + 2OH^- \qquad (6.C.R6)$$

The alkalinity yield is 2 moles HCO_3^- per mole of SO_4^{2-} reduced to H_2S. Some H_2S precipitates as insoluble iron sulfide in deep ocean sediments and some escapes the ocean into the atmosphere. Aerobic chemolithotrophic bacteria in the marine water column increase the likelihood H_2S will be reoxidized to SO_4^{2-} but despite these losses marine chemists (Chen and Wang, 1999; Thomas et al., 2009) consider sulfate respiration the major alkalinity source in the ocean with the potential to neutralize steadily increasing $CO_2(aq)$ levels from fossil fuel combustion.

6.D SELECTIVITY OF (Na^+, Ca^{2+}) EXCHANGE

Oster et al. (1980) and Sposito et al. (1983a,b) published (Na^+, Ca^{2+}) exchange on two closely related smectite clay specimens (American Petroleum Institute specimen API-25, Upton, Wyoming and Clay Minerals Society specimen SWy-1, smectite, Crook County, Wyoming).

Changes in clay chemistry appear when Na^+ ions occupy a significant fraction of the exchange complex. Recall the equivalent fraction-dependent selectivity coefficient for (Mg^{2+}, Ca^{2+}) exchange on the Libby vermiculite (see Appendix 4.E). In that case selectivity favors Ca^{2+} ions at the Ca^{2+}-rich limit of the exchange isotherm, while favoring Mg^{2+} ions at the Mg^{2+}-rich limit of the exchange isotherm. The shift in ion selectivity away from Ca^{2+} occurs when the equivalent fraction of exchangeable Mg^{2+} ions increases from trace levels to chemically significant levels, altering the character of the interlayer. Similarly, as the equivalent fraction of exchangeable Na^+ ions increases during (Na^+, Ca^{2+}) exchange, the character of the interlayer gradually changes from an environment limited to crystalline swelling at the Ca^{2+}-rich limit to an interlayer capable of free swelling at the Na^+-rich limit.

Fig. 6.D.1 plots the exchange isotherm reported by Sposito et al. (1983a,b) while the exchange isotherm from (Oster et al., 1980) appear in Fig. 6.D.2. Despite employing a significantly higher electrolyte concentration (Oster et al., 1980), the selectivity coefficient derived from fitting data from Sposito et al. (1983a,b) provides an excellent fit of data reported by

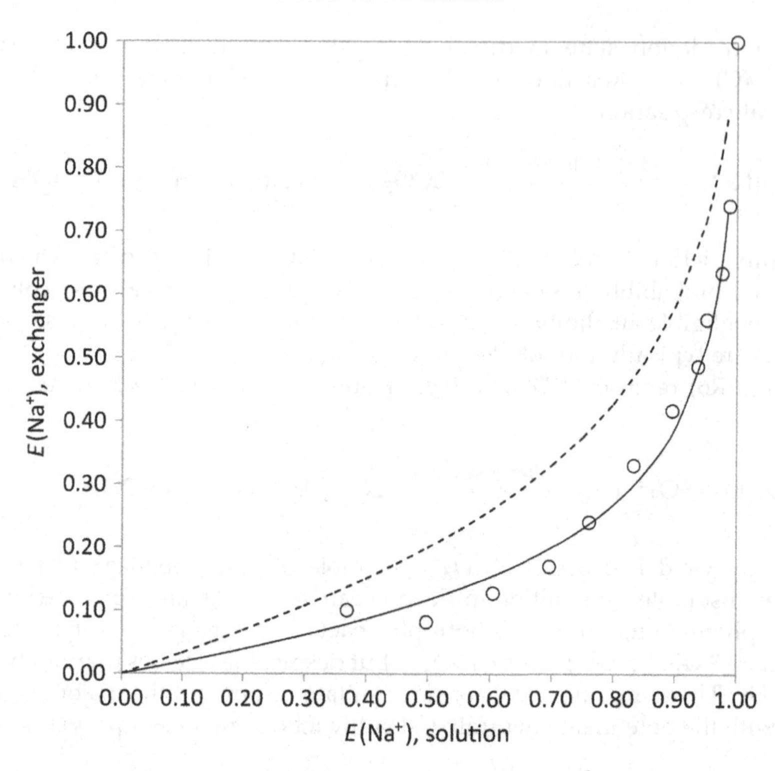

FIG. 6.D.1 The experimental exchange isotherms for asymmetric (Na^+, Ca^{2+}) exchange for Wyoming smectite SWy-1 (Sposito et al., 1983a,b). The exchange isotherm representing selectivity coefficient $K^{\diamond}_{Na/Ca} = 0.30$ is plotted as a *smooth line*, and the nonselective exchange isotherm appears as a *dashed line*.

Oster et al. (1980). The shift in position of the exchange isotherm going from Figs. 6.D.1 to 6.D.2 results from the different salt concentrations used in the two studies.

Modest increases in the ESP reduces the *plasticity index* PI of smectite clays, significantly lowering the activity *A* of soils containing smectite clays (Fig. 7.9). The reduction of plasticity lowers the mechanical strength of the soil fabric and leads to deterioration of soil structure. Soil structure degradation caused by high *ESP* increases the soil air-water permeability ratio, lowering hydraulic conductivity and water infiltration rates leading to increasing runoff. This reduces both soil moisture recharge efficiency and internal drainage. Internal soil drainage controls the leaching of excess salts from the soil profile.

6.E FITTING THE ASYMMETRIC (Al^{3+}, Ca^{2+}) ION-EXCHANGE ISOTHERM

Finding the one real root of the asymmetric (Al^{3+}, Ca^{2+}) exchange isotherm third-order polynomial brings us to a mathematical decision point. One option involved the analytic solution to third-order polynomials, similar in spirit but significantly more complex than the quadratic equation needed to find the positive root for asymmetric (Na^+, Ca^{2+}) exchange

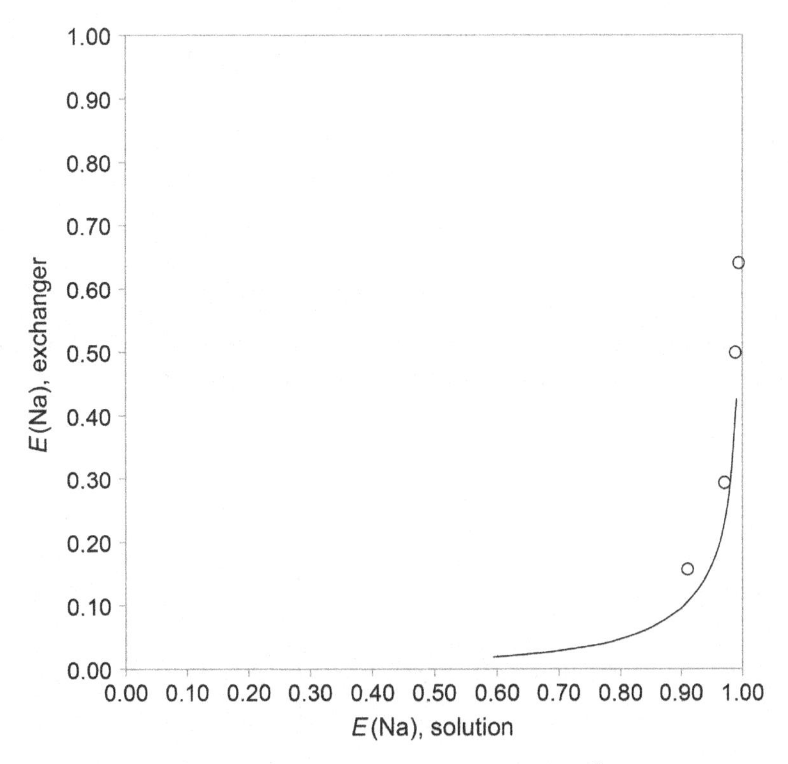

FIG. 6.D.2 The experimental exchange isotherms for asymmetric (Na^+, Ca^{2+}) exchange for Wyoming smectite API-25 (Oster et al., 1980). The exchange isotherm representing selectivity coefficient $K^{\ominus}_{Na/Ca} = 0.30$ is plotted as a *smooth line*.

isotherm (a second-order polynomial). The first edition of this book explained the analytic solution to third-order polynomials.

Adding to the sheer complexity of the analytic solution is an additional requirement: each point defining the exchange isotherm requires a distinct solution. The analytic solution described in the first edition (Patel, 1980; Delano, 1998) parsed the solution into two different forms that depended on whether a key parameter was positive or negative.

Inevitably the point on the exchange isotherm where the solution is parsed depends on the selectivity coefficient $K^{\ominus}_{Al/Ca}$, the solution ion-charge concentration parameter N_0, and the solution equivalent fraction $E_{\overline{Al^{3+}}}$. Implementing the analytic method requires a logical IF(logical-test, value-if-true, value-if-false) to test the parsing parameter and assign the appropriate solution.

Chapter 8 will discuss a special site-limited adsorption isotherm based on symmetric and asymmetric exchange isotherms (cf. Chapter 4). These ion-exchange based site-limited adsorption isotherms are all four-order and higher-order polynomials that require a numerical method to determine the real root. In this appendix we will derive the third-order polynomial for the asymmetric (Al^{3+}, Ca^{2+}) exchange isotherm and apply the relatively simple, and extremely flexible, numerical solution known as the *Newton-Rapson method*.

$$2Al^{3+}(aq) + 3\overline{Ca^{2+}} \longleftrightarrow 2\overline{Al^{3+}} + 3Ca^{2+}(aq) \quad (6.E.R1)$$

Applying Gaines-Thomas equivalent-fraction convention, the selectivity quotient expression for the asymmetric (Al^{3+}, Ca^{2+}) exchange reaction (6.E.R1) is given by expression (6.E.1).

$$K^{\circ}_{Al/Ca} = \frac{E^2_{\overline{Al^{3+}}} \cdot a^3_{Ca^{2+}}}{a^2_{Al^{3+}} \cdot E^3_{\overline{Ca^{2+}}}} = \frac{E^2_{\overline{Al^{3+}}} \cdot \left(\gamma^3_{2\pm} \cdot c^3_{Ca^{2+}}\right)}{\left(\gamma^2_{3\pm} \cdot c^2_{Al^{3+}}\right) \cdot E^3_{\overline{Ca^{2+}}}} \quad (6.E.1)$$

The solution ion-charge concentration parameter N_0 required for all asymmetric ion-exchange isotherm (Eq. 6.E.2) is used to define the solution equivalent fractions (e.g., $E_{Al^{3+}}$).

$$N_0 = 3 \cdot c_{Ca^{2+}} + 2 \cdot c_{Al^{3+}} \quad (6.E.2)$$

$$E_{Al^{3+}} = \frac{3 \cdot c_{Al^{3+}}}{N_0} = \left(1 - E_{Ca^{2+}}\right)$$

The modified selectivity quotient expression lists all ions using only equivalent fractions (Eq. 6.E.3), where $\gamma^6_{1\pm} = (\gamma^2_{3\pm}/\gamma^3_{2\pm})$.

$$\left(\frac{\gamma^2_{3\pm} \cdot K^{\circ}_{Al/Ca}}{\gamma^3_{2\pm}}\right) = \left(\frac{9 \cdot N_0}{8}\right) \cdot \left(\frac{E^2_{\overline{Al^{3+}}} \cdot E^3_{Ca^{2+}}}{E^2_{Al^{3+}} \cdot E^3_{\overline{Ca^{2+}}}}\right)$$

$$\left(\frac{8 \cdot \gamma^6_{1\pm} \cdot K^{\circ}_{Al/Ca}}{9 \cdot N_0}\right) = \left(\frac{E^2_{\overline{Al^{3+}}} \cdot E^3_{Ca^{2+}}}{E^2_{Al^{3+}} \cdot E^3_{\overline{Ca^{2+}}}}\right) \quad (6.E.3)$$

Replacing Ca^{2+} equivalent fractions yields the desired third-order polynomial.

$$\left(\frac{8 \cdot K^{c}_{Al/Ca}}{9 \cdot N_0}\right) = \left(\frac{E^2_{\overline{Al^{3+}}} \cdot \left(1 - E_{Al^{3+}}\right)^3}{E^2_{Al^{3+}} \cdot \left(1 - E_{\overline{Al^{3+}}}\right)^3}\right)$$

$$\left(\frac{8 \cdot K^{c}_{Al/Ca} \cdot E^2_{Al^{3+}}}{9 \cdot N_0 \cdot \left(1 - E_{Al^{3+}}\right)^3}\right) = \left(\frac{E^2_{\overline{Al^{3+}}}}{1 - 3 \cdot E_{\overline{Al^{3+}}} + 3 \cdot E^2_{\overline{Al^{3+}}} - E^3_{\overline{Al^{3+}}}}\right)$$

Rearranging the term and defining the *cubic exchange parameter* β_3 (Eq. 6.E.4) results in a simplified third-order polynomial expression (6.E.5) for the asymmetric (Al^{3+}, Ca^{2+}) exchange isotherm.

$$\beta_3 \equiv \left(\frac{9 \cdot N_0 \cdot \left(1 - E_{Al^{3+}}\right)^3}{8 \cdot K^{c}_{Al/Ca} \cdot E^2_{Al^{3+}}}\right) - 3 \quad (6.E.4)$$

Define: $r \equiv E_{\overline{Al^{3+}}}$

$$p(r) \equiv r^3 + \beta_3 \cdot r^2 + 3 \cdot r - 1 = 0 \quad (6.E.5)$$

The first step in the *Newton-Rapson method* begin with an initial estimate of a real root of the polynomial function: represented by r_0. Since we are tracing the exchange isotherm the initial estimate of any exchanger equivalent fraction can be simply the independent variable (i.e., the *solution equivalent fraction* $r_0 = E_{Al^{3+}}$).

A second improved approximation r_1 (Eq. 6.E.6) is the sum the initial estimate r_0 and the value of the first derivative of the polynomial evaluated at r_0 (Eq. 6.E.7).

$$r_1 = r_0 + p'(r_0) \quad (6.E.6)$$

$$p'(r_0) \equiv 3 \cdot r_0^2 + (2 \cdot \beta_3) \cdot r_0 + 3 \quad (6.E.7)$$

The process is repeated as Eq. (6.E.8) until the estimate converges (i.e., $r_{n+1} \approx r_n$).

$$r_{n+1} = r_n + p'(r_n) \quad (6.E.8)$$

The *Newton-Rapson* numerical solution to a polynomial asymmetric-exchange isotherm can yield a selectivity coefficient estimate using a single point from an asymmetric exchange isotherm or fit the entire isotherm by minimizing the sum-of-squares error (cf. Appendix 4.B).

A nonselective asymmetric isotherm is found by setting the selectivity coefficient β_3 in expression (6.E.4) equal to unity and finding the real root for points sampling the dependent variable—the *solution* equivalent fraction of one ion—on the interval $0 \le r \le 1$ and finding the real root (i.e., the dependent variable being the corresponding *exchanger* equivalent fraction) for each point.

The polynomial (6.E.5) with cubic exchange parameter β_3 result from asymmetric exchange reaction (6.E.R1). The asymmetric (Na^+, Al^{3+}) exchange reaction would yield a different polynomial with a different set of coefficients.

References

Allison, L.E., Bernstein, L., Bower, C.A., Brown, J.W., Fireman, M., Hatcher, J.T., Hayward, H.E., Pearson, G.A., Reeve, R.C., Richards, L.A., Wilcox, L.V., 1954. Diagnosis and Improvement of Saline and Alkali Soils. Revised ed. Agriculture Handbook, 60. United States Salinity Laboratory, Washington, DC, 160 pp.

Altshuller, A.P., 1973. Atmospheric sulfur dioxide and sulfate. distribution of concentration at urban and nonurban sites in united states. Environ. Sci. Technol. 7 (8), 709–712.

Arrhenius, S., 1887. Über die dissociation der in wasser gelösten stoff. Z. Phys. Chem. 1, 631–648.

Arrhenius, S., 1888. Über den gefrierpunkt verdünnter wässeriger lösungen. Z. Phys. Chem. 2, 491–505.

Ayers, R.S., Westcot, D.W., 1977. Water quality for agriculture. In: Proceedings of the International Conference on Managing Saline Water Irrigation. Int. Cent. Arid Semi-Arid Land Stud.

Ayers, R.S., Westcot, D.W., 1985. Water quality for agriculture In: Proceedings of the International Conference on Managing Saline Water Irrigation. Int. Cent. Arid Semi-Arid Land Stud., http://www.fao.org/DOCREP/003/T0234E/T0234E00.htm.

Bandura, A.V., Lvov, S.N., 2006. The ionization constant of water over wide ranges of temperature and density. J. Phys. Chem. Ref. Data 35 (1), 15–30.

Bates, T.S., Lamb, B.K., Guenther, A., Dignon, J., Stoiber, R.E., 1992. Sulfur emissions to the atmosphere from natural sources. J. Atmos. Chem. 14 (1–4), 315–337.

Benkovitz, C.M., Scholtz, M.T., Pacyna, J., Tarrason, L., Dignon, J., Voldner, E.C., Spiro, P.A., Logan, J.A., Graedel, T.E., 1996. Global gridded inventories of anthropogenic emissions of sulfur and nitrogen. J. Geophys. Res. [Atmos.] 101 (D22), 29239–29253.

Bhattacharya, P., Hasan, M.A., Sracek, O., Smith, E., Ahmed, K.M., Von Brömssen, M., Huq, S.I., Naidu, R., 2009. Groundwater chemistry and arsenic mobilization in the holocene flood plains in south-central Bangladesh. Environ. Geochem Health 31 (1), 23–43.

Bjerrum, L., 1954. Geotechnical properties of Norwegian marine clays. Geotechnique, 4, 49–69.

Blatt, H., Tracy, R., Owens, B., 2006. Petrology: Igneous, Sedimentary, and Metamorphic. W. H. Freeman, New York.

Bower, C.A., Wilcox, L.V., Akin, G.W., Keyes, M.G., 1965. An index of the tendency of CACO3 to precipitate from irrigation waters. Soil Sci. Soc. Am. Proc. 29 (1), 91–92.

Bower, C.A., Ogata, G., Tucker, J.M., 1968. Sodium hazard of irrigation waters as influenced by leaching fraction and by precipitation or solution of calcium carbonate. Soil Sci. 106 (1), 29–34.

Brønsted, J.N., 1923. Einige bemerkungen über den begriff der säuren und basen. Recl. Trav. Chim. Pays-Bas. 42 (8), 718–728.

Chen, C.T.A., Wang, S.L., 1999. Carbon, alkalinity and nutrient budgets on the East China Sea continental shelf. J. Geophys. Res., [Oceans] 104 (C9), 20675–20686.

Choi, J., Hulseapple, S., Conklin, M., Harvey, J., 1998. Modeling CO_2 degassing and ph in a stream-aquifer system. J. Hydrol. 209 (1), 297–310.

Cohen, E.R., 2007. Quantities, Units and Symbols in Physical Chemistry, third ed. Royal Society of Chemistry, London, United Kingdom.

Coulter, B.S., Talibudeen, O., 1968. Calcium: aluminum exchange equilibrium in clay minerals and acid soils. J. Soil Sci. 19 (2), 237–250.

Crandall, C.A., Katz, B.G., Hirten, J.J., 1999. Hydrochemical evidence for mixing of river water and groundwater during high-flow conditions, lower Suwannee River basin, Florida, USA. Hydrogeol. J. 7 (5), 454–467. http://dx.doi.org/10.1007/s100400050218.

Davis, L.E., 1945. Kinetic theory of ionic exchange for ions of unequal charge. J. Phys. Chem. 49, 473–479.

De Gregorio, S., Camarda, M., Longo, M., Cappuzzo, S., Giudice, G., Gurrieri, S., 2011. Long-term continuous monitoring of the dissolved CO_2 performed by using a new device in groundwater of the Mt. Etna (Southern Italy). Water Res. 45 (9), 3005–3011.

Debye, P., Hückel, E., 1923. Zur Theorie der Elektrolyte. I. Gefrierpunktserniedrigung und verwandte Erscheinungen. Phys. Z. 24, 185–206.

Deffeyes, K.S., 1965. Carbonate equilibria: A graphic and algebraic approach. Limnol. Oceanogr. 10 (3), 412–426.

Delano, A., 1998. Design analysis of the Einstein refrigeration cycle. Appendix A. Analytical solution of a cubic equation. http://www-old.me.gatech.edu/energy/andy_phd/appA.

Delany, J.M., Wolery, T.J., 1984. Fixed fugacity option for the EQ6 geochemical reaction path code UCRL-53598; Order No. DE85016345.

Dixon, J.B., Jackson, M.L., 1962. Properties of intergradient chlorite-expansible layer silicates of soils. Soil Sci. Soc. Am. Proc. 26 (4), 358–362.

Eilers, J.M., Brakke, D.F., Henriksen, A., 1992. The inapplicability of the gibbs model of world water chemistry of dilute lakes (comments). Limnol. Oceanogr. 37 (6), 1335–1337.

Feth, J.H., Gibbs, R.J., 1971. Mechanisms controlling world water chemistry: evaporation-crystallization process. Science 172 (3985), 870–872.

Frost, B.R., 1991. Introduction to oxygen fugacity and its petrologic importance. Rev. Mineral. 25, 1–9.

Fuoss, R.M., 1959. The velocity field in electrolytic solutions. J. Phys. Chem. 63 (4), 633–636.

Fuoss, R.M., Onsager, L., 1957. Conductance of unassociated electrolytes. J. Phys. Chem. 61 (5), 668–682.

Fuoss, R.M., Onsager, L., 1958. The kinetic term in electrolytic conductance. J. Phys. Chem. 62 (10), 1339–1340.

Gapon, E.N., 1933a. Theory of exchange adsorption in soils. I. J. Gen. Chem. (USSR) 3, 144–163.

Gapon, E.N., 1933b. Theory of exchange adsorption. IV. J. Gen. Chem. (USSR) 3, 660–666.

Garrels, R.M., Mackenzie, F.T., 1971. Evolution of Sedimentary Rocks, first ed. Norton, New York.

George, G.N., Gorbaty, M.L., 1989. Sulfur K-edge x-ray absorption spectroscopy of petroleum asphaltenes and model compounds. J. Am. Chem. Soc. 111 (9), 3182–3186.

Gibbs, R.J., 1970. Mechanisms controlling world water chemistry. Science 170 (3962), 1088–1090.

Gibbs, R.J., 1971. Mechanisms controlling world water chemistry: evaporation-crystallization process. reply to comments. Science 172 (3985), 871–872.

Gibbs, R.J., 1972. Water chemistry of the Amazon river. Geochim. Cosmochim. Acta 36 (9), 1061–1066.

Gibbs, R.J., 1992. The inapplicability of the Gibbs model of world water chemistry for dilute lakes. reply to comments. Limnol. Oceanogr. 37 (6), 1338–1339.

Gorbaty, M.L., George, G.N., Kelemen, S.R., 1989. Direct determination and quantification of sulfur forms in heavy petroleum and coals. Part II: the sulfur K edge x-ray absorption spectroscopy approach. Am. Chem. Soc. 34 (3), 738–744.

Griffin, R.A., Jurinak, J.J., 1973. Estimation of activity coefficients from the electrical conductivity of natural aquatic systems and soil extracts. Soil Sci. 116 (1), 26–30.

Haggerty, S.E., 1976. Opaque mineral oxides in terrestrial igneous rocks. Rev. Mineral. 3, Hg101–Hg300.

Halmer, M.M., Schmincke, H.U., Graf, H.F., 2002. The annual volcanic gas input into the atmosphere, in particular into the stratosphere: a global data set for the past 100 years. J. Volcanol. Geotherm. Res. 115 (3), 511–528.

Hasan, M.A., Ahmed, K.M., Sracek, O., Bhattacharya, P., Von Broemssen, M., Broms, S., Fogelström, J., Mazumder, M.L., Jacks, G., 2007. Arsenic in shallow groundwater of Bangladesh: investigations from three different physiographic settings. Hydrogeol. J. 15 (8), 1507–1522.

Hegg, D.A., Hobbs, P.V., 1978. Oxidation of sulfur dioxide in aqueous systems with particular reference to the atmosphere. Atmos. Environ. 12 (1–3), 241–253.

Helgeson, H.C., Garrels, R.M., Mackenzie, F.T., 1969. Evaluation of irreversible reactions in geochemical processes involving minerals and aqueous solutions. II. Applications. Geochim. Cosmochim. Acta 33 (4), 455–481.

Hopkins, C., Knox, W., Pettit, H.J., 1903. A quantitative method for determining the acidity of soils. In: Proceedings of the 19th Annual Convention, Bureau of Chemistry Bulletin, 74. Government Printing Office, Washington, DC, pp. 114–119.

Hsu, P.H., Bates, T.F., 1964. Fixation of hydroxy-aluminum polymers by vermiculite. Soil Sci. Soc. Am. Proc. 28 (6), 763–769.

Jackson, M.L., Tyler, S.A., Willis, A.L., Bourbeau, G.A., Pennington, R.P., 1948. Weathering sequence of clay-size minerals in soils and sediments. I. Fundamental generalizations. J. Phys. Colloid Chem. 52, 1237–1260.

Jackson, M.L., Hseung, Y., Corey, R.B., Evans, E.J., Heuvel, R.C.V., 1952. Weathering sequence of clay-size minerals in soils and sediments. II. Chemical weathering of layer silicates. Soil Sci. Soc. Am. Proc. 16, 3–6.

Jaeschke, W., 2013. Chemistry of Multiphase Atmospheric Systems. Springer-Verlag, Berlin, New York.

Kaewmano, C., Kheoruenromne, I., Suddhiprakarn, A., Gilkes, R., 2009. Aggregate stability of salt-affected kaolinitic soils on the North-east Plateau, Thailand. Soil Res. 47 (7), 697–706.

Kilham, P., 1990. Mechanisms controlling the chemical composition of lakes and rivers: data from Africa. Limnol. Oceanogr. 35 (1), 80–83.

Krešić, N., 2007. Hydrogeology and Groundwater Modeling, second ed. CRC Press, Boca Raton.

Langelier, W.F., 1936. The analytical control of anticorrosion water treatment. J. Am. Water Works Assoc. 28, 1500–1521.

Lebron, I., Suarez, D.L., Alberto, F., 1994. Stability of a calcareous saline-sodic soil during reclamation. Soil Sci. Soc. Am. J. 58 (6), 1753–1762.

Lessard, G.J.P., 1981. Biogeochemical phenomena in quick clays and their effects on engineering properties. American Chemical Society, 362 pp.

Li, S., Matthews, J., Sinha, A., 2008. Atmospheric hydroxyl radical production from electronically excited NO_2 and H_2O. Science (Washington, DC) 319 (5870), 1657–1660.

Li, S., Matthews, J., Sinha, A., 2009. Response to comment on "atmospheric hydroxyl radical production from electronically excited NO_2 and H_2O". Science (Washington, DC) 324 (5925), 336.

Lide, D.R., (Ed.) 2005. Sectilon 14: geophysics, astronomy, and acoustics. In: Abundance of Elements in the Earth's Crust and in the Sea. Internet version ed. CRC Press, Boca Raton, FL, p. 17.

Livingstone, D.A., 1963. Data for geochemistry. Chemical composition of rivers and lakes. U.S.G.S. Professional Paper No. 440-G, 64 pp.

Lowry, T., 1923. The uniqueness of hydrogen. J. Soc. Chem. Ind. 42 (3), 43–47.

Margitan, J.J., 1984. Mechanism of the atmospheric oxidation of sulfur dioxide. catalysis by hydroxyl radicals. J. Phys. Chem. 88 (15), 3314–3318.

Matthess, G., 1982. The Properties of Groundwater. Wiley, New York.

Mattson, S., Wiklander, L., 1940. The laws of soil colloidal behavior. XXI. A. The amphoteric points, the pH and the Donnan equilibrium. Soil Sci. 49, 109–134.

Merrill, T., Lane, R.W., Pisigan, R.A., Richards, G.G., Rossum, J.R., Scanlan, L.P., Singley, J.E., Montgomery, J.M., Whitney, G.R., 1990. Suggested methods for calculating and interpreting calcium carbonate saturation indexes. J. Am. Water Works Assoc. 82 (7), 71–77.

Morel, F.M.M., Hering, J.G., 1993. Principles and Applications of Aquatic Chemistry. John Wiley & Sons, New York.

NCSS, 2014. Kellogg Soil Survey Laboratory Methods Manual, version 5 ed. Soil Survey Investigations Report No. 42. Department of Agriculture, Natural Resources Conservation Service, Kellogg Soil Survey Laboratory, Lincoln, NE.

Nordstrom, D.K., Ball, J.W., McCleskey, R.B., 2005. Ground water to surface water: chemistry of thermal outflows in Yellowstone National Park. In: Inskeep, W.P., McDermott, T.R. (Eds.), Geothermal Biology and Geochemistry in Yellowstone National Park, Proceeding of the Thermal Biology Institute Workshop. Montana State University, pp. 73–94.

Oster, J.D., Shainberg, I., Wood, J.D., 1980. Flocculation value and gel structure of sodium/calcium montmorillonite and illite suspensions. Soil Sci. Soc. Am. J. 44 (5), 955–959.

Ostwald, W.F., 1888. Zur theorie der lösungen. Z. Phys. Chem. Stoechiom. Verwandtschafts. 2, 36–37.

Ostwald, W.F., 1889. Zur theorie der lösungen. Z. Phys. Chem. Stoechiom. Verwandtschafts. 3, 588–602.

Patel, N.C., 1980. The calculation of thermodynamic properties and phase equilibria using a new cubic equation of state. Ph.D. thesis, Loughborough University of Technology, Leicestershire, UK.

Persily, A.K., Gorfain, J., 2008. Analysis of ventilation data from the US Environmental Protection Agency building assessment survey and evaluation (base) study. NISTIR 7145 (revised). National Institute of Standards and Technology, Building and Fire Research Laboratory.

Pisigan Jr., RA., Singley, J.E., 1985. Calculating the pH of calcium carbonate saturation. J. Am. Water Works Assoc. 77 (10), 83–91.

Resendiz, D., 1977. Relevance of atterberg limits in evaluating piping and breaching potential. ASTM Spec. Tech. Publ. 341–353.

Rhoades, J., Merrill, S., 1976. Assessing the suitability of water for irrigation: theoretical and empirical approaches. In: Prognosis of Salinity and Alkalinity, vol. 31. Food and Agriculture Organization of the United Nations, Rome, pp. 69–110.

Rhoades, J.D., Krueger, D.B., Reed, M.J., 1968. The effect of soil-mineral weathering on the sodium hazard of irrigation waters. Soil Sci. Soc. Am. Proc. 32 (5), 643–647.

Schofield, R.K., 1947. A ratio law governing the equilibrium of cations in the soil solution. Proceedings of the 11th International Congress Pure Applied Chemistry, London, Vol. 3, pp. 257–261.

Scott, W.D., Hobbs, P.V., 1967. Formation of sulfate in water droplets. J. Atmos. Sci. 24 (1), 54–57.

Shainberg, I., Oster, J.D., Wood, J.D., 1980. Sodium/calcium exchange in montmorillonite and illite suspensions. Soil Sci. Soc. Am. J. 44 (5), 960–964.

Sikdar, P., Chakraborty, S., 2008. Genesis of arsenic in groundwater of north bengal plain using PCA: a case study of English Bazar block, Malda District, West Bengal, India. Hydrol. Process. 22 (12), 1796–1809.

Skempton, A.W., 1953. The colloidal activity of clays. Proceedings of the Third International Conference on Soil Mechanics and Foundation Engineering, Vol. 1, pp. 57–61.

Snyder, L.R., 1970. Petroleum nitrogen compounds and oxygen compounds. Acc. Chem. Res. 3 (9), 290–299.

Sørensen, S.P.L., 1909. Enzymstudien ii: Über die messung und die bedeutung der wasserstoffionenkonzentration bei enzymatischen prozessen. Biochem. Z. 21, 131–200.

Sposito, G., Mattigod, S.V., 1977. On the chemical foundation of the sodium adsorption ratio. Soil Sci. Soc. Am. J. 41 (2), 323–329.

Sposito, G., Holtzclaw, K.M., Charlet, L., Jouany, C., Page, A.L., 1983a. Sodium-calcium and sodium-magnesium exchange on Wyoming bentonite in perchlorate and chloride background ionic media. Soil Sci. Soc. Am. J. 47 (1), 51–56.

Sposito, G., Holtzclaw, K.M., Jouany, C., Charlet, L., 1983b. Cation selectivity in sodium-calcium, sodium-magnesium, and calcium-magnesium exchange on Wyoming bentonite at 298 K. Soil Sci. Soc. Am. J. 47 (5), 917–921.

Stumm, W., Morgan, J.J., 2013. Aquatic Chemistry: Chemical Equilibria and Rates in Natural Waters, third ed. John Wiley & Sons, New York.

Suarez, D.L., 1981. Relation between PHC and sodium adsorption ratio (SAR) and an alternative method of estimating SAR of soil or drainage waters. Soil Sci. Soc. Am. J. 45 (3), 469–475.

Tans, P., Keeling, R., 2015, Trends in atmospheric carbon dioxide—Global Greenhouse Gas Reference Network. ESRL CO2 Trends RSS, http://www.esrl.noaa.gov/gmd/ccgg/trends/.

Thomas, G.W., 1977. Historical developments in soil chemistry: ion exchange. Soil Sci. Soc. Am. J. 41 (2), 230–238.

Thomas, H., Schiettecatte, L.S., Suykens, K., Koné, Y.J.M., Shadwick, E.H., Prowe, F., Bozec, Y., de Baar, H.J.W., Borges, A.V., 2009. Enhanced ocean carbon storage from anaerobic alkalinity generation in coastal sediments. Biogeosciences 6 (2), 267–274.

Van Denburgh, A.S., Feth, J.H., 1965. Solute erosion and chloride balance in selected river basins of the western conterminous united states. Water Resour. Res. 1 (4), 537–541.

van't Hoff, J.H., 1887. Die rolle osmotischen druckes in der analogie zwichen lösungen und gasen. Z. Phys. Chem. 1, 481–508.

van't Hoff, J.H., Reicher, L.T., 1888. Über die dissociationstheorie der electrolyte. Z. Phys. Chem. Stoechiom. Verwandtschafts. 2, 77–81.

van't Hoff, J.H., Reicher, L.T., 1889. Beziehung zwischen osmotischer druck, gefrierpunktserniedrigung und elektrischer leitfähigkeit. Z. Phys. Chem. Stoechiom. Verwandtschafts. 3, 198–202.

Veitch, F.P., 1902. The estimation of soil acidity and the lime requirements of soils. J. Am. Chem. Soc. 24 (11), 1120–1128. http://dx.doi.org/10.1021/ja02025a018.

Veitch, F.P., 1904. Comparison of methods for the estimation of soil acidity. J. Am. Chem. Soc. 26 (6), 637–662. http://dx.doi.org/10.1021/ja01996a005.

Waldo, G.S., Carlson, R.M.K., Moldowan, J.M., Peters, K.E., Penner-Hahn, J.E., 1991. Sulfur speciation in heavy petroleums: Information from x-ray absorption near-edge structure. Geochim. Cosmochim. Acta 55 (3), 801–814.

Waldo, G.S., Mullins, O.C., Penner-Hahn, J.E., Cramer, S.P., 1992. Determination of the chemical environment of sulfur in petroleum asphaltenes by x-ray absorption spectroscopy. Fuel 71 (1), 53–57.

White, D.E., Hem, J.D., Waring, G.A., 1963. Data for geochemistry. Chemical composition of subsurface waters. U.S.G.S. Professional Paper, No. 440-F, 44 pp.

Whittig, L.D., Janitzky, P., 1963. Mechanisms of formation of sodium carbonate in soils. I. Manifestations of biological conversions. J. Soil Sci. 14 (2), 322–333.

Wongpokhom, N., Kheoruenromne, I., Suddhiprakarn, A., Gilkes, R.J., 2008a. Micromorphological properties of salt affected soils in northeast Thailand. Geoderma 144 (1), 158–170.

Wongpokhom, N., Kheoruenromne, I., Suddhiprakarn, A., Smirk, M., Gilkes, R.J., 2008b. Geochemistry of salt-affected aqualfs in northeast Thailand. Soil Sci. 173 (2), 143–167.

Yoo, J.M., Lee, Y.R., Kim, D., Jeong, M.J., Stockwell, W.R., Kundu, P.K., Oh, S.M., Shin, D.B., Lee, S.J., 2014. New indices for wet scavenging of air pollutants (O_3, CO, NO_2, SO_2, and PM10) by summertime rain [erratum to document cited in CA160:728919]. Atmos. Environ. 91, 178.

CHAPTER

7

Natural Organic Matter

OUTLINE

7.1 Introduction 333

7.1.1 Organic Matter Extraction 334

7.1.2 Organic Matter Fractionation 335

7.2 Biological Attributes of Natural Organic Matter 337

7.2.1 Stoichiometric Composition of Natural Organic Matter 337

7.2.2 Organic Matter as Substrate for Microbial Growth 341

7.2.3 Exocellular Bioorganic Compounds 344

7.3 Organic Carbon Turnover and the Terrestrial Carbon Cycle 349

7.3.1 Carbon Fixation 350

7.3.2 Carbon Mineralization 351

7.3.3 Oxidation of Organic Compounds by Dioxygen 352

7.3.4 Organic-Matter Turnover Models 353

7.3.5 Soil Carbon Pools 362

7.4 Chemical Properties of Natural Organic Matter 363

7.4.1 Oxygen Functional Groups 363

7.4.2 Nitrogen Functional Groups 367

7.4.3 Phosphorus Functional Groups 370

7.4.4 Sulfur Functional Groups 372

7.4.5 Carbon Moieties and Functional Groups 375

7.4.6 Colloidal Properties 376

7.5 Summary 376

Appendices 377

7.A Limitations of Carbon K-Edge NEXAFS and STXM 377

7.B Spin Conservation and Spin Forbidden Reactions 378

References 380

7.1 INTRODUCTION

The path leading from net primary production, through microbial biomass to natural organic matter takes tens to hundreds of years, a path subject to considerable speculation but little verification. Our focus will be the biological and chemical properties of natural organic matter that are, in turn, intimately linked to their composition and chemical functionality.

Soil and Environmental Chemistry
http://dx.doi.org/10.1016/B978-0-12-804178-9.00007-0

The chemically active components lie at the extreme ends of the carbon transformation trajectory: identifiable biomolecules and unidentifiable organic substances are the product of untold microbial mineralization cycles.

The terms *humus* and *humic substances* have been synonymous with *natural organic matter* from the earliest organic matter research (Box 7.1). This chapter largely employs the term *natural organic matter* in an effort to remain noncommittal on specifics of the *humification* process.

BOX 7.1

THE CONTENTIOUS DEBATE OVER SEMANTICS

Many terms used by scientists studying natural organic matter still in use today originated during the late 1700s and early 1800s. For example, Waksman (1936) attributes the term *humus*[1] to the doctoral thesis (written in Latin) of Wallerius (1761). The noun *humification*, derived from humus, is taken to mean the process that transforms plant remains into humus.

Waksman (1936) also discussed several issues that continue to vex present day humus chemists (cf. Lehmann and Kleber, 2015). Among the issues raised by Waksman (1936) are: whether or not humic substances are chemically refractory (cf. de Saussure, 1804; Hoppe-Seyler, 1889), the impact of chemical extractions on organic-matter properties, and the usefulness of postextraction fraction. Scientists intent on fractionation or promotion of the carbon-pool paradigm (cf. Section 7.3.5) have forged a reductionist perception of natural organic matter. Waksman (1936) writes:

> *No other phase of chemistry has been so much confused as that of humus, as a result of which it frequently becomes necessary to lay considerable emphasis upon the proper definition of the terms used. Vague generalizations ... have been used for specific processes of decomposition; the great complexity of the numerous "humic acids", all of which designate not definite chemical compounds but mere preparations, and the unjustified comparisons between the natural humus compounds, formed in soils, composts, or bogs, with artificial preparations produced in the laboratory by the action of strong mineral acids on carbohydrates, served not to advance the subject of humus, but rather to confuse it.*

[1] Latin: *humus*, n. "ground, soil, earth, land, country."

7.1.1 Organic Matter Extraction

The stoichiometric composition, chemical functionality, and colloidal properties of natural organic matter are virtually impossible to determine in situ because natural organic matter entrains mineral colloids (Mikutta et al., 2006). This section describes the most widely used extraction protocol, separating organic matter from entrained mineral colloids; a protocol designed to minimize chemical alteration.

Oden (1919) published an extended paper on humus chemistry that included a review of humus research from 1833 to 1903. By the late 1800s scientists knew aqueous strong base extracted most, but not all, of the organic carbon in mineral soils and peat (cf. Achard, 1786 cited by Waksman, 1936). Natural organic matter isolated by strong-base extraction was, and still is, widely identified as *humic substances* (German: *Humusstoffe*). The extraction protocol described by Oden (1919) is strikingly similar to the original protocol of Achard (1786) and the modern protocol endorsed by the *International Humic Substances Society (IHSS)* (cf. Box 7.2).

BOX 7.2

THE INTERNATIONAL HUMIC SUBSTANCES SOCIETY EXTRACTION PROTOCOL

The protocol begins with an initial 1-h extraction with 0.1 M HCl to dissolve an acid-soluble "fulvic acid" fraction. The second step employs a 4-h extraction with 0.1 M NaOH to dissolve base-soluble organic matter, taking care to minimize oxidation at high pH by excluding O_2. The NaOH extract, which typically contains a considerable quantity of entrained minerals, is acidified by adding 6 M HCl and allowed to stand for 12–16 h.[2] The supernatant—the second acid-soluble "fulvic acid" fraction—is combined with the first acid-soluble extract while the flocculated organic matter is identified as the "humic acid" fraction.

The acid-flocculated "humic acid" fraction is dispersed a second time using 0.1 M KOH and sufficient KCl to make the final $K^+(aq)$ concentration 0.2 M (this step is designed to flocculate much of the remaining entrained mineral colloids). The "humic acid" fraction is flocculated a second time using a combination of 0.1 M HCl and 0.3 M HF and allowed to sit for 12 h (this step is designed to dissolve any remaining silicate clays entrained with the "humic acid" fraction).

The final cleanup involves adjusting to pH 7 and removing excess salts.[3] The organic matter that cannot be extracted by this protocol is commonly designated *humin*.

[2] The standard method for protein hydrolysis refluxing 24 h at 110°C in 6 M HCl.

[3] Originally the *IHSS* protocol called for dialysis but since the late 1990s electrolytes are removed by reverse osmosis.

Chemists studying humic substances during this early period also evaluated other extraction solvents (e.g., alcohol; Hoppe-Seyler, 1889) and coined names for the soil organic matter extracted by other solvents. Nonetheless, strong base dissolves the largest proportion of the total organic carbon and, eventually, other extraction protocols fell out of common use.

7.1.2 Organic Matter Fractionation

Chemists relied heavily on qualitative methods during the 1800s. It was only natural for them to evaluate the effect of various chemical reagents on the stability of organic matter extracts; most notable strong acids, neutral salts, and various transition-metal salts. Solubility is not the best term when referring to the strong-base organic-matter extract. Oden and others (cf. Oden, 1919, and references therein) realized the organic matter extracted from soil, sediment, or peat by strong base is a *colloidal dispersion*. Colloid chemists immediately recognize the effect of the aforementioned reagents on the *stability* of most colloidal dispersions (cf. Section 7.4.6).

Oden (1919) was the first to use the term *yellow acids* (German: *Fulvosäuren*; cf. Box 7.3) to identify the soluble fraction separated by acidifying a strong-base organic-matter extract. Oden (1919) found that adding 1% HCl causes most of the dark-brown organic matter to flocculate, leaving a supernatant solution containing colorless inorganic salts and yellow acid-soluble organic matter. The dark-brown base-soluble organic-matter fraction flocculated by strong acid was identified as *humic acids* (German: *Humussäuren*).

Although the chemical distinction between "humic acids" and "fulvic acids" (i.e., between *Humussäuren* and *Fulvosäuren* organic-matter fractions) is operational, the colloidal distinction remains relevant: the so-called humic acids are best understood as the colloidal component susceptible to flocculation. The so-called fulvic acids are the noncolloidal component, the component that resists acid flocculation.

The *IHSS* extraction protocol outlined in Box 7.2 was designed to reduce the ash content of the organic-matter extract. The protocol unavoidably leads to the separation of acid-soluble and acid-flocculated fractions. The question confronting chemists studying natural organic matter is whether or not their research or application emphasizes colloidal or purely chemical properties.

BOX 7.3

FULVOSÄUREN: TANNINS OR ACID-SOLUBLE HUMIC SUBSTANCES

Oden (1919) introduced a scientific term for the soluble *yellow acids*[4] in water draining from bogs: *Fulvosäuren*.

> *water soluble humus, yellow to amber colored, is found in bog waters along with colloidal humus. Filtered bog water is a pale-yellow to golden-yellow solution.... The hue is always yellowish, not brown or black brown.... An appropriate name for this soluble humus is missing, which is why I suggest, because of its characteristic yellow color, the name of yellow acids.*[5]

Reddish-yellow to amber-hued aquatic organic matter is common in the lakes and rivers draining boreal (taiga) peat lands from Scandinavia and Russia to Alaska and Canada. The water-soluble organic matter in these settings is often identified as *tannins*. The so-called *blackwater* rivers draining sub-tropical and tropical forested wetlands in the Amazon Basin, the Orinoco Basin, Indonesia, and Malaya also contain tannins.

Given their connotation it appears the terms *tannins* and *yellow acids* (Oden, 1919) both refer to the same group of water soluble organic compounds. Neither term implies a definitive chemical substance; both refer to water-soluble substances produced when plant residue decomposes in water-logged wetlands.

BOX 7.3 *(cont'd)*

Oden (1919) appears to have associated the distinctive yellow hue and water solubility of bog waters with the soluble organic-matter fraction separated by acidifying a strong-based extract of peat, soil, or sediment. It is doubtful any chemist studying natural organic matter would conclude—based solely on hue and solubility—that the acid-soluble organic-matter fraction isolated by coagulating strong-base extracts of peat, soil, or sediment are identical to the *yellow acids* characteristic of bog water.

[4] Latin: *fulvus* adj. "reddish-yellow, tawny, amber-colored."

[5] "der Humusstoffe geht schon durch Behandlung mit Wasser in Lösung, wobei leicht diffundierbare echte Lösungen von gelber bis gelbbrauner Farbe entstehen. In der Natur kommen sie in Moorwässern vor, jedoch stets durch die Kolloide der folgenden Stoffgruppen verunreinigt. Filtriert man solche Moorwässer durch Porzellanfilter, so erhalt man blassgelbe bis goldgelbe Lösungen.... Der Farbenton its aber stets überwiegend gelblich, nicht braun order schwarzbraun.... Ein zweckmaßiger Sammelname fehlt, weshalb ich hier, eben der charakteristischen gelben Farbe wegen, für dieser Stoffgruppe den Namen Fulvosäuren vorschlage."

7.2 BIOLOGICAL ATTRIBUTES OF NATURAL ORGANIC MATTER

7.2.1 Stoichiometric Composition of Natural Organic Matter

Rice and MacCarthy (1991) published the stoichiometric composition of organic matter extracted from numerous soil, peat, aquatic, and marine specimens. They report composition probability distributions for five elements (C, H, O, N, and S) from a population of 215 "humic acid" specimens, 127 "fulvic acid" specimens, and 26 "humin" specimens. Table 7.1 lists the mean composition and standard deviation for "humic acids," "fulvic acids," and "humin" isolated from soils. The concentration probability distributions of most organic matter elemental components generally follow a normal distribution.[6] As a consequence, the arithmetic mean and standard deviation are suitable distribution statistics for most sample populations.

Rice and MacCarthy (1991) found remarkably low variation in stoichiometric composition within each population and no significant difference between "humic acid" and "humin" fractions. The only statistically significant composition difference between populations was the composition of the "fulvic acid" fraction relative to "humic acid" and "humin" fractions.

[6] The statistics for sulfur are geometric; all others are arithmetic. The sample populations for the sulfur analyses are consistently smaller because the sulfur content is rather low: "humic acids" ($N = 67$), "fulvic acids" ($N = 45$), and "humin" ($N = 16$).

TABLE 7.1 Mean Mass Fraction and Standard Deviation of Soil "Humic Acids" ($N = 215$), Soil "Fulvic Acids" ($N = 127$), and Soil "Humin" ($N = 26$)

	$w(C)$ ($g\,g^{-1}$)	$w(O)$ ($g\,g^{-1}$)	$w(H)$ ($g\,g^{-1}$)	$w(N)$ ($g\,g^{-1}$)	$w(S)$ ($g\,g^{-1}$)
Humic acids					
Mean	0.554	0.048	0.360	0.036	0.008
Standard deviation	0.038	0.010	0.037	0.013	0.006
Fulvic acids					
Mean	0.453	0.050	0.462	0.026	0.013
Standard deviation	0.054	0.010	0.052	0.013	0.011
Humin					
Mean	0.561	0.055	0.347	0.037	0.004
Standard deviation	0.026	0.010	0.034	0.013	0.003

Most scientists reporting stoichiometric composition quote mass fractions. The mole fractions in Table 7.2, computed from the data in Table 7.1, are more useful because elements combine on a molar basis. The oxygen-to-nitrogen mole ratio in "fulvic acids" (≈ 16) is twice the ratio in "humic acids" and "humin" (≈ 8). The oxygen-to-carbon mole ratio in "fulvic acids" (≈ 0.8) is nearly twice the ratio in "humic acids" and "humin" (≈ 0.5). These results clearly indicate that the "fulvic acid" fraction has significantly higher oxygen content than the "humic acid" and "humin" fractions.

Using the results from Table 7.2 (columns 1–6), we can estimate the mean carbon oxidation state $\overline{C}$ in each soil organic-matter fraction (Table 7.2, column 7). Chemical bonding details are irrelevant provided nitrogen is assigned an oxidation state of $\overline{N} = -3$ (cf. Example 7.1). It makes no difference whether hydrogen is assigned to H-O bonds or H-C bonds because, ultimately, the formal oxidation state of oxygen is $O = -2$. Example 7.1 illustrates how this estimate is made.

TABLE 7.2 The Mean Mole Fraction $\overline{x}_E$ for Each of the Five Most Abundant Elements in Soil "Humic Acids," Soil "Fulvic Acids," and Soil "Humin"

Fraction	$\overline{x}_C$	$\overline{x}_O$	$\overline{x}_H$	$\overline{x}_N$	$\overline{x}_S$	$\overline{C}$
Humic acids	0.387	0.400	0.189	0.022	0.002	+0.111
Fulvic acids	0.318	0.419	0.244	0.016	0.003	+0.364
Humin	0.371	0.434	0.172	0.021	0.001	−0.070

EXAMPLE 7.1

Example Permalink

http://soilenvirochem.net/7McnPY

Estimate the nominal molecular formula and mean carbon oxidation state $\overline{\text{C}}$ for an organic-matter specimen based on its stoichiometric composition.

The specimen in this example is: Elliot soil "humic acid," *IHSS* specimen 1S102H.

Part 1. Convert the stoichiometric mass fraction $w(\text{E})$ to a mole-per-mass fraction $b(\text{E})$ for each element.

Table 7.3 lists the stoichiometric composition of specimen 1S102H (Elliot soil "humic acid") in column 2 and atomic masses in column 3. The mole-per-mass composition is found by dividing the mass fraction $w(\text{E})$ of each element by its atomic mass (expression 7.1). The mole-per-mass compositions appear in Table 7.3 column 4.

$$b(\text{E}) = \frac{w(\text{E})}{\overline{m}_a(\text{E})} \tag{7.1}$$

TABLE 7.3 The Stoichiometric Mass-Fraction $w(\text{E})$ and Mole-per-Mass $b(\text{E})$ Fraction of *IHSS* Specimen 1S10-2H (Elliot Soil "Humic Acid" Standard)

Element	$w(\text{E})$ (g g^{-1})	$\overline{m}_a(\text{E})$ (g mol^{-1})	$b(\text{E})$ (mol g^{-1})
C	0.5813	12.01	0.0484
H	0.0368	1.01	0.0364
O	0.3408	16.00	0.0213
N	0.0414	14.01	0.0030
S	0.0044	32.07	0.0001
P	0.0024	30.97	0.0001

$$b(\text{C}) = \frac{0.5813}{12.01} = 0.0484$$

Part 2. Convert the mole-per-mass fraction $b(\text{E})$ to a mole fraction x_E and compute the stoichiometric number ν_E for each element.

The total moles per gram is computed by summing the mole-per-mass composition $b(\text{E})$ of each element (expression 7.2). Divide the mole-per-mass fraction of each element by the total moles per gram to estimate the mole fraction of each element (expression 7.3).

$$b = \sum_{\text{E}} b(\text{E}) \tag{7.2}$$

$$x_\text{E} = \frac{b(\text{E})}{b} \tag{7.3}$$

$$x_C = \frac{0.0484}{0.1093} = 0.4428$$

Table 7.4 lists the stoichiometric mole fractions x_E for specimen 1S102H (Elliot soil "humic acid") in column 2. The stoichiometric numbers listed in Table 7.4 column 3 are found by normalizing (expression 7.4) the mole fractions listed in column 2 by the carbon mole fraction $x_C = 0.4428$.

$$\nu_E = \frac{x_E}{x_C} \tag{7.4}$$

$$\nu_H = \frac{x_H}{x_C} = \frac{0.3333}{0.4428} = 0.7528$$

TABLE 7.4 The Stoichiometric Mole-Fraction Composition x_E and Stoichiometric Numbers ν_E for *IHSS* Specimen 1S102H (Elliot Soil "Humic Acid" Standard)

Element	x_E (mol mol^{-1})	ν_E (unitless)
C	0.4428	1.0000
H	0.3333	0.7528
O	0.1949	0.4401
N	0.0270	0.0611
S	0.0013	0.0028
P	0.0007	0.0016

The nominal molecular formula for *IHSS* specimen 1S102H (Elliot soil "humic acid" standard) is: $CH_{0.7528}O_{0.4401}N_{0.0611}S_{0.0028}P_{0.0016}$.

Part 3. Estimate the mean carbon oxidation state $\overline{C}$ for *IHSS* specimen 1S102H (Elliot soil "humic acid" standard) using the stoichiometric numbers from Part 2.

The mean carbon oxidation state $\overline{C}$ is relatively insensitive to the oxidation states of trace elements, therefore this estimate neglects the oxidation states of sulfur and phosphorus.

General chemistry textbooks assign the following oxidation state values to oxygen (expression 7.5) and hydrogen (expression 7.6) in most compounds.[7]

$$O \equiv -2 \tag{7.5}$$

$$H \equiv +1 \tag{7.6}$$

Nuclear magnetic resonance spectroscopy ^{15}N reveals the nitrogen oxidation state in organic matter, regardless of origin, is $\overline{N} = -3$ (cf. Section 7.4.2).

Using the oxidation states of hydrogen, oxygen, and nitrogen quoted earlier and the stoichiometric numbers listed in Table 7.4 (column 3) we can estimate the mean carbon oxidation state $\overline{C}$ in specimen 1S102H (Elliot soil "humic acid" standard) using expression (7.7).

$$\nu_C \cdot \overline{C} + \nu_H \cdot H + \nu_O \cdot O + \nu_N \cdot \overline{N} = 0 \tag{7.7}$$

$$\overline{\text{C}} = -\frac{\nu_{\text{H}}}{\nu_{\text{C}}} \cdot (+1) - \frac{\nu_{\text{O}}}{\nu_{\text{C}}} \cdot (-2) - \frac{\nu_{\text{H}}}{\nu_{\text{C}}} \cdot (-3)$$

$$\overline{\text{C}} = -0.7528 \cdot (+1) - 0.4401 \cdot (-2) - 0.0611 \cdot (-3) = 0.31$$

The mean carbon oxidation state in natural organic matter is: $\overline{\text{C}} \approx 0$. Given to the higher oxygen content relative to "humic acid" and "humin" fractions, the mean carbon oxidation state of the "fulvic acid" fraction with be more positive. That being said, the increase is negligible.

[7] Exceptions do not apply to natural products such as organic matter because peroxide and hydride are not encountered.

7.2.2 Organic Matter as Substrate for Microbial Growth

7.2.2.1 Carbon Use Efficiency

The carbon turnover rate depends on the amount of microbial biomass which, in turn, depends on *carbon-use efficiency (CUE)*. The focus in this section is organic matter as substrate, not plant senescence residue (cf. "slow carbon" and "passive carbon" in Fig. 7.13).

In global terms, *CUE* is the ratio of *microbial biomass growth (G)* to the microbial *carbon uptake (U)* (expression 7.8) as illustrated in Fig. 7.1.

$$CUE \equiv \frac{G}{U} \tag{7.8}$$

Fig. 7.1 shows, among other things, that *CUE* depends on the carbon-to-nitrogen ratio of substrate and biomass and the flow of substrate carbon into microbial respiration. What Fig. 7.1 does not show is the constraint imposed by the mean carbon oxidation state $\overline{\text{C}}$ of the substrate and microbial biomass.

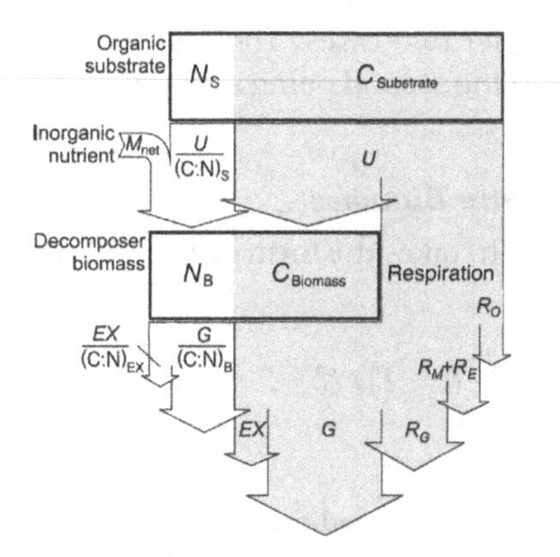

Inorganic nutrients *E*
$U/(C{:}N)_S$: substrate C-to-N ratio
$U/(C{:}N)_B$: biomass C-to-N ratio

Biomass carbon
U : carbon uptake
EX : carbon excretion
G : growth (biomass)

Respiration carbon
R_G : growth
R_M : maintenance
R_E : enzyme production
R_O : carbon overflow

Manzoni et al. (2012) *New Phytologist*, **196**, 79

FIG. 7.1 Schematic illustration of the mass balances of microbial biomass carbon C_B and the essential element nitrogen N_B, in relation to the stoichiometry of the substrate C_S and the substrate nutrient content N_S. *Gray arrows* and *boxes* represent the carbon C fluxes and pools, *open arrows* and *boxes* refer to the fluxes of nitrogen N. *Modified from Manzoni, S., Taylor, P., Richter, A., Porporato, A., Ågren, G.I., 2012. Environmental and stoichiometric controls on microbial carbon-use efficiency in soils. New Phytol. 196 (1), 79–91.*

7.2.2.2 Mean Carbon Reduction

Microbiologists typically use *carbon reduction* (expression 7.9) rather than carbon oxidation state $\overline{C}$ to evaluate growth yield on substrate. Carbon reduction ranges from a maximum of $\Gamma = 8$ in CH_4 to a minimum of $\Gamma = 0$ in carbon dioxide CO_2.

$$\Gamma \equiv 4 - \overline{C} \tag{7.9}$$

Specifically, the mean carbon reduction $\overline{\Gamma}$ of an organic compound $CH_aO_bN_c$ is determined by its major element stoichiometry (expression 7.10).

$$\overline{\Gamma}_S = 4 + a - 2 \cdot b - 3 \cdot c \tag{7.10}$$

Using expression (7.10) and stoichiometric compositions from various sources (Rice and MacCarthy, 1991; Sun et al., 1997; Schaul et al., 1997; Vodyanitskii, 2000), the mean carbon reduction in natural organic matter is estimated to be: $\overline{\Gamma}_{OM} = 3.8$. The estimated mean molar carbon-to-nitrogen ratio in organic matter is: $\overline{C:N}_{OM} = 19.6$.

Bacterial biomass compositions from several sources (Hoover and Porges, 1952; Sawyer, 1956; Symons and McKinney, 1958; Speece and McCarty, 1964; Heijnen and Van Dijken, 1992; Schaul et al., 1997; Von Stockar and Liu, 1999; VanBriesen, 2002; Mouginot et al., 2014) result in a somewhat higher estimate of the mean carbon reduction in microbial biomass: $\overline{\Gamma}_B = 4.2$. The estimated mean molar carbon-to-nitrogen ratio of microbial biomass is significantly lower than organic matter: $\overline{C:N}_B = 5.3$.

Since the $\overline{C:N}_B > \overline{C:N}_{OM}$ microbes growing on natural organic matter will not require inorganic nitrogen (cf. M_{net} in Fig. 7.1). For the present discussion, because the difference in carbon reduction is a mere 10%, we will assume: $\overline{\Gamma}_{OM} \approx \overline{\Gamma}_B$.

The relative stoichiometry of natural organic matter and microbial biomass means that natural organic matter is an ideal substrate for microbial growth. Microbes need to mineralize additional organic-matter carbon to supply the respiratory demand for growth R_G, maintenance R_M, and enzyme production R_E.[8]

What effect does substrate-carbon reduction $\overline{\Gamma}_S$ have on *CUE*? To illustrate the importance of substrate-carbon reduction $\overline{\Gamma}_S$ we will consider two cases: The first being substrates with carbon more oxidized than biomass carbon, and the second being substrates with carbon more reduced than biomass carbon.

7.2.2.3 Case I. Substrate Is Less Reduced Than Biomass

If $\overline{\Gamma}_S < \overline{\Gamma}_B$ then biomass formation (i.e., growth) takes the form of a *dis proportionate* reaction (7.R2) with CO_2 appearing as a product.[9]

$$\chi \equiv ((a-d) - 3 \cdot (c-f))/2$$

[8] Carbon overflow respiration R_O occurs under nitrogen limiting conditions, which clearly do not apply when organic matter is the growth substrate.

[9] It is possible, in general, to replace CO_2 with any metabolite less reduced than the substrate but the metabolite cannot be counted as biomass. The sign of the stoichiometric coefficients for NH_3 and H_2O determine which side of reaction (7.R2) these two components appear. If $(\nu_{NH_3} = (f-c) < 0)$ then NH_3 appears on the product or right side. If $(\nu_{H_2O} = \chi < 0)$ then H_2O appears on the educt or left side.

$$CH_aO_bN_c + (f - c)NH_3 \longrightarrow \alpha CH_dO_eN_f + (1 - \alpha)CO_2 + \chi H_2O \qquad (7.R2)$$

The microbial biomass stoichiometric coefficient α in reaction (7.R2) is determined by the ratio of substrate carbon reduction to biomass carbon reduction (cf. Fig. 7.2 and expressions 7.11, 7.12).

$$\alpha = \frac{\overline{\Gamma}_S}{\overline{\Gamma}_B} \qquad (7.11)$$

$$\overline{\Gamma}_S = \alpha \cdot \overline{\Gamma}_B + (1 - \alpha) \cdot \Gamma_{CO_2} \qquad (7.12)$$

Stoichiometric or maximum growth yield Y_{max} (expression 7.13) approaches zero as $\overline{\Gamma}_S \rightarrow 0$ (expression 7.14). Stated differently, as $\overline{\Gamma}_S \rightarrow 0$ less substrate carbon can be incorporated as biomass carbon and *CUE* approaches zero.

$$Y_{max} = \frac{w(C)_B}{w(C)_S} = \alpha \qquad (7.13)$$

$$\lim_{\alpha \rightarrow 0} Y_{max}(\alpha) = 0 \qquad (7.14)$$

7.2.2.4 *Case II. Substrate Is More Reduced Than Biomass*

If $\overline{\Gamma}_S > \overline{\Gamma}_B$ then biomass formation requires the transfer of electrons from substrate carbon to an electron acceptor (e.g., O_2, NO_3^-, etc.) to oxidize substrate carbon to the mean carbon reduction of microbial biomass $\overline{\Gamma}_B$.

$$\kappa \equiv \left(\overline{\Gamma}_S - \overline{\Gamma}_B\right)/4$$

$$\chi \equiv \left((a - d) - 3 \cdot \left(c - f\right)\right)/2$$

$$CH_aO_bN_c + (f{-}c)NH_3 + \kappa O_2 \longrightarrow CH_dO_eN_f + \chi H_2O \qquad (7.R3)$$

Reaction (7.R3) reveals that *CUE* reaches its maximum when $\overline{\Gamma}_S = \overline{\Gamma}_B$. Substrates with $\overline{\Gamma}_S > \overline{\Gamma}_B$ offer no growth advantage over substrates with $\overline{\Gamma}_S = \overline{\Gamma}_B$.

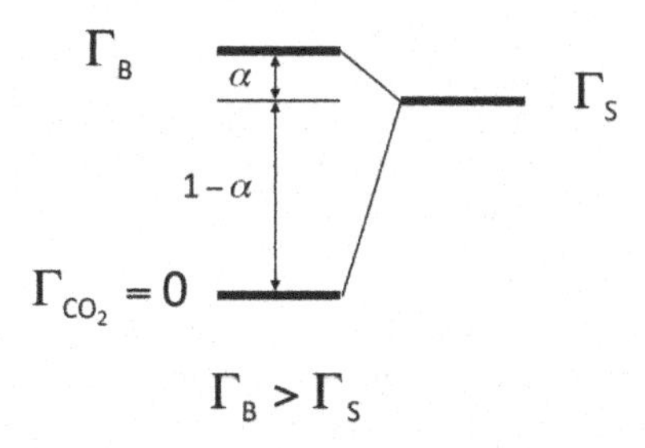

FIG. 7.2 Diagram illustrating the disproportionation (cf. reaction (7.R2)) of a substrate with a mean carbon reduction less than the mean carbon reduction of microbial biomass: $\overline{\Gamma}_S < \overline{\Gamma}_B$.

7.2.3 Exocellular Bioorganic Compounds

Dissolved organic matter (DOC) is organic matter extracted solely by water.[10] While DOC contains a substantial proportion of unknown substances, many known bioorganic compounds can be found in and isolated from DOC. This section identifies some of the bioorganic compounds dissolved in soil pore water. The bioorganic compounds dissolved in soil pore water are extremely labile, characterized by soil half-lives on the order of hours to days. Their short half-life does not diminish their importance because they are extremely reactive and continually replenished by soil organisms.

7.2.3.1 *Organic Acids*

Soil pore water, particularly in the *rhizosphere*,[11] contains a large variety of low molecular weight organic acids. Certain plant species actively secrete weak acids from their roots to increase nutrient availability (e.g., citric acid (Gardner et al., 1982); malic acid, fumaric acid, and oxalic acid). Other weak acids (e.g., the lignin precursors, Fig. 7.3) leak from plant roots during cell-wall biosynthesis. Bacteria and fungi release a variety of organic acids as fermentation by-products: succinic acid, butanoic acid, propionic acid, and lactic acid.

Organic acids dissolved in soil pore water are on the order of 10^{-3} to 10^{-2} $mmol_c\ dm^{-3}$ (Quideau and Bockheim, 1997). Proton dissociation constants and formation constants for metal-weak acid complexes for many of the common organic acids are included in

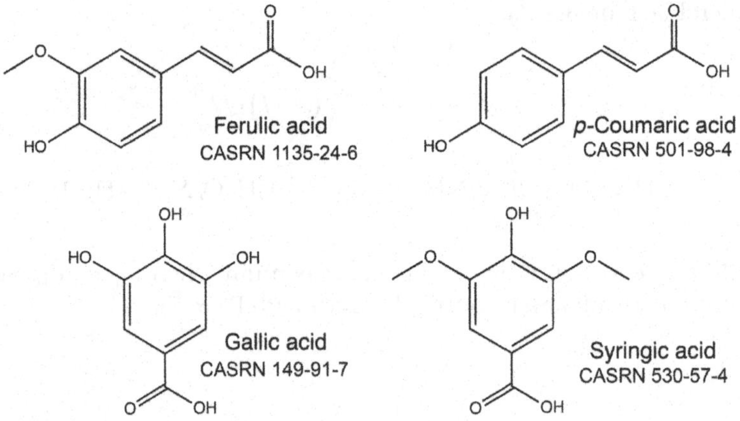

Ferulic acid
CASRN 1135-24-6
p-Coumaric acid
CASRN 501-98-4
Gallic acid
CASRN 149-91-7
Syringic acid
CASRN 530-57-4

FIG. 7.3 Lignin precursor molecules: ferulic acid (3-(4-hydroxy-3-methoxyphenyl)-2-propenoic acid, CAS registry number 1135-24-6), p-courmaric acid ((2*E*)-3-(4-hydroxyphenyl)-2-propenoic acid, CAS registry number 501-98-4), gallic acid (3,4,5-trihydroxybenzoic acid, CAS registry number 149-91-7), and syringic acid (4-hydroxy-3, 5-dimethoxybenzoic acid, CAS registry number 530-57-4).

[10] Some authors use the term "water extractable organic matter" (WEOM) interchangeably with DOM (Chantigny, 2003), both encompass all major elements (carbon, oxygen, nitrogen, etc.), while dissolved organic carbon DOC or dissolved DON quantify carbon (or nitrogen) only.

[11] "The rhizosphere is the narrow region of soil directly around roots. It is teeming with bacteria that feed on sloughed-off plant cells and the proteins and sugars released by roots" (Ingham et al., 2000).

all widely used computer-based water chemistry models (cf. Chapter 5) or can be found in comprehensive chemistry databases (e.g., *Critically Selected Stability Constants of Metal Complexes*; Martell et al., 2004).

7.2.3.2 Amino Acids

Free amino acids in soil pore water are on the order of 10^{-5} mol dm^{-3} (Jones et al., 2002, 2004; Hannam and Prescott, 2003) and represent about 1–10% of the DON. The remaining DON is unidentifiable. The solution chemistry of free amino acids is best understood in the context of computer-based water chemistry models (cf. Chapter 5).

7.2.3.3 Microbial Growth Factors

Bacteria and fungi cannot grow in simple media containing inorganic salts and a carbon source (Lochhead and Chase, 1943; Bécard and Piché, 1989). Soil extracts, particularly, extracts of *rhizosphere*[12] soil are essential for microbial growth. Familiar vitamins (cf. Fig. 7.4) are among the growth factors that have been identified. Other growth factors remain unidentified.

7.2.3.4 Exocellular Enzymes

Some enzymes remain active after the death of plant and microbial cells, associated with ruptured (lysed) cell debris or adsorbed to soil particles. Soil biochemists refer to these as *exocellular* enzymes. Many endocellular enzymes require cofactors, but active exocellular

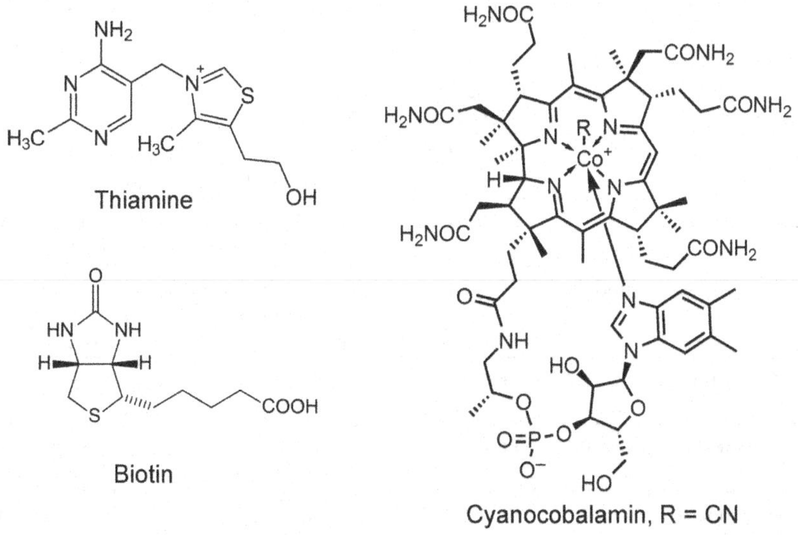

FIG. 7.4 Microbial growth factors identified from soil rhizosphere extracts include the following vitamins secreted by plant roots: vitamin B_1 (thiamine, CAS registry number 59-43-8), vitamin B_7 (biotin, CAS registry number 58-85-5), and vitamin B_{12} (cyanocobalamin, CAS registry number 68-19-9).

[12] The rhizosphere encompasses the soil indirect contact with plant roots, a soil microdomain rich in photosynthate carbon and plant root secretions. Microbial biomass, activity, and diversity are particularly high in the rhizosphere.

enzymes cannot be cofactor dependent (Pietramellara et al., 2002). Enzymes, regardless of whether they are endo- or exocellular, tend to lose activity outside a relatively narrow pH range because tertiary enzyme structure is pH sensitive. Apparently the adsorption of exocellular enzymes to soil particles stabilizes tertiary structure, at least active-site tertiary structure, extending both the active pH range and the lifetime in soil.

A review assayed exocellular enzyme activity in soil specimens collected at 40 US sites (Sinsabaugh et al., 2008). The assays evaluated the activity of seven enzyme classes (Table 7.5). The activity of glucosidase and cellobiohydrolase shows no apparent change between pH 4.0 and 8.5. The activity of N-acetylglucosaminidase and phosphatase decreases in the same pH range, while the activity of the remaining exocellular soil enzymes increases throughout the range.

Soil biochemists believe that exocellular enzymes might explain soil organic-matter turnover rates exceeding predictions based strictly on soil microbial biomass and respiration. There is no question that exocellular phosphatase enzymes are essential for phosphate release in soils (cf. Section 7.4.3).

7.2.3.5 Siderophores

Plants, fungi, and bacteria face identical challenges to meet their iron nutritional requirements. Kosman (2003) provides an excellent summary of these chemical challenges in the typical pH range of aerated soils: (1) Fe^{2+} spontaneously oxidizes to Fe^{3+}; (2) Fe^{3+} undergoes hydrolysis to form insoluble ferric oxyhydroxide minerals[13]; and (3) Fe^{3+} forms extremely stable complexes that are kinetically inert.[14] Kosman (2003) places particular emphasis on the

TABLE 7.5 A Partial Listing of Active Exocellular Enzymes Identified in Soils

Enzyme	Classification	Reaction
β-1, 4-Glucosidase	EC 3.2.1.21	Hydrolysis of terminal β-1,4-glucosyl residues, releasing β-D-glucose
Cellobiohydrolase	EC 3.2.1.91	Hydrolysis of 1,4-β-D-glucosidic linkages in cellulose, releasing cellobiose
β-N-acetylglucosaminidase	EC 3.2.1.14	Random hydrolysis of *N*-acetyl-β-D-glucosaminide (1 → 4)-β-linkages in chitin
Aminopeptidase	EC 3.4.11.1	Hydrolysis of *N*-terminal amide linkages, releasing amino acids
Acid or alkaline phosphatase	EC 3.1.3.1	Hydrolysis of phosphate monoester, releasing phosphate
Phenol oxidase	EC 1.10.3.2	Hydroxylation of diphenols to semiquinones, releasing water
Peroxidase	EC 1.11.1.7	Oxidation (wide specificity) using H_2O_2

Sinsabaugh, R.L., Lauber, C.L., Weintraub, M.N., Ahmed, B., Allison, S.D., Crenshaw, C., Contosta, A.R., Cusack, D., Frey, S., Gallo, M.E., et al., 2008. Stoichiometry of soil enzyme activity at global scale. Ecol. Lett. 11 (11), 1252–1264.

[13] Ferrous cation Fe^{2+} has little tendency to hydrolyze, thereby explaining its higher inherent solubility.

[14] Ferrous cation Fe^{2+} forms weak complexes that are very labile to ligand exchange.

final chemical property; even if Fe^{3+} were not susceptible to hydrolysis (i.e., insoluble), the kinetic stability of its complexes would render them biologically inert. Plants, bacteria, and fungi rely on elaborate biological systems to solubilize, transport, and absorb ferric iron from soil pore water; central to this system is a family of compounds known as siderophores.

Fungi (Kosman, 2003; Philpott, 2006) rely on three uptake pathways to balance their iron nutritional requirements against iron cytotoxicity: a nonreductive siderophore uptake pathway, a reductive iron uptake pathway, and a direct Fe^{2+} uptake pathway. Plants, regardless of whether they secrete siderophores, employ two of these pathways: a reductive iron uptake pathway and a direct ferrous uptake pathway. Bacteria, regardless of whether they secrete siderophores, employ two of these pathways: a nonreductive siderophore uptake and a direct ferrous uptake pathway. The direct Fe^{2+} uptake pathways (plants, fungi, and bacteria)—which are not operative under aerobic conditions when iron solubility is at its lowest—employ low-affinity transmembrane ion channels (or porins).

Poaceae use phytosiderophores to transport Fe^{3+} ions to root surfaces, where ferric reductase enzymes reduce Fe^{3+} to Fe^{2+} (Chaney et al., 1972; Römheld and Marschner, 1981). The labile ferrous complex releases Fe^{2+} to ion channels that transport it into the cytoplasm. The fungal reductive iron uptake pathway is functionally equivalent to the plant reductive iron uptake pathway.

The nonreductive siderophore uptake pathways in both bacteria and fungi rely on outer membrane receptors that bind the ferri-siderophore complex (Ferguson and Deisenhofer, 2002; Kosman, 2003; Philpott, 2006). Once the ferri-siderophore reaches the cytoplasm, it is reduced, and the labile ferrous complex releases Fe^{2+} to a cytoplasmic chaperone. As noted by Kosman (2003), Fe^{2+} is cytotoxic, making iron uptake and homeostasis a tightly regulated process in all living organisms.

The *dissolved iron* pore-water concentration of neutral aerated soils consistently falls in the 1–10 μM range (Fuller et al., 1988; Grieve, 1990; Ammari and Mengel, 2006), well above the saturation concentration of even the most soluble secondary ferric oxide minerals but within the range required to meet the nutritional requirements of plants, fungi, and bacteria. Weak organic acids do not form complexes sufficiently stable to account for the observed iron solubility, providing solid evidence that *siderophore* soil pore-water concentrations fall in the 1–10 μM range.

7.2.3.6 Biosurfactants

Microbial biosurfactants in soil pore water and surface water are secreted as either exocellular agents (Table 7.6, Fig. 7.5) with specific functions (Desai and Banat, 1997; Lang, 2002; Wosten et al., 1999; Wosten, 2001) or cell envelope constituents released by the autolysis of dying cells. Cell membrane biosurfactants are believed to influence bacterial attachment to environmental surfaces by altering cell-envelope properties.

TABLE 7.6 Exocellular Bacterial Biosurfactants

Type	Examples
Glycolipids	Rhamnolipids, mycolic acids
Lipoproteins	Surfactin, visconsin, putisolvin
Proteins	Hydrophobins

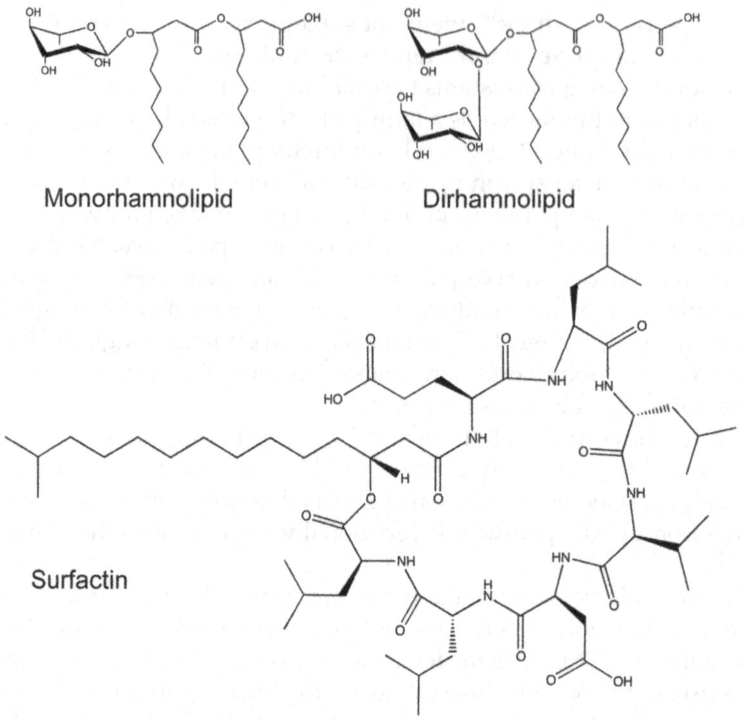

FIG. 7.5 The biosurfactants monorhamnolipid and dirhamnolipid (CAS registry number 4348-76-9) are secreted by *Pseudomonas aeruginosa* and surfactin (CAS registry number 24730-31-2) secreted by *Bacillus subtilis*.

The molecular properties of microbial biosurfactants arise from their amphiphilic[15] nature: limited water solubility (a result of their hydrophobic molecular segment) and a tendency to form micelles above a characteristic critical micelle concentration CMC.

An intriguing consequence of microbial biosurfactants is the enhanced solubility and bioavailability of hydrophobic or sparingly soluble organic contaminants in soil pore water (Cameotra and Bollag, 2003; Mulligan, 2005). This behavior is intimately dependent on their adsorption at mineral-water interfaces to form *hemimicelles*.

7.2.3.7 DNA

Every sector of the environment—soils, sediments, fresh, and marine water bodies—contain DNA sequences. By one estimate (Vogel et al., 2009) each gram of soil contains 10^{12} base pairs.[16] The vast majority of the DNA sequences in environmental media derive from microbial genomes.

[15] Amphiphilic molecules have a primary structure that contains at least one significant segment that is strongly hydrophilic or polar and at least one significant segment that is strongly hydrophobic or nonpolar.

[16] The human genome contains an estimated 3×10^9 base pairs.

DNA has unusually high chemical stability in the environment and sizable intact genetic segments are readily extracted, amplified, and sequenced. Two major international projects, the *Global Ocean Survey* (Fuhrman, 2003; Venter et al., 2004) and the *Terragenome Project* (Vogel et al., 2009) employ a method known as *shotgun sequencing* (Tyson et al., 2004) to recover meaningful genetic information from environmental DNA specimens.

Briefly, *shotgun sequencing* randomly breaks a DNA specimen into numerous small segments; each segment is sequenced using the chain-termination method. The DNA specimen is processed repeatedly: shotgun fragmentation followed by chain-termination sequencing, yielding multiple overlapping DNA sequences. Finally, these sequences are analyzed by computer programs to identify overlapping ends from different small sequences to assemble them into a continuous sequence.

Because DNA specimens extracted from environmental media is a sampling of the genomes of uncounted organisms, the "continuous sequence" is know as a *metagenome*. Schloss and Handelsman (2007) published a metaphorical account of shotgun sequencing and its use in reconstructing microbial communities. Before metagenomics, environmental microbiologists needed to isolate and culture each microbe in order to understand its biology. Unfortunately, only a small number of the microbial species in environmental media can be grown in culture (Handelsman, 2005).

Metagenomics allows microbiologists to study microbial communities without having to culture each specie. That being said, after a "continuous sequence" of an environmental metagenome is assembled microbiologists search the sequence looking for matches to known microbial genes. Metagenomic interpretation is still tethered to traditional isolation-culturing methods because this remains the only approach that allows microbiologists to identify genes and their function. As the database of isolated and cultured microbial genomes grows, scientists studying metagenomics have more opportunities to identify potential gene sequences in environmental metagenomes.

7.3 ORGANIC CARBON TURNOVER AND THE TERRESTRIAL CARBON CYCLE

The global carbon cycle (Fig. 7.6) encompasses both marine and terrestrial environments, but here we focus on a key component of the terrestrial carbon cycle.[17] The cycle begins with biological carbon fixation: the conversion of CO_2 to biomass carbon by photolithotrophic (i.e., photosynthetic) eukaryotes and prokaryots and, to a much lesser degree, chemolithotrophic microorganisms. The cycle ends with the mineralization of organic carbon into CO_2 through a process called oxidative metabolism. Oxidative metabolism releases the chemical energy stored during carbon fixation.

[17] The freshwater and marine carbon cycles involve similar carbon transformations and yield organic-matter products with chemical properties similar to soil organic matter.

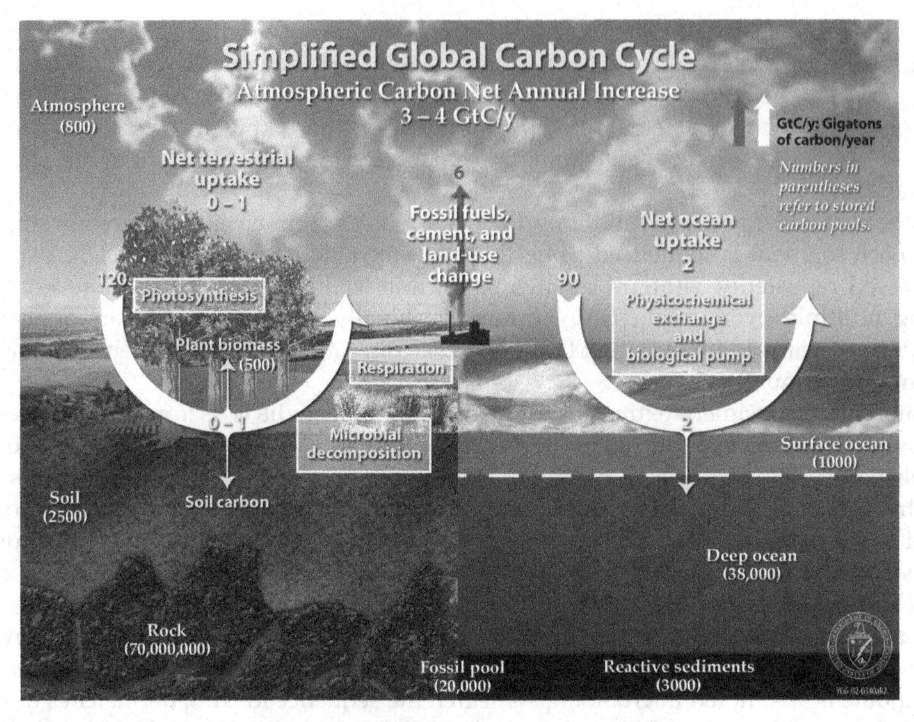

FIG. 7.6 Simplified representation of the global carbon cycle (Genomics: GTL, 2005).

7.3.1 Carbon Fixation

Photosynthetic carbon fixation converts light energy into chemical energy. Photosynthesis reduces the carbon in carbon dioxide from OSC = +4 to OSC = +1 in the terminal carbon in glyceraldehyde-3-phosphate, the feedstock for simple sugars, amino acids, and lipids.[18] The process (Fig. 7.7) involves four steps in what is known as the *Calvin cycle*.

In the first step, a single CO_2 molecule attaches to carbon-2 of D-ribose 1,5-bisphosphate (CAS registry number 14689-84-0). The intermediate—2-carboxy-3-keto-arabinitol-1,5-bisphosphate—is an extremely unstable compound. Hydrolysis of the bond between carbons 2 and 3 in 2-carboxy-3-ketoarabinitol-1,5-bisphosphate occurs during the second step, yielding two molecules of 3-phospho-D-glycerate (CAS registry number 3443-58-1). The net effect of forming the C–C bond in reaction 7.R2 is the reduction of the carbon in CO_2 from OSC = +4 to OSC = +3.

The third step is the phosphorylation of 3-phospho-D-glycerate by adenosine-5′-triphosphate ATP to yield 1,3-bisphospho-D-glycerate (CAS registry number 38168-82-0) and adenosine-5′-diphosphate ADP. The fourth step is the reduction of 1,3-bisphospho-D-glycerate

[18] Carbon oxidation states range from C = +1 to C = −1 in carbohydrates and peptides and from C = −2 to C = −3 in lipids. The average OSC ≈ 0 in biomass.

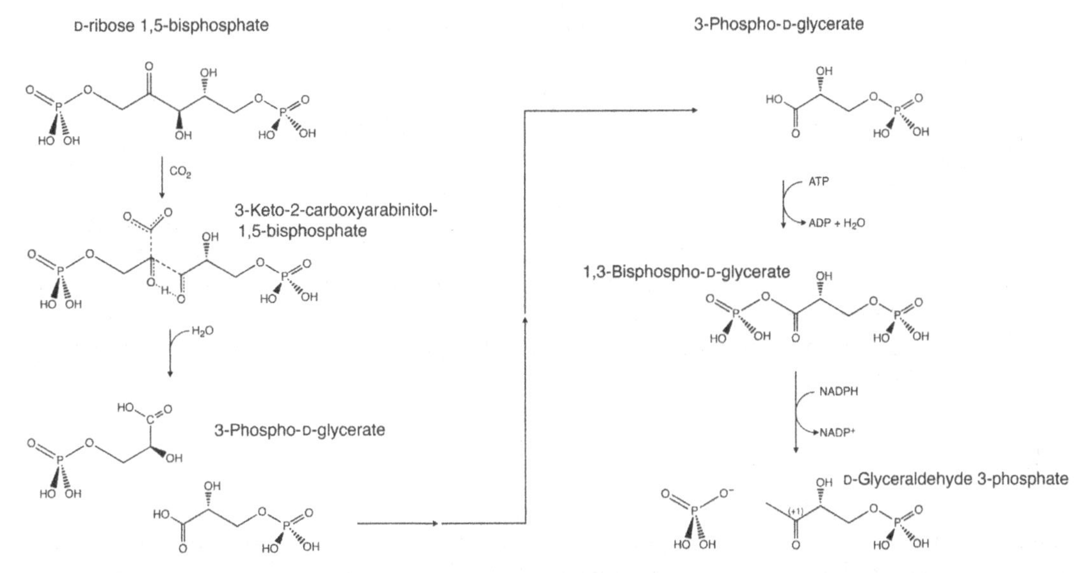

FIG. 7.7 Biological fixation of CO_2 involving: (1) D-ribose 1,5-bisphosphate (CAS registry number 14689-84-0), (2) 3-keto-2-carboxyarabinitol-1,5-bisphosphate (CAS registry number 82334-97-2), (3) 3-phospho-D-glycerate (CAS registry number 3443-58-1), (4) 1,3-bisphospho-D-glycerate (CAS registry number 38168-82-0), and D-glyceraldehyde 3-phosphate (CAS registry number 591-57-1).

to D-glyceraldehyde 3-phosphate (CAS registry number 591-57-1) by nicotinamide adenine dinucleotide phosphate (NADPH). NADPH transfers two electrons to carbon 1, reducing it from OSC = +3 to OSC = +1.

The final product of the Calvin cycle, D-glyceraldehyde 3-phosphate, is a simple 3-carbon sugar that serves as the premier feedstock for all bioorganic compounds. The net chemical reaction (7.R4) fixes 6 molecules of CO_2 as carbon 1 in 6 molecules of D-glyceraldehyde 3-phosphate. The absorption of photoenergy $E = h\nu$ by chlorophyll (reaction 7.R5) generates the chemical energy (i.e., ATP) and reducing capacity (i.e., NADPH) required by reactions 7.R4 and 7.R5.

$$6\,CO_2 + 6\,\text{ribulose-1,5-bisphosphate} + 12\,\text{ATP} + 12\,\text{NADPH} \longrightarrow 12\,\text{glyceraldehyde-3-phosphate} \quad (7.R4)$$

$$12\,H_2O + 12\,\text{NADP}^+ + 18\,\text{ADP}^{3-} + 18\,H_2PO_4^- \underset{E=h\nu}{\longrightarrow} 6\,O_2 + 12\,\text{NADPH} + 12\,H^+ + 18\,\text{ATP}^{4-} \quad (7.R5)$$

7.3.2 Carbon Mineralization

The mineralization of reduced carbon in biomass and senescence residue to CO_2 releases chemical energy for the growth and maintenance of both plants and soil organisms. The plants that fix virtually all carbon end up mineralizing a considerable fraction for their growth and maintenance. A fraction of standing plant biomass is consumed directly by herbivores, while the remainder becomes senescence residue supporting a diverse community of decomposing chemoorganotrophic organisms ranging from arthropods to bacteria and fungi.

In Chapter 9 we examine the reduction-oxidation chemistry of organic matter mineralization in greater detail. The diverse metabolic pathways termed *catabolism* deconstruct large organic molecules into smaller molecules. The chemical energy released during catabolism is transferred to a small number of electron carriers such as the nicotinamide adenine dinucleotide phosphate ($NADP^+$) appearing in reaction (7.R5). These electron carriers serve as the electron donors for the *electron transport chain*. The chemical energy stored in biomass is ultimately transferred to adenosine-5′-triphosphate (ATP), the primary energy currency for all metabolic processes in living organisms.

7.3.3 Oxidation of Organic Compounds by Dioxygen

An often overlooked characteristic of the carbon cycle is the slow abiotic oxidation of biomass and senescence residue by dioxygen (O_2). Make no mistake, the oxidation of most organic compounds by O_2 is thermodynamically favorable, releasing considerable free energy. For example, the standard free energy of reaction for the oxidation of glucose ($C_6H_{12}O_5$) by O_2 (reaction 7.R6) equals $\Delta_r G^{\circ} = -470.82$ kJ per mole of carbon. The standard free energy of reaction for the oxidation of methane (CH_4) by O_2 (reaction 7.R7) equals $\Delta_r G^{\circ} = -826.44$ kJ per mole of carbon.

$$C_6H_{12}O_6(aq) + 6\,O_2(g) \longrightarrow 6\,CO_2(aq) + 6\,H_2O(l) \qquad (7.R6)$$

$$CH_4(aq) + O_2(g) \longrightarrow CO_2(aq) + H_2O(l) \qquad (7.R7)$$

O_2 is the primary oxidant for the oxidative degradation of natural products, completing the global carbon cycle that begins with the biological fixation of CO_2. The mechanism of carbon oxidation in the carbon cycle, however, is almost exclusively biological; catalyzed by a host of microbial enzymes.

The carbon cycle is, for all practical purposes, a biologically driven cycle; oxidative degradation is no less biologically dependent than carbon fixation. A fundamental principle of organic chemistry and biochemistry, with direct relevance to the oxidative degradation of natural products in the carbon cycle, can be simply stated: the direct oxidation of organic molecules by O_2 is kinetically sluggish.[19]

Organic chemists, inorganic chemists, and biochemists studying reaction mechanisms eventually discovered why the direct oxidation of organic molecules by O_2 is kinetically unfavorable despite being thermodynamically highly favorable. The explanation relates to the *spin-state* of reactants and products and, in particular, the difference between the spin-state of O_2 compared to the spin-state of most organic molecules.

[19] The rate limiting step in the direct oxidation of organic molecules by ground-state O_2 is 6-orders of magnitude slower than typical reactions involving bond cleavage and formation (Lipscomb et al., 1988).

Electrons have an intrinsic angular momentum known as *spin* which gives rise to the *electron magnetic moment*. Paired electrons have their spins aligned opposite one another ⟨↑↓⟩ resulting in a zero net electron magnetic moment. Most organic compounds and natural products have spin-paired electrons in every occupied molecular orbital and, therefore, a zero net electron magnetic moment. Furthermore, the ultimate oxidation products CO_2 and H_2O have paired electrons in every molecular orbital. Chemists use the term *closed-shell* compounds when referring to compounds with paired electrons in every occupied molecular orbital.

Reactions are generally quite rapid if the reactants and products are all *closed-shell* compounds. Dioxygen—specifically ground-state O_2—is an *open-shell* molecule (cf. Box 7.4). The direct oxidation of organic molecules (including most natural products) by *triplet* O_2 is *spin forbidden* (cf. Appendix 7.B, *Spin Conservation and Spin Forbidden Reactions*).

The rationale behind electron-spin conservation and its implications for the direct reaction between O_2 and natural products may seem difficult to grasp, but this restriction has enormous implications for the carbon cycle. Bacteria, fungi, and other decomposers use enzyme catalysis to bypass the energy barrier blocking direct oxidation (i.e., allow the reaction to follow an alternate pathway that conserves electron-spin). Most of the chemical energy stored in biosphere organic compounds is released by enzyme-catalyzed biological oxidation reactions. Organic compounds represent kinetically stable chemical energy that organisms release at a rate matched to their metabolic needs.

7.3.4 Organic-Matter Turnover Models

7.3.4.1 Early Soil Organic Nitrogen Turnover Models

Below-ground carbon decreases with increasing depth within the soil profile; relatively little carbon is found below the plant root zone in the intermediate vadose and saturated zones. This distribution reflects the photosynthetic carbon fixation by the plant community. Litter fall from standing plant biomass accumulates at the surface and may, especially in forest ecosystems, form an organic soil horizon (>18% organic carbon by weight). Below-ground net primary production comes from the root system that supports a diverse decomposer community that, along with root biomass, represents total below-ground biomass.

BOX 7.4

GROUND AND EXCITED STATES OF DIOXYGEN

The *highest occupied molecular orbitals (HOMOs)* in ground-state O_2 are degenerate (i.e., equal energy but different symmetry): (π_x^*, π_y^*). Given the total number of electrons from two oxygen atoms and the number of molecular orbitals, each of the two *HOMOs* is occupied by a single electron. The total electron angular momentum $S = 1$, one-half spin for each electron: $\pi_x^*\langle\uparrow\rangle\ \pi_y^*\langle\uparrow\rangle$). The O_2 ground-state is a 3-fold degenerate or *triplet* spin-state (Fig. 7.8, $^3\Sigma_g^-$ lower left).

Continued

BOX 7.4 *(cont'd)*

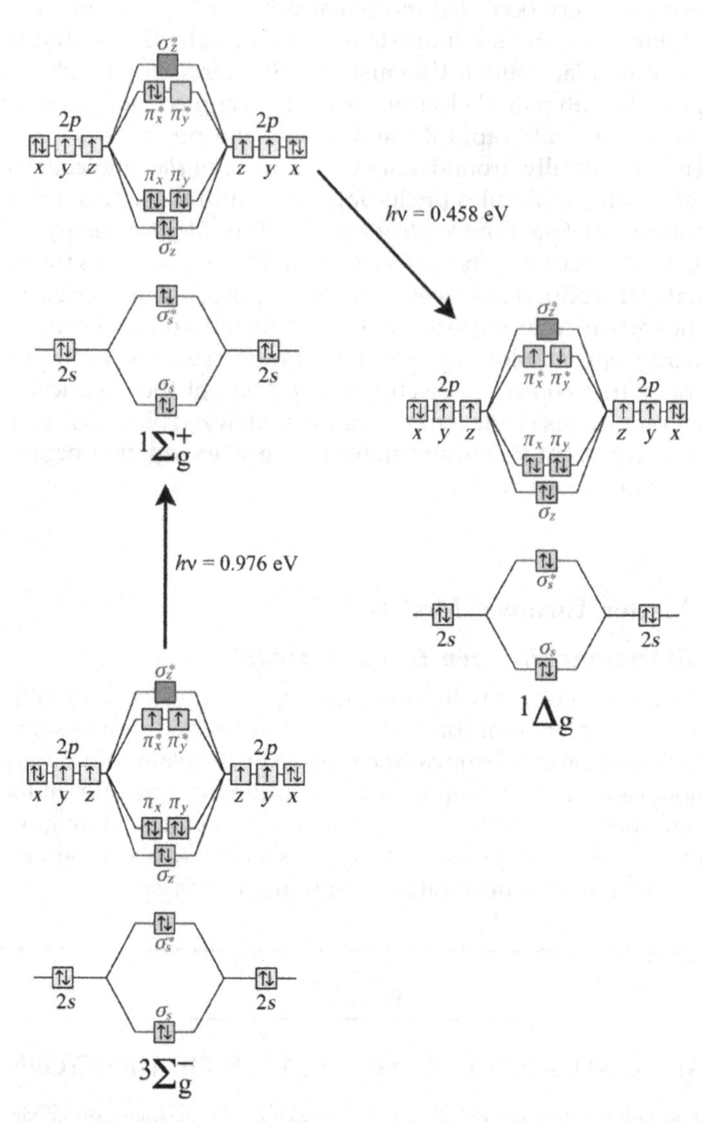

FIG. 7.8 Molecular orbital diagrams of dioxygen O_2 showing electron-spin configurations. The triplet ground-state $^3\Sigma_g^-$ appears at lower left with an electron-spin configuration of $(\pi_x^*\langle\uparrow\rangle\ \pi_y^*\langle\uparrow\rangle)$ in the highest occupied molecular orbitals. The most unstable excited state (upper left) $^1\Sigma_g^+$ lies 0.976 eV (equivalent to 158 kJ mol^{-1}) above the triplet ground-state. The unstable excited singlet state $^1\Sigma_g^+$ rapidly relaxes—emitting a 0.458 eV photon—to a more stable excited singlet state $^1\Delta_g$ (middle right) with an electron-spin configuration of $(\pi_x^*\langle\uparrow\rangle\ \pi_y^*\langle\downarrow\rangle)$.

BOX 7.4 *(cont'd)*

Immediately above the triplet ground-state lie two excited states (Fig. 7.8). The lowest excited state,[20] lying an estimated 94.3 kJ mol^{-1} above the triplet ground-state, is a stable singlet state (Fig. 7.8, $^1\Delta_g$ middle right) with two unpaired electrons in antialignment: $(\pi_x^*\langle\uparrow\rangle\ \pi_y^*\langle\downarrow\rangle)$.

The O_2 excited singlet state with two unpaired, antialigned electron spins $^1\Delta_g$ is the excited singlet-state responsible for the sluggish kinetics associated with the direct oxidation of organic compounds by dioxygen.

[20] The second excited state (Fig. 7.8, upper left) is an unstable, singlet state which quickly relaxes to a lower-lying, stable excited state (Fig. 7.8, middle right), which can persist for 72 min in gas-phase O_2 (Wilkinson et al., 1995).

Soil, by definition, must pass through a 2 mm sieve. This applies to all mineral grains, organic residue, and biomass. This definition excludes many arthropods, earthworms, and larger plant roots—to name a few—but does include much of the decomposer community. Soil biomass, regardless of whether it is from the plant community or the soil decomposer community supported by net primary production, is the starting point of the carbon trajectory that ultimately leads to the complex product known as humus or natural organic matter.

Net primary production—the total carbon fixed by photosynthesis during the growing season—ends up in one of three forms: carbon dioxide, biomass, and senescence residue. Respiration and biosynthesis derived from photosynthetic carbon represent investments in building and maintaining plant biomass. Senescence residue, the third component, is the second stage of the soil carbon trajectory.

Ecologists have developed models to describe the turnover of senescence residue: its mineralization to carbon dioxide, assimilation into decomposer biomass, and sequestration as below-ground organic matter. Jenny (1941) proposed one of the simplest models (expression 7.15) of organic-carbon turnover in soils. Jenny originally applied his turnover model to soil organic nitrogen dynamics (Fig. 7.9).

The Jenny soil organic nitrogen turnover model (expression 7.15) is a differential rate law that combines a zero-order term—the annual biological nitrogen fixation k_f [$\text{kg}_{\text{on}}\,\text{kg}_{\text{s}}^{-1}$ year^{-1}]—and a first-order term that represents the mass fraction of soil organic nitrogen $w(\text{N}_\text{s})$ [$\text{kg}_{\text{on}}\,\text{kg}_{\text{s}}^{-1}$]. Parameter k_m [year^{-1}] is first-order rate constant for the soil organic matter mineralization. Jenny also defined a new parameter $w(\text{N}_\text{s})_{\text{SS}} \equiv k_f/k_\text{m}$ representing the equilibrium or steady-state soil organic nitrogen mass fraction.

$$\frac{\Delta w(\text{N}_\text{s})}{\Delta t} = k_f - k_\text{m}\cdot w(\text{N}_\text{s}) = k_\text{m}\cdot\left(w(\text{N}_\text{s})_{\text{SS}} - w(\text{N}_\text{s})\right) \qquad (7.15)$$

Table 7.7 lists data from four soil organic-matter turnover studies (Salter and Green, 1933; Jenny, 1941; Myers et al., 1943; Haynes and Thatcher, 1955). The organic carbon mineralization rate constant (k_m) and half-life ($t_{1/2}$) for soil organic carbon for these midwestern US agricultural soils are very similar.

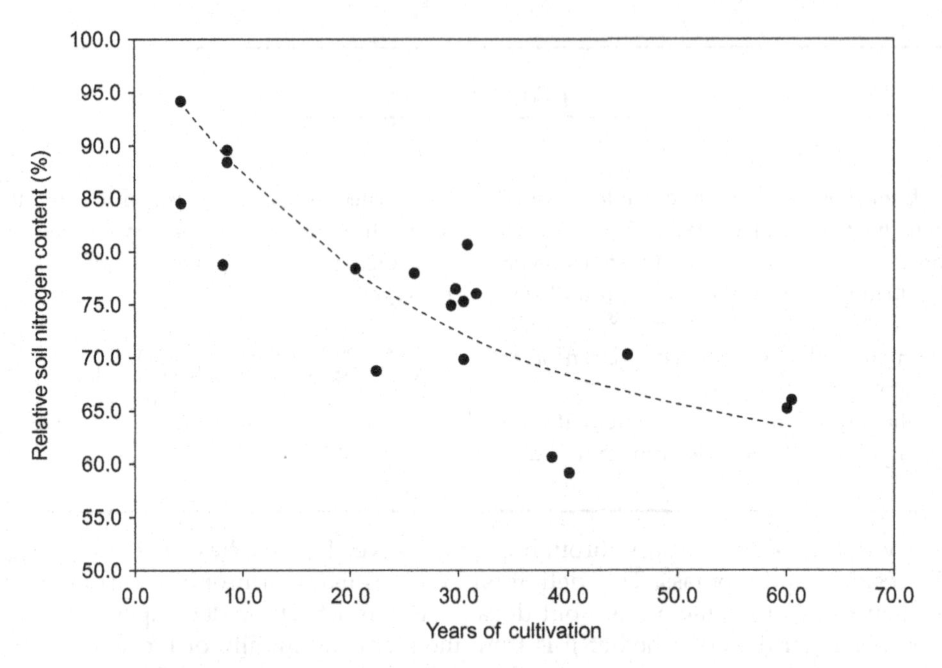

FIG. 7.9 The accelerated mineralization of soil organic nitrogen during 60 years of agricultural cultivation resulted in a net 35% loss of soil organic nitrogen in US prairie soils. *Reproduced with permission from Jenny, H., 1941. Factors of Soil Formation: A System of Quantitative Pedology. McGraw-Hill, New York/London.*

TABLE 7.7 Soil Organic Matter Mineralization Rate Constants and Half-Lives for Agricultural Soil

Rate constant k_m [year^{-1}]	Half-life $t_{1/2}$ [year]	Location	Reference
0.0608	11	Ohio (USA)	Jenny (1941)
0.052	13	Ohio (USA)	Salter and Green (1933)
0.072	10	Ohio (USA)	Haynes and Thatcher (1955)
0.07	10	Kansas (USA)	Myers et al. (1943)

Hénin and Dupuis (1945) proposed a slightly different model (expression 7.16) that defines the *isohumic coefficient* (A_{iso}), representing a constant fraction of net primary production that will be transformed into soil organic nitrogen each year.

$$\frac{\Delta w(N_s)}{\Delta t} = A_{iso} - k_m \cdot w(N_s) \tag{7.16}$$

Turnover rate laws (expressions 7.15, 7.16) appear in differential form. The integral forms (expressions 7.17, 7.18) are better suited to fitting measured changes in soil organic nitrogen content as a function of time $w(\mathrm{N_s})_t$. The soil organic nitrogen content at the beginning of the turnover experiment is $w(\mathrm{N_s})_0$.

$$w(\mathrm{N_s})_t = w(\mathrm{N_s})_{\mathrm{SS}} - (w(\mathrm{N_s})_{\mathrm{SS}} - w(\mathrm{N_s})_0) \cdot \mathrm{e}^{-k_{\mathrm{m}} \cdot t} \tag{7.17}$$

$$w(\mathrm{N_s})_t = \left(\frac{A_{\mathrm{iso}}}{k_{\mathrm{m}}}\right) - \left(\frac{A_{\mathrm{iso}}}{k_{\mathrm{m}}} - w(\mathrm{N_s})_0\right) \cdot \mathrm{e}^{-k_{\mathrm{m}} \cdot t} \tag{7.18}$$

The soil organic nitrogen content will converge on a steady-state content $w(\mathrm{N_s})_{\mathrm{SS}}$ after a time interval that is long relative to the mineralization rate constant k_{m}. Expression (7.19) reveals that the models by Jenny (1941) and Hénin and Dupuis (1945) are mathematically equivalent, differing only in the parameters used.

$$\lim_{t \to \infty} (w(N_s)_t) = w(\mathrm{N_s})_{\mathrm{SS}}$$

$$w(\mathrm{N_s})_{\mathrm{SS}} = \left(\frac{A_{\mathrm{iso}}}{k_{\mathrm{m}}}\right) \tag{7.19}$$

Woodruff (1950) elaborated the Jenny-Henin-Dupuis model by subdividing soil organic matter into several pools i—each with a characteristic isohumic coefficient A_i, steady-state organic-nitrogen mass fraction $w(\mathrm{N_s})_i$, and mineralization rate constant k_i. The rationale for elaborating the soil organic nitrogen model was assertion that soil organic matter is heterogeneous. Expression (7.20) is an operational definition of organic matter heterogeneity insofar as heterogeneity means each organic nitrogen pool has distinct turnover parameters.

$$w(\mathrm{N_s})_t = \sum_i^n \left[\left(\frac{A_{\mathrm{i}}}{k_i}\right) - \left(\frac{A_{\mathrm{i}}}{k_i} - w(\mathrm{N_s})_i\right) \cdot \mathrm{e}^{-k_i \cdot t}\right] \tag{7.20}$$

The Jenny-Henin-Dupuis turnover model (expressions 7.18, 7.19) recognizes two organic carbon pools: net primary production (i.e., senescence residue) and soil organic carbon. The Woodruff (1950) elaboration (expression 7.20) does not indicate the number of soil organic matter pools n.

In conclusion, although biological nitrogen fixation and biological carbon fixation are two very different processes, to say nothing of the relative quantities of nitrogen and carbon fixed annually, the same rate law applies to both processes. While nitrogen mineralization fulfills the nutritional demand for building soil microbial biomass, carbon mineralization must fulfill the carbon demand to build *and* sustain microbial biomass. Soil microbes will mineralize organic nitrogen in excess of nitrogen nutritional requirements simply to meet metabolic carbon demand.

7.3.4.2 Modern Soil Organic Carbon Turnover Models: RothC and CENTURY

The original soil organic-matter turnover models (Jenny, 1941; Hénin and Dupuis, 1945; Woodruff, 1950) addressed the impact of agricultural crop production on soil organic nitrogen levels. Scientists realized soil nitrogen, an essential major crop nutrient, declines under crop production. Modern soil organic-carbon turnover models shifted their focus from nitrogen to carbon because sustainable agriculture and crop production depends on a range of soil properties, besides fertility (e.g., moisture retention, aeration, erodibility), depend on soil organic carbon levels.

Global climate models rely on an accurate terrestrial and marine carbon budgets. Simulating the impact of climate on net primary production is relatively straight forward. Modeling soil carbon dynamics has proven a more daunting challenge. Modern soil organic-carbon turnover models, originally designed to reveal the impact of agricultural practices and land management on soil carbon levels in North American and Northern Europe, now play a critical role in global carbon cycle models.

RothC, the Rothamsted carbon-turnover model (Jenkinson, 1977; Jenkinson and Rayner, 1977; Jenkinson et al., 1990), builds upon and refines the organic-carbon pool concept introduced by Woodruff (1950). Plant material is subdivided into two forms—decomposable plant material (DPM) and resistant plant material (RPM)—each plant senescence residue form has a characteristic rate constant for its conversion into soil organic carbon.

Data from a relatively short-term experiment measuring the mineralization of a one-time addition of ^{14}C-labeled rye grass appear in Fig. 7.10 clearly showing the precipitous loss of DPM during the first year followed by a more gradual loss of RPM over the remainder of the study. The soil organic^{14}C data $w(^{14}C_s)$ in Fig. 7.10 are fitted with expression (7.21).

$$w(^{14}C_s)_t = w(^{14}C_s) \cdot \left[f_{dpm} \cdot e^{-k_{dpm} \cdot t}\right] + \left[f_{rpm} \cdot e^{-k_{rpm} \cdot t}\right] \tag{7.21}$$

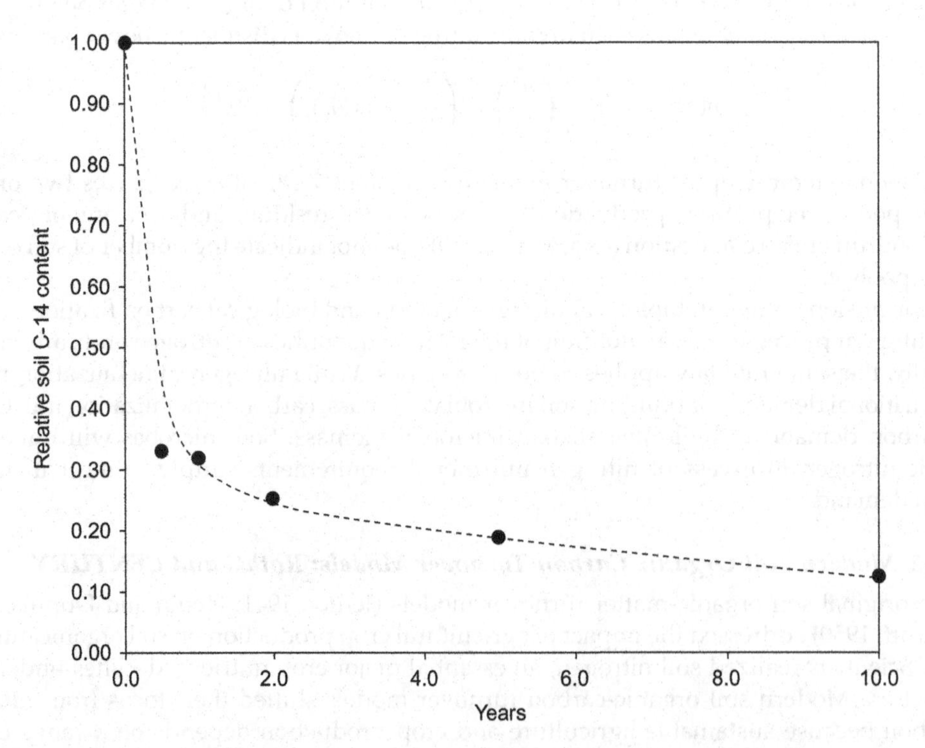

FIG. 7.10 Decomposition of ^{14}C-labeled rye-grass under field conditions (Jenkinson, 1977; Jenkinson and Rayner, 1977) is fitted to expression (7.21).

RothC divides soil organic carbon into two actively cycling pools: microbial biomass BIO and humus HUM. A third soil carbon pool designated *inert organic matter* (IOM) does not turnover, representing what some soil scientists call *char* or *block carbon*. All biologically fixed carbon (net primary production (NPP)) partitions (expression 7.21) between a rapidly decomposing senescence residue fraction f_{dpm} and a slowly decomposing senescence residue fraction f_{rpm}. *RothC* partitions carbon from each the senescence residue fraction into three pathways (Fig. 7.11), arrows): mineralization into CO_2, assimilation into chemoorganotrophic biomass (BIO), or transformation into soil humus (HUM).

Scientists studying carbon turnover understood several factors besides NPP controlled mineralization rate constants: soil temperature, soil moisture, plant growth stage, and soil clay content. *RothC* represents the effect of these factors as rate modifiers. Expression (7.22) illustrates this for the soil temperature rate modifier a, the soil moisture rate modifier b, and the plant-retention[21] modifier c.

$$\begin{aligned} w(^{14}\mathrm{C_s})_t = {} & w(^{14}\mathrm{C_s}) \cdot \left\{ \left(f_{\mathrm{dpm}} \cdot \mathrm{e}^{-a \cdot b \cdot c \cdot k_{\mathrm{dpm}} \cdot t} \right) + \left(f_{\mathrm{rpm}} \cdot \mathrm{e}^{-a \cdot b \cdot c \cdot k_{\mathrm{rpm}} \cdot t} \right) \right\} \\ & + w(^{14}\mathrm{C_s}) \cdot \left\{ \left(f_{bio} \cdot \mathrm{e}^{-a \cdot b \cdot c \cdot k_{bio} \cdot t} \right) + \left(f_{hum} \cdot \mathrm{e}^{-a \cdot b \cdot c \cdot k_{hum} \cdot t} \right) \right\} \end{aligned} \tag{7.22}$$

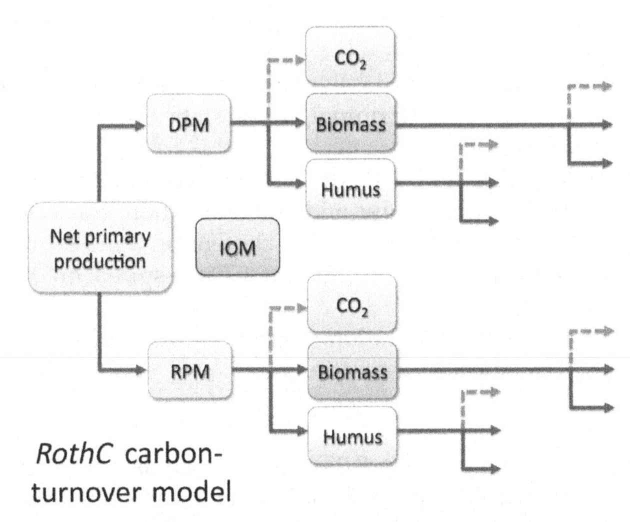

FIG. 7.11 The carbon transformation pathways and carbon pools of the Rothamsted soil organic carbon turnover model (*RothC*): net primary production (NPP), decomposable (DPM), and resistant (RPM), mineralization CO_2, assimilation (BIO), and transformation into soil humus (HUM). Inert organic matter (IOM) does not actively turnover. Each pathway has its characteristic transformation rate constant. *Reproduced with permission from Jenkinson, D.S., Andrew, S.P.S., et al., 1990. The turnover of organic carbon and nitrogen in soil. Philos. Trans. R. Soc. Lond. B Biol. Sci. 329 (1255), 361–368.*

[21] *RothC* assumes plant senescence residue accounts for 60% of NPP when plants are actively growing and 100% when plants are no longer growing.

RothC models the impact of soil texture on mineralization rate constants though its effect on the ratio of $^{14}CO_2$ evolved to ^{14}C in soil organic carbon (expression 7.23), assumed to be an undefined combination of microbial biomass and *fresh* (i.e., ^{14}C-labeled soil organic matter). The operative soil texture parameter is soil cation exchange capacity (CEC) which correlates with the soil clay content.

$$\frac{w(^{14}CO_2)}{w(^{14}C_s)} = 1.21 + 2.24 \cdot e^{-0.085 \cdot CEC} \tag{7.23}$$

RothC was developed and validated at the Rothamsted Experiment Station located near Harpenden in southern England (51.8175°N, −0.3524°E). A second, prominent carbon turnover model—the *CENTURY* model (Parton et al., 1987)—was developed and validated in the Great Plains of the United States. Unlike *RothC, CENTURY* was developed to model soil organic carbon dynamics in response to both range management and crop production. Regardless of their details, these two models are fundamentally equivalent and have been validated in climate zones markedly different from the settings where they were originally developed and validated (Jenkinson et al., 1999; Parton et al., 1993).

CENTURY distinguishes resistant and decomposable plant senescence carbon as *structural carbon* represented by plant cell wall lignin and *metabolic carbon* representing plant cytoplasmic material. *CENTURY* partitions soil organic carbon into four carbon pools, each with a different mean residence time: microbial biomass, dissolved organic carbon, slow carbon, and passive carbon. Dissolved organic carbon can leach out of the soil profile, slow carbon can either sustain microbial biomass or advance to passive carbon. Passive carbon—the soil carbon pool with the longest residence time—is the terminal carbon state whose ultimate fate is either CO_2 or sustaining microbial biomass.

The flow charts in Figs. 7.11 and 7.12 do not, indeed cannot, properly represent the role of microbial biomass in carbon flow and mineralization. Since microbial biomass is typically represented as X, the microbial biomass content on a mass-fraction basis is $w(C_X)$. Respiration is typically represented as R, the net CO_2 respired on a mass-fraction basis becomes $w(C_R) \equiv w(CO_2)$.

Expression (7.24) gives the net respiration (soil mass-fraction basis) as a function of the microbial biomass $w(C_X)$, the specific respiration rate r [s^{-1}], and time.

$$\Delta w(C_R) = w(C_X) \cdot r \cdot t \tag{7.24}$$

Growth yield on substrate Y (expression 7.25), also known as *carbon-use efficiency* (CUE), is the ratio of net growth ΔX to assimilation, the latter being carbon invested in growth ΔX (cf. Section 7.2.2) and carbon mineralized during respiration ΔR.

$$Y = \frac{\Delta X}{\Delta X + \Delta R} = \frac{\Delta w(C_X)}{\Delta w(C_X) + \Delta w(C_R)} \tag{7.25}$$

Fig. 7.13 shows the role of microbial biomass as the driver for carbon flow from one soil carbon pool (litter to "slow" carbon and "slow" carbon to "passive" carbon) where net respiration ΔR is determined by the total microbial biomass and the specific respiration rate of the microbial biomass.

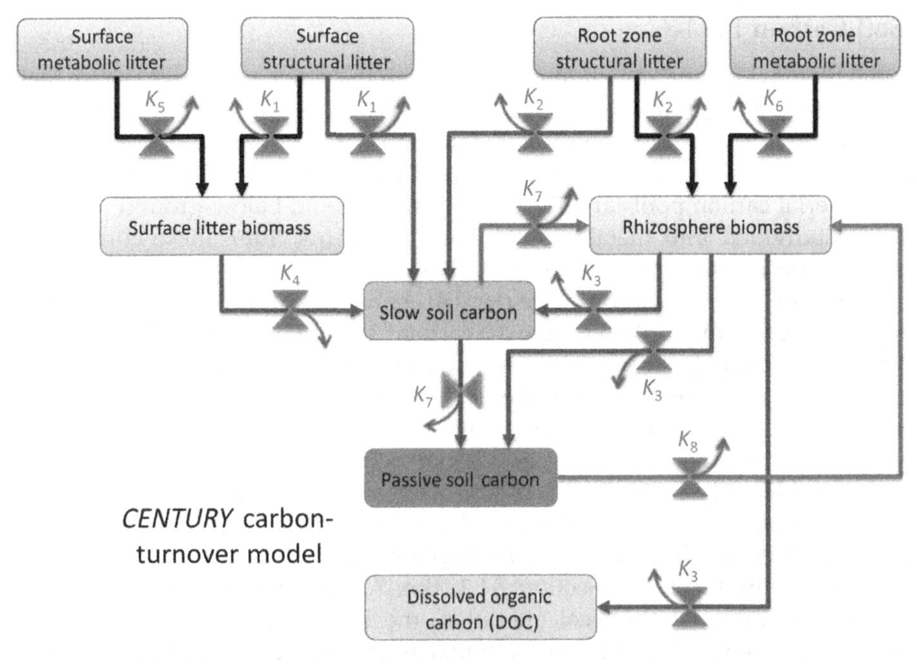

FIG. 7.12 The *CENTURY* carbon mineralization flow chart. As carbon flows from one carbon pool to the next a fraction is mineralized to CO_2, depicted as *arrows*. A universal valve symbol restricts carbon flow along each pathway with the mineralization rate constant listed above each valve symbol.

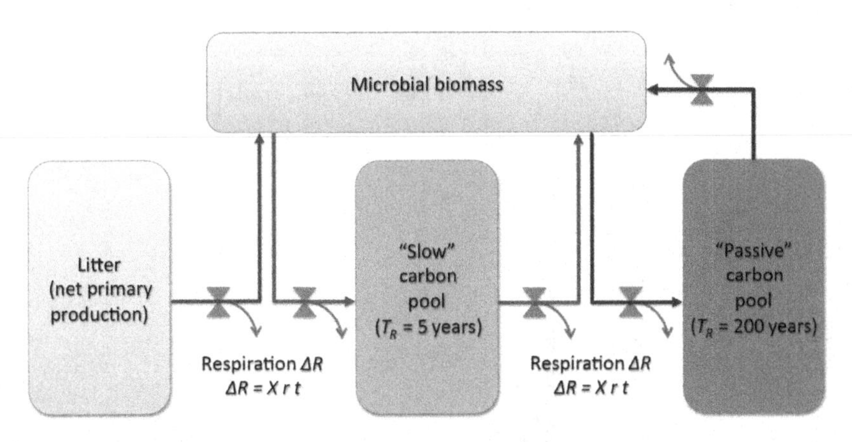

FIG. 7.13 Net respiration ΔR is a product of microbial biomass X, the specific respiration rate r, and time t. All soil carbon pools (litter, "slow," and "passive") are substrate supporting microbial biomass growth ΔX.

7.3.5 Soil Carbon Pools

Microbial biomass is fundamentally different from the other soil carbon pools included in all soil carbon-turnover models (cf. Fig. 7.1). *RothC* recognizes four soil carbon pools (besides microbial biomass): (1) decomposable plant material, (2) resistant plant material, (3) "humus," and (4) "inert organic matter" (Fig. 7.11). *CENTURY* recognizes six soil carbon pools (Fig. 7.12). The plant material carbon pools are divided into surface and below-ground pools and these are further subdivided into metabolic litter and structural litter, these correspond to the decomposable and resistent pools of *RothC*. *CENTURY* identifies two other soil carbon pools "slow carbon" and "passive carbon." The parallels between "humus" and "slow carbon" and between "inert organic matter" and "passive carbon" are apparent if not entirely equivalent.

Scientists studying natural organic matter in soils and sediments have used numerous extraction methods to isolate organic matter from entrained mineral colloids. Some extraction treatments release soluble carbon or CO_2 without regard to chemical alteration, merely to quantify organic carbon or isotope ratios. Carbon released by mild chemical hydrolysis or oxidation is considered labile to mineralization while carbon released by aggressive chemical hydrolysis or oxidation is deemed either physically protected or chemically refractory (Trumbore et al., 1996; Six et al., 2002; Kaiser and Guggenberger, 2003; Zimmermann et al., 2006).

Some extraction protocols (cf. Section 7.1.1, Box 7.2) are designed to extract natural organic matter with minimal chemical alteration; the intent being to expose stoichiometric composition, chemical functionality, and colloidal properties without undue interference from entrained mineral colloids. Do the soil organic matter pools identified by these and other soil organic carbon turnover models actually exist? The short answer: these distinctions are operational and, therefore, artificial. Bolker et al. (1998) demonstrated that both *RothC* and *CENTURY* are "pure decay" models. An example of "pure decay" can be found in Fig. 7.14 which plots the contribution of individual years to the soil organic carbon percentage of a Vertisol from Kenya (1.333333° S, 36.833333° E).

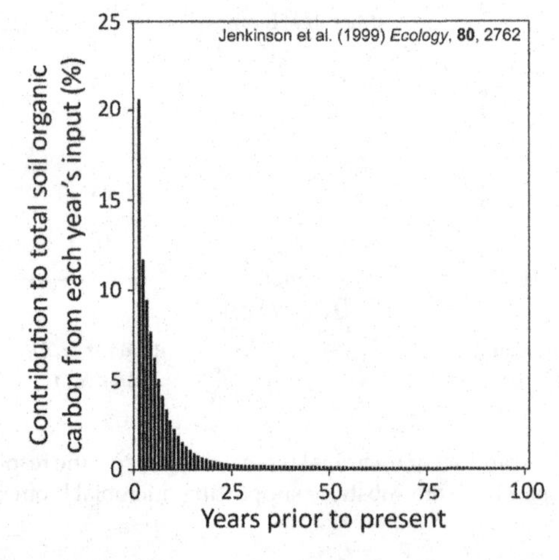

FIG. 7.14 The contribution of organic carbon inputs from individual years to the total soil organic carbon stock. The simulation was made by ROTHC-26.3 and refer to the 0–15 cm layer of a Kenyan vertisol (Jenkinson et al., 1999), assuming this soil has attained steady-state conditions.

Whether one wishes to define an arbitrary series of soil carbon pools whose residence time increase as their proportion of total soil carbon decreases or imagine a continuum where increasing residence time correlates with decreasing fraction of the total soil carbon, the final result is the same for carbon turnover dynamics.

7.4 CHEMICAL PROPERTIES OF NATURAL ORGANIC MATTER

Biochemists employ a hierarchy to describe the structure of complex biological molecules. The *primary structure* of a peptide is its amino acid sequence, while the primary structure of a glycan (or polysaccharide) is its saccharide sequence. *Secondary structure* refers to three-dimensional geometry of a particular segment of a biomolecule (i.e., local conformation) strongly influenced by hydrogen bonding, while *tertiary structure* refers to the three-dimensional geometry of the entire biomolecule.

Despite considerable research, the current understanding of molecular structure in the context of natural organic-matter specimens is limited to cataloging chemical functional groups.[22] How identifiable functional groups are linked together to form the primary structure of natural organic-matter molecules remains a mystery. In short, the primary structure of organic-matter molecules is best viewed as is a complex assembly of identifiable functional groups; the specimen as a whole is a mixture where no two molecules are alike.

The following sections summarize experimental results that identify specific chemical functional groups and quantify their abundance. An account of the physicochemical basis for the spectroscopic method employed and details justifying their interpretation are beyond the scope of this book.

7.4.1 Oxygen Functional Groups

7.4.1.1 Titratable Weak Acid Groups

Acidometric titration is the most common method for quantifying the weak-acid content of organic-matter specimens. Acidometric titration curves for two organic-matter specimens appear in Fig. 7.15: Suwannee River (Georgia, USA) aquatic "fulvic acid" and Summit Hill (New Zealand) soil "humic acid."

The displacement of the "fulvic acid" titration curve in Fig. 7.15 (*open circles*) reveals the higher weak acid content of the "fulvic acid" specimen relative to the "humic acid" specimen (*open triangles*), consistent with the higher oxygen content of "fulvic acid" fractions relative to "humic acid" fractions (cf. Table 7.7).

$$Q_{TOT} = \frac{Q_1}{1 + \left(K_{a1} \cdot a_{H^+}\right)^{1/n_1}} + \frac{Q_2}{1 + \left(K_{a2} \cdot a_{H^+}\right)^{1/n_2}} \tag{7.26}$$

Expression (7.26) represents the total acidity of organic-matter specimen as the sum of a weak acid population with hydrolysis constant $K_{a1} \approx 10^{-4}$ (nominally a carboxyl functional group) and a $K_{a2} \approx 10^{-10}$ (nominally a phenol functional group). Each acidometric titration

[22] A functional group is a well-defined group of atoms which display similar chemical reactivity in most molecules that contain that group.

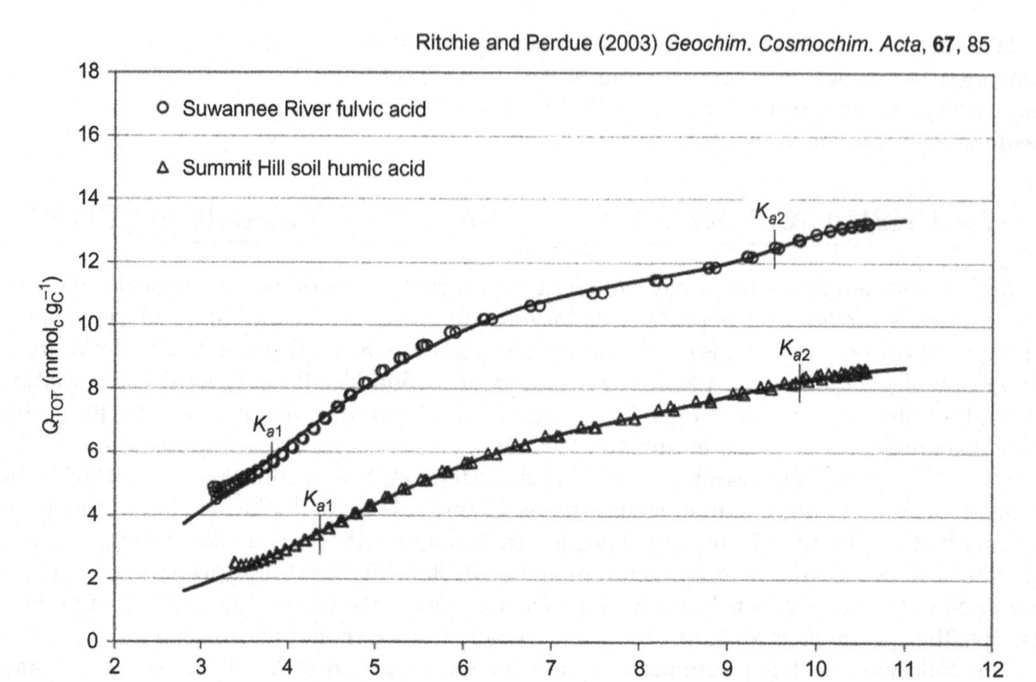

FIG. 7.15 Both acidometric titration curves reveal two weak-acid populations: a population with mean hydrolysis constant $K_{a1} \approx 10^{-4}$ and a population with mean hydrolysis constant $K_{a2} \approx 10^{-10}$. *Reproduced with permission from Ritchie, J.D., Perdue, E.M., 2003. Proton-binding study of standard and reference fulvic acids, humic acids, and natural organic matter. Geochim. Cosmochim. Acta 67 (1), 85–96.*

curve in Fig. 7.15 is fitted using expression (7.26), a six (6) parameter model (Katchalsky and Spitnik, 1947). The amount of base added during the titration Q_{TOT} [mmol$_c$ g$_C^{-1}$] is normalized by the mass of dissolved organic carbon.

The hydrolysis constants (K_{a1}, K_{a2}) in expression (7.26) determine the pH of the equivalence points, as indicated in Fig. 7.15. The amount of each weak acid—Q_1 [mmol$_c$ g$_C^{-1}$] and Q_2 [mmol$_c$ g$_C^{-1}$], respectively—determines the position of the two equivalence points along the Q_{TOT} axis.

Organic-matter titration curves of are significantly broadened compared to the titration curve typical of solutions composed of low molecular-weight compounds with a single weak-acid functional group.[23] The two remaining parameters (n_1, n_2) in expression (7.26) account for the broadening effect.

Katchalsky and Spitnik (1947) attribute a diffuse equivalence point to electrostatic interactions between closely spaced weak acid groups. The electrostatic effect, in its simplest terms, reduces the tendency of protons to dissociate from weak acid groups when surrounded by

[23] Katchalsky and Spitnik (1947) report this behavior for polymeric acids.

chemically identical weak acids that have already dissociated. In essence, the proximity of negatively charged (i.e., dissociated weak acid) sites exerts a collective attractive electrostatic force on protons, lowering the tendency of remaining weak acid functional groups to release their protons. Some organic-matter chemists attribute equivalence-point broadening to chemical heterogeneity of the two weak-acid populations.

7.4.1.2 Other Oxygen-Containing Functional Groups

Acidometric titration typically identifies two weak organic acid populations in organic-matter extracts based on two equivalence points in the titration curves (cf. Fig. 7.15). Carboxyl functional groups estimated using NMR (Ritchie and Perdue, 2008) overestimate titratable carboxyl acidity by roughly 40–50%. NMR spectroscopy fails to distinguish carboxyl functional groups from carboxyl ester and amide functional groups, accounting for this overestimate. Phenol functional groups cannot be reliably resolved using either H-1 or C-13 NMR.

Example 7.2 illustrates how the oxygen and weak acid content can be used to estimate the abundance of nontitratable oxygen functional groups. Regardless of whether the specimen is a "fulvic acid" or a "humic acid," a considerable fraction of the oxygen in organic-matter specimens cannot be assigned to weak acid functional groups.

EXAMPLE 7.2

Example Permalink

http://soilenvirochem.net/dh1061

Assign the oxygen content of an organic-matter specimen to titratable weak-acid functional groups and nontitratable oxygen functional groups.

The specimen in this example is the same specimen as in Example 7.2: Elliot soil "humic acid," *IHSS* specimen 1S102H. Required composition data appear in Table 7.8.

TABLE 7.8 The Carbon Mass-Fraction $w(C)$, Oxygen Mass-Fraction $w(O)$, and Titratable Acidity of *IHSS* Specimen 1S102H (Elliot Soil "Humic Acid" Standard)

$w(C)$ (g_C, g^{-1})	$w(O)$ (g_O, g^{-1})	"Carboxyl" groups ($mmol_c$, g_C^{-1})	"Phenol" groups ($mmol_c$ g_C^{-1})
0.5813	0.3408	8.28	1.87

Part 1 Convert moles of titratable acidity to grams of oxygen.

IHSS reports titratable weak acidity using millimoles-of-charge per gram carbon: $mmol_c\ g_C^{-1}$. The oxygen mass fraction $w(O)$, however, reported as grams of oxygen per gram of organic matter (dry weight): $g_O\ g_{OM}^{-1}$.

This change of units requires the oxygen atomic mass $\overline{m}_a(O)$ and the moles of oxygen in each of the weak-acid functional groups assigned to the equivalence points in the titration curve (cf. Fig. 7.15). These weak-acid functional groups appear in Fig. 7.16.

Each mole-of-charge assigned as "carboxyl" is equivalent to two moles of oxygen while each mole-of-charge assigned as "phenol" is equivalent to one moles of oxygen. Use the "carbonyl" content (Table 7.8, column 3) and the "phenol" content (Table 7.8, column 4).

Carboxyl functional group Phenol functional group

FIG. 7.16 Weak acid functional groups assigned to organic-matter acidometric titration data: carboxyl (*left*) and phenol (*right*). The weak-acid population with an equivalence point $K_{a1} \approx 10^{-4}$ is assigned to "carboxyl" groups. The weak-acid population with an equivalence point $K_{a2} \approx 10^{-10}$ is assigned to "phenol" groups.

carboxyl oxygen

$$8.28\ \text{mmol}_c\ \text{g}_C^{-1} \cdot \left(\frac{2\ \text{mol}_O}{1000\ \text{mmol}_c\ \text{mol}^{-1}} \right) = 1.656 \times 10^{-2}\ \text{mol}_O^{-1}\ \text{g}_C$$

phenol oxygen

$$1.87\ \text{mmol}_c\ \text{g}_C^{-1} \cdot \left(\frac{1\ \text{mol}_O}{1000\ \text{mmol}_c\ \text{mol}^{-1}} \right) = 1.87 \times 10^{-3}\ \text{mol}_O\ \text{g}_C^{-1}$$

Part 2. Convert the molar weak-acid oxygen content from Part 1 using the oxygen atomic mass: $\overline{m}_a(\text{O}) = 16.00\,\text{g}\,\text{mol}^{-1}$.

Carboxyl oxygen

$$\left(1.656 \times 10^{-2}\ \text{mol}_O\ \text{g}_C^{-1}\right) \cdot \left(16.00\,\text{g}\,\text{mol}_O^{-1}\right) = 0.265\,\text{g}_O\,\text{g}_C^{-1}$$

Phenol oxygen

$$\left(1.87 \times 10^{-3}\ \text{mol}_O\ \text{g}_C^{-1}\right) \cdot \left(16.00\,\text{g}\,\text{mol}_O^{-1}\right) = 0.0299\,\text{g}_O\,\text{g}_C^{-1}$$

Titratable weak-acid oxygen

$$0.295\ \text{g}_O\ \text{g}_C^{-1} = \left(0.265\ \text{g}_O\ \text{g}_C^{-1}\right) + \left(0.0299\ \text{g}_O\ \text{g}_C^{-1}\right)$$

Part 3. Convert the weak-acid content from Part 2 the same units as the oxygen mass fraction $w(\text{O})$.

IHSS reports titratable weak acidity as millimoles-of-charge per gram carbon: $\text{mol}_c\ \text{g}_C^{-1}$. The oxygen mass fraction $w(\text{O})$, however, is normalized by the organic-matter total mass: $\text{g}_O\ \text{g}_{OM}^{-1}$.

Using carbon mass fraction $w(\text{C})$ listed in Table 7.8 column 1, renormalize the weak-acid content from Part 2.

$$0.295\,\text{g}_O\,\text{g}_C^{-1} \cdot 0.5813\,\text{g}_C\,\text{g}^{-1} = 0.171\,\text{g}_O\,\text{g}^{-1}$$

The titratable weak-acidity, assigned as "carboxyl" and "phenol," accounts for 50.3% of the total oxygen content of this particular organic-matter specimen.

$$\frac{0.171}{0.341} = 0.503$$

The oxygen assignment illustrated in Example 7.2 is generally applicable to natural organic matter, regardless of origin. Titratable acidity accounts for about half of the total oxygen content.

The remaining organic oxygen is not titratable: enol (R–OH), keto (R–(C=O)–R′), amide (cf. Section 7.4.2), and ester (cf. Section 7.4.3) moieties. The assignment of nontitratable organic oxygen is also indirect, relying on assignments made to ^{1}H and ^{13}C NMR spectra (cf. Section 7.4.5.1).

7.4.2 Nitrogen Functional Groups

What little we know about nitrogen in organic matter comes from N-15 NMR spectroscopy. NMR spectroscopy is an isotope-specific method. Although both stable nitrogen isotopes have a nuclear magnetic moment essential for NMR, each isotope presents distinct challenges when studying organic matter.

Nitrogen isotope ^{14}N is very abundant (99.63%), but the large electric quadrupole moment of its nucleus causes profound signal broadening, rendering the most abundant nitrogen isotope of little practical use (Lambert et al., 1964). Nitrogen isotope ^{15}N, though not susceptible to electric-quadrupole signal broadening, produces a relatively weak signal in natural organic-matter specimens because both its low natural abundance (0.37%) (Lambert et al., 1964) and small relative sensitivity (cf. Table 7.9, columns 2 and 4).

Thorn and Cox (2009) recorded natural abundance N-15 NMR spectra of "fulvic" and "humic acids," shown in Fig. 7.17. The recording of these spectra required 56–560 h, reflecting the low relative sensitivity of natural abundance N-15 NMR. Thorn and Cox (2009) assign the 119–120 ppm resonance peak to amide nitrogen typical of peptides (Fig. 7.18, upper left) and N-acetylated amino-polysaccharides (Fig. 7.18, upper right and lower center) and the 31–36 ppm to amine nitrogen found in amino sugars and terminal amino acids. Amide nitrogen

TABLE 7.9 Nuclear Magnetic Properties of the Most Abundant Elements in Natural Organic Matter

Isotope	Natural abundance (%)	Nuclear Spin (unitless)	Relative sensitivity (unitless)
$^{1}_{1}H$	99.985	$+\frac{1}{2}$	$1.00 \times 10^{+0}$
$^{2}_{1}H$	0.015	+1	5.43×10^{-7}
$^{13}_{6}C$	1.10	$+\frac{1}{2}$	1.69×10^{-4}
$^{14}_{7}N$	99.63	+1	2.05×10^{-5}
$^{15}_{7}N$	0.37	$+\frac{1}{2}$	2.11×10^{-7}
$^{17}_{8}O$	0.038	$-\frac{5}{2}$	4.48×10^{-7}
$^{31}_{15}P$	100.	$+\frac{1}{2}$	1.39×10^{-4}
$^{33}_{16}S$	0.75	$+\frac{3}{2}$	1.43×10^{-8}

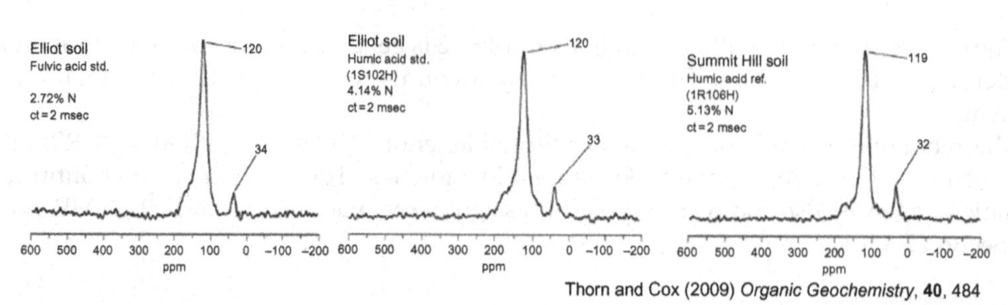

FIG. 7.17 Nitrogen-15 NMR spectra of soil "fulvic acid" and "humic acid" from the *IHSS* specimen collection. Chemical shifts are relative to the single N-15 resonance in the amino acid glycine. *Reproduced with permission from Thorn, K.A., Cox, L.G., 2009. N-15 NMR spectra of naturally abundant nitrogen in soil and aquatic natural organic-matter specimens of the International Humic Substances Society. Org. Geochem. 40, 484–499.*

typically accounts for 75–85% of the total organic nitrogen, terminal amine 6–8%, and unidentified nitrogen accounts for the remainder.

The survival of amide nitrogen in organic matter extracted by the *IHSS* extraction protocol (cf. Box 7.2) is a significant concern. The acid phase of the *IHSS* protocol represents conditions suitable for the acid-hydrolysis of labile amide linkages. Aluwihare et al. (2005) found that mild acid hydrolysis[24] converted about 24% of amide-N into amine-N via deacetylation hydrolysis. They suggest that most of the amide in marine organic nitrogen is acetylated amino sugars.

Peptidoglycan (Fig. 7.18, lower center)—the primary cell wall polymer in bacteria—is cross-linked by tetrapeptide segments besides containing acetylated amino sugars. Peptide

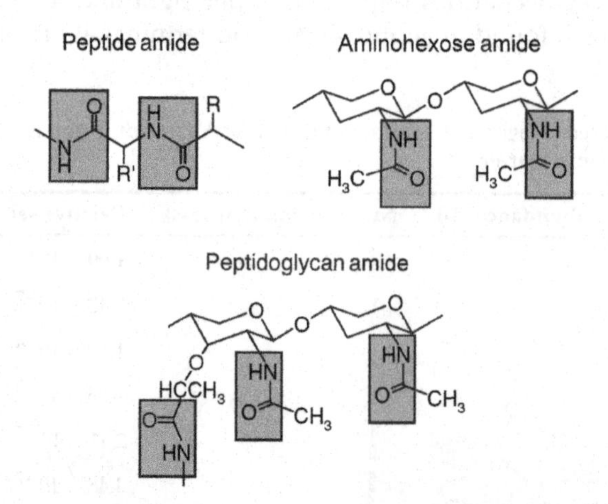

FIG. 7.18 Amide moieties in peptides, glycoproteins (aminohexose), and peptidoglycan.

[24] 10-h reflux at 90°C in 1 M HCl.

hydrolysis releases free amino acids, while deacetylation releases acetic acid from chitin and lactic acid from peptidoglycan; both hydrolysis reactions convert amide-N to amine-N.

Although the significance of amide-N as the major form of humic nitrogen remains a topic of scientific discussion, it strongly suggests that microbial cell wall residue is a prominent secondary source of organic matter. The assimilation of plant carbon (and nitrogen) as a soil microbial biomass is consistent with all major soil carbon turnover model (cf. Figs. 7.11–7.13) and represents an evolving understanding of the respective roles of plant and microbial biomass in the formation of organic matter.

Natural abundance nitrogen-15 NMR clearly identifies two nitrogen-containing chemical functional groups in organic matter: amine and amide. The chemical significance is two-fold. First, amide functional groups have low affinity for bonding to trace metals, making amine functional groups the only significant nitrogen-containing trace-metal bonding site in organic matter (Example 7.3). Second, amide functional groups occur in peptides and amino sugars (Fig. 7.18). Peptide amides are labile to both exocellular soil protease enzymes and the *IHSS* extraction protocol, suggesting acetylated amino sugars may explain the slow rate of organic nitrogen turnover: soil organic nitrogen half-life on the order of 50 years (cf. Fig. 7.9).

EXAMPLE 7.3

Example Permalink

http://soilenvirochem.net/XlChzb

Estimate the potential metal-binding capacity of Elliot soil organic matter at the location where the *IHSS* specimen was collected: Joliet Army Ammunition Plant, Will County, IL, USA. Table 7.10 lists the required data to compute the metal-binding capacity of organic matter. The specimen in this example is the same specimen as in Example 7.2: Elliot soil "humic acid," *IHSS* specimen 1S102H.

TABLE 7.10 The Carbon Mass-Fraction $w(C)$, Nitrogen Mass-Fraction $w(N)$, and Total Titratable Acidity of *IHSS* Specimen 1S102H (Elliot Soil "Humic Acid" Standard)

w(C) (g_C, g^{-1})	w(N) (g_N, g^{-1})	"Carboxyl" groups ($mmol_c$, g_C^{-1})	"Phenol" groups ($mmol_c$ g_C^{-1})
0.5813	0.0414	8.28	1.87

Part 1. Estimate the terminal amine content of soil organic matter using the organic nitrogen mass fraction and the N-15 NMR results in Fig. 7.18.

Assume 15% of the organic nitrogen in the Elliot soil is terminal amine and the remaining 85% is amide. The molar amine content is the function of the organic nitrogen mass fraction (Table 7.10, column 2) and the mole fraction of total organic nitrogen as amine divided by the atomic mass of nitrogen: $\overline{m}_a(N) = 14.00\,g\,mol^{-1}$.

Amine-binding sites

$$\frac{\left(0.0414\ g_N\ g_{SOM}^{-1}\right) \cdot \left(0.15\ mol\ mol^{-1}\right)}{\left(14.00\ g\ mol^{-1}\right) \cdot \left(0.5813\ g_C\ g_{SOM}^{-1}\right)} = 7.63 \times 10^{-4}\ mol_{amine}\ g_C^{-1}$$

Part 2. Estimate the combined molar metal-binding capacity of the soil organic-matter specimen.

The soil pH, measured using a 1-to-2 soil-to-solution (0.01 M $CaCl_2(aq)$) mass ratio, is: pH 5.9. Amine nitrogen and carboxyl in soil organic matter are potential metal-binding sites at pH 5.9; phenol functional groups hydrolyze in a pH range higher than found in most soils.

The molar "carboxyl" content in Table 7.10 is based on the carbon content, the same units as the "amine" content estimated in Part 1. The soil organic matter metal-binding capacity is simply the sum of the two functional group populations.

Total binding-sites

$$\left(8.28 \times 10^{-3}\ \mathrm{mol_c}\ \mathrm{g_C^{-1}}\right) + \left(7.63 \times 10^{-4}\ \mathrm{mol_{amine}}\ \mathrm{g_C^{-1}}\right) = 9.04 \times 10^{-3}\ \mathrm{mol_{sites}}\ \mathrm{g_C^{-1}}$$

The Elliot soil in Will County, IL, USA, contains 4.15% organic carbon.[25] Multiplying the number of sites per unit mass of organic carbon by the soil organic carbon content yields the total binding sites, however this value is difficult to evaluate.

Elliot soil metal-binding sites

$$\left(9.04 \times 10^{-3}\ \mathrm{mol_{sites}}\ \mathrm{g_C^{-1}}\right) \cdot \left(0.0415\ \mathrm{g_C}\ \mathrm{g_{soil}^{-1}}\right) = 3.75 \times 10^{-4}\ \mathrm{mol_{sites}}\ \mathrm{g_{soil}^{-1}}$$

Consider the above result in a different light: multiply the moles of binding sites per gram of soil by the atomic mass of copper $\overline{m}_a(\mathrm{Cu}) = 63.55\ \mathrm{g\,mol^{-1}}$. The mean copper content in Histosols is significantly higher than its mean content in mineral soils, unlike other elements, because the Cu^{2+} cation has a particular affinity for the amine functional group in soil and peat organic matter (Helmke, 2000).

Elliot soil Cu^{2+} binding-capacity

$$\left(3.75 \times 10^{-4}\ \mathrm{mol_{sites\,g_{soil}^{-1}}}\right) \cdot \left(63.55\ \mathrm{g\,mol^{-1}}\right) = 2.38 \times 10^{-2}\ \mathrm{g\,g_{soil}^{-1}}$$

Organic matter in the Elliot soil has the capacity to bind 24,000 $\mathrm{mg_{Cu^{2+}}\,kg^{-1}}$. It is a simple matter to compare the soil Cu^{2+} binding-capacity with the mean Cu content of soils worldwide (Helmke, 2000): 20 $\mathrm{mg_{Cu}\,kg^{-1}}$.

This result is typical: the metal binding content for transition and posttransition metal cations is about 1000-fold higher than the *natural abundance* of these elements.

[25] National Cooperative Soil Survey Pedon ID 1948IL197004, 0–18 cm.

7.4.3 Phosphorus Functional Groups

Phosphorus-31 NMR results (e.g., Fig. 7.19) indicate phosphorus in organic-matter specimens takes the form of phosphate mono- and diesters (Fig. 7.20). The cluster of resonance peaks in the range +3.5 ppm to +5.2 ppm is assigned to various monoesters, while the broad peak in the range −0.5 to −0.9 ppm is assigned to diesters. The resonance peaks at +5.6 ppm and −5.0 ppm are orthophosphate and pyrophosphate, respectively. Both inorganic phosphate species would be absent if the sample were thoroughly dialyzed to remove all small ions.

The final step in the *IHSS* base extraction protocol relies on dialysis or reverse osmosis to remove the excess salt that accumulates during the various extraction treatments. Dialysis confines dissolved and dispersed organic matter, allowing small ions to diffuse through the pores of the dialysis membrane (cf. Fig. 8.9). An indeterminate amount of the phosphorus in the initial extract is lost during dialysis as orthophosphate $PO_4^{3-}(aq)$ anion. Some of orthophosphate in the dialysate derives from mineral colloids dispersed and ultimately dissolved during the extraction procedure, the remaining orthophosphate derives from the hydrolysis

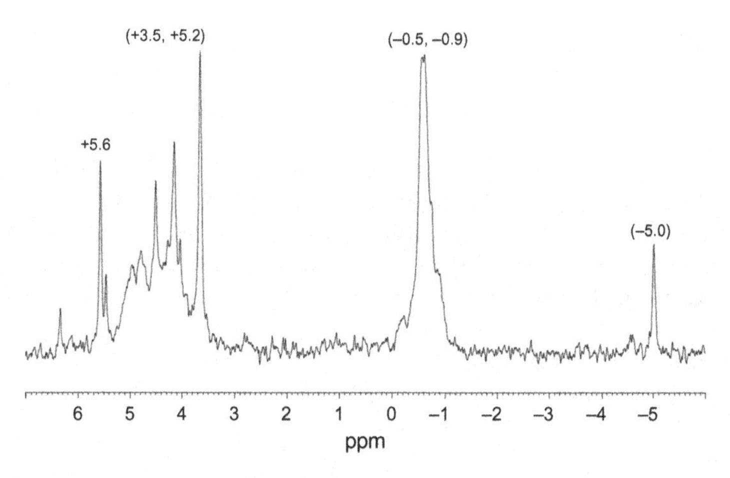

FIG. 7.19 Phosphorus-31 NMR spectra of a soil "humic acid" specimen collected from the Bankhead National Forest (Alabama). Chemical shifts are relative to the single P-31 resonance in phosphoric acid $H_3PO_4(aq)$, an external reference. *Courtesy of R.W. Taylor and T. Ranatunga.*

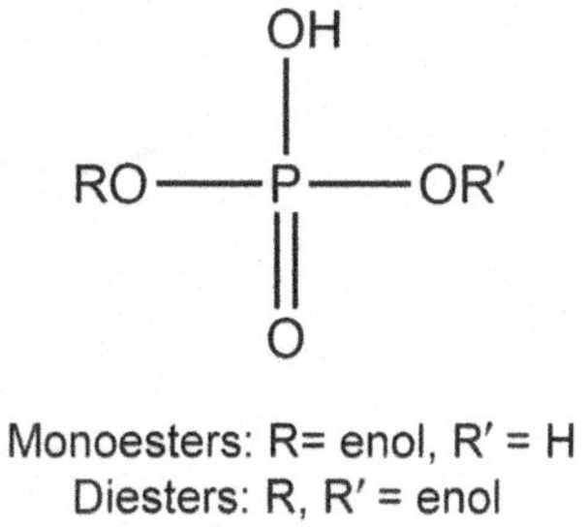

OH

RO—P—OR′

O

Monoesters: R= enol, R′ = H
Diesters: R, R′ = enol

FIG. 7.20 Phosphate esters for when one or more orthophosphate PO_4^{3-} oxygen binds to enol functional groups (R–OH). The monoester is attached to a single enol group while a diester is attached to two enol groups.

of phosphate mono- and diesters. Some chemists studying organic matter phosphorus modify their extraction protocol to specifically minimize phosphate-ester hydrolysis.

It is unclear whether the phosphate esters detected by P-31 NMR (Fig. 7.19) derive from microbial biomass and dispersed plant residue particles or from esters that form during carbon turnover.

Natural abundance phosphorus-31 NMR identifies phosphate mono- and diesters as the dominant phosphorus moiety in organic matter. Phosphate esters (Fig. 7.20) represent a significant source of phosphorus fertility in natural ecosystems. Exocellular phosphatase enzymes secreted by plant roots, soil bacteria, and soil fungi (Dodd et al., 1987; Kim et al., 1998) hydrolyze phosphate ester bonds, releasing orthophosphate ions into soil solution for plant and microbial uptake.

7.4.4 Sulfur Functional Groups

Our understanding of organosulfur in organic matter draws heavily on X-ray absorption spectroscopy. In many respects the level of our understanding of organosulfur is comparable to our understanding of organonitrogen, despite differences in chemistry and spectroscopic method.

Sulfur absorbs X-ray photons in a relatively narrow energy range greater than 2472.0 eV, the sulfur K X-ray absorption edge (or K-edge). X-ray absorption by sulfur at the K-edge excites an electron in the lowest energy atomic orbital—an electron in a 1s atomic orbital—to one of the lowest unoccupied molecular orbitals (LUMOs). The LUMO energy distribution mirrors the bonding interactions that determine the energy of each unoccupied molecular orbital (cf. Fig. 7.8).

Fig. 7.21 displays the normalized X-ray absorption spectrum for thiol (R–SH) sulfur in the amino acid L-cysteine. Peaks and oscillations near the edge—the X-ray absorption

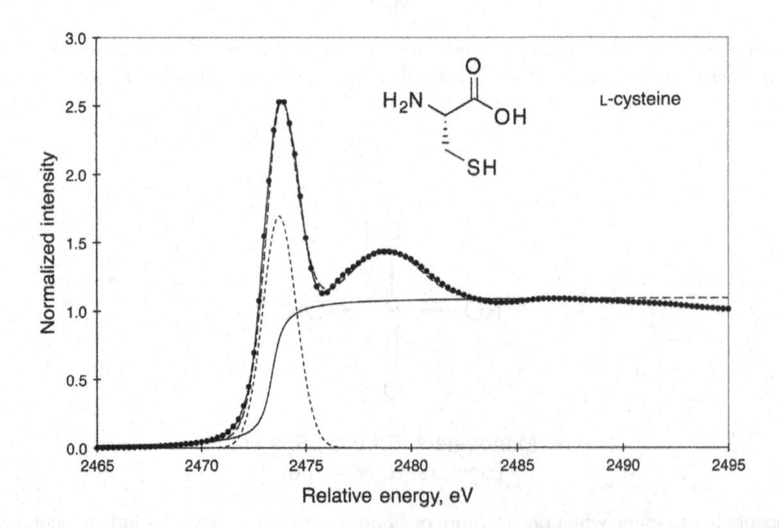

FIG. 7.21 Normalized sulfur K-edge X-ray absorption spectrum of the thiol (R–SH) sulfur in L-cysteine shows negligible X-ray absorption below the absorption edge (2472.5 eV) but considerable absorption above the edge.

near-edge structure (XANES) (or near-edge X-ray absorption fine structure (NEXAFS))—obscure the precise position of the X-ray absorption edge. Two key components of the sulfur K-edge XANES spectrum in Fig. 7.21 are the absorption edge (fitted to an arctangent step function plotted as a *solid line*) and the *white-line* peak (fitted to a Gaussian function plotted as a *dashed line*) centered on the absorption edge.

The position of the edge step and both the position and intensity of the *white-line* peak are sensitive to the effective sulfur oxidation state. By comparing the sulfur K-edge XANES spectra of natural specimens with the XANES spectra of known organosulfur compounds, we can infer the presence of different sulfur oxidation states and their relative abundance in the organic-matter specimens. The peat "fulvic acid" specimen shown in Fig. 7.22 contains what appear to be six (6) distinct sulfur oxidation states. The estimated relative abundance of each sulfur oxidation state and its chemical name appear in Table 7.11.

The effective sulfur oxidation states listed in Fig. 7.22 do not necessarily match the formal oxidation states of representative organosulfur species listed in Table 7.11. The sulfur in disulfide (R–SS–R′) (Fig. 7.23, upper left), thioester (Fig. 7.23, upper right), and thioether (R–S–R′) functional groups all have a formal oxidation state OS(S) = 0, assuming that the difference in Pauling electronegativity for carbon $\chi_C = 3.55$ and sulfur $\chi_S = 3.58$ is negligible.

Thiol (R–SH) has a formal oxidation state of OS(S) = −1, a result of the difference in Pauling electronegativity for hydrogen $\chi_H = 2.20$ and sulfur $\chi_S = 3.58$. The formal oxidation state OS(S) = +2 assigned to thiosulfate (Fig. 7.23, center left) is the average of two sulfur atoms with different oxidation states. The sulfone sulfur (Fig. 7.23, center right) is bonded to two oxygen atoms and two carbon atoms, resulting in a formal oxidation state of OS(S) = +4. Sulfonate sulfur (Fig. 7.23, lower left) is identical to the OS(S) = +5 sulfur in thiosulfate, while sulfate esters (Fig. 7.23, lower right) register a formal oxidation state of: OS(S) = +6.

Chemical heterogeneity prevents scientists from distinguishing subtle variations in the chemical bonding of each sulfur oxidation state in organic matter. Sulfur K-edge XANES

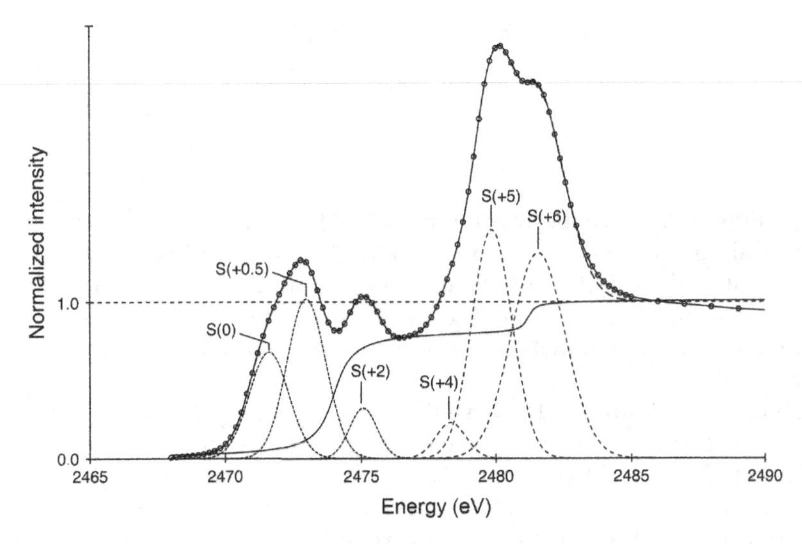

FIG. 7.22 Normalized sulfur K-edge X-ray absorption spectra of sulfur in a peat "fulvic acid" specimen (Loxely peat, Marcell Experimental Forest, Grand Rapids, MN). *Sample courtesy of P.R. Bloom.*

TABLE 7.11 Formal Oxidation States of Organosulfur Species and Their Relative Abundance in the Peat "Fulvic Acid" Specimen Appearing in Fig. 7.22

OS(S)	Relative *white-line* position (eV)	Relative abundance (%)
0.0	−0.4	24
$+\frac{1}{2}$	+1.0	24
+2	+3.5	27
+4	+5.5	5
+5	+8.5	3
+6	+10.5	24

White-line position is quoted relative to the sulfur K-edge at 2472.5 eV.

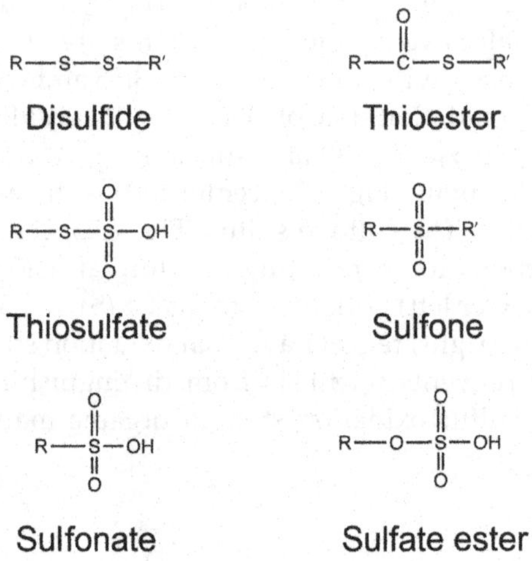

Disulfide

Thioester

Thiosulfate

Sulfone

Sulfonate

Sulfate ester

FIG. 7.23 Representative organosulfur functional groups corresponding to the sulfur oxidation states OS(S) appearing in Table 7.11.

distinguishes little difference between thiol (R–SH), disulfide (R–SS–R′), and thioether (R–S–R′) when all are present in a complex natural sample. Sulfur K-edge XANES allows chemists to quantify the effective oxidation states of sulfur in organic matter and to infer the identity of organosulfur functional groups. What little we know about the chemical bonding of organosulfur functional groups in organic matter requires supplemental chemical information.

Natural abundance sulfur K-edge XANES identifies multiple sulfur chemical functional groups in organic matter representing a considerable range of formal oxidation states (cf. Fig. 7.22). Exocellular soil sulfatase enzymes hydrolyze sulfate esters in a process resembling phosphate release by phosphatase enzymes in soil. Chemists have no explanation for the occurrence of sulfur functional groups registering oxidation states in the range

$+2 < OS(S) < +6$ because these intermediate forms do not correspond to the accepted mechanism for dissimilatory sulfate reduction (cf. Chapter 9).

Reduced (i.e., $OS(S) \leq 0$) organosulfur functional groups with have two-fold importance: reducing agents and functional groups with a binding affinity for *chalcophilic* (sulfur-loving) trace metals. Szulczewski et al. (2001) found that organosulfur species in soil organic matter could reduce chromate CrO_4^{2+} to Cr^{3+}. Several scientists studying the binding of methylmercury CH_3Hg^+ by organic matter observed the formation of Hg–S bonds that originate from organic matter thiol groups (Amirbahman et al., 2002; Qian et al., 2002; Yoon et al., 2005). Roughly one-third of the reduced sulfur in the organic-matter specimens appears to be capable of binding methylmercury CH_3Hg^+ cations. The remaining, unreactive reduced sulfur is most likely thioether, disulfide, and thioester functional groups.

7.4.5 Carbon Moieties and Functional Groups

7.4.5.1 Carbon-13 and Hydrogen-1 NMR Spectroscopy

The assignment of proton or H-1 NMR spectra derives largely from the carbon to which the protons are bonded, providing essentially the same chemical information found in C-13 NMR spectra. The most practical distinction between H-1 NMR spectra and C-13 NMR spectra of organic matter is found in Table 7.9; natural abundance C-13 NMR spectroscopy is roughly 10,000-fold less sensitive than H-1 NMR.

The sensitivity advantage enjoyed by H-1 NMR enhances a particular C-13 NMR experiment: cross-polarization $^1H \rightarrow {}^{13}C$ solid-state NMR (Pines et al., 1973; Schaefer and Stejskal, 1976). Hydrogen-1 nuclei—rather than carbon-13 nuclei—absorb radiation in a cross-polarization $^1H \rightarrow {}^{13}C$ solid-state NMR experiment and are then coaxed to transfer their magnetic alignment to a dilute population of C-13 nuclei before the spectrum is recorded. Technical issues beyond the scope of this book make H-1 NMR the method of choice if the organic-matter specimen is dissolved, while cross-polarization $^1H \rightarrow {}^{13}C$ solid-state NMR is the method of choice for solid organic-matter specimens.

Regardless of the method, H-1 and C-13 NMR spectra of organic matter contain resonance peaks assigned to the following carbon moieties: aromatic, aliphatic, enols (alcohols or polysaccharides), and ketones (including carboxylic groups). Keeping in mind that these assignments pertain to the primary structure of organic-matter molecules, NMR spectroscopy has yielded little reliable information on the secondary structure of organic-matter molecules. Of course, absent a meaningful primary structure it is pointless to speculate on the secondary molecular structure. In simple terms, we know the type of carbon moieties found in organic matter but are uncertain of their precise relative abundance[26] or how these simple structures combine to form an entire molecule. Since organic matter is a mixture composed of molecules, each with a different primary and secondary structure, it is doubtful we will ever be able

[26] NMR spectra carry little meaningful quantitative results because each different bonding environment relaxes at a different rate. Trace amounts of Fe^{3+} can cause considerable line broadening because its magnetic moment induces rapid relaxation of nearby H-1 and C-13 nuclei.

to draw the structure of representative organic-matter molecule whose tertiary structure and molecular properties match the collective properties of a natural organic-matter specimen.

7.4.5.2 Carbon K-Edge X-Ray Absorption Spectroscopy

Some chemists have recently used X-ray absorption spectroscopy at the carbon K-edge to collect chemical data from organic matter in situ without resorting to conventional strong-base extraction. Appendix 7.A provides a brief review of recent studies reporting carbon K-edge X-ray absorption spectra of natural organic matter. The review identifies and evaluate each major limitation to carbon K-edge X-ray absorption spectroscopy of natural organic matter.

Appendix 7.A concludes that carbon K-edge NEXAFS spectra (cf. Fig. 7.1) supply comparable chemical information content as H-1 or C-13 NMR spectra. Chemical functional groups, identified by carbon K-edge NEXAFS spectroscopy (Solomon et al., 2005) will not reveal chemical functionality that chemists have not already collected by other methods.

7.4.6 Colloidal Properties

Early chemists studying natural organic matter (Oden, 1919) were probably more sensitive to the colloidal behavior of natural organic matter than most present-day chemists, as evidenced by the relative emphasis on molecular rather than colloidal characteristics. When colloid chemist are presented with a colloidal dispersion, such as a strong-base organic-matter extract, the initial evaluation would explore the effect of pH and electrolytes on the stability of the colloidal dispersion.

Acidifying a strong-base organic-matter extract does, in fact, trigger the flocculation of a substantial fraction of the extract (Oden, 1919). Following Oden (1919), the flocculated product (identified as "humic acids") is the colloidal component of the extract. The acid-soluble, "yellow acids" (also identified as "fulvic acids") can be considered the molecular component with weak or nonexistent colloidal properties.

Simpson (2002) used solution-state ^{1}H NMR to estimate the molecular weight of "humic" and "fulvic acids" specimens and reports the following.

> Diffusion studies of ["fulvic acid" specimens] indicate that little if any aggregation occurs in solution and that, on average, the components display diffusivities that are consistent with molecules of relatively low molecular weights of [about] 1000 Da . . . In addition to a range of small molecular components, . . . larger molecular weight components exist within the mixtures.

Colloidal organic matter dispersed in water exhibits amphiphilic behavior similar to microbial bioamphiphiles (cf. Section 7.2.3.6. The colloidal behavior of natural organic matter is deferred to Chapter 8. There we will present evidence for the molecular-association and colloidal behavior of natural organic matter.

7.5 SUMMARY

Natural organic matter is a component of the terrestrial carbon cycle, representing both an important environmental substance and a vital component of the biological energy cycle. Photolithotrophs convert light energy into chemical energy during photosynthesis, storing this chemical energy in a host of bioorganic compounds comprising living biomass. The reduced carbon in bioorganic compounds is kinetically stable in the presence of ground state (triplet) O_2, a consequence of the peculiar electron configuration of the O_2 molecule. The release of the chemical energy stored in natural organic matter requires enzymatic catalysis, firmly placing carbon turnover in the hands of a complex community of soil organisms.

The mathematical models quantifying organic carbon (and organic nitrogen) turnover have changed little in the past 70 years. The decomposition of soil organic carbon appears to follow a first-order rate law—a rate law familiar from general chemistry. Ecologists studying carbon turnover distinguish two or three carbon pools that turn over at different rates.

A substantial fraction of plant residue has a residence time of less than a year but contains many chemically and biologically active compounds. Though much of the dissolved organic carbon in pore water cannot be identified, environmental chemists have identified organic acids, free amino acids, active exocellular enzymes, siderophores, and biosurfactants (to say nothing of vitamins, growth factors, DNA fragments, etc.).

With the exception of the biologically active components mentioned in the previous paragraph, organic matter is a colloidal substance intimately associated with mineral colloids. Environmental chemists routinely extract organic matter for chemical analysis and characterization. The stoichiometric composition of strong-base extracts is remarkably invariant. Unfortunately, we know little about the primary or secondary structure of organic-matter molecules beyond the identity of chemical moieties formed by the principle elements: carbon, oxygen, nitrogen, phosphorus, and sulfur. Our most comprehensive knowledge relates to molecular and colloidal properties rather than molecular structure.

Organic matter has a substantial capacity to adsorb ionic species from solution by ion exchange and complexation of metal ions to specific chemical moieties, a chemical property similar to mineral colloids in its effect on solution chemistry but arising from profoundly different chemical mechanisms. The affinity of organic matter for nonpolar organic compounds is a colloidal property that has no parallel in mineral colloids (cf. Chapter 8). Organic matter in situ should be considered self-organized association of amphiphilic molecules.

The stoichiometric analysis (cf. Section 7.2.2) demonstrates a key characteristic of natural organic matter: stoichiometric composition makes natural organic matter the optimal substrate for microbial growth because $\overline{C:N}_{OM} \geq \overline{C:N}_{B}$ and $\overline{\Gamma}_{OM} \approx \overline{\Gamma}_{B}$. This conclusion raises a profound question that our current understanding cannot resolve: what prevents microbes from completely mineralizing natural organic matter? If natural organic matter is not chemically refractory, and there is no evidence that it is (Lehmann and Kleber, 2015), then perhaps a biological or ecological imperative limits organic matter mineralization (cf. Von Stockar and Liu, 1999).

APPENDICES

7.A LIMITATIONS OF CARBON K-EDGE NEXAFS AND STXM

This appendix is a critical review of recent carbon K-edge X-ray absorption studies of *in situ* natural organic matter. While *in situ* X-ray adsorption spectroscopy eliminates the need for chemical extraction, scientists studying natural organic matter must weight this particular advantage against limitations that apply to all forms of X-ray absorption spectroscopy in addition to specific limitations encountered when relying on soft X-ray[27] absorption spectroscopy.

First, consider limitations imposed by soft X-rays in the carbon K-edge vicinity. X-ray attenuation by dry air at sea level reduce X-ray intensity by 95% after traveling a mere 68.7 cm. Much of the early carbon K-edge research required a ultra-high vacuum between the X-ray source and the detector. Even then, X-ray absorption by trace levels of carbon that accumulate on the monochrometer over time contaminate spectra (Jokic et al., 2003).

Boese et al. (1997) describes an alternative X-ray absorption configuration; the synchrotron source produces a high X-ray flux which passes through a helium-filled chamber enclosing a zone plate that collects and focuses X-rays and then enters a second helium-filled chamber where it passes through a 0.2 μm thick sample film before entering the X-ray detector. Silicon nitride windows allow X-rays in the vicinity of the carbon K-edge to pass through the monochrometer and sample chambers without undue absorption. The sample absorption coefficient $\mu(E)$, however, is very sensitive to sample-film uniformity and thickness (Boese et al., 1997).

Second, the fine structure resolution of any NEXAFS spectrum is constrained by the resolution of the X-ray monochrometer. The resolution of the monochrometer used to collect the carbon K-edge NEXAFS in Fig. 7.A.1 is $\Delta E = 0.3\,\text{eV}$ while the *full-width at half-maximum FWHM* for the aromatic-carbon component is $FWHM = 0.45\,\text{eV}$, slightly greater than the monochromater energy resolution.

Third, the chemical resolution of a NEXAFS spectrum is also constrained by the chemical heterogeneity of the specimen. The monochrometer energy resolution ΔE quoted earlier is broader than the energy difference between individual antibonding orbitals in many simple organic compounds. Compare the carbon K-edge NEXAFS spectra of the amino acid glycine and human serum fibrinogen (Gordon et al., 2003). Glycine, at 5%, is the *seventh* most abundant amino acid in human serum fibrinogen (Triantaphyllopoulos and Triantaphyllopoulos, 1967) yet the carbon K-edge NEXAFS spectra of the amino acid glycine and the protein fibrinogen are nearly indistinguishable.

Scanning transmission X-ray microscopy (STXM) provides the opportunity to collect microscopic images of soil organic carbon in situ, however this also comes at a steep cost: the resolution of individual image elements is significantly poorer than spectra from films prepared using organic-matter extracts (Heymann et al., 2014).

7.B SPIN CONSERVATION AND SPIN FORBIDDEN REACTIONS

The conservation of orbital symmetry and electron-spin determine the pathway followed when educts combine to form products. The *Woodward-Hoffman Rules* (Woodward and Hoffmann, 1965) describes the conservation of orbital symmetry on reaction pathway: any reaction following a concerted pathway (i.e., existing bonds break as incipient bonds form) demands

[27] Physicists define X-rays as photons with energies in the range: $1\,\text{keV} \le E \le 100\,\text{keV}$. The carbon X-ray absorption K-edge = 0.282 keV, well below the 10 keV threshold for soft X-rays.

Solomon et al. (2005) Soil Sci. Soc. Am. J., 69, 107

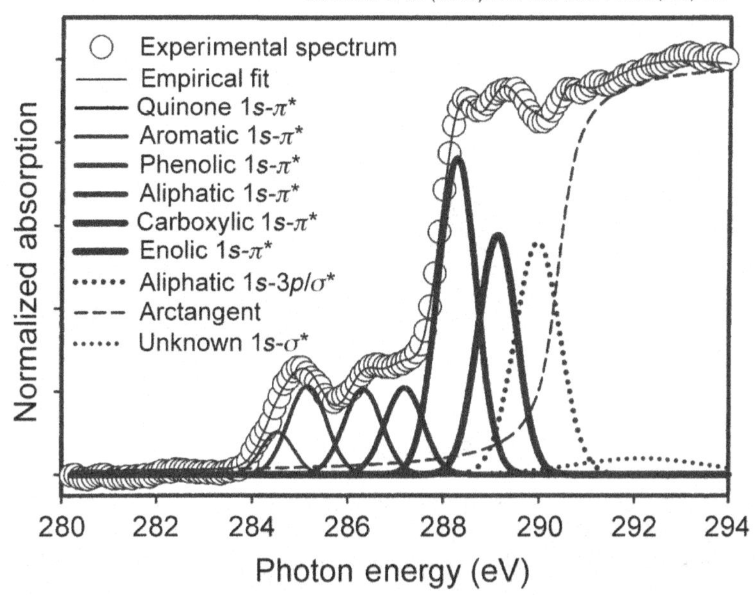

FIG. 7.A.1 Typical C K-edge NEXAFS spectra deconvolution showing the main $1s-\pi^*$ transitions, two $1s-\sigma^*$ transitions and the arctangent step function. The specimen is organic matter extracted from soil collected near Munesa, Ethopia (7.583°N, 38.750°E). *Solomon, D., Lehmann, J., Kinyangi, J., Liang, B., Schäfer, T., 2005. Carbon K-edge NEXAFS and FTIR-ATR spectroscopic investigation of organic carbon speciation in soils. Soil Sci. Soc. Am. J. 69 (1), 107–119.*

that all occupied molecular orbitals in the reactant must correlate with occupied molecular orbitals of the *same symmetry* in the product. The reaction is *symmetry forbidden* if orbital symmetry is not conserved; a direct low-energy pathway is unavailable to the reaction. To identify a reaction as *symmetry forbidden* does not mean the reaction cannot take place at all, instead the reaction must follow a *different pathway* with *different symmetry*. The alternative, higher-energy pathway follows a different orbital-symmetry correlation pattern from reactants to products.

Spin conservation operates in much the same way as orbital-symmetry conservation. In fact, orbital-symmetry conservation may *allow* a reaction pathway but the pathway is *spin forbidden* because the electron spin-states of the reactants do not correlate with the spin-states of the product(s). The direct oxidation of most organic molecules—which are singlet in the ground-state—by ground-state triplet O_2 (cf. Box 7.4) is *spin forbidden* because the electron-spin configuration in most organic molecules and their oxidation products, regardless of whether oxidation is complete (yielding CO_2 and H_2O) or partial (yielding somewhat more oxidized organic products), is singlet in the ground-state.

The following quote from Swart and Costas (2016) explains the activation barrier imposed on *spin-forbidden* reactions that results in unusually slow reaction kinetics.

> ... for light atoms the change in the spin state is accompanied by a large barrier (...an example being the slow reaction of organic molecules with triplet oxygen, as compared with the fast spin-allowed reaction with singlet oxygen) ...

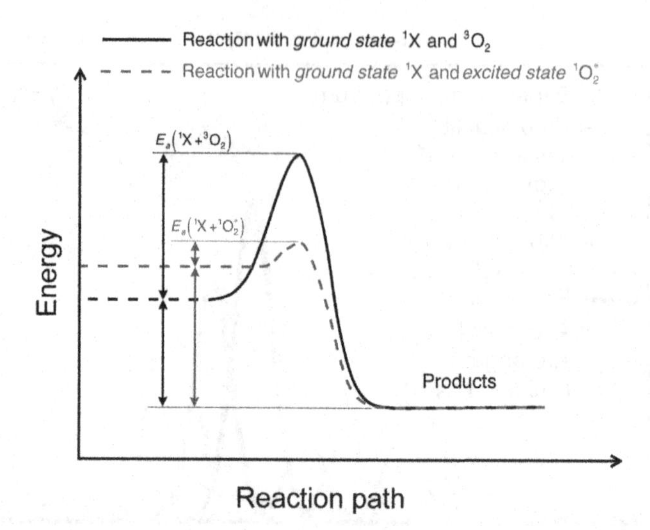

FIG. 7.B.1 Activation energy barriers for spin-forbidden oxidation of singlet organic molecules ^1X by triple dioxygen $^3O_2$ and spin-allowed oxidation of singlet organic molecules ^1X by excited-state singlet dioxygen $^1O_2^*$. Spin conservation allows ^1X and $^1O_2^*$ to follow a lower-energy reaction pathway than the pathway available to ^1X and $^3O_2$.

When O_2 reacts with an organic molecule, the organic molecule transfers two electrons to O_2. Since the two electrons are spin paired in the electron donor (i.e., the organic molecule), one electron spin must invert before it can pair with the unpaired electron already occupying a triplet O_2 molecular orbital. Alternatively, ground-state O_2 can absorb energy to enter an excited O_2 singlet-state (Fig. 7.B.1). The large activation energy required to flip the electron spin accounts for the kinetic sluggishness of the direct reaction between ground-state O_2 and most organic molecules.

References

Achard, F.K., 1786. Chemische untersuchung des torfs. Crell's Chem. Ann. 2, 391–403.

Aluwihare, L.I., Repeta, D.J., Pantoja, S., Johnson, C.G., 2005. Two chemically distinct pools of organic nitrogen accumulate in the ocean. Science 308 (5724), 1007–1010.

Amirbahman, A., Reid, A.L., Haines, T.A., Kahl, J.S., Arnold, C., 2002. Association of methylmercury with dissolved humic acids. Environ. Sci. Technol. 36 (4), 690–695.

Ammari, T., Mengel, K., 2006. Total soluble Fe in soil solutions of chemically different soils. Geoderma 136 (3–4), 876–885.

Bécard, G., Piché, Y., 1989. Fungal growth stimulation by CO_2 and root exudates in vesicular-arbuscular mycorrhizal symbiosis. Appl. Environ. Microbiol. 55 (9), 2320–2325.

Boese, J., Osanna, A., Jacobsen, C., Kirz, J., 1997. Carbon edge XANES spectroscopy of amino acids and peptides. J. Electron Spectrosc. Relat. Phenom. 85 (1), 9–15.

Bolker, B.M., Pacala, S.W., Parton Jr., W.J., 1998. Linear analysis of soil decomposition: insights from the century model. Ecol. Appl. 8 (2), 425–439.

Cameotra, S.S., Bollag, J.-M., 2003. Biosurfactant-enhanced bioremediation of polycyclic aromatic hydrocarbons. Crit. Rev. Environ. Sci. Technol. 33 (2), 111–126.

Chaney, R.L., Brown, J.C., Tiffin, L.O., 1972. Obligatory reduction of ferric chelates in iron uptake by soybeans. Plant Physiol. 50 (2), 208–213.
Chantigny, M.H., 2003. Dissolved and water-extractable organic matter in soils: a review on the influence of land use and management practices. Geoderma 113 (3–4), 357–380.
de Saussure, T., 1804. Recherches chimiques sur la végétation. Nyon, Paris.
Desai, J.D., Banat, I.M., 1997. Microbial production of surfactants and their commercial potential. Microbiol. Mol. Biol. Rev. 61 (1), 47–64.
Dodd, J.C., Burton, C.C., Burns, R.G., Jeffries, P., 1987. Phosphatase activity associated with the roots and the rhizosphere of plants infected with vesicular-arbuscular mycorrhizal fungi. New Phytol. 107 (1), 163–172.
Ferguson, A.D., Deisenhofer, J., 2002. TonB-dependent receptors—structural perspectives. Biochim. Biophys. Acta Biomembr. 1565 (2), 318–332.
Fuhrman, J., 2003. Genome sequences from the sea. Nature 424 (6952), 1001–1002.
Fuller, R.D., Simone, D.M., Driscoll, C.T., 1988. Forest clearcutting effects on trace metal concentrations: spatial patterns in soil solutions and streams. Water Air Soil Pollut. 40 (1–2), 185–195.
Gardner, W.K., Parbery, D.G., Barber, D.A., 1982. The acquisition of phosphorus by Lupinus albus L. I. Some characteristics of the soil/root interface. Plant Soil 68 (1), 19–32.
Gordon, M.L., Cooper, G., Morin, C., Araki, T., Turci, C.C., Kaznatcheev, K., Hitchcock, A.P., 2003. Inner-shell excitation spectroscopy of the peptide bond: comparison of the C 1s, N 1s, and O 1s spectra of glycine, glycyl-glycine, and glycyl-glycyl-glycine. J. Phys. Chem. A 107 (32), 6144–6159.
Grieve, I.C., 1990. Variations in chemical composition of the soil solution over a four-year period at an upland site in southwest Scotland. Geoderma 46 (4), 351–362.
Handelsman, J., 2005. Metagenomics: application of genomics to uncultured microorganisms. Microbiol. Mol. Biol. Rev. 69 (1), 195–195.
Hannam, K.D., Prescott, C.E., 2003. Soluble organic nitrogen in forests and adjacent clearcuts in British Columbia, Canada. Can. J. For. Res. 33 (9), 1709–1718.
Haynes, J.L., Thatcher, J.S., 1955. Crop rotation and soil nitrogen. Soil Sci. Soc. Am. Proc. 19, 324–327.
Heijnen, J.J., Van Dijken, J.P., 1992. In search of a thermodynamic description of biomass yields for the chemotrophic growth of microorganisms. Biotechnol. Bioeng. 39 (8), 833–858.
Helmke, P.A., 2000. The chemical composition of soils. In: Sumner, M.E. (Ed.), Handbook of Soil Science. CRC Press, London, pp. B3–B24.
Hénin, S., Dupuis, M., 1945. Essai de bilan de la matière organique du sol. Ann. Agron. 15 (1), 17–29.
Heymann, K., Lehmann, J., Solomon, D., Liang, B., Neves, E., Wirick, S., 2014. Can functional group composition of alkaline isolates from black carbon-rich soils be identified on a sub-100 nm scale?. Geoderma 235, 163–169.
Hoover, S.R., Porges, N., 1952. Assimilation of dairy wastes by activated sludge. II. The equation of synthesis and rate of oxygen utilization. Sewage Ind. Waste. 24, 306–312.
Hoppe-Seyler, F., 1889. Über Huminsubstanzen, ihre Entstehung und ihre Eigenschaften. Z. Physiol. Chem. 13, 66–121.
Ingham, E.R., Moldenke, A.R., Edwards, C.A., 2000. In: Tugel, A.J., Lewandowski, A.M., Happe-vonArb, D. (Eds.), Soil Biology Primer. Soil and Water Conservation Society, Ankeny, IA, http://www.nrcs.usda.gov/wps/portal/nrcs/main/soils/health/biology/.
Jenkinson, D.S., 1977. Studies on the decomposition of plant material in soil. V. The effects of plant cover and soil type on the loss of carbon from ^{14}C labelled ryegrass decomposing under field conditions. J. Soil Sci. 28 (3), 424–434.
Jenkinson, D.S., Andrew, S.P.S., et al., 1990. The turnover of organic carbon and nitrogen in soil. Philos. Trans. R. Soc. Lond. B Biol. Sci. 329 (1255), 361–368.
Jenkinson, D.S., Rayner, J.H., 1977. The turnover of soil organic matter in some of the Rothamsted classical experiments. Soil Sci. 123 (5), 298–305.
Jenkinson, D.S., Meredith, J., Kinyamario, J.I., Warren, G.P., Wong, M.T.F., Harkness, D.D., Bol, R., Coleman, K., 1999. Estimating net primary production from measurements made on soil organic matter. Ecology 80 (8), 2762–2773.
Jenny, H., 1941. Factors of Soil Formation: A System of Quantitative Pedology. McGraw-Hill, New York/London.
Jokic, A., Cutler, J.N., Ponomarenko, E., van der Kamp, G., Anderson, D.W., 2003. Organic carbon and sulphur compounds in wetland soils: insights on structure and transformation processes using K-edge XANES and NMR spectroscopy. Geochim. Cosmochim. Acta 67 (14), 2585–2597.

Jones, D.L., Owen, A.G., Farrar, J.F., 2002. Simple method to enable the high resolution determination of total free amino acids in soil solutions and soil extracts. Soil Biol. Biochem. 34 (12), 1893–1902.

Jones, D.L., Shannon, D., Murphy, D.V., Farrar, J., 2004. Role of dissolved organic nitrogen (DON) in soil N cycling in grassland soils. Soil Biol. Biochem. 36 (5), 749–756.

Kaiser, K., Guggenberger, G., 2003. Mineral surfaces and soil organic matter. Eur. J. Soil Sci. 54 (2), 219–236.

Katchalsky, A., Spitnik, P., 1947. Potentiometric titrations of polymethacrylic acid. J. Polym. Sci. 2, 432–446.

Kim, K.Y., Jordan, D., McDonald, G.A., 1998. Effect of phosphate-solubilizing bacteria and vesicular-arbuscular mycorrhizae on tomato growth and soil microbial activity. Biol. Fertil. Soils 26 (2), 79–87.

Kosman, D.J., 2003. Molecular mechanisms of iron uptake in fungi. Mol. Microbiol. 47 (5), 1185–1197.

Lambert, J.B., Binsch, G., Roberts, J.D., 1964. ^{15}N magnetic resonance spectroscopy. I. Chemical shifts. Proc. Natl. Acad. Sci. U. S. A. 51 (5), 735–737.

Lang, S., 2002. Biological amphiphiles (microbial biosurfactants). Curr. Opin. Colloid Interface Sci. 7 (1–2), 12–20.

Lehmann, J., Kleber, M., 2015. The contentious nature of soil organic matter. Nature (London, U. K.) 528 (7580), 60–68.

Lipscomb, J.D., Whittaker, J.W., Arciero, D.M., Orville, A.M., Wolgel, S.A., 1988. Mechanisms of catechol dioxygenases. Microbial Metabolism and the Carbon Cycle. Harwood Academic, New York, pp. 259–282.

Lochhead, A.G., Chase, F.E., 1943. Qualitative studies of soil microorganisms: V. Nutritional requirements of the predominant bacterial flora. Soil Sci. 55 (2), 185–196.

Martell, A.E., Smith, R.M., Motekaitis, R.J., 2004. NIST Critically Selected Stability Constants of Metal Complexes, NIST Standard Reference Database 46.6.0. United States Department of Commerce, Gaithersburg, MD.

Mikutta, R., Kleber, M., Torn, M.S., Jahn, R., 2006. Stabilization of soil organic matter: association with minerals or chemical recalcitrance?. Biogeochemistry 77 (1), 25–56.

Mouginot, C., Kawamura, R., Matulich, K.L., Berlemont, R., Allison, S.D., Amend, A.S., Martiny, A.C., 2014. Elemental stoichiometry of Fungi and Bacteria strains from grassland leaf litter. Soil Biol. Biochem. 76, 278–285.

Mulligan, C.N., 2005. Environmental applications for biosurfactants. Environ. Pollut. 133 (2), 183–198.

Myers, H.E., Hallsted, A.L., Kuska, J.B., Haas, H.J., 1943. Nitrogen and Carbon Changes in Soils Under Low Rainfall as Influenced by Cropping Systems and Soil Treatment No. 56. Kansas Agricultural Experiment Station, Manhattan, KS, 47 pp..

Oden, S., 1919. Die Huminsäuren. Chemische, physikalische und bodenkundliche Forschungen. Kolloidchem. Beih. 11, 75–260.

Parton, W.J., Schimel, D.S., Cole, C.V., Ojima, D.S., 1987. Analysis of factors controlling soil organic matter levels in Great Plains grasslands. Soil Sci. Soc. Am. J. 51 (5), 1173–1179.

Parton, W.J., Scurlock, J.M.O., Ojima, D.S., Gilmanov, T.G., Scholes, R.J., Schimel, D.S., Kirchner, T., Menaut, J.-C., Seastedt, T., Garcia Moya, E., et al., 1993. Observations and modeling of biomass and soil organic matter dynamics for the grassland biome worldwide. Global Biogeochem. Cycles 7 (4), 785–809.

Philpott, C.C., 2006. Iron uptake in fungi: a system for every source. Biochim. Biophys. Acta, Mol. Cell Res. 1763 (7), 636–645.

Pietramellara, G., Ascher, J., Ceccherini, M.T., Renella, G., 2002. Soil as a biological system. Ann. Microbiol. 52 (2), 119–132.

Pines, A., Gibby, M.G., Waugh, J.S., 1973. Proton-enhanced NMR of dilute spins in solids. J. Chem. Phys. 59 (2), 569–590.

Qian, J., Skyllberg, U., Frech, W., Bleam, W.F., Bloom, P.R., Petit, P.E., 2002. Bonding of methyl mercury to reduced sulfur groups in soil and stream organic matter as determined by X-ray absorption spectroscopy and binding affinity studies. Geochim. Cosmochim. Acta 66 (22), 3873–3885.

Quideau, S.A., Bockheim, J.G., 1997. Biogeochemical cycling following planting to red pine on a sandy prairie soil. J. Environ. Qual. 26 (4), 1167–1175.

Rice, J.A., MacCarthy, P., 1991. Statistical evaluation of the elemental composition of humic substances. Org. Geochem. 17 (5), 635–648.

Ritchie, J.D., Perdue, E.M., 2008. Analytical constraints on acidic functional groups in humic substances. Org. Geochem. 39 (6), 783–799.

Römheld, V., Marschner, H., 1981. Rhythmic iron stress reactions in sunflower at suboptimal iron supply. Physiol. Plant. 53 (3), 347–353.

Salter, R.M., Green, T.C., 1933. Factors affecting the accumulation and loss of nitrogen and organic carbon in cropped soils. J. Am. Soc. Agron. 25, 622–630.

Sawyer, C.N., 1956. Bacterial nutrition and synthesis. In: McCabe, J., Eckenfelder Jr., W.W. (Eds.), Biological Treatment of Sewage and Industrial Wastes, vol. 1. Reinhold, New York, pp. 3–17.

Schaefer, J., Stejskal, E.O., 1976. Carbon-13 nuclear magnetic resonance of polymers spinning at the magic angle. J. Am. Chem. Soc. 98 (4), 1031–1032.

Schaul, R.C., Duplay, J., Tardy, Y., 1997. Propriétés thermodynamiques, température et solubilité de composés organiques du sol. C. R. Acad. Sci. Ser. IIA Earth Planet. Sci. 325 (1), 27–33.

Schloss, P.D., Handelsman, J., 2007. The last word: books as a statistical metaphor for microbial communities. Annu. Rev. Microbiol. 61, 23–34.

Simpson, A.J., 2002. Determining the molecular weight, aggregation, structures and interactions of natural organic matter using diffusion ordered spectroscopy. Magn. Reson. Chem. 40 (13), S72–S82.

Sinsabaugh, R.L., Lauber, C.L., Weintraub, M.N., Ahmed, B., Allison, S.D., Crenshaw, C., Contosta, A.R., Cusack, D., Frey, S., Gallo, M.E., et al., 2008. Stoichiometry of soil enzyme activity at global scale. Ecol. Lett. 11 (11), 1252–1264.

Six, J., Callewaert, P., Lenders, S., De Gryze, S., Morris, S.J., Gregorich, E.G., Paul, E.A., Paustian, K., 2002. Measuring and understanding carbon storage in afforested soils by physical fractionation. Soil Sci. Soc. Am. J. 66 (6), 1981–1987.

Solomon, D., Lehmann, J., Kinyangi, J., Liang, B., Schäfer, T., 2005. Carbon K-edge NEXAFS and FTIR-ATR spectroscopic investigation of organic carbon speciation in soils. Soil Sci. Soc. Am. J. 69 (1), 107–119.

Speece, R.E., McCarty, P.L., 1964. Nutrient requirements and biological solids accumulation in anaerobic digestion. Adv. Water Pollut. Res. 2, 305–322.

Sun, L., Perdue, E.M., Meyer, J.L., Weis, J., 1997. Use of elemental composition to predict bioavailability of dissolved organic matter in a Georgia river. Limnol. Oceanogr. 42 (4), 714–721.

Swart, M., Costas, M., 2016. Spin States in Biochemistry and Inorganic Chemistry: Influence on Structure and Reactivity. John Wiley, Chichester, UK.

Symons, J.M., McKinney, R.E., 1958. The biochemistry of nitrogen in the synthesis of activated sludge. Sewage Ind. Waste. 30 (7), 874–890.

Szulczewski, M.D., Helmke, P.A., Bleam, W.F., 2001. XANES spectroscopy studies of Cr(VI) reduction by thiols in organosulfur compounds and humic substances. Environ. Sci. Technol. 35 (6), 1134–1141.

Thorn, K.A., Cox, L.G., 2009. N-15 NMR spectra of naturally abundant nitrogen in soil and aquatic natural organic matter samples of the International Humic Substances Society. Org. Geochem. 40 (4), 484–499.

Triantaphyllopoulos, E., Triantaphyllopoulos, D.C., 1967. Amino acid composition of human fibrinogen and anticoagulant derivatives. Biochem. J. 105 (1), 393–400.

Trumbore, S.E., Chadwick, O.A., Amundson, R., 1996. Rapid exchange between soil carbon and atmospheric carbon dioxide driven by temperature change. Science (Washington, D. C.) 272 (5260), 393–396.

Tyson, G.W., Chapman, J., Hugenholtz, P., Allen, E.E., Ram, R.J., Richardson, P.M., Solovyev, V.V., Rubin, E.M., Rokhsar, D.S., Banfield, J.F., 2004. Community structure and metabolism through reconstruction of microbial genomes from the environment. Nature 428 (6978), 37–43.

VanBriesen, J.M., 2002. Evaluation of methods to predict bacterial yield using thermodynamics. Biodegradation 13 (3), 171–190.

Venter, J.C., Remington, K., Heidelberg, J.F., Halpern, A.L., Rusch, D., Eisen, J.A., Wu, D., Paulsen, I., Nelson, K.E., Nelson, W., et al., 2004. Environmental genome shotgun sequencing of the Sargasso Sea. Science 304 (5667), 66–74.

Vodyanitskii, Y.N., 2000. Application of thermodynamic characteristics to the description of humus acids in soils. Eurasian Soil Sci. 33 (1), 43–48.

Vogel, T.M., Simonet, P., Jansson, J.K., Hirsch, P.R., Tiedje, J.M., van Elsas, J.D., Bailey, M.J., Nalin, R., Philippot, L., 2009. TerraGenome: a consortium for the sequencing of a soil metagenome. Nat. Rev. Microbiol. 7 (4), 252.

Von Stockar, U., Liu, J.-S., 1999. Does microbial life always feed on negative entropy? Thermodynamic analysis of microbial growth. Biochim. Biophys. Acta Bioenerg. 1412 (3), 191–211.

Waksman, S.A., 1936. Humus: Origin, Chemical Composition, and Importance in Nature. Williams & Wilkins, Baltimore.

Wallerius, J.G., 1761. Agriculturae Fundamenta Chemica. Doctoral Thesis, Upsaliae.

Wilkinson, F., Helman, W.P., Ross, A.B., 1995. Rate constants for the decay and reactions of the lowest electronically excited singlet state of molecular oxygen in solution. An expanded and revised compilation. J. Phys. Chem. Ref. Data 24 (2), 663–677.

Woodruff, C.M., 1950. Estimating the nitrogen delivery of soil from the organic matter determination as reflected by Sanborn Field. Soil Sci. Soc. Am. Proc. 14, 208–212.

Woodward, R.B., Hoffmann, R., 1965. Stereochemistry of electrocyclic reactions. J. Am. Chem. Soc. 87 (2), 395–397.

Wosten, H.A.B., 2001. Hydrophobins: multipurpose proteins. Annu. Rev. Microbiol. 55, 625–646.

Wosten, H.A.B., Van Wetter, M.A., Lugones, L.G., Van der Mei, H.C., Busscher, H.J., Wessels, J.G.H., 1999. How a fungus escapes the water to grow into the air. Curr. Biol. 9 (2), 85–88.

Yoon, S.-J., Diener, L.M., Bloom, P.R., Nater, E.A., Bleam, W.F., 2005. X-ray absorption studies of $CH_3 Hg^+$-binding sites in humic substances. Geochim. Cosmochim. Acta 69 (5), 1111–1121.

Zimmermann, M., Leifeld, J., Fuhrer, J., 2006. Quantifying soil organic carbon fractions by infrared-spectroscopy. Soil Biol. Biochem. 39 (1), 224–231.

CHAPTER

8

Surface Chemistry and Adsorption

OUTLINE

8.1 Introduction 385

8.2 Mineral and Organic Colloids as Environmental Adsorbents 386

8.3 Adsorption Isotherm Experiments 389
- *8.3.1 Area-Based Adsorption Isotherms* 389
- *8.3.2 Mass-Based Adsorption Isotherms* 392
- *8.3.3 Chemisorption and Physisorption* 392
- *8.3.4 Langmuir Isotherm Model* 394
- *8.3.5 Ion-Exchange Isotherm Model* 399
- *8.3.6 Partitioning Isotherm Model* 401
- *8.3.7 Freundlich Adsorption Isotherm* 411

8.4 pH-Dependent Surface Charge 413
- *8.4.1 Weak Acid-Conjugate Base Proton Adsorption Model* 414
- *8.4.2 Crystallographic Proton Adsorption Models* 415

8.5 The Adsorption Envelope Experiment 423
- *8.5.1 Adsorption Edges* 424
- *8.5.2 Interpreting Adsorption Envelopes* 426
- *8.5.3 The Structure of Adsorption Complexes* 429
- *8.5.4 Overview of Surface Complexation Models* 430

8.6 Summary 432

Appendices 434

8.A Hydrophilic Colloids 434

8.B Bond-Valence Model 436

8.C Valence-Bond Proton Adsorption Model: Goethite (100) Surface 436

8.D Using the PBT Profiler 439

References 440

8.1 INTRODUCTION

An earlier chapter (cf. Chapter 4) has already introduced one type of adsorption reaction; ion-exchange adsorption involves the exchange of one or more solution cations (or anions) with one or more cations (or anions) bound to charged sites on the surface of mineral particles or macromolecules. Hereafter the solute being adsorbed is referred to as the *adsorptive* and the material to which it binds as the *adsorbent*.

Soil and Environmental Chemistry
http://dx.doi.org/10.1016/B978-0-12-804178-9.00008-2

Ion exchange is but a special case of a more general class of adsorption reactions. Adsorption and adhesion share the same root, and both imply binding to a surface. An adsorption reaction involves the transfer of the adsorptive from the dissolved state to a bound state associated with an adsorbent. The adsorptive does not have to be an ion and adsorption does not require the displacement or exchange of a molecule or ion bound to the adsorbent. In fact, adsorption does not necessarily result in an *adsorption complex* confined the interface separating the solution and the adsorbent.

Ion-exchange adsorption is a reaction—enforced by the charge-neutrality condition—that confines *counter-ions* to the vicinity of oppositely charged surface sites.[1] Adsorption, being a more general chemical process, retains molecules or ions through the chemical bonding, hydrogen-bonding, and intermolecular (i.e., van der Waals[2]) forces besides *Coulomb* forces.

Protons, for instance, are bound to H_2O molecules and OH^- anions in solution and to adsorbents through covalent bonding. Protons also form *hydrogen bonds* between absorptive molecules and adsorbents, the absorptive and the adsorbent acting as a *hydrogen-bond donor-receptor pair* whose roles are interchangeable between absorptive and adsorbent.

As with hydrogen bonds, the absorptive and the adsorbent constitute an *electron-pair donor* (i.e., ligand or *Lewis base*) and an *electron-pair receptor* (i.e., metal ion or *Lewis acid*), whose roles are interchangeable between absorptive and adsorbent. Weak dipolar forces also bind absorptive molecules to adsorbents. Dipolar forces act in concert, becoming the dominant binding mechanism when the absorptive is a large neutral molecule.

8.2 MINERAL AND ORGANIC COLLOIDS AS ENVIRONMENTAL ADSORBENTS

Adsorption processes, though ubiquitous, become apparent only when the surface-to-volume ratio of the adsorbent is extremely high. This condition is commonly satisfied in environmental systems where clay-size mineral particles (Fig. 8.1) or natural organic matter are present in significant amounts.

Table 8.1 lists the size range and specific surface area for the three particle-size classes (sand, silt, and clay). Computational details appear in Example 8.1, and Table 8.2 lists the specific surface area of several soil textural classes.

Tables 8.1 and 8.2 demonstrate the importance of clay-size mineral particles to the specific surface area of surficial material: soils, sediments, and aquifers. A similar case could be made for organic matter that, in the context of this chapter, is considered the organic colloidal fraction in soils and sediments.

[1] In the case of cation-exchange adsorption by a smectite adsorbent, the smectite layers are negatively charged. The cations are *counter-ions* bound to negatively charged smectite layers by an electrostatic (i.e., *Coulomb*) potential energy E_{coul}. Solution anions in any cation-exchange adsorption reaction are identified as *co-ions*. Anions are counter-ions and cation are co-ions in anion-exchange adsorption reactions.

[2] The term *van der Waals* force is applied to three distinct electrostatic dipole-dipole forces: (1) the *Keeson* force between permanent molecular dipoles, (2) the *Debye* force between permanent molecular dipoles and induced dipoles, and (3) the *London dispersion* force between instantaneously induced dipoles.

FIG. 8.1 These electron micrographs depict fine grains of the minerals kaolinite $Al_4^{vi}Si_2^{iv}O_{10}(OH)_8(s)$ (*left*; Roe, 2016) and hematite $Fe_2O_3(s)$ (*right*; Contributors, 2008). *The images are courtesy of the Clay Minerals Group of the Mineralogical Society.*

TABLE 8.1 Specific Surface Area a_s of Spherical $SiO_2(s)$ Particles With Density $\rho_{SiO_2(s)} = 2.65$ Mg m^{-3}

Particle Class	Diameter (mm)	Specific Surface Area (m^2 g^{-1})
Coarse sand (CSa)	2.0000	1.13×10^{-3}
Medium sand (MSa)	0.6300	3.50×10^{-3}
Fine sand (FSa)	0.0200	1.13×10^{-2}
Coarse silt (CSi)	0.0630	3.59×10^{-2}
Medium silt (MSi)	0.0200	1.13×10^{-1}
Fine silt (FSi)	0.0063	3.59×10^{-1}
Clay (Cl)	0.0020	$1.13 \times 10^{+0}$

One further example demonstrates how solute adsorption by the clay-size fraction can significantly alter pore-water composition. Given a typical 25% volumetric water content of soils at field capacity and a typical bulk density of $\rho_{bulk} \approx 1.5$ Mg m^{-3}, a cubic meter of soil will hold about 0.250 m^3 of water.

$$\theta_{FC} \approx 0.25 \text{ m}^3 \text{ m}^{-3}$$

From Tables 8.1 and 8.2 we can see that the typical specific surface area[3] of soils and sediments is on the order of $a_s \approx 0.3$ m^2 g^{-1} (cf. column 5, Table 8.2). A cubic meter of soilwill

[3] This chapter uses the official *International Union of Pure and Applied Chemistry* colloid and surface chemistry symbols and terminology (Everett, 1972).

TABLE 8.2 Specific Surface Area of USDA Texture Classes (Grain Density: $\rho_{SiO_2(s)} = 2.65$ Mg m^{-3}) and Percentage of the Specific Surface Area a_s From the Clay-Size Fraction

Texture Class	Sand (%)	Silt (%)	Clay (%)	a_s (Total) (m^2 g^{-1})	a_s (Clay, %)
Clay	20	20	60	0.70	96.7
Loam	40	40	20	0.27	82.9
Silt	20	70	10	0.19	58.6
Sandy loam	70	20	10	0.14	81.8

have a dry mass of $m_{soil} \approx 1.5 \times 10^{+6}$ g with a total surface area of $A_s \approx 0.45 \times 10^{+6}$ m^2, mostly from the clay fraction (cf. column 6, Table 8.2).

$$A_s = a_s \cdot \rho_{bulk} \cdot V_{soil}$$

$$A_s = \left(0.3\ \mathrm{m^2\ g^{-1}}\right) \cdot \left(1.5 \times 10^{+6}\ \mathrm{g\ m^{-3}}\right) \cdot \left(1\ \mathrm{m^3}\right)$$

$$A_s = 0.45 \times 10^{+6}\ \mathrm{m^2}$$

Later in this chapter we will show the typical capacity of oxide minerals to adsorb a small ionic component i is on the order of $\Gamma_i \approx 10^{-6}$ mol m^{-2}. One cubic meter of soil can remove through adsorption onto mineral surfaces about 2×10^{-3} mol dm^{-3}, a substantial and readily measured change in solution concentration.

$$\Delta c_i = \frac{A_s \cdot \Gamma_i}{\theta_{FC}}$$

$$\Delta c_i = \frac{\left(0.45 \times 10^{+6}\ \mathrm{m^2}\right) \cdot \left(10^{-6}\ \mathrm{mol\ m^{-2}}\right)}{\left(0.25\ \mathrm{m^3}\right) \cdot \left(10^{+3}\ \mathrm{dm^3\ m^{-3}}\right)} = 1.8 \times 10^{-3}\ \mathrm{mol\ dm^{-3}}$$

The capacity of adsorption processes to alter pore-water solute concentrations Δc is a direct result of the extremely high specific surface area a_s of the soil clay fraction. To grasp these differences, we must first understand how an adsorption experiment is performed and the general appearance of the experimental results.

EXAMPLE 8.1

Example Permalink

http://soilenvirochem.net/ct9t1A

Compute the specific surface area a_s [m^2 g^{-1}] of mineral particles given their effective spherical diameter d_{grain} [m] and density ρ_{grain} [Mg m^{-3}]

Step 1. Determine the surface area A_S [m^2] and mass of a single particle of medium silt ($d_{MSl} = 2.0 \times 10^{-5}$ m).

The surface area and volume of a single spherical particle is founding using simple geometry.

$$A_S = 4\pi \cdot \left(\frac{d_{MSl}}{2}\right)^2 = 4\pi \cdot \left(\frac{2.0 \times 10^{-5}\ \text{m}}{2}\right)^2 = 1.26 \times 10^{-9}\ \text{m}^2$$

$$V_{MSl} = \tfrac{4}{3}\pi \cdot \left(\frac{d_{MSl}}{3}\right)^2 = \tfrac{4}{3}\pi \cdot \left(\frac{2.0 \times 10^{-5}\ \text{m}}{3}\right)^2 = 4.91 \times 10^{-15}\ \text{m}^3$$

$$m_{MSl} = V_{MSl} \cdot \rho_{\text{quartz}}$$

$$m_{MSl} = \left(4.19 \times 10^{-15}\ \text{m}^3\right) \cdot \left(2.65 \times 10^6\ \text{g m}^{-3}\right) = 1.11 \times 10^{-8}\ \text{g}$$

Step 2. Determine the specific surface area a_S [$m^2\ g^{-1}$] of medium silt.

The specific surface area a_S is found by dividing the area of a single spherical particle by its mass.

$$a_S = \frac{A_S}{m} = \frac{1.26 \times 10^{-9}\ \text{m}^2}{1.11 \times 10^{-8}\ \text{g}} = 1.13 \times 10^{-1}\ \text{m}^2\ \text{g}^{-1}$$

8.3 ADSORPTION ISOTHERM EXPERIMENTS

The *International Union of Pure and Applied Chemistry* defines *adsorption isotherm* in the following statement (Everett, 1972).

> Adsorption isotherm is the relation between the quantity adsorbed (suitably defined) and the composition of the bulk phase …under equilibrium conditions at constant temperature. Equilibrium between a bulk fluid and an interfacial layer may be established with respect to neutral species or to ionic species.

Adsorption and ion exchange are fundamentally different chemical processes as the following statement makes clear (Everett, 1972).

> If the adsorption of one or several ionic species is accompanied by the *simultaneous desorption* (displacement) of an equivalent amount of one or more other ionic species this process is called *ion exchange*.

Physical chemists, surface chemists, and soil chemists have repeatedly tried to reconcile adsorption and ion exchange into simple adsorption isotherm model. We will return to this presently.

8.3.1 Area-Based Adsorption Isotherms

In this section we describe *adsorption isotherm* experiments involving solid/liquid (S/L) interfaces. The experiment measures adsorption at the S/L interface as a change in solution (i.e., liquid-phase) concentration Δc_i of a single *adsorptive i*. The solution contains a specified solid adsorbent mass m suspended in a specified liquid volume V. The independent variable (i.e., treatment) is the initial solution-phase absorptive concentration c'_i.

$$\Delta c_i = \left(c'_i - c_i\right) \tag{8.1}$$

In many cases the *adsorbent-specific surface area* a_s [$m^2\ g^{-1}$] is known. If this is the case, the total S/L area A_s [m^2] is proportional to the total adsorbent mass m suspended in the liquid phase.

$$A_s = m \cdot a_s \tag{8.2}$$

The *surface excess amount* n^{σ_i} [mol] identifies the absorptive by subscript i. Superscript σ associates the amount n_i with a two-dimensional surface or interface (Everett, 1972). The solution volume V [dm^3] as the difference between the total volume of the suspension and the volume of the suspended solid. The *surface excess concentration* Γ_i [$mol\ m^{-2}$] is simply the surface excess amount n_i^σ normalized by the total adsorbent surface area A_s in the suspension.

$$n_i^\sigma = \Gamma_i \cdot A_s = \Delta c_i \cdot V \tag{8.3}$$

Fig. 8.2A depicts the initial state of a suspension containing a clay-sized mineral adsorbent. During the equilibration stage the suspension is agitated to ensure ample opportunity for the dissolved absorptive to contact the adsorbent (Fig. 8.2B). The duration of the equilibration stage depends on how rapidly the absorptive binds to the mineral surface, usually a few minutes to several hours or days.

The final stage (Fig. 8.2C) requires separation of suspended clay-sized mineral particles from the solution for each treatment in the series, usually by filtration or accelerated sedimentation in a centrifuge (cf. Appendix 3.C). The equilibrium supernatant solutions contain varying final concentrations c_i of the component i, the absorptive. The amount of component i adsorbed is given as the surface excess concentration (expression 8.4).

$$\Gamma_i = \frac{n_i^\sigma}{m \cdot a_s} = \frac{\Delta c_i \cdot V}{m \cdot a_s} \tag{8.4}$$

Table 8.3 lists equilibrium results from a representative adsorption isotherm; the data are plotted in Fig. 8.3. The adsorbent is the ferric oxyhydroxide mineral goethite α-FeO(OH)(s). The adsorptive HPO_4^{2-} is a divalent molecular anion. The adsorption isotherm displays a steep initial slope that decreases steadily toward what appears to be a zero slope representing an adsorption maximum. Adsorption isotherms showing this behavior are called *site-limited* adsorption isotherms.

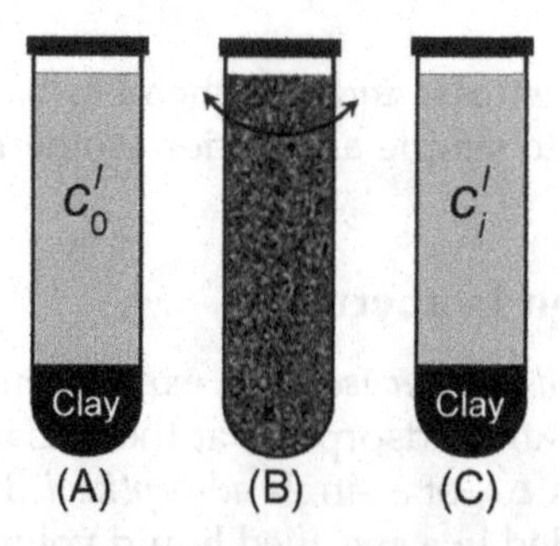

FIG. 8.2 Measuring an adsorption isotherm requires a series of solutions containing varying concentrations c_i^l (A) of an adsorptive i and the solid or molecular adsorbent. The suspension is agitated (B) during the adsorption reaction. After reaching adsorption equilibrium the adsorbent is separated from the solution (C) for analysis of c_i.

TABLE 8.3 Equilibrium Adsorption Isotherm Data for Adsorptive HPO_4^{2-} and Adsorbent Goethite α-FeOOH

Dissolved Phosphate, $c_{HPO_4^{2-}}$ (mol dm^{-3})	Adsorbed Phosphate, $\Gamma_{HPO_4^{2-}}$ (mol m^{-2})
4.23E−07	1.30E−06
2.33E−06	1.77E−06
1.27E−05	1.91E−06
2.07E−05	2.09E−06
5.94E−05	2.22E−06
7.32E−05	2.28E−06
9.99E−05	2.37E−06
1.24E−04	2.52E−06

Notes: The surface excess concentration $\Gamma_{HPO_4^{2-}}$ [mol m^{-2}] *is plotted on the vertical axis. The system has the following properties:* 25°C, pH 4, $c_{KNO_3(aq)}$ = 0.01 mol dm^{-3}.
From Antelo, J., Avena, M., Fiol, S., Lopez, R., Arce, F., 2005. Effects of pH and ionic strength on the adsorption of phosphate and arsenate at the goethite-water interface. J. Colloid Interface Sci. 285 (2), 476–486.

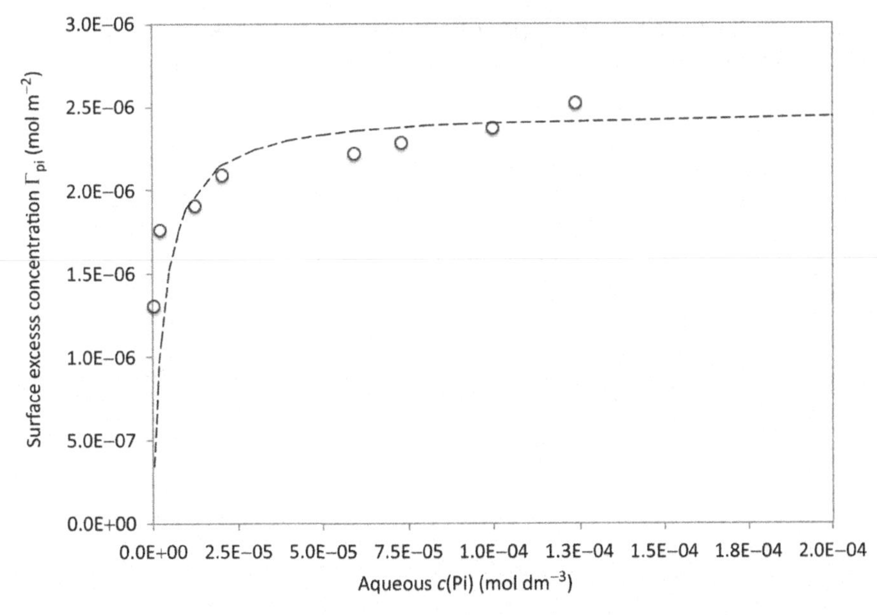

FIG. 8.3 The adsorption isotherm for absorptive HPO_4^{2-} and adsorbent goethite α-FeOOH at pH 4. *From Antelo, J., Avena, M., Fiol, S., Lopez, R., Arce, F., 2005. Effects of pH and ionic strength on the adsorption of phosphate and arsenate at the goethite-water interface. J. Colloid Interface Sci. 285 (2), 476–486.*

8.3.2 Mass-Based Adsorption Isotherms

In some cases adsorbent-specific surface area a_s is not an appropriate or logical adsorption parameter. For example, the natural organic-matter in soils, sediments, and aquifers will take up nonpolar organic absorptives in a process known as *physisorption by partitioning*. Physisorption by partitioning is best represented by the *specific surface excess* (n_i^σ/m) defined by expression (8.5). Note, all *specific* parameters (e.g., *specific* surface area a_s and *specific* surface excess (n_i^σ/m)) are normalized by adsorbent mass m.

$$\frac{n_i^\sigma}{m} = \frac{\Delta c_i \cdot V}{m} \tag{8.5}$$

Lambert (1968) reported isotherms for adsorption of the insecticide *Sapona* (chlorfenvinphos, CAS registry number 470-90-6) two California soil series: Yolo silty clay loam and Hanford (formerly Ripperdan) fine sandy loam. The adsorption isotherms for chlorfenvinphos by these two soils appear in Fig. 8.4.

Lambert (1968) relied on chlorfenvinphos labeled with radioisotope ${}^{14}_{6}C$; adsorptive solution-phase concentration and specific surface excess (n_i^σ/m) are both reported in radioactive units: becquerel Bq s^{-1}.

The adsorption isotherm in Fig. 8.4 is linear throughout the entire solution-phase concentration range c_{Sapona}. Adsorption isotherms showing this behavior are called *partition* isotherms.

8.3.3 Chemisorption and Physisorption

Figs. 8.3 and 8.4 illustrate the two types of adsorption isotherms most commonly encountered by environmental chemists. Fig. 8.3 is a nonlinear site-limited isotherm exhibiting an

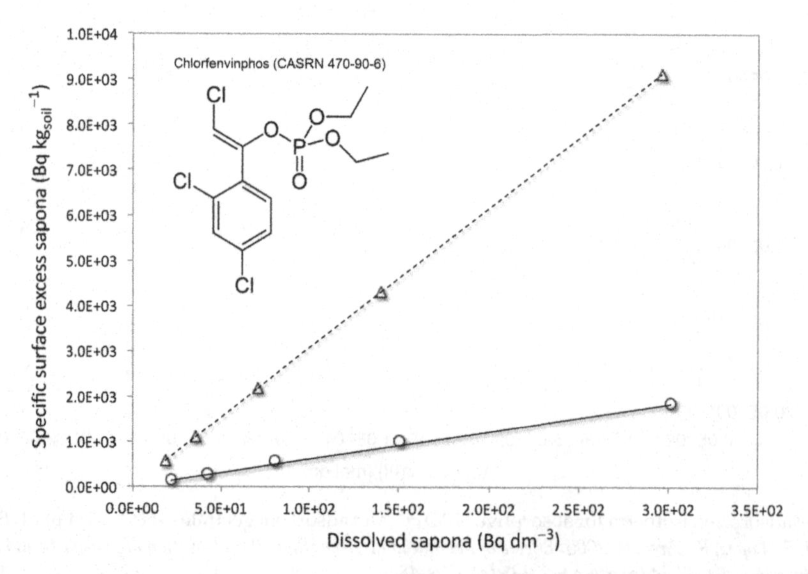

FIG. 8.4 The adsorption isotherm for absorptive *Sapona* (chlorfenvinphos, CAS registry number 470-90-6) and two whole soil adsorbents: Yolo soil (*open triangles*) and Hanford soil (*open circles*). The specific surface excess n_A^σ/m values are normalized by whole-soil mass m_{soil}. *From Lambert, S.M., 1968. Omega (Ω), a useful index of soil sorption equilibria. J. Agric. Food Chem. 16 (2), 340–343.*

apparent adsorption maximum. Fig. 8.4 is a linear partitioning isotherm with no apparent adsorption maximum.

Surface chemists usually associate site-limited adsorption with *chemisorption reactions* (reaction 8.R1). Chemisorption occurs when an absorptive molecule A bonds with an *adsorption site* (*) to form an *adsorption complex* A*.

General Chemisorption Reaction

$$A(aq) + {}^{*} \longleftrightarrow A^{*} \tag{8.R1}$$

There is general agreement that chemisorption requires chemical bonding between the adsorptive and the adsorption sites. Given the nature of mineral surfaces and the ionic nature of many adsorptives studied by soil and environmental chemists, we will include ion exchange as a chemisorption reaction. The adsorption maximum, characteristic of site-limited isotherms, represents the saturation of a limited number of adsorption sites.

Linear adsorption isotherms are based on partitioning equilibrium.[4] Reaction (8.R2) is the partitioning of substance A between aqueous phase (aq) and an immiscible organic liquid phase (org).

General Partition Reaction

$$A(aq) \longleftrightarrow A(org) \tag{8.R2}$$

Environmental, biological, and medical chemists typically quote partition constants where the organic phase is 1-octanol (CAS registry number 111-87-5), using the alternative partition constant notation $K^{\ominus}_{o/w}$ appearing in expression (8.6) for partitioning reaction (8.R2). Henceforth we adopt the partition constant notation used by the environmental science community rather than the notation recommended by the International Union of Pure and Applied Chemistry.[5]

Octanol-Water Partition

$$K^{\ominus}_{o/w} = \frac{c_{A(oct)}}{c_{A(aq)}} \tag{8.6}$$

Surface chemists associate *partitioning* adsorption with *physisorption*. Physisorption is adsorption in which the forces involved are the same intermolecular forces[6]) responsible for vapor condensation to the liquid state. Physisorption does not result in significant changes in the electronic structure (i.e., molecular orbitals) of the adsorptive molecule.

General Physisorption Reaction

$$A(aq) \longleftrightarrow A(ads) \tag{8.R3}$$

[4] Some sources identify Henry's law as the basis for linear adsorption isotherms by analogy with the dimensionless "water-air partitioning coefficient" $H^{cc} = c_a/c_g$.

[5] The International Union of Pure and Applied Chemistry recommends the following partition constant notation $(K^{\ominus}_D)_A$, which assumes the aqueous phase solute A(aq) is the educt and the organic phase solute A(org) is the product.

[6] The term van der Waals adsorption is synonymous with physical adsorption, but the International Union of Pure and Applied Chemistry does not recommended its use.

$$K^{\ominus}_{\text{adsorbent/w}} = \frac{n^{\sigma}_{\text{A}}/m}{c_{\text{A(aq)}}} \tag{8.7}$$

Adsorption processes that produce linear partition isotherms involve organic adsorptive compounds that adsorb by *partitioning into solvent-like domains* found in molecular-association adsorbents.

In summary, adsorption isotherms reveal the fundamental nature of the adsorption process, determined by the dominant interaction between the absorptive and the adsorbent. The following sections present models for both types of adsorption isotherms. Nonlinear site-limited adsorption isotherms are generally represented by a particular two-parameter function. Simple linear partition isotherms require a single *partition ratio* parameter but, as we will see, molecular association may occur in two stages requiring two partition ratios.

8.3.4 Langmuir Isotherm Model

Langmuir (1916) derived the model widely used to represent site-limited adsorption. The assumptions upon which this model was originally based (Eq. 8.4) are idealized and generally do not apply to site-limited adsorption at S/L interfaces. The use of the Langmuir model generated considerable debate within the environmental chemistry community, particularly the appropriateness of using a model whose foundational assumptions are not satisfied. We can apply this model to parameterize nonlinear adsorption isotherms, provided that we carefully avoid assigning certain physical interpretations of the site-limited adsorption isotherm parameters appearing in Table 8.4.

TABLE 8.4 Assumptions Used by Langmuir to Derive the Nonlinear Site-Limited Gas Adsorption Isotherm Expression

1. The interface is "a plane surface having only one kind of elementary space."
2. "If the surface is that of a crystal, there will be a definite number of spaces, N_0, on each sq. cm. of surface capable of holding adsorbed gas molecules."
3. "...the rate of condensation of the gas on the crystal surface is [proportional to] the fraction of the surface which is bare."
4. "...each space can hold only one adsorbed molecule".
5. "Since evaporation and condensation are in general thermodynamically reversible phenomena, the mechanism of evaporation must be the exact reverse of that of condensation, even down to the smallest detail."
6. "The forces acting between two layers of gas molecules will usually be very much less than those between the crystal surface and the first layer of molecules. The rate of evaporation in the second layer will, therefore, generally be so much more rapid than in the first, that the number of molecules in the second layer will be negligible."
7. The rate constant for desorption is independent of the fraction of sites occupied.[a]

[a] *This assumption is implicit in the equilibrium state expression* (8.11).

Langmuir (1916, 1918) based his theory of site-limited adsorption on a kinetic model. The rate adsorptive A desorbs from a solid surface $(r_{des})_A$, by his reasoning, is proportional to the fraction of occupied sites θ (expression 8.8). The proportionality constant $(k_{des})_A$ is the rate constant for desorption.

$$(r_{des})_A = (k_{des})_A \cdot \theta \tag{8.8}$$

The adsorption rate $(r_{ads})_A$ is proportional to the fraction of unoccupied states $(1-\theta)$ and the absorptive concentration c_A when we apply the Langmuir model to adsorption from solution. The proportionality constant for adsorption $(k_{ads})_A$ (expression 8.9) is the rate constant for adsorption from solution. Langmuir (1916) developed the theory for gas molecule adsorption using the partial pressure of the adsorbent instead of the concentration in solution.

$$(r_{ads})_A = (k_{ads})_A \cdot (1-\theta) \cdot c_A \tag{8.9}$$

Provided adsorption involves a *reversible, one-step mechanism* (cf. Assumption 5, Table 8.4) then adsorption equilibrium is the state where the rates of adsorption and desorption are equal (Eq. 8.10).

$$(k_{des})_A \cdot \theta = (k_{ads})_A \cdot (1-\theta) \cdot c_A \tag{8.10}$$

Assumption 5 (Eq. 8.4) permits the kinetic definition of an equilibrium constant for reversible adsorption K_A° as the rate constant ratio (8.11).

$$K_A^{\circ} \equiv \frac{(k_{ads})_A}{(k_{des})_A} \tag{8.11}$$

Solving expression (8.10) for θ using rate constant ratio expression (8.11) yields the familiar Langmuir adsorption isotherm expression (8.12).

$$\theta = \frac{K_A^{\circ} \cdot c_A}{1 + (K_A^{\circ} \cdot c_A)} \tag{8.12}$$

Plotting θ as a function of the absorptive concentration c_A illustrates an important characteristic of the Langmuir adsorption isotherm: adsorption approaches a plateau as the fraction of occupied sites θ approaches unity (cf. Fig. 8.3).

The fraction of occupied sites θ can be represented using either *surface excess concentration* ratios or *specific surface excess* ratios (expression 8.13). Expression (8.14) is derived from the ratio of surface excess concentration Γ_A to the *maximum* surface excess concentration $\Gamma_{A(m)}$ when all adsorption sites are occupied. Expression (8.15) is derived from the ratio of specific surface excess (n_A^{σ}/m) to the *maximum* specific surface excess $(n_{A(m)}^{\sigma}/m)$.

$$\theta = \frac{\Gamma_A}{\Gamma_{A(m)}} = \frac{(n_A^{\sigma}/m)}{(n_{A(m)}^{\sigma}/m)} \tag{8.13}$$

$$\Gamma_A = \Gamma_{A(m)} \cdot \left(\frac{K_A^{\circ} \cdot c_A}{1 + (K_A^{\circ} \cdot c_A)} \right) \tag{8.14}$$

$$\frac{n_A^{\sigma}}{m} = \left(\frac{n_{A(m)}^{\sigma}}{m} \right) \cdot \left(\frac{K_A^{\circ} \cdot c_A}{1 + (K_A^{\circ} \cdot c_A)} \right) \tag{8.15}$$

For *solid/gas* (S/G) interfaces of the type Langmuir studied adsorption-site saturation represents monolayer coverage m. Adsorption-site saturation of the S/L interface generally falls short of a close-packed monolayer of adsorptive molecules.

As the absorptive concentration becomes large—specifically when $(K_A^{\circ} \cdot c_A) \gg 1$—then $\theta \approx 1$ in expression (8.13) and $\Gamma_A \approx \Gamma_{A(m)}$ in expression (8.14) or $(n_A^{\sigma}/m) \approx (n_{A(m)}^{\sigma}/m)$ in expression (8.15).

At the limit where the absorptive concentration becomes very small—specifically when $(K_A^{\circ} \cdot c_A) \ll 1$—then θ and either the surface excess concentration Γ_A (expression 8.14) or the specific surface excess (n_A^{σ}/m) in expression (8.15) appear to be proportional to the absorptive concentration c_A.

Two parameters determine site-limited adsorption isotherms: the adsorption equilibrium constant K_A° and the *maximum* surface excess concentration $\Gamma_{A(m)}$ (or the *maximum* specific surface excess $(n_{A(m)}^{\sigma}/m)$). The significance of *maximum* surface excess concentration or specific surface excess in site-limited adsorption isotherms is self-evident, but the adsorption equilibrium constant K_A° requires further explanation.

Langmuir (1916) developed his kinetic adsorption model subject to stringent conditions (cf. Table 8.4), enabling him to associate the rate-constant ratio with the kinetic definition of an equilibrium constant for a reversible reaction. This association provides a thermodynamic basis for the adsorption equilibrium constant K_A° under the conditions imposed by Langmuir (1916, 1918).

In general, the *Langmuir conditions* (cf. Table 8.4) are not satisfied for site-limited S/L adsorption encountered by soil and environmental chemists (Harter and Baker, 1977). Veith and Sposito (1977) showed that solute loss from solution by precipitation could be fit to expression (8.12). In the following section we will show how site-limited adsorption isotherms can also be derived from ion-exchange isotherms where, once again, the *Langmuir conditions* are not satisfied.

Recognizing both the value of a simple, empirical site-limited isotherm and the failure to satisfy the *Langmuir conditions* (8.4) we will henceforth use an *empirical* two-parameter site-limited isotherm defined by expression (8.16) to fit experimental adsorption isotherms exhibiting site-limited behavior. Empirical parameter k_L quantifies adsorbate affinity for absorptive A without any claim it represents a thermodynamic adsorption equilibrium constant K_A° with explicit ties to the Gibbs energy of adsorption ΔG_{ads}°. Example 8.2 demonstrates parameter fitting using a linearized form of expression (8.16).

$$\theta = \frac{k_L \cdot c_A}{1 + (k_L \cdot c_A)} \tag{8.16}$$

EXAMPLE 8.2

Example Permalink

http://soilenvirochem.net/Iqv47B

Fit the HPO_4^{2-} adsorption isotherm data listed in Table 8.3 and plotted in Fig. 8.3 using the empirical site-limited adsorption isotherm (expression 8.16).

This example will fit the experimental HPO_4^{2-} adsorption isotherm data by transforming expression (8.16) into a quasi-linear form. The experimental data require replacement of θ in expression (8.16) with the surface excess concentration ratio in expression (8.13).

$$\Gamma_A = \Gamma_{A(m)} \cdot \left(\frac{k_L \cdot c_A}{1 + (k_L \cdot c_A)} \right)$$

Step 1. Convert expression (8.16) into a quasi-linear form by inverting both sides of the expression, canceling duplicate terms and otherwise simplifying the linearized expression.

$$\frac{1}{\Gamma_A} = \frac{1 + (k_L \cdot c_A)}{\Gamma_{A(m)} \cdot k_L \cdot c_A}$$

$$\frac{1}{\Gamma_A} = \frac{1}{\Gamma_{A(m)} \cdot k_L \cdot c_A} + \frac{(k_L \cdot c_A)}{\Gamma_{A(m)} \cdot k_L \cdot c_A}$$

$$\frac{1}{\Gamma_A} = \frac{1}{\Gamma_{A(m)} \cdot k_L \cdot c_A} + \frac{1}{\Gamma_{A(m)}}$$

Multiply both sides of the inverted and simplified expression by c_A.

$$\frac{c_A}{\Gamma_A} = \frac{c_A}{\Gamma_{A(m)} \cdot k_L \cdot c_A} + \frac{c_A}{\Gamma_{A(m)}}$$

$$\frac{c_A}{\Gamma_A} = \left(\frac{1}{\Gamma_{A(m)} \cdot k_L} \right) + \left(\frac{1}{\Gamma_{A(m)}} \right) \cdot c_A$$

Step 2. Transform the original isotherm dataset listed in Table 8.3 by generating a transformed dependent variable c_A / Γ_A (cf. column 2, Table 8.5).

TABLE 8.5 Equilibrium Adsorption Isotherm Data for Adsorptive HPO_4^{2-} and Adsorbent Goethite α-FeOOH

Dissolved Phosphate, $c_{HPO_4^{2-}}$ (mol dm^{-3})	Transformed Variable, $c_{HPO_4^{2-}} / \Gamma_{HPO_4^{2-}}$ (m^2 dm^{-3})
4.23E−07	0.324
2.33E−06	1.318
1.27E−05	6.654
2.07E−05	9.910
5.94E−05	26.781
7.32E−05	32.054
9.99E−05	42.156
1.24E−04	49.103

Notes: The second column lists the transformed dependent variable: $c_{HPO_4^{2-}} / \Gamma_{HPO_4^{2-}}$ [m^2 dm^{-3}]. *The original data appear in Table 8.3.*
From Antelo, J., Avena, M., Fiol, S., Lopez, R., Arce, F., 2005. Effects of pH and ionic strength on the adsorption of phosphate and arsenate at the goethite-water interface. J. Colloid Interface Sci. 285 (2), 476–486.

A plot (Fig. 8.5) of the transformed variable c_A/Γ_A as a function of c_A confirms the transformed dataset is linearized.

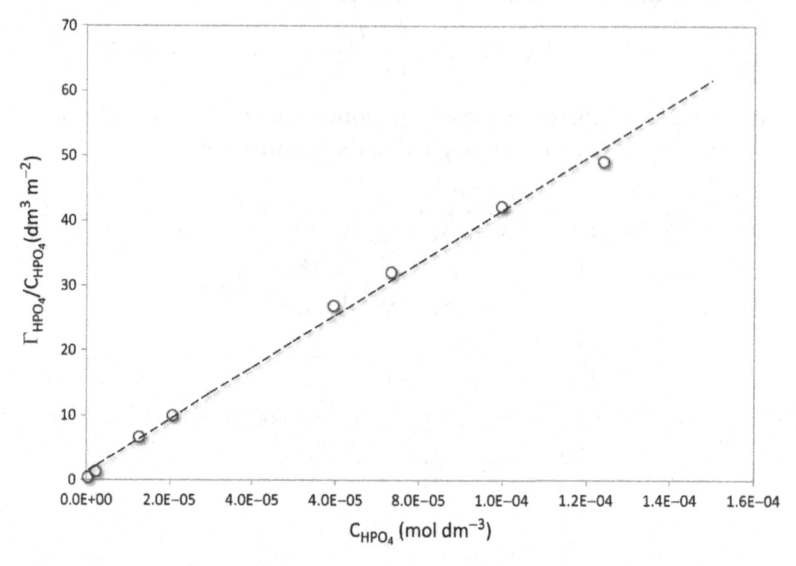

FIG. 8.5 The linearized adsorption isotherm for absorptive HPO_4^{2-} and adsorbent goethite α-FeOOH at pH 4. The transformed data are listed in Table 8.5. *From Antelo, J., Avena, M., Fiol, S., Lopez, R., Arce, F., 2005. Effects of pH and ionic strength on the adsorption of phosphate and arsenate at the goethite-water interface. J. Colloid Interface Sci. 285 (2), 476–486.*

Step 3. Perform a linear regression of the data in Table 8.5 to generate the intercept parameter α and the slope parameter β.

$$x' \equiv c_A$$

$$y' \equiv c_A/\Gamma_A$$

$$\hat{y} = \alpha + \beta \cdot x'$$

$$\beta \equiv \Gamma_{A(m)}^{-1} = 4.03 \times 10^5\ m^2\ mol^{-1}$$

$$\alpha \equiv \left(\Gamma_{A(m)} \cdot k_L\right)^{-1} = \frac{\beta}{k_L} = 1.26\ m^2\ dm^{-3}$$

The two parameters $\Gamma_{HPO_4{}^{2-}(m)}$ and k_L are computed from the intercept parameter α and the slope parameter β listed earlier.

$$\Gamma_{A(m)} = \beta^{-1} = \left(4.03 \times 10^5\ m^2\ mol^{-1}\right)^{-1} = 2.48 \times 10^{-6}\ mol\ m^{-2}$$

$$k_L = \frac{\beta}{\alpha} = \frac{1.26\ m^2\ dm^{-3}}{4.03 \times 10^5\ m^2\ mol^{-1}} = 3.19 \times 10^5\ dm^3\ mol^{-1}$$

Expression (8.16) is plotted in Fig. 8.3 using the parameters $\left\{\Gamma_{HPO_4{}^{2-}(m)}, k_L\right\}$ calculated earlier.

8.3.5 Ion-Exchange Isotherm Model

Chemists studying ion exchange have repeatedly attempted to derive site-limited adsorption isotherm expressions despite the fact that ion exchange violates the *Langmuir conditions* (cf. Table 8.4). The typical route represents ion exchange as competitive adsorption while attempting to cleave as closely as possible to the original *Langmuir conditions.*

Taylor (1931) derived a competitive isotherm expression built on the original *Langmuir conditions* involving two adsorptive gases *A* and *B*.

$$(k_{des})_A \cdot \theta_A = (k_{ads})_A \cdot (1 - \theta_A - \theta_B) \cdot c_A$$

$$(k_{des})_B \cdot \theta_B = (k_{ads})_B \cdot (1 - \theta_A - \theta_B) \cdot c_B$$

$$K_A^{\ominus} \equiv \frac{(k_{ads})_A}{(k_{des})_A}$$

$$K_B^{\ominus} \equiv \frac{(k_{ads})_B}{(k_{des})_B}$$

$$\theta_A = \frac{K_A^{\ominus} \cdot c_A}{1 + (K_A^{\ominus} \cdot c_A) + (K_B^{\ominus} \cdot c_B)} \tag{8.17}$$

$$\theta_B = \frac{K_B^{\ominus} \cdot c_B}{1 + (K_A^{\ominus} \cdot c_A) + (K_B^{\ominus} \cdot c_B)} \tag{8.18}$$

Boyd et al. (1947) attempted to derive a Langmuir-type isotherm expression by adopting the Taylor competitive adsorption isotherms. Boyd et al. (1947) simplified expressions (8.17), (8.18) by deleting the unit term in the denominator, justified by the following assumption.

> The number of unsaturated anionic exchanging groups (i. e., amount of bare surface in the Langmuir picture) must always be very small, for otherwise an appreciable free negative surface charge would be created leading to particles of adsorbent bearing a high negative charge, which is not the case. Accordingly, to a good approximation the quantity unity in the denominator ...can be neglected relative to the [remaining terms].

While this assumption may appear justified, eliminating the unit term means the competitive adsorption isotherms (8.16), (8.17) fail to converge on the appropriate single-adsorptive isotherm at the limit of either adsorptive concentration approaches zero.

As we will see, complex thermodynamic derivations (e.g. Sposito, 1979; Elprince and Sposito, 1981; Polzer et al., 1992; Misak, 1995) are unnecessary. A thermodynamically valid, site-limited adsorption isotherm can be derived directly from ion-exchange isotherm expressions. These derived adsorption isotherms are polynomial expressions easily solved using the Newton-Rapson method described in Appendix 6.E.

A simple symmetric (Na^+, K^+) cation-exchange reaction (8.R4) and the corresponding *equilibrium quotient expression* appears below.

$$Na^+(aq) + \overline{K^+} \longleftrightarrow \overline{Na^+} + K^+(aq) \tag{8.R4}$$

$$K_{Na/K}^{\ominus} = \frac{\left(n_{\overline{Na^+}}/m\right) \cdot \gamma_{1\pm} \cdot c_{K^+}}{\gamma_{1\pm} \cdot c_{Na^+} \cdot \left(n_{\overline{K^+}}/m\right)} \tag{8.19}$$

Notice the concentration and specific surface excess ratios in expression (8.19) reduce to molar ratios (expression 8.20).

$$\frac{c_{K^+}}{c_{Na^+}} = \frac{\left(n_{K^+}/V\right)}{\left(n_{Na^+}/V\right)} = \frac{n_{K^+}}{n_{Na^+}} \tag{8.20}$$

$$\frac{\left(n_{\overline{Na^+}}/m\right)}{\left(n_{\overline{K^+}}/m\right)} = \frac{n_{\overline{Na^+}}}{n_{\overline{K^+}}} \tag{8.21}$$

$$K^{\circ}_{Na/K} = \frac{n_{\overline{Na^+}} \cdot n_{K^+}}{n_{Na^+} \cdot n_{\overline{K^+}}} \tag{8.22}$$

The adsorption experiment is represented as the titration of a K^+-saturated cation exchanger by a $Na^+(aq)$ solution. The reaction quotient expression is prepared using an RICE table (Table 2). The initial K^+-specific surface excess adsorbed to the cation exchanger equals the *cation-exchange capacity*: $(n^0_{\overline{K^+}}/m) = CEC$.

$$K^{\circ}_{Na/K} = \frac{x \cdot x}{\left(n^0_{Na^+} - x\right) \cdot \left((CEC \cdot m) - x\right)}$$

$$x^2 + \left(\frac{\left(n^0_{Na^+} + (CEC \cdot m)\right) \cdot K^{\circ}_{Na/K}}{K^{\circ}_{Na/K} - 1}\right) \cdot x + \left(\frac{n^0_{Na^+} \cdot (CEC \cdot m) \cdot K^{\circ}_{Na/K}}{K^{\circ}_{Na/K} - 1}\right) = 0$$

The roots of this second-order polynomial are found using the quadratic equation with parameter $a \equiv 1$ and parameters b and c defines as follows.

$$b \equiv \left(\frac{\left(n^0_{Na^+} + (CEC \cdot m)\right) \cdot K^{\circ}_{Na/K}}{K^{\circ}_{Na/K} - 1}\right)$$

$$c \equiv \left(\frac{n^0_{Na^+} \cdot (CEC \cdot m) \cdot K^{\circ}_{Na/K}}{K^{\circ}_{Na/K} - 1}\right)$$

The site-limited adsorption isotherm for symmetric cation exchange is found by entering appropriate values for (Na^+, K^+) cation-exchange selectivity $K^{\circ}_{Na/K}$, water volume V, exchanger mass m, and cation-exchange capacity CEC then finding the polynomial roots using the quadratic equation for a series of $c^0_{Na^+}$ values representing the titration.

Fig. 8.6 plots a simulated adsorption isotherm for adsorptive Na^+ by K^+-saturated bentonite. The symmetric (Na^+, K^+) exchange-selectivity constant $K^{\circ}_{Na/K} = 0.278$ comes from Vanselow (1932). Each data point (open circles in Fig. 8.6) represents the root of a second-order polynomial generated using Table 8.6.

The solid line plotted in Fig. 8.6 is the optimal fit of expression (8.16) to the adsorption isotherm explicitly based on the ion-exchange equilibrium quotient. Both curves plot two-parameter isotherms but one can clearly see a two-parameter site-limited isotherm does not have to have the form of expression (8.16).

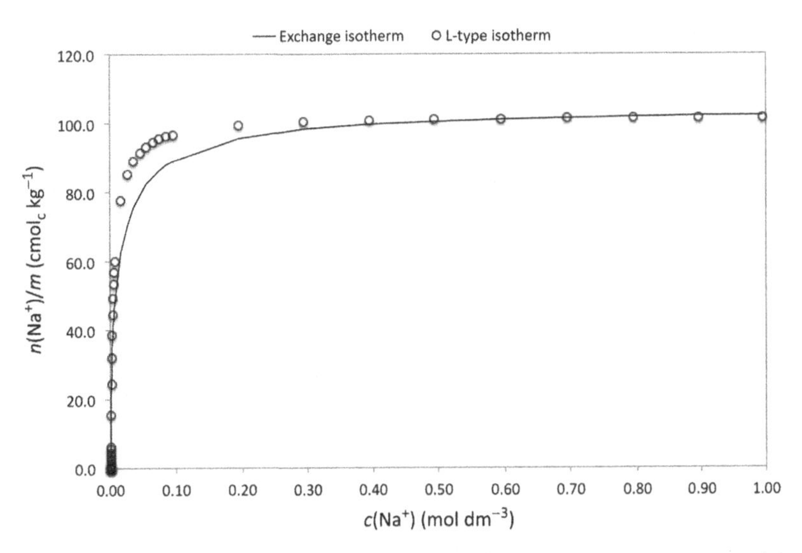

FIG. 8.6 A simulated site-limited adsorption isotherm (*open circles*) for the adsorption of Na^+ by K^+-saturated bentonite based on (Na^+, K^+) cation-exchange data from Vanselow (1932) with $K^{\circ}_{Na/K} = 0.278$. A Langmuir-type site-limited isotherm is also plotted (*solid line*) based on expression (8.16) where $(n^{\sigma}_{Na^+(m)}/m) = 104.1\ cmol_c\ kg^{-1}$ at the limit where $\theta \to 1$.

TABLE 8.6 *RICE* Tables for Symmetric (Na^+, K^+) Cation-Exchange Reaction

Reaction → (mol_c)	$Na^+(aq)$ (mol_c)	$\overline{K^+}$	↔ (mol_c)	$\overline{Na^+}$ (mol_c)	$K^+(aq)$
Initial	$n^0_{Na^+}$	$n^0_{\overline{K^+}}$		0	0
Reaction	$-x$	$-x$		$+x$	$+x$
Equilibrium	$(n^0_{Na^+} - x)$	$((CEC \cdot m) - x)$		x	x

8.3.6 Partitioning Isotherm Model

The linear partition isotherms are reminiscent of solute partitioning between two immiscible solvents: water and an organic solvent. The adsorption isotherm of nonpolar pesticides led Lambert et al. (1965) to hypothesize the adsorption of organic compounds by soil "*corresponds to partition of the compound into the organic matter.*" The Yolo soil clearly adsorbs much more of the insecticide than the Hanford soil (Fig. 8.4), but if the quantity adsorbed is normalized by the organic carbon content m_{OC} of the soil (expression 8.24), the adsorption isotherms for the two soils are essentially identical (Fig. 8.7).

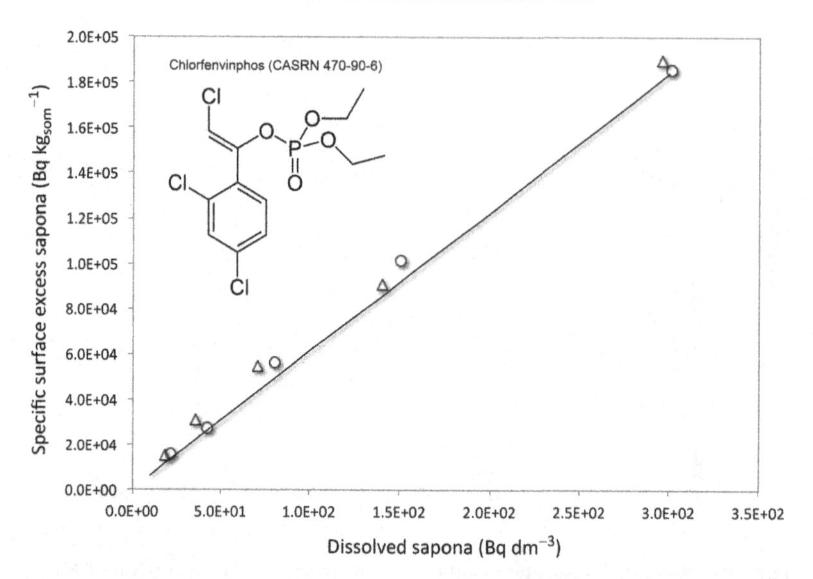

FIG. 8.7 The adsorption isotherm for absorptive *Sapona* (chlorfenvinphos, CAS registry number 470-90-6) and two whole soil adsorbents: Yolo soil (*open triangles*) and Hanford soil (*open circles*). The adsorbent is soil organic matter and the specific surface excess n_A^σ/m is normalized using the soil organic matter mass m_{SOM} of each soil. *From Lambert, S.M., 1968. Omega (Ω), a useful index of soil sorption equilibria. J. Agric. Food Chem. 16 (2), 340–343.*

$$\left(\frac{n_A}{m_{soil}}\right) \equiv K_{s/w}^{\circ} \cdot c_A \tag{8.23}$$

$$\left(\frac{n_A}{m_{oc}}\right) \equiv K_{oc/w}^{\circ} \cdot c_A \tag{8.24}$$

The (*s*) appearing in partition constant $K_{s/w}^{\circ}$ (expression 8.23) indicates the adsorbent (cf. reaction 8.R3) is *whole soil or sediment*. The (*oc*) appearing in partition constant $K_{oc/w}^{\circ}$ (expression 8.24) indicates the adsorbent (cf. reaction 8.R3) is *organic carbon* fraction from a soil or sediment specimen. Example 8.3 demonstrates parameter fitting using expression (8.23).

EXAMPLE 8.3

Example Permalink

http://soilenvirochem.net/Rp3hrH

Fit the *Sapona* adsorption isotherm data plotted in Fig. 8.4 using linear regression to determine the organic-carbon/water partition constant $K_{oc/w}^{\circ}$ for the soil *organic carbon* fraction.

This example will fit experimental *Sapona* adsorption isotherm data listed in Table 8.7 using partition-isotherm expression (8.23).

TABLE 8.7 Equilibrium Adsorption Isotherm Data for Adsorptive *Sapona* (Chlorfenvinphos, CAS Registry Number 470-90-6) With a Whole-Soil Specimen of the Yolo Soil Series as the Adsorbent

Dissolved Concentration, c_{Sapona} (Bq dm^{-3})	Specific Surface Excess, $n^{\sigma}_{Sapona}/m_{soil}$ (Bq kg$^{-1}_{soil}$)
1.86E+01	7.63E+02
3.57E+01	1.55E+03
7.07E+01	2.72E+03
1.40E+02	4.53E+03
2.96E+02	9.50E+03
2.14E+01	1.55E+02
4.21E+01	2.73E+02
8.07E+01	5.61E+02
1.51E+02	1.01E+03
3.01E+02	1.86E+03

From Lambert, S.M., 1968. Omega (Ω), a useful index of soil sorption equilibria. J. Agric. Food Chem. 16 (2), 340–343.

Step 1. Perform a linear regression of the data in Table 8.5 to generate the intercept parameter α and the slope parameter β.

$$x' \equiv c_{Sapona}$$
$$y' \equiv n^{\sigma}_{Sapona}/m_{soil}$$
$$\hat{y} = \alpha + \beta \cdot x'$$
$$\beta = 30.8 \text{ dm}^3 \text{ kg}^{-1}_{soil}$$
$$\alpha = 353 \text{ Bq kg}^{-1}_{soil}$$

The soil-water partition constant for *Sapona* adsorption by the Yolo soil series is equal to the slope from the linear regression: $K^{\ominus}_{s/w} = 30.8 \text{ dm}^3 \text{ kg}^{-1}_{soil}$.

Step 2. Estimate the partition constant $K^{\ominus}_{oc/w}$ for *Sapona* adsorption by the soil *organic carbon* fraction.

The partition constant $K^{\ominus}_{oc/w}$ is estimated from the whole-soil partition constant $K^{\ominus}_{s/w}$ and the organic carbon content of the Yolo soil series. Lambert (1968) does not give the location where the soil specimen was collected and quotes the organic matter content rather than the organic carbon content. Here we use data for a Yolo soil specimen from the University of California-Davis campus (National Soil Survey Center site ID *S1965CA113002*, 38.533225° N 121.784378° W): $f_{oc} = 0.0134 \text{ kg}_{oc} \text{ kg}_{soil}$.

$$K^{\ominus}_{oc/w} = \frac{K^{\ominus}_{s/w}}{f_{oc}}$$

$$K^{\ominus}_{oc/w} = \frac{30.8 \text{ dm}^3 \text{ kg}^{-1}_{soil}}{0.0134 \text{ kg}_{oc} \text{ kg}^{-1}_{soil}} = 2.30 \times 10^3 \text{ dm}^3 \text{ kg}^{-1}_{oc}$$

The Lambert hypothesis has been widely adopted by environmental chemists studying the adsorption of neutral organic compounds by soils and sediments because normalizing the adsorption isotherms for the same compound on many different soils and sediments typically results in a plot similar to that in Fig. 8.7. Lambert (1968) states that the adsorption behavior displayed in Figs. 8.4 and 8.7 "may be construed as evidence that the organic matter of soils from widely differing locations and of widely differing composition is so similar that for all practical purposes it can be considered the same." The "similarity" pertains to the adsorption behavior of neutral organic compounds by natural organic matter and should not be construed to apply to other properties of soil organic matter.

As just mentioned, linear partition isotherms result from adsorption processes where the interaction between absorptive and adsorbent does not involve binding to a limited number of sites. Instead, the adsorbent behaves similar to a solvent, a characteristic of hydrophilic molecular-association adsorbents. The absorptive is believed to *partition* from water into nonpolar solvent-like domains found within the molecular-association colloid.

8.3.6.1 Association Behavior of Natural Organic Matter Molecules

Amphiphilic organic matter molecules aggregate in aqueous solution to form *association colloids* typically identified as *micelles*.[7]

The *critical micelle concentration* c_M of organic matter solutions appears as an inflection point when the air/water surface tension γ (Wershaw et al., 1969; Hayase and Tsubota, 1983; Yonebayashi and Hattori, 1987; Shinozuka and Lee, 1991; Guetzloff and Rice, 1994) or the *apparent solubility* of nonpolar organic compounds (Wershaw et al., 1969; Shinozuka and Lee, 1991; Guetzloff and Rice, 1994; Quagliotto et al., 2006) are plotted as a function of organic matter mass concentration. Since molecules of organic matter are ionic amphiphiles, the critical micelle concentration c_M also appears as an inflection point when solution electric conductance G is plotted as a function of amphiphile concentration (Fishman and Elrich, 1975). Fig. 8.8 plots the specific conductance G of organic matter extracted from soil. Similar results appear in Quagliotto et al. (2006).

Micelles, supramolecular assemblies that reach colloidal dimensions, are dispersed in the liquid phase.[8] A dynamic equilibrium couples amphiphile solution phase solubility, amphiphile adsorption at the air/water interface and—above the c_M—supramolecular assembly of amphiphile molecules. Amphiphiles also form association colloids at S/L interfaces such as

[7] Some humic chemists object to use of *micelle* or *micellar solution* when referring to organic matter solutions. Surfactant chemists, however, use *micelle* to describe all *association colloids* that spontaneous aggregate as a characteristic concentration, known as the *critical micelle concentration* c_M.

[8] If the micelles are dispersed in water they are *hydrophilic* colloids. If micelles composed of the same amphiphile are dispersed in a nonpolar fluid their exposed surface is hydrophobic with a hydrophilic core, making them *hydrophobic* colloids.

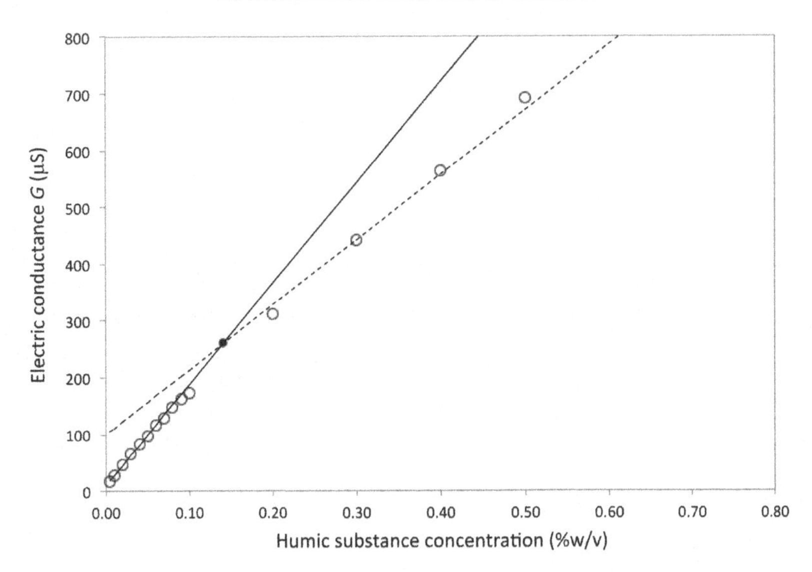

FIG. 8.8 The electric conductance *G* (*filled circles*) of Aldrich sodium humate (an alkaline extract from brown coal, Sigma-Aldrich Corp.) solutions. The *critical micelle concentration* appears as an inflection point (*filled circle*, $c_M = 0.14\%$ (w/v)). The electric conductance *G* inflection point for "humic acids" extracted from three Wisconsin soil series (Houghton, Sparta, and Wacousta) fall in a narrow range: 0.13% (w/v) $< c_M < 0.15\%$ (w/v).

mineral/water interfaces. *Admicelles* are supramolecular association colloids attached to S/L interfaces.

Natural organic matter is a heterogeneous molecular mixture therefore all organic matter association colloids should be regarded as *mixed micelles* composed of uncounted organic matter molecules, each with a unique molecular weight, elemental composition, molecular structure, and amphiphilic properties (MacCarthy, 2001; Piccolo, 2001; Sutton and Sposito, 2005).

Unlike most synthetic amphiphiles, there is good evidence organic matter molecules associate at concentrations well below the c_M (Piccolo, 2001; Sutton and Sposito, 2005). Empirical estimates of the mean molecular weight of natural organic matter specimens by a variety of methods range from 10^{+3} to 10^{+6} Da (Cameron et al., 1972). Reexamination of these results, however, indicate a clear concentration dependence (Simpson, 2002).

8.3.6.2 Adsorption by Partitioning in Molecular and Micellar Solutions

Aqueous micellar solutions will solubilize insoluble nonpolar organic compounds. Partitioning into the hydrophobic core at the center of hydrophilic micelles creates what can become a microemulsion. A environmental remediation application is solubilizing nonpolar compounds in contaminated soil or aquifers using synthetic or biological amphiphiles (i.e., surfactants). Scientists studying adsorption of nonpolar organic compounds to dissolved organic matter rely on dialysis cells to separate the dissolved adsorptive from a molecular adsorbent even if associative colloids do not form.

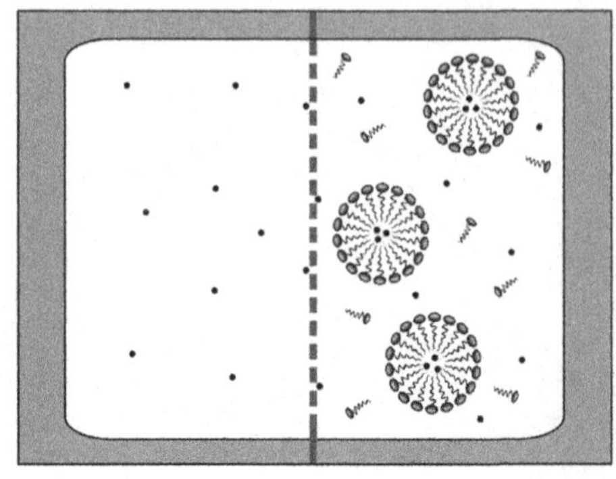

FIG. 8.9 A dialysis cell with two cells (labeled permeate and retentate) separated by a dialysis membrane (*dashed line, center*). The membrane confines micelles and amphiphiles from diffusing from the retentate cell (*right chamber*) into the permeate cell (*left chamber*). Adsorptive molecules (*filled circles*) are free to diffuse through the membrane and occupy either cell.

Fig. 8.9 depicts the equilibrium state in a dialysis cell used to measure sorption isotherms in molecular or micellar solutions. If the adsorbent is a molecular adsorbent such as organic matter (see Section 6.5.1), it is impossible to separate the solution from the adsorbent by filtration or centrifugal sedimentation. Separation is achieved by equilibrium dialysis.

The appropriate dialysis membrane is permeable to the absorptive but impermeable to the molecular adsorbent. A series of dialysis cells (Fig. 8.9) are prepared, each containing the same total liquid volume V^i, the same mass of molecular adsorbent m—confined to retentate half of the cell by a dialysis membrane—and varying initial concentrations absorptive. The *dissolved* absorptive concentration is the same on both sides of the dialysis membrane because it can freely diffuse across the membrane, but the *total* concentration—dissolved plus adsorbed—is higher in the retentate half containing the adsorbent.

Numerous studies (Carter and Suffet, 1982; Chiou et al., 1986, 1987; Kile and Chiou, 1989; Reid et al., 1991; Engebretson and von Wandruszka, 1994) demonstrate dilute organic matter solutions are capable of *molecular adsorption* well below the inflection-point c_M (cf. Fig. 8.8).

Fig. 8.10 plots results from two molecular adsorption experiments (Carter and Suffet, 1982; Chiou et al., 1987) performed using dialysis membranes to restrict organic matter molecules to the retentate cell but permit access of the absorptive to both retentate and permeate cells. The adsorptive was DDT (1,1,1-trichloro-2,2-bis(4-chlorophenyl)ethane, CAS registry number 50-29-3). Example 8.4 demonstrates parameter fitting of data plotted Fig. 8.10.

The apparent solubility of many hydrophobic, nonpolar compounds increases linearly as the organic matter concentration ranges from a few milligrams per cubic decimeter to a thousand milligrams per cubic decimeter. An inflection point in apparent solubility occurs at the same organic matter concentration where the electric conductance G inflection point (cf. Fig. 8.8) and where the air/water surface tension γ inflection point occur.

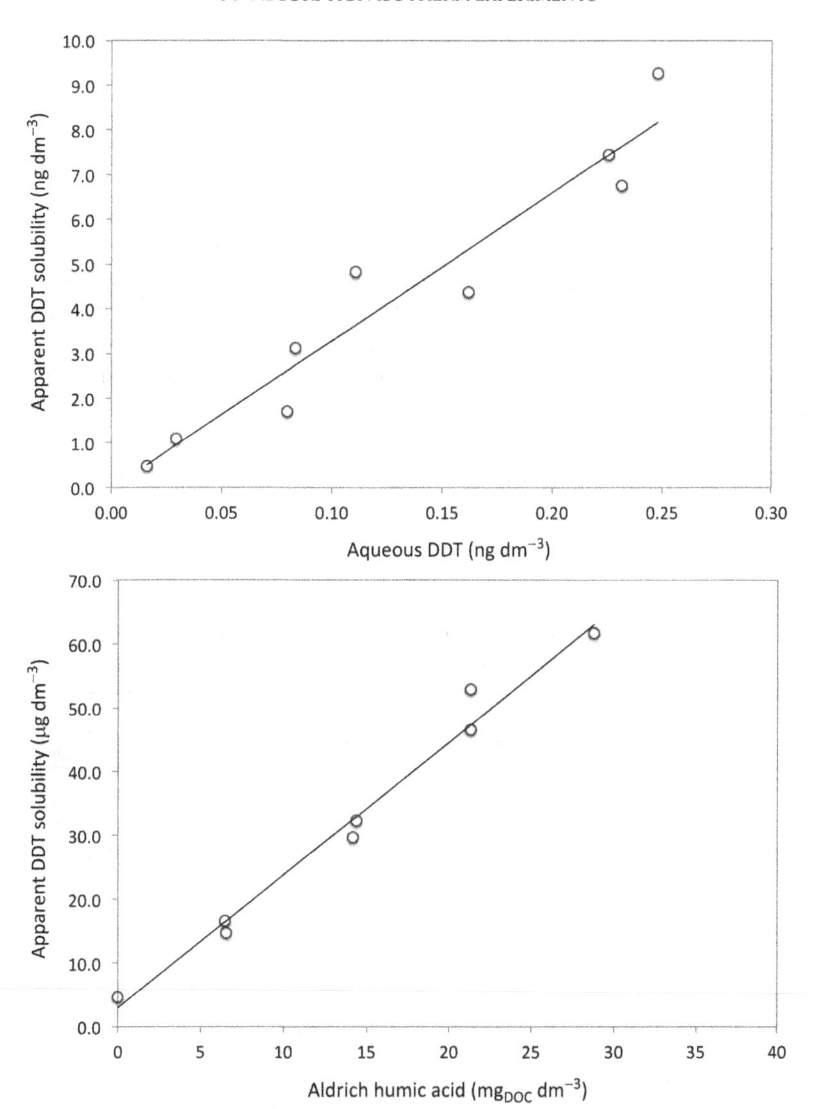

FIG. 8.10 These graphs plot DDT (1,1,1-trichloro-2,2-bis(4-chlorophenyl)ethane, CAS registry number 50-29-3) *molecular adsorption* by "humic acids" (Aldrich, sodium humate: AHA) in dilute aqueous solution. Aqueous DDT solubility is $s_{DDT} = 5.5 \times 10^{-6}$ g dm^{-3} at 25°C. *Upper graph* (Carter and Suffet, 1982) plots DDT apparent solubility in solutions containing $(m_{AHA}/V) = 15.4$ mg dm^{-3} "humic acid" (pH = 8.3) as a function of aqueous DDT concentration (permeate cell): $c_{DDT} < 5.5 \times 10^{-6}$ g dm^{-3}. *Lower graph* (Chiou et al., 1987) plots DDT apparent solubility in solutions containing $(m_{AHA}/V) < 35$ mg dm^{-3} "humic acid" (pH = 6.5) where aqueous DDT remains saturated (permeate cell): $s_{DDT} = 5.5 \times 10^{-6}$ g dm^{-3}.

EXAMPLE 8.4

Example Permalink

http://soilenvirochem.net/T396b7

Fit the DDT adsorption isotherm data plotted in Fig. 8.10 using linear regression to determine the organic-carbon partition constant $K^{\ominus}_{oc/w}$ for a molecular "humic acid" solution.

The adsorptive is DDT (1,1,1-trichloro-2,2-bis(4-chlorophenyl)ethane, CAS registry number 50-29-3), a nonpolar organic insecticide.[9] The adsorbent is a molecular "humic acid" solution. Supramolecular associations of individual "humic acid" molecules create hydrophobic domains suitable for DDT adsorption-by-partitioning. The experimental adsorption data appear in Table 8.8.

DDT mass ratios from the permeate cell $m_{DDT(p)}/m_W$ appear in column 1, Table 8.8. Retentate DDT mass ratios normalized by retentate-cell water mass $m_{DDT(p)}/m_W$ appear in column 2. The retentate cell in each experiment contained $c_{humate} = 1.54 \times 10^{-2}\ g_{humate}\ dm^{-3}$. The DDT mass ratios listed in column 3 are normalized by the "humic acid" mass concentration the retentate cell $m_{DDT(p)}/m_{humate}$.

TABLE 8.8 Equilibrium Adsorptive DDT Concentrations in the Permeate and Retentate Cells of a Dialysis Chamber

Permeate, $m_{DDT(p)}/m_W$ (ng g_W^{-1})	Retentate, $m_{DDT(r)}/m_W$ (ng g_W^{-1})	Retentate, $n_{DDT(r)}/m_{humate}$ (ng g_{humate}^{-1})
0.016	0.048	3.12E+03
0.030	0.113	7.35E+03
0.080	0.178	1.16E+04
0.084	0.320	2.08E+04
0.110	0.495	3.21E+04
0.162	0.453	2.94E+04
0.226	0.767	4.98E+04
0.231	0.699	4.54E+04
0.248	0.950	6.17E+04

Notes: The adsorptive DDT is radiolabeled with $^{14}_{6}C$. The molecular adsorbent—Aldrich "humic acid," an alkaline extract of a soft brown coal—is confined to the retentate cell.
From Carter, C.W., Suffet, I.H., 1982. Binding of DDT to dissolved humic materials. Environ. Sci. Technol. 16 (11), 735–740.

Step 1. Normalize the retentate-cell apparent solubility values listed in column 2 by the humate mass ratio to generate the values listed in column 3.

$$\frac{n_{DDT(r)}}{m_{humate}} = \frac{n_{DDT(r)}/m_W}{m_{humate}/m_W} = \frac{n_{DDT(r)}/m_W}{1.52 \times 10^{-2}\ g_{humate}\ g_W}$$

Step 2. Performed a linear regression of the data in columns 1 and 3, Table 8.8 to estimate the DDT partition constant $K^{\ominus}_{\text{humate/w}}$.

$$x' \equiv m_{\text{DDT}(p)}/m_{\text{w}}$$
$$y' \equiv m_{\text{DDT}(r)}/m_{\text{humate}}$$
$$\hat{y} = K^{\ominus}_{\text{humate/w}} \cdot x'$$
$$K^{\ominus}_{\text{humate/w}} = 2.22 \times 10^5$$
$$\log_{10}\left(K^{\ominus}_{\text{humate/w}}\right) = 5.35$$

Using the organic carbon content of Aldrich "humic acid" ($f_{\text{oc}} = 0.549\ \text{g}_{\text{oc}}\ \text{g}^{-1}_{\text{humate}}$), compute the $K^{\ominus}_{\text{oc/w}}$.

$$K^{\ominus}_{\text{oc/w}} = \frac{K^{\ominus}_{\text{humate/w}}}{f_{\text{oc}}} = \frac{2.22 \times 10^5}{0.549} = 4.04 \times 10^5$$
$$\log_{10}\left(K^{\ominus}_{\text{oc/w}}\right) = 5.61$$

Carter and Suffet (1982, Fig. 2 and Table II, row 8, columns 6 and 7) quote both partition constants.

[9] The US Environmental Protection Agency banned DDT in 1972.

8.3.6.3 *Chemical Basis for Partitioning Adsorption*

The slope of the linear partition isotherm (cf. Figs. 8.4, 8.7, and 8.10) is the equilibrium *partition constant* $K^{\ominus}_{\text{oc/w}}$ (expression 8.23) for adsorptive A partitioning into a organic matter molecular-association adsorbent from aqueous solution.[10]

Substances that have a very low affinity for organic matter molecular-association adsorbents have partition constant near zero while substances with very high partition constant $K^{\ominus}_{\text{oc/w}}$ tend to be highly nonpolar organic compounds with very low water solubility.

As pointed out earlier, Langmuir-type site-limited adsorption isotherms (expression 8.16) at low absorptive concentration c_{A} range (i.e., $k_{\text{L}} \cdot c_{\text{A}} \ll 1$) and ion-exchange adsorption isotherms (cf. Fig. 8.6) approach a linear limit. Linear partitioning isotherms involve a specific adsorbent—amphiphilic organic matter micellar colloids or subcolloidal amphiphilic organic matter molecular-aggregates—and a particular type of absorptive: nonpolar organic molecules. True linear partitioning isotherm represents what humic chemists believe is fundamentally a partitioning process; nonpolar organic adsorptives partition between two solvents: the aqueous solvent and hydrophobic *solvent-like* domains formed when organic matter

[10] The International Union of Pure and Applied Chemistry considers *partition coefficient* an obsolete term and does not recommend use of *distribution coefficient*. The *distribution ratio D*, reserved for empirical ratios representative of solvent extraction procedures, does not meet the criteria for an equilibrium constant.

amphiphiles aggregate in solution (Fig. 8.10) or on the surfaces of the clay-size fraction in soils and sediments (Figs. 8.4, 8.7, and 8.10).

Small nonpolar compounds typically have a high vapor pressure and readily evaporate from water rather than adsorb to molecular-association colloids. Vapor pressure decreases as molecular weight increases, increasing the tendency of nonpolar organics to adsorb to organic matter colloids. Water solubility correlates with the overall polarity of organic compounds and is a major factor determining the partitioning between aqueous solution and hydrophobic domains of organic matter colloids.

Lambert et al. (1965) cited earlier studies of pesticide adsorption by soil with varying organic matter contents that confirm the prevalence of linear partition isotherms for nonpolar pesticide adsorption by whole soils. While Lambert's simple partitioning model identifies *organic carbon* as the *organic phase* implied by partition constant for nonpolar adsorptive A (expression 8.23), but does not provide a basis for relating the partition constant for adsorptive A to its molecular structure.

Lambert (1967, 1968) applied the *linear free energy principle* to predict changes in $K^{\ominus}_{oc/w}$ from altering molecular volume by adding or replacing substituents.[11] Lambert's linear free energy relationship for predicting $K^{\ominus}_{oc/w}$ is based the *principle of independent surface action* introduced by Langmuir (1925).

> In the case of the molecules of organic substances of non-polar type the so-called physical properties are roughly additive. For example, the addition of each $-CH_2-$ to a hydrocarbon chain in most compounds containing such chains increases the [molar] volume, raises the boiling point, and alters the solubilities in approximately the same way. It is reasonable to assume, therefore, that the field of force about any particular group or radical in a large organic molecule is characteristic of that group and, as a first approximation, is independent of the nature of the rest of the molecule. For convenience we shall refer to this as the *principle of independent surface action*.

Verbruggen et al. (2001) demonstrate the linear dependence of aqueous solubility s, the logarithm of the octanol-water partition constant $\log_{10}(K^{\ominus}_{o/w})$, and vapor pressure p on molar volume V_m in their review of the physicochemical properties of nonaromatic hydrocarbon. Consequently, any change in molecular structure that increases the *nonpolar molecular volume* decreases water solubility, increasing partitioning from water into an immiscible organic phase (e.g., 1-octanol). Similarly, any change in molecular structure that increases the *nonpolar molecular surface area* will have a similar effect: decreased water solubility and increased water-to-organic carbon partition constant $K^{\ominus}_{oc/w}$.

This model for predicting the organic carbon partition constant anticipates increasing water solubility and decreasing $K^{\ominus}_{oc/w}$ from adding polar fragment to a molecular structure. Adding polar groups to a nonpolar molecule reduces the energy cost of creating a solute cavity in water. Lambert (1967, 1968) found a linear correlation between the *parachor* P of similar nonpolar organic adsorptives and the organic carbon partition constant $K^{\ominus}_{oc/w}$.

[11] Lambert (1967, 1968) quantified molecular volume using *parachor* P. Parachor P is a function of a compound's molar volume V_m and the compound's gas/liquid surface tension $\gamma_{G/L}$: $P = V_m \cdot \gamma_{G/L}^{1/4}$.

BOX 8.1

BIOMAGNIFICATION AND BIOACCUMULATION

The Office of Chemical Safety and Pollution Prevention (US Environmental Protection Agency) maintains an internet site called the *PBT Profiler* (USEPA, 2012) that estimates a variety of important environmental properties of organic compounds relating to environmental persistence, bioaccumulation, and toxicity. The Office of Chemical Safety and Pollution Prevention defines biomagnification as "the concentration of a chemical to a level that exceeds that resulting from its diet." Biomagnification is quantified (expression 8.25) using the *biomagnification factor* (BMF), where c_B indicates the concentration in an organism (e.g., zooplankton, fish, etc.) and c_A the total concentration in the organism's diet.

$$BMF \equiv \frac{c_B}{c_A} \tag{8.25}$$

Bioaccumulation encompasses both *bioconcentration* and biomagnification, where the former relates to uptake from all sources and the latter specifically relates to diet. Bioconcentration is quantified (expression 8.26) using the *bioconcentration factor* (BCF), where c_W is the total concentration in water.

$$BCF \equiv \frac{c_B}{c_W} \tag{8.26}$$

Mackay and Fraser (2000) reviewed bioaccumulation models and methods for estimating *BCF*. A popular approach draws on the extensive octanol-water partition constant $K^{\ominus}_{o/w}$ database for organic chemicals (Mackay, 1982).

These principles underlying linear partition isotherms for nonpolar organic adsorptives by organic matter colloids were subsequently confirmed and extended by Chiou et al. (1979), Doucette and Andren (1987, 1988), and Shiu et al. (1988). Some of these later studies correlate the molecular volume and total molecular surface area of nonpolar organic compounds with the octanol-water partition constant $K^{\ominus}_{o/w}$. Octanol-water partition constants interest environmental chemists because $K^{\ominus}_{o/w}$ constants correlates with fat solubility of a compound and the tendency for biomagnification (Box 8.1).

8.3.7 Freundlich Adsorption Isotherm

Freundlich (1909) published an empirical, two-parameter isotherm resembling the simple partition isotherm discussed earlier (expression 8.27). Unlike the two-parameter Langmuir isotherm (expression 8.16), the *Freundlich* isotherm (expression 8.27) is strictly empirical and has no apparent adsorption maximum.

$$\frac{n^{\sigma}_{A}}{m} = k_F \cdot c_A^{1/n} \tag{8.27}$$

Scientists studying the adsorption of nonpolar organic adsorptives by whole soils and sediments containing natural organic matter report instances where the adsorption isotherm is not linear. Deviations from linear partitioning isotherms take different forms.

Karickhoff et al. (1979) reported a consistent deviation as the adsorptive aqueous concentration approached its solubility limit. The deviation, described in the following quote, was attributed to the possible precipitation of a crystalline phase of the nonpolar adsorptive rather than a change in the adsorption mechanism.

> Deviations from linearity (least squares fitted) approximated deviations in replicate determinations for individual isotherm points. As the sorbate water concentration approached 60-70% of the sorbate aqueous solubility, the isotherms typically bent upward, indicative of increased sorption.

Karickhoff (1981) revised the nonlinearity issue in a later review paper that specifically mentions widespread use of the *Freundlich* isotherm (expression 8.26) to fit nonlinear adsorption isotherms. Karickhoff (1981) argued that these deviations from nonlinear adsorption result from nonideal solute behavior requiring activity coefficients (expression 8.28).

$$K^{\ominus}_{o/w} = \frac{a_{A(org)}}{a_{A(aq)}} = \frac{\gamma_{c,\,A(org)} \cdot c_{A(org)}}{\gamma_{c,\,A(aq)} \cdot c_{A(aq)}} \tag{8.28}$$

$$K^{c}_{o/w} = \frac{c_{A(org)}}{c_{A(aq)}} \approx K^{\ominus}_{o/w} \cdot \left(\frac{\gamma_{c,\,A(aq)}}{\gamma_{c,\,A(org)}}\right) \tag{8.29}$$

Karickhoff (1981) noted that adsorption isotherms were linear at the low solute concentration limit where activity coefficients in both aqueous and adsorbed phases approach unity. Deviations from linearity, implying a concentration-dependent partition ratio $K^{c}_{o/w}$ (expression 8.29) appear only at the high solute concentration limit.

One alternative explanation for nonlinear adsorption isotherms arose from evidence that the adsorption isotherms for nonpolar organic adsorptives by whole soil or sediment adsorbents changed gradually over time. The *Distributed-Reactivity* model (Weber et al., 1992) assumes the nonpolar organic adsorptives by whole soil or sediment is best described by the generic site-limited adsorption isotherm (8.16). Soils and sediments contain a heterogeneous mixture of many adsorbents, each with its characteristic isotherm (hence, *distributed reactivity*). The observed adsorption isotherm, according to the *Distributed-Reactivity* model, is a linear combination of site-limited isotherms whose composite isotherm is approximated by the empirical Freundlich isotherm (expression 8.27).

The *Dual-Domain* model (Xing and Pignatello, 1996; Xing et al., 1996; Xing and Pignatello, 1997) assumes nonlinear behavior results from the site-limited adsorption at the surface of "holes" within the organic matter film coating mineral particles. Accardi-Dey and Gschwend (2002, 2003) assume solid particles of *black carbon* (i.e., carbon soot) mixed in with organic matter molecules provide the solid surface required for site-limited adsorption. The net isotherm, in both cases, is a linear combination of the partition isotherm (expression 8.23) and the generic site-limited isotherm (expression 8.16).

The *Distributed-Reactivity* and the *Dual-Domain* models seek to explain two phenomena: nonlinearity in the adsorption isotherms for nonpolar organic adsorptives by whole soil or sediment and time-dependent change in Freundlich isotherm parameters $\{k_F, n\}$ fitted to the data. Regardless, the adsorption isotherms for nonpolar organic adsorptives by dissolved organic matter and organic matter colloidal dispersions are consistently linear. While there are insufficient results to decide between the competing models, the contribution of *black carbon* remains a credible explanation for nonlinearity.

8.4 pH-DEPENDENT SURFACE CHARGE

Environmental acidity is a complex phenomenon discussed in Chapter 6. Suffice it to say at this point, pore water pH in soils, sediments, and aquifers covers a relatively large range. Natural biological and geochemical processes result in pore water acidity as low as pH 4 and basicity as high as pH 9. Chemists find ion adsorption by environmental colloids varies significantly with pore water pH. Some of this could be explained by the hydrolysis of weak acid groups in natural organic matter, but weak acid hydrolysis of organic matter could not explain all of the variability that the chemists measured. Clay-size oxide minerals account for much of the observed pH-dependent adsorption in soil and sediments.

One method of measuring pH-dependent surface charge relies on recording solution pH while titrating an aqueous adsorbent suspension with a standard strong acid or strong base.[12] The *proton surface-charge* experiment measures the relative adsorption of hydrogen ions H^+ or hydroxyl ions OH^- by the adsorbent over a specified solution pH range.

The strong acid concentration c_{acid} and base concentration c_{base} of the titrating solutions must be precisely measured because the mineral adsorbents may adsorb hydrogen ions H^+ over a characteristic pH interval then desorb hydrogen ions H^+ over another interval. Furthermore, this type of adsorption experiment cannot distinguish between hydrogen ion H^+ *adsorption* and hydroxyl ion OH^- *desorption*.

Parks and De Brnyn (1962) defined of proton surface charge Γ_{H^+} $[\text{mol}_c\ \text{m}^{-2}]$ using expression (8.30), where a_s is the specific surface area of a mineral adsorbent and m/V is the adsorbent mass concentration.[13]

$$\Gamma_{H^+} = \frac{\left(c_{acid} - 10^{-pH}\right) + \left(c_{base} - 10^{14-pH}\right)}{a_s \cdot (m/V)} \tag{8.30}$$

Zhou and Gunter (1992) measured the pH-dependent proton surface charge Γ_{H^+} of kaolinite $Al_4Si_4O_{10}(OH)_8(s)$, comparing their results with earlier pH-dependent proton surface charge Γ_{H^+} measurements of silica ($SiO_2 \cdot nH_2O(s)$) (Bolt, 1957) and alumina ($Al_2O_3 \cdot nH_2O(s)$) (Sprycha, 1989). Kaolinite is a clay mineral with layers composed of silica and alumina (cf. Figs. 3.9 and 3.10) and develops surface charge at its edge surface. Fig. 8.11 reveals a considerably greater surface charge than its component oxides. The kaolinite pH-dependent proton surface charge Γ_{H^+} cannot be considered the additive sum of the two components.

[12] Measuring pH-dependent surface charge is fundamentally a pH-dependent ion adsorption.

[13] Parks and De Brnyn (1962) used σ_{H^+} $[\text{mol}_c\ \text{m}^{-2}]$ to represent proton surface charge. Expression (8.30) uses the recommended surface excess concentration.

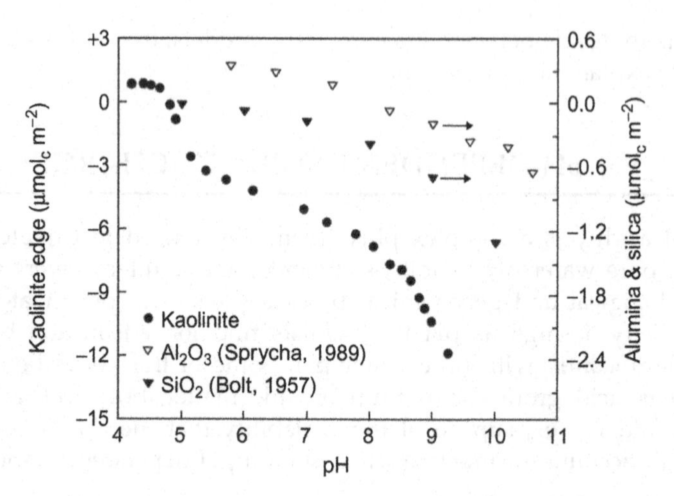

FIG. 8.11 The pH-dependent proton surface charge Γ_{H^+} of the mineral kaolinite $Al_4Si_4O_{10}(OH)_8(s)$ from Zhou and Gunter (1992), with additional data published previously by Bolt (1957) and Sprycha (1989).

8.4.1 Weak Acid-Conjugate Base Proton Adsorption Model

Parks and De Brnyn (1962) were the first to suggest that hydrogen ion H^+ adsorption by mineral surfaces could be described by a hydrolysis reaction with a characteristic equilibrium constant. They believed a charge-neutral mineral surface occurs at a characteristic pH depending on mineral chemical composition. The charge-neutral mineral surface develops a *positive* surface charge through hydrogen ion H^+ adsorption as illustrated in reaction (8.R5), with the neutral amine RNH_2 designated as the conjugate base.

$$\underset{\text{Weak base}}{RNH_2} + H_3O^+(aq) \longleftrightarrow \underset{\text{Conjugate acid}}{RNH_3^+} + H_2O(l) \tag{8.R5}$$

The apparent desorption of hydroxyl ions OH^- from a weak base surface sites (reaction 8.R6) is equivalent to hydrogen ion H^+ adsorption by a weak base surface site (reaction 8.R5).

$$\underset{\text{Weak base}}{RNH_2} + H_2O(l) \longleftrightarrow \underset{\text{Conjugate acid}}{RNH_3{}^+} + OH^-(aq) \tag{8.R6}$$

The charge-neutral mineral surface develops a *negative* surface charge through hydrogen ion H^+ desorption as illustrated by reaction (8.R7), with the neutral carboxyl RCOOH acting as the weak acid.

$$\underset{\text{Weak acid}}{RCOOH} + H_2O(l) \longleftrightarrow \underset{\text{Conjugate base}}{RCOO^-} + H_3O^+(aq) \tag{8.R7}$$

In this case the apparent hydroxyl ions OH^- adsorption to weak acid surface sites (reaction 8.R8) is equivalent to hydrogen ion H^+ desorption by a weak acid surface site (reaction 8.R7).

$$\underset{\text{Weak acid}}{RCOOH} + OH^-(aq) \longleftrightarrow \underset{\text{Conjugate base}}{RCOO^-} + H_2O(l) \tag{8.R8}$$

The model proposed by Parks and de Bruyn, and later Stumm and Morgan (1970), is shown in reactions (8.R9), (8.R10).

The positively charged oxide surface site[14]—representing the oxide mineral S/L interface at low pH—is a diprotic weak acid MOH_2^+. Acid hydrolysis reaction (8.R9) releases a single hydrogen ion H^+ to generate a neutral site MOH^0. The neutral site MOH^0 is, in fact, the *monobasic* conjugate base of the positively charged diprotic weak-acid site MOH_2^+. The acid hydrolysis equilibrium constant for reaction (8.R9) is K°_{a1}.

$$\underset{\text{Diprotic weak acid}}{MOH_2^+} + H_2O(l) \longleftrightarrow \underset{\text{Conjugate base}}{MOH_2^{\ 0}} + H_3O^+(aq) \quad (8.R9)$$

Acid hydrolysis reaction (8.R10) releases a second hydrogen ion H^+ to generate a negatively charged site MO^-. The negatively charged site MO^-—representing the oxide mineral S/L interface at high pH—is the dibasic conjugate base of the positively charged diprotic weak acid site MOH_2^+. The acid hydrolysis equilibrium constant for reaction (8.R10) is K°_{a2}.

$$\underset{\text{Conjugate base}}{MOH^0} + H_2O(l) \longleftrightarrow \underset{\text{Conjugate base}}{MO^-} + H_3O^+(aq) \quad (8.R10)$$

Reactions (8.R9), (8.R10) rationalize surface-site charge and its relation to hydrogen ion H^+ desorption when an acidic oxide mineral suspension is titrated by a strong base. Conversely, reversal of reactions (8.R9), (8.R10) relate surface-site charge to hydrogen ion H^+ adsorption when a basic oxide mineral suspension is titrated by a strong acid. It implies positively charged oxide mineral surfaces are populated by water molecules under acidic conditions and that negatively charged oxide surfaces are populated by oxygen anions under basic conditions.

The following two sections describe the experiments that measure pH-dependent ion adsorption by mineral colloids and pH-dependent surface charge, the latter being a major determinant of pH-dependent ion adsorption.

8.4.2 Crystallographic Proton Adsorption Models

Chemical weathering transforms igneous minerals into secondary minerals. Of particular interest to surface chemistry and adsorption are the mineral colloids of the clay-size fraction. Having discussed the structure and cation-exchange properties of the clay mineral group (cf. Chapters 3 and 4), we turn our attention to the S/L interface of colloidal oxide minerals.

The outermost atomic layer at the mineral/water interface of all oxide and silicate minerals consists of oxygen ions at various stages of protonation. Hence, hydrogen ion H^+ adsorption and desorption reactions (cf. reactions 8.R9, 8.R10) are a reasonable representation of the surface charging process.

[14] The diprotic weak-acid oxide surface site MOH_2^+ develops a positive charge through hydrogen ion H^+ adsorption, the reverse of reaction (8.R9).

BOX 8.2

PROTON SURFACE SITE DENSITY: CLOSE-PACKED OH^- SPHERES

The nature of chemisorption surface sites in the Langmuir model (Langmuir, 1916) and the pH-dependent surface charge model (e.g., Stumm and Morgan, 1970) are fundamentally quite similar. An implicit assumption of all chemisorption models is the adsorption complex occupies a finite "footprint" at the S/L interface that ultimately determines the surface excess concentration Γ_A at the adsorption maximum.

If the Parks and de Bruyn pH-dependent surface charge model represented in reactions (8.R9), (8.R10) is chemically reasonable, then the maximum oxide proton surface charge Γ_{H^+} should approximate the molar area of close-packed hydroxide OH^- ions.

Expression (8.31) gives the relationship between the *molar surface area* A_m of a monolayer of close-packed hydroxide OH^- ions; N_A is the Avogardo constant and a_m is the area occupied by a single OH^- ion in a closed-packed monolayer (Fig. 8.12).

$$A_m = N_A \cdot a_m \quad (8.31)$$

The Pauling ionic radius of a OH^- ion is $R_{OH^-} = 140$ pm, resulting in $a_m \approx 8.32 \times 10^{+4}$ pm^2 and a molar area $A_m \approx 5.01 \times 10^4$ $m^2\ mol^{-1}$.

$$a_m = \left(\pi \cdot R^2_{OH^-}\right) \cdot \left(3 \cdot \sqrt{2}/\pi\right)$$

$$a_m = \left(6.16 \times 10^4\ pm^2\right) \cdot (1.35)$$

$$= 8.32 \times 10^4\ pm^2$$

$$A_m = \left(6.022 \times 10^{23}\ mol^{-1}\right) \cdot \left(8.32 \times 10^4\ pm^2\right)$$

A monolayer of close-packed OH^- ions results in a surface excess concentration $\Gamma_{H^+(m)} \approx 20\ \mu mol\ m^{-2}$.

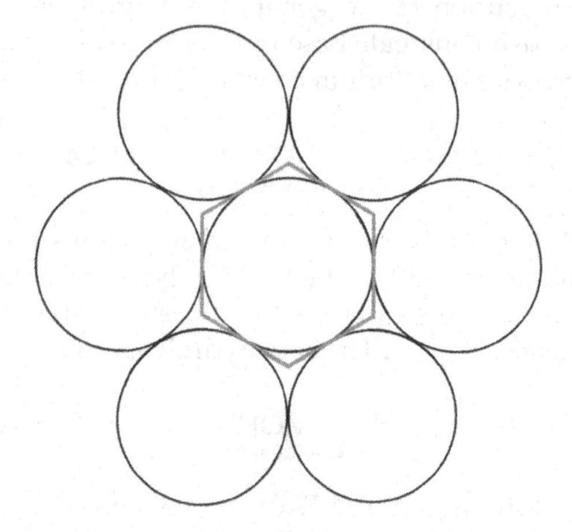

FIG. 8.12 The effective area occupied by a single sphere in a monolayer of close-packed spheres is $\left(3 \cdot \sqrt{2}/\pi\right) \approx 1.35$ times the cross-sectional area of the sphere.

$$\Gamma_{H^+(m)} = A_m^{-1} = 2.00 \times 10^{-5}\ mol\ m^{-2}$$

$$= 20\ \mu mol\ m^{-2}$$

Schindler and Kamber (1968) estimate the area density of proton adsorption sites on silica gel (colloidal silica $SiO_2(s)$) to be $\Gamma_{H^+(m)} \approx 4\ \mu mol_c\ m^2$. The limiting proton surface charge of kaolinite (cf. Fig. 8.11) is $\Gamma_{H^+(m)} \approx 12\ \mu mol_c\ m^2$ at pH 9.

The following section will using oxide crystal structures to show the maximum surface site density is the area occupied by two O^{2-} (or a hydroxyl OH^-) ions, reducing the maximum H^+ surface excess concentration to $\Gamma_{H^+(m)} \approx 10\ \mu mol\ m^2$.

8.4.2.1 Valence-Bond Proton Adsorption Model: Gibbsite (100) Surface

The crystal structure of gibbsite $Al(OH)_3(s)$ consists of atomic planes composed of oxygen O^{2-} anions in a close-packed arrangement. A plane of cations lies between each oxygen O^{2-} anion plane, a plane of H^+ cations alternates with a plane of Al^{3+} cations (Fig. 8.13).

The Al^{3+} cations occupy two-thirds of the octahedral sites formed by the close-packed O^{2-} anion plane on either side (cf. Figs. 8.12 and 8.13). The gibbsite $Al(OH)_3(s)$ unit cell contains two octahedral sheets and a total of eight formula units, four formula units per sheet.

The high-resolution electron micrograph (Fig. 8.14, left) reveals the common hexagonal gibbsite crystal habit. The gibbsite crystal structure (Fig. 8.14, right) delineates the three symmetry-equivalent crystal planes (100), (110), and $(1\bar{1}0)$ that appear in the micrograph (Fig. 8.14, left).

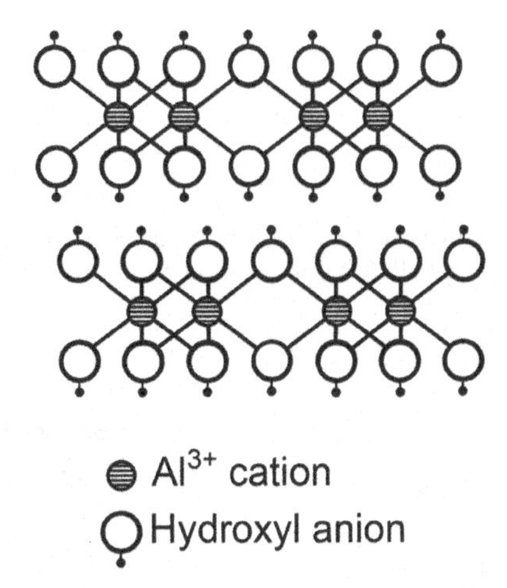

FIG. 8.13 Idealized gibbsite $Al(OH)_3(s)$ crystal structure showing octahedral Al^{3+} coordination and hydroxyl OH^- ions. The O–H vector and the relative position of each later in the actual crystal structure allow each hydroxyl OH^- ion to donate and receive a hydrogen bond.

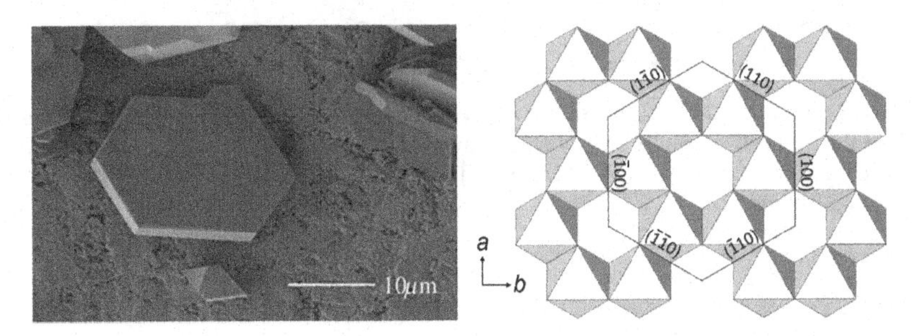

FIG. 8.14 Well crystallized gibbsite $Al(OH)_3(s)$ will often appear as hexagonal plates (*left*). The top and bottom of the plates is the (001) surface while the hexagonal faces are (100) and (110) surfaces (*right*). The hexagonal surfaces are equivalent by symmetry.

The choice of unit cell boundary is somewhat arbitrary but crystallographers prefer boundaries that emphasize crystal structure symmetry. Symmetry preserving boundaries are often drawn through atoms. When this happens the unit cell stoichiometry counts half of each atom lying on a plane.[15] Cleavage through a single gibbsite layer to form a conceptual model of the (100) surface is illustrated in Fig. 8.15.

Half of the hydroxyl OH^- ions lying in the gibbsite (100) plane are lost creating one 5-coordinate Al^{3+} ion in each surface unit cell. The adsorption of a single H_2O molecule to each surface unit cell restores sixfold coordination to all Al^{3+} ions at the gibbsite (100) surface. The surface unit cell formula becomes $Al_4(OH)_{12} \cdot H_2O$.

The conceptual model of the gibbsite (100) surface (Fig. 8.15, left) is the simplest model of the gibbsite (100) surface and suggests a charge-neutral surface is populated by an equal number of adsorbed H_2O molecules and OH^- ions. We can evaluate the stability of this crystallographic model using Pauling's second rule (Pauling, 1929): the *electrostatic valence principle*.

> Let (ze) be the electric charge of a cation and ν its coordination number. Then the strength of the *electrostatic valence bond* going to each corner of the polyhedron of anions about it is defined as:
>
> $$s \equiv \frac{z}{\nu} \tag{8.32}$$
>
> Let $(-\zeta e)$ be the charge of the anion located at a corner shared among several polyhedra. We now postulate the following *electrostatic valence principle*: II. In a stable coordination structure the electric charge of each anion

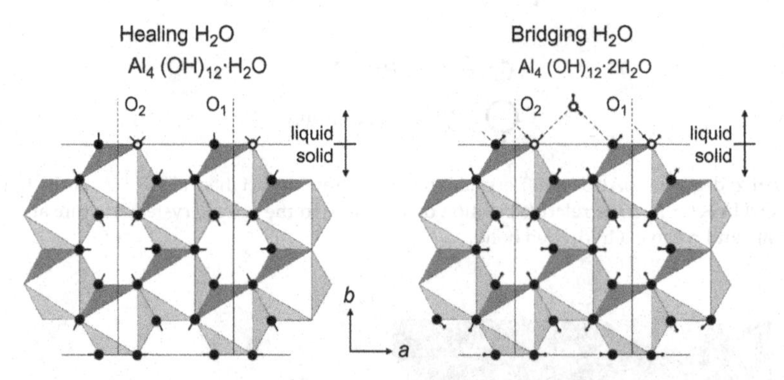

FIG. 8.15 Two conceptual models of the gibbsite $Al(OH)_3(s)$ (100) surface. The diagrams show one choice of unit cell boundary and the unit cell composition of a single-layer $Al_4(OH)_{12}$. The *filled circles* are hydroxyl OH^- ions, a *short bar* to represent the orientation of each O–H bond. The *open circle with two short bars* represent adsorbed H_2O molecules. See text for details.

[15] If the atom lies on the intersection of two planes the stoichiometry counts one-fourth. If the atom lies at the intersection of three planes the stoichiometry counts one-eighth.

tends to compensate the strength of the electrostatic valence bonds reaching to it from the cations at the centers of the polyhedra of which it forms a corner; that is, for each anion:

$$\zeta = \sum_i \left(\frac{z_i}{\nu_i} \right) \tag{8.33}$$

The *electrostatic valence bond* of each Al–O bond in gibbsite is $s_{Al-O} = (3/6)$ while that of each H–O bond $s_{H-O} = (1/1)$. The *electrostatic valence principle* is satisfied in gibbsite because $\zeta_O = 2 \cdot s_{Al-O} + s_{H-O} = 2$. This is not the case for the two oxygen ions coordinating Al^{3+} ions at the gibbsite (100) surface. The surface hydroxyl ion OH_1^- is coordinate by one Al^{3+} ion and one H^+ ion while the surface water molecule $OH_2{}^0$ is coordinate by one Al^{3+} ion and two H^+ ions.

$$\zeta_{O_1} = s_{Al-O} + s_{H-O} = 1\tfrac{1}{2}$$
$$\zeta_{O_2} = s_{Al-O} + 2 \cdot (s_{H-O}) = 2\tfrac{1}{2}$$

Pauling (1929) stated: "It is not to be anticipated that [the *electrostatic valence principle*] will be rigorously satisfied by all crystals. It should, however, be always satisfied approximately." The deviations for the surface hydroxyl ion OH^- and water molecule O_2H at the gibbsite (100) are substantial. What is missing is the impact of *hydrogen bonding*, which is illustrated by the second conceptual model appearing at the right in Fig. 8.15 where a water molecule bridges the two surface oxygen ions through hydrogen bonds.

Pauling's second principle has proven very powerful but has deficiencies. It assumes that the cation remains at the center of its coordination polyhedron rather than being displaced to some off-center position. The best example of this is the hydrogen bond O–H···O.

Most hydrogen bonds are asymmetric: the donor O–H bond length is much shorter than the receptor H···O. The *bond-valence theory* of I. David Brown (Brown and Shannon, 1973; Brown and Wu, 1976; Brown, 1976; Brown and Altermatt, 1985) provides a much improved explanation of hydrogen bonding.

Brown (1976) published a detailed analysis of hydrogen-bond geometry in crystals that documented the asymmetric tendency of many hydrogen bonds O–H···O. Off-center displacement of H^+ lengthens the nonbonded O···O distance and shifts the H^+ off the nonbonded O···O axis resulting in a bent hydrogen bond.

> The asymmetry of hydrogen bonds arises from the repulsion between the O atoms forming the bonds. A bond-valence analysis of the repulsion leads to the conclusion that strong and weak hydrogen bonds are different in kind, the stronger ones (O···O less than 2.7 Å) involve strain and are *linear* while the weaker ones (O···O greater than 2.7 Å) have an extra degree of freedom and are generally bent.

Bond-valence analysis of the gibbsite (100) showing hydrogen bonding and a bridging H_2O (Fig. 8.15, right) illustrates how hydrogen bonding with water at the interface will lead to a more stable bonding environment for the ions a the mineral surface. The hydroxide OH^- ion (O_1) will donate a hydrogen bond to the bridging H_2O water molecule and receive a hydrogen bond from another water molecule in the vicinity.

Hydroxide OH^- ion (O_1) will form two hydrogen bonds, one will be asymmetric the other will be symmetric. The asymmetric hydrogen bond results in a *short* O–H donor bond with a bond valence approaching unity. The second symmetric hydrogen bond results in a *short* H···O receptor bond with a bond valence less than 1/2. The hydrogen bonding configuration for hydroxide OH^- ion (O_1) will *increase* the O_1 bond-valence sum.

$$\zeta_{O_1} > 1\tfrac{1}{2}$$

Water molecule H_2O (O_2) will also form a two hydrogen bonds; both will be symmetric. Symmetric hydrogen bonds result in *long* O–H donor bonds with bond valences approaching 1/2. The hydrogen bonding configuration for water molecule H_2O (O_2) will *decrease* the O_2 bond-valence sum.

$$\zeta_{O_2} < 2\tfrac{1}{2}$$

The bridging water molecule H_2O (Fig. 8.14, right)—spanning surface-bound OH^- and H_2O—determines site dimensions on the gibbsite (100) surface. The effective surface-site area—unit cell dimension a times the thickness ($c/2 \cdot \sin\beta$) of a single gibbsite layer—is about three times greater than the area of a single hydroxide ion (cf. Box 8.2).

$$a_{\mathrm{m}} = a \cdot (c/2) \cdot \sin\beta \tag{8.34}$$

$$a_{\mathrm{m}} = 507.8\ \mathrm{pm} \cdot 486.8\ \mathrm{pm} \cdot (0.997) = 2.46 \times 10^5\ \mathrm{pm}^2$$

$$A_{\mathrm{m}} = \left(2.46 \times 10^5\ \mathrm{pm}^2\right) \cdot 6.022 \times 10^{23}\ \mathrm{mol}^{-1} = 1.48 \times 10^5\ \mathrm{m}^2\ \mathrm{mol}^{-1}$$

$$\Gamma_{\mathrm{H}^+(m)} = A_{\mathrm{m}}^{-1} = 6.74 \times 10^{-6}\ \mathrm{mol\ m}^{-2} = 6.74\ \mu\mathrm{mol\ m}^{-2} \tag{8.35}$$

The third, and final, consequence of the gibbsite (100) surface site is an entirely different perspective of surface-site hydrolysis. The neutral surface (illustrated in Fig. 8.15, right and Fig. 8.16, center) consists of an equal number of surface-bound OH^- ions and H_2O molecules.

Imagine the adsorption of a strong acid HCl to the neutral gibbsite (100) surface site. The H^+ adsorbs to the surface-bound OH^- anion forming a surface-bound H_2O molecule (Fig. 8.16, lower right). In acid-base terms, the surface-bound OH^- anion is a weak base and the surface-bound H_2O molecule is its conjugate acid.

In surface-chemistry terms, the surface complex formed when a H^+ ion adsorbs to the surface-bound OH^- anion is a surface-bound H_2O molecule. The pair of surface-bound H_2O molecules constitute a positively charged surface site on the gibbsite (100) surface formed at low pH with the Cl^- co-ion taking the place of the bridging water molecule.

Now, imagine the surface hydrolysis induced strong base NaOH to the neutral gibbsite (100) surface site (Fig. 8.14, center). The OH^- ion accepts a H^+ from a surface-bound H_2O molecule resulting in two surface-bound OH^- (Fig. 8.16, lower left). In acid-base terms, the surface-bound H_2O molecule is a weak acid and the surface-bound OH^- is its conjugate base. The pair of surface-bound OH^- ions are the negatively charged surface site formed at high pH with the Na^+ taking the place of the bridging water molecule.

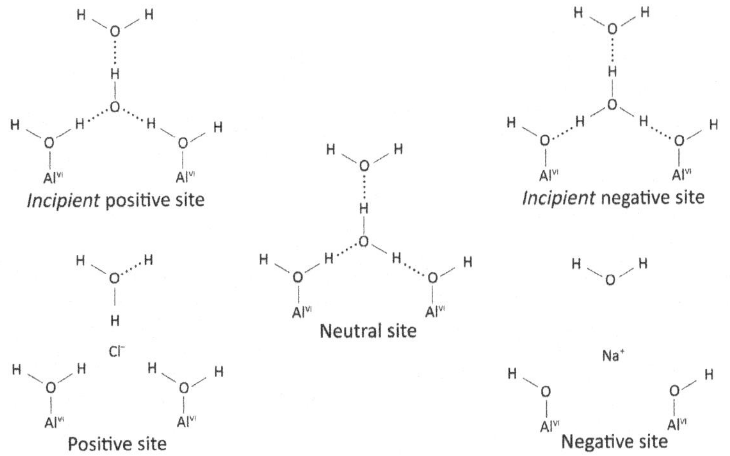

FIG. 8.16 Hydrolysis of surface-bound H_2O and surface-bound OH^- to yield positively charged site at low pH and negatively charged site at high pH. The two diagrams show how the *redistribution* of donor O–H and receptor O···H bonds at the neutral site generate: (1) a OH^- ion hydrogen-bonded to two surface-bound H_2O of an *incipient positive site* (*upper left*) or (2) a H_3O^+ ion hydrogen-bonded to two surface-bound OH^- of an *incipient negative site* (*upper right*). Chemisorption of a strong acid H^+Cl^- transforms a neutral site into a positive site with Cl^- occupying an anion-exchange site (*lower left*). Chemisorption of a strong base Na^+OH^- transforms a neutral site into a negative site with Na^+ occupying a cation-exchange site (*lower right*).

8.4.2.2 Significance of Crystallographic Proton-Adsorption Models for Chemisorption at Oxide Mineral Surfaces

Appendix 8.A presents the valence-bond proton adsorption model for the goethite α-FeO(OH)(s) (100) surface. Hydrogen bonding plays a greater role in the goethite crystal structure than is the case for gibbsite. The goethite (100) is populated by three types of oxygen but only two have altered coordination when exposed at the surface. As in the gibbsite case, hydrogen bonding plays an important role stabilizing the surface structure.

Although the crystallographic structure of the gibbsite (100) and goethite (100) surfaces (Appendix 8.A) are considerably different they have one common feature: the neutral surface has an equal number of surface-bound H_2O and surface-bound OH^- species (the three-coordinate site O_3 at the goehite (100) retains the bonding configuration of the bulk crystal and, therefore, should not be included in this analysis).

Positively charged surface sites form on both oxide surfaces at low pH when H^+ ions adsorb to conjugate base sites—the surface-bound OH^-—present on both surfaces. The adsorption complex in both cases is a weak acid site—the surface-bound H_2O molecule. Negatively charged surface sites form at high pH when the weak-acid sites—surface-bound H_2O molecules—hydrolyze to release H^+ ions into solution leaving behind conjugate base sites—surface-bound OH^-. There is no adsorption complex in the latter case because the hydrolysis of the surface-bound weak acid is effectively the desorption of a H^+ ion.

The first and most obvious significance of strong acid chemisorption by an oxide surface is formation of positively charged anion-exchange sites (cf. Fig. 8.16, lower left). Likewise strong base chemisorption by an oxide surface results in a negatively charged cation-exchange sites (cf. Fig. 8.16, lower right). Chapter 4 and Section 8.3.5 provide ample coverage of ion-exchange adsorption and the role of the equilibrium ion-exchange quotient, ion-exchange selectivity constants and factors influencing ion-exchange isotherms.

The crystallographic H^+ adsorption model carry a deeper significance implied by the Parks and De Brnyn (1962) surface hydrolysis model (reactions 8.R9, 8.R10). The positively charged surface at low pH is populated by Brønsted weak-acid sites (nominally surface-bound H_2O) while the negatively charged surface at high pH is populated by Brønsted conjugate-base sites (nominally surface-bound OH^-).

The adsorption of H^+ cations by an oxide surface is a site-limited chemisorption reaction whose acid-base equivalent is the reaction of a strong acid with a limited number of Brønsted conjugate-base sites. At some pH value the Brønsted conjugate-base sites are saturated; Γ_{H^+} has reached its adsorption maximum $\Gamma_{H^+(m)}$. Likewise, the adsorption of OH^- anions by an oxide surface is a site-limited chemisorption reaction whose acid-base equivalent is the reaction of a strong base with a limited number of Brønsted weak-acid sites. At some pH value the Brønsted weak-acid sites are completely hydrolyzed; Γ_{OH^-} has reached its adsorption maximum $\Gamma_{OH^-(m)}$.

In short, the oxide/solution interface at low pH is populated by Brønsted weak-acid sites in the form of surface-bound H_2O. The strength of these Brønsted weak-acid sites depends on the cation (or cations) of the oxide mineral. The same oxide/solution interface as high pH is populated by Brønsted conjugate-base sites in the form of surface-bound OH^-. The strength of these Brønsted conjugate-base sites also depends on the cation (or cations) of the oxide mineral. The chemisorption of adsorptives other than H^+ and OH^- will be influenced by the relative *abundance* and *strength* of Brønsted weak-acid and Brønsted conjugate-base sites present the oxide/solution interface.

Finally, the crystallographic H^+ adsorption model has a significance beyond the simple surface hydrolysis model of Parks and De Brnyn (1962) tied to the *Lewis acid-base theory*. Lewis (1923) developed a valence-electron theory based on the sharing and exchange of electron pairs. A *Lewis base* donates an electron pair to form a *dative* covalent bond while a *Lewis acid* accepts an electron pair.

A Lewis acid is defined as any substance (e.g., Al^{3+} or Fe^{3+} cations in coordination polyhedra at the S/L interface of gibbsite and goethite, respectively; cf. Figs. 8.15 and 8.C.1) that can accept a nonbonding electron pair. A Lewis base is defined as any substance (e.g., the OH^- ion) that can donate a nonbonding electron pair. The Lewis acid-base theory expands the number of acids and therefore the number of acid-base reactions.

The reaction between Al^{3+} cations and H_2O molecules in aqueous solution to form the hexaaquaaluminum(III) complex cation (CAS registry number 15453-67-5) can be represented as a *Lewis acid-base* reaction.

$$Al^{3+}(aq) + 6H_2O(l) \longleftrightarrow Al(H_2O)_6^{3+}(aq)$$

The Lewis structure of the H_2O molecule places two nonbonding (i.e., *lone*) valence electron pairs on the oxygen; the H_2O molecule is a *Lewis base*.

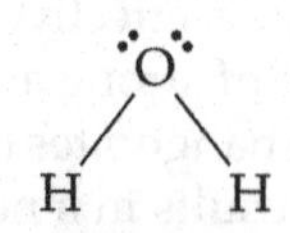

The electron configuration of the Al^{3+} cation reveals nine (9) empty valence orbitals,[16] each of which can accept one nonbonding electron pair donated by H_2O molecules; the Al^{3+} cation is a *Lewis acid.*

$$Al^{3+}: (1s)^2\ (2s)^2\ (2p)^6\ (3s)^0\ (3p)^0\ (3d)^0$$

The hexaaquaaluminum(III) cation is the product of a Lewis acid-base reaction (8.R11); the Lewis-acid Al^{3+} accepts lone-electron pairs, one each from six Lewis-base H_2O molecules, to yield a *Lewis acid-base complex.*

$$\underset{\text{Lewis acid}}{Al^{3+}(aq)} + \underset{\text{Lewis base}}{6\ H_2O(l)} \longleftrightarrow \underset{\text{Lewis complex}}{Al(H_2O)_6^{3+}(aq)} \qquad (8.R11)$$

The cation M at the center of each coordination polyhedron exposed at the oxide/solution interface (cf. Figs. 8.15, 8.C.1, and 8.C.2) is a Lewis-acid surface-site that can accept a nonbonding electron pair from Lewis-base adsorptive molecules or ions. The M–OH_2 bond anchoring a surface-bound H_2O molecules at the oxide/water interface is weaker than the M–OH bond anchoring a surface-bound OH^- at the same interface. The surface excess of reactive Lewis-acid sites increases at low pH.

The desorption of H^+ cations from the oxide/water interface exposes a nonbonding electron pair on each surface-bound OH^- ion; the surface excess of Lewis base sites increases at high pH. The nonbonding electron pair of surface-bound OH^- Lewis-base sites react with Lewis-acid adsorptive molecules or ions. The strength of surface-bound OH^- Lewis-base sites is influenced by the chemical composition of the oxide mineral.

The pH-dependent chemical properties of the oxide/solution interface involve much more than surface charging and adsorption by ion exchange. The surface excess of Brønsted-acid and Brønsted-base sites is pH dependent as is the surface excess of Lewis-acid and Lewis-base sites. We should anticipate a complex pH-dependent transformation of the oxide/solution interface and, as a consequence, a transformation of the chemical affinity of adsorptive molecules and ions for the interface.

8.5 THE ADSORPTION ENVELOPE EXPERIMENT

Adsorption-envelope experiments are pH-dependent adsorption experiments that reveal the impact of H^+ ion adsorption and desorption by the mineral/solution interface on surface charge, the relative abundance of surface sites—Brøsted weak-acid sites, Brøsted conjugate-base sites, Lewis-acid sites, and Lewis-base sites—at the mineral/solution interface.

The real dilemma facing scientists studying adsorption at the mineral/solution interface is how to model adsorption. Adsorption envelopes are generally very simple curves that can fit using two-parameter empirical models. The simplicity of the adsorption envelope does not justify use of more than two parameters. Furthermore, the adsorption envelope provides few if any clues about the adsorption complex(es) that form at the mineral/solution interface.

[16] Valence atomic orbital (3s) can accept one electron-pair, orbitals (3p) can accept three electron-pairs, and orbitals (3d) can accept five electron-pairs.

The design of any adsorption edge experiment must assess the possibility that precipitation instead of adsorption accounts for the loss of inorganic adsorptives from solution. Water chemistry simulations using the initial adsorptive concentration can identify the pH range where the solution becomes super-saturated, setting limits on the pH-range of the adsorption-edge experiment.

8.5.1 Adsorption Edges

Typical cation adsorption envelopes appear in Fig. 8.17, which shows a rapid increase in the percent adsorbed in a narrow pH range, typically two pH units from negligible to nearly complete adsorption from solution. Adsorption envelopes such as those in Fig. 8.17 can be fit using a simple, two-parameter $\left\{pH_{edge}, w\right\}$ empirical model (cf. expressions 8.36, 8.37). Parameter pH_{edge} is the pH at 50% adsorption and parameter w is the dispersion or width of the edge.

Expression (8.36) is typical of *cation adsorptives*: maximum adsorption $n^{\sigma}_{A(m)}$ occurs in the range $pH \gg pH_{edge}$.

$$\theta_{cation} = \frac{\left(n^{\sigma}_{A}\right)}{\left(n^{\sigma}_{A(m)}\right)} = \frac{1}{1 + \left(10^{(pH_{edge} - pH)/w}\right)} \quad (8.36)$$

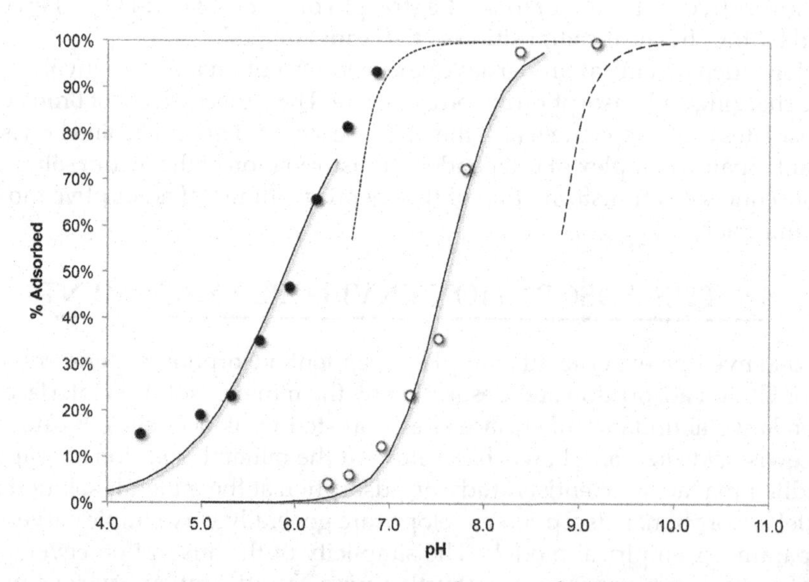

FIG. 8.17 The graph plots pH-dependent adsorption of Co^{2+} and Cu^{2+} by imogolite $Al_2SiO_3(OH)_4(s)$ (Clark and McBride, 1984). The initial adsorptive concentration was 2.50×10^{-4} mol dm^3, the suspensions contained 5 g of synthetic imogolite per cubic decimeter and the ionic strength $I_c = 0.15$ as $Ca(NO_3)_2(aq)$. *Filled circles* plot the Cu^{2+} adsorption envelope, *open circles* plot the Co^{2+} adsorption envelope. The *solid lines* plot the adsorption envelope up to the pH where the solution becomes saturated by the hydroxide solid phase. The *dashed lines* trace metal cation removal by precipitation.

Expression (8.37) is typical of *anion adsorptives*: maximum adsorption $n^{\sigma}_{A(m)}$ occurs in the range $pH \ll pH_{edge}$.

$$\theta_{anion} = \frac{(n^{\sigma}_{A})}{(n^{\sigma}_{A(m)})} = \frac{1}{1 + \left(10^{(pH - pH_{edge})/w}\right)} \tag{8.37}$$

Parameter w used to fit edge width deserves additional comment. Ritchie and Perdue (2008) used a modified Henderson-Hasselbach function (expression 8.38) to model acidometric titration curves of natural organic matter.

$$c_{OH^-} = \frac{c_{acid-1}}{1 + (a_{H^+}/K_1)^{w_1}} + \frac{c_{acid-2}}{1 + (a_{H^+}/K_2)^{w_2}} \tag{8.38}$$

Expression (8.38) is a four-parameter $\{c_{acid-1}, c_{acid-2}, K_1, K_2, w_1, w_2\}$ empirical model. Expression (8.38) parameters K_1 and K_2 correspond to $10^{-pH_{edge}}$ in expressions (8.36), (8.37) because they locate the pH at the center of two equivalence points. Expression (8.38) parameters w_1 and w_2 correspond to w in expressions (8.36), (8.37), representing "width parameters that describe the distribution of [hydrolysis constant] values" for the two weak acid sites.

The empirical *width* parameters w_1 and w_2 are an essential representation of weak-acid chemical heterogeneity in natural organic matter. Likewise, we must anticipate surface-site chemical heterogeneity at the mineral/solution interface. The reader will note the four adsorption envelopes in Figs. 8.17 and 8.18 are not a uniform width.

If the experiment is properly designed the adsorptive will not precipitate in the pH range of the experiment. Under these conditions maximum adsorption $n^{\sigma}_{A(m)}$ is determined by the adsorption-site (*) specific surface excess n^{σ}_{*} relative to the adsorptive concentration in the system (expressions 8.39, 8.40, respectively).

$$n^{\sigma}_{A(m)} = n^{\sigma}_{*} \quad \text{if: } (c_A \cdot V) > n^{\sigma}_{*} \tag{8.39}$$

$$n^{\sigma}_{A(m)} = c_A \cdot V \quad \text{if: } (c_A \cdot V) < n^{\sigma}_{*} \tag{8.40}$$

The Cu^{2+} adsorption edge is $pH_{edge} = 5.9$ (cf. Fig. 8.17) with an edge width[17] of $\Delta pH_{edge} = 7.0$. The Co^{2+} adsorption edge is $pH_{edge} = 7.6$, one-and-a-half pH units higher than the Cu^{2+} edge and the Co^{2+} edge width $\Delta pH_{edge} = 3.8$ is about half the Cu^{2+} edge.

Typical anion adsorption envelopes appear in Fig. 8.18, which shows a rapid increase in the percent adsorbed with decreasing pH, the adsorption transition covers two pH units. While cation adsorption is negligible when the solution pH is acidic, anion adsorption is negligible when the solution pH is basic. The selenate SeO_4^{2-} adsorption edge is $pH_{edge} = 5.3$ with an edge width of $\Delta pH_{edge} = 3.6$. The selenite SeO_3^{2-} adsorption edge is $pH_{edge} = 9.9$, nearly

[17] The edge width being quoted here is the pH range from 10% adsorption to 90% adsorption and not the fitting parameter w in expression (8.36).

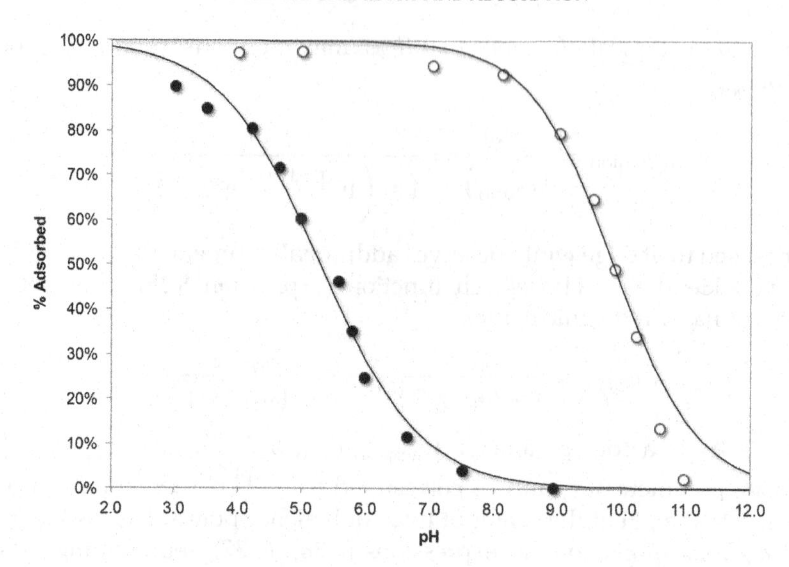

FIG. 8.18 The graph plots pH-dependent adsorption of selenate SeO_4^{2-} and selenite SeO_3^{2-} by goethite γ-FeOOH(s) (Su and Suarez, 2000). The initial adsorptive concentration was 1.0×10^{-4} mol dm^3, the suspensions contained 4 g of goethite per cubic decimeter and the ionic strength $I_c = 0.1$ as NaCl(aq). *Filled circles* plot the selenate SeO_4^{2-} adsorption envelope, *open circles* plot the selenite SeO_3^{2-} adsorption envelope.

five pH units higher than the SeO_4^{2-}, and SeO_3^{2-} edge width $\Delta pH_{edge} = 2.8$ is significantly more narrow than the SeO_4^{2-} edge. Example 8.5 demonstrates the fitting of data plotted in Fig. 8.18 using the anion adsorption-edge model (8.37).

8.5.2 Interpreting Adsorption Envelopes

Fig. 8.17 plots pH-dependent adsorption of Co^{2+} and Cu^{2+} by imogolite, a tubular mineral believed to be similar to halloysite. The distinguishing solution chemistry characteristic of these two components is the solubility of the respective hydroxide solid phases. The solution containing Cu^{2+} becomes saturated by $Cu(OH)_2(s)$ at pH 6.43 (cf. dotted line extending curved through filled circles, Fig. 8.17). Adsorption of Co^{2+} is restricted to pH < 6.43. The solution containing Co^{2+} becomes saturated by $Co(OH)_2(s)$ at pH 8.62 (cf. dashed line extending curved through open circles Fig. 8.17). Adsorption of Co^{2+} is restricted to pH < 8.62.

The explanation for the displacement of the Co^{2+} adsorption envelope to a higher pH range cannot rest on a cation-exchange adsorption mechanism, after all Ca^{2+} remains over 10-fold higher than either Cu^{2+} or Co^{2+} throughout the experimental pH range. The adsorption mechanism must involve chemisorption of Lewis-acid adsorptive cations (Cu^{2+} or Co^{2+}) reacting with Lewis-base sites at the imogolite/solution interface.

Both Cu^{2+} and Co^{2+} form solutions complexes with the OH^- anion but Cu^{2+} begins to form OH^- at a lower pH value than Co^{2+}. At the Cu^{2+} adsorption edge $pH_{edge} = 5.9$, 16.7%

of the solution Cu^{2+} species are OH^- complexes while only 3.6% of the solution Co^{2+} species are OH^- complexes at the Co^{2+} adsorption edge $pH_{edge} = 7.6$.

Fig. 8.18 plots the pH-dependent adsorption of selenate $SeO_4{}^{2-}$ and selenite $SeO_3{}^{2-}$ by goethite. Both adsorptive anions are highly soluble with no tendency to precipitate within the experimental pH range. The solution chemistry of these two components is also quite different. Selenate exists predominantly as the divalent anion $SeO_4{}^{2-}(aq)$ throughout the entire pH range. Selenite exists predominantly as the neutral species $H_2SeO_4{}^0(aq)$ below pH 2.65, the univalent anion $HSeO_4{}^-$ predominates over the pH range 2.65–8.54, and the divalent anion $SeO_4{}^{2-}$ becomes the major solution species at pH 8.54. The hydrolysis behavior identifies $SeO_3{}^{2-}$ as the stronger Lewis base.

EXAMPLE 8.5

Example Permalink

http://soilenvirochem.net/AjC360

Fit the selenite $SeO_3{}^{2-}$ adsorption envelope data plotted in Fig. 8.17 using expression (8.37) to determine parameters $\{pH_{edge}, w\}$.

The data plotted in Fig. 8.17 are from a study by Su and Suarez (2000). The adsorptive is selenite $SeO_3{}^{2-}$ and the adsorbent is goethite γ-FeOOH(s), an iron oxyhydroxide mineral. The experimental adsorption envelope data appear in Table 8.9.

Expression (8.37), reproduced following, is a two-parameter $\{pH_{edge}, w\}$ model for a simple adsorption envelope of the type appearing in Fig. 8.17.

TABLE 8.9 Fraction of Selenite $SeO_3{}^{2-}$ Adsorbed by Goethite as a Function of pH: $\theta = c_{SO3^{2-}}/c'_{SO3^{2-}}$

pH′	θ′ (%)
4.00	97.2
5.00	97.6
7.04	94.5
8.13	92.7
9.03	79.3
9.56	64.6
9.88	49.0
10.22	33.9
10.60	13.6
10.97	2.2

Notes: The initial concentration was $c'_{SeO_3{}^{2-}} = 1.0 \times 10^{-4}$ mol dm^{-3}. The solution ionic strength $I_c = 0.1$ remained constant throughout the entire pH range.
From Su, C., Suarez, D.L., 2000. Selenate and selenite sorption on iron oxides: an infrared and electrophoretic study. Soil Sci. Soc. Am. J. 64 (1), 101–111.

The best-fit parameters $\{pH_{edge}, w\}$ are those that minimize the *residual square sum RSS* where pH and θ are the experimental concentration ratios listed in Table 8.9.

$$\hat{\theta}_{anion} = \frac{1}{1 + \left(10^{(pH - pH_{edge})/w}\right)}$$

$$RSS = \sum_i \left(\hat{\theta} - \theta\right)^2$$

$$pH_{edge} = 9.868$$

$$w = 1.498$$

Parameter $pH_{edge} = 9.868$ corresponds to the center of the adsorption envelope curve where $\theta = 0.5$. The edge width $\Delta pH_{edge} = 2.8$ for the SeO_3^{2-} adsorption envelope is the pH range from 10% adsorption to 90% adsorption, not the fitting parameter w in expression (8.37).

While the decrease in selenate adsorption with increasing pH appears to correlate with the decrease in negative surface charge (Su and Suarez, 2000), nearly all of the selenite in solution adsorbs to a negatively charged goethite surface while the dominant solution species is neutral $H_2SeO_4^0(aq)$ and substantial selenite adsorption occurs in the pH range 9–11, where the goethite surface either has a net zero surface charge or is negatively charged and the dominant selenite solution species is a divalent anion SeO_4^{2-}. Selenite adsorption cannot be explained by anion-exchange adsorption, it clearly involves Lewis acid-base reaction with Lewis acid sites at the goethite/solution interface. Most likely selenite displaced surface-bound H_2O and to form an adsorption complex anchored by a Fe^{3+}–O–Se^{4+} bond.

Numerous studies report similar results where the adsorption of solution species by mineral surfaces does not correlate with surface charge. This type of adsorption behavior is clearly not ion-exchange adsorption, though ion-exchange and electrostatic forces may contribute to the overall adsorption process. Apparently the adsorption affinity for sites on a particular mineral surface for a given adsorptive ion is determined by the Brønsted acid-base and Lewis acid-base properties of both reactants—surface site and absorptive. Surface chemists have yet to develop a predictive model for the adsorption of inorganic components (neutral or charged species) by mineral surfaces based on the chemical properties of both reactants.

The chemical state and properties of a mineral surface site are more difficult to characterize than solution species. Measurements of proton surface charge (cf. Fig. 8.11) have proven insufficient to characterize adsorption affinity because it does not provide a direct measure of surface-site Lewis acid-base strength. In the past few decades, most adsorption research by environmental chemists has focused on characterizing the chemical structure of the absorptive-surface site complex—the chemical product that forms when absorptive binds to the surface of an adsorbent.

8.5.3 The Structure of Adsorption Complexes

In much the same way that earlier proton adsorption models depict the binding of protons to the surface by forming a complex with a surface site. Ion may adsorb to charged sites at the mineral/solution interface through ion exchange (cf. Fig. 8.16) or they may form a Lewis acid-base adsorption complex. A Lewis acid-base reaction occurs when a surface-bond H_2O molecule or OH^- ion displaces one or more H_2O molecules coordinating a metal cation in solution to form an inner-sphere adsorption complex.

The bridging bi-dentate surface complex (Fig. 8.19, structure **1**) represents the probable structure of Pb^{2+} adsorption complexes formed at the α-$Al_2O_3(s)$ mineral/solution interface at the relatively low surface excess concentration of $\Gamma_{Pb^{2+}} = 1.5\ \mu mol\ m^{-2}$ (Bargar et al., 1997). The mono-dentate surface complex (Fig. 8.19, structure **2**) appears at a surface excess concentration of $\Gamma_{Pb^{2+}} = 2.0\ \mu mol\ m^{-2}$ and represents the coordination of about half of the adsorbed Pb^{2+} under those conditions. Bargar et al. (1997) estimate the bond length *d*(Pb–O) and the nonbonded internuclear distance *d*(Pb···Al) using scattering data from X-ray absorption spectroscopy.

Bargar et al. (1997) assumed threefold coordination for the observed *d*(Pb–O) bond distance of 232 pm. Lead Pb^{2+} ion coordination in the mineral litharge PbO(s) is fourfold, and the mean *d*(Pb–O) bond distance is 231 pm. Structures 1 and 2 (Fig. 8.19) have been modified to show the most probable Pb^{2+} coordination environment; the nonbonded internuclear distance *d*(Pb···Al) is not affected by Pb^{2+} coordination as long as the *d*(Pb–O) bond distance remains the same.

X-ray absorption methods offer two independent estimates of coordination number: the number of oxygen atoms directly bonded to Pb^{2+}, which is a direct estimate based on the intensity of the scattering peak (Fig. 8.20) for the first atomic shell, and an indirect estimate based on the bond valence (Brown and Altermatt, 1985), which is calculated from the *d*(Pb–O) bond distance. Bond distance measurements from X-ray absorption methods, based on the position of the scattering peaks (Fig. 8.20), are consistently more accurate than coordination number measurements, making the latter approach for estimating coordination number the most reliable.

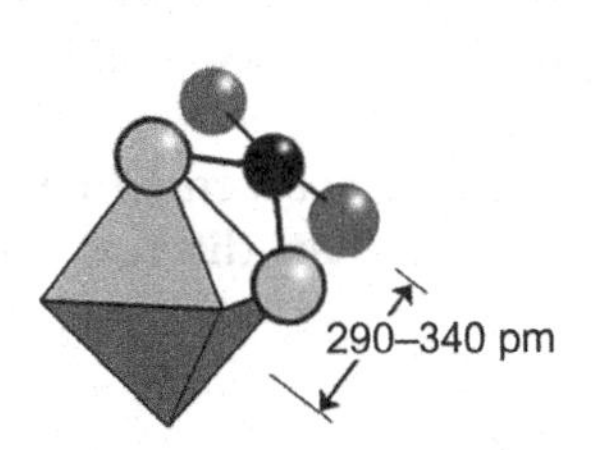

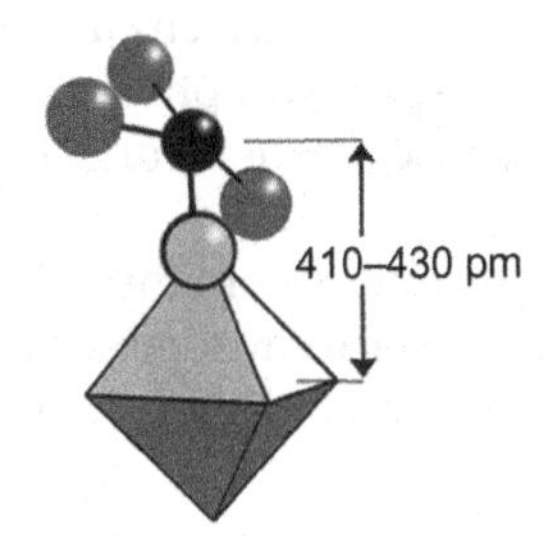

FIG. 8.19 Structure **1**: bi-dentate adsorption complex. Structure **2**: mono-dentate adsorption complex. *Reproduced with permission from Bargar, J.R., Brown Jr., G.E., Parks, G.A., 1997. Surface complexation of Pb(II) at oxide-water interfaces: I. XAFS and bond-valence determination of mononuclear and polynuclear Pb(II) sorption products on aluminum oxides. Geochim. Cosmochim. Acta 61 (13), 2617–2637.*

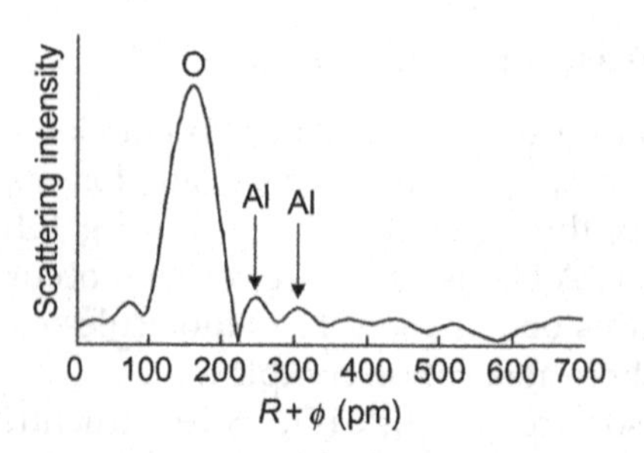

FIG. 8.20 This graph plots the X-ray absorption scattering intensity at the Pb L_{III} absorption edge from 2.0 μmol m^2 Pb^{2+} adsorbed to α-Al_2O_3(s) at pH 7. *From Bargar, J.R., Brown Jr., G.E., Parks, G.A., 1997. Surface complexation of Pb(II) at oxide-water interfaces: I. XAFS and bond-valence determination of mononuclear and polynuclear Pb(II) sorption products on aluminum oxides. Geochim. Cosmochim. Acta 61 (13), 2617–2637.*

The X-ray absorption scattering curve (Fig. 8.20) is equivalent to a radial distribution function, showing the intensity of X-ray scattering by atoms surrounding the atom that absorbs the X-ray radiation. Since the scattering curve in Fig. 8.20 is from X-ray absorption at the Pb L_{III} absorption edge (i.e., 13,055 eV), the origin is occupied by Pb^{2+}. The first scattering peak—labeled O for oxygen atoms in the first atomic shell—is the most intense (the distance R in the scattering curve is not the actual bond distance because X-ray scattering introduces a *phase shift* ϕ that depends on the scattering atom). Two other scattering peaks—labeled Al for aluminum atoms in the second atomic shell—also appear in Fig. 8.20. The first corresponds to the shorter bi-dentate surface complex (Fig. 8.19, structure **1**) and the second to the longer mono-dentate surface complex (Fig. 8.19, structure **2**).

Surface chemists using X-ray absorption spectroscopy and other methods have also collected evidence of surface precipitates forming at the oxide/solution interface under conditions in which water chemistry simulations fail to predict precipitation. The X-ray scattering curve in Fig. 8.21 represents a surface complex between Dy^{3+}, a lanthanide element whose chemistry resembles trivalent actinide elements, and surface-bound phosphate ions on the hydrous aluminum oxide boehmite γ-AlO(OH)(s).

Phosphate ions adsorb to all mineral surfaces in nature, reaching their maximum surface excess $\Gamma_{H_xPO_4(m)}$ under acid conditions. The presence of surface-bound phosphate—itself an adsorption complex—dramatically increases the affinity of the boehmite surface for Dy^{3+}, producing a surface precipitate whose structure is very similar to the structure of known crystalline dysprosium phosphate solids (Yoon et al., 2002). Surface-bound phosphate ions represent a new class of strong Lewis-base surface sites not found on pristine oxide surface.

This is but one example of many where the surface product formed when inorganic ions adsorb at mineral surfaces resemble a precipitate rather than discrete surface complexes illustrated in Figs. 8.15 and 8.19.

8.5.4 Overview of Surface Complexation Models

Several *surface complexation* models emerged in the early to mid-1970s, applying *diffuse double-layer theory* to ion-exchange and Lewis acid-base adsorption reactions. Diffuse double-layer theory has many parallels with the pioneering work of Debye and Hückel (1923).

Chapter 5 discussed the powerful and successful Debye and Hückel (1923) method to compute ion-activity coefficients. The Debye and Hückel (1923) method recognized an electric potential ψ arises from electrolyte ions dissolved in solution, generating a significant electric potential energy U_E acting on all ions in solution. The electric potential energy U_E was not included in the standard Gibbs energy $\Delta_f G^{\circ}$, which is largely the Gibbs energy of ion hydration at the limit of an infinitely dilute electrolyte.

Diffuse double-layer theory represents the solution counter-ion distribution in the vicinity of a uniformly charged surface as *diffuse*. The counter-ion distribution perpendicular to the charged surface represents a balance between the kinetic energy of the ions in solution and their electrostatic potential energy U_E. In the diffuse-double layer case the electric potential ψ accounts for both the uniformly charge surface and the ions within the diffuse layer near the surface.

Eriksson (1952) used diffuse double-layer theory to model cation exchange by smectite clay minerals in an attempt to explain exchange selectivity. The electric potential used by Eriksson (1952) was the diffuse double-layer potential at the mid-plane of the smectite interlayer. The Eriksson (1952) diffuse double-layer model of ion-exchange adsorption failed to account for the effect of ion hydration on the placement of exchangeable counter-ions. Furthermore, a diffuse double layer is not a valid representation of the cation distribution in crystalline hydrates limited to two-layer or two-layer hydrates.

Despite the obvious limitations of diffuse double-layer theory as a model for ion-exchange adsorption, 20 years later several research groups derived a new set of diffuse double-layer models to represent ion-exchange and Lewis acid-base adsorption by oxide minerals. The reader is referred to the excellent review by Westall and Hohl (1980) covering all major surface-complexation models from which following statement is quoted,

> It is found that a *wide range of parameter values yield optimal fit*, i.e., that it is very difficult to separate adsorption energy unambiguously into electrostatic and chemical components. All models can represent the experimental data equally well, but *the values of corresponding parameters from different models are not the same.* Hence, the models must be viewed as being of the correct mathematical form to represent the data, but not necessarily an accurate physical description of the interface.

The surface-complexation models examined by Westall and Hohl (1980) share a common flaw that is entirely independent of diffuse double-layer theory or the importance of a surface electric potential energy U_E for the standard Gibbs energy of adsorption $\Delta_f G^{\circ}_{ads}$. Figs. 8.17 and 8.18 clearly demonstrate that no adsorption-envelope model can justify more than two fitting parameters. The Debye and Hückel (1923) ion-activity function contains *zero adjustable parameters*; all parameters in the Debye and Hückel (1923) ion-activity function are physical parameters.

An excess of adjustable parameters, beyond the minimum required, generally does not lower a model's performance representing experimental data. A model user would expect all models sharing the same fundamental assumptions would reach a consensus on critical parameters (e.g., equilibrium constants for the formation of adsorption complexes) but clearly these major surface-complexation models fail to converge on a consensus representation of adsorption complexes.

Fig. 8.22 plots data from Hohl and Stumm (1976) for the pH-dependent adsorption of Pb^{2+} by α-$Al_2O_3(s)$, the same system studied by Bargar et al. (1997) (cf. Figs. 8.19 and 8.20). The

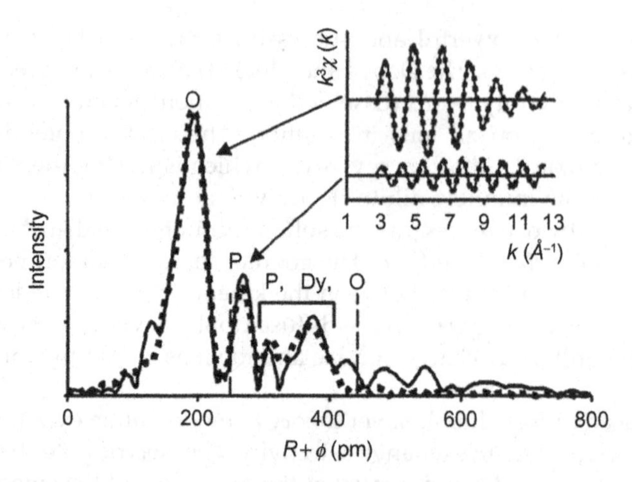

FIG. 8.21 This graph plots the X-ray absorption scattering intensity at the Dy L_{III} absorption edge from 1.0 μmol m^2 Dy^{3+} adsorbed to γ-AlO(OH)(s) at pH 7. *From Yoon, S.-J., Helmke, P.A., Amonette, J.E., Bleam, W.F., 2002. X-ray absorption and magnetic studies of trivalent lanthanide ions sorbed on pristine and phosphate-modified boehmite surfaces. Langmuir 18 (26), 10128–10136.*

maximum surface excess $\Gamma_{Pb^{2+}(m)}$ is nearly identical in both adsorption envelopes plotted in Fig. 8.22: (filled circles: 0.215 μmol m^2, open circles: 0.263 μmol m^2).

Hohl and Stumm (1976) fit the two adsorption envelope curves in Fig. 8.22 by assuming the formation of two adsorption complexes: a Pb^{2+} cation-exchange adsorption complex and a bi-dentate Pb^{2+} Lewis acid-base adsorption complex. There is nothing in the adsorption data or the adsorption envelopes to justify two sets of parameters, one for each hypothetical adsorption complex. Furthermore, the authors (Hohl and Stumm, 1976) admit "bidentate [Lewis acid-base] complexes are formed to a very small extent in comparison to the 1: 1 complexes." By postulating two adsorption complexes the authors have at their disposal sufficient adjustable parameters to simultaneously fit both curves.

Detailed analysis of the Hohl and Stumm (1976) study, one of the earliest and most cited of the surface-complexation models, should be sufficient motivation to take the warning of Westall and Hohl (1980) quoted above seriously.

8.6 SUMMARY

This chapter began with a description of mineral and organic colloidal adsorbents commonly found in soils, sediments, aquifers, and suspended surface waters. These colloidal adsorbents have sufficient capacity, as a direct consequence of their high surface-to-volume ratio, to adsorb compounds dissolved in water to significantly alter pore water chemistry.

Environmental chemists typically encounter two types of adsorption processes: site-limited and partitioning. The two processes have distinctly different adsorption isotherms: the empirical Langmuir isotherm model (expression 8.16) and the partition isotherm model (expressions 8.23, 8.24), respectively. Site-limited adsorption isotherms characterize the adsorption of most compounds by mineral colloids or cation adsorption by organic matter colloids, but partitioning isotherms are restricted to the adsorption of neutral organic compounds by organic matter colloids and other molecular-association colloids. Neutral organic compounds partition into nonpolar domains within molecular-association colloids, giving rise to their distinctive adsorption isotherm.

Both the capacity and the affinity of mineral colloids, with the exception of layer silicate clays, to adsorb inorganic and organic ions depend on pH. The pH-dependent adsorbing behavior of mineral colloids results from the adsorption and desorption of H^+, leading to the development of surface charge Γ_{H^+}. The effect of pH-dependent surface charge on the adsorbing behavior of mineral colloids is well documented but poorly understood because surface-site hydrolysis creates ion-exchange sites, Brønsted weak-acid and conjugate-base sites, and Lewis-acid and Lewis-base sites, each with its attendant chemical characteristics.

The current state of our understanding of Lewis acid-base adsorption complexes comes from molecular spectroscopy studies that determine the structure and composition of the adsorption complex that forms when inorganic and organic ions bind to mineral surfaces. In some cases the chemical structure of the adsorption complex resembles the simple model postulated for proton binding to oxygen atoms at the mineral surface (cf. Figs. 8.15, 8.19, and 8.C.2), but in other cases the chemical structure of the surface complex has a composition that is incompatible with the simple site-binding model (cf. Fig. 8.22).

The ultimate significance of surface chemistry and adsorption for environmental chemists can be traced back to the larger question of biological availability and toxicity. Plants

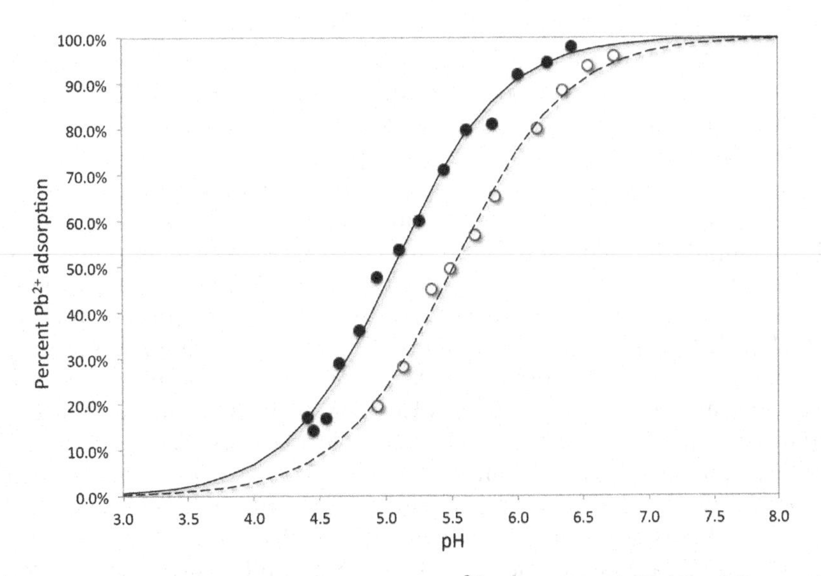

FIG. 8.22 The graph plots pH-dependent adsorption of Pb^{2+} by α-Al_2O_3(s) (Hohl and Stumm, 1976) for two suspensions. *Filled circles* plot the adsorption envelope for an initial Pb^{2+} concentration of 2.94×10^{-4} mol dm^3 in a suspensions containing 11.72 g of α-Al_2O_3(s) per cubic decimeter. *Open circles* plot the adsorption envelope for an initial Pb^{2+} concentration of 9.8×10^{-5} mol dm^3 in a suspensions containing 3.18 g of α-Al_2O_3(s) per cubic decimeter. The ionic strength for both experiments was $I_c = 0.1$ as $NaClO_4$(aq).

and animals take up compounds dissolved in water much more readily than compounds that are precipitated as insoluble solids or adsorbed to the surfaces of environmental colloids. Adsorption processes immobilize potential toxicants, lowering their uptake by living organisms and their movement with groundwater. Adsorption also ensures that toxicants adhering to environmental colloids pose a continuing health risk because it represents a reservoir that continually releases toxicants into pore water for biological uptake.

APPENDICES

8.A HYDROPHILIC COLLOIDS

Stokes' Law (cf. Appendix 3.C) predicts the settling rate of particles suspended in a liquid. Below a certain diameter, the colloid threshold, the particles remain suspended indefinitely: a *stable colloidal dispersion*. The constant thermal motion of liquid-phase molecules (i.e., *Brownian motion*) is sufficient to overcome the effects of buoyancy and gravity on the suspended particles. Molecular interactions between the liquid and the colloidal particles are a vital component in dispersion stability. The colloids found in soils, clays and organic matter, are *hydrophilic* meaning water molecules strongly hydrate natural soil colloids.

Stable Na^+-saturated smectite clays will gel at water contents between the liquid limit $w_L \approx 5\ g_{water}\ g_{clay}^{-1}$ and $w/c \approx 50\ g_{water}\ g_{clay}^{-1}$ (i.e., about 2% smectite by mass). Dispersed clay particles remain in constant contact up to the smectite sol-gel limit, which explains gelation. The Na^+-saturated smectite dispersion enters the sol state beyond this threshold.

Smectite clays form stable colloidal dispersions only if the clays are Na^+- or Li^+-saturated,[18] meaning free swelling. As shown earlier (cf. Example 3.3) over 90% of the water in a clay gel is interlayer water which hydrates the interlayer cation rather than the mineral surface itself.

Organic matter molecules exhibit amphiphilic properties similar to simple amphiphiles such as phospholipids (cf. Chapter 7, *Natural Organic Matter*). It is unlikely, though not impossible, that organic matter molecules are *bipolar* amphiphiles, having a single polar end and a single nonpolar end. A more probable structure would resemble an amphiphilic block-copolymer, a flexible linear molecule with alternating polar and nonpolar segments.

Amphiphilic block-copolymer molecules have low water solubility owing to their amphiphilic structure and escape solution by adsorbing at the air-water interface the same as simple bipolar amphiphiles (Fig. 8.A.1). Once a molecular monolayer completely occupies the air-water interface the only way amphiphilic block-copolymer molecules can enter the aqueous phase is by molecular folding or aggregation, presenting polar segments at the aqueous contact and burying nonpolar segments within the interior.

Bacteria and fungi secrete a variety of bioamphiphiles (e.g., glycolipids, lipopeptides, hydrophobins). Most of the microbial bioamphiphiles are bipolar, however, fungal

[18] Vermiculite can free swell to the gel and sol state only if Li^+-saturated.

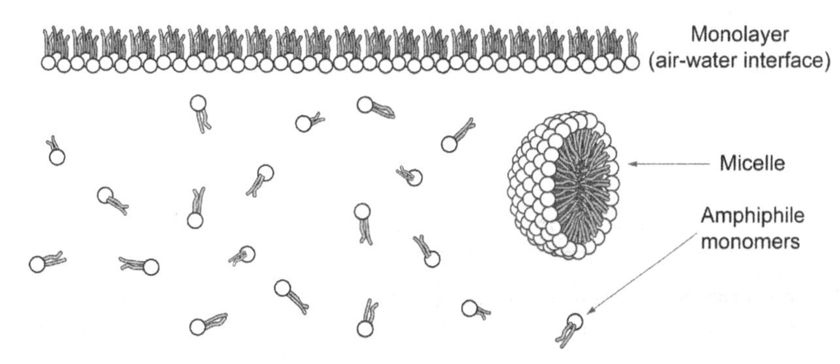

FIG. 8.A.1 Surfactant molecules self-organize, forming films at the air-water interface and colloidal aggregates (micelles) dispersed in water. *Illustration adapted from Villarreal, M.R., 2007. Cross Section of the Different Structures that Phospholipids can Take in a Aqueous Solution. The Circles are the Hydrophilic Heads and the Wavy Lines are the Fatty Acyl Side Chains. Wikimedia Commons. https://upload.wikimedia.org/wikipedia/commons/c/c6/Phospholipids_aqueous_solution_structures.svg.*

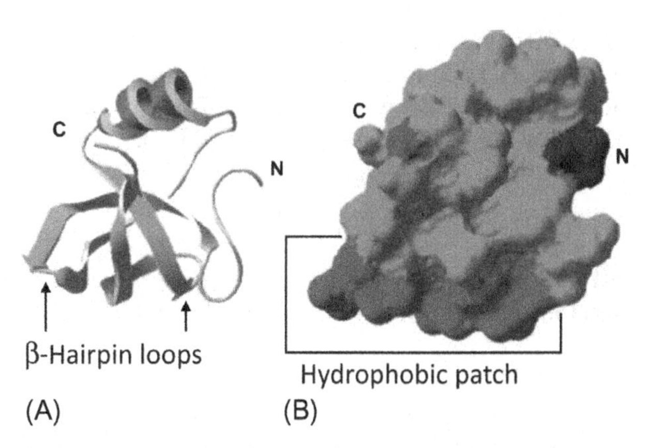

FIG. 8.A.2 The structure of the *Trichoderma reesei* HFBII hydrophobin shows an amphiphilic molecule with one hydrophilic and one hydrophobic part: A) ribbon diagram showing *α-helix*: spiral at top and *β-sheet* (hairpin loops) at bottom, B) space-filling or *calotte* diagram with hydrophilic surface as top and hydrophobic patch at bottom. *Image source: Linder, M.B., Szilvay, G.R., Nakari-Setaelae, T., Penttilae, M.E., 2005. Hydrophobins: the protein-amphiphiles of filamentous fungi. FEMS Microbiol. Rev. 29 (5), 877–896.*

hydrophobins are amphiphilic block-polypeptides (Fig. 8.A.2). Bioamphiphiles play a variety of roles, including increasing the biological availability of nonpolar organic contaminants for degradation.

Organic matter amphiphiles have little tendency to form solution *micellar aggregates*, except when concentrated Na^+-saturated organic matter base-extract solutions are prepared in the laboratory.[19] Otherwise, organic matter molecules form self-organized aggregate at mineral

[19] If organic matter molecules are saturated by divalent Group 2 cations the molecules coagulate.

surfaces, folding in such a manner as to present their polar segments to either water or the mineral surface and withdrawing their nonpolar segments into the aggregate interior to create nonpolar domains much like the interior of the micelles and colloidal films pictured in Fig. 8.A.1.

8.B BOND-VALENCE MODEL

Bond-valence theory assumes a functional relationship between bond length R_{ij} and bond valence s_{ij}. Brown and co-workers parameterized a series of three empirical expressions by fitting bond lengths drawn from crystallographic refinements.

The first empirical bond-valence expression (8.B.1) was composed of three parameters (Brown and Shannon, 1973). Each M–O bond type was fitted to a unique set of parameters $\{s_0, R_0, N\}$.

$$s_{ij} = s_0 \cdot \left(\frac{R_{ij}}{R_0}\right)^N \tag{8.B.1}$$

Brown and Shannon (1973) also fitted bond lengths to a second two-parameter empirical bond-valence expression (8.B.1). Each of the 22 M–O bond types, once again, was fitted using a unique set of parameters $\{R_1, N_1\}$. Brown and Wu (1976) substantially expanded the M–O bond types parameterized for bond-valence expression (8.B.1).

$$s_{ij} = \left(\frac{R_{ij}}{R_1}\right)^{N_1} \tag{8.B.2}$$

Brown and Altermatt (1985) ultimately fitted bond length data to simplified, two-parameter bond-valence expression (8.B.2) containing one universal parameter B and a single bond-type parameter R_2.[20]

$$s_{ij} = e^{\left(R_2 - R_{ij}\right)/B} \tag{8.B.3}$$

A consequence of each of these three bond-valence expressions, regardless of the number of parameters used, is that the mean bond length R_{M-O} of a coordination polyhedron increases if the cation M moves off center leading to the lengthening of some bonds and a shortening of other bonds.

8.C VALENCE-BOND PROTON ADSORPTION MODEL: GOETHITE (100) SURFACE

The goethite α-FeO(OH)(s) crystal structure (Fig. 8.C.1) contains two distinct O^{2-} ions (Szytula et al., 1968; Yang et al., 2006): O_h is a four-coordinate OH^- ion and O is three-

[20] Brown and co-workers used a different bond length parameter for each bond-valence expression: R_0 for expression (8.B.1), R_1 for expression (8.B.2), and R_2 for expression (8.B.3).

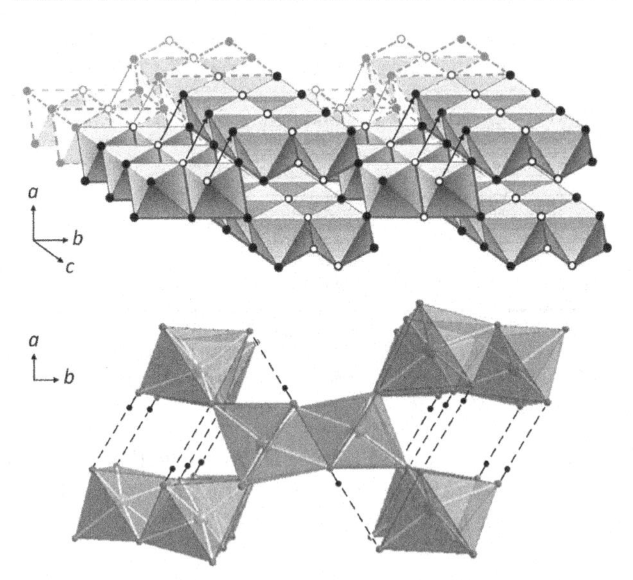

FIG. 8.C.1 The goethite α-FeO(OH)(s) crystal structure can be visualized as *double rows* of edge-sharing Fe^{3+} octahedra aligned parallel to the *c*-axis (*top perspective view*). Corner-sharing O^{2-} cross-link Fe^{3+} octahedral chains to form what appear to be channels, in the polyhedral rendering used here, running parallel to the *c*-axis (*bottom*). Oxygen ions O^{2-} appear as *filled circles* in the top illustration while hydroxide ions OH^- are *open circles*. The *arrows* in both illustrations represent the orientation of O–H bonds. Notice every OH^- ion is a hydrogen-bond donor while every O^{2-} ion is a hydrogen bond receptor. See text for details. *Modified from Carvalho-e-Silva, M.L., Ramos, A.Y., Tolentino, H.C.N., Enzweiler, J., Netto, S.M., Alves, M.D.C.M., 2003. Incorporation of Ni into natural goethite: an investigation by X-ray absorption spectroscopy. Am. Mineral. 88 (5–6), 876–882 and Casey, W.H., Rustad, J.R., Spiccia, L., 2009. Minerals as molecules: use of aqueous oxide and hydroxide clusters to understand geochemical reactions. Chem. A Eur. J. 15 (18), 4496–4515.*

coordinate O^{2-} ion. The goethite structure fails to satisfy the Pauling (1929) *electrostatic valence principle*; the hydroxyl ions OH^- are "over bonded" while the oxygen ions O^{2-} are "under bonded."

$$\zeta_{O_h} = 3 \cdot (s_{Fe-O}) + s_{H-O} = 2\tfrac{1}{2}$$
$$\zeta_{O} = 3 \cdot (s_{Fe-O}) = 1\tfrac{1}{2}$$

Unlike gibbsite, hydrogen bonding cannot be ignored when analyzing cation coordination in the goethite crystal structure. The crystal structure refinement (cf. Table 8.C.1 and Yang et al., 2006) finds Fe–OH bonds are significantly longer than Fe–O bonds: 210.5 and 193.9 pm, respectively.

Table 8.C.1 lists individual bond-valence values for the two oxygen ions in goethite, computed using the empirical bond valence function of Brown and Altermatt (1985) and the goethite crystal data of Yang et al. (2006). The bond-valence sums (expressions 8.C.1, 8.C.2) indicate a more stable bonding configuration than implied by bond-valence sums computed using the simplistic electrostatic valance principle.

$$\zeta_{O_h} = 3 \cdot \left(s_{Fe-O_h}\right) + s_{H-O_h} = 3 \cdot (0.39) + 0.99 = 1.84 \tag{8.C.1}$$

$$\zeta_{O} = 3 \cdot (s_{Fe-O}) + s_{H\cdots O} = 3 \cdot (0.61) + 0.02 = 2.18 \tag{8.C.2}$$

TABLE 8.C.1 Bond Lengths and Bond Valences for the Oxide Mineral Goethite α-FeO(OH)(s)

Bond	Length (pm)	Bond Parameter (pm)	Valence
Fe–O_h	210.3	175.9	0.39
Fe–O	194.4	175.9	0.61
H–O_h	88.2	88.5	0.99
H···O	231.1	88.5	0.02

Notes: Bond-valance constant B = 0.37 in expression (8.B.3).
From Yang, H., Lu, R., Downs, R.T., Costin, G., 2006. Goethite, α-FeO(OH), from single-crystal data. Acta Crystallogr., Sect. E: Struct. Rep. Online 62 (12), i250–i252.

The goethite (100) surface configuration proposed by Russell et al. (1974) assigns OH^- ions to all three oxygen positions illustrated in Fig. 8.C.2. The index corresponds to the number of Fe^{3+} ions each oxygen is bonded to at the surface.

> [Three-coordinate O_3 hydroxyls are] "exposed on the (100) surface together with two new types of hydroxyls (labelled A [O_1 hydroxyls] and C [O_2 hydroxyls]) which arise from protonation of the oxide ions along the edges of the strips. Of these surface hydroxyls, types [O_3] and [O_2] cannot form hydrogen bonds within the surface, since all adjacent hydroxyls are coordinated to a common Fe^{3+} ion, but type [O_1] hydroxyls, each coordinated to only one Fe^{3+}, can form bonds [300 pm] long with each other."

The crystallographic structure in Fig. 8.C.2 illustrates a slightly different bonding configuration at the goethite (100) surface, the difference being the assignment of a water molecule H_2O molecule to the one-coordinate O_1. Adsorption of this H_2O molecule from the aqueous phase completes the sixfold coordination of Fe^{3+} at the surface. Site O_2 is a hydroxyl OH^- ion bonded to two Fe^{3+} ions at the surface and site O_3 is hydroxyl OH^- ion bonded to three Fe^{3+} ions at the surface, the latter two site assignments are identical to the Russell et al. (1974) assignments.

Although the goethite (100) exposes three distinct surface-bound oxygen atoms, only two (O_1 and O_2) have a bonding configuration different from the bulk structure. Although oxygen

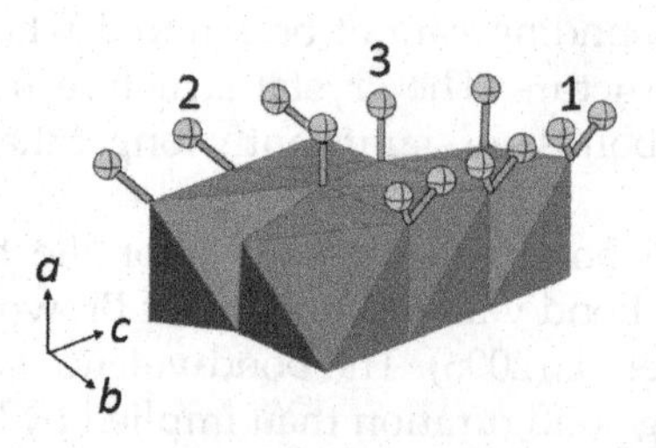

FIG. 8.C.2 This crystallographic structure illustrates the termination of goethite α-FeO(OH)(s) by (100) crystal plane. Three crystallographically distinct oxygen ions lie at the (100) surface, the numbering coincides with the number of Fe–O bonds linked to each surface oxygen. See text for details. *Modified from Rustad, J.R., Zarzycki, P., 2008. Calculation of site-specific carbon-isotope fractionation in pedogenic oxide minerals. Proc. Natl. Acad. Sci. U. S. A. 105 (30), 10297–10301.*

TABLE 8.C.2 Bond Lengths and Bond Valences for the (100) Surface of Goethite α-FeO(OH)(s)

Bond	Length (pm)	Bond Parameter (pm)	Valence
$Fe–O_1$	210.0	175.9	0.40
$H–O_1$	96.5	88.5	0.80
$Fe–O_2$	201.5	175.9	0.50
$H–O_2$	88.5	88.5	0.99

Notes: Bond-valance constant B = 0.37 in expression (8.B.3). The three-coordinate OH^- is not listed but would have the valence bond sum in expression (8.C.2).
From Yang, H., Lu, R., Downs, R.T., Costin, G., 2006. Goethite, α-FeO(OH), from single-crystal data. Acta Crystallogr., Sect. E: Struct. Rep. Online 62 (12), i250–i252.

atom O_3 is exposed at the goethite (010) surface its bonding configuration is unchanged from an oxygen O^{2-} ion occupying the same position in the bulk structure.

Applying the electrostatic valence principle to the bonding of surface-bound H_2O molecule O_1 and surface-bound OH^- ion O_2 reveals an unstable bonding configuration at H_2O molecule O_1 and a stable bonding configuration at OH^- ion O_2.

Electrostatic Valence Principle

$$\zeta_{O_1} = s_{Fe-O} + 2 \cdot s_{H-O} = 2\tfrac{1}{2}$$

$$\zeta_{O_2} = 2 \cdot s_{Fe-O} + (s_{H-O}) = 2$$

If hydrogen bonding is taken into account a more stable bonding configuration at site O_1 would have the water molecule donating two relatively weak asymmetric hydrogen bonds and lengthening the Fe–O bond (rows 1 and 2 in Table 8.C.2, and expression 8.C.3). Site O_2 would have the hydroxide donating a very weak hydrogen bond (rows 3 and 4 in Table 8.C.2 and expression 8.C.4).

Bond Valence Theory

$$\zeta_{O_1} = s_{Fe-O_1} + 2 \cdot \left(s_{H-O_1}\right) = 0.40 + 2 \cdot (0.80) = 2.00 \quad (8.C.3)$$

$$\zeta_{O_2} = 2 \cdot \left(s_{Fe-O_2}\right) + s_{H-O_2} = 2 \cdot (0.50) + 0.99 = 1.99 \quad (8.C.4)$$

8.D USING THE PBT PROFILER

The Office of Chemical Safety and Pollution Prevention (US Environmental Protection Agency) maintains an internet site called the *PBT Profiler* (USEPA, 2012).

PBT Profiler

http://www.pbtprofiler.net/

The chemical and biological data available from the *PBT Profiler* are restricted to organic compounds and include a variety of important environmental properties relating to environmental persistence, bioaccumulation, and toxicity. The *PBT Profiler* is a hybrid data source because some of the data are consensus experimental results while others are simulated from the molecular structure.[21]

The user can search the database using the *Chemical Abstracts Service CAS registry number* if a registry number has been assigned to the organic compound of interest. If the query is for a novel organic compound lacking a registry number, the *PBT Profiler* will compute chemical and biological parameters based on either a molecular drawing or *simplified molecular-input line-entry system SMILES* notation.

The *PBT Profiler* will list water solubility and the octanol-water partition constant $K^{\ominus}_{o/w}$. The *soil adsorption coefficient* corresponds to the soil-water partition constant $K^{\ominus}_{s/w}$ with an unspecified mean organic carbon content. The term *sediment* is not restricted to riverine and lacustrine sediments.

soil is considered aerobic (oxygenated) and sediment is consider anaerobic (free of oxygen)

As a general rule, a organic carbon-water partition constant $K^{\ominus}_{oc/w}$ is more useful than an undocumented soil-water partition constant $K^{\ominus}_{s/w}$.

References

Accardi-Dey, A.M., Gschwend, P.M., 2002. Assessing the combined roles of natural organic matter and black carbon as sorbents in sediments. Environ. Sci. Technol. 36 (1), 21–29.

Accardi-Dey, A., Gschwend, P.M., 2003. Reinterpreting literature sorption data considering both absorption into organic carbon and adsorption onto black carbon. Environ. Sci. Technol. 37 (1), 99–106.

Bargar, J.R., Brown Jr., G.E., Parks, G.A., 1997. Surface complexation of Pb(II) at oxide-water interfaces: I. XAFS and bond-valence determination of mononuclear and polynuclear Pb(II) sorption products on aluminum oxides. Geochim. Cosmochim. Acta 61 (13), 2617–2637.

Bolt, G.H., 1957. Determination of the charge density of silica sols. J. Phys. Chem. 61, 1166–1169.

Boyd, G.E., Schubert, J., Adamson, A.W., 1947. The exchange adsorption of ions from aqueous solutions by organic zeolites. I. Ion-exchange equilibria. J. Am. Chem. Soc. 69, 2818–2829.

Brown, I.D., 1976. Bond valence theory. Part II, examples. J. Chem. Educ. 53 (4), 231–232.

Brown, I.D., Altermatt, D., 1985. Bond-valence parameters obtained from a systematic analysis of the inorganic crystal structure database. Acta Crystallogr., Sect. B: Struct. Sci. B41 (4), 244–247.

Brown, I.D., Shannon, R.D., 1973. Empirical bond-strength-bond-length curves for oxides. Acta Crystallogr. Sect. A: Cryst. Phys. Diffr. Theor. Gen. Crystallogr. 29 (Pt. 3), 266–282.

Brown, I.D., Wu, K.K., 1976. Empirical parameters for calculating cation-oxygen bond valences. Acta Crystallogr. Sect. B: Struct. Crystallogr. Cryst. Chem. B32 (7), 1957–1959.

Cameron, R.S., Thornton, B.K., Swift, R.S., Posner, A.M., 1972. Molecular weight and shape of humic acid from sedimentation and diffusion measurements on fractionated extracts. J. Soil Sci. 23 (4), 394–408.

Carter, C.W., Suffet, I.H., 1982. Binding of DDT to dissolved humic materials. Environ. Sci. Technol. 16 (11), 735–740.

Chiou, C.T., Peters, L.J., Freed, V.H., 1979. A physical concept of soil-water equilibria for nonionic organic compounds. Science (New York, N.Y.) 206 (4420), 831–832.

[21] Follow the *Methodology* weblink listed on the *PBT Profiler* home page.

Chiou, C.T., Malcolm, R.L., Brinton, T.I., Kile, D.E., 1986. Water solubility enhancement of some organic pollutants and pesticides by dissolved humic and fulvic acids. Environ. Sci. Technol. 20 (5), 502–508.
Chiou, C.T., Kile, D.E., Brinton, T.I., Malcolm, R.L., Leenheer, J.A., MacCarthy, P., 1987. A comparison of water solubility enhancements of organic solutes by aquatic humic materials and commercial humic acids. Environ. Sci. Technol. 21 (12), 1231–1234.
Clark, C.J., McBride, M.B., 1984. Chemisorption of copper(II) and cobalt(II) on allophane and imogolite. Clays Clay Miner. 32 (4), 300–310.
Contributors, 2008. Hematite in Scanning Electron Microscope, Magnification 100×. Wikimedia Commons. https://commons.wikimedia.org/wiki/File:Hematite_in_Scanning_Electron_Microscope,_magnification_100x.GIF.
Debye, P., Hückel, E., 1923. Zur Theorie der Elektrolyte. I. Gefrierpunktserniedrigung und verwandte Erscheinungen. Phys. Z. 24, 185–206.
Doucette, W.J., Andren, A.W., 1987. Correlation of octanol/water partition coefficients and total molecular surface area for highly hydrophobic aromatic compounds. Environ. Sci. Technol. 21 (8), 821–824.
Doucette, W.J., Andren, A.W., 1988. Estimation of octanol/water partition coefficients: evaluation of six methods for highly hydrophobic aromatic hydrocarbons. Chemosphere 17 (2), 345–359.
Elprince, A.M., Sposito, G., 1981. Thermodynamic derivation of equations of the Langmuir type for ion equilibriums in soils. Soil Sci. Soc. Am. J. 45 (2), 277–282.
Engebretson, R.R., von Wandruszka, R., 1994. Micro-organization in dissolved humic acids. Environ. Sci. Technol. 28 (11), 1934–1941.
Eriksson, E., 1952. Cation-exchange equilibria on clay minerals. Soil Sci. 74, 103–113.
Everett, D.H., 1972. Manual of symbols and terminology for physicochemical quantitites and units. Appendix II. Definitions, terminology and symbols in colloid and surface chemistry. Pure Appl. Chem. 31 (4), 577–638.
Fishman, M.L., Elrich, F.R., 1975. Interactions of aqueous poly(*N*-vinylpyrrolidone) with sodium dodecyl sulfate. II. Correlation of electric conductance and viscosity measurements with equilibrium dialysis measurements. J. Phys. Chem. 79 (25), 2740–2744.
Freundlich, H.M.F., 1909. Kapillarchemie, eine Darstellung der Chemie der Kolloide und verwandter Gebiete. Akademische Verlagsgesellschaft M.B.H., Leipzig, Germany
Guetzloff, T.F., Rice, J.A., 1994. Does humic acid form a micelle?. Sci. Total Environ. 152 (1), 31–35.
Harter, R.D., Baker, D.E., 1977. Applications and misapplications of the Langmuir equation to soil adsorption phenomena. Soil Sci. Soc. Am. J. 41 (6), 1077–1080.
Hayase, K., Tsubota, H., 1983. Sedimentary humic acid and fulvic acid as surface active substances. Geochim. Cosmochim. Acta 47 (5), 947–952.
Hohl, H., Stumm, W., 1976. Interaction of Pb^{2+} with hydrous γ-Al_2O_3. J. Colloid Interface Sci. 55 (2), 281–288.
Karickhoff, S.W., 1981. Semi-empirical estimation of sorption of hydrophobic pollutants on natural sediments and soils. Chemosphere 10 (8), 833–846.
Karickhoff, S.W., Brown, D.S., Scott, T.A., 1979. Sorption of hydrophobic pollutants on natural sediments. Water Res. 13 (3), 241–248.
Kile, D.E., Chiou, C.T., 1989. Water solubility enhancements of DDT and trichlorobenzene by some surfactants below and above the critical micelle concentration. Environ. Sci. Technol. 23 (7), 832–838.
Lambert, S.M., 1967. Functional relation between sorption in soil and chemical structure. J. Agric. Food Chem. 15 (4), 572–576.
Lambert, S.M., 1968. Omega (Ω), a useful index of soil sorption equilibria. J. Agric. Food Chem. 16 (2), 340–343.
Lambert, S.M., Porter, P.E., Schieferstein, R.H., 1965. Movement and sorption of chemicals applied to the soil. Weeds 13 (3), 185–190.
Langmuir, I., 1916. Constitution and fundamental properties of solids and liquids. I. Solids. J. Am. Chem. Soc. 38, 2221–2295.
Langmuir, I., 1918. The adsorption of gases on plane surfaces of glass, mica and platinum. J. Am. Chem. Soc. 40, 1361–1402.
Langmuir, I., 1925. The distribution and orientation of molecules. Colloid Symp. Monogr. 48–75.
Lewis, G.N., 1923. Valence and the structure of atoms and molecules. American Chemical Society Monograph Series. The Chemical Catalog Company, Inc., New York.
MacCarthy, P., 2001. The principles of humic substances. Soil Sci. 166 (11), 738–751.

Mackay, D., 1982. Correlation of bioconcentration factors. Environ. Sci. Technol. 16 (5), 274–278.
Mackay, D., Fraser, A., 2000. Bioaccumulation of persistent organic chemicals: mechanisms and models. Environ. Pollut. (Oxford, U. K.) 110 (3), 375–391.
Misak, N.Z., 1995. Adsorption isotherms in ion exchange reactions. Further treatments and remarks on the application of the Langmuir isotherm. Colloids Surf. A: Physicochem. Eng. Asp. 97 (2), 129–140.
Parks, G.A., De Brnyn, P.L., 1962. Zero point of charge of oxides. J. Phys. Chem. 66, 967–973.
Pauling, L., 1929. The principles determining the structure of complex ionic crystals. J. Am. Chem. Soc. 51, 1010–1026.
Piccolo, A., 2001. The supramolecular structure of humic substances. Soil Sci. 166 (11), 810–832.
Polzer, W.L., Rao, M.G., Fuentes, H.R., Beckman, R.J., 1992. Thermodynamically derived relationships between the modified Langmuir isotherm and experimental parameters. Environ. Sci. Technol. 26 (9), 1780–1786.
Quagliotto, P., Montoneri, E., Tambone, F., Adani, F., Gobetto, R., Viscardi, G., 2006. Chemicals from wastes: compost-derived humic acid-like matter as surfactant. Environ. Sci. Technol. 40 (5), 1686–1692.
Reid, P.M., Wilkinson, A.E., Tipping, E., Jones, M.N., 1991. Aggregation of humic substances in aqueous media as determined by light-scattering methods. J. Soil Sci. 42 (2), 259–270.
Ritchie, J.D., Perdue, E.M., 2008. Analytical constraints on acidic functional groups in humic substances. Org. Geochem. 39 (6), 783–799.
Roe, M., 2016. Well-Crystallized Kaolinite from the Keokuk Geode, USA. Mineralogical Society of Great Britain & Ireland & The Clay Minerals Society. http://www.minersoc.org/photo.php?id=95.
Russell, J.D., Parfitt, R.L., Fraser, A.R., Farmer, V.C., 1974. Surface structures of gibbsite goethite and phosphated goethite. Nature (London, U. K.) 248 (5445), 220–221.
Schindler, P., Kamber, H.R., 1968. Die Acidität von Silanolgruppen. Helv. Chim. Acta 51 (7), 1781–1786.
Shinozuka, N., Lee, C., 1991. Aggregate formation of humic acids from marine sediments. Mar. Chem. 33 (3), 229–241.
Shiu, W.Y., Doucette, W., Gobas, F.A.P.C., Andren, A., Mackay, D., 1988. Physical-chemical properties of chlorinated dibenzo-*p*-dioxins. Environ. Sci. Technol. 22 (6), 651–658.
Simpson, A.J., 2002. Determining the molecular weight, aggregation, structures and interactions of natural organic matter using diffusion ordered spectroscopy. Magn. Reson. Chem. 40 (Spec. Issue), S72–S82.
Sposito, G., 1979. Derivation of the Langmuir equation for ion-exchange reactions in soils. Soil Sci. Soc. Am. J. 43 (1), 197–198.
Sprycha, R., 1989. Electrical double layer at alumina/electrolyte interface. I. Surface charge and zeta potential. J. Colloid Interface Sci. 127 (1), 1–11.
Stumm, W., Morgan, J.A., 1970. Aquatic Chemistry: An Introduction Emphasizing Chemical Equilibria in Natural Waters. Wiley-Interscience, New York.
Stumm, W., Huang, C.P., Jenkins, S.R., 1970. Specific chemical interaction affecting the stability of dispersed systems. Croat. Chem. Acta 42, 223–245.
Su, C., Suarez, D.L., 2000. Selenate and selenite sorption on iron oxides: an infrared and electrophoretic study. Soil Sci. Soc. Am. J. 64 (1), 101–111.
Sutton, R., Sposito, G., 2005. Molecular structure in soil humic substances: the new view. Environ. Sci. Technol. 39 (23), 9009–9015.
Szytula, A., Burewicz, A., Dimitrijevic, Z., Krasnicki, S., Rzany, H., Todorovic, J., Wanic, A., Wolski, W., 1968. Neutron diffraction studies of goethite. Phys. Status Solidi 26 (2), 429–434.
Taylor, H.S., 1931. Reaction velocity in heterogeneous systems. In: Taylor, H.S. (Ed.), A Treatise on Physical Chemistry: A Co-operative Effort by a Group of Physical Chemists, second ed. D. van Nostrand, New York, pp. 1019–1102.
USEPA, 2012. PBT Profiler: Profiler: Persistent, Bioaccumulative, and Toxic Profiles Estimated for Organic Chemicals. Office of Chemical Safety and Pollution Prevention, U.S. Environmental Protection Agency. http://www.pbtprofiler.net/.
Vanselow, A.P., 1932. Equilibria of the base-exchange reactions of bentonites, permutites, soil colloids and zeolites. Soil Sci. 33, 95–113.
Veith, J.A., Sposito, G., 1977. On the use of the Langmuir equation in the interpretation of "adsorption" phenomena. Soil Sci. Soc. Am. J. 41 (4), 697–702.
Verbruggen, E.M.J., Hermens, J.L.M., Tolls, J., 2001. Physicochemical properties of higher nonaromatic hydrocarbons: a literature study. J. Phys. Chem. Ref. Data 29 (6), 1435–1446.
Weber Jr., W.J., McGinley, P.M., Katz, L.E., 1992. A distributed reactivity model for sorption by soils and sediments. 1. Conceptual basis and equilibrium assessments. Environ. Sci. Technol. 26 (10), 1955–1962.

Wershaw, R.L., Burcar, P.J., Goldberg, M.C., 1969. Interaction of pesticides with natural organic material. Environ. Sci. Technol. 3 (3), 271–273.

Westall, J., Hohl, H., 1980. A comparison of electrostatic models for the oxide/solution interface. Adv. Colloid Interface Sci. 12 (4), 265–294.

Xing, B., Pignatello, J.J., 1996. Time-dependent isotherm shape of organic compounds in soil organic matter: implications for sorption mechanism. Environ. Toxicol. Chem. 15 (8), 1282–1288.

Xing, B., Pignatello, J.J., 1997. Dual-mode sorption of low-polarity compounds in glassy poly (vinyl chloride) and soil organic matter. Environ. Sci. Technol. 31 (3), 792–799.

Xing, B., Pignatello, J.J., Gigliotti, B., 1996. Competitive sorption between atrazine and other organic compounds in soils and model sorbents. Environ. Sci. Technol. 30 (8), 2432–2440.

Yang, H., Lu, R., Downs, R.T., Costin, G., 2006. Goethite, α-FeO(OH), from single-crystal data. Acta Crystallogr., Sect. E: Struct. Rep. Online 62 (12), i250–i252.

Yonebayashi, K., Hattori, T., 1987. Surface active properties of soil humic acids. Sci. Total Environ. 62, 55–64.

Yoon, S.-J., Helmke, P.A., Amonette, J.E., Bleam, W.F., 2002. X-ray absorption and magnetic studies of trivalent lanthanide ions sorbed on pristine and phosphate-modified boehmite surfaces. Langmuir 18 (26), 10128–10136.

Zhou, Z., Gunter, W.D., 1992. The nature of the surface charge of kaolinite. Clays Clay Miner. 40 (3), 365–368.

CHAPTER

9

Reduction-Oxidation Chemistry

OUTLINE

9.1 Introduction 445

9.2 Electrochemical Principles 446
- 9.2.1 *Formal Oxidation States* 446
- 9.2.2 *Balancing Reduction Half-Reactions* 450
- 9.2.3 *Standard Electrochemical Potentials* 453
- 9.2.4 *The Nernst Equation* 453
- 9.2.5 *Standard Biological Potentials* 455

9.3 Measurement and Interpretation of Electrochemical Potentials in Soils and Sediments 457
- 9.3.1 *Electrochemical Stability Diagrams* 457
- 9.3.2 *Pourbaix Electrochemical Stability Diagrams* 459

9.4 Microbial Respiration 470
- 9.4.1 *Catabolism and Electron Transport Chains* 473
- 9.4.2 *Microbial Electron Transport Chains* 475
- 9.4.3 *Impact of Respiratory Efficiency on Carbon Use Efficiency* 481

9.5 Methanogenesis 482

9.6 Summary 482

Appendices 483

9.A Reduction-Oxidation Reactions Without Electron Transfer 483

9.B Limitation of Platinum Oxidation-Reduction Electrodes 484

9.C Standard and Biological Electrochemical Potentials for Environmental and Biological Half-Reactions (Tables 9.C.1 and 9.C.2) 486

9.D Facultative and Obligate Anaerobes 486

9.E Fermentative Anaerobic Bacteria 487

References 488

9.1 INTRODUCTION

Reduction-oxidation chemistry in the environmental context blends biochemistry and geochemistry in ways that can make the transition from what you learned in general chemistry challenging. The driving force behind reduction-oxidation reactions in natural settings is

http://dx.doi.org/10.1016/B978-0-12-804178-9.00009-4

microbial respiration, a complex series of electron-transfer reactions that ultimately couple cellular metabolism to the environment, consuming electron acceptors ranging from molecular oxygen to carbon dioxide in order to release the chemical energy stored in reduced carbon compounds.

Environmental reduction-oxidation chemistry has at its source the microbial oxidation of biomolecules, organic compounds, and organic matter, relying on a variety of electron acceptors in the absence of O_2. This means that the locus of most environmental reduction-oxidation reactions is the zone of biological activity, the zone of organic carbon accumulation.

The progress of environmental reduction-oxidation reactions and development of zones where certain reduction-oxidation processes dominate (Lovley and Goodwin, 1988) are both governed by the level of biological activity. The seeming absence of molecular oxygen—*anoxia* being the condition where the $O_2(aq)$ concentration is very low—is necessary but not sufficient for reducing conditions. The development of anoxia leads to changes in the active microbial population from communities that rely on aerobic respiration, where O_2 serves as the terminal electron acceptor, to anaerobic respiration, where other electron acceptors replace O_2 as the terminal electron acceptor supporting respiration.

This transition from one microbial community to another—from one type of respiration to another—sets the stage for the chemical reduction of the environment that will ultimately couple the reduction-oxidation reactions required for biological respiration to a host of reduction-oxidation reactions that occur simply because the electrochemical potential is drawn down by anaerobic respiration.

This chapter is organized into three major sections. The first is a reprise of reduction-oxidation chemistry fundamentals designed to bridge the gap between general chemistry and environmental chemistry. The second develops the methods used by geochemists to quantify and interpret reduction-oxidation conditions as they occur in the environment. The final section provides the mechanism that generates reducing conditions in soils and groundwater: *anaerobiosis*—microbial respiration in the absence of molecular oxygen.

9.2 ELECTROCHEMICAL PRINCIPLES

9.2.1 Formal Oxidation States

Reduction-oxidation reactions are chemical reactions involving a change in oxidation state in an educt and a product through the transfer of one or more electrons from an electron donor (the reducing agent) to an electron acceptor (the oxidizing agent). The assignment of oxidation states is essential because it will verify whether electron transfer has occurred and identify electron donors and acceptors. Formal oxidation states are also used for electron accounting when balancing reduction-oxidation reactions.

The oxidation state of an element in a compound is usually a poor estimate of its effective charge.[1] Regardless, an electron transfer reaction always results in the loss of one or more

[1] The following is an extended quote from Vitz (2002): "Oxidation states are not required to have anything to do with charges or electrons, as emphasized by the *IUPAC* recommendation that they be written, for example, '+1', with sign first and no dimensions, to distinguish them from charge numbers, which are denoted for example as '1+', and which would be associated with a dimension (esu or coulomb)."

electrons from the electron donor compound and the acquisition of an equal number of electrons by the electron acceptor compound. The assignment of oxidation states demonstrates whether or not electron transfer has occurred in a reaction and, if electron transfer has occurred, identifies electron donors and acceptors.

Table 9.1 lists the chemical rules for assigning oxidation states to elements in ions and compounds. There are exceptions, but they are relatively inconsequential for environmental chemistry. Examples 9.1 and 9.2 demonstrate the assignment of oxidation states.

TABLE 9.1 Rules for Assigning Formal Oxidation States to Elements

1.	The oxidation state of the element in its uncombined state is always zero, regardless of physical state: $OS(E) = 0$
2.	The oxidation state of a monatomic ion equals the charge of the ion. **Example**: $OS(Ca^{2+}) = +2$, $OS(F^-) = -1$
3.	The oxidation state of hydrogen in a compound is usually: $OS(H) = +1$. **Exceptions**: The oxidation state of hydrogen is $OS(H) = -1$ in Groups 1 and 2 hydrides (e.g., $NaH(s)$ or $CaH_2(s)$) because the electronegativity of these elements are less than that of hydrogen (Fig. 9.1)
4.	The oxidation state of oxygen in a compound is usually: $OS(O) = -2$. **Exceptions**: Oxygen-oxygen bonds in peroxides (e.g., hydrogen peroxide H_2O_2) lowers the oxidation state of oxygen by one unit: $OS(O) = -1$
5.	The oxidation state of Group 1 elements $E(G_1)$ in a compound is: $OS(E(G1)) = +1$. **Example**: $OS(K) = +1$
6.	The oxidation state of Group 2 elements $E(G_2)$ in a compound is: $OS(E(G2)) = +2$. **Example**: $OS(Mg) = +2$
7.	The oxidation state of Group 17 (halogen) elements $E(G_{17})$ in a compound is: $OS(E(G17)) = -1$. **Exception**: Since oxygen is more electronegative than chlorine (Fig. 9.1), its oxidation state takes precedence. The oxidation state of Cl in HOCl is: $OS(Cl) = +1$
8.	The sum of the oxidation states of all of the atoms in a neutral compound is zero. **Example**: The oxidation state sum for the ascorbic acid $C_6H_8O_6$ equals: $(6 \cdot OS(C)) + (8 \cdot OS(H)) + (6 \cdot OS(O)) = 0$
9.	The sum of the oxidation states in a polyatomic ion equals ion charge. **Example**: The oxidation state sum for the sulfate anion $SO_4{}^{2-}$ equals: $OS(S) + (4 \cdot OS(O)) = -2$

EXAMPLE 9.1

Example Permalink

http://soilenvirochem.net/QzSWph

Assigning oxidation states using the rules in Table 9.1.

Rules 3 and 4 in Table 9.1 assign the oxidation states for hydrogen and oxygen. This example will apply rule 9 to assign oxidation states for carbon.

Group (vertical) / Period (horizontal)	1	2	3	4	5	6	7	8	9	10	11	12	13	14	15	16	17	18
1	H 2.20																	He
2	Li 0.98	Be 1.57											B 2.04	C 2.55	N 3.04	O 3.44	F 3.98	Ne
3	Na 0.93	Mg 1.31											Al 1.61	Si 1.90	P 2.19	S 2.58	Cl 3.16	Ar
4	K 0.82	Ca 1.00	Sc 1.36	Ti 1.54	V 1.63	Cr 1.66	Mn 1.55	Fe 1.83	Co 1.88	Ni 1.91	Cu 1.90	Zn 1.65	Ga 1.81	Ge 2.01	As 2.18	Se 2.55	Br 2.96	Kr 3.00
5	Rb 0.82	Sr 0.95	Y 1.22	Zr 1.33	Nb 1.6	Mo 2.16	Tc 1.9	Ru 2.2	Rh 2.28	Pd 2.20	Ag 1.93	Cd 1.69	In 1.78	Sn 1.96	Sb 2.05	Te 2.1	I 2.66	Xe 2.60
6	Cs 0.79	Ba 0.89	*	Hf 1.3	Ta 1.5	W 2.36	Re 1.9	Os 2.2	Ir 2.20	Pt 2.28	Au 2.54	Hg 2.00	Tl 1.62	Pb 2.33	Bi 2.02	Po 2.0	At 2.2	Rn 2.2
7	Fr 0.7	Ra 0.9	**	Rf	Db	Sg	Bh	Hs	Mt	Ds	Rg	Uub	Uut	Uuq	Uup	Uuh	Uus	Uuo
Lanthanides	*	La 1.1	Ce 1.12	Pr 1.13	Nd 1.14	Pm 1.13	Sm 1.17	Eu 1.2	Gd 1.2	Tb 1.1	Dy 1.22	Ho 1.23	Er 1.24	Tm 1.25	Yb 1.1	Lu 1.27		
Actinides	**	Ac 1.1	Th 1.3	Pa 1.5	U 1.38	Np 1.36	Pu 1.28	Am 1.13	Cm 1.28	Bk 1.3	Cf 1.3	Es 1.3	Fm 1.3	Md 1.3	No 1.3	Lr 1.291		

FIG. 9.1 Periodic table of Pauling electronegativity values.

Part 1. Determine the number of unique carbon bonding sites in phenol.

Phenol contains two unique carbon bonding sites: carbon C_1 bonded to carbon and oxygen. The remaining five carbon atoms are bonded to carbon and hydrogen (Fig. 9.2).

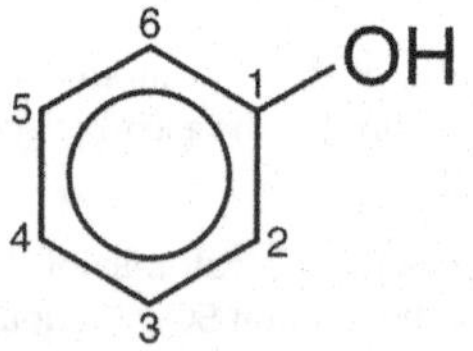

6 1 OH 5 4 2 3

FIG. 9.2 Molecular structure of phenol (CAS registry number 108-95-2).

Part 2. Assign the oxidation-state contribution of each unique carbon in phenol.

Table 9.2 lists each bond for the two unique carbon atoms in phenol. Each carbon-carbon bond is connected by a double arrow ⟷, indicating the electron pair assigned to this bond type are equally shared.

Since carbon is less electronegative than oxygen (Fig. 9.1), the bond-arrow points from carbon to oxygen, indicating oxygen is assigned the bonding electron pair. The electron transfer in each carbon-oxygen bond adds one unit to the carbon C_1 oxidation state.

Since carbon is more electronegative than hydrogen (Fig. 9.1), the bond-arrow points from hydrogen to carbon, indicating carbon is assigned the bonding electron pair. The electron transfer in each carbon-hydrogen bond subtracts one unit from the carbon C_2 oxidation state.

TABLE 9.2 Bond Oxidation State Assignments for Phenol

Bonding	OS(C1)
$C_1 \longleftrightarrow C_2$	0
$C_1 \longleftrightarrow C_6$	0
$C_1 \longrightarrow O$	+1
Bonding	**OS(C2)**
$C_2 \longleftrightarrow C_1$	0
$C_2 \longleftrightarrow C_3$	0
$C_2 \longleftarrow H$	−1

Part 3. Sum the oxidation-state bond contributions for each unique carbon to determine the net oxidation state of each unique carbon.

The net oxidation state of phenol carbon C_1 equals bond contributions summed over Table 9.2 rows 1–3: OS(C1) = +1.

The net oxidation state of phenol carbon C_2 equals bond contributions summed over Table 9.2 rows 4–6: OS(C2) = −1.

EXAMPLE 9.2

Example Permalink

http://soilenvirochem.net/wRKRkM

Assigning oxidation states using the rules in Table 9.1.

Rules 3 and 4 in Table 9.1 assign the oxidation states for hydrogen and oxygen. This example will apply rule 9 to assign oxidation states for carbon.

Part 1. Determine the number of unique carbon bonding sites in pyridine.

Pyridine contains two unique carbon bonding sites: carbon C_2 bonded to nitrogen, carbon, and hydrogen. The remaining three carbon atoms are bonded to carbon and hydrogen (Fig. 9.3).

Part 2. Assign the oxidation-state contribution of each unique carbon in pyridine.

Table 9.3 lists each bond for the two unique carbon atoms in pyridine. Each carbon-carbon bond is connected by a double arrow $\longleftrightarrow$, indicating the electron pair assigned to this bond type are equally shared.

Since carbon is less electronegative than nitrogen (Fig. 9.1), the bond-arrow points from carbon to nitrogen, indicating nitrogen is assigned the bonding electron pair. The electron transfer in each carbon-nitrogen bond adds one unit to the carbon C_2 oxidation state.

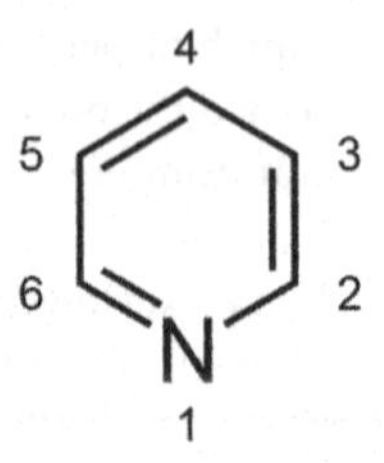

FIG. 9.3 Molecular structure of pyridine (CAS registry number 110-86-1).

Carbon is more electronegative than hydrogen (Fig. 9.1) therefore the bond-arrow points from hydrogen to carbon, indicating carbon is assigned the bonding electron pair. The electron transfer in each carbon-hydrogen bond subtracts one unit from the carbon C_3 oxidation state.

TABLE 9.3 Bond Oxidation State Assignments for Pyridine

Bonding	**OS(C2)**
$C_2 \longleftrightarrow C_3$	0
$C_2 \longrightarrow N$	+1
Bonding	**OS(C3)**
$C_3 \longleftrightarrow C_2$	0
$C_3 \longleftarrow H$	−1

Part 3. Sum the oxidation-state bond contributions for each unique carbon to determine the net oxidation state of each unique carbon.

The net oxidation state of pyridine carbon C_2 equals bond contributions summed over Table 9.3 rows 1–2: OS(C2) = +1.

The net oxidation state of pyridine carbon C_3 equals bond contributions summed over Table 9.3 rows 3–4: OS(C3) = −1.

9.2.2 Balancing Reduction Half-Reactions

The primary utility of oxidation states comes when balancing reduction-oxidation reactions: verifying that electron transfer occurs, identifying electron donor and acceptor atoms, and determining electron stoichiometry. General chemistry introduced the *reduction half-reaction* concept.

$$\nu_O \mathrm{Ox}^{o\pm} + m\mathrm{H}^+(aq) + ne^- \longrightarrow \nu_R \mathrm{Red}^{r\pm} \tag{9.R1}$$

$$(\pm o \cdot \nu_O) + (m - n) = (\pm r \cdot \nu_R) \tag{9.1}$$

Table 9.4 lists the steps you should follow when balancing a reduction half-reaction (9.R1 and Box 9.1). The steps must be executed in the order given to avoid confusion and error. Example 9.3 demonstrates how to balance a half-reaction.

TABLE 9.4 Steps for Balancing Reduction Half-Reactions

1. Assign formal oxidation states to identify the electron-acceptor specie (the educt in a reduction half-cell reaction) and the electron-donor (the product in a reduction half-reaction). **Example**: $\overset{OS(N)=+1}{N_2O(aq)} \longrightarrow \overset{OS(N)=-3}{NH_4^+(aq)}$
2. Balance the electron-acceptor and electron-donor stoichiometry. **Example**: $\overset{OS(N)=+1}{N_2O(aq)} \longrightarrow \overset{OS(N)=-3}{2NH_4^+(aq)}$
3. Determine the change in oxidation state and assign that integral number to the number of electrons n on the electron acceptor (or educt) side of the reaction equation. **Example**: $N_2O(aq) + 8e^- \longrightarrow 2NH_4^+(aq)$
4. Balance the net charge (*include both ion charges and electron charges*) by adding m H^+ ions to the educt side of the reaction (or an appropriate number of OH^- ions to the product side). **Example**: $N_2O(aq) + 10H^+(aq) + 8e^- \longrightarrow 2NH_4^+(aq)$
5. Balance proton (or hydroxyl) stoichiometry by adding an appropriate number of H_2O. **Example**: $N_2O(aq) + 10H^+(aq) + 8e^- \longrightarrow 2NH_4^+(aq) + H_2O(l)$

BOX 9.1

REDUCTION HALF-REACTIONS

Every balanced reduction-oxidation reaction can be factored into two half-reactions.[2] Factoring reaction (9.R4) results in half-reactions (9.R2), (9.R3).

$$Cu^{2+}(aq) + 2e^- \longrightarrow Cu(s) \qquad (9.R2)$$

$$2H+(aq) + 2e^- \longrightarrow H_2(g) \qquad (9.R3)$$

The *International Union of Pure and Applied Chemistry* endorses a formalism that writes all half-reactions as *reduction* half-reactions, listing the electron acceptor as an educt and the electron donor as a product.

Another convention appears in reactions (9.R1), (9.R2), (9.R3): the explicit listing of electron stoichiometry n. Although e^- appears as an educt in reduction half-reactions it is not a chemical specie. The appearance of electron stoichiometry n and e^- in half-reactions greatly simplifies the balancing of reduction-oxidation reactions (cf. Example 9.3).

[2] In Chapter 6 the hydronium cation $H_3O^+(aq)$ is used to represent the hydrated proton because water self-ionization is an integral part of the Brønsted-Lowery acid-base model. In the context of reduction-oxidation reactions, however, the abbreviated proton symbol H^+ simplifies the balancing of half-reactions.

EXAMPLE 9.3

Example Permalink

http://soilenvirochem.net/P2p7eA

Balancing reduction half-reactions.

Balancing the reduction half-reaction involving ascorbic acid (CAS registry number 50-81-7) and dehydroascorbic acid (CAS registry number 490-83-5) (Fig. 9.4) follows the steps appearing in Table 9.4.

Part 1. Identity the electron-acceptor and electron-donor species.

This requires calculating the oxidation state of each atom whose bonding environment changes in the reaction because acceptor and donor species may not be immediately evident from inspection.

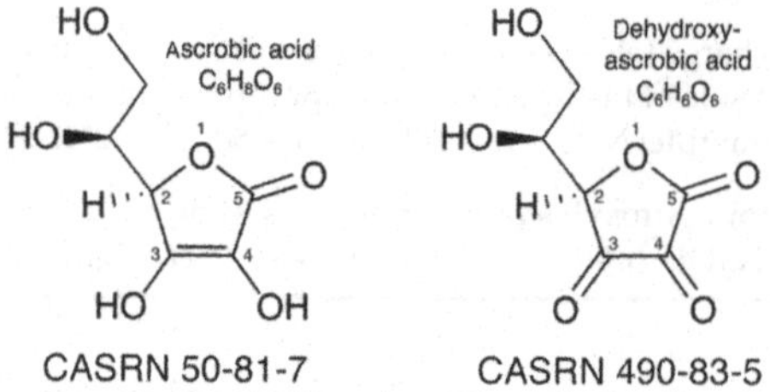

Ascorbic acid $C_6H_8O_6$ — CASRN 50-81-7; Dehydroxyascrobic acid $C_6H_6O_6$ — CASRN 490-83-5

FIG. 9.4 Molecular structures of ascorbic acid (CAS registry number 50-81-7) and dehydroascorbic acid (CAS registry number 490-83-5).

These two compounds differ at carbon C_3 (and C_4). The oxidation states of carbon C_3 are OS(C3) in ascorbic acid and OS(C3) in dehydroascorbic acid. This identifies dehydroascorbic acid as the electron acceptor and ascorbic acid as the electron donor.

The reduction half-reaction places electron-acceptor dehydroascorbic acid on the left-hand (educt) side and electron-donor ascorbic acid on the right-hand (product) side.

$$\underset{\text{Dehydroascorbic acid}}{C_6H_6O_6(aq)} \longrightarrow \underset{\text{Ascorbic acid}}{C_6H_8O_6(aq)}$$

Part 2. Balance electron-acceptor and electron-donor elements whose oxidation states change.

The electron-acceptor stoichiometry number and the electron-donor stoichiometry number, in this case, both equal unity.

Part 3. Determine the electron stoichiometry number, balancing the oxidation states of the electron acceptor and electron donor.

The net change in oxidation state summed over both carbons is −2. The required electron stoichiometry number *n* is 2.

$$\underset{\text{Dehydroascorbic acid}}{C_6H_6O_6(aq)} + 2e^- \longrightarrow \underset{\text{Ascorbic acid}}{C_6H_8O_6(aq)}$$

Part 4. Determine the proton stoichiometry number *m*, balancing both species and electron charge.

Charge balance requires a proton stoichiometry number *m* of 2, balancing the electron charge only since ascorbic acid and dehydroascorbic acid are both neutral molecules.

$$\underset{\text{Dehydroascorbic acid}}{C_6H_6O_6(aq)} + 2H^+ + 2e^- \longrightarrow \underset{\text{Ascorbic acid}}{C_6H_8O_6(aq)}$$

The proton mass balance does not require any additional water molecules; the reaction above is the fully balanced reduction half-reaction.

9.2.3 Standard Electrochemical Potentials

A standard galvanic cell appears in Fig. 9.5, a familiar illustration from general chemistry. Electrons enter the external circuit at the anode, an inert platinum electrode where $H_2(g)$ is oxidized to $H^+(aq)$ ions. These electrons reduce $Cu^{2+}(aq)$ ions to Cu(s) at the cathode. A salt bridge permits the diffusion of spectator ions to maintain charge neutrality during electron transfer (reaction 9.R4). A high impedance[3] voltmeter measures the cell potential. The $H_2(g)$ gas pressures are 1 atmosphere and the ion activities are 1 molar (cf. Section 5.2.1).

$$Cu^{2+}(aq) + H_2(g) \longrightarrow Cu(s) + 2H^+(aq) \tag{9.R4}$$

The standard electrochemical potential for a hydrogen half-cell is, by definition, equal to zero volts: $E^{\circ} = 0$ mV. Consequently the cell potential recorded in Fig. 9.5 is equal to the standard electrochemical potential of the other half-cell: $E^{\circ} = +337$ mV.

Most reduction-oxidation reactions encountered in biology, geochemistry, and environmental chemistry cannot be replicated using galvanic cells of the type shown in Fig. 9.5. Regardless, it is possible to use Gibbs energy relations to determine standard electrode half-reaction potentials for all relevant reduction-oxidation reactions. We will explain this in the following sections.

Standard potentials serve the same purpose as the standard conditions discussed in Chapter 5: the efficient exchange of essential chemical data in a form that all chemists understand. Standard potentials are not intended to serve any other purpose. The following section discusses how standard potentials can be adjusted to reflect prevailing conditions.

9.2.4 The Nernst Equation

The Nernst equation, which you first encountered in general chemistry, allows you to calculate a half-reaction electrochemical potential under any nonstandard condition. It is derived from the Gibbs energy expression (9.2) (cf. Chapter 5) and an expression relating Gibbs energy to the electric potential (9.3).

[3] A high impedance voltmeter requires a very small current to record an electric potential.

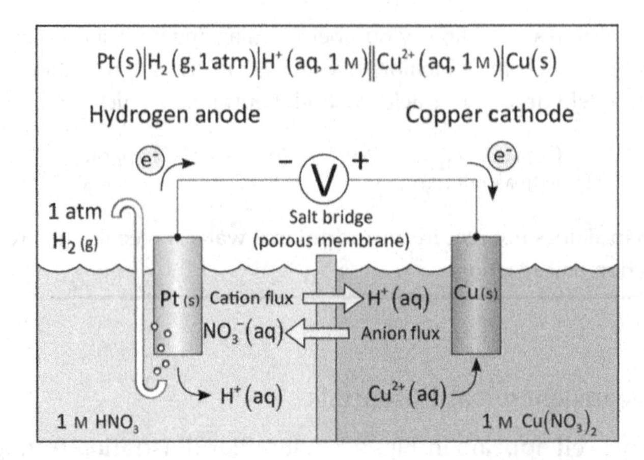

FIG. 9.5 A complete galvanic cell consists of a cathode and an anode. The cathode is a standard cupric half-cell, and the anode is a standard hydrogen half-cell. Shorthand notation for the galvanic cell in appears at the top (*double vertical lines* indicate salt bridge).

Expression (9.2) represents the Gibbs energy change $\Delta_r G$ in any chemical reaction. The symbols in expression (9.2) are: the reaction quotient Q, the *gas constant* $R = 8.314472 \times 10^{-3}$ kJ mol^{-1} K^{-1}, and the *absolute temperature* T.

$$\Delta_r G = \Delta_r G^{\circ} + (R \cdot T) \cdot \ln\left(\frac{a_{\mathrm{Red}^{r\pm}}^{\nu_R}}{a_{\mathrm{H}^+}^{m} \cdot a_{\mathrm{Ox}^{o\pm}}^{\nu_O}}\right) = \Delta_r G^{\circ} + (R \cdot T) \cdot \ln Q \tag{9.2}$$

Expression (9.3) represents the Gibbs energy change $\Delta_r G$ as electrons move from one *electrostatic potential* to another. The initial and final electrostatic potentials are those existing in the electron acceptor and donor compounds. The symbols in Eq. (9.3) are: the electron stoichiometry number n and the *Faraday constant* $F = 9.6485 \times 10^{-2}$ kJ mol^{-1} mV^{-1}.

$$\Delta_r G = -n \cdot F \cdot E \tag{9.3}$$

Every reduction-oxidation reaction can be factored into half-reactions (cf. Box 9.1). The electrochemical potential for a balanced reduction-oxidation reaction and the Gibbs energy change are denoted E and $\Delta_r G$, respectively, regardless of whether it is a complete or half-reaction.

Setting expression (9.2) equal to Eq. (9.3) results in the seminal Nernst equation (9.4). Solving expression (9.4) for half-reaction potential E yields the Nernst equation in its most familiar form (expression 9.5).

$$(-n \cdot F \cdot E) = \left(-n \cdot F \cdot E^{\circ}\right) + (R \cdot T) \cdot \ln\left(\frac{a_{\mathrm{Red}^{r\pm}}^{\nu_R}}{a_{\mathrm{H}^+}^{m} \cdot a_{\mathrm{Ox}^{o\pm}}^{\nu_O}}\right) \tag{9.4}$$

$$E = E^{\circ} - \left(\frac{R \cdot T}{n \cdot F}\right) \cdot \ln\left(\frac{a_{\mathrm{Red}^{r\pm}}^{\nu_R}}{a_{\mathrm{H}^+}^{m} \cdot a_{\mathrm{Ox}^{o\pm}}^{\nu_O}}\right) \tag{9.5}$$

Usually the natural logarithm in expression (9.5) is replaced with the base-10 logarithm (expression 9.6).[4]

$$E = E^{\circ} - \left(\frac{R \cdot T \cdot \ln(10)}{n \cdot F}\right) \cdot \log_{10}\left(\frac{a_{\mathrm{Red}^{r\pm}}^{\nu_R}}{a_{\mathrm{H}^+}^{m} \cdot a_{\mathrm{Ox}^{o\pm}}^{\nu_O}}\right) \tag{9.6}$$

At 298.15 K (25°C),[5] the bracketed term listing temperature T (expression 9.6) reduces to an expression (9.7) dependent solely on electron stoichiometry n.

$$\left(\frac{R \cdot (298.15\ \mathrm{K}) \cdot \ln(10)}{n \cdot F}\right) = \left(\frac{59.160\ \mathrm{mV}}{n}\right) \tag{9.7}$$

$$E(298.15\ \mathrm{K}) = E^{\circ}(298.15\ \mathrm{K}) - \left(\frac{59.160\ \mathrm{mV}}{n}\right) \cdot \log_{10}\left(\frac{a_{\mathrm{Red}^{r\pm}}^{\nu_R}}{a_{\mathrm{H}^+}^{m} \cdot a_{\mathrm{Ox}^{o\pm}}^{\nu_O}}\right) \tag{9.8}$$

9.2.5 Standard Biological Potentials

Biochemists tabulate *standard biological*[6] *potentials* $E^{\oplus}$ that approximate physiological conditions. Some environmental chemists also quote *standard environmental potentials* which are identical to standard biological potentials $E^{\oplus}$.

Rearranging expression (9.8) yields expression (9.9), which explicitly lists proton m and electron n stoichiometry numbers.

$$\begin{aligned} E(298.15\ \mathrm{K}) = {} & \left(E^{\circ}(298.15\ \mathrm{K}) - \left(\frac{m}{n}\right) \cdot (59.160\ \mathrm{mV}) \cdot \mathrm{pH}\right) \\ & - \left(\frac{59.160\ \mathrm{mV}}{n}\right) \cdot \log_{10}\left(\frac{a_{\mathrm{Red}^{r\pm}}^{\nu_R}}{a_{\mathrm{Ox}^{o\pm}}^{\nu_O}}\right) \end{aligned} \tag{9.9}$$

The first and second terms in expression (9.9) define the standard biological electrochemical potential $E^{\oplus}$ (expression 9.10), regardless of the values assigned to the remaining terms in the reaction quotient Q. Expression (9.11) allows you to calculate a nonstandard electrochemical potential using $E^{\oplus}$ instead of E°.

[4] $\ln x = \ln 10 \cdot \log_{10} x \approx 2.3026 \cdot \log_{10} x$.

[5] Conversion from degrees Celsius to absolute temperature Kelvin: 0°C = 273.15 K.

[6] The *International Union of Pure and Applied Chemistry* does not recognize the *standard biological state* and does not endorse a symbol for the standard biological electrochemical potential. The biochemistry community, however, has adopted the *standard biological state* as defined here. The standard biological electrochemical potential notation adopted here is that used by Atkins and de Paula (2005).

$$E^{\oplus}(298.15\text{ K}) \equiv E^{\circ}(298.15\text{ K}) - \left(\left(\frac{m}{n}\right) \cdot (59.160\text{ mV}) \cdot 7\right) \tag{9.10}$$

$$E(298.15\text{ K}) = E^{\oplus}(298.15\text{ K}) - \left(\frac{59.160\text{ mV}}{n}\right) \cdot \log_{10}\left(\frac{a_{\text{Red}^{r\pm}}^{\nu_R}}{a_{\text{Ox}^{o\pm}}^{\nu_O}}\right) \tag{9.11}$$

Example 9.4 demonstrates the calculation of the standard *biological* potential $E^{\oplus}$ using the Nernst equation (9.9). A standard biological electrochemical potential $E^{\oplus}$ requires that all components—other than $H^+(aq)$—have activities or partial pressures equal to unity.

EXAMPLE 9.4

Example Permalink

http://soilenvirochem.net/jTAHAt

Calculate the standard *biological* electrochemical potential $E^{\oplus}$ for the following reduction half-reaction.

$$O_2(g) + 4H^+(aq) + 4e^- \xrightarrow{E^{\circ} = 1229\text{ mV}} 2H_2O(l)$$

The standard electrochemical potential for the O_2 reduction half-reaction is: $E^{\circ} = 1229$ mV. Under standard conditions the proton activity $a_{H^+} = 1$ mol dm^{-3} (cf. Appendix 5.A).

Part 1. Apply the standard state condition: $p_{O_2} = 1$ atm.

$$E^{\oplus} = E^{\circ} - \left(\frac{R \cdot T \cdot \ln(10)}{4 \cdot F}\right) \cdot \log_{10}\frac{1}{a_{H^+}^4 \cdot p_{O_2}}$$

$$E^{\oplus}(298.15\text{ K}) = (1229\text{ mV}) - \left(\frac{59.16\text{ mV}}{4}\right) \cdot \log_{10}\frac{1}{a_{H^+}^4}$$

$$E^{\oplus}(298.15\text{ K}) = (1229\text{ mV}) + \left(\frac{59.16\text{ mV}}{4}\right) \cdot 4 \cdot \log_{10} a_{H^+}$$

$$E^{\oplus}(298.15\text{ K}) = (1229\text{ mV}) + (59.16\text{ mV}) \cdot \log_{10} a_{H^+}$$

Part 2. Apply the standard *biological* state condition: $a_{H^+} = 10^{-7}$ mol dm^{-3}.
Using the definition: $\text{pH} = -\log_{10} a_{H^+}$.

$$E^{\oplus} = (1229\text{ mV}) - 7 \cdot (59.16\text{ mV})$$

$$E^{\oplus} = (1229\text{ mV}) - (414.1\text{ mV}) = 814.9\text{ mV}$$

9.3 MEASUREMENT AND INTERPRETATION OF ELECTROCHEMICAL POTENTIALS IN SOILS AND SEDIMENTS

One of the most important tasks facing an environmental chemist is assessing and interpreting chemical conditions at a site. Chapter 5 discussed the use of water chemistry simulations to identify minerals that potentially control water chemistry in environmental samples. Chapters 4 and 8 discuss the use of adsorption and exchange isotherms to quantify an insoluble but chemically-active reserve that sustains solute concentrations without the involvement of precipitation reactions. The assessment and interpretation of reduction-oxidation conditions require chemical coordinates that characterize reduction-oxidation conditions and a means of interpreting the significance of those chemical coordinates.

9.3.1 Electrochemical Stability Diagrams

The two chemical coordinates that chemists use to characterize reduction-oxidation conditions are the measured electrochemical potential E and pH. Electrochemist Marcel Pourbaix (1945) was the first to prepare stability diagrams to interpret reduction-oxidation conditions using (E, pH) coordinates. Corrosion scientists were the first to employ electrochemical (E, pH) diagrams, but environmental chemists, soil chemists, and geochemists quickly adopted them to interpret geochemical and environmental reduction-oxidation conditions.

Baas-Becking et al. (1960) published a electrochemical diagram plotting a host of experimental (E, pH) coordinates from scientific papers reporting environmental reduction-oxidation conditions (Fig. 9.6 and Box 9.2).

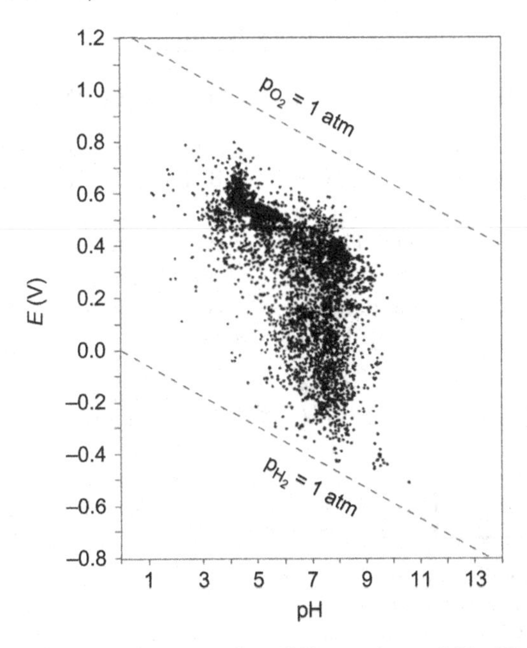

FIG. 9.6 This electrochemical diagram plots more than 6200 experimental (E, pH) coordinates representative of the reduction-oxidation conditions found in the environment. *Reproduced with permission from Baas-Becking, L.G.M., Kaplan, I.R., Moore, D., 1960. Limits of the natural environment in terms of pH and oxidation-reduction potentials. J. Geol. 68, 243–284.*

Few of the experimental (*E*, pH) points fall outside the pH range 4–8. The pH 4 limit represents the pH of maximum buffering by exchangeable Al^{3+} ions (cf. Chapter 6). Carbonate buffering at pH 8 resists further pH increases.

Baas-Becking et al. (1960) attribute the significant increase in the range of environmental reduction-oxidation conditions centered on the alkaline pH limit to microbial sulfate respiration; however, there is no significant change in sulfate solubility in this pH range. An alternative interpretation would assign the dramatic expansion of reducing conditions at the alkaline pH limit to microbial carbonate respiration. Carbonate solubility increases dramatically above pH 6 (see Chapter 7), increasing both buffering capacity and the availability of the terminal electron donor required for methanogenesis (cf. Section 9.5).

$$HCO_3^-(aq) + 7H^+(aq) + 8e^- \xrightarrow{E^{\oplus}=-245\ \mathrm{mV}} CH_4(g) + 2H_2O(l) \qquad (9.R5)$$

Regardless of why the range of reducing conditions recorded in soils expands with increasing pH, Fig. 9.6 confirms the profound importance of anaerobic microbial respiration in defining environmental reduction-oxidation conditions.

BOX 9.2

THE PLATINUM OXIDATION-REDUCTION ELECTRODE

Environmental reduction potentials are measured using some variation of the platinum combination oxidation-reduction potential (ORP) electrode, illustrated in Fig. 9.7. The electrode is a compact version of the galvanic cell in Fig. 9.5. The standard hydrogen half-cell (anode, Fig. 9.5) is not practical for environmental measurements and is replaced by a reference half-cell such as the calomel half-cell with reduction half-reaction (9.R6). The voltmeter employs a high-impedance design that requires a very small current to record an accurate potential.

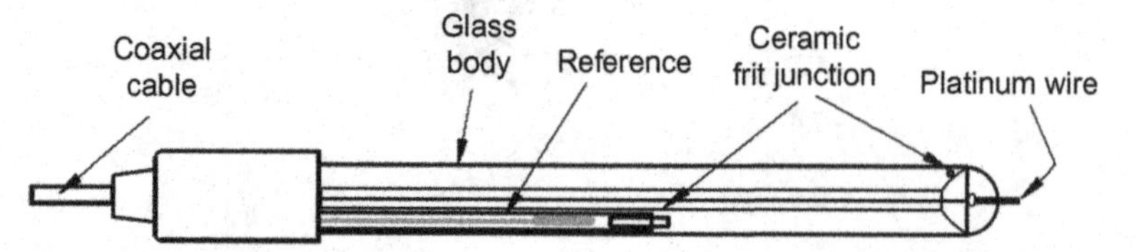

FIG. 9.7 A typical platinum oxidation-reduction potential (ORP) electrode is a combination electrode. The electrochemical potential of the reference half cell in this case is based on the calomel electrochemical potential (reaction 9.R6, $E^{\oplus} = +244.4$ mV). The reduction potential is measured using a platinum wire as the second electrode.

BOX 9.2 *(cont'd)*

$$Hg_2Cl_2(s) + 2e^- \xrightarrow{E^\circ=+244.4\ mV} 2Hg(l) + 2Cl^- \quad (9.R6)$$

The platinum ORP combination electrode includes two porous ceramic junctions that function as salt bridges, allowing a spectator ion current to flow between reference electrode and the external solution being measured. The electron current passes through a voltmeter in the circuit connecting the platinum sensing wire and the liquid mercury of the reference cell. Nordstrom and Wilde (1998) outline the limitations of measuring environmental reduction potentials using platinum ORP electrodes. These limitations are discussed in Appendix 9.B.

The true electrochemical potential of a soil that is rich in organic matter is highly reducing, but the reducing potential of natural organic matter is blocked by an activation barrier that prevents the transfer of electrons from organic matter to dioxygen (cf. Chapter 7). Although biological respiration (i.e., catabolism and the electron transport chain) eliminates the activation barrier blocking carbon oxidation by O_2, scientists are unable to quantify environmental electrochemical potentials using platinum ORP electrodes.

9.3.2 Pourbaix Electrochemical Stability Diagrams

This section describes the components in a Fe–O–C–H Pourbaix stability diagram to illustrate the essential features found in all Pourbaix stability diagrams. The preparation of accurate Pourbaix stability diagrams is beyond the scope of this book, since it requires an advanced understanding of chemical thermodynamics and reduction-oxidation chemistry. This section, however, will provide the foundation that will allow you to interpret virtually any Pourbaix stability diagram you will encounter, however complex.

Table 9.5 lists the features found in any complete Pourbaix stability diagram. Interpretation begins with identifying each of these features. Most of these features will be readily apparent, but some may be listed in the caption or text accompanying the diagram. The absence of these features makes interpretation impossible. Once you have correctly identified all of the essential features, the interpretation is surprisingly simple because Pourbaix stability diagrams provide a limited representation of the chemical system.

TABLE 9.5 The Essential Features Needed to Interpret a Pourbaix Diagram

1.	The elemental components whose chemical stability the diagram is designed to portray
2.	Total dissolved concentration of all solutes and partial pressure of all gases influencing the chemical stability of the defined system
3.	The nature of each stability field: gas, solute, or precipitate

9.3.2.1 Water Stability Limits

The two-boundaries in Fig. 9.6 delineate the stability domain of water (Delahay et al., 1950). These boundaries are derived from the two reduction half-reactions (9.R7), (9.R8).

$$O_2(g) + 4H^+(aq) + 4e^- \xrightarrow{E^\circ=+1229\,\mathrm{mV}} 2H_2O(l) \tag{9.R7}$$

$$2H^+(aq) + 2e^- \xrightarrow{E^\circ=0\,\mathrm{mV}} H_2(g) \tag{9.R8}$$

Expression (9.12) is the Nernst equation for half-reaction (9.R7) defining the boundary labeled $p_{O_2} = 1$ atm in Fig. 9.8.

$$E(298.15\ \mathrm{K}) = E^\circ(298.15\ \mathrm{K}) - \left(\frac{59.160\ \mathrm{mV}}{4}\right) \cdot \log_{10}\left(\frac{1}{a_{H^+}^4 \cdot p_{O_2}}\right) \tag{9.12}$$

Factoring the second term on the right of expression (9.12) results in expression (9.13), containing three terms on the right: the standard electrochemical potential $E^\circ(298.15\ \mathrm{K})$, the dependence on the O_2 partial pressure (cf. Box 9.3), and the dependence on proton activity. The electron stoichiometry number n and proton stoichiometry number m are identical, canceling in the third term (expression 9.14).

The $O_2(g)|H_2O(l)$ stability boundary (expression 9.14) is usually drawn assuming: $p_{O_2} = 1$ atm. Expression (9.14) is the equation of a line with a slope equal to 1229 mV and an intercept equal to 59.16 mV.

BOX 9.3

ANOXIA

The $O_2(g)|H_2O(l)$ stability boundary is usually drawn assuming (expression 9.14): $p_{O_2} = 1$ atm. To demonstrate the effects of anoxia substitute $p_{O_2} = 10^{-6}$ atm into expression (9.13) and combine terms.

$$E(298.15\ \mathrm{K}) = E^\circ(298.15\ \mathrm{K}) + \left(\frac{(59.160\ \mathrm{mV}) \cdot \log_{10} 10^{-6}}{4}\right) - (59.160\ \mathrm{mV}) \cdot \mathrm{pH}$$

$$E(298.15\ \mathrm{K}) = 1141\ \mathrm{mV} - (59.160\ \mathrm{mV}) \cdot \mathrm{pH}$$

The pH = 0 intercept is shifted downward a mere 88 mV relative to the intercept of expression (9.14), a relatively minor shift considering the 700–800 mV potential recorded by a typical platinum ORP electrode E^{Pt} in aerated water (cf. Fig. 9.6 and Appendix 9.B).

Reducing conditions develop when anaerobiosis (i.e., anaerobic microbial respiration) releases byproducts such as nitrite $NO_2^-(aq)$, ferrous iron $Fe^{2+}(aq)$, hydrogen sulfite $HSO_3^-(aq)$, hydrogen sulfide $HS^-(aq)$, and the like into solution.

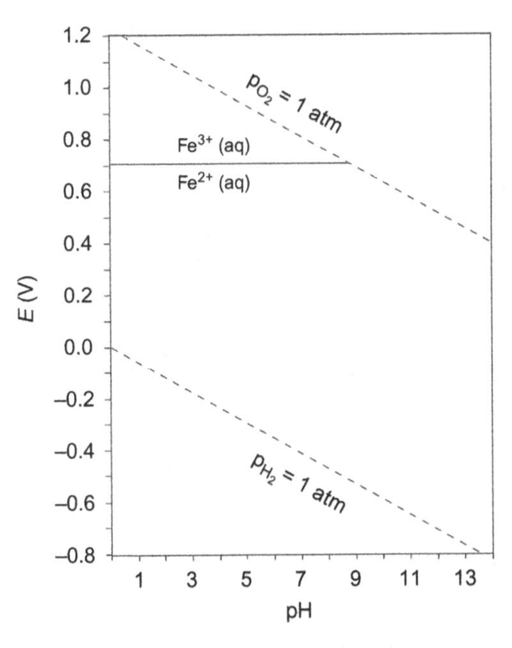

FIG. 9.8 An incomplete Pourbaix stability diagram plotting the two water stability boundaries, expressions (9.14), (9.17), and one solute-solute reduction boundary expression (9.19). The positions of the water stability boundaries are determined by $p_{O_2} = 1$ atm and $p_{H_2} = 1$ atm, as reported in the diagram.

$$E(298.15\ \text{K}) = E^{\circ}(298.15\ \text{K}) + \left(\frac{59.160\ \text{mV}}{4}\right) \cdot \log_{10} p_{O_2} - (59.160\ \text{mV}) \cdot \text{pH} \tag{9.13}$$

$$E(298.15\ \text{K}) = 1229\ \text{mV} - (59.160\ \text{mV}) \cdot \text{pH} \tag{9.14}$$

Expression (9.15) is the Nernst equation half-reaction (9.R8) defining the boundary labeled $p_{H_2} = 1$ atm in Fig. 9.8.

$$E(298.15\ \text{K}) = E^{\circ}(298.15\ \text{K}) - \left(\frac{59.160\ \text{mV}}{2}\right) \cdot \log_{10}\left(\frac{p_{H_2}}{a_{H^+}^2}\right) \tag{9.15}$$

Factoring the second term on the right of expression (9.15) results in expression (9.16), containing three terms on the right: the standard electrochemical potential E°, the dependence on the $H_2(g)$ partial pressure, and the dependence on proton activity. The electron stoichiometry number n and proton stoichiometry number m are identical and cancel in the third term. The $H^+(aq)|H_2(g)$ stability boundary (expression 9.17) is usually drawn assuming: $p_{H_2} = 1$ atm.

$$E(298.15\ \text{K}) = E^{\circ}(298.15\ \text{K}) - \left(\frac{59.160\ \text{mV}}{2}\right) \cdot \log_{10} p_{H_2} - (59.160\ \text{mV}) \cdot \text{pH} \tag{9.16}$$

$$E(298.15\ \text{K}) = 0\ \text{mV} - (59.160\ \text{mV}) \cdot \text{pH} \tag{9.17}$$

Boundary expressions (9.14), (9.17) are parallel and differ only in their intercept, which at $pH = 0$ are the standard electrochemical potentials E° for the two reduction half-reactions (9.R7) and (9.R8) defining the electrochemical limits of water stability.

9.3.2.2 *Solute-Solute Reduction Boundary*

The first boundary unique to the Fe–O–C–H Pourbaix stability diagram we are preparing is the pH-independent equilibrium boundary for half-reaction (9.R9).

$$Fe^{3+}(aq) + e^{-} \xrightarrow{E^{\circ} = +771\ \text{mV}} Fe^{2+}(aq) \tag{9.R9}$$

The Nernst equation for half-reaction (9.R9) is Eq. (9.18). Since this half-reaction electrochemical potential is pH independent, we can anticipate the boundary runs perpendicular to the E axis of our Pourbaix stability diagram.

$$E(298.15\ \text{K}) = E^{\circ}(298.15\ \text{K}) - \left(\frac{59.160\ \text{mV}}{1}\right) \cdot \log_{10} \frac{a_{Fe^{2+}}}{a_{Fe^{3+}}} \tag{9.18}$$

The convention used to define a solute-solute reduction equilibrium boundary is to draw the boundary where the activities of the two solutes are equal: $a_{Fe^{2+}} = a_{Fe^{3+}}$. Expression (9.18) reduces to solute-solute reduction equilibrium boundary expression (9.19).

$$E(298.15\ \text{K}) = E^{\circ}(298.15\ \text{K}) = 771\ \text{mV} \tag{9.19}$$

Solute $Fe^{3+}(aq)$ undergoes hydrolysis reactions (9.R10), (9.R11) as the pH increases.

$$Fe^{3+}(aq) + H_2O(l) \xleftrightarrow{K_{a1}^{\circ} = 10^{-2.19}} Fe(OH)^{2+}(aq) + H^{+}(aq) \tag{9.R10}$$

$$Fe(OH)^{2+}(aq) + H_2O(l) \xleftrightarrow{K_{a2}^{\circ} = 10^{-3.50}} Fe(OH)_2{}^{+}(aq) + H^{+}(aq) \tag{9.R11}$$

Solution specie $Fe(OH)^{2+}(aq)$ appearing in both Eqs. (9.R10), (9.R11) never reaches an activity where it accounts for 50% of the soluble Fe. Hydrolysis equilibrium involving the two dominant solute species is defined by reaction (9.R12), which combines reactions (9.R10), (9.R11).

$$Fe^{3+}(aq) + 2H_2O(l) \longleftrightarrow Fe(OH)_2{}^{+}(aq) + 2H^{+}(aq) \tag{9.R12}$$

The equilibrium expression for hydrolysis reaction (9.R12) is given by expression (9.20).

$$K_{a1}^{\circ} \cdot K_{a2}^{\circ} = \frac{a_{Fe(OH)_2{}^{+}} \cdot a_{H^{+}}^{2}}{a_{Fe^{3+}}} \tag{9.20}$$

Since electron acceptor $Fe(OH)_2{}^{+}(aq)$ will exceed 50% of the total soluble Fe in the range $pH > 2.84$, we require a solute-solute reduction equilibrium boundary. This boundary is found by adding reduction half-reaction (9.R9) and hydrolysis reaction (9.R12). The combined reaction is given by (9.R13).

$$Fe(OH)_2{}^{+}(aq) + 2H^{+}(aq) + e^{-} \longrightarrow Fe^{2+}(aq) + 2H_2O(l) \qquad (9.R13)$$

The electrochemical potential for this half-reaction is calculated from the standard Gibbs energy for the reduction half-reaction (9.R13) using the standard Gibbs energy of formation $\Delta_f G^{\circ}$ for each educt and product (expression 9.21).[7]

$$\begin{aligned} \Delta_r G^{\circ} &= \left(\Delta_f G^{\circ}\left(Fe^{2+}, aq\right) + 2 \cdot \Delta_f G^{\circ}\left(H_2O, l\right)\right)_{products} \\ &\quad - \left(\Delta_f G^{\circ}\left(Fe(OH)_2{}^{+}, aq\right) + 2 \cdot \Delta_f G^{\circ}\left(H^{+}, aq\right)\right)_{educts} \qquad (9.21) \\ \Delta_r G^{\circ}(298.15\text{ K}) &= \left(\left(-91.21\text{ kJ mol}^{-1}\right) + 2 \cdot \left(-237.2\text{ kJ mol}^{-1}\right)\right) \\ &\quad - \left(\left(-458.7\text{ kJ mol}^{-1}\right) + 2 \cdot \left(0\text{ kJ mol}^{-1}\right)\right) = -106.9\text{ kJ mol}^{-1} \end{aligned}$$

The electrochemical potential expression (9.22) is calculated by substituting the standard Gibbs energy $\Delta_r G^{\circ}$ for reduction half-reaction (9.R13) into expression (9.3).

$$\begin{aligned} E^{\circ}(298.15\text{ K}) &= \frac{-\Delta_r G^{\circ}(298.15\text{ K})}{n \cdot F} \\ E^{\circ}(298.15\text{ K}) &= \frac{-\left(-106.9\text{ kJ mol}^{-1}\right)}{1 \cdot \left(9.6485 \times 10^{-2}\text{ kJ mol}^{-1}\text{ mV}^{-1}\right)} = -1108\text{ mV} \qquad (9.22) \end{aligned}$$

The Nernst equation for half-reaction (9.R13) is Eq. (9.23).

$$\begin{aligned} E(298.15\text{ K}) &= E^{\circ}(298.15\text{ K}) - \left(\frac{59.160\text{ mV}}{1}\right) \cdot \log_{10} \frac{a_{Fe^{2+}}}{a_{Fe(OH)_2{}^{+}} \cdot a_{H^{+}}^{2}} \\ E(298.15\text{ K}) &= (1108\text{ mV}) - (59.160\text{ mV}) \cdot \log_{10} \frac{a_{Fe^{2+}}}{a_{Fe(OH)_2{}^{+}}} - 2 \cdot (59.160\text{ mV}) \cdot \text{pH} \qquad (9.23) \end{aligned}$$

The convention used to define solute-solute reduction equilibrium boundaries is to draw the boundary where the activities of the two solutes are equal: $a_{Fe^{2+}} = a_{Fe(OH)_2{}^{+}}$. Expression (9.23) reduces to boundary expression (9.24).

$$E(298.15\text{ K}) = (1108\text{ mV}) - (118.3\text{ mV}) \cdot \text{pH} \qquad (9.24)$$

[7] The standard Gibbs energy of formation $H^{+}(aq)$ is defined as: $\Delta_f G^{\circ} \equiv 0\text{ kJ mol}^{-1}$. The source for standard Gibbs energy of formation $\Delta_f G^{\circ}$ values and various equilibrium constants K° is Lindsay (1979) and Dick (2008).

9.3.2.3 *Solute-Solute Hydrolysis Boundary*

Solute hydrolysis is a common equilibrium included in Pourbaix stability diagrams. Solute $Fe^{3+}(aq)$ undergoes hydrolysis reactions (9.R10), (9.R11), as noted in the preceding section. Solving expression (9.20) for the proton activity results in expression (9.25).

$$a^2_{H^+} = K^{\ominus}_{a1} \cdot K^{\ominus}_{a2} \cdot \frac{a_{Fe^{3+}}}{a_{Fe(OH)_2^+}} = 10^{-5.69} \cdot \frac{a_{Fe^{3+}}}{a_{Fe(OH)_2^+}}$$

$$a_{H^+} = 10^{-2.84} \cdot \left(\frac{a_{Fe^{3+}}}{a_{Fe(OH)_2^+}}\right)^{1/2} \qquad (9.25)$$

Following the convention used to derive (expression 9.19), we will draw the solute-solute hydrolysis equilibrium boundary where the activities of the two solutes are equal: $a_{Fe^{3+}} = a_{Fe(OH)_2^+}$. Expression (9.25) reduces to boundary expression (9.26). Since this hydrolysis reaction is independent of *E*, boundary (9.26) runs perpendicular to the pH axis of our Pourbaix stability diagram.

$$pH = 2.84 \qquad (9.26)$$

9.3.2.4 *Solute-Precipitate Solubility Boundary*

Solute precipitation is a common equilibrium included in Pourbaix stability diagrams. Solute $Fe^{3+}(aq)$ undergoes hydrolysis as the pH increases until precipitation occurs in a pH range where $Fe(OH)_3^0(aq)$ is the most abundant ferric solute (reaction 9.R14). Solute-precipitate equilibrium involving the dominant specie $Fe(OH)_2^+(aq)$ is defined by reaction (9.R15), which combines hydrolysis reaction (9.R12) with solubility reaction (9.R14).

$$Fe(OH)_3(s) + 3H^+(aq) \xleftrightarrow{K^{\ominus}_{s0}\cdot(K^{\ominus}_{w})^{-3}=10^{+3.54}} Fe^{3+}(aq) + 3H_2O(l) \qquad (9.R14)$$

$$Fe(OH)_3(s) + H^+(aq) \overset{K^{\ominus}_{s2}}{\longleftrightarrow} Fe(OH)_2^+(aq) + H_2O(l) \qquad (9.R15)$$

Combining the equilibrium constants and quotients for reactions (9.R12), (9.R14) and solving for H^+ activity yields boundary expression (9.27).

$$K^{\ominus}_{s2} = \left(\frac{K^{\ominus}_{s0}}{K^{\ominus}_{w}}\right) \cdot \left(K^{\ominus}_{a1} \cdot K^{\ominus}_{a2}\right) = 10^{+3.54} \times 10^{-5.69} = 10^{-2.15}$$

$$K^{\ominus}_{s2} = \left(\frac{a_{Fe^{3+}}}{a^3_{H^+}}\right) \cdot \left(\frac{a_{Fe(OH)_2^+} \cdot a^2_{H^+}}{a_{Fe^{3+}}}\right) = \frac{a_{Fe(OH)_2^+}}{a_{H^+}}$$

$$a_{H^+} = \frac{a_{Fe(OH)_2^+}}{K^{\ominus}_{s2}} = \frac{a_{Fe(OH)_2^+}}{10^{-2.15}} \qquad (9.27)$$

The position of the solute-precipitate equilibrium boundary depends on the activity of solute $Fe(OH)_2^+(aq)$ appearing in the numerator of expression (9.27). The selection of this boundary condition relates to the second item in Table 9.5. *The total soluble concentration of all solutes must be identified before you can interpret any Pourbaix diagram*. In this example, the total soluble iron concentration is chosen to be: $c_{Fe} = 10^{-7}$ mol dm^{-3}. The solute-precipitate equilibrium boundary for this choice is expression (9.28). The soluble iron boundary condition is reported by labeling the Pourbaix stability diagram itself (cf. Fig. 9.9).

$$a_{H^+} = \frac{10^{-7.00}}{10^{-2.15}} = 10^{-4.85} \tag{9.28}$$

Carbon dioxide gas dissolved in water will cause solute $Fe^{2+}(aq)$ to precipitate as siderite $FeCO_3(s)$ at high pH. This solute-precipitate equilibrium boundary depends on the carbon dioxide partial pressure, as shown in solubility reaction (9.R16) and carbonate hydrolysis reactions (9.R17) and (9.R18).

$$\underset{\text{Siderite}}{FeCO_3(s)} \xleftrightarrow{K^{\ominus}_{s0}=10^{-10.89}} Fe^{2+}(aq) + CO_3{}^{2-}(aq) \tag{9.R16}$$

$$CO_2(g) + H_2O(l) \xleftrightarrow{K^{\ominus}_{a1}=10^{-6.37}} H^+(aq) + HCO_3{}^-(aq) \tag{9.R17}$$

$$HCO_3{}^-(aq) + H_2O(l) \xleftrightarrow{K^{\ominus}_{a2}=10^{-10.33}} H^+(aq) + CO_3{}^{2-}(aq) \tag{9.R18}$$

The base-10 logarithm of solubility expression (9.29) is rearranged to obtain boundary expression (9.30).

$$\frac{K^{\ominus}_{s0}}{K^{\ominus}_{a1} \cdot K^{\ominus}_{a2}} = 10^{+5.81} = \frac{a_{Fe^{2+}} \cdot p_{CO_2(g)}}{a^2_{H^+}}$$

$$a^2_{H^+} = \frac{a_{Fe^{2+}} \cdot p_{CO_2(g)}}{10^{+5.81}} \tag{9.29}$$

$$2 \cdot \log_{10} a_{H^+} = \log_{10} a_{Fe^{2+}} + \log_{10} p_{CO_2(g)} - 5.81$$

$$pH = \tfrac{1}{2} \cdot (-\log_{10} a_{Fe^{2+}} - \log_{10} p_{CO_2(g)} + 5.81) \tag{9.30}$$

The solute-precipitate equilibrium boundary (expression 9.31) is found by applying our earlier condition on total soluble $c_{Fe} = 10^{-7}$ mol dm^{-3} and choosing $p_{CO_2} = 10^{-3.45}$ atm (the atmospheric carbon dioxide partial pressure at mean sea level).

$$pH = \tfrac{1}{2} \cdot (7.00 + 3.45 + 5.81) = 9.91 \tag{9.31}$$

9.3.2.5 Solute-Precipitate Reduction Boundary

Adding reduction half-reaction (9.R9) to solubility reaction (9.R14) generates a reduction half-reaction (9.R19) where the educt is a solid (amorphous $Fe(OH)_3(s)$) and the product is the solute $Fe^{2+}(aq)$. Reduction half-reactions of this type are common in environmental systems.

$$Fe(OH)_3(s) + 3H^+(aq) + e^- \longrightarrow Fe^{2+}(aq) + 3H_2O(l) \tag{9.R19}$$

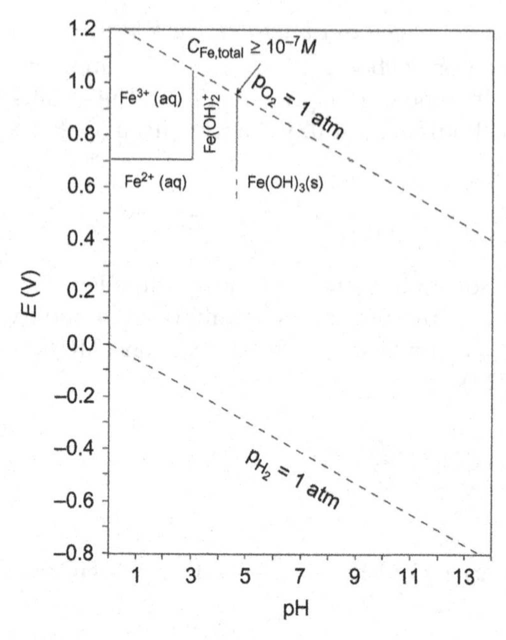

FIG. 9.9 An incomplete Pourbaix stability diagram displaying two water stability boundary lines, one solute-solute reduction equilibrium boundary (9.19), a solute-solute hydrolysis equilibrium boundary (9.26), and a solute-precipitate equilibrium boundary (9.28).

You will not find the reduction potential for reaction (9.R19) tabulated in any database but it is readily computed using the same approach applied to reaction (9.R13). The electrochemical potential for this half-reaction is calculated from the standard Gibbs energy $\Delta_r G^\circ$ for reduction half-reaction (9.R19) using the standard Gibbs energy of formation $\Delta_f G^\circ$ for each educt and product (9.32). The standard electrochemical potential E° is given by Eq. (9.33).

$$\Delta_r G^\circ = \left(\Delta_f G^\circ\left(Fe^{2+}, aq\right) + 3 \cdot \Delta_f G^\circ\left(H_2O, l\right)\right)_{products} - \left(\Delta_f G^\circ\left(Fe(OH)_3, s\right) + 3 \cdot \Delta_f G^\circ\left(H^+, aq\right)\right)_{educts} \tag{9.32}$$

$$\Delta_r G^\circ(298.15\ K) = \left(\left(-91.21\ kJ\ mol^{-1}\right) + 3 \cdot \left(-237.2\ kJ\ mol^{-1}\right)\right) - \left(\left(-708.1\ kJ\ mol^{-1}\right) + 3 \cdot \left(0\ kJ\ mol^{-1}\right)\right) = -94.60\ kJ\ mol^{-1}$$

$$E^\circ(298.15\ K) = \frac{-\left(-94.60\ kJ\ mol^{-1}\right)}{1 \cdot \left(9.6485 \times 10^{-2}\ kJ\ mol^{-1}\ mV^{-1}\right)} = +981.0\ mV \tag{9.33}$$

The Nernst equation for half-reaction (9.R19) is Eq. (9.34). Substituting the standard electrochemical potential E° (9.33) into Eq. (9.34) and separating terms for the contributions from Fe^{2+}(aq) and H^+(aq) yield expression (9.35).

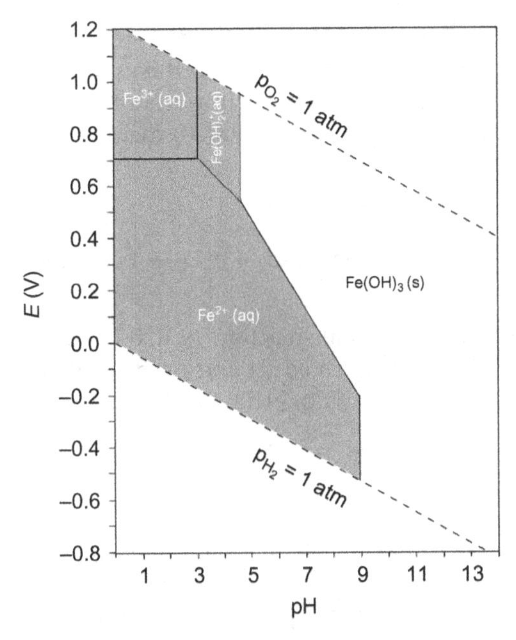

FIG. 9.10 An incomplete Pourbaix stability diagram plotting the following equilibrium boundaries: two solute-solute reduction equilibrium boundaries (expressions 9.19, 9.24), solute-solute hydrolysis equilibrium boundary (expression 9.26), two solute-precipitate equilibrium boundaries (expressions 9.28, 9.31), and solute-precipitate reduction equilibrium boundary (expression 9.36). The domains where $c_{Fe} \geq 10^{-7}\ \text{mol dm}^{-3}$ are *shaded gray*.

$$E(298.15\ \text{K}) = E^{\circ}(298.15\ \text{K}) - \left(\frac{59.160\ \text{mV}}{1}\right) \cdot \log_{10} \frac{a_{Fe^{2+}}}{a^3_{H^+}} \tag{9.34}$$

$$E(298.15\ \text{K}) = (981\ \text{mV}) - (59.160\ \text{mV}) \cdot \log_{10} a_{Fe^{2+}} - 3 \cdot (59.160\ \text{mV}) \cdot \text{pH} \tag{9.35}$$

Every Pourbaix stability diagram that includes reduction half-reactions where an educt or product is a solid will result in an expression much like Eq. (9.35). The position of the solute/precipitate reduction equilibrium boundary (9.35) depends on the activity of solute Fe^{2+}(aq), the second term on the right.

Earlier, when drawing the solute-precipitate equilibrium boundary (expression 9.28), the following condition was adopted: $c_{Fe} = 10^{-7}\ \text{mol dm}^{-3}$. This condition applies to Eq. (9.35): $a_{Fe^{2+}} = 10^{-7}\ \text{mol dm}^{-3}$ or $\log_{10} a_{Fe^{2+}} = -7$. The solute-precipitate reduction equilibrium boundary for this choice is expression (9.36) and is plotted in Fig. 9.10.

$$E(298.15\ \text{K}) = (981.0\ \text{mV}) - (59.16\ \text{mV}) \cdot (-7.00) - (177.5\ \text{mV}) \cdot \text{pH}$$

$$E(298.15\ \text{K}) = (1395\ \text{mV}) - (177.5\ \text{mV}) \cdot \text{pH} \tag{9.36}$$

9.3.2.6 *Precipitate-Precipitate Reduction Boundary*

Reduction equilibrium can involve two solids. In this case the electron acceptor is amorphous ferric hydroxide $Fe(OH)_3(s)$, and the electron donor is the ferrous carbonate mineral siderite $FeCO_3(s)$. The reduction half-reaction (9.R20) includes $CO_2(g)$ as an educt, making this reaction dependent on p_{CO_2}.

$$Fe(OH)_3(s) + CO_2(g) + H^+(aq) + e^- \longrightarrow FeCO_3(s) + 2H_2O(l) \tag{9.R20}$$

The electrochemical potential for half-reaction (9.R20) is calculated from the standard Gibbs energy of formation $\Delta_f G^\circ$ for each educt and product (expression 9.37). The standard electrochemical potential E° is given by Eq. (9.38).

$$\begin{aligned}\Delta_r G^\circ &= \left(\Delta_f G^\circ\,(FeCO_3, s) + 2\cdot\Delta_f G^\circ\,(H_2O, l)\right)_{\text{products}}\\ &\quad - \left(\Delta_f G^\circ\,(Fe(OH)_3, s) + \Delta_f G^\circ\,(CO_2, g) + \Delta_f G^\circ\,(H^+, aq)\right)_{\text{educts}}\end{aligned} \tag{9.37}$$

$$\begin{aligned}\Delta_r G^\circ(298.15\ \text{K}) &= \left(\left(-677.6\ \text{kJ mol}^{-1}\right) + 2\cdot\left(-237.2\ \text{kJ mol}^{-1}\right)\right)\\ &\quad - \left(\left(-708.1\ \text{kJ mol}^{-1}\right) + \left(-394.4\ \text{kJ mol}^{-1}\right) + \left(0\ \text{kJ mol}^{-1}\right)\right)\\ &= -\,49.43\ \text{kJ mol}^{-1}\end{aligned}$$

$$E^\circ(298.15\ \text{K}) = \frac{-\left(-49.43\ \text{kJ mol}^{-1}\right)}{1\cdot\left(9.6485\times 10^{-2}\ \text{kJ mol}^{-1}\ \text{mV}^{-1}\right)} = +512.3\ \text{mV} \tag{9.38}$$

The Nernst equation for half-reaction (9.R20) is Eq. (9.39).

$$E(298.15\ \text{K}) = E^\circ(298.15\ \text{K}) - \left(\frac{59.160\ \text{mV}}{1}\right)\cdot\log_{10}\left(\frac{1}{a_{H^+}\cdot p_{CO_2}}\right) \tag{9.39}$$

Earlier, when drawing the solute-precipitate equilibrium boundary (9.31), the following condition was adopted: $p_{CO_2} = 10^{-3.45}$ atm. This condition also applies to expression (9.39).

Substituting the standard reduction potential E° (9.38) into Eq. (9.39) and separating terms for the contributions from $CO_2(g)$ and $H^+(aq)$ yields stability boundary expression (9.40) for precipitate-precipitate reduction equilibrium between electron donor $FeCO_3(s)$ and electron acceptor $Fe(OH)_3(s)$.

$$\begin{aligned}E(298.15\ \text{K}) &= (512.3\ \text{mV}) - (59.16\ \text{mV})\cdot\log_{10} p_{CO_2} - (59.16\ \text{mV})\cdot\text{pH}\\ E(298.15\ \text{K}) &= (308.2\ \text{mV}) - (59.16\ \text{mV})\cdot\text{pH}\end{aligned} \tag{9.40}$$

9.3.2.7 *Simple Rules for Interpreting Pourbaix Stability Diagrams*

The complete Pourbaix stability diagram (Fig. 9.11) was earlier identified as a Fe–O–C–H Pourbaix stability diagram because these four elements are the only components displayed in the diagram (Table 9.5, item 1). Adding other components will introduce new solutes and solids, altering the appearance of the diagram. A Fe–O–C–H Pourbaix stability diagram has no value when interpreting Fe chemistry for a system containing components besides carbon, assuming Pourbaix stability diagrams will always contain water components.

The position of numerous boundary lines depends on solute concentrations and gas partial pressures: boundary expression (9.14) depends on p_{O_2}, boundary expression (9.17) depends on p_{H_2}, boundaries (9.31) and boundary expression (9.40) depend on p_{CO_2}, and boundaries (9.28), (9.31), (9.36) depend on the total soluble iron concentration.

Experimental (E, pH) coordinates plotting to the left of boundaries (9.28), (9.31), (9.36) encompass conditions where $c_{Fe} > 10^{-7}$ mol dm^{-3}. Experimental (E, pH) coordinates plotting to the right of these same boundaries cover conditions where $c_{Fe} < 10^{-7}$ mol dm^{-3}.

Notice that domains where the solute concentration exceeds the defined limit (Fig. 9.11) are labeled as solutes: Fe^{3+}(aq), $Fe(OH)_2{}^+$(aq), and Fe^{2+}(aq). Domains where the solute concentration is less than the defined limit (Fig. 9.11) are labeled as solids: $Fe(OH)_3$(s) and $FeCO_3$(s). This labeling convention (Table 9.5, item 3) is a reminder regarding the total solute concentration. Each boundary represents a specific type of equilibrium: solute-solute reduction, solute-solute hydrolysis, solute-precipitate reduction, and precipitate-precipitate reduction.

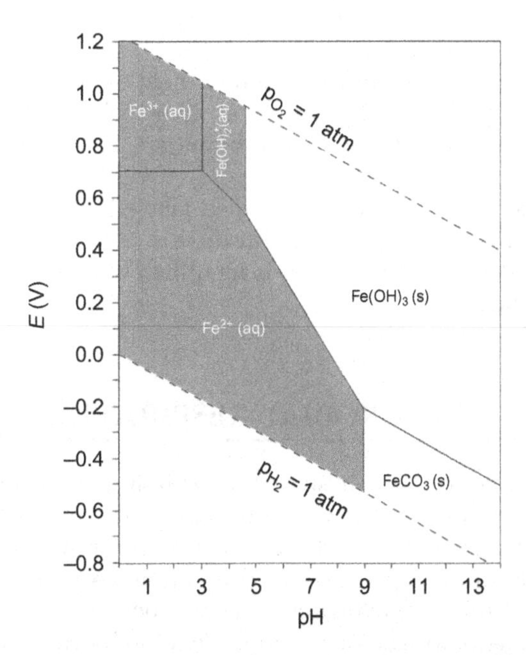

FIG. 9.11 A complete Pourbaix stability diagram plotting the following equilibrium boundaries: two solute-solute reduction equilibrium boundaries (expressions 9.19, 9.24), solute-solute hydrolysis equilibrium boundary (expression 9.26), two solute-precipitate equilibrium boundaries (expressions 9.28, 9.31), a solute-precipitate reduction equilibrium boundary (expression 9.36), and a precipitate-precipitate reduction equilibrium boundary (expression 9.40). The domains where $c_{Fe} \geq 10^{-7}$ mol dm^{-3} are *shaded gray*.

Experimental (*E*, pH) coordinates plotting along solute-solute reduction boundaries indicate that the activities of the electron-donor specie and the electron-acceptor specie are *equal to* each other. Similarly, (*E*, pH) coordinates plotting along solute-solute hydrolysis boundaries indicate that the activities of the two solute species separated by the boundary are *equal to* each other.

Experimental (*E*, pH) coordinates plotting along solute-precipitate boundaries indicate that the activity of the solute species is *equal to* the solubility boundary condition and the bordering solid controls solubility. For example, (*E*, pH) coordinates along boundary (9.36) indicate that $Fe(OH)_3(s)$ is the solid controlling iron solubility, iron solubility is at the designated limit, and the most abundant solute is $Fe(OH)_2^+(aq)$.

Experimental (*E*, pH) coordinates plotting along precipitate-precipitate reduction boundaries indicate that solute activities are *less than* the solubility boundary condition and the two solids separated by the boundary *coexist*. When (*E*, pH) coordinates plot on either side of the boundary, one of the two solids must transform into the other by reduction or oxidation (depending on which side of the boundary the (*E*, pH) coordinate lies).

For example, (*E*, pH) coordinates along boundary (9.40) represent conditions where $Fe(OH)_3(s)$ and $FeCO_3(s)$ coexist and iron solubility is below the designated limit. If experimental (*E*, pH) coordinates plot on the reducing side of boundary (9.40) (i.e., below), then $Fe(OH)_3(s)$ is unstable and will transform into siderite $FeCO_3(s)$ under prevailing conditions ($p_{CO_2} = 10^{-3.45}$ atm). If experimental $Fe(OH)_3(s)$ coordinates plot on the oxidizing side of boundary (9.40) (i.e., above), then $FeCO_3(s)$ is unstable and will transform into $Fe(OH)_3(s)$ under prevailing conditions.

If (*E*, pH) coordinates plot within a stability field labeled with a solid phase the interpretation will identify the solid-phase controlling solubility and recognize the total solute concentration is *less than* the solubility boundary condition. Solute species remain unidentified.

If (*E*, pH) coordinates plot within a stability field labeled with a solution specie the interpretation will recognize the total solute concentration is *greater than* the solubility boundary condition. The *most abundant* solution species is identified but any solid phases that might be present remain unidentified.

9.4 MICROBIAL RESPIRATION

Section 9.3 provides the chemical basis for interpreting reduction-oxidation conditions in soils and the saturated zone, but it does not supply the mechanism and driving force behind the development of reducing conditions in these environments. Experimental (*E*, pH) coordinates reflect the profound impact of biological activity. Absent biological respiration, the reducing potential of natural organic matter remains blocked (cf. Section 7.3.3). Respiration, particularly microbial anaerobiosis, removes the activation barriers blocking the oxidation of organic matter and releases reduced solutes that a platinum oxidation-reduction potential (ORP) electrode can detect as a voltage displacement (cf. Fig. 9.5 and Fig. 9.7).

Planet Earth as a whole (metallic core, mantle, and crust) is chemically reduced. Even though oxygen is the most abundant element in crustal rocks, the most abundant transition elements are not completely oxidized. The 10 most abundant elements in Earth's (cf. Table 1.4) crust account for 99.48% of the mass and, based on the listed formal oxidation numbers, 99.86% of the charge. The mean formal oxidation state for iron is $\overline{OS(Fe)} \approx +1.8$ (assuming iron in crust and mantle determined by fayalite-magnetite-quartz redox buffer (Haggerty, 1976) and core is metallic Fe(s)), indicating that the formation of the crust resulted in the partial oxidation of iron from the elemental state found in the core.

The present atmosphere (mean sea level) has the following volume percent composition (cf. Table 6.8, Chapter 6, *Acid-Base Chemistry*): 78.084% $N_2(g)$, 20.946% $O_2(g)$, 0.9340% Ar(g), and 0.0387% $CO_2(g)$. The primordial atmosphere contained an estimated 0.2% $O_2(g)$, most probably the product of abiotic water photolysis (Berkner and Marshall, 1965). Primordial $O_2(g)$ levels were believed to be sufficient for aerobic respiration as early as the beginning of the Paleozoic Era 542 million years ago. Estimated atmospheric $CO_2(g)$ levels early in the Paleozoic Era were in the range 0.42–0.60% or 11- to 15-fold higher than today.

Present $O_2(g)$ levels are the result of photosynthesis. The reduction half-reactions for carbon dioxide (9.R21) and oxygen (9.R22) are combined to yield the net reduction-oxidation reaction for biological carbon fixation to form glucose (9.R22) (Example 9.5). Example 9.5 demonstrates the calculation of the standard biological potential $E^{\oplus}$ for reaction (9.R21) using expression (9.10).

$$6CO_2(g) + 24H^+(aq) + 24e^- \xrightarrow{E^{\oplus}=-404.1\ \text{mV}} \underset{\text{Glucose}}{C_6H_{12}O_6(aq)} + 6H_2O(l) \quad (9.R21)$$

$$6O_2(g) + 24H^+(aq) + 24e^- \xrightarrow{E^{\oplus}=+814.9\ \text{mV}} 6H_2O(l) \quad (9.R22)$$

$$6CO_2(g) + 6H_2O(l) \xrightarrow{E^{\oplus}=-1219\ \text{mV}} \underset{\text{Glucose}}{C_6H_{12}O_6(aq)} + 6O_2(g) \quad (9.R23)$$

EXAMPLE 9.5

Example Permalink

http://soilenvirochem.net/3F0J4d

Calculate the standard *biological* electrochemical potential $E^{\oplus}$ for reduction half-reaction (9.R21).

The carbon dioxide-to-glucose reduction half-reaction (9.R21) list a biological standard electrochemical potential of $E^{\oplus} = -404.1$ mV. It would be impossible to construct a galvanic cell similar to the one shown in Fig. 9.5 to measure this potential, but we can calculate the electrochemical potential using Gibbs energy expression (9.3).

Part 1. Calculate the change in Gibbs energy for reduction half-reaction (9.R21) under standard conditions $\Delta_r G^{\oplus}$.

The change in Gibbs energy for reduction half-reaction (9.R21) under standard conditions $\Delta_r G^{\circ}$ is the sum of the standard Gibbs energy of formation for each product minus the sum of the standard Gibbs energy of formation for each educt.

$$\Delta_r G^{\circ} = \left(\Delta_f G^{\circ}(\text{glucose, aq}) + 6 \cdot \Delta_f G^{\circ}(H_2O, l)\right)_{products} - \left(6 \cdot \Delta_f G^{\circ}(CO_2, g) + 24 \cdot \Delta_f G^{\circ}(H^+, aq)\right)_{educts}$$

$$\Delta_r G^{\circ}(298.15\ \text{K}) = \left(\left(-915.9\ \text{kJ mol}^{-1}\right) + 2 \cdot \left(-237.2\ \text{kJ mol}^{-1}\right)\right) - \left(\left(-386.0\ \text{kJ mol}^{-1}\right) + \left(0\ \text{kJ mol}^{-1}\right)\right) = -23.10\ \text{kJ mol}^{-1}$$

Part 2. Calculate the standard electrochemical potential E° for reduction half-reaction (9.R21).
Substitute half-reaction $\Delta_r G^{\circ}$ into expression (9.3).

$$E^{\circ}(298.15\ \text{K}) = \frac{-\left(-23.10\ \text{kJ mol}^{-1}\right)}{1 \cdot \left(9.6485 \times 10^{-2}\ \text{kJ mol}^{-1}\ \text{mV}^{-1}\right)} = +9.976\ \text{mV}$$

Part 3. Calculate the standard biological electrochemical potential $E^{\oplus}$ for reduction half-reaction (9.R21).

Substitute E° from Part 2 into expression (9.10) to calculate the standard biological electrochemical potential for reaction (9.R21).

$$E^{\oplus}(298.15\ \text{K}) = (+9.976\ \text{mV}) - \left(\frac{m}{n}\right) \cdot (59.160\ \text{mV}) \cdot (7)$$

$$E^{\oplus}(298.15\ \text{K}) = (+9.976\ \text{mV}) - (414.1\ \text{mV}) = -404.1\ \text{mV}$$

Carbon dioxide fixation occurs when the enzyme rubisco (ribulose 1,5-bisphosphate carboxylase) catalyzes the conversion (Table 9.6, reaction 1) of RuBP (ribulose 1,5-bisphosphate, CAS registry number 14689-84-0) and CO_2 into G3P (glyceraldehyde 3-phosphate, CAS registry number 591-59-3). The energy (adenosine triphosphate [ATP]) and reducing equivalents (NADPH) required for this process is provided by the capture of light energy in earlier steps of the photosynthetic process.

Ten of 12 G3P formed by carbon fixation (Table 9.6, reaction 1) regenerate the original RuBP consumed during carbon fixation (Table 9.6, reaction 2) while the remaining 2 G3P combine to form glucose (Table 9.6, reaction 3). The chemical energy captured during photosynthesis is released during respiration, the reverse of reaction (9.R23).

TABLE 9.6 Calvin Cycle Balance Sheet

1.	$6CO_2 + 6RuBP + 12ATP + 12NADPH \longrightarrow 12G3P$
2.	$10G3P + 6ATP \longrightarrow 6RuBP$
3.	$2G3P \longrightarrow$ glucose

TABLE 9.7 Classification of Living Organisms: Energy and Carbon Requirements

Carbon Source	Light Energy	Chemical Oxidation
Carbon dioxide	Photolithotrophs	Chemolithotrophs
Organic carbon	Photoorganotrophs	Chemoorganotrophs

Table 9.7 lists a classification of living organisms based on their carbon and energy requirements. *Photolithotrophs* such as plants, algae, and certain bacteria derive their metabolic energy from photosynthesis and assimilate carbon dioxide for biosynthesis. *Chemolithotrophic* bacteria derive metabolic energy by oxidizing a variety of inorganic electron donors (thiosulfate $S_2O_3^-$, trithionate $S_3O_6^{2-}$, elemental sulfur S(s), ammonium NH_4^+, and dihydrogen H_2) and fix carbon dioxide for biosynthesis. *Chemolithotrophic* bacteria make an insignificant contribution to net carbon dioxide fixation but play a prominent role in environmental reduction-oxidation cycling.

Organotrophic organisms cannot assimilate carbon dioxide for biosynthesis and, therefore, must assimilate organic compounds produced by lithotrophic organisms. Chemoorganotrophs (all eukaryotic organisms that are not photolithotrophs and most prokaryotes) derive their metabolic energy from the oxidation of organic compounds.

When lithotrophs assimilate CO_2, they generate small organic molecules (cf. G3P, Table 9.6) used to synthesize complex biomolecules that comprise living biomass while simultaneously capturing the chemical energy required by all metabolic processes. Both lithotrophs and organotrophs release stored chemical energy by catabolic pathways that disassemble large molecules into fragments and oxidize the fragments.

9.4.1 Catabolism and Electron Transport Chains

The release of chemical energy stored in organic compounds occurs in two stages: catabolism and respiration. Reduction leading to CO_2 release occurs during catabolism: the electron donor is the organic compound being catabolized, while the electron acceptor is the oxidized form of nicotinamide adenine dinucleotide NAD^+ (Fig. 9.12) or a similar compound (e.g., flavin adenine dinucleotide FAD or nicotinamide adenine dinucleotide phosphate[8] $NADP^+$).

$$NAD^+(aq) + H^+(aq) + 2e^- \longrightarrow NADH(aq) \qquad (9.R24)$$

For example, glucose catabolism involves three processes: glycolysis, pyruvate oxidation, and the citric acid cycle. Reduction half-reaction (9.R21) represents glucose catabolism by this pathway.

Combining reduction half-reactions (9.R21), (9.R24) illustrates glucose catabolism to carbon dioxide (9.R25). Notice that the electron acceptor is NAD^+ (Fig. 9.13), not O_2.

[8] cf. NADPH, Table 9.6

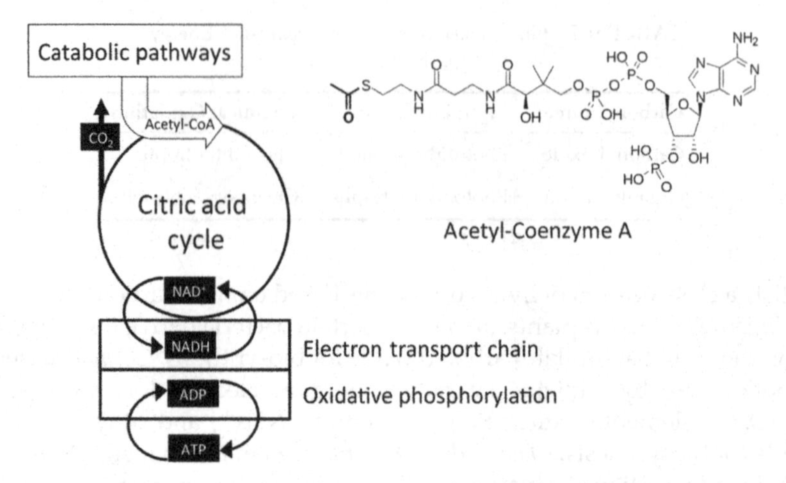

FIG. 9.12 Diverse catabolic pathways deliver acetyl-Coenzyme A (CAS registry number 72-89-9) to the *Citric Acid Cycle* where the acetyl group (terminal $CH_3(C = O)$– bonded to the thiol R–SH) acting as electron donor reduces NAD^+ to NADH. NADH is the electron donor for the *electron transport chain* which produces the energy carrier ATP through *oxidative phosphorylation*.

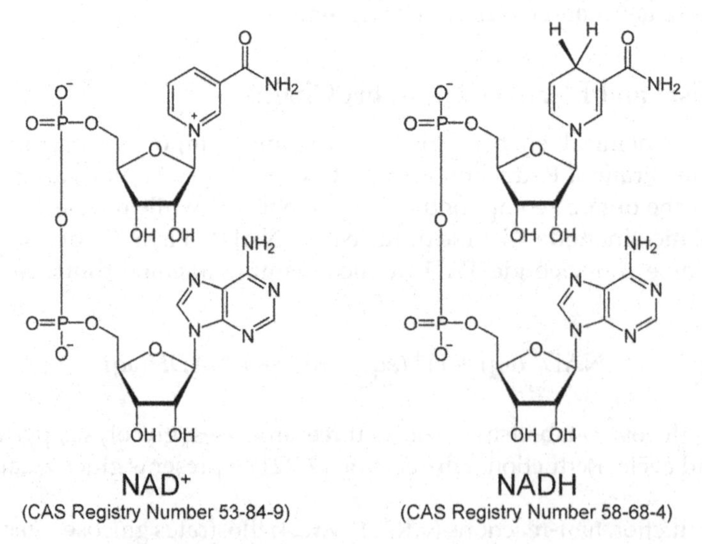

FIG. 9.13 Molecular structures of nicotinamide adenine dinucleotide in its oxidized NAD^+ (CAS registry number 53-84-9) and reduced NADH (CAS registry number 58-68-4) forms.

$$\underset{\text{Glucose}}{C_6H_{12}O_6(aq)} + 12NAD^+(aq) \longrightarrow CO_2(g) + 12NADH + 12H^+ \qquad (9.R25)$$

The electron-donor NADH functions as a reduction-oxidation cofactor; an electron carrier that shuttles electrons from various catabolic pathways to the electron transport chain (Fig. 9.13). While all organisms employ numerous catabolic pathways, bacteria have evolved a broader array of catabolic pathways than other organisms, making them important players in the environmental degradation of organic contaminants. Bacteria also draw upon a more flexible electron transport chain, providing them with respiratory alternatives absent in eukaryotes. Consequently, many bacteria continue to respire in the absence of O_2 (a process called *anaerobic respiration* or *anaerobiosis*) by utilizing a variety of terminal electron acceptors as the need arises. *The electron transport chain, not the catabolic pathway, defines the terminal electron acceptor.*

9.4.2 Microbial Electron Transport Chains

Aerobic respiration uses an electron transport chain whose electron donor is NADH with O_2 acting as terminal electron acceptor. The net reduction-oxidation reaction for the aerobic electron transport chain (9.R26) combines reduction half-reactions for the electron donor (9.R24) and the terminal electron acceptor (9.R7).

$$6O_2(g) + 12NADH(aq) + 12H^+ \longrightarrow 12NAD^+ + 12H_2O(l) \qquad (9.R26)$$

The net reduction-oxidation reaction in aerobic respiration (9.R27) combines reduction reactions for glucose catabolism (9.R25) and aerobic electron transport chain (9.R26).

$$\underset{\text{Glucose}}{C_6H_{12}O_6(aq)} + 6O_2(g) \longrightarrow 6CO_2(g) + 12H_2O(l) \qquad (9.R27)$$

The $NAD^+(aq)|NADH(aq)$ couple (9.R24) does not appear in reaction (9.R27) because it is recycled between the catabolic pathway and electron transport chain (Fig. 9.12).

9.4.2.1 Electron Transport Chain of Aerobic Bacteria

The reduction half-reactions listed in Table 9.C.2 are components of the electron transport chain typical of aerobic bacteria and prokaryotes. The electron donor for the sequence is typically NADH, and the terminal electron acceptor is O_2.

The half-reactions listed in Table 9.C.2 combine to yield the reduction-oxidation reaction sequence of the aerobic electron transport chain, where NADH serves as the electron donor. These reactions release the chemical energy carried to the electron transport chain by the electron donor in relatively small increments.

Electrons enter a reduction-oxidation cascade from the electron donor NADH and release energy as they cascade over each reduction step until they reach the terminal electron acceptor, which defines the floor of the cascade. The floor of the aerobic electron transport cascade is the $O_2(g)|H_2O(l)$ couple.

Although the reduction potential of each step in the reduction-oxidation cascade is more positive than the preceding step, using the Gibbs energy expression (9.3) requires each step in the cascade to lie at a more negative Gibbs energy than the preceding step. The complete reduction-oxidation reaction represented by the aerobic electron transport chain is reaction (9.R26), and the total Gibbs energy change associated with the complete cascade of electrons from NADH to O_2 is: $\Delta_r G^{\oplus} = -436.2\ \text{kJ mol}^{-1}$ or $\Delta_r G^{\oplus} = -218.1\ \text{kJ mol}^{-1}$ per NADH.

The aerobic *ETC* is coupled to three complexes (Fig. 9.14) that generate the *proton motive force* (PMF)[9] required for oxidative phosphorylation. Oxidative phosphorylation produces the energy carrier ATP used by virtually all biochemical reactions in living organisms. The Gibbs energy released by the complexes illustrated in Fig. 9.14 transfers protons across a cytoplasmic or periplasmic membrane to generate the proton motive force that drives the oxidative phosphorylation of ADP^{3-} to yield ATP^{4-}.

$$ADP^{3-}(aq) + H_2PO_4^- \longrightarrow ATP^{4-} + H_2O(l) \tag{9.R28}$$

Complex I is the NADH–CoQ reductase complex that transfers electrons from NADH to ubiquinone (Coenzyme CoQ). *Complex III* is the $CoQH_2$-cytochrome *c* reductase complex that

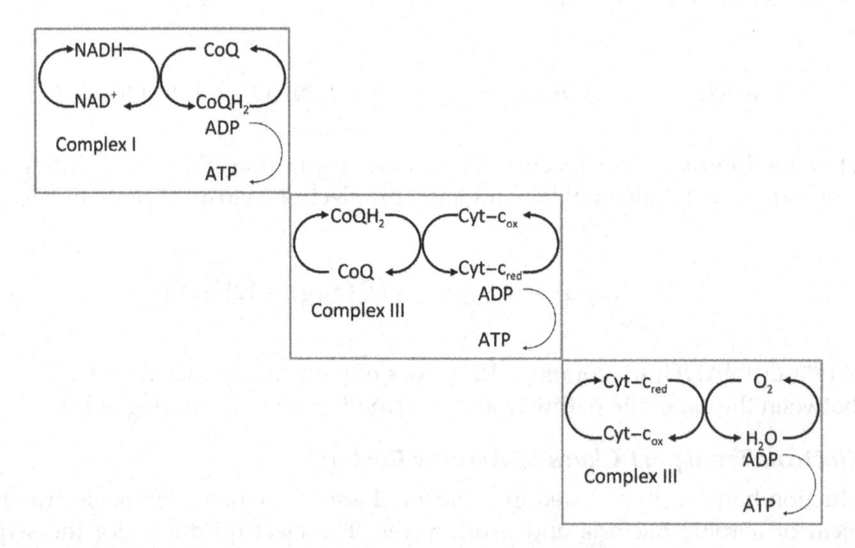

FIG. 9.14 The aerobic electron transport chain is coupled to the synthesis of energy carrier ATP (in a process called oxidative phosphorylation) at three points: Complexes I, III, and IV.

[9] Protons H^+ are pumped across the cytoplasmic membrane by the complexes illustrated in Fig. 9.14 to generate an electrochemical potential gradient known as the *PMF*. Although the potential gradient arises from H^+ export across the membrane the *PMF* is not determined by the a_{H^+} on either side of the membrane. In particular, H^+ pumped across the cytoplasmic membrane of *Gram-positive* bacteria enter the surroundings yet the pH of the surroundings does not determine the *PMF*.

transfers electrons from ubquinol $CoQH_2$ to cytochrome c. *Complex IV* is the cytochrome *c* oxidase complex that transfers electrons from the reduced form of cytochrome *c* to the terminal electron acceptor O_2 to yield H_2O.

The mineralization of one glucose $C_6H_{12}O_6$ molecule to CO_2 and H_2O yields a total of 33 ATP by oxidative phosphorylation.[10]

9.4.2.2 Dissimilatory Nitrate Reduction: Denitrifying Bacteria

The electron transport chain of denitrifying bacteria utilizes most of the aerobic electron transport chain, diverting electrons from cytochrome *c* to nitrate and eliminating reactions 6 through 9 in Table 9.C.2. The use of nitrate as the terminal electron acceptor, however, requires an enzyme system that transfers electrons from cytochrome *c* to nitrate. The reduction of nitrate to N_2 involves the transfer of five electrons per nitrogen. Since biological reduction-oxidation reactions typically are one- and two-electron reductions, the complete reduction of nitrate to N_2 cannot occur in a single reduction-oxidation reaction (Table 9.C.1).

Fig. 9.15 illustrates the components of the nitrate reduction system in *denitrifying* bacteria. The first step—the reduction of nitrate to nitrite—is catalyzed by a nitrate reductase (NAR) bound to plasma membrane. The nitrite reductase (NIR)—reducing nitrite to nitric oxide—is located in the periplasm because nitric oxide (NO) is cytotoxic. The remaining steps occur in the periplasm: NO reduction to nitrous oxide (N_2O) and nitrous oxide reduction to N_2. The Gibbs energy change for electron transfer from NADH to N_2 is: $\Delta_r G^{\oplus} = -205.0 \text{ kJ mol}^{-1}$ per NADH. The respiratory efficiency of anaerobic denitrifying bacteria is comparable to aerobic respiration (aerobic respiration: $\Delta_r G^{\oplus} = -218.1 \text{ kJ mol}^{-1}$ per NADH). Many microbiologists believe the aerobic electron transport chain originated with the electron transport chain used by ancient denitrifying bacteria.

$$2NO_3^-(aq) + 5NADH(aq) + 7H^+(aq) \longrightarrow N_2(g) + 5NAD^+ + 6H_2O(l) \quad (9.R29)$$

9.4.2.3 Dissimilatory Nitrate Reduction: Nitrate-Reducing Bacteria

The terminal reductase in the electron transport chain used by nitrate-reducing bacteria appears to be a nitrate reductase (NAR). The electron donor is ubiquinone and the product is nitrite NO_2^-. The truncated electron transport chain of nitrate-reducing bacteria effectively eliminates oxidative phosphorylation Complex IV (cf. reactions 6 through 9, Table 9.C.2).

If aerobic glucose mineralization yields 37 ATP per glucose molecule (4 ATP from substrate-level phosphorylation and 33 from oxidative phosphorylation using all three complexes) then denitrifying bacteria will produce about 26 ATP per glucose molecule: $26 = 4 + (2/3) \cdot 33$. This estimate assumes two-thirds efficiency when only Complexes I and III yield ATP. Stated differently, nitrate-reducing bacteria would need to mineralize 142% as much glucose to generate the same amount of ATP as aerobic bacteria.

[10] This does not count an additional 4 ATP from substrate-level phosphorylation independent of the *ETC*.

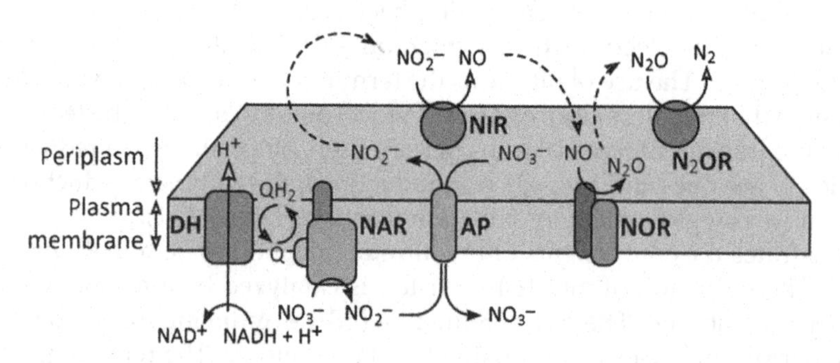

FIG. 9.15 Components of the nitrate dissimilatory reduction system of denitrifying bacteria are located in the inner or plasma membrane and the periplasm. *Reproduced with permission from Zumft, W.G., 1997. Cell biology and molecular basis of denitrification. Microbiol. Mol. Biol. Rev. 61 (4), 533–616.*

9.4.2.4 Dissimilatory Iron(III) Reduction

Most iron-reducing bacteria are capable of using several terminal acceptors besides ferric iron (Ruebush et al., 2006), but their distinguishing characteristic is their ability to grow under anaerobic conditions where the sole electron acceptor is either Fe^{3+} or Mn^{4+}. The relative natural abundance of iron and manganese means Fe^{3+} is by far the more abundant terminal electron acceptor in the environment.

The iron respiratory chain faces a singular challenge: the terminal electron acceptor is insoluble and cannot be taken up into the cytoplasm like $NO_3^-(aq)$, $SO_4^{2-}(aq)$, or $HCO_3^-(aq)$. There is considerable evidence for membrane-bound terminal ferric reductase enzymes (Lovley, 1993) that overcome the need for a soluble terminal electron acceptor, relying on direct contact between cell surface and insoluble ferric minerals. Some studies (Lovley et al., 2004) suggest that some iron-reducing species (e.g., *Shewanella* and *Geothrix*) may secrete soluble electron shuttles capable of reducing ferric oxide minerals without coming into direct physical contact.

Microbiologists studying the iron respiratory chain have identified several components of an electron transport chain that couples to terminal ferric reductase enzymes embedded in the outer-membrane (Fig. 9.16). Identification of soluble electron shuttles, however, remains

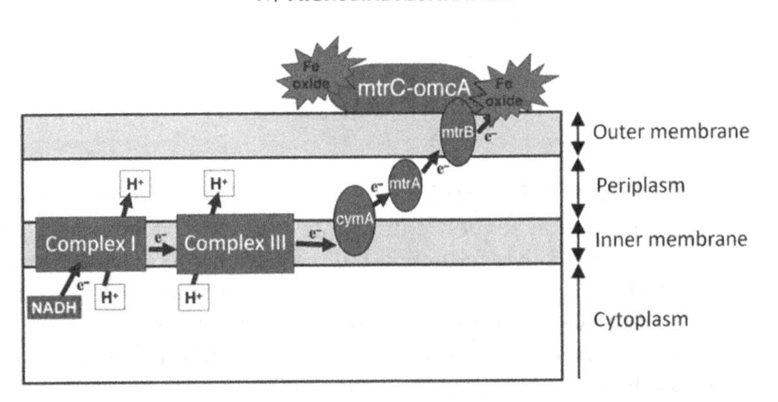

FIG. 9.16 The terminal reductase of dissimilatory iron-reducing *Gram-negative* bacteria is located in the outer membrane while the other components are located in the cytoplasmic membrane. *From Fredrickson, J.K., Zachara, J.M. 2008. Electron transfer at the microbe–mineral interface: a grand challenge in biogeochemistry. Geobiology, 6, 245–253.*

elusive. One of the more unusual features of the terminal ferric reductase is its location. The position of reductase enzymes in the nitrate respiratory chain (cf. Fig. 9.15) in the inner or cytoplasmic membrane and the periplasm[11] also applies to the reductase enzymes of the sulfate and carbonate respiratory chains. The placement of the terminal ferric reductase, however, is on the outer surface of the outer membrane (DiChristina et al., 2002; Myers and Myers, 2002).

The complete reduction-oxidation reaction represented by the electron transport chain of iron-reducing bacteria is reaction (9.R30). The total Gibbs energy change associated with the complete cascade of electrons from NADH to $Fe^{2+}(aq)$ is: $\Delta_r G^{\oplus} = -46.51 \text{ kJ mol}^{-1}_{NADH}$. The Gibbs energy change for reaction (9.R30), however, does not represent the relative respiratory efficiency of iron-reducing bacteria.

Fig. 9.16 shows how the truncated electron transport chain of iron-reducing bacteria eliminates oxidative phosphorylation Complex IV (cf. reactions 6 through 9, Table 9.C.2). In fact, electron transfer to exocellular ferric oxides dissipates the electrochemical gradient generated by Complex III (Fig. 9.16) leaving iron-reducing reliant solely on Complex I to generate ATP.

If aerobic glucose mineralization yields 37 ATP per glucose molecule (4 ATP from substrate-level phosphorylation and 33 from oxidative phosphorylation using all three complexes) then iron-reducing bacteria will produce about 15 ATP per glucose molecule: $15 = 4+(1/3)\cdot 33$. This estimate assumes one-thirds efficiency when only Complex I yields ATP. Stated differently, iron-reducing bacteria would need to mineralize 247% as much glucose to generate the same amount of ATP as aerobic bacteria.

[11] The cell wall of *Gram-negative* bacteria lies between the inner and outer membranes. This zone between the inner and outer membranes is called the periplasm.

$$2Fe(OH)_3(s) + NADH(aq) + 5H^+(aq) \longrightarrow$$
$$2Fe^{2+}(aq) + NAD^+ + 6H_2O(l) \qquad (9.R30)$$

9.4.2.5 Dissimilatory Sulfate Reduction

The proposed SO_4^{2-} reduction sequence in sulfate-reducing bacteria appears in Fig. 9.17 (Bradley et al., 2011). Table 9.C.1 lists reduction half-reactions and electrochemical potentials of reduction half-reactions appearing in Fig. 9.17. The appearance of the intermediates in Fig. 9.17—hydrogen sulfite HSO_3^-, thiosulfate $S_2O_3^{2-}$ and trithionate $S_3O_6^{2-}$—caused considerable confusion as scientists worked out the sulfate reduction pathway.

The reduction of sulfate to hydrogen sulfite is the first reaction in the sequence appearing in Fig. 9.17. The electrochemical potential for sulfate SO_4^{2-} reduction to hydrogen sulfite HSO_3^- (reaction 2, Table 9.C.1) is: $E^\oplus = -519$ mV. The Gibbs energy for reaction (9.R31) is positive; reaction (9.R31) is not spontaneous under physiological conditions.

$$SO_4^{2-}(aq) + NADH(aq) + 2H^+(aq) \longrightarrow HSO_3^-(aq) + NAD^+ + H_2O(l) \qquad (9.R31)$$

All sulfate-reducing bacteria activate sulfate for assimilatory and dissimilatory reduction by forming adenosine 5′-phosphosulfate (CAS 485-84-7, Fig. 9.18) from adenosine 5′-triphosphate

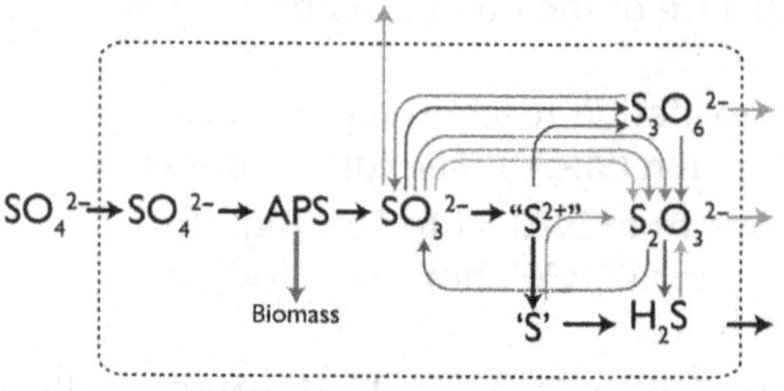

FIG. 9.17 Components of the sulfate reduction system of dissimilatory sulfate-reducing bacteria are located in the inner or cytoplasmic membrane. *From Bradley, A.S., Leavitt, W.D., Johnston, D.T., 2011. Revisiting the dissimilatory sulfate reduction pathway. Geobiology, 9, 446–457.*

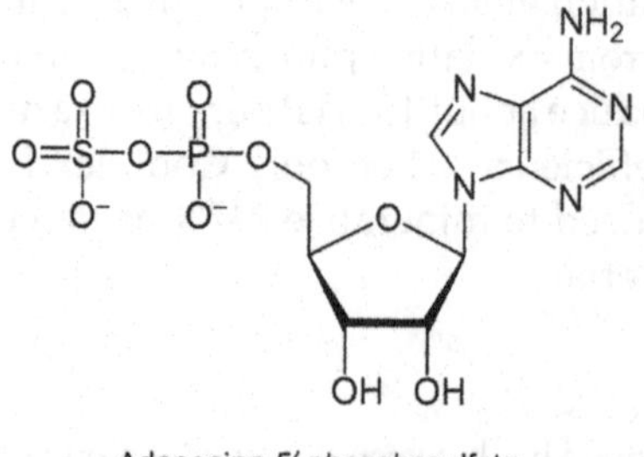

FIG. 9.18 Molecular structure of adenosine 5′-phosphosulfate (CAS registry number 485-84-7).

ATP^{4-} and sulfate in reaction (9.R32). APS^{3-} synthesis is coupled to the exergonic hydrolysis of guanosine 5′-triphosphate GTP^{4-} (Liu et al., 1994).

$$ATP^{4-}(aq) + SO_4{}^{2-}(aq) \xrightarrow{\text{ATP sulfurylase}} APS^{3-}(aq) + HP_2O_7{}^{3-}(aq) \quad (9.R32)$$

$$GTP^{4-}(aq) + H_2O(l) \longrightarrow GDP^{3-} + H_2PO_4{}^{-} \quad (9.R33)$$

Activation of the sulfate lowers the potential for the reduction of sulfate to hydrogen sulfite to $E^{\oplus} = -60$ mV (reaction 3, Table 9.C.1) and allows the remaining reduction steps to follow. The Gibbs energy change $\Delta_r G^{\circ}$ for electron transfer from NADH to HS^- is: $\Delta_r G^{\oplus} = -24.73$ kJ mol^{-1}_{NADH} (counting the Gibbs energy of activation). This Gibbs energy change for reaction (9.R33), however, does not represent the relative respiratory efficiency of sulfate-reducing bacteria.

Sulfate-reducing bacteria lose two-thirds of the complexes available to aerobic bacteria for oxidative phosphorylation (reactions 4–9, Table 9.C.2). Based solely reliance on Complex I it would appear iron-reducing and sulfate-reducing bacteria would have similar respiratory efficiencies; however, this neglects the ATP invested in activating sulfate to APS.

Once again, aerobic glucose mineralization yields 37 ATP per glucose molecule (4 ATP from substrate-level phosphorylation and 33 from oxidative phosphorylation using all three complexes). Sulfate-reducing bacteria will produce about 12 ATP per glucose molecule: $12 = 4 - 3 + (1/3) \cdot 33$, assuming one-thirds efficiency when only Complex I yields ATP and counting the 3 ATP invested to activate sulfate during glucose mineralization. Sulfate-reducing bacteria would need to mineralize 308% as much glucose to generate the same amount of ATP as aerobic bacteria. Sulfate activation significantly reduces the respiratory efficiency of sulfate-reducing bacteria relative to iron-reducing bacteria despite sole reliance of both groups on ATP generated by Complex I.

9.4.3 Impact of Respiratory Efficiency on Carbon Use Efficiency

Section 7.2.2 discussed how carbon reduction influences carbon use efficiency (CUE). Briefly, if the mean carbon reduction of the substrate $\overline{\Gamma}_S$ is equal or greater than the mean carbon reduction of the biomass $\overline{\Gamma}_B$ then *CUE* has reached its maximum value.

The mean carbon reduction of natural organic matter is roughly equal to the mean carbon reduction of microbial biomass and, furthermore, the carbon-to-nitrogen ratio of natural organic matter is higher than the carbon-to-nitrogen ratio of microbial biomass. As a consequence we can conclude: natural organic matter has the ideal composition for forming microbial biomass.

CUE also depends on respiratory efficiency. There is no reason to assume the energy requirements for biosynthesis and maintenance metabolism are different for aerobic and anaerobic organisms. The *CUE* for aerobic bacteria is defined by expression (9.41).

$$CUE_{max} = \frac{G}{U} = \frac{G}{G + R_{O_2}} \quad (9.41)$$

The preceding sections have demonstrated that ATP yield declines when anaerobic respiration eliminates oxidative phosphorylation by first Complex IV and then Complex III. As a result, we can estimate the relative *CUE* of anaerobic bacteria relative to aerobic bacteria growing on the same substrate. The metabolic activity of facultative anaerobic bacteria will be lower than that of aerobic bacteria and the activity of obligate anaerobic bacteria is the lowest of all.

$$CUE_{\text{nitrate}} = \frac{G}{G + R_{NO_3^-}} = \frac{G}{G + (1.42 \cdot R_{O_2})}$$

$$CUE_{\text{ferric}} = \frac{G}{G + R_{Fe^{3+}}} = \frac{G}{G + (2.47 \cdot R_{O_2})}$$

$$CUE_{\text{sulfate}} = \frac{G}{G + R_{SO_4^{2-}}} = \frac{G}{G + (3.08 \cdot R_{O_2})}$$

9.5 METHANOGENESIS

Methanogenic organisms belong to the domain archaea (Thauer, 1998). Methanoarchaea derive metabolic energy by synthesizing CH_4 from a variety of one- and two-carbon compounds (reactions 9.R34–9.R36).

$$\underset{\text{Acetic acid}}{C_2H_4O_2(aq)} \longrightarrow CO_2(g) + CH_4(g) \qquad (9.R34)$$

$$\underset{\text{Formic acid}}{4CH_2O_2(aq)} \longrightarrow 3CO_2(g) + CH_4(g) + 2H_2O(l) \qquad (9.R35)$$

$$\underset{\text{Methyl amine}}{4\,CH_3NH_2} + 2H_2O(l) \longrightarrow CO_2(g) + 3CH_4(g) + 4NH_3(aq) \qquad (9.R36)$$

Reactions (9.R34)–(9.R36) are all *disproportionation* reactions. They are not true reduction-oxidation reactions because the reactions cannot be factored into half-reactions with distinct electron-donor and electron-acceptor species. For example, reaction (9.R34) involves the transfer of one electron from the carboxylic carbon to the methyl group of acetic acid to produce CO_2 and CH_4. These disproportionation reactions are enzymatically catalyzed.

9.6 SUMMARY

This chapter began with a review of reduction-oxidation principles from general chemistry: formal oxidation states, balancing reduction-oxidation reactions, and the Nernst equation. These are the basic tools needed to calculate standard biological (or environmental) electrochemical potentials $E^{\oplus}$ and to understand biological oxidation-reduction reactions.

Section 9.3 focused on the measurement and interpretation of reduction-oxidation conditions by plotting experimental (E, pH) coordinates on Pourbaix stability diagrams.

Table 9.5 summarizes the essential features common to Pourbaix stability diagrams and the foundation for interpreting reducing conditions using coordinates (*E*, pH).

Reducing conditions in the environment result from a key biological process: bacterial respiration (Baas-Becking et al., 1960). Section 9.4 explained how catabolism and electron transport are fundamentally chemical reduction processes that release the chemical energy stored in bioorganic compounds. Numerous catabolic pathways feed electrons into electron transport chains that harvest the chemical energy released during catabolism. Reducing conditions in the environment typically result from anaerobiosis. Biologically generated reducing conditions develop in water-saturated soil and sediments when there is abundance of carbon, a suitable terminal electron acceptor, and temperatures favoring biological activity.

Anoxia develops as aerobic organisms deplete pore water of dissolved O_2, setting the stage for the sequential rise and decline in nitrate-reducing bacteria, iron-reducing, sulfate-reducing, and, ultimately, methanoarchaea. Each succeeding population drives the reducing potential lower as each electron acceptor pool is depleted and a new terminal electron acceptor is sought. With each new terminal electron acceptor, a new population of anaerobic bacteria becomes active.

Since biological activity requires an energy (i.e., reduced carbon) source and suitable temperatures, reducing conditions develop when and where there is a significant biological activity. Anaerobic respiration effectively pumps reactive electron donors into the environment, depending on the type of anaerobic respiration. These electron donors—especially thiosulfate, hydrogen sulfide, and ferrous iron—transfer electrons to a host of electron acceptors through abiotic reduction reactions. The mobility and biological availability of many inorganic contaminants are sensitive to prevailing reduction-oxidation conditions. Solubility increases under reducing conditions for some contaminants and decreases for others. Regardless, prevailing reduction-oxidation conditions determine environmental risk for many contaminants.

The fate of nitrate in groundwater illustrates the role of anaerobic respiration on nitrate levels. Nitrate readily leaches through soil because ion-exchange capacity of soils is overwhelmingly cation-exchange capacity. Denitrifying bacteria in water-saturated soil reduce nitrate, but the likelihood of that happening drops dramatically once the nitrate leaches into the subsoil and underlying surficial materials, where organic carbon levels and biological activity tend to be quite low. Nitrate in the groundwater persists until it enters a surface water recharge (or *hyporheic*) zone in a wetland or sediment zone of a stream or lake. Biological activity in these zones is much higher because these zones are carbon rich. Dissolved nitrate rapidly disappears as nitrate-reducing bacteria in the recharge zone utilize the nitrate to metabolize available carbon.

APPENDICES

9.A REDUCTION-OXIDATION REACTIONS WITHOUT ELECTRON TRANSFER

Reduction-oxidation reactions are chemical reactions in which the oxidation state of one (or more) product atoms differ from their oxidation state in the educt. Although many

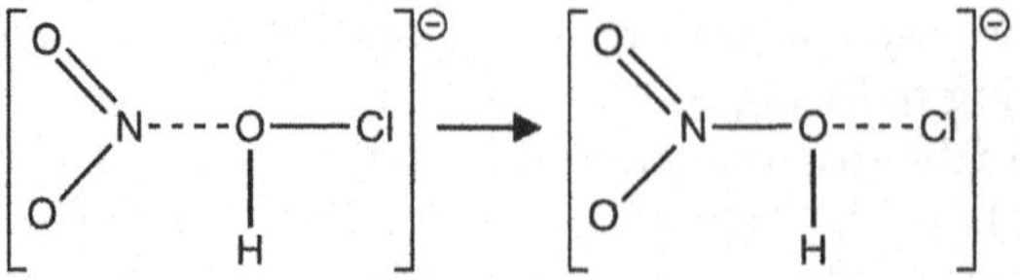

FIG. 9.A.1 Nucleophilic S_N2 displacement of Cl^- by NO_2^- in the oxidation of nitrate by hychlorous acid.

reduction-oxidation reactions do involve electron-transfer between educts, many involve atom-transfer mechanisms (Vitz, 2002).

The oxidation of nitrite NO_2^- by hypochlorous acid HClO (reaction 9.A.R1)—which appears to be an electron transfer reaction—can be factored into an oxidation half-reaction (9.A.R2) and a reduction half-reaction (9.A.R3).

$$NO_2^-(aq) + HClO(aq) \longleftrightarrow NO_3^-(aq) + H^+(aq) + Cl^-(aq) \quad (9.A.R1)$$

$$NO_2^-(aq) + H_2O(l) \longleftrightarrow NO_3^-(aq) + 2H^+(aq) + 2e^- \quad (9.A.R2)$$

$$HClO(aq) + 2H^+(aq) + 2e^- \longleftrightarrow H^+(aq) + Cl^-(aq) + H_2O(l) \quad (9.A.R3)$$

The S_N2 reaction mechanism[12] involves nucleophilic attack by the NO_2^- nitrogen atom on the HClO oxygen atom in resulting in displacement of product Cl^- (Fig. 9.A.1).

> This reaction is certainly a [reduction-oxidation] reaction because the oxidation state of the N changes from +3 to +5, even though no electrons are transferred from the NO_2^- to HClO. The non-bonding pair of electrons on N is replaced by a bonding pair, increasing the formal oxidation state. Similarly, the bonding pair on Cl is replaced by a non-bonding pair in Cl^-, lowering the formal oxidation state. …Transfer of an oxygen atom [from HClO] to NO_2^- is formally equivalent to adding water and removing two protons and two electrons, [but] the resulting change in oxidation state cannot be taken to imply either mechanism (Vitz, 2002).

9.B LIMITATION OF PLATINUM OXIDATION-REDUCTION ELECTRODES

Aqueous solutions saturated with O_2 under standard conditions ($p_{O_2} = 1$ atm, pH = 0) typically result in platinum ORP electrode-potentials in the range 700 mV $< E^{Pt} <$ 800 mV (Garrels and Christ, 1965; Stumm and Morgan, 1970) rather than the expected $E^{\circ} = 1229$ mV (cf. Table 9.C.1). The discrepancy results from the detailed electron transfer mechanism between the platinum electrode and O_2 molecules adsorbed on its

[12] The designation S_N2 is short-hand for *substitution nucleophilic (bi-molecular)*.

surface and the slow kinetics of the electron transfer process. A thin layer of platinum oxide forms when O_2 is being reduced by electrons at the electrode surface (Liang and Juliard, 1965a,b). This oxide layer passivates the platinum metal, inhibiting electron transfer by forming an insulating layer and altering the experimental potential. The net effect of the reduction mechanism and kinetics is a zero electrode current over a considerable potential range (Stumm and Morgan, 1970).

Numerous electrochemically active reduction-oxidation couples are dissolved in any environmental sample. The kinetics, mechanism, and reversibility of electron transfer and the extent to which each couple deviates from its equilibrium state ensure that any platinum ORP electrode-potential E^{Pt} represents a composite (i.e., mixed) electron transfer potential between all of these species and the platinum electrode surface (Table 9.B.1).

TABLE 9.B.1 Limitations on Experimental Platinum ORP Electrode Measurements of Reduction Potentials in Environmental Samples

1.	Many electrochemical species exhibit no tendency to transfer electrons at the platinum metal electrode surface (e.g., dinitrogen gas N_2, sulfate $SO_4{}^{3-}$, and methane CH_4) and, therefore, fail to contribute to the platinum ORP electrode-potential E^{Pt}
2.	A platinum ORP electrode will record the potential of a couple only if the electron transfer reaction is reversible. Rosca and Koper (2005) and de Vooys et al. (2001a,b) discovered at least two pathways for nitric oxide NO reduction at a platinum electrode, one yielding nitrous oxide N_2O and the other yielding ammonia NH_3
3.	The electron transfer rate at the platinum metal electrode surface is a function of solution activity for reduction-oxidation couples where one or both of the electron donor and acceptor species are solutes. Solutes such as $Fe^{3+}(aq)$ and $Mn^{4+}(aq)$, and their hydrolysis products, are extremely insoluble over a considerable pH range, rendering the platinum ORP electrode-potential E^{Pt} insensitive to significant variations in the electrochemical status of these couples
4.	Many reduction-oxidation couples do not attain equilibrium under environmental conditions—regardless of the reversibility or kinetics of the half-reaction at the platinum metal electrode surface—because of slow electron transfer kinetics in situ. The platinum ORP electrode-potential E^{Pt} fails to represent chemical equilibrium for these couples

9.C STANDARD AND BIOLOGICAL ELECTROCHEMICAL POTENTIALS FOR ENVIRONMENTAL AND BIOLOGICAL HALF-REACTIONS (TABLES 9.C.1 AND 9.C.2)

TABLE 9.C.1 Environmental Reduction Half-Reactions and Electrochemical Potentials

Half-Reaction	Potential Standard (E° mV)	Biological ($E^{\oplus}$ mV)
$2H^+(aq) + 2e^- \longrightarrow H_2(g)$	0	−414
$SO_4^{2-}(aq) + 3H^+(aq) + 2e^- \longrightarrow HSO_3^- + H_2O(l)$	+102	−519
$APS^{2-}(aq) + H^+(aq) + 2e^- \longrightarrow HSO_3^-(aq) + AMP^{2-}(aq)$	+147	−60
$CO_2(g) + 8H^+(aq) + 8e^- \longrightarrow CH_4(g) + 2H_2O(l)$	+169	−245
$2SO_4^{2-}(aq) + 10H^+(aq) + 8e^- \longrightarrow S_2O_3^{2-} + 5H_2O(l)$	+284	−233
$N_2(g) + 8H^+(aq) + 6e^- \longrightarrow 2NH_4^+(aq)$	+306	−246
$SO_4^{2-}(aq) + 8H^+(aq) + 6e^- \longrightarrow S(s) + 4H_2O(l)$	+353	−199
$HSO_3^-(aq) + 6H^+(aq) + 6e^- \longrightarrow HS^-(aq) + 3H_2O(l)$	+357	−140
$FeOOH(s) + 3H^+ + e^- \longrightarrow Fe^{2+}(aq) + 2H_2O(l)$	+770	−472
$Fe^{3+}(aq) + e^- \longrightarrow Fe^{2+}(aq)$	+771	+771
$Fe_2O_3(s) + 6H^+ + 2e^- \longrightarrow 2Fe^{2+}(aq) + 3H_2O(l)$	+773	−469
$NO_3^-(aq) + 2H^+(aq) + 2e^- \longrightarrow NO_2^-(aq) + H_2O(l)$	+838	+434
$NO_3^-(aq) + 10H^+(aq) + 8e^- \longrightarrow NH_4^+(aq) + 3H_2O(l)$	+881	+363
$Fe(OH)_3(s) + 3H^+ + e^- \longrightarrow Fe^{2+}(aq) + 3H_2O(l)$	+1059	−183
$NO_2^-(aq) + 2H^+(aq) + e^- \longrightarrow NO(g) + H_2O(l)$	+1170	+342
$O_2(g) + 4H^+ + 4e^- \longrightarrow 2H_2O(l)$	+1229	+815.8
$2NO(g) + 2H^+ + 2e^- \longrightarrow N_2O(g) + H_2O(l)$	+1588	+1174
$N_2O(g) + 2H^+ + 2e^- \longrightarrow N^2(g) + H_2O(l)$	+1754	+1340

9.D FACULTATIVE AND OBLIGATE ANAEROBES

Reaction 3 in the aerobic electron transport chain (Table 9.C.2) is particularly problematic for aerobic organisms because electrons divert prematurely to O_2 at this stage. The normal electron transfer process begins with the oxidation of ubiquinol $CoQH_2$, delivering one electron to two different one-electron acceptors, one being cytochrome *b* (reaction 4, Table 9.C.2) in a bifurcated pathway. Disruption occurs when the one-electron transfer to cytochrome *b* fails, leaving an unstable ubisemiquinone radical $CoQH^{\bullet-}$. The ubisemiquinone radical $CoQH^{\bullet-}$

TABLE 9.C.2 Half-Reactions and Potentials of the Aerobic Electron Transport Chain

Complex	Half-Reaction	Potential ($E^{\oplus}$ mV)
I	$NAD^+ + H^+(aq) + 2e^- \longrightarrow NADH$	−320
I	$FMN + 2H^+(aq) + 2e^- \longrightarrow FMNH_2$	−212
I	$\text{Ubiquinone} + 2H^+(aq) + 2e^- \longrightarrow \text{Ubquinol}$	+100
III	$\underset{Fe^{3+}}{\text{Cytochrome } b} + e^- \longrightarrow \underset{Fe^{2+}}{\text{Cytochrome } b}$	+120
III	$\underset{Fe^{3+}}{\text{Cytochrome } c_1} + e^- \longrightarrow \underset{Fe^{2+}}{\text{Cytochrome } c_1}$	+220
IV	$\underset{Fe^{3+}}{\text{Cytochrome } c} + e^- \longrightarrow \underset{Fe^{2+}}{\text{Cytochrome } c}$	+250
IV	$\underset{Fe^{3+}}{\text{Cytochrome } a} + e^- \longrightarrow \underset{Fe^{2+}}{\text{Cytochrome } a}$	+290
IV	$\underset{Fe^{3+}}{\text{Cytochrome } a_3} + e^- \longrightarrow \underset{Fe^{2+}}{\text{Cytochrome } a_3}$	+350
IV	$O_2(g) + 4H^+ + 4e^- \longrightarrow 2H_2O(l)$	+816

reacts with O_2 to form superoxide anion radical $O_2^{\bullet -}$. The occurrence of this electron-transport disruption has resulted in the evolution of two key enzymes in aerobic organisms: superoxide dismutase (SOD)—an enzyme that swiftly and safely converts the superoxide anion radical into hydrogen peroxide $H_2O_2(aq)$ (reaction 9.D.R1)—and catalase—an enzyme that disproportionates hydrogen peroxide into water and O_2 (reaction 9.D.R2).

$$2O_2^{\bullet -} + 2H^+ \xrightarrow{\text{SOD}} O_2(aq) + H_2O_2(aq) \qquad (9.D.R1)$$

$$2H_2O_2(aq) \xrightarrow{\text{Catalase}} O_2(aq) + 2H_2O(l) \qquad (9.D.R2)$$

Facultative anaerobic bacteria possess superoxide dismutase, enabling them to tolerate disruption of the aerobic electron transport chain in reaction (9.D.R1) by decomposing the reactive oxygen species it produces. Obligate anaerobic bacteria lack superoxide dismutase and, as a consequence, cannot tolerate significant levels of O_2.

9.E FERMENTATIVE ANAEROBIC BACTERIA

One anaerobic respiratory process is fermentation. Fermentation, unlike other anaerobic respiratory processes, does not use the electron transport chain. Organisms capable of fermentation (certain fungi and many bacteria) are facultative anaerobes. During fermentation the electron transport chain is inactive; oxidation occurs during catabolism.

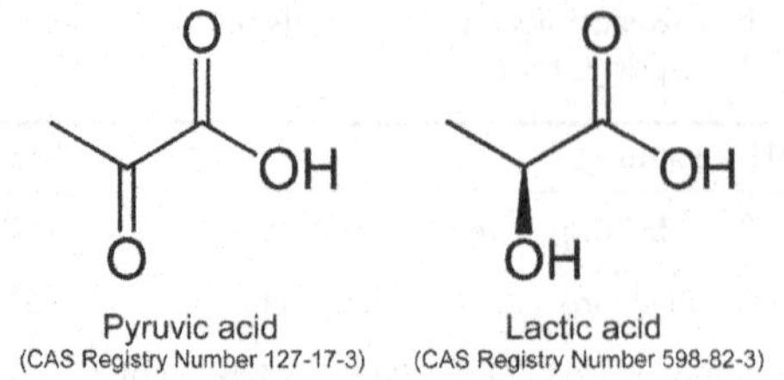

Pyruvic acid
(CAS Registry Number 127-17-3)

Lactic acid
(CAS Registry Number 598-82-3)

FIG. 9.E.1 Molecular structure of pyruvic acid (CAS registry number 127-17-3) and lactic acid (CAS registry number 598-82-3).

The first stage of glucose catabolism (glycolysis) produces two molecules of pyruvic acid (CAS registry number 127-17-3, Fig. 9.E.1). Lactic acid fermentation (reaction 9.E.R1) uses NADH as an electron donor and pyruvic acid as an electron acceptor to produce lactic acid (CAS registry number 598-82-3, Fig. 9.E.1).

$$\underset{\text{Pyruvic acid}}{2C_3H_4O_3(aq)} + NADH(aq) + H^+ \longrightarrow \underset{\text{Lactic acid}}{2C_3H_6O_3(aq)} + NAD^+(aq) \qquad (9.E.R1)$$

Alcohol fermentation begins with the disproportionation of pyruvic acid into acetaldehyde ($CH_3(C=O)H$, CAS 75-07-0) and CO_2 by pyruvate decarboxylase (reaction 9.E.R2).

$$\underset{\text{Pyruvic acid}}{C_3H_4O_3(aq)} \longrightarrow \underset{\text{Acetaldehyde}}{C_2H_4O(aq)} + CO_2(g) \qquad (9.E.R2)$$

The final product, ethanol (CH_3CH_2OH, CAS 64-17-5), is formed when electron donor NADH reduces acetaldehyde in reaction (9.E.R3).

$$\underset{\text{Ethanol}}{C_2H_6O(aq)} + NADH(aq) + H^+ \longrightarrow \underset{\text{Acetaldehyde}}{2C_2H_4O(aq)} + NAD^+(aq) \qquad (9.E.R3)$$

Bacteria use other fermentation pathways to produce a variety of low molecular weight organic acids (also known as volatile fatty acids) found in the pore water of soils and sediments.

References

Atkins, P., de Paula, J., 2005. Physical Chemistry for the Life Sciences. W.H. Freeman and Company, New York.

Baas-Becking, L.G.M., Kaplan, I.R., Moore, D., 1960. Limits of the natural environment in terms of pH and oxidation-reduction potentials. J. Geol. 68, 243–284.

Berkner, L.V., Marshall, L.C., 1965. History of major atmospheric components. Proc. Natl. Acad. Sci. 53 (6), 1215–1226.

Bradley, A.S., Leavitt, W.D., Johnston, D.T., 2011. Revisiting the dissimilatory sulfate reduction pathway. Geobiology 9 (5), 446–457.

de Vooys, A.C.A., Koper, M.T.M., Van Santen, R.A., Van Veen, J.A.R., 2001a. Mechanistic study of the nitric oxide reduction on a polycrystalline platinum electrode. Electrochim. Acta 46 (6), 923–930.

de Vooys, A.C.A., Koper, M.T.M., Van Santen, R.A., Van Veen, J.A.R., 2001b. Mechanistic study on the electrocatalytic reduction of nitric oxide on transition-metal electrodes. J. Catal. 202 (2), 387–394.

Delahay, P., Pourbaix, M., Van Rysselberghe, P., 1950. Potential-pH diagrams. J. Chem. Educ. 27 (12), 683–688.

DiChristina, T.J., Moore, C.M., Haller, C.A., 2002. Dissimilatory Fe(III) and Mn(IV) reduction by Shewanella putrefaciens requires ferE, a homolog of the pulE (gspE) type II protein secretion gene. J. Bacteriol. 184 (1), 142–151.

Dick, J.M., 2008. Calculation of the relative metastabilities of proteins using the CHNOSZ software package. Geochem. Trans. 9, 1–17.

Garrels, R.M., Christ, C.L., 1965. Solutions, Minerals, and Equilibria. Harper and Row, New York.

Haggerty, S.E., 1976. Opaque mineral oxides in terrestrial igneous rocks. Rev. Mineral. 3 (Oxide Minerals), Hg101–Hg300.

Liang, C.C., Juliard, A.L., 1965a. The overpotential of oxygen reduction at platinum electrodes. J. Electroanal. Chem. 9 (5–6), 390–394.

Liang, C.C., Juliard, A.L., 1965b. Reduction of oxygen at the platinum electrode. Nature (London, U. K.) 207 (4997), 629–630.

Lindsay, W.L., 1979. Equilibria in Soils. John Wiley, New York.

Liu, C., Suo, Y., Leyh, T.S., 1994. The energetic linkage of GTP hydrolysis and the synthesis of activated sulfate. Biochemistry 33 (23), 7309–7314.

Lovley, D.R., 1993. Dissimilatory metal reduction. Ann. Rev. Microbiol. 47 (1), 263–290.

Lovley, D.R., Goodwin, S., 1988. Hydrogen concentrations as an indicator of the predominant terminal electron-accepting reactions in aquatic sediments. Geochim. Cosmochim. Acta 52 (12), 2993–3003.

Lovley, D.R., Holmes, D.E., Nevin, K.P., 2004. Dissimilatory Fe(III) and Mn(IV) reduction. Adv. Microb. Physiol. 49, 219–286.

Myers, J.M., Myers, C.R., 2002. Genetic complementation of an outer membrane cytochrome omcB mutant of Shewanella putrefaciens MR-1 requires omcB plus downstream DNA. Appl. Environ. Microbiol. 68 (6), 2781–2793.

Nordstrom, D.K., Wilde, F.D., 1998. Reduction-oxidation potential (electrode method). Geological Survey Techniques of Water-Resources Investigations, vol. 9. United States Geological Survey, Reston, VA, pp. 1–20 (chap. 6.5).

Pourbaix, M.J.N., 1945. Thermodynamique des solutions aqueuses diluées : représentation graphique du rôle du pH et du potentiel (Ph.D. thesis). Technische Hogeschool Delft, Delft, Netherlands.

Rosca, V., Koper, M.T.M., 2005. Mechanism of electrocatalytic reduction of nitric oxide on Pt (100). J. Phys. Chem. B 109 (35), 16750–16759.

Ruebush, S.S., Brantley, S.L., Tien, M., 2006. Reduction of soluble and insoluble iron forms by membrane fractions of Shewanella oneidensis grown under aerobic and anaerobic conditions. Appl. Environ. Microbiol. 72 (4), 2925–2935.

Stumm, W., Morgan, J.J., 1970. Aquatic Chemistry: An Introduction Emphasizing Chemical Equilibria in Natural Waters, first ed. Wiley-Interscience, New York.

Thauer, R.K., 1998. Biochemistry of methanogenesis: a tribute to Marjory Stephenson: 1998 Marjory Stephenson Prize Lecture. Microbiology 144 (9), 2377–2406.

Vitz, E., 2002. Redox redux: recommendations for improving textbook and IUPAC definitions. J. Chem. Educ. 79 (3), 397–400.

CHAPTER

10

Human Health and Ecological Risk Analysis

OUTLINE

10.1 Introduction 492

10.2 The Federal Risk Assessment Paradigm 493
- 10.2.1 *Risk Assessment* 493
- 10.2.2 *Risk Management and Mitigation* 493

10.3 Dose-Response Assessment 493
- 10.3.1 *Dose-Response Functions* 494
- 10.3.2 *Low-Dose Extrapolation* 498

10.4 Exposure Pathway Assessment 502
- 10.4.1 *Receptors* 503
- 10.4.2 *Exposure Routes* 504
- 10.4.3 *Exposure Points* 505
- 10.4.4 *Fate and Transport* 506
- 10.4.5 *Primary and Secondary Sources* 507
- 10.4.6 *Exposure Pathway Assessment* 508
- 10.4.7 *Exposure Factors* 509

10.5 Intake Estimates 511

10.6 Risk Characterization 512
- 10.6.1 *Target Cancer Risk* 512
- 10.6.2 *Cumulative Target Risk* 513
- 10.6.3 *Hazard Quotient* 514
- 10.6.4 *Cumulative Risk: Hazard Index* 515

10.7 Exposure Mitigation 515

10.8 Risk-Based Screening Levels 516

10.9 Ecological Risk Assessment 519
- 10.9.1 *Wildlife Risk Model* 519
- 10.9.2 *Ecological Soil Screening Levels* 522

10.10 Summary 523

Appendices 524

10.A Factors Affecting Contaminant Transport by Surface Water 524

10.B Factors Affecting Contaminant Transport by Groundwater 526

10.C Factors Affecting Contaminant Transport by Soils or Sediments 528

10.D Water Ingestion Equation 529

10.E Soil Ingestion Equation 530

10.F Food Ingestion Equation 531

10.G Air Inhalation Equation 532

References 532

Soil and Environmental Chemistry
http://dx.doi.org/10.1016/B978-0-12-804178-9.00010-0

10.1 INTRODUCTION

A major impetus for the growth of environmental chemistry dates from the jarring discovery in the 1960s and 1970s of human health hazards caused by environmental pollution. Prior to the modern environmental movement, key amendments to the Federal Food, Drug and Cosmetic Act (United States Congress, 1938)—the Pesticide Amendments of 1954 (United States Congress, 1954), the Food Additives Amendment (United States Congress, 1958), and the Color Additives Amendment (United States Congress, 1960)—were enacted to mitigate exposure to potential carcinogens in food. This formed a basis for developing the risk assessment paradigm employed by the US federal government (Fig. 10.1) that expanded risk assessment beyond food to include contaminants in air, water, and soil.

Chemical pollution of the environment results from the release of compounds that are toxic to humans. Chronic human toxicity can take many forms (Table 10.1) that invariably originate with the impairment of normal biochemical processes, a loss of fertility, birth defects or cancer.

The following section reviews the paradigm employed by the US Federal government to assess human health hazards resulting from chemical pollution of the environment and protocols to mitigate adverse health effects resulting from chemical pollution.

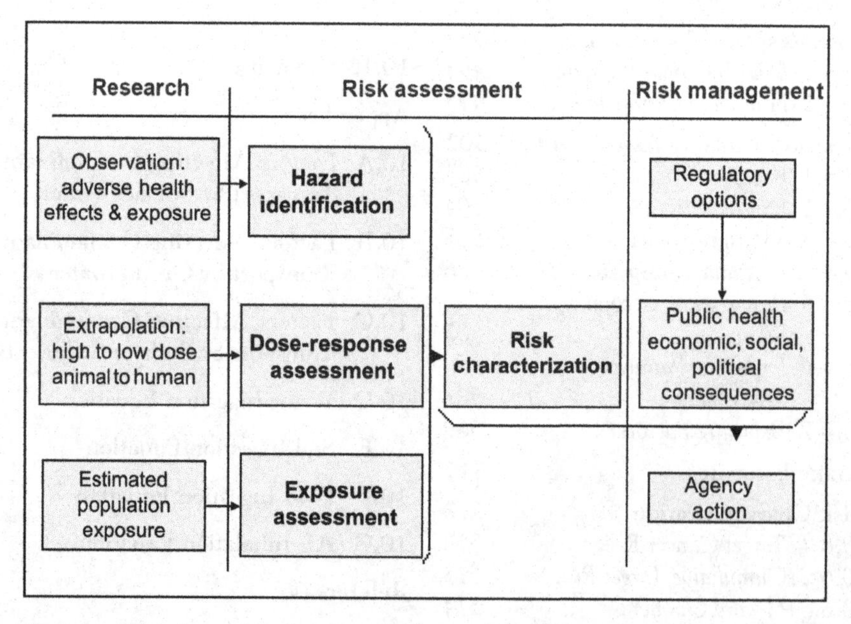

FIG. 10.1 The Federal Risk Assessment Paradigm. *Reproduced with permission from National Research Council, 1983. Risk Assessment in the Federal Government: Managing the Process. National Academies Press, Washington, DC.*

TABLE 10.1 A Classification of Chronic Toxicities and Their Human Health Effects

Toxicity	Adverse Health Effect
Germ cell mutagenicity	Genetic aberrations of adult germ cells resulting in heritable abnormal traits.
Carcinogenicity	Genetic aberrations resulting in the development of malignant tumors
Reproductive toxicity (teratogenicity)	Loss of adult fertility and abnormal development of fetus or children
Target organ toxicity	Loss of organ function

10.2 THE FEDERAL RISK ASSESSMENT PARADIGM

10.2.1 Risk Assessment

Scientific research supports human health risk assessment in three ways (cf. Fig. 10.1): identifying hazardous chemical compounds, developing dose-response functions necessary to predict adverse effects based on low-level chronic doses delivered by the environment, and estimating uptake from various routes by exposed human populations. Risk assessment draws from each of these scientific fields to characterize specific health risks, whether that is an increased cancer risk or the probability of an adverse effect in a particular population.

10.2.2 Risk Management and Mitigation

Risk characterization is essentially a quantitative prediction of the adverse effect on human health that integrates our scientific understanding of toxicology and environmental exposure. Federal, state, and local government agencies assume ultimate responsibility for mitigating health hazards by establishing emission standards or action levels, both designed to reduce contaminant levels at various exposure points (air, water, and soil). Mitigation may involve removal and secure disposal of hazardous waste or excessively contaminated soil or, in many cases, management of contaminated sites by imposing institutional controls designed to limit human contact with contaminants that are impractical to remove. As indicated in Fig. 10.1, risk management seeks to balance the risk posed by pollution against the economic, social, and political costs of reducing risk. Not surprisingly, marginal costs increase substantially as risk declines.

10.3 DOSE-RESPONSE ASSESSMENT

From the outset, risk assessment follows different routes depending on whether a chemical compound is identified as a known or potential carcinogen. The fundamental distinction arises from very different toxicity paradigms.

The adverse physiological effects of noncarcinogens are presumed to manifest only during exposure. Adverse effects cease to manifest once intake is suspended and the toxicant is eliminated. Stated differently, the adverse physiological effects of noncarcinogens are assumed to be reversible. This is, of course, an over simplification. The neurological effects of chronic lead exposure in children can persist into adulthood (ATSDR, 2010). Repeated episodes of chronic lead exposure can cause irreversible renal damage (chronic lead tubulo-interstitial nephritis).

Cancer mortality has a distinctive characteristic; it increases with age (Knudson, 2001), the result of accumulated "hits" or mutations.

> "A conspicuous feature of the epidemiology of common cancers is that their incidence increases with age, so the notion of multiple mutations was invoked by way of explanation. If r successive mutations occur in some cells at constant rates ..., if the size of the target-cell population remains constant, and if cells with an intermediate number of mutations have no growth advantage, the age-specific incidence would be $I = k \cdot t^{r-1}$. ...Many cancers show this relationship, and r has been estimated for numerous cancers; for example, $r = 6$ for colon cancer. This, of course, would be the number of rate-limiting events that produce a recognizable cancer."

10.3.1 Dose-Response Functions

Toxicologists fit dose-response data to numerous mathematical response functions (e.g., Fig. 10.2). The response is usually *binary*—either there is an adverse effect or there is no adverse effect—and the response function reports the fraction of responders manifesting the adverse effect.

The dose-response functions plotted in Fig. 10.2 are all deterministic responses. In general the response includes both stochastic and deterministic effects (Fig. 10.3), the former being uncorrelated with dose while the latter reports a positive response to dose.

10.3.1.1 Normal Binary-Response Function

The normal-distribution response (or *cumulative distribution function*, CDF) appearing at the upper-left quadrant in Fig. 10.2 is centered on the arithmetic mean μ of a normal probability distribution, the width determined by the arithmetic standard deviation σ. The normal response function in closed form uses the *error function* (expression 10.1) which is plotted in Fig. 10.2 (upper left quadrant).

$$P_N(d) = \frac{1}{2} \cdot \left(1 + \text{erf}\left(\frac{d\mu}{2 \cdot \sigma}\right)\right) \quad (10.1)$$

10.3.1.2 Weibull Binary-Response Function

Two Weibull response functions are plotted in the upper-right quadrant in Fig. 10.2. The curve with no threshold (*dashed line*) is plotted using $\alpha = 1$ and $\beta = 1$. The curve with a threshold (*solid line*) is plotted using $\alpha = 5$ and $\beta = 5$. The Weibull response function is given by expression (10.2).

$$P_W(d) = 1 - e^{(d/\beta)^{\alpha}} \quad (10.2)$$

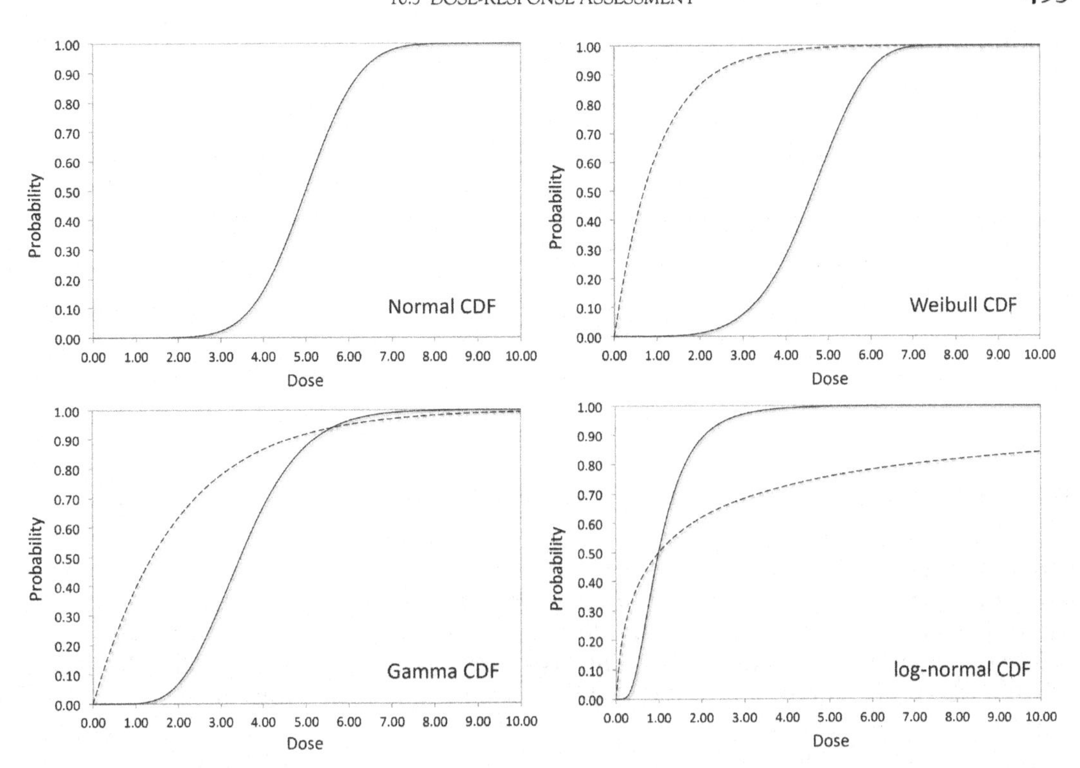

FIG. 10.2 The four cumulative probability distribution functions (CDF) appearing in this figure are widely-used 2-parameter, binary dose-response functions. The response function for a normal distribution appears at upper left. The Weibull response function appears at upper right. The gamma response function appears at lower left. The logarithmic-normal response appears at lower right. The Weibull, gamma and log-normal graphs plot two curves to illustrate the effect of function parameters on the presence or absence of a threshold.

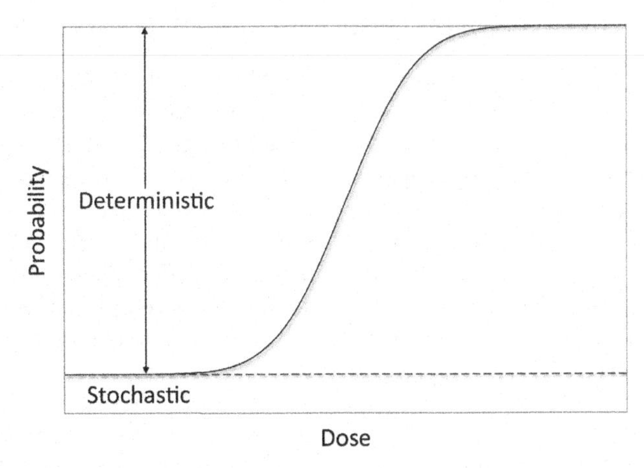

FIG. 10.3 Typical response functions include a stochastic range where dose and response are not correlated and a deterministic range where response correlates with dose.

10.3.1.3 *Gamma Binary-Response Function*

Two gamma response functions are plotted in the lower left quadrant in Fig. 10.2. The curve with no threshold (*dashed line*) is plotted using $k = 1$ and $\theta = 2$. The curve with a threshold (*solid line*) is plotted using $k = 9$ and $\theta = 0.4$. The gamma response function is given by expression (10.3) where $\gamma(k, x/\theta)$ and $\Gamma(k)$ are the *lower incomplete* gamma function and *complete* gamma functions, respectively. Most current spreadsheet applications include a gamma CDF function.

$$P_G(d) = \frac{\gamma(k, x/\theta)}{\Gamma(k)} \tag{10.3}$$

10.3.1.4 *Logarithmic-Normal Binary-Response Function*

The logarithmic normal-distribution response functions appearing at the lower-right quadrant in Fig. 10.2 is centered on the *geometric* mean $\log(\mu_G)$ of a log-normal probability distribution, the width determined by the *geometric* standard deviation $\log(\sigma_G)$. The logarithmic-normal response function in closed form uses the *error function* (expression 10.4).

The curve with no threshold (*dashed line*) is plotted using $\log(\mu_G) = 0.0$ and $\log(\sigma_G) = 1.0$ (Box 10.1). The curve with a threshold (*solid line*) is plotted using $\log(\mu_G) = 0$ and $\log(\sigma_G) = 0.25$.

$$P_{LN}(d) = \frac{1}{2} \cdot \left(1 + \text{erf}\left(\frac{\log(d) \log(\mu_G)}{2 \cdot \log(\sigma_G)}\right)\right) \tag{10.4}$$

BOX 10.1

CARCINOGENESIS: THE KNUDSON HYPOTHESIS

A handful of important milestones mark the emergence of the modern understanding of carcinogenesis. First, geneticist Boveri (1914) identified the link between mutagenesis and carcinogenesis. Second, Nordling (1953) used statistical analysis to propose a multi-stage model for increased cancer mortality with age; cancer mortality among males of European descent appears to require about six genetic mutations (10.5).[1]

$$I(t) = k \cdot t^{r-1} \tag{10.5}$$

This interpretation of cumulative cancer mortality failed to recognize, however, the high mutation rates required by a six-mutation model, the nature and dynamic state of target cells, and the distinctive age distributions of childhood cancers.

Target cells are, in all probability, actively-dividing stem cells rather than quiescent differentiated cells. Taking into account stem-cell growth kinetics and their age-adjusted abundance, age-specific cancer mortality (encompassing both childhood and adulthood) can be explained by a single age-specific cancer mortality model with two (2) rather than six (6) rate-limiting mutations (Moolgavkar and Knudson, 1981).

BOX 10.1 (*cont'd*)

Muller (1950a,b) and Nordling (1952) recognized mutations could eliminate genetic control of normal cell growth, resulting in malignant cells with a distinct a growth advantage. Huebner and Todaro (1969) coined the term *oncogene* to identify aberrant genes[2] responsible for tumor-cell proliferation. Weinberg and others (Shilo and Weinberg, 1981; Shih and Weinberg, 1982) were able to demonstrate the oncogenes isolated from cancer cells were remarkably similar to *proto-oncogenes* found in normal cells. Finally, Knudson (1986) proposed, in a wide-ranging review of carcinogenic research, a number of mechanisms that could explain the role of two gene classes in carcinogenesis.

> The two classes of cancer genes are different in several ways. The classical oncogene is active; its product must be present and abnormal either in structure or in amount. The antioncogene is inactive; cancer results when no normal copy is present ...Oncogene abnormality therefore seems to be a common, perhaps even universal, feature of cancer ...Primary antioncogene change operates in many other cancers. One might even predict that most cancers are caused by the latter mechanism.

Tumor-suppressor gene has replaced *antioncogene* in common usage but the conclusions drawn by Knudson (1986) remain intact. Most human cancers involve loss-of-function in tumor-suppressor genes. Loss-of-function usually involves chromosomal mutations arising from double-strand breaks: translocation, deletion and, less commonly, inversion of large chromosomal segments. The *position effect*[3] explains the profound impact of chromosomal translocation on loss-of-function. Chemicals that induce double-strand breaks (e.g., benzene, free radicals, and reactive oxygen species) increase the likelihood of chromosomal mutation.

Knudson (1971) studied the incidence of retinal cancer in children from birth to 4 years. The incidence-rate model required only two disabling mutations, one for each allele of a tumor-suppressor gene. Two mutations is sufficient to initiate tumor growth in individuals with no hereditary susceptibility. If the individual inherits a single inactivated tumor-suppressor allele (the results of germ cell DNA mutations or chromosomal aberrations) then a disabling mutation in the second, active allele would be sufficient to initiate carcinogenesis. Heritable carcinogenic susceptibility in this instance manifests as single inactivated or silenced tumor-suppressor allele (cf. Moolgavkar and Knudson, 1981).

Gene inactivation can occurs without genetic mutation. Hypermethylation targeting the gene promoter region inactivates the gene while leaving the genetic sequence intact. Aberrant hypermethylation (e.g., silencing tumor-suppressor genes (Chen and Baylin, 2004; Esteller, 2007; Baylin and Jones, 2011)) has generated considerable interest because it represents a heritable epigenetic aberration that does not require mutation.

[1] The number of rate-limiting steps or mutations r in expression (10.5) is: $r \approx 6$.

[2] Oncogenes were at that time assumed to be of viral origin.

[3] A change in gene expression linked to gene location in a chromosome is know as the position effect.

10.3.2 Low-Dose Extrapolation

Low-dose extrapolation is necessary because the acceptable adverse effects from chronic exposure implied by the Federal risk assessment paradigm (National Research Council, 1983) typically requires extrapolation beyond the *lowest observable adverse effect level*. Furthermore, fundamental differences between reversible intoxication by anticarcinogen and irreversible intoxication by carcinogens (i.e., genetic mutations or chromosomal aberrations) demand different assumptions governing low-dose extrapolation.

Low-dose extrapolation seeks to identify the dose threshold separating deterministic and stochastic adverse effects (cf. Fig. 10.3), the former correlated to dose and the latter uncorrelated to dose. The presence of a threshold response implied by Fig. 10.3 is consistent with reversible intoxication by noncarcinogens.

In contrast, a fundamental assumption of carcinogenesis is that "malignant transformation of a single cell is sufficient to give rise to a tumor" (Moolgavkar and Knudson, 1981). The concept of a threshold dose simply does not apply to carcinogens. A second complicating factor—relevant to low-dose extrapolation in carcinogenesis—is the absence of characteristics distinguishing mutations uncorrelated to environmental exposure from mutations arising from carcinogen exposure. The following quote paraphrases an excerpt from *Health Effects of Exposure to Low Levels of Ionizing Radiation* (BEIR, 1996).

> Estimates of the risk of cancer, therefore, must rely largely on observations of the numbers of cancers of different kinds that arise in [exposed populations]. Since nearly 20% of all deaths in the United States result from cancer, the estimated number of cancers attributable to [chemical exposure or ionizing radiation] is only a small fraction of the total number that occur. Furthermore, the cancers that result from [exposure to carcinogens] have no special features by which they can be distinguished from those produced by other causes. Thus the probability that cancer will result from a small dose can be estimated only by extrapolation from the increased rates of cancer that have been observed after larger doses, based on assumptions about the dose-incidence relationship at low doses.

10.3.2.1 Noncarcinogenic Agents

Adverse effects from chronic intoxication are a function of the steady-state concentration in the body—also known as the *body-burden*. The steady-state body-burden is a function of the intake and elimination rates from the body and depend on the biological half-life $t_{1/2}$ for toxicant retention in the body (cf. Box 10.2).

BOX 10.2

BIOLOGICAL HALF-LIFE AND STEADY-STATE BODY-BURDEN

A steady-state body-burden exists when the toxicant elimination rate equal the toxicant ingestion rate. For this example the toxicant is dissolved in water being ingested orally and the toxicant is eliminated in urine.

$$C_u \cdot ER = C_W \cdot IR \tag{10.6}$$

$$C_u = C_W \cdot \frac{IR\ [\mathrm{L\ day^{-1}}]}{ER\ [\mathrm{L\ day^{-1}}]} \tag{10.7}$$

BOX 10.2 *(cont'd)*

Toxicant elimination in urine follows a first-order rate law, expression (10.8) gives the differential first-order rate law and expression (10.9) gives the integral first-order rate law. The daily mean urine volume is ER and the volume over which the toxicant distributes within the body is V_d while $C_{u,0}$ and $C_{u,t}$ are the urine concentrations at time zero and time t, respectively.

$$\frac{dC_u}{dt} = -k \cdot C_u = -\left(\frac{ER}{V_d}\right) \cdot C_u \quad (10.8)$$

$$\ln\left(\frac{C_{u,t}}{C_{u,0}}\right) = -\left(\frac{ER}{V_d}\right) \cdot t \quad (10.9)$$

Using the kinetic half-life definition, expression (10.9) can be rearranged, solved for ER, and substituted into expression (10.7) to yield expression (10.10)

$$C_u = \left(\frac{t_{1/2} \cdot IR \cdot C_W}{\ln 2 \cdot V_d}\right) \quad (10.10)$$

Chronic exposure disregards the time interval required to reach a steady-state body-burden following the onset of exposure and the time interval required to completely eliminate the toxicant once exposure is suspended.

The dose plotted in a noncarcinogenic dose-response curve is the steady-state body burden [mg kg^{-1}] prorated on a daily basis: the *average daily dose* (ADD) [mg kg^{-1} day^{-1}]. Low-dose extrapolation using the *benchmark–dose* method relies on fitting dose-response data using one or more cumulative distribution functions. The *benchmark dose* (BMD) is determined by the empirical dose-response function used to fit the experimental data and the *benchmark response* (BMR) chosen to represent the stochastic response absent exposure. The cumulative dose ingested throughout the exposure duration is considered irrelevant.

Fig. 10.4 illustrates the low-dose extrapolation method for noncarcinogenic dose-response functions. Two sets of toxicological end-points appear in Fig. 10.4, one based whether or not a dose results in statistically significant deterministic response and the other based on the *benchmark dose method* (Davis et al., 2011).

The *lowest observable adverse effect level* (LOAEL) in Fig. 10.4 represents the lowest dose that has a statistically significant deterministic response. The *no observable adverse effect level* (NOAEL) in Fig. 10.4 represents the highest dose that has no statistically significant deterministic response. The deterministic threshold lies somewhere between the NOAEL and the LOAEL. The low-dose extrapolation method employing (NOAEL, LOAEL) limits does not rely on fitting dose-response data a cumulative distribution function.

Fig. 10.4 also illustrates low-dose extrapolation using the *benchmark dose method*. In this case the dose-response data are fitted to a cumulative distribution function (cf. the *solid line* representing the central estimate in Fig. 10.4). Based on this empirical fit the toxicologist chooses a *benchmark response*—a low deterministic incidence rate greater than the stochastic

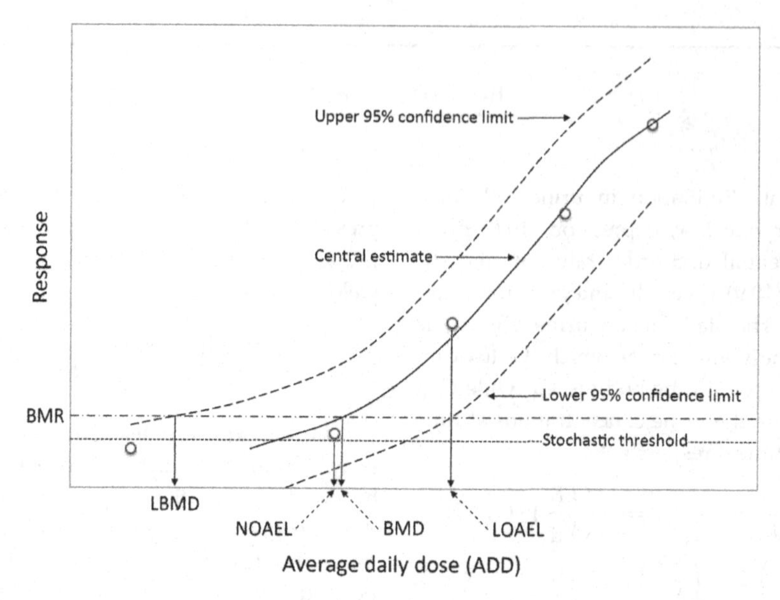

FIG. 10.4 The graph plots data points (*open circles*), a central estimate (*solid line*) from an empirical fit of the data points and the 95% confidence limits (*dashed lines*) for an idealized dose-response curve. The *solid line* parallel to the horizontal axis is the estimated stochastic threshold (cf. Fig. 10.4). The data point labeled NOAEL is the highest dose that has no statistically significant deterministic response while the point labeled LOAEL is the lowest dose that has a statistically significant deterministic response. The *dashed-dotted line* parallel to the horizontal axis is the *benchmark response BMR*. The BMR intersects the upper 95% confidence limit at the *lower benchmark dose LBMD* and intersects the central estimate at the *benchmark dose* (BMD).

incidence rate. This *benchmark response* intersects the upper 95% confidence limit for the central estimate as the *lower benchmark dose* (LBMD).

The *reference dose RfD*, the maximum acceptable dose for chronic human intoxication, is an extrapolated toxicological end-point defined relative to either the NOAEL or the LBMD. Box 10.3 summarizes the options for determining the *reference dose* (RfD).

BOX 10.3

REFERENCE DOSE RFD

The reference dose RfD is an extrapolated toxicological end-point limited to noncarcinogenic toxicants. The reference dose RfD has the same units ($mg\ kg^{-1}\ day^{-1}$) as the *average daily dose ADD* appearing in Fig. 10.4.

$$\text{RfD} = \frac{\text{LOAEL}}{10 \cdot \text{UF}} = \frac{\text{LOAEL}}{10 \times 10^n} \qquad (10.11)$$

$$\text{RfD} = \frac{\text{NOAEL}}{\text{UF}} = \frac{\text{NOAEL}}{10^n} \qquad (10.12)$$

BOX 10.3 (*cont'd*)

The RfD can be defined relative to either a NOAEL or a LOAEL. A RfD estimate extrapolated from a LOAEL (expression 10.11) includes an additional factor of 10 in the denominator that does not appear when the estimate is extrapolated from a NOAEL (expression 10.12).

The LBMD toxicological end-point defines a tolerable daily intake roughly equivalent to RfD. The LBMD differs from RfD because tolerance is defined by the benchmark response[4] BMR and uncertainty is defined by the upper confidence limit which the BMR intersects. This approach, however, does not provide for extrapolation from dose-response studies involving animals to adverse effects in humans.

The RfD estimate (expressions 10.11, 10.12) include an *uncertainty factor* (UF) that depends on the type of the dose-response study. If the dose-response study involves humans the uncertainty factor is: $UF = 10^1$. Dose-response studies employing animals requires animal-to-human extrapolation, which calls for a larger uncertainty factor: $UF = 10^2$. Short-term dose-response studies, typically involving mice or rats, call for another factor of 10: $UF = 10^3$.

[4] The reader must keep in mind the deterministic incidence rate chosen for this example is not a standard value and can range from 1% to 10%. Risk management considerations (cf. Fig. 10.1) determine whether higher or lower incidence rates are employed.

10.3.2.2 Carcinogenic Agents

Epidemiological studies suggest roughly 80% of all cancers can be traced to environmental causes such as diet, work, and chemical exposure (Benigni et al., 2013). The irreversibility of genetic mutations underlies the correlation between cancer mortality and age. DNA reactive carcinogens and ionizing radiation damage DNA and cause chromosomal aberrations (i.e., genetic mutations) that ultimately result in malignant stem cells. Oncologists believe malignancy can result from DNA or chromosomal damage to a single cell. In the context of low-dose extrapolation this eliminates the notion of a threshold dose.

Fig. 10.5 illustrates the low-dose extrapolation method for carcinogenic dose-response curves. A line parallel to the horizontal dose axis and passing through the lowest data point defines the response used for low-dose extrapolation: the *incremental life-time cancer risk* (IELCR). As with the BMR line in Fig. 10.4, the IELCR line passes through the upper 95% confidence limit and the central estimate of the dose-response curve. The former intersection corresponds to the *lower effective dose* (LED) while the latter intersections corresponds to the *effective dose* (ED).

Linear low-dose extrapolation extends through a POD to the origin. The coordinates of the POD are (IELCR, LED). The slope of the linear low-dose extrapolation is identified as the *slope factor* (SF). Later we will use the SF to quantify carcinogenic risk.

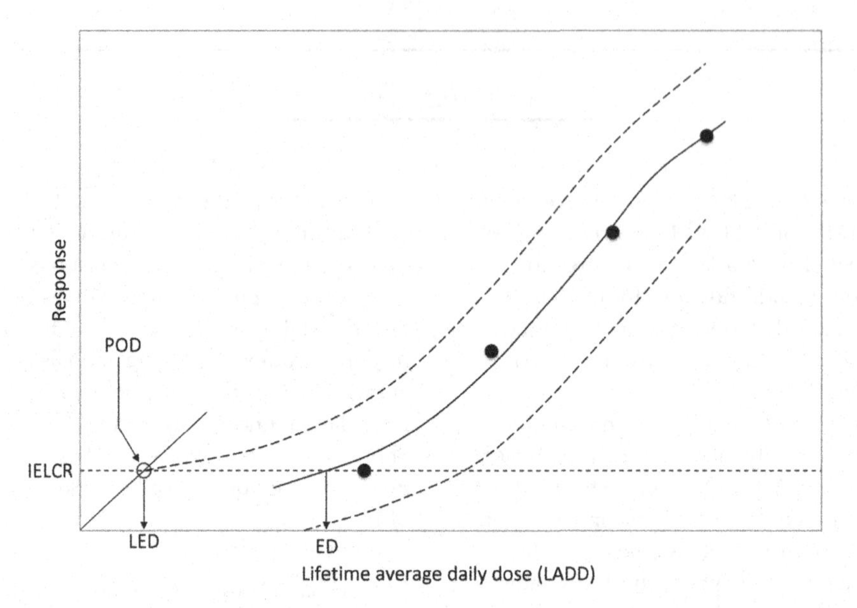

FIG. 10.5 The graph plots data points (*filled circles*), a central estimate (*solid line*) from an empirical fit of the data points and the 95% confidence limits (*dashed lines*) for an idealized dose-response curve. The *dashed line* parallel to the horizontal axis is the *incremental life-time cancer risk* (IELCR) corresponding to the lowest dose in the dose-response curve. Notice *effective dose* (ED) for this IELCR is defined by the intersection of the *dashed line* and the central estimate of the dose-response curve. The *point of departure* (POD) (*open circle*) is defined by the intersection of the IELCR and the upper 95% confidence limit for the dose-response which also defines the *lower effective dose* (LED).

The incremental life-time cancer risk IELCR is best understood by returning to an earlier quote from *Health Effects of Exposure to Low Levels of Ionizing Radiation* (BEIR, 1996).

> Since nearly 20% of all deaths in the United States result from cancer, the estimated number of cancers attributable to [chemical exposure or ionizing radiation] is only a small fraction of the total number that occur.

This means roughly 20,000 deaths per 100,000 result from cancer. Moreover, 16,000 deaths per 100,000 are cancers that arguably arise from environmental causes (Benigni et al., 2013). The response metric used to quantify risk from exposure to carcinogenic chemicals (or ionizing radiation) is the incremental (increased) life-time cancer risk IELCR.

The *benchmark dose method* defines a benchmark response BMR which is an incremental (typically 1–10%) increased adverse effect in the exposed population. The IELCR is a similar response metric.

10.4 EXPOSURE PATHWAY ASSESSMENT

Fig. 10.6 illustrates the elements of exposure pathway assessment. An exposed population receives a contaminant dose provided that all elements in the exposure pathway are present.

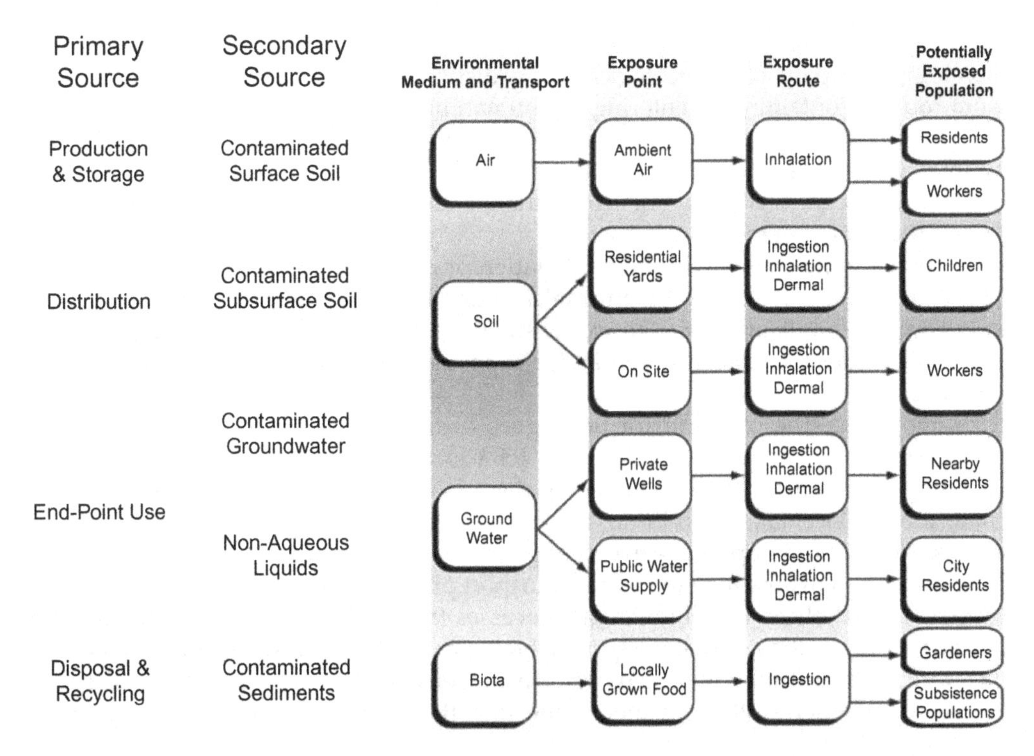

FIG. 10.6 A complete exposure pathway begins with a contamination source and a mode of environmental transport and reaches an exposed population or receptors through some exposure point—contaminated environmental media—and an exposure (uptake) route. *Reproduced with permission from ATSDR, 2005. Public Health Assessment Guidance Manual, ATSDR-208. U.S. Department of Health and Human Services, Public Health Service, Atlanta, GA.*

10.4.1 Receptors

The exposure pathway terminates with a specific human population—the *receptor* in risk assessment terminology—that receives a contaminant dose from the environment (Fig. 10.6). The exposed or potentially-exposed population is distinguished from the general population by some characteristic that brings this sub-population into contact with contaminated environmental media. For instance, urban air pollution affects ambient air throughout the urban environment, placing the entire residential urban population in contact with contaminant gases and particulates, while certain manufacturing processes pollute a work site, exposing only workers at that site. One can further distinguish high-risk populations: a particular age group in a residential urban population or workers performing specific tasks at a work site.

The *Exposure Factors Handbook* (USEPA, 2011) provides numerous examples of activity patterns that influence the type and duration of contact experienced by different sub-populations. Day et al. (2007), for instance, reported on the exposure pathway assessment at a factory producing copper-beryllium wire that distinguishes work area sub-populations.

10.4.2 Exposure Routes

Toxicologists employ pharmacological drug disposition models (Fig. 10.7) to represent exposure routes. Contaminants entering the human body ultimately elicit an adverse effect by acting on a target organ. The magnitude of the adverse effect is most closely linked to the contaminant concentration or dose at the target organ—the *biologically effective dose*. A complex internal pathway connects the point where the contaminant enters the body—the *exposure point*—and the target organ.

The exposure route begins with oral, inhalation or dermal intake that delivers a potential dose once the contaminant crosses an intake boundary—the lining of the sinus passages or lungs, various points along the gastrointestinal tract, or the skin.[5] The potential dose is typically lower than the concentration at the exposure point because intake is usually less than 100% efficient.

Following intake, the contaminant undergoes distribution throughout some portion of the human body. The distribution volume V_d (cf. expressions 10.8–10.10) and the possibility for clearance (i.e., elimination) reduce the applied dose that arrives at the biological barrier (vascular, renal, placental, blood-brain, etc.) that protects the organ from potentially toxic substances. Contaminant uptake is the passage of the compound across a biological barrier, resulting in an internal dose lowered by the transport properties of the barrier from the applied dose. Finally, metabolism and elimination processes (reverse transport across the biological barrier) further lower the internal dose to the biologically effective dose just mentioned.

Eukaryotes—organisms whose cells have a nucleus—rely on selective ion transporters to regulate uptake across the cell membrane into the cytoplasm. The selectivity of these

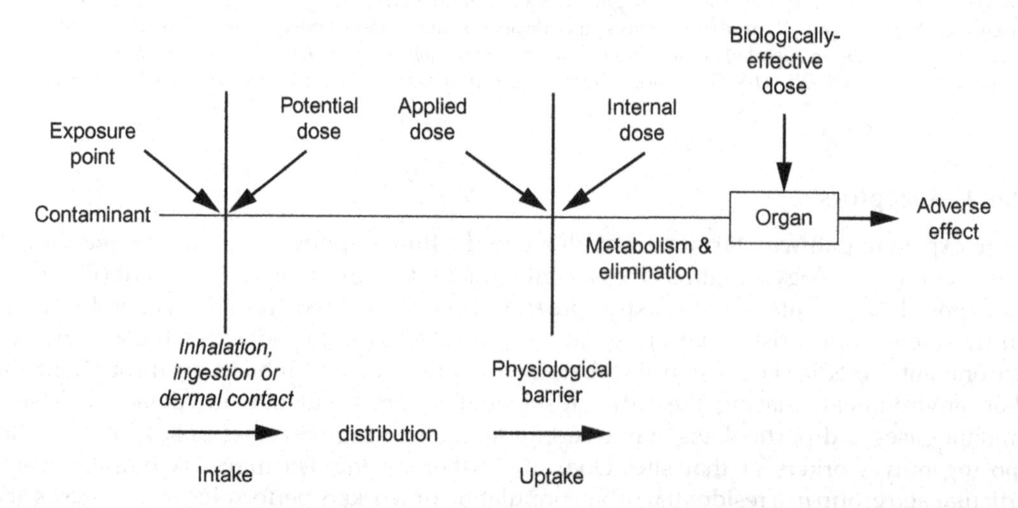

FIG. 10.7 The generalized exposure route model is adopted from pharmacology drug disposition models. *Reproduced with permission from USEPA, 2011. Exposure Factors Handbook, National Center for Environmental Assessment Report EPA/600/R-090/052F. U.S. Environmental Protection Agency.*

[5] Skin and mucous membranes are considered nonspecific barriers.

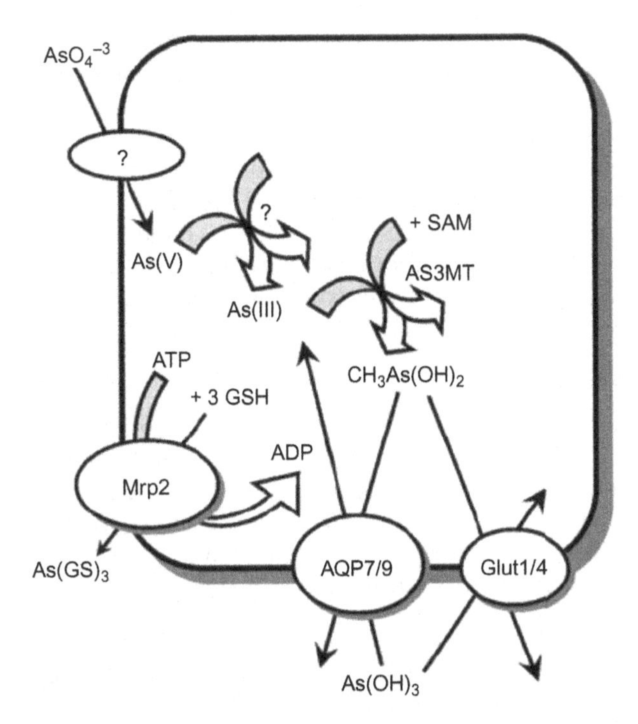

FIG. 10.8 The generalized cellular uptake and efflux pathways of arsenic in mammals include both the electrochemical reduction of As(V) to As(III) and transporters that export either the arsenic(III)-gluthathione complex $As(GS)_3$ or arsenous acid $As(OH)_3$. Arsenic uptake is believed to occur via phosphate ion transporters. *Reproduced with permission from Rosen, B.P., Liu, Z., 2009. Transport pathways for arsenic and selenium: a minireview. Environ. Int. 35, 512–515.*

ion transporters serves as a biological barrier to the uptake of inorganic toxic substances. Zinc homeostasis in mammalian cells is maintained by a combination of import and export transporters (Liuzzi and Cousins, 2004). There is no evidence that cadmium ions are exported from mammalian cells by any of the known zinc efflux transporters ZnT (Martelli et al., 2006), but cadmium export is possible by alternative routes: efflux by transporters of other divalent metal cations (e.g., Mn^{2+}) or export of cadmium complexes with ligands such as cysteine or glutathione.

Fig. 10.8 illustrates the cytoplasmic processes initiating the elimination of arsenate by mammalian cells; the intracellular reduction of As(V) followed by the methylation and export of As(III).

10.4.3 Exposure Points

Potential exposed populations come into contact with media containing contaminants, known as the exposure point (cf. Figs. 10.6 and 10.7). An exposure point can be a wide range of media (USEPA, 2011) ranging from ambient air, water, sediments, and soil to food and consumer products. Our primary concern here is with environmental media—air, water, sediments, and soil—that deliver contaminants to humans by inhalation, ingestion, and dermal contact (cf. Fig. 10.6).

Table 10.2 lists possible exposure points. Risk analysts distinguish exposure points from environmental media; an exposure point is the setting where a receptor encounters contaminated media. Each exposure point implies potential exposure routes and exposed populations, the latter depending on specific activities.

Establishing exposure points requires detailed knowledge of both the nature and extent of contamination through environmental testing and biological monitoring. Environmental testing analyzes all of the environmental media listed in Table 10.2 to identify and quantify potential contaminants. Biological monitoring collects and analyzes biological fluid and tissue samples from potential exposed populations to confirm and quantify exposure.

10.4.4 Fate and Transport

Continuing to work our way toward the contaminant source, the next component in the exposure pathway (cf. Fig. 10.6) is contaminant fate and transport in environmental media. This is environmental chemistry's primary focus and the principal topic of this book. Most hazardous substances undergo profound biological and chemical transformation upon entering soils or aquifers, the atmosphere, or water; often the transformations are quite complex and depend on prevailing conditions.

TABLE 10.2 Possible Exposure Points for Contaminated Environmental Media

Environmental Media	Exposure Points
Soil	Surface soil at work sites, residential sites, and parks; subsurface soil exposed during excavation or other disturbance of the surface; excavated and transported soil used as fill
Sediment	Submerged or exposed stream and lake sediment, sandbars, overbank flood deposits, and drainage ditches; excavated and transported sediments used as fill or the result of ditch, drainage channel, canal, and watercourses maintenance
Groundwater	Wells and springs supplying water for municipal, domestic, industrial, agricultural, and recreational purposes
Surface Water	Lakes, ponds, streams, and storm drainage supplying water for municipal, domestic, industrial, agricultural, and recreational purposes
Air	Ambient air downwind of a contaminated site, indoor air containing migrating soil gases—e.g., flammable (CH_4) and asphyxiating (CO_2) gas entering buildings on or adjacent to landfills
Biota	Plants (cultivated and gathered fruits and vegetables, plants used for medicinal purposes), animals (livestock, game, etc.), or other natural products that have contacted contaminated soil, sediment, waste materials, groundwater, surface water, or air
Other	Contaminated materials at commercial or industrial sites may provide a direct point of contact for on-site workers, visitors, or trespassers

Biological and chemical transformations acting in concert may degrade a hazardous substance, eliminating or reducing toxicity, or they can activate a substance, substantially enhancing toxicity. Examples of both are well known. Often biological and chemical transformations of inorganic substances that lower toxicity under prevailing conditions are reversed when chemical and biological conditions change over time and location.

Adsorption lowers toxicity by reducing biological availability (i.e., water solubility) and, in the process, leads to a substance's accumulation in soils, sediments, and aquifers. Contaminant storage and buildup in environmental solids contributes to legacy pollution, whose long-term risk is often difficult to assess.

Mobile environmental media—water (above or below ground), solvents, and gases (above and below ground)—disperse contaminants from their primary and secondary sources, expanding the area and volume of contaminated environmental media and increasing the opportunity for human exposure. Water percolating through soil, sediments, and aquifers transports contaminants slowly and over rather short distances below ground. Surface water flow and atmospheric currents disperse suspended particles rapidly and, in some cases, over considerable distances. The cost of environmental mitigation and the scope of the exposed population can increase dramatically with environmental transport.

Appendixes 10.A–10.C list chemical-specific and site-specific factors that influence the fate and transport of contaminants in the environment. Though lacking chemical and mechanistic detail, the tables in Appendixes 10.A–10.C give a broad perspective of the way risk analysts understand fate and transport processes.

10.4.5 Primary and Secondary Sources

Every hazardous substance that contaminates the environment has a history. The history begins with a production process that utilizes the compound of concern or that generates the compound as either a primary product or byproduct. The history terminates with the compound of concern entering a waste stream that may or may not involve recovery or recycling. The agriculture, mining, manufacturing, and chemical industries in the United States are undergoing a transformation designed to minimize the use and production of hazardous substances, but this transformation will not entirely eliminate the release of hazardous substances into the environment at production and storage facilities.

Opportunities for environmental pollution occur during the distribution and end-point use. Often distributors or end-users fail to properly store or handle products that are hazardous. Waste management, especially when recovery and recycling are costly or inefficient, creates further opportunities for environmental contamination during waste storage, transport, and disposal.

Hazardous substances released into the environment from any of the primary sources just mentioned will contaminate soils, surface water bodies, groundwater, sediments, and aquifers. Contaminated environmental media become secondary sources upon the removal of the primary source— dismantling production, storage or distribution facilities, termination of endpoint use, or closing of waste disposal sites—often leaving secondary sources as a continuing pollution legacy.

10.4.6 Exposure Pathway Assessment

The ultimate goal of exposure pathway assessment is identifying pathways of concern linking a contaminant source to an exposed population. A key factor in this process is exposure pathway elimination. If all components of an exposure pathway are present, then risk analysts identify this as a completed exposure pathway—a pathway of concern because a specific population is exposed via the pathway to a contamination source. If one or more elements of an exposure pathway are absent, then risk analysts identify this as an eliminated exposure pathway. A potential exposure pathway would be one where the presence or absence of one or more pathway elements is unverified. Box 10.4 illustrates this process with a real case.

BOX 10.4

EXPOSURE PATHWAY ELIMINATION

In 1974, over opposition by surrounding landowners, the Dane County (Wisconsin, USA) Zoning Board granted DeBeck permission to open and operate a landfill in an abandoned limestone quarry near Black Earth Creek. The Board did not require DeBeck to install a secure liner or monitoring wells. DeBeck operated the Refuse Hideaway landfill from 1974 through 1987 in the face of continued opposition by landowners whose properties bordered the landfill. At one point homeowners filed a lawsuit that temporarily closed the landfill for 15 months.

In March 1988, the Wisconsin Department of Natural Resources WDNR detected cleaning solvents from the landfill in two domestic wells and contaminated runoff in a drainage ditch that emptied into Black Earth Creek. The WDNR asked the Dane County Zoning Board to rescind the landfill permit issued to DeBeck. After threatening to file a lawsuit, the Wisconsin attorney general, Hanaway, reached an agreement with DeBeck to stop accepting waste by May 16, 1988, and to install a clay cap and methane vents at the landfill. DeBeck filed for bankruptcy a year later, and in October 1992, the US Environmental Protection Agency added Refuse Hideaway to the National Priorities List, designating it a Superfund site.

Table 10.3 lists the exposure pathway components and the exposure time frame for Refuse Hideaway. The exposure point (drinking water from private wells), the exposure route (ingestion of drinking water), and a potentially exposed population (landowners relying on private wells for drinking water) existed prior to the opening of Refuse Hideaway, but the exposure pathway is eliminated because a contaminant source and transport did not exist during that time frame.

The presence of a potential exposure pathway linking Refuse Hideaway to domestic drinking water wells motivated the 53 landowners in the vicinity to oppose the opening and operation of the landfill. It provided a basis for the court judgment temporarily closing the landfill and for early efforts by the Wisconsin attorney general and the WDNR to terminate operations. Legal efforts to close the landfill were unsuccessful until the detection of solvents in domestic wells completed the exposure pathway.

BOX 10.4 (*cont'd*)

TABLE 10.3 Exposure Pathway Analysis

	Exposure Time Frame		
Pathway Component	**Before 1974**	**1974–1987**	**1988–Present**
Contaminant source	No	Yes	Yes
Exposure point	Yes	Yes	Yes
Exposure route	Yes	Yes	Yes
Potentially exposed population	Yes	Yes	Yes
Fate and transport	No	Unknown	Yes
Pathway	Eliminated	Potential	Completed

10.4.7 Exposure Factors

Key parameters influencing intake dose and average daily dose estimates relate to the behavior and physical characteristics of each potential exposed population: the ingestion rate IR associated with a particular exposure route, body mass BM and exposure duration ED. Earlier we discussed the importance of probability distributions on statistical estimates of central tendency and variation in any population. Toxicologists studying exposure factors rely on statistical methods to estimate the most likely exposure MLE and the reasonable maximum exposure RME, based on the mode and the 95th percentile for the potential exposed population, respectively.

For example, the *Exposure Factors Handbook* (USEPA, 2011) distinguishes several potential exposed populations, each with distinctive IR and BM: adults and children, males and females, lactating and pregnant females, and highly active or sedentary individuals. Population studies also distinguish IR and ED for residents, industrial workers, commercial workers, agricultural workers, and recreational activity.

10.4.7.1 Ingestion and Inhalation Rates

Table 10.4 lists standard inhalation rates where the key variable is receptor behavioral characteristics. The resident adult RME daily inhalation rate of 15 $m^3\ day^{-1}$ is less than the RME inhalation rate for workers because the relative activity levels of the two groups.

Soil ingestion includes actual "outdoor" soil and indoor dust. Soil ingestion covers both purposeful soil consumption (a behavior known as *pica*) and incidental ingestion through hand-to-mouth contact or particulates entering the mouth or nostrils while breathing. Children younger than 6 years old ingest significantly more soil than other age groups, another population behavior characteristic.

TABLE 10.4 Ingestion and Inhalation Rates for Selected Exposure Routes and Populations

Intake Route	Reasonable Maximum Intake Rate
Water ingestion	Child: 0.6 $dm^3 day^{-1}$
	Adult: 2 dm^3 day^{-1}
Soil ingestion	Child: 200 mg day^{-1}
	Adult: 100 mg day^{-1}
Air inhalation	Child: 10 m^3 day^{-1}
	Resident: adult 15 m^3 day^{-1}
	Worker: 20 m^3 day^{-1}
Produce ingestion	Fruit: 42 g day^{-1}
	Vegetables: 80 g day^{-1}
Fish ingestion	132 g day^{-1}

Child: *1–6 years old.*
Adult: *more than 6 years old.*
Produce Ingestion: *subsistence, homegrown.*
Fish Ingestion: *subsistence, wild-caught.*

Ingestion rates for certain foods identify sub-populations or food sources associated with exposure pathways of concern: homegrown fruits and vegetables grown in contaminated soil, a sub-population that subsists on fish caught in contaminated lakes or streams, or nursing infants subsisting on breast milk. Intake equations for water, soil, and food ingestion and air inhalation, along with examples of each, appear in Appendixes 10.D–10.G.

10.4.7.2 Exposure Duration

The resident ED assumes this population is absent from their residential water supply about 15 days per year. Potable water ingestion in the workplace is assumed to be 50% of the total water IR, a population behavior characteristic. Workplace exposure must account for annual work days and distinguish between populations based on their work activity.

10.4.7.3 Averaging Time

OSWER Directive 9285.6-03 (USEPA, 1991) provides the following guidance for setting the *averaging time* for intake equations.

> Averaging time (AT) for exposure to noncarcinogenic compounds is always equal to [exposure duration] ED, whereas, for carcinogens, a 70-year AT is still used in order to compare to Agency slope factors typically based on that value.

$$\text{Average Daily Dose} = \frac{\text{Intake Dose}}{\text{BM} \cdot \text{AT}} \tag{10.13}$$

$$\text{Intake Dose} = \text{C} \cdot \text{IR} \cdot \text{ED} \tag{10.14}$$

Noncarcinogens

$$\mathrm{AT} = \mathrm{ED} \tag{10.15}$$

The cumulative mutation risk increases over time, a result of the correlation between cancer mortality and age. The daily intake of carcinogens are prorated differently than noncarcinogens.

> Carcinogen risk assessment models have generally been based on the premise that risk is proportional to cumulative lifetime dose. For lifetime human exposure scenarios, therefore, the exposure metric used for carcinogenic risk assessment has been the *lifetime average daily dose*. **USEPA (2005)**

Carcinogens

$$\mathrm{AT} = (70\ \mathrm{years}) \cdot \left(365\ \mathrm{days\ years}^{-1}\right) = 2.56 \times 10^{4}\ \mathrm{days} \tag{10.16}$$

10.5 INTAKE ESTIMATES

The general exposure route model (see Fig. 10.7) begins with intake of contaminated media. This section discusses intake estimates for specific exposure routes. Intake equations (see Appendixes 10.D–10.G) estimate the dose delivered to an exposed population. The steady-state body burdens computed for noncarcinogens are called the *average daily dose* (ADD) (Example 10.1 and 10.2).

EXAMPLE 10.1

Example Permalink

http://soilenvirochem.net/2pp6zh

Estimate the *average daily dose ADD* for a child exposed to toluene (CAS Registry Number 108-88-3) in residential soil. The USEPA lists toluene as a noncarcinogen (USEPA, 2016b).

The resident child ingests soil containing 10 mg kg^{-1} toluene. The child's age is irrelevant because exposure duration is not considered (i.e., $AT = ED$).

$$ADD = \frac{C_S \cdot CF \cdot IR \cdot ED}{BM \cdot AT}$$

$$ADD = \frac{\left(10\ \mathrm{mg\,kg}^{-1}\right) \cdot \left(10^{-6}\ \mathrm{kg\,mg}^{-1}\right) \cdot \left(200\ \mathrm{mg\,day}^{-1}\right)}{15\ \mathrm{kg}}$$

$$ADD = 1.33 \times 10^{-4}\ \mathrm{mg\,kg\,day}^{-1}$$

EXAMPLE 10.2

Example Permalink

http://soilenvirochem.net/18temM

Estimate the *lifetime average daily dose* (LADD) for an adult exposed to benzene (CAS Registry Number 71-43-2) in drinking water. The USEPA lists benzene as carcinogen (USEPA, 2016a).

The receptor is a 30-year-old resident adult ingesting water containing 0.05 mg dm^{-3} benzene.

$$\mathrm{LADD} = \frac{C_W \cdot \mathrm{IR} \cdot \mathrm{ED}}{\mathrm{BM} \cdot \mathrm{AT}}$$

$$\mathrm{ED} = \left(350\ \mathrm{day\ year^{-1}} \cdot 30\ \mathrm{year}\right) = 1.05 \times 10^4\ \mathrm{days}$$

$$\mathrm{LADD} = \frac{\left(0.05\ \mathrm{mg\ dm^{-3}}\right) \cdot \left(2\ \mathrm{dm^3\ day^{-1}}\right) \cdot \left(1.05 \times 10^4\ \mathrm{days}\right)}{(70\ \mathrm{kg}) \cdot (2.56 \times 10^4\ \mathrm{day})}$$

$$\mathrm{LADD} = 5.87 \times 10^{-4}\ \mathrm{mg\ kg\ day^{-1}}$$

10.6 RISK CHARACTERIZATION

Armed with equations for estimating contaminant intake based on concentrations at the exposure point (Appendixes 10.D–10.G), we are now prepared to quantify risk. Risk is a function of intake and toxicity. Quantifying risk from exposure to a carcinogen is based on a definite statistical probability: the incremental excess lifetime cancer risk IELCR. The risk factor for a noncarcinogen, on the other hand, is not directly linked to the probability of an adverse effect on a population.

10.6.1 Target Cancer Risk

The IELCR is the probability of developing cancer as the result of exposure to a specific carcinogen and appears as an incremental increase in cancer cases in the exposed population over what would occur in the absence of exposure. An acceptable IELCR can vary but is often 1–10%, depending on variability in the population study.[6] The probability of developing cancer, based on the number of cancer cases in the population—regardless of age—is the lifetime cancer risk. Carcinogen toxicity is quantified by the cancer slope factor SF of each

[6] Variability in any population study of carcinogenesis ultimately determines the minimal IELCR detectable with statistical confidence. A population study with low variability permits IELCR detection as low as 1% with reasonable statistical confidence. Inherent variability typically limits IELCR detection with reasonable statistical confidence to 5% or 10%. The corresponding LADD is a toxicological end-point dose symbolized, respectively, as ED_{01}, ED_{05}, or ED_{10}.

substance (expression 10.17), whose units are the inverse of the lifetime average daily dose LADD because the ratio is a probability (i.e., unitless) (Example 10.3).

$$\text{Risk} = \text{Dose} \cdot \text{Toxicity}$$
$$\text{IELCR} = \text{LADD} \cdot \text{SF} \tag{10.17}$$

EXAMPLE 10.3

Example Permalink

http://soilenvirochem.net/Fc4oDC

Estimate the IELCR for an adult exposed to the carcinogen benzene (CAS Registry Number 71-43-2) in drinking water. The USEPA lists benzene as carcinogen (USEPA, 2016a).

The receptor and exposure in this case is identical to those described in Example 10.2.

$$\text{LADD} = 5.87 \times 10^{-4}\ \text{mg kg day}^{-1}$$

The US Environmental Protection Agency *Integrated Risk Information System* quotes a range for the benzene cancer slope factor: $1.5 \times 10^{-2}\ \left(\text{mg kg}^{-1}\ \text{day}^{-1}\right)^{-1} \leq \text{SF} \leq 5.5 \times 10^{-2}\ \left(\text{mg kg}^{-1}\ \text{day}^{-1}\right)^{-1}$.

The IELCR for this example is simply the product of the benzene LADD and the benzene cancer slope factor SF (expression 10.17).

$$\text{IELCR} = 5.87 \times 10^{-4}\ \left(\text{mg kg day}^{-1}\right) \cdot 5.5 \times 10^{-2}\ (\text{mg kg day})^{-1}$$
$$\text{IELCR} = 3.2 \times 10^{-5}$$

The IELCR resulting from benzene exposure in this example amounts to about 3 excess cancer cases per one-hundred thousand. The stochastic cancer rate from all causes is 20,000 cancer cases per one-hundred thousand in the United States.

10.6.2 Cumulative Target Risk

The *cumulative target risk* (CTR) (expression 10.18) estimates the IELCR resulting from exposure to two or more carcinogens.

$$\text{CTR} = \sum_i (\text{IELCR}_i) \tag{10.18}$$

Table 10.5 lists two known carcinogens, each with a different lifetime average daily dose LADD based on exposure to untreated well water samples taken near the Refuse Hideaway landfill in 1987 (benzene 24 $\mu g \cdot dm^{-3}$; vinyl chloride 20 $\mu g \cdot dm^{-3}$). The exposure duration is calculated based on continuous exposure from 1987 to 2010. The cancer SF applies to all forms of cancer. The cumulative target risk CTR is the sum of the IELCR values listed in column 4 of Table 10.5: $\text{CTR} = 1.9 \times 10^{-4}$.

TABLE 10.5 Cumulative Target Risk for Exposure to Benzene (CAS Registry Number 71-43-2) and Vinyl Chloride (CAS Registry Number 75-01-4)

Carcinogen	LADD ($mg\ kg^{-1}\ day^{-1}$)	SF$\left(\left(mg\ kg^{-1}\ day^{-1}\right)^{-1}\right)$	IELCR
Benzene	2.3×10^{-4}	5.5×10^{-2}	1.3×10^{-5}
Vinyl chloride	2.5×10^{-4}	7.2×10^{-1}	1.8×10^{-4}

The excess cancer risk resulting from vinyl chloride exposure is predicted to be 18 cases per one-hundred thousand, while about 1 excess case result from benzene exposure in drinking water contaminated by Refuse Hideaway landfill, resulting in a CTR of about 19 excess cancer cases per one-hundred thousand in the exposed population.

Although tolerable risk is influenced by social, economic and political factors (cf. Fig. 10.1), tolerable risk also depends on the statistical uncertainty in incremental excess lifetime cancer risk estimates.

10.6.3 Hazard Quotient

The *hazard quotient* (HQ)—the risk metric for noncarcinogens—relates the dose delivered at the exposure point ADD to the tolerable dose: the RfD (expression 10.19).

$$HQ \equiv \frac{ADD}{RfD} = ADD \cdot (RfD)^{-1} \tag{10.19}$$

Expressions for HQ and IELCR (expression 10.17) may initially appear different. Notice, however, the units for SF are the inverse of the units for daily dose which is effectively the same as multiplying ADD by the inverse of RfD (expression 10.19)

An acceptable HQ is less than 1, meaning the ADD is less than the tolerable dose RfD (Example 10.4).

EXAMPLE 10.4

Example Permalink

http://soilenvirochem.net/KBlddL

Estimate the HQ for a child exposed to soil contaminated by toluene (CAS Registry Number 108-88-3). The USEPA lists toluene as noncarcinogen (USEPA, 2016b).

The receptor and exposure in this case is identical to those described in Example 10.1.

$$ADD = 1.33 \times 10^{-4}\ mg\,kg\,day^{-1}$$

The US Environmental Protection Agency *Integrated Risk Information System* quotes a tolerable oral dose for chronic toluene exposure: RfD = $0.08\ mg\,kg^{-1}\ day^{-1}$.

The HQ for this example is simply the toluene ADD divided by the toluene oral reference dose RfD (expression 10.19).

$$HQ = \frac{1.33 \times 10^{-4}\ \text{mg kg}^{-1}\ \text{day}^{-1}}{8.0 \times 10^{-2}\ \text{mg kg}^{-1}\ \text{day}^{-1}}$$

$$HQ = 1.7 \times 10^{-3} \ll 1$$

10.6.4 Cumulative Risk: Hazard Index

Simultaneous exposure at sub-threshold levels to several toxicants can result in adverse health effects. The *hazard index* (HI) (expression 10.20) estimates the cumulative risk from exposure to multiple noncarcinogens. Acceptable risk is defined as $HI < 1$.

$$HI \equiv \sum_i (HQ_i) \tag{10.20}$$

Table 10.6 lists the ADD for three noncarcinogenic elements assuming water consumption by a child. The concentration of each element is at the 75th percentile for US groundwater (Newcomb and Rimstidt, 2002). Table 10.6 lists the HQ for each substance in column 4. The hazard index exceeds a tolerable value: $HI = 1.18 > 1$.

10.7 EXPOSURE MITIGATION

Federal and state agencies employ several practices to mitigate exposure. Mitigation typically involves specific actions taken to eliminate exposure pathways and to reduce risk to acceptable levels at exposure points. The following example illustrates actions taken by the Wisconsin Department of Natural Resources to mitigate exposure from contamination at Refuse Hideaway. The actions represent the selected remedy announced in the official Record of Decision (USEPA, 1995a,b) for Refuse Hideaway landfill.

The Source Control actions in Table 10.7 address contamination at the source of the exposure pathway and are designed to restrict access, eliminate primary, and secondary contamination sources at the site, and restrict contaminant transport off-site. The Groundwater Treatment actions focus on off-site mitigation, primarily treating the groundwater plume to prevent

TABLE 10.6 Hazard Index for the Average Daily Dose of a Child Ingesting Water Containing Cadmium, Arsenic and Lead at the 75th Percentile for US Groundwater

Carcinogen	ADD ($mg\,kg^{-1}\,day^{-1}$)	RfD ($mg\,kg^{-1}\,day^{-1}$)	HQ
Cadmium	3×10^{-4}	1×10^{-3}	0.30
Arsenic	2.4×10^{-4}	3×10^{-4}	0.80
Lead	4×10^{-5}	5×10^{-4}	0.08

TABLE 10.7 Selected Remedies for Risk Mitigation at Refuse Hideaway Landfill

Source control	Deed restrictions and zoning modifications
	Warning signs posted around the perimeter of the property
	Maintenance of the existing single barrier (clay) cap, vegetation, and surface runoff controls
	Operation and maintenance of the existing landfill gas extraction and destruction system and leachate extraction and off-site treatment and disposal system
	Groundwater monitoring of selected monitoring wells and private home wells
Groundwater treatment	Extraction of the most highly contaminated groundwater (greater than 200 ppb total [volatile organic compounds]) in the vicinity of the landfill and treatment of groundwater to meet applicable groundwater discharge standards
	Injection of the treated water into the aquifer up-gradient of the landfill to stimulate in situ biodegradation of degradable components of the contamination
	Monitoring and evaluating the effectiveness of the groundwater extraction, treatment, and re-injection system in achieving progress toward cleanup standards
Water supply	Supply a point-of-entry treatment system for any private well exhibiting contaminants originating at the Refuse Hideaway landfill with concentrations exceeding NR 140 Enforcement Standards (Wis. Stat. §35.93, NR 140, 2015) or that are believed by the [Wisconsin Department of Natural Resources] and [USEPA] to be imminently at risk for exceeding those standards
	Construct a community water supply well if the number of homes requiring replacement water supplies makes [construction of a community well] cost effective

further contaminant migration off-site. The Water Supply actions address the exposure point: domestic wells.

Mitigation also relies on setting and enforcing standards designed to limit risk in the exposed population. The selected remedies at Refuse Hideaway landfill (Table 10.7) refer to standards twice: extraction of groundwater when volatile organic compounds are "greater than 200 ppb" and installation of treatment systems for any private well "with concentrations exceeding NR 140 Enforcement Standards." The role of risk assessment in setting enforcement standards is the topic of the following section.

10.8 RISK-BASED SCREENING LEVELS

You will encounter numerous acronyms used to identify risk-based standards; a few appear in Table 10.8. *Risk-based screening levels*, associated with a particular HQ or IELCR,

TABLE 10.8 Acronyms Used to Identify Air, Water, Soil, and Food Standards

Acronym	Explanation
MCL	A maximum contaminant level subject to enforcement
MCLG	A maximum contaminant level goal that serves as a nonenforceable target for remediation
MOE	A margin of exposure (*NOAEL* ÷ *ADD*) related to HQ (cf. expression 10.19)

are not enforceable standards but merely guidelines for setting enforceable standards. *Risk-based screening levels* would appear in the box labeled *Regulatory Options* in the Federal Risk Assessment Paradigm (cf. Fig. 10.1). The setting of a nonenforceable *maximum contaminant level goal* (MCLG) or an enforceable *maximum contaminant levels* (MCL) occurs at the level of *Agency Action*, representing a regulatory consensus designed to provide acceptable protection of public health while taking into account attendant economic, social, and political constraints.

A *risk-based screening level* always relates to a particular exposure route because it estimates the contaminant concentration at the exposure point associated with a quantifiable risk. Example 10.5 estimates the *risk-based screening level* for a carcinogen dissolved in water, the exposure point being ground or surface water and the exposure route water ingestion. Expression (10.21) is the LADD equation for water ingestion. Expression (10.22) quantifies LADD using the target IELCR.

Oral Intake Estimate

$$\text{LADD} = \frac{C_W \cdot \text{IR} \cdot \text{ED}}{\text{BM} \cdot \text{AT}} \tag{10.21}$$

Risk Quantification

$$\text{LADD} = \frac{\text{Target IELCR}}{\text{SF}} \tag{10.22}$$

The *risk-based screening level* for a carcinogen dissolved in water C_W^{RB} is found by equating the two preceding expressions and solving for the concentration in water.

$$C_W^{RB} = \left(\frac{\text{Target IELCR}}{\text{SF}}\right) \cdot \left(\frac{\text{BM} \cdot \text{AT}}{\text{IR} \cdot \text{ED}}\right) \tag{10.23}$$

EXAMPLE 10.5

Example Permalink

http://soilenvirochem.net/0Ewye6

Compute the risk-based screening level for the water ingestion of benzene by adults assuming lifetime exposure. Use expression (10.23) to compute the risk-based benzene concentration C_W^{RB}.

The target IELCR for lifetime exposure assumes an adult population that drinks benzene-containing water for 70 years. In this example a conservative target risk is selected: IELCR $= 10^{-5}$.

The cancer slope factor for benzene is $SF = 5.5 \times 10^{-2}\ \left(\text{mg kg}^{-1}\ \text{day}^{-1}\right)^{-1}$.

$$C_{w}^{RB} = \left(\frac{10^{-5}}{5.5 \times 10^{-2}\ \mathrm{mg\,kg^{-1}\,day^{-1}}}\right) \cdot \left(\frac{70\ \mathrm{kg} \cdot 2.56 \times 10^{4}\ \mathrm{day}}{2\ \mathrm{dm^{3}\,day^{-1}} \cdot 2.56 \times 10^{4}\ \mathrm{day}}\right)$$

$$C_{w}^{RB} = 6.4 \times 10^{-3}\ \mathrm{mg\,dm^{-3}}$$

The State of Wisconsin has adopted an enforcement standard (Wis. Stat. §35.93, NR 140, 2015) for benzene dissolved in groundwater of 5 parts per billion. The IELCR for lifetime adult exposure to MCL = 5×10^{-3} mg dm^{-3} benzene is close to the target risk used in this example.

Example 10.6 estimates the *risk-based screening level* for a noncarcinogen dissolved in water, the exposure point being ground or surface water and the exposure route water ingestion. Expression (10.24) is the ADD equation for water ingestion. Expression (10.25) quantifies ADD using the a hazard quotient HQ of unity.

Oral Intake Estimate

$$\mathrm{ADD} = \frac{C_{W} \cdot \mathrm{IR} \cdot \mathrm{ED}}{\mathrm{BM} \cdot \mathrm{AT}} = \frac{C_{W} \cdot \mathrm{IR}}{\mathrm{BM}} \tag{10.24}$$

Risk Quantification

$$\mathrm{ADD} = \mathrm{HQ} \cdot \mathrm{RfD} = \mathrm{RfD} \tag{10.25}$$

The *risk-based screening level* for a noncarcinogen dissolved in water C_{w}^{RB} is found by equating the two preceding expressions and solving for the concentration in water.

$$C_{w}^{RB} = \mathrm{RfD} \cdot \left(\frac{\mathrm{BM}}{\mathrm{IR}}\right) \tag{10.26}$$

EXAMPLE 10.6

Example Permalink

http://soilenvirochem.net/s6qp29

Compute the risk-based screening level for the water ingestion of toluene by adults. Use expression (10.26) to compute the risk-based toluene concentration C_{w}^{RB}.

The target risk for a noncarcinogen such as toluene yields a hazard quotient HQ of unity (expression 10.19). The oral reference dose for toluene is RfD = 8×10^{-2} mg kg^{-1} day^{-1}.

$$C_{w}^{RB} = \left(8 \times 10^{-2}\ \mathrm{mg\,kg^{-1}\,day^{-1}}\right) \cdot \left(\frac{70\ \mathrm{kg}}{2\ \mathrm{dm^{3}\,day^{-1}}}\right)$$

$$C_{w}^{RB} = 2.8\ \mathrm{mg\,dm^{-3}}$$

The State of Wisconsin has adopted an enforcement standard (Wis. Stat. §35.93, NR 140, 2015) for toluene dissolved in groundwater of MCL = 0.8 mg dm^{-3}.

10.9 ECOLOGICAL RISK ASSESSMENT

The USEPA issued OSWER Directive 9285.7-55 (USEPA, 2003) to provide guidance to the Agency, as well as other state and federal agencies, involved in ecological risk assessment. The focus of ecological risk assessment is the impact of ecological toxicity on plants and animals at contaminated sites not human health risk assessment.

OSWER Directive 9285.7-55 (USEPA, 2003) outlines the methods and models for setting risk-based soil screening levels that are "protective of the terrestrial environment." Identification of *contaminants of potential concern* (COPCs) is an essential part of any site evaluation leading to toxicity risk analysis. Ecological soil screening levels identify COPCs for in-depth ecological risk assessment.

The following discussion will reveal numerous parallels between human-health risk assessment and ecological risk assessment. Perhaps the impost important distinction being the abandonment of cancer as an adverse effect in animal receptors.

10.9.1 Wildlife Risk Model

The wildlife risk model adopted by OSWER Directive 9285.7-55 (USEPA, 2003) assesses ecotoxicity using a hazard quotient similar to expression (10.19) for human health risk. The biota hazard quotient employs a biota-specific *toxic reference value* (TRV) for each contaminant.

$$\mathrm{HQ} = \frac{\text{Exposure Dose}}{\text{Toxic Reference Value}}$$

By analogy with expression (10.20), the wildlife risk model also defines a overall hazard index for exposure to two or more contaminants (expression 10.27) where ADD_j represents the average daily dose for exposure to contaminant j.

$$\mathrm{HI} = \sum_j \left(\mathrm{HQ}_j\right) = \sum_j \left(\frac{\mathrm{ADD}_j}{\mathrm{TRV}_j}\right) \leq 1 \tag{10.27}$$

10.9.1.1 Biota Toxicity Reference Value TRV

Ecological dose-response assessment begins with identification of the maximum acceptable toxicant concentration MATC, which is the threshold for deterministic adverse effects from exposure (cf. Fig. 10.3). The MATC threshold lies above an empirical no observable adverse effect concentration NOAEC and below an empirical lowest observable adverse effect concentration LOAEC. Often the MATC is computed as the geometric mean of NOAEC and LOAEC (expression 10.28).

$$\mathrm{MATC} = (\mathrm{NOAEC} \cdot \mathrm{LOAEC})^{1/2} \tag{10.28}$$

The USEPA *PBT Profiler* (http://www.pbtprofiler.net/) quotes the *fish chronic value* (ChV) as the aquatic toxicity metric, computed using expression (10.28). *Fish ChV* quantifies larval survival and growth of Fathead minnow. Table 10.9 lists test species used to quantify chronic (toxicity) values ChV for aquatic organisms.

OSWER Directive 9285.7-55 (USEPA, 2003) assesses ecotoxicity arising from the exposure of terrestrial biota to contaminated soil. Table 10.9 lists test species used to quantify the terrestrial ecotoxicity metric: toxicity reference value TRV. The ChV and TRV are the ecological equivalent of the reference dose RfD or lower benchmark dose LBMD in human health risk assessment.

10.9.1.2 Soil-to-Biota Bioaccumulation Estimates

Terrestrial plants are assigned a TRV (cf. Table 10.9) because they serve as food source for soil invertebrates, mammalian herbivores, and avian grainivors. For terrestrial plants soil-to-foliage and soil-to-seed bioaccumulation becomes an important exposure route. Fig. 10.9 plots the cadmium soil-to-foliage and soil-to-seed bioaccumulation data from OSWER Directive 9285.7-55 (USEPA, 2003).

The central estimate from data such as those plotted in Fig. 10.9 yield a series of soil-to-plant bioaccumulation expressions, one for each contaminant. For instance, the cadmium

TABLE 10.9 Aquatic Chronic Toxicity Values and Terrestrial Toxic Reference Values

Chronic Toxicity Value ChV	Test Specie
Fish	Fathead minnow, *Pimephales promelas*
Invertebrate	Daphnia, *Daphnia magna*
Aquatic plant	Green algae, *Raphidocelis subcapitata*
Toxic Reference Value TRV	**Test Specie**
Plant	Various[a]
Invertebrate	Earthworm, *Eisenia fetida*
Mammalian herbivore	Meadow vole, *Microtus pennsylvanicus*
Mammalian insectivore	Short-tailed shrew, *Blarina brevicauda*
Mammalian carnivore	Long-tailed weasel, *Mustela frenata*
Avian grainivore	Mourning dove *Zenaida macroura*
Avian insectivore	American woodcock, *Scolopax minor*
Avian carnivore	Red-tailed hawk, *Buteo jamaicensis*

[a] *ASTM (2014) Method E1963-09.*

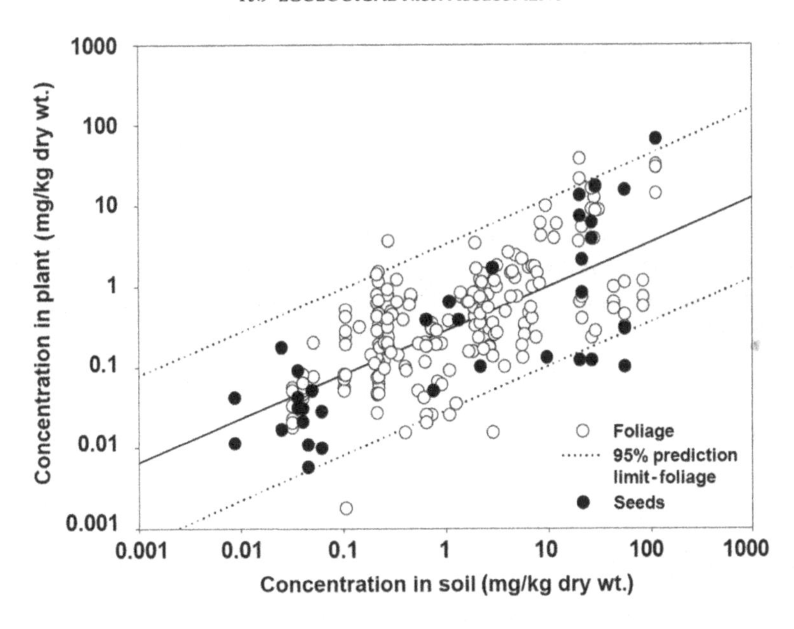

FIG. 10.9 Soil-to-foilage (*open circles*) and soil-to-seed (*filled circles*) bioaccumulation for cadmium. *USEPA, 2003. Guidance for Developing Ecological Soil Screening Levels, OSWER Directive 9285.7-55. U.S. Environmental Protection Agency, Washington.*

concentration in plants derived from soil uptake $B_{plant,\,\mathrm{Cd}}$ is given by the following expression where S_{Cd} represents the soil cadmium concentration.

$$\ln\left(B_{plant,\,\mathrm{Cd}}\right) = 0.546 \cdot \ln\left(S_{\mathrm{Cd}}\right) - 0.475$$

Expression (10.29) is the general soil-to-biota expression where i represents one of three biota (plant, invertebrate or animal) and j represents one of twenty four (24) contaminants. A soil-to-animal bioaccumulation expression is required because oral soil ingestion is a significant exposure route for all terrestrial animals.

$$\ln\left(B_{i,j}\right) = a_i \cdot \ln\left(S_j\right) + b_i \tag{10.29}$$

10.9.1.3 Biota Intake Estimates

Contaminated soil is the sole exposure point for plants and soil invertebrates. Terrestrial mammals and birds have two exposure points: soil and food source. Table 10.9 distinguishes three food-source biota for terrestrial animals: plants, invertebrates, and animals. Each terrestrial animal test specie requires a soil-to-biota bioaccumulation expression (10.29) for its principal food-source exposure point.

Expression (10.30) gives the exposure dose ADD_j for an animal consuming food-source biota i containing contaminant j.[7]

$$ADD_j = (S_j \cdot P_s \cdot FIR) + \sum_i (B_{i,j} \cdot P_i \cdot FIR) \tag{10.30}$$

The daily food intake rate is FIR is subdivided into the fraction of the daily intake rate as direct soil consumption P_s and the fraction of the daily intake P_i for the principal food-source biota i. Table 10.10 lists the food intake rate FIR and the soil ingestion parameter P_s appearing in expression (10.30).

10.9.2 Ecological Soil Screening Levels

The risk-based ecological soil screening level S_j^{RB} for contaminant j is defined by expression (10.31). Expression (10.31) is derived by substituting expression (10.29) into expression (10.30) then substituting expression (10.30) into expression (10.27) for each contaminant j.

$$\frac{\left(S_j^{RB} \cdot P_s \cdot FIR\right) + \sum_i \left(B_{i,j}^{RB} \cdot P_i \cdot FIR\right)}{TRV_j} \equiv 1 \tag{10.31}$$

Expression (10.31) is not solved analytically. The ecological soil screening level S_j^{RB} is found by adjusting S_j^{RB} to satisfy the condition given by (10.31). The reader should note: ecological

TABLE 10.10 Parameterization of the Wildlife Exposure Model

Receptor Specie	Food Ingestion Rate ($kg_{dw}\ kg_{bw}^{-1}\ day^{-1}$)	Soil Ingestion (P_s)	Assumed Diet
Meadow vole	0.0875	0.032	100% plant foliage
Short-tailed shrew	0.209	0.030	100% earthworms
Long-tailed weasel	0.130	0.043	100% small mammals
Mourning dove	0.190	0.139	100% plant seeds
American woodcock	0.164	0.032	100% earthworms
Red-tailed hawk	0.0353	0.057	100% small mammals[a]

[a] *Small mammals diet 100% earthworms/*

[7] Expression (10.30) is a simplified form of the wildlife risk model that omits factors that are typically set equal to unity: the absorbed fraction of contaminant i from soil $AF_{S,j}$ and the absorbed fraction of contaminant i from food-source biota j $AF_{i,j}$.

soil screening level S_j^{RB} vary from site to site because the hazard index HI for exposure to multiple contaminants at each site must satisfy expression (10.27).

10.10 SUMMARY

Chemical contaminant risk assessment in the United States follows a paradigm (cf. Fig. 10.1) that begins with hazard identification and dose-response assessment and reaches its completion with exposure assessment and risk characterization.

Risk assessment results provide a scientific basis for risk management and mitigation. Dose-response assessment follows separate tracks for carcinogens and noncarcinogens, the principal difference being the absence of a threshold dose and response based on cumulative rather than steady-state dose when predicting carcinogenesis. Distinctions between carcinogens and noncarcinogens also determine the method for low-dose extrapolation and, ultimately, risk characterization.

A key element in all risk assessment studies of contaminants in the environment is exposure assessment, more specifically exposure pathway assessment. Exposure pathway assessment establishes links between the contamination source and potential exposed populations, evaluating all potential transport pathways, exposure points, and exposure routes that could connect a potential exposed population to a contamination source.

Risk characterization quantifies risk in exposed populations where an exposure pathway is completed. Risk characterization employs intake equations for each potential exposure route to estimate the potential dose. Risk characterization allows the setting of risk-based screening levels for each contaminant and exposure route. The enforcement standards—maximum contaminant levels MCLs—adopted by federal and state environmental agencies derive from risk-based screening levels. The environmental chemist is a key participant in the risk assessment process, whether through research studies of fate and transport or collecting field data during environmental monitoring studies.

APPENDICES

10.A FACTORS AFFECTING CONTAMINANT TRANSPORT BY SURFACE WATER

Tables 10.A.1 and 10.A.2 and lists the transport mechanisms and the chemical-specific and site-specific factors affecting contaminant fate and transport by surface water. It is excerpted from Appendix E of *Public Health Assessment Guidance Manual* (ATSDR, 2005).

TABLE 10.A.1 Chemical- and Site-Specific Factors Affecting Contaminant Transport: Surface Water

	Factors Affecting Transport	
Transport Mechanism	**Chemical-Specific**	**Site-Specific**
Overland flow	• Water solubility • Soil organic-carbon partition constant[a]	• Precipitation • Infiltration rate • Topography • Vegetative cover and land use • Soil/sediment type and chemistry • Use as water supply intake areas • Location, width, and depth of channel; velocity; dilution factors; direction of flow • Floodplains • Point and nonpoint source discharge areas
Volatilization	• Water solubility • Vapor pressure • Henry's Law gas-solubility constant[b]	• Climatic conditions • Surface area • Contaminant concentration
Hydrologic connection between surface water and groundwater	• Henry's Law gas-solubility constant	• Groundwater/surface water recharge and discharge • Stream bed permeability • Soil type and chemistry • Surface geology (especially karst)

[a] *Soil organic-carbon partition constant:* $K_{oc/w}$.
[b] *Soil Henry's Law gas-solubility constant:* H_X^{cp}.

TABLE 10.A.2 Chemical- and Site-Specific Factors Affecting Contaminant Transport: Surface Water (continued)

	Factors Affecting Transport	
Transport Mechanism	**Chemical-Specific**	**Site-Specific**
Adsorption to suspended soil particles and sedimentation	• Water solubility • Soil organic-carbon partition constant[a] • Octanol-water partition constant[b]	• Sediment-particle size and density • Geochemistry (soils and sediments) • Organic carbon content (soils and sediments)
Biologic uptake	• Octanol-water partition constant • Bioconcentration factor[c]	• Chemical concentration (water and soil) • Presence of fish, plants, and other animals

[a] *Soil organic-carbon partition constant:* $K_{oc/w}$.

[b] *Octanol-water partition constant:* $K_{o/w}$.

[c] *Bioconcentration factor:* $BCF = C_{biota}/C_{water}$.

10.B FACTORS AFFECTING CONTAMINANT TRANSPORT BY GROUNDWATER

Tables 10.B.1 and 10.B.1 lists the transport mechanisms and the chemical-specific and site-specific factors affecting contaminant fate and transport by groundwater. It is excerpted from Appendix E of *Public Health Assessment Guidance Manual* (ATSDR, 2005).

TABLE 10.B.1 Chemical- and Site-Specific Factors Affecting Contaminant Transport: Groundwater

	Factors Affecting Transport	
Transport Mechanism	**Chemical-Specific**	**Site-Specific**
Movement within and across aquifers or discharge to surface water	• Density (relative to water) • Water solubility • Soil organic-carbon partition constant[a]	• Hydrogeology • Precipitation • Infiltration rate • Porosity • Hydraulic conductivity • Groundwater flow direction • Depth to aquifer • Groundwater/surface water recharge and discharge zones • Aquifer geochemistry • Soil properties • Presence and condition of wells, conduits and sewers
Volatilization	• Water solubility • u0240Vapor pressure • u0245Henry's Law gas-solubility constant[b] • u0250Gas diffusion coefficient	• Depth to water table • Soil properties and vegetative cover • Climate conditions • Contaminant concentration • Porosity and gas permeability of soils and shallow geologic materials

[a] *Soil organic-carbon partition constant:* $K_{oc/w}$.

[b] *Soil Henry's Law gas-solubility constant:* H_X^{cp}.

TABLE 10.B.2 Chemical- and Site-Specific Factors Affecting Contaminant Transport: Groundwater (continued)

	Factors Affecting Transport	
Transport Mechanism	**Chemical-Specific**	**Site-Specific**
Adsorption to soils and aquifers	• Water solubility • Soil organic-carbon partition constant • Octanol-water partition constant[a]	• Geochemistry (soils and aquifers) • Organic carbon content (soils and aquifers)
Biologic uptake	• Octanol-water partition constant • Bioconcentration factor[b]	• Groundwater use for irrigation and livestock watering

[a] *Octanol-water partition constant:* $K_{o/w}$.

[b] *Bioconcentration factor:* $BCF = C_{biota}/C_{water}$.

10.C FACTORS AFFECTING CONTAMINANT TRANSPORT BY SOILS OR SEDIMENTS

Tables 10.C.1 and 10.C.2 lists the transport mechanisms and the chemical-specific and site-specific factors affecting contaminant fate and transport by soils or sediments. It is excerpted from Appendix E of *Public Health Assessment Guidance Manual* (ATSDR, 2005).

TABLE 10.C.1 Chemical- and Site-Specific Factors Affecting Contaminant Transport: Soil or Sediment

	Factors Affecting Transport	
Transport Mechanism	**Chemical-Specific**	**Site-Specific**
Runoff (soil erosion)	• Water solubility • Soil organic carbon partition constant[a]	• Presence of plants • Soil type • Precipitation rate • Land surface configuration • Land surface condition
leaching	• Water solubility • Soil organic carbon partition constant[b]	• Soil type • Soil porosity and permeability • Soil chemistry (especially acid-base chemistry) • Cation exchange capacity • Organic carbon content

[a] *Soil organic-carbon partition constant:* $K_{oc/w}$.
[b] *Soil organic-carbon partition constant:* $K_{oc/w}$.

TABLE 10.C.2 Chemical- and Site-Specific Factors Affecting Contaminant Transport: Soil and Sediment (continued)

	Factors Affecting Transport	
Transport Mechanism	**Chemical-Specific**	**Site-Specific**
Volatilization	• Vapor pressure • Henry's Law gas solubility constant[a]	• Soil physical properties • Soil chemical properties • Climate conditions
Biologic uptake	• Biological availability • Bioconcentration factor[b]	• Soil properties • Contaminant concentration

[a] *soil Henry's Law gas solubility constant:* H_X^{cp}
[b] *bioconcentration factor:* $BCF = C_{biota}/C_{water}$

10.D WATER INGESTION EQUATION

The water ingestion equation is a function of six variables listed in Table 10.D.1. One parameter is the contaminant water concentration C_W at the exposure point. The averaging time AT depends on the nature of the contaminant, noncarcinogen (expression 10.D.1) or carcinogen (expression 10.D.2). The remaining parameters depend on the physical and behavioral characteristics of the potential exposed population.

Noncarcinogen

$$ADD(C_W) = \left(\frac{C_W \cdot IR_W}{BM}\right) \tag{10.D.1}$$

Carcinogen

$$ADD(C_W) = \left(\frac{C_W \cdot IR_W}{BM}\right) \cdot \left(\frac{ED \cdot EF}{AT}\right) \tag{10.D.2}$$

TABLE 10.D.1 Water Ingestion Equation Parameters

Parameter	Units	Symbol
Water concentration	$mg\ dm^{-3}$	C_W
Ingestion rate	$dm^3\ day^{-1}$	IR_W
Exposure duration	years	ED
Exposure frequency	$days\ year^{-1}$	EF
Body mass	kg	BM
Averaging time	days	AT

10.E SOIL INGESTION EQUATION

The soil ingestion equation is a function of six variables listed in Table 10.E.1. One parameter is the contaminant soil concentration C_S at the exposure point. The averaging time AT depends on the nature of the contaminant, noncarcinogen (expression 10.E.1) or carcinogen (expression 10.E.2). The remaining parameters depend on the physical and behavioral characteristics of the potential exposed population.

noncarcinogen

$$\mathrm{ADD}(C_S) = \left(\frac{C_S \cdot \mathrm{CF} \cdot \mathrm{IR}_S}{\mathrm{BM}}\right) \tag{10.E.1}$$

carcinogen

$$\mathrm{ADD}(C_S) = \left(\frac{C_S \cdot \mathrm{CF} \cdot \mathrm{IR}_S}{\mathrm{BM}}\right) \cdot \left(\frac{\mathrm{ED} \cdot \mathrm{EF}}{\mathrm{AT}}\right) \tag{10.E.2}$$

TABLE 10.E.1 Soil Ingestion Equation Parameters

Parameter	Units	Symbol
Soil concentration	$\mathrm{mg\,kg^{-1}}$	C_S
Unit conversion Factor	$\mathrm{kg\,mg^{-1}}$	CF
Ingestion rate	$\mathrm{mg\,day^{-1}}$	IR_S
Exposure duration	years	ED
Exposure frequency	$\mathrm{days\,year^{-1}}$	EF
Body mass	kg	BM
Averaging time	days	AT

10.F FOOD INGESTION EQUATION

The food ingestion equation is a function of six variables listed in Table 10.F.1. One parameter is the contaminant food concentration C_F at the exposure point. The averaging time AT depends on the nature of the contaminant, noncarcinogen (expression 10.F.1) or carcinogen (expression 10.F.2). The remaining parameters depend on the physical and behavioral characteristics of the potential exposed population.

Noncarcinogen

$$ADD(C_F) = \left(\frac{C_F \cdot CF \cdot IR_F}{BM}\right) \tag{10.F.1}$$

Carcinogen

$$ADD(C_F) = \left(\frac{C_F \cdot CF \cdot IR_F}{BM}\right) \cdot \left(\frac{ED \cdot EF}{AT}\right) \tag{10.F.2}$$

TABLE 10.F.1 Food Ingestion Equation Parameters

Parameter	Units	Symbol
Food concentration	$mg\,kg^{-1}$	C_F
Unit conversion Factor	$kg\,mg^{-1}$	CF
Ingestion rate	$mg\,day^{-1}$	IR_F
Exposure duration	years	ED
Exposure frequency	$days\,year^{-1}$	EF
Body mass	kg	BM
Averaging time	*days*	AT

10.G AIR INHALATION EQUATION

The air inhalation equation is a function of six variables listed in Table 10.G.1. One parameter is the contaminant air concentration C_A at the exposure point. The averaging time AT depends on the nature of the contaminant, noncarcinogen (expression 10.G.1), or carcinogen (expression 10.G.2). The remaining parameters depend on the physical and behavioral characteristics of the potential exposed population.

Noncarcinogen

$$ADD(C_A) = \left(\frac{C_A \cdot CF \cdot IR_A}{BM}\right) \tag{10.G.1}$$

Carcinogen

$$ADD(C_A) = \left(\frac{C_A \cdot CF \cdot IR_A}{BM}\right) \cdot \left(\frac{ED \cdot EF}{AT}\right) \tag{10.G.2}$$

TABLE 10.G.1 Air Inhalation Equation Parameters

Parameter	Units	Symbol
Air concentration	$mg\,m^{-3}$	C_A
Ingestion rate	$m^3\,day^{-1}$	IR_A
Exposure duration	years	ED
Exposure frequency	days $year^{-1}$	EF
Body mass	kg	BM
Averaging time	days	AT

References

ASTM, 2014. Standard Guide for Conducting Terrestrial Plant Toxicity Tests, E1963-09, vol. 11.06. American Society for Testing and Materials, p. 21.

ATSDR, 2005. Public Health Assessment Guidance Manual, ATSDR-208. U.S. Department of Health and Human Services, Public Health Service, Atlanta, GA.

ATSDR, 2010. Case Studies in Environmental Medicine. Lead Toxicity, WB 1105. U.S. Department of Health and Human Services, Washington, DC.

Baylin, S.B., Jones, P.A., 2011. A decade of exploring the cancer epigenome–biological and translational implications. Nat. Rev. Cancer 11 (10), 726–734.

BEIR, 1996. Health Effects of Exposure to Low Levels of Ionizing Radiation: Beir V, Biological Effects of Ionizing Radiations. National Academy Press, Washington, DC.

Benigni, R., Bossa, C., Tcheremenskaia, O., 2013. Nongenotoxic carcinogenicity of chemicals: mechanisms of action and early recognition through a new set of structural alerts. Chem. Rev. 113 (5), 2940–2957.

Boveri, T., 1914. Zur Frage der Entstehung Maligner Tumoren. Gustav Fisher Verlag, Jena, Germany.

Chen, W.Y., Baylin, S.B., 2004. Inactivation of tumor suppressor genes. Choice between genetic and epigenetic routes. Cell Cycle 4 (1), 10–12.

Davis, J.A., Gift, J.S., Zhao, Q.J., 2011. Introduction to benchmark dose methods and US EPA's benchmark dose software (BMDS) version 2.1. 1. Toxicol. Appl. Pharmacol. 254 (2), 181–191.

Day, G.A., Dufresne, A., Stefaniak, A.B., Schuler, C.R., Stanton, M.L., Miller, W.E., Kent, M.S., Deubner, D.C., Kreiss, K., Hoover, M.D., 2007. Exposure pathway assessment at a copper-beryllium alloy facility. Ann. Occup. Hyg. 51 (1), 67–80.

Esteller, M., 2007. Epigenetic gene silencing in cancer: the DNA hypermethylome. Hum. Mol. Genet. 16 (Rev. Iss. 1), R50–R59.

Huebner, R.J., Todaro, G.J., 1969. Oncogenes of RNA tumor viruses as determinants of cancer. Proc. Natl. Acad. Sci. U. S. A. 64 (3), 1087–1094.

Knudson, A.G.J., 1971. Mutation and cancer: statistical study of retinoblastoma. Proc. Natl. Acad. Sci. U. S. A. 68 (4), 820–823.

Knudson, A.G., 1986. Genetics of human cancer. J. Cell. Physiol. 129 (S4), 7–11.

Knudson, A.G., 2001. Two genetic hits (more or less) to cancer. Nat. Rev. Cancer 1 (2), 157–162.

Liuzzi, J.P., Cousins, R.J., 2004. Mammalian zinc transporters. Annu. Rev. Nutr. 24, 151–172.

Martelli, A., Rousselet, E., Dycke, C., Bouron, A., Moulis, J.-M., 2006. Cadmium toxicity in animal cells by interference with essential metals. Biochimie 88 (11), 1807–1814.

Moolgavkar, S.H., Knudson, A.G., 1981. Mutation and cancer: a model for human carcinogenesis. J. Natl. Cancer Inst. 66 (6), 1037–1052.

Muller, H.J., 1950a. Radiation damage to the genetic material. Am. Sci. 38 (1), 33–59.

Muller, H.J., 1950b. Radiation damage to the genetic material. II. Effects manifested mainly in the exposed individuals. Am. Sci. 38 (3), 399–425.

National Research Council, 1983. Risk Assessment in the Federal Government: Managing the Process. National Academies Press, Washington, DC.

Newcomb, W.D., Rimstidt, J.D., 2002. Trace element distribution in US groundwaters: a probabilistic assessment using public domain data. Appl. Geochem. 17 (1), 49–57.

Nordling, C.O., 1952. Theories and statistics of cancer. Nord. Med. 47 (24), 817–820.

Nordling, C.O., 1953. A new theory on the cancer-inducing mechanism. Br. J. Cancer 7 (1), 68.

Shih, C., Weinberg, R.A., 1982. Isolation of a transforming sequence from a human bladder carcinoma cell line. Cell (Cambridge, MA, U. S.) 29 (1), 161–169.

Shilo, B.Z., Weinberg, R.A., 1981. Unique transforming gene in carcinogen-transformed mouse cells. Nature (London, U. K.) 289 (5798), 607–609.

United States Congress, 1938. Federal Food, Drug, and Cosmetic Act, 75th Congress, 3rd Session, Public Law 75-717, vol. 52 Statute. Government Printing Office, Washington, pp. 1040–1059.

United States Congress, 1954. Pesticide Amendments to Food, Drug and Cosmetic Act, 83rd Congress, 2nd Session, Public Law 83-518, vol. 68 Statute. Government Printing Office, Washington, pp. 511–517.

United States Congress, 1958. Food Additives Amendment of 1958, 85th Congress, 2nd Session, Public Law 85-929, vol. 72 Statute. Government Printing Office, Washington, pp. 1784–1789.

United States Congress, 1960. Color Additives Amendment of 1960, 86th Congress, 2nd Session, Public Law 86-618, vol. 74 Statute. Government Printing Office, Washington, pp. 397–407.

USEPA, 1991. Risk Assessment Guidance for Superfund: Volume I, Human Health Evaluation Manual, Supplemental Guidance, OSWER Directive 9285.6-03. U.S. Environmental Protection Agency, Washington.

USEPA, 1995a, Refuse Hideaway Landfill, Superfund Record of Decision EPA/ROD/R05-95/281. U.S. Environmental Protection Agency, Chicago, IL.

USEPA, 1995b, U.S. EPA Announces Changes in Refuse Hideaway Superfund Site Cleanup Plan. U.S. Environmental Protection Agency, Chicago, IL.

USEPA, 2003. Guidance for Developing Ecological Soil Screening Levels, OSWER Directive 9285.7-55. U.S. Environmental Protection Agency, Washington.

USEPA, 2005. Guidelines for Carcinogen Risk Assessment, Risk Assessment Forum EPA/630/P-03/001F. U.S. Environmental Protection Agency, Washington.

USEPA, 2011. Exposure Factors Handbook, National Center for Environmental Assessment Report EPA/600/R-090/052F. U.S. Environmental Protection Agency, Washington, DC.

USEPA, 2016a, Benzene; CASRN 71-43-2. United States Environmental Protection Agency, Integrated Risk Information System. https://cfpub.epa.gov/ncea/iris2/chemicalLanding.cfm?substance_nmbr=276.

USEPA, 2016b, Toluene; CASRN 108-88-3. United States Environmental Protection Agency, Integrated Risk Information System. https://cfpub.epa.gov/ncea/iris2/chemicalLanding.cfm?substance_nmbr=118.

Wisconsin Legislature, 2015. Groundwater Quality, Wisconsin Statutes § 35.93, NR 140, vol. 715. Wisconsin Legislative Reference Bureau, Madison, WI, 323–336.

Soil and Environmental Chemistry: Exercises

OUTLINE

1 Elemental Abundance 535
2 Chemical Hydrology 537
3 Clay Mineralogy and Chemistry 539
4 Ion Exchange 541
5 Water Chemistry 542
6 Acid-Base Chemistry 545
7 Natural Organic Matter 549
8 Surface Chemistry and Adsorption 551
9 Reduction-Oxidation Chemistry 555
10 Risk Analysis 558
References 561

1 ELEMENTAL ABUNDANCE

(1) Calculate the binding energy per nucleon E_b/A in MeV for the carbon isotope ${}^{12}_{6}C$. Proton rest mass $m_p = 1.00727646688\,u$, and the neutron rest mass is $m_n = 1.00866491588\,u$.

The mass-energy equivalent is: $1\,u = 931.4940954$ MeV. You can obtain a complete list of isotope rest masses from the National Institute of Standards and Technology (www.physics.nist.gov/).

(2) List the number of stable isotopes for the period 3 elements (Na through Ar). Explain why some elements have more stable isotopes than others. Isotope relative abundances can be found at websites maintained by either the US *National Institute of Standards and Technology (NIST)* (www.physics.nist.gov) or *National Physical Laboratory: Kaye & Laby Online* (www.kayelaby.npl.co.uk).

(3) Fig. 2 plots the relative abundance of several elements in the solar system, Earth's crust, and Earth's soil. Explain the factors that determine the relative abundance of the following elements: cobalt ${}_{27}Co$, nickel ${}_{28}Ni$, and copper ${}_{29}Cu$.

(4) List the half-life of the longest-lived isotope for each element from Po to Th ($84 < Z < 90$). A complete *Table of the Isotopes* is available on the *Abundances of the Elements* (Kaye & Laby Tables of Physical and Chemical Constants, National Physics Laboratory; www.kayelaby.npl.co.uk).

(5) A reactor accident occurred in Chernobyl, Ukraine in 1986 that deposited $^{137}_{55}Cs$ and other fission products in a large region of Ukraine, Belarus, and Russia. A significant fraction of the area was contaminated by Cs-137 at levels of 37–555 GBq km^{-2}, equivalent to a radioactivity of 0.055-0.833 Bq g^{-1} in the top 1 m of soil.

Cs-137 is a radioactive nuclide with a half-life of $t_{1/2} \approx 30$ years. Estimate the length of time necessary for soil Cs-137 levels in this zone to reach 0.018 Bq g^{-1}, a level of radioactivity associated with a lifetime cancer risk of 3 per 100,000 persons living in the area.

(6) The Earth's crust contains thorium and uranium although no stable isotope exists for either element. Explain why these unstable elements are present in the Earth's crust.

(7) The Earth's crust is depleted of certain elements relative to the overall composition of the Solar System. Explain why the following elements (helium, neon, argon, krypton, xenon, silver, gold, platinum) are depleted in the crust.

(8) Explain the significance of the following geochemical terms: atmophilic, lithophilic, siderophilic, and chalcophilic.

(9) The following data given in Table 1 are the vanadium content of samples of Canadian granite (Ahrens, 1954). Determine the geometric mean and geometric standard deviation from this data set.

TABLE 1 The Vanadium Content in 27 Canadian Granite Specimens

Sample	w(V) (mg kg^{-1})	Sample	w(V) (mg kg^{-1})
G-1	21	KB-1	75
KB-2	43	KB-3	200
KB-4	50	KB-5	7.5
KB-6	30	KB-7	33
KB-8	27	KB-9	34
KB-10	42	KB-11	10
KB-12	52	KB-13	51
KB-14	182	KB-15	630
KB-16	144	KB-17	94
KB-18	94	KB-19	29
48-63	8	48-115	11
48-158	5.5	48-118	23
48-485	19	48-489	7
48-490	5.6		

Ahrens, L.H., 1954. The lognormal distribution of the elements. Geochim. Cosmochim. Acta 5, 49–73.

(10) Data given in Table 2 is the nitrogen content of soil specimens collected in Black Hawk County, Iowa (Professor M.E. Thompson, Iowa State University). Determine the arithmetic mean and geometric mean soil nitrogen content of this data set.

TABLE 2 The Organic Nitrogen Content of Soil Specimens From Black Hawk, County, Iowa

Bulk Density ($Mg\,m^{-3}$)	Organic Nitrogen ($mg_N\,g^{-1}$)	Bulk Density ($Mg\,m^{-3}$)	Organic Nitrogen ($mg_N\,g^{-1}$)
1.37	1.010	1.36	1.988
1.39	1.495	1.34	2.091
1.30	1.496	1.32	2.272
1.39	1.510	1.30	2.445
1.20	1.555	1.21	2.882
1.31	1.679	1.31	2.898
1.31	1.723	1.31	3.054
1.32	1.774	1.37	3.655
1.37	1.786	1.31	3.763
1.31	1.936	1.28	5.289
1.28	1.955		

2 CHEMICAL HYDROLOGY

(1) A cylindrical soil core ($L = 20\,cm$, $r = 3\,cm$) was collected from a field site in Madison County, Iowa, The weight of the empty metal cylinder is: $m_1 = 225\,g$. The metal cylinder containing field-moist soil is weighed ($m_2 = 1217.8\,g$) then dried in a 100°C oven until it ceases to lose weight. The weight of the cylinder containing oven-dried soil is: $m_3 = 977.1\,g$.

Calculate: (1) the mass water content $w_{fm}[Mg\,Mg^{-1}]$ of the field moist soil and (2) the *moist* soil bulk density $\rho_t[Mg\,m^{-3}]$.

(2) A cylindrical soil core ($L = 20\,cm$, $r = 3\,cm$) was collected from a field site in Madison County, Iowa. The weight of the empty metal cylinder is: $m_1 = 225\,g$. The metal cylinder containing field-moist soil is weighed ($m_2 = 1217.8\,g$) then dried in a 100°C oven until it ceases to lose weight. The weight of the cylinder containing oven-dried soil is: $m_3 = 977.1\,g$.

Calculate: (1) the *moist* porosity ϕ_t and (2) the *volumetric* water content $\theta_{fm}[m^3\,m^{-3}]$.

(3) Table 3 lists the physical and water retention data for the Macksburg soil series from Madison County, Iowa. The Macksburg soil is located on nearly level (0–2% slopes) upland sites, has moderately high permeability and is considered well drained.

Determine the volumetric plant-available water-holding capacity θ_{AWC} of the Macksburg soil profile, reporting the water-holding capacity as centimeters of water for entire the 60 cm soil depth.

TABLE 3 Moist Bulk Density ρ_t and Mass Water Contents w of the Macksburg Soil From Madison County, Iowa.

Horizon (cm)	ρ_t(Mg m^{-3})	w_{1500}(Mg Mg^{-1})	w_{1500}(Mg Mg^{-1})
0–18	1.32	0.29	0.14
18–30	1.38	0.39	0.25
30–42	1.45	0.38	0.22
42–60	1.42	0.28	0.16

Notes: Mass water content at $p_{tension} = -10$ *kPa is* w_{10} *and mass water content at* $p_{tension} = -1500$ *kPa is* w_{1500}.

(4) The Robbs soil from Johnson County, Illinois has a moist bulk density of $\rho_t = 1.34\,\text{Mg m}^{-3}$ and a field-capacity volumetric water content of $\theta_{fc} = 0.38\,\text{cm cm}^{-1}$. The soil-water partition coefficient for the herbicide Cyanazine (CAS Registry Number 21725-46-2) in the Robbs soil is: $K^{\circ}_{s/w} = 2.2\,\text{m}^3\,\text{Mg}^{-1}$.

Calculate the Cyanazine retardation coefficient R_f for the Robbs soil at field capacity.

(5) The Macksbury soil (Madison County, Iowa) has a retardation coefficient of $R_f = 42.2$ for the herbicide Atrazine (CAS Registry Number 1912-24-9). In early spring the Macksbury soil is at field capacity and the depth to water table is 45 cm.

Estimate how deep Atrazine applied at the surface of Macksbury soil will migrate after a 7.6 cm rainfall under prevailing conditions.

(6) The Antigo soil (Langlade County, Wisconsin) has a retardation coefficient of $R_f = 3.90$ for the organophosphate insecticide Phosmet (CAS Registry Number 732-11-6). The Antigo soil has a field capacity water content of $\theta_{fc} = 0.37\,\text{cm cm}^{-1}$.

Estimate how deep Phosmet applied at the surface of Antigo soil will migrate after a 3.27 cm rainfall. The soil water content as rain begins to fall is: $\theta_{fm} = 0.20\,\text{cm cm}^{-1}$.

(7) The volumetric moisture content of a fine-sand soil is $\theta = 0.12\,\text{m}^3\,\text{m}^{-3}$.

Use the empirical Clapp-Hornberger water retension function in Appendix 2.D to estimate the tension head $h_{tension}$ at this water content.

(8) Use the empirical Clapp-Hornberger unsaturated hydraulic conductivity function in Appendix 2.D to estimate the hydraulic conductivity K_D of a loam-texture soil at field capacity.

(9) Use the Thornthwaite potential evapotranspiration model to estimate the mean evapotranspiration water loss during May 2016 in Dane County, Wisconsin. Identify the weather station you use to make your estimate.

(10) The estimated evapotranspiration water loss during June 2008 in a Dane County, Wisconsin watershed averaged 4.9 mm of water per day. The average soil depth in the watershed is 100 cm and has a total water storage capacity of: $\theta = 0.400\,\text{cm cm}^{-1}$. At the beginning of June 2008 the soil moisture content in the watershed averaged $\theta_{initial} = 0.370\,\text{cm cm}^{-1}$.

Predict stream discharge from this watershed at the end of June 2008 during which 167.6 mm of precipitation fell.

3 CLAY MINERALOGY AND CHEMISTRY

(1) Identify the Jackson Weathering Stage where smectite minerals first appear.
(2) Identify any crystallographic feature that would permit the solid-state transformation of feldspar minerals into layer silicate minerals.
(3) Describe the importance of plasticity in the field identification of clay content and identify the minerals typically found in the clay-size fraction that exhibit plasticity.
(4) Apply Pauling's *Radius Ratio rule* to determine the preferred coordination of the cation listed in Table 4. The ionic radius of oxygen is: $r_{O^{2-}} = 0.140\,nm$.

TABLE 4 Ionic Radii r_c of the Eight Most Abundant Cations in the Earth's Outer Crust

Abundance Rank	Cation	Ionic Radius r_c (nm)
2	Si^{4+}	0.034
3	Al^{3+}	0.053
4	Fe^{2+}	0.077
4	Fe^{3+}	0.065
5	Ca^{2+}	0.100
6	K^{+}	0.138
7	Na^{+}	0.102
8	Mg^{2+}	0.072
9	Ti^{4+}	0.069

(5) Four layer silicate minerals are shown in Fig. 1.
Associate the following with each structure: (1) a specific mineral name, (2) the Jackson chemical weathering stage dominated by each mineral, and (3) whether the mineral is capable of *crystalline* swelling when hydrated.
(6) The specific surface area of kaolinite specimens from the clay (<0.2 μm) fraction is: $a_s \approx 10\,m^2\,g^{-1}$. The specific surface area of smectite specimens from the clay (<0.2 μm) fraction is: $a_s \approx 700\,m^2\,g^{-1}$.
Explain the dramatic difference in surface area despite the fact that both specimens have the same particle radius.
(7) Consider a landscape underlain by granite that weathers in situ to residuum. The x-ray diffraction pattern from the silt-size fraction of the residuum collected near the contact with the underlying granite bedrock clearly shows the presence of fine-grained muscovite. The x-ray diffraction pattern from the clay-size fraction of the residuum collected from the soil profile at the land surface lacks the characteristic diffraction lines of muscovite; instead, the dominant mineral is kaolinite.
Explain the significance of these findings.
(8) Clay mineral layer charge influences a variety of chemical and physical properties: swelling behavior, surface area, and the exchangeability of interlayer ions.

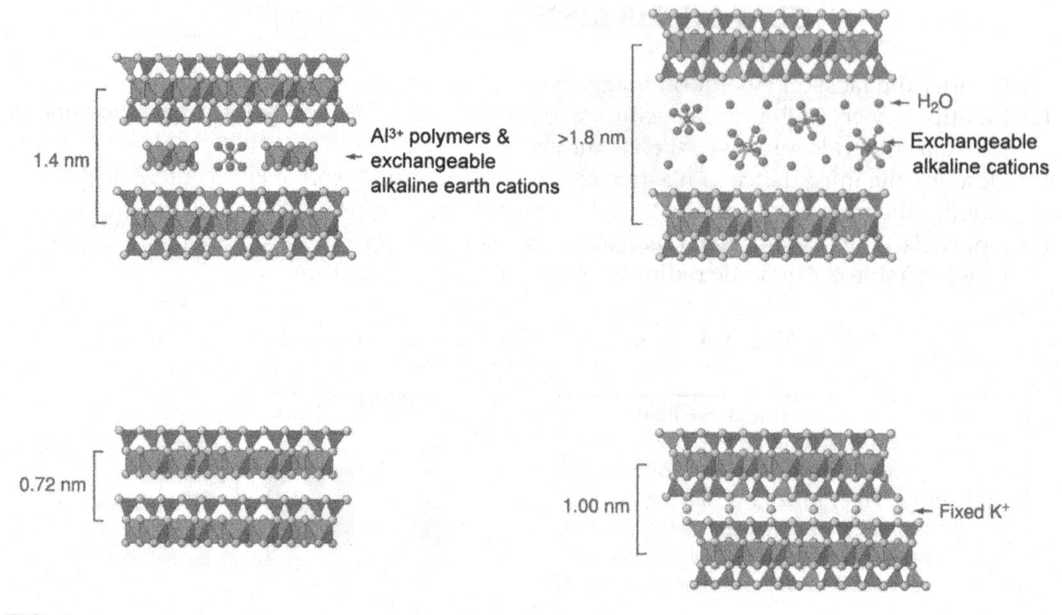

FIG. 1 Layer silicate structures showing tetrahedral sheets, octahedral sheets, and, where appropriate, interlayer components.

Explain the relationship between layer charge and the chemical and physical properties listed above.

(9) Describe the special characteristics of *Tschermak* cation substitution and its significance for cation substitution in layer silicates.

(10) Estimate the unit cell composition of a Wyoming montmorillonite specimen "Volclay" (American Colloid Company) using the oxide mass-fraction data listed in Table 5.

TABLE 5 The Chemical Composition of a Na^+-Saturated Volclay Montmorillonite Specimen, Expressed in Oxide Mass-Fraction Units

Oxide	w(Oxide)($g\,g^{-1}$)
SiO_2(s)	0.613
Al_2O_3(s)	0.224
Fe_2O_3(s)	0.0355
FeO(s)	0.0039
TiO_2(s)	0.0011
MgO(s)	0.0281
Na_2O(s)	0.0271

(11) Compute the osmotic head $h_{osmotic}$ of a $CaSO_4(aq)$ solution:
$c_{CaSO_4} = 1.6 \times 10^{-2}\,mol\,dm^{-3}$.

(12) Fig. 3.16A plots the crystalline swelling of a Na^+-saturate smectite specimen.
Estimate the number of water layers in the interlayer of this smectite at each crystalline swelling state.

(13) Fig. 3.24 shows the tension head $h_{tension}$ of a 13.5% suspension of Cheto montmorillonite.
Compute the actual height of the clay gel above pure water in an osmotic cell equivalent to the osmotic head: $h_{osmotic} = 13.4\,cm$.

4 ION EXCHANGE

(1) The clay fraction of the Twotop soil series (Crook County, Wyoming) is predominantly smectite with a reported cation exchange capacity of $n_+^\sigma/m = 76.4\,cmol_c\,kg^{-1}$. An asymmetric (Ca^{2+}, K^+) cation exchange experiment reports the following results: $c_{K^+} = 10.17\,mmol\,dm^{-3}$, $c_{Ca^{2+}} = 0.565\,mmol\,dm^{-3}$, $n_{K^+}^\sigma/m = 38.2\,cmol_c\,kg^{-1}$, and $n_{K^+}^\sigma/m = 38.2\,cmol_c\,kg^{-1}$.
Compute the selectivity coefficient $K_{Ca/K}^c$ for $(Ca^{2+}, K^+$ exchange in this soil clay specimen.

(2) The Ness soil series (Hodgeman County, Kansas) has a cation exchange capacity of $n_+^\sigma/m = 38.0\,cmol_c\,kg^{-1}$ balanced by $n_{Ca^{2+}}^\sigma/m = 33.1\,mol\,kg^{-1}$ of exchangeable Ca^{2+} and $n_{K^+}^\sigma/m = 4.9\,mol\,kg^{-1}$ exchangeable Na^+. The concentrations of Ca^{2+} and Na^+ in the soil saturated paste extract are $c_{Ca^{2+}} = 10\,mmol\,dm^{-3}$ and $c_{Na^+} = 81\,mmol\,dm^{-3}$, respectively.
Determine which cation—Na^+ or Ca^{2+}—is selectively enriched during this ion exchange reaction.

(3) The Maxfield soil series (Olmsted County, Minnesota) has a cation exchange capacity of $n_+^\sigma/m = 41.0\,cmol_c\,kg^{-1}$ balanced by $n_{Ca^{2+}}^\sigma/m = 39.5\,mol\,kg^{-1}$ of exchangeable Ca^{2+} and $n_{Na^+}^\sigma/m = 1.5\,mol\,kg^{-1}$ exchangeable Na^+. The concentrations of Ca^{2+} and Na^+ in the soil saturated paste extract are $c_{Ca^{2+}} = 4.49\,mmol\,dm^{-3}$ and $c_{Na^+} = 4.58\,mmol\,dm^{-3}$, respectively. Calculate the equilibrium concentrations Ca^{2+} and Na^+ in soil pore water concentrated through water loss by evaporation to the wilting point (concentration factor Fc = 3:31). The following are soil physical properties of the Maxfield soil: moist (i.e., $p_{tension} = -33\,kPa$) bulk density $\rho_t = 1.30\,Mg \cdot_m^{-3}$, volumetric water content at field capacity $\theta_{fc} = 0.255\,m^3 \cdot_m^{-3}$ and the wilting point $\theta_{wp} = 0.176\,m^3 \cdot_m^{-3}$.

(4) The Bssyz2 horizon of the Duke soil series (Fine, mixed, active, thermic Sodic Haplusterts; Harmon County, OK) is 64.7% clay, has the following chemical properties (NCSS Pedon ID: S2003OK-057-006): pH = 7.7, CEC7 = $24.2\,cmol_c \cdot kg^{-1}$, and a saturated-paste sodium adsorption ratio SAR = 19.4. Determine the conditional selectivity coefficient $K_{Na/K}^c$ for (Na^+, K^+) exchange using the ammonium-acetate and saturated-paste extract data from the National Cooperative Soil Survey for the Duke soil series (Hamon County, OK; NCSS Pedon ID: S2003OK-057-006). The ion exchange reaction between each pair of cations is independent of the exchange reactions between all other cation pairs.

TABLE 6 Empirical Results for (Na^+, Ca^{2+}) Exchange on the Clay Fraction From Sawyers-I Field, Rothamsted Experiment Station (Quirk and Schofield, 1955)

$c_{NaCl(aq)}$ (mol dm^{-3})	$c_{CaCl_2(aq)}$ (mol dm^{-3})	E_{Na^+} (Fraction)
2.25E−03	4.00E−04	0.058
4.48E−03	4.00E−04	0.089
1.13E−02	4.00E−04	0.210
2.27E−02	4.00E−04	0.352

(5) Quirk and Schofield (1955) measured (Na^+, Ca^{2+}) exchange by a vermiculite-dominated soil clay fraction collected at Rothamsted Experiment Station (Harpenden, UK). The results are listed in Table 6. Use the least sum-square method to determine the selectivity coefficient $K^c_{Na/Ca}$ for (Na^+, Ca^{2+}) exchange on this soil vermiculite clay.

5 WATER CHEMISTRY

(1) The *ChemEQL* database Solid Phase Library (i.e., mineral species formation reactions and equilibrium solubility constants) contains the following entry for the mineral CALCITE: $CaCO_3(s)$

$$Ca^{2+}(aq) + HCO_3^-(aq) \xleftrightarrow{K^{\circ}=10^{-1.85}} H^+ + \underset{\text{Calcite}}{CaCO_3(s)}$$

Using the equilibrium reaction and constant above, derive the linear, base-10 logarithmic solubility expression plotted in Fig. 5.7 and representing calcite solubility in groundwater without contact with the above-ground atmosphere.

(2) The *ChemEQL* database Solid Phase Library (i.e., mineral species formation reactions and equilibrium solubility constants) contains the following entries for the carbonate mineral DOLOMITE: $CaMg(CO_3)_3(s)$

$$Ca^{2+}(aq) + Mg^{2+}(aq) + CO_2(aq) \longleftrightarrow 4H^+(aq) + \underset{\text{Dolomite}}{CaMg(CO_3)_3(s)}$$

$$K^{\circ} = 10^{-16.30} = \frac{a^4_{H^+}}{a_{Ca^{2+}} \cdot a_{Mg^{2+}} \cdot p_{CO_2}}$$

Fig. 5.8 plots activity $a_{Mg^{2+}}$ as a line on a logarithmic activity diagram. Using the equilibrium reactions and constants for CALCITE and DOLOMITE solubility, derive the logarithmic solubility expression plotted in Fig. 5.8.

(3) Install *ChemEQL* and simulate the chemistry of a 10^{-5} mol dm^{-3} phthalic acid solution at pH 5 (cf. *ChemEQL Manual*, example simulation of 10^{-3} mol dm^{-3} acetic acid solution at pH 4).

Save the input file[1] (`.txt`) and the output file[2] (`.xls`) and perform a charge-balance validation to determine whether this condition is satisfied.

(4) Install *ChemEQL* and simulate the chemistry of a 10^{-5} mol dm^{-3} phthalic acid solution at pH 5 *without setting* mode(H+) = free.

Save the input matrix file (`.txt`) and the output file (`.xls`) and perform a charge-balance validation to determine whether the condition is satisfied when additional components are included to simulate pH 4 without setting mode(H+) = free.

(5) The standard free energies of formation $\Delta_f G^{\circ}$ (298.15 K) for each component in the FLUORAPATITE: $Ca_5(PO_4)_3F(s)$ solubility reaction appear in Table 7 (Source: CHNOSZ database; Dick, 2008)

$$\underset{\text{Fluorapatite}}{Ca_5(PO_4)_3F(s)} \longleftrightarrow 5Ca^{2+}(aq) + 3PO_4{}^{3-}(aq) + F^-(aq)$$

TABLE 7 Gibbs Energy of Formation for Fluorapatite and Its Components

Component	$\Delta_f G^{\circ}$ (298.15 K) (kJ mol^{-1})
$Ca^{2+}(aq)$	−553.16
$PO_4{}^{3-}(aq)$	−1019.49
$F^-(aq)$	−281.94
$Ca_5(PO_4)_3F(s)$	−6494.04

Calculate the standard Gibbs energy $\Delta_r G^{\circ}$ (298.15 K) and the equilibrium constant K_{s0}° for the FLUORAPATITE solubility reaction at 298.15 K (25°C)—the appropriate ideal gas constant R has units: kJ mol^{-1} K^{-1}.

(6) **Variation 1.** Using the *R.I.C.E.* table method, calculate the pH of a solution containing $c_0 = 3.80 \cdot 10^{-4}$ mol · dm^{-3} dihydrogen sulfide $H_2S(aq)$.

The following reactions give the hydrolysis of dihydrogen sulfide $H_2S(aq)$ to hydrogen sulfide HS^- and the dissociation of water $H_2O(l)$.

$$H_2S(aq) \longleftrightarrow [K_{a1}^{\circ} = 9.22 \cdot 10^{-8}] HS^-(aq) + H^+(aq) \tag{0.R1}$$

$$H_2O(l) \longleftrightarrow [K_{W}^{\circ} = 1.023 \cdot 10^{-14}] H^+(aq) + OH^-(aq) \tag{0.R2}$$

[1] Early versions of *ChemEQL* generated input files with one of two extensions (`.mat` or `.cql`). The latest *ChemEQL* version generates files with extension `.txt` when compiling an input file.

[2] *ChemEQL* output files are `.txt` format but are saved with the Excel file extension `.xls`. See Appendix 5.B for further details.

Variation 2. Using the *R.I.C.E.* table method, calculate the pH of a solution containing $c_0 = 3.80 \cdot 10^{-4}$ mol · dm^{-3} hydrogen sulfide $HS^-(aq)$.

The following reactions give the hydrolysis of dihydrogen sulfide HS^- to sulfide S^{2-} and the dissociation of water $H_2O(l)$.

$$HS^-(aq) + H_2O(l) \longleftrightarrow [K^{\circ}_{b2} = 1.05 \cdot 10^{-7}]H_2S(aq) + OH^-(aq) \quad (0.R3)$$

$$H_2O(l) \longleftrightarrow [K^{\circ}_{w} = 1.023 \cdot 10^{-14}]H^+(aq) + OH^-(aq) \quad (0.R4)$$

Variation 3. Using the *R.I.C.E.* table method, calculate the pH of a solution containing $c_0 = 3.80 \cdot 10^{-4}$mol · dm^{-3} sulfide $S^{2-}(aq)$.

The following reactions give the hydrolysis of sulfide S^{2-} to hydrogen sulfide S^{2-} and the dissociation of water $H_2O(l)$.

$$S^{2-}(aq) + H_2O(l) \longleftrightarrow [K^{\circ}_{b1} = 7.94 \cdot 10^{-1}]HS^-(aq) + OH^-(aq) \quad (0.R5)$$

$$H_2O(l) \longleftrightarrow [K^{\circ}_{w} = 1.023 \cdot 10^{-14}]H^+(aq) + OH^-(aq) \quad (0.R6)$$

(7) Variation 1. Using the *R.I.C.E.* table method, calculate the activities of all hydrogen sulfide species for a solution containing $c_0 = 3.80 \cdot 10^{-4}$ mol · dm^{-3} dihydrogen sulfide $H_2S(aq)$.

The following reactions give the hydrolysis of dihydrogen sulfide $H_2S(aq)$ to hydrogen sulfide HS^- and the dissociation of water $H_2O(l)$.

$$H_2S(aq) \longleftrightarrow [K^{\circ}_{a1} = 9.22 \cdot 10^{-8}]HS^-(aq) + H^+(aq) \quad (0.R7)$$

$$H_2O(l) \longleftrightarrow [K^{\circ}_{w} = 1.023 \cdot 10^{-14}]H^+(aq) + OH^-(aq) \quad (0.R8)$$

Variation 2. Using the *R.I.C.E.* table method, calculate the activities of all hydrogen sulfide species for a solution containing $c_0 = 3.80 \cdot 10^{-4}$ mol · dm^{-3} hydrogen sulfide $HS^-(aq)$.

The following reactions give the hydrolysis of dihydrogen sulfide HS^- to sulfide S^{2-} and the dissociation of water $H_2O(l)$.

$$HS^-(aq) + H_2O(l) \longleftrightarrow [K^{\circ}_{b2} = 1.05 \cdot 10^{-7}]H_2S(aq) + OH^-(aq) \quad (0.R9)$$

$$H_2O(l) \longleftrightarrow [K^{\circ}_{w} = 1.023 \cdot 10^{-14}]H^+(aq) + OH^-(aq) \quad (0.R10)$$

Variation 3. Using the *R.I.C.E.* table method, calculate the activities of all hydrogen sulfide species for a solution containing $c_0 = 3.80 \cdot 10^{-4}\ mol \cdot dm^{-3}$ sulfide $S^{2-}(aq)$.

The following reactions give the hydrolysis of sulfide S^{2-} to hydrogen sulfide S^{2-} and the dissociation of water $H_2O(l)$.

$$S^{2-}(aq) + H_2O(l) \longleftrightarrow [K^{\circ}_{b1} = 7.94 \cdot 10^{-1}]HS^{-}(aq) + OH^{-}(aq) \quad (0.R11)$$

$$H_2O(l) \longleftrightarrow [K^{\circ}_{w} = 1.023 \cdot 10^{-14}]H^{+}(aq) + OH^{-}(aq) \quad (0.R12)$$

(8) Simulate the dissolution of GYPSUM: $CaSO_4 \cdot 2H_2O(s)$, saving both the input matrix file (`.txt`) and the output file (`.xls`).
Validate the ion activity coefficients $\gamma_{2\pm}$ and the gypsum *ion activity product (IAP)* using results in the output file.

(9) Simulate HYDROXYAPATITE: $Ca_5(PO_4)_3OH(s)$ solubility in the pH range: $4 \leq pH \leq 9$.
Save both the input file (`.txt`) and the output file (`.xls`) for the simulation.
Plot the total solution phosphate concentration and the concentrations of the four most abundant solution phosphate species over the specified pH range.

(10) Simulate TROILITE: FeS(s) solubility in the pH range: $4 \leq pH \leq 9$.
Save both the input file (`.txt`) and the output file (`.xls`) for the simulation.

(11) Simulate total soluble lead (Pb) in a 3-phase system consisting of an aqueous solution in equilibrium with the phosphate minerals HYDROXYAPATITE: $Ca_5(PO_4)_3OH(s)$ and PYROMORPHITE: $Pb_5(PO_4)_3OH(s)$.
Save both the input file (`.txt`) and the output file (`.xls`) for the simulation.

(12) Simulate the 4-phase system consisting of an aqueous solution in solubility equilibrium with the following three minerals: CALCITE: $CaCO_3(s)$, HYDROXYAPATITE: $Ca_5(PO_4)_3OH(s)$, and PYROMORPHITE: $Pb_5(PO_4)_3OH(s)$.
Save both the input file (`.txt`) and the output file (`.xls`) for the simulation.

(13) The *ChemEQL* Solid Phase Library contains the following entry for the calcium phosphate mineral FLUORAPATITE: $Ca_5(PO_4)_3F(s)$

$$5Ca^{2+}(aq) + 3PO_4^{3-}(aq) + F^{-}(aq) \xleftrightarrow{K^{\circ}_{s0}=10^{+59.60}} \underset{\text{Fluorapatite}}{Ca_5(PO_4)_3F(s)}$$

Calculate the equilibrium coefficient K^c for this reaction using the Debye-Hückel empirical activity coefficient expressions for an ionic strength $I_c = 2 \times 10^{-3}\ mol\,dm^{-3}$.

6 ACID-BASE CHEMISTRY

(1) Crandall et al. (1999) published a study of groundwater and river water mixing, listing dissolved salts in groundwater from a number of wells and sinkholes and surface water from the Suwannee and Little Springs rivers in Suwannee County, Florida. Table 8 lists the water analysis for one groundwater sample under low-flow (i.e., low rainfall) conditions. The water analysis pH is 7.55 and alkalinity is reported as bicarbonate.

Use ChemEQL to simulate the alkalinity of this groundwater sample using the water analysis pH value. Save both the input file (`.txt`) and the output file (`.xls`) for the simulation. List the assumptions you make to set up your input file.

TABLE 8 Groundwater Composition Under Low-Flow Conditions, Suwannee County, Florida

Component	Dissolved Concentration (mg dm^{-3})
Ca	40
Mg	5.2
Na	3.3
K	0.9
Bicarbonate	190
Cl	5.9
Sulfate	8.1
$SiO_2(aq)$	6.3

Crandall, C.A., Katz, B.G., Hirten, J.J., 1999. Hydrochemical evidence for mixing of river water and groundwater during high-flow conditions, lower Suwannee River basin, Florida, USA. Hydrogeol. J. 7 (5), 454–467.

(2) Table 9 lists the chemical analysis of a groundwater sample from Green Bay, Wisconsin. The analytic pH is 7.7.

TABLE 9 Groundwater Composition Under Green Bay, Wisconsin

Component	Dissolved Concentration (mg dm^{-3})
Ca	151
Mg	45
Na	113
K	14
Bicarbonate	167.8
Cl	8
Sulfate	333

Use ChemEQL to simulate the pH of this groundwater sample. Save both the input file (`.txt`) and the output file (`.xls`) for the simulation. List the assumptions you make to set up your input file.

(3) Calculate the amount of $CaCO_3(s)$ required to neutralize the acidity in a year's worth—500 mm—of acid precipitation (mean pH = 4) falling on a 1-hectare area.

(4) Dai et al. (1998) studied the effect of acid rain on soils from the provinces of Hunan and Guangxi in southern China. The soil chemical properties relevant to acidity of the Hongmaochong Inceptisols (surface 0–10 cm) are: pH (1M KCl) = 3.91, $CEC7 = 22.5\,\mathrm{cmol_c\,kg^{-1}}$, exchangeable $Al^{3+} = 4.45\,\mathrm{cmol_c\,kg^{-1}}$. Exchangeable spectator (i.e., base) cations occupy 9.12% of the *CEC7*.

Calculate the amount of limestone ($w(CaCO_3)$ $[\mathrm{g\,kg_{soil}^{-1}}]$) necessary to neutralize the acidity from exchangeable aluminum in this soil. Estimate the nonexchangeable aluminum content of this soil and discuss the contribution of nonexchangeable aluminum to soil acidity.

(5) Porębska et al. (2008) studied the effect of acid rain on soils in Poland. The soil chemical properties relevant to acidity of a spodosol near Gubin in western Poland: pH (1M KCl) = 3.2, $CEC7 = 34.2\,\mathrm{cmol_c\,kg^{-1}}$, exchangeable $Al^{3+} = 26.0\,\mathrm{cmol_c\,kg^{-1}}$. Exchangeable spectator (i.e., base) cations occupy 23.9% of the *CEC7*.

Calculate the amount of limestone ($w(CaCO_3)$ $[\mathrm{g\,kg_{soil}^{-1}}]$) necessary to neutralize the acidity from exchangeable aluminum in this soil. Estimate the nonexchangeable aluminum content of this soil and discuss the contribution of nonexchangeable aluminum to soil acidity.

(6) The subsurface soil chemical properties of a Rhodiudult from Guangxi Province (China) (Dai et al., 1998) are: pH (1M KCl) = 4.0, $CEC7 = 29.8\,\mathrm{cmol_c\,kg^{-1}}$.

Using Figs. 6.15 and 6.16, estimate the exchangeable Al^{3+} $[\mathrm{cmol_c\,kg^{-1}}]$ content of this soil.

(7) Table 10 lists the chemical analysis of a groundwater sample from Waukesha, Wisconsin. The electrical (specific) conductivity of the water is $EC = 1.05\,\mathrm{dS\,m^{-1}}$ and the analytic pH is 8.0.

Calculate the following from the water analysis: carbonate alkalinity, ionic strength, and the *SAR*.

(8) Table 11 lists the chemical analysis of a groundwater sample from Waukesha, Wisconsin. The electrical (specific) conductivity of the water is $EC = 1.05\,\mathrm{dS\,m^{-1}}$ and the analytic pH is 8.0.

Simulate the chemistry of this solution assuming a 3-phase system: solution, CALCITE $CaCO_3(s)$, and $CO_2(g)$. Justify your choice of p_{CO_2}. Save both the input file (`.txt`) and the output file (`.xls`) for the simulation.

Calculate the adjusted *SAR* using the total soluble calcium from the simulation.

(9) The Byz2 horizon of the Duke soil series (fine, mixed, active, thermic Sodic Haplusterts; Harmon County, OK) is 47% smectite clay. The soil displays a "slight effervescence" when treated with dilute hydrochloric acid (indicating $CaCO_3(s)$). The *National Cooperative Soil Survey* analysis (Pedon ID: S2003OK-057-006) reports the following chemical properties: pH 7.7 and $CEC7 = 24.2\,\mathrm{cmol_c\,kg^{-1}}$. Analysis of the saturated paste extract yields the following: $c_{Ca} = 32.0\,\mathrm{mmol_c\,dm^{-3}}$ and $c_{HCO_3^-} = 2.3\,\mathrm{mmol_c\,dm^{-3}}$.

TABLE 10 Groundwater Composition Under Waukesha, Wisconsin

Component	Dissolved Concentration ($mg\ dm^{-3}$)
Ca	78.0
Mg	38.0
Na	79.0
K	5.0
HCO_3	250.7
CO_3	0
Cl	190.0
SO_4	85.0
NO_3	0.04
PO_4	0.01

TABLE 11 Groundwater Composition Under Waukesha, Wisconsin

Component	Dissolved Concentration ($mg\ dm^{-3}$)
Ca	78.0
Mg	38.0
Na	79.0
K	5.0
HCO_3	250.7
CO_3	0
Cl	190.0
SO_4	85.0
NO_3	0.04
PO_4	0.01

Simulate soil pore water chemistry assuming $CaCO_3(s)$ controls calcium solubility (i.e., a 3-phase system containing aqueous solution, gas phase containing CO_2 and the mineral CALCITE). Justify your choice of $CO_2(g)$ partial pressure.

Compare the soil analysis pH and HCO_3^- concentration with the 3-phase simulation.

(10) The Byz2 horizon of the Duke soil series (fine, mixed, active, thermic Sodic Haplusterts; Harmon County, Oklahoma) is 47% smectite clay. The soil displays a "slight effervescence"Ì when treated with dilute hydrochloric acid (indicating $CaCO_3(s)$). The *National Cooperative Soil Survey* analysis (Pedon ID: S2003OK-057-006) reports the following chemical properties: pH 7.7 and $CEC7 = 24.2\,\text{cmol}_c\,\text{kg}^{-1}$. Analysis of the saturated paste extract yields the following: $c_{Ca} = 32.0\,\text{mmol}_c\,\text{dm}^{-3}$, $c_{Mg} = 15.8$ $\text{mmol}_c\,\text{dm}^{-3}$ and $c_{Na} = 94.9\,\text{mmol}_c\,\text{dm}^{-3}$, which results in a $SAR = 19.4$.

Simulate soil pore water chemistry assuming a 2-phase system containing aqueous solution and gas phase containing CO_2. Justify your choice of $CO_2(g)$ partial pressure.

Determine whether the water analysis supports the presence of $CaCO_3(s)$ in the Duke soil.

7 NATURAL ORGANIC MATTER

(1) The International Humic Substances Society supplies reference samples to researchers (IHSS Products link at the website www.ihss.gatech.edu/). The Chemical Properties link leads you to data under the Elemental Composition and Acidic Functional Groups.

Using the oxygen content from the Elemental Composition page and the combined carboxyl and phenol content from the Acidic Functional Groups page, determine the oxygen percentage in the Summit Hill Soil Reference Humic Acid (IHSS sample 1R106H) attributable to carboxyl and phenol groups. What might be the chemical nature of the remaining oxygen?

(2) The International Humic Substances Society supplies reference samples to researchers (IHSS Products link at the website www.ihss.gatech.edu/). The Chemical Properties link leads you to data on the Elemental Composition and Acidic Functional Groups.

Using the oxygen content from the Elemental Composition page and the combined carboxyl and phenol content from the Acidic Functional Groups page, determine the percentage of oxygen content in the Pahokee Reference Humic Acid (IHSS sample 1R103H) attributable to carboxyl and phenol groups.

(3) Table 1.2 in The Chemical Composition of Soils (Helmke, 2000) lists the geometric mean Cu content of world soils: $20\,\text{mg}\,\text{g}^{-1}$. The soil organic carbon content of the Summit Hill (Christchurch, New Zealand) A-horizon is 4.3% ($f_{oc} = 0.043\,\text{g}_{oc}\,\text{g}^{-1}$). Assume the metal binding capacity is the weak acid content of the Summit Hill Soil Reference Humic Acid.

Compare the metal binding capacity of the A-horizon with the Cu content of the soil, assuming one Cu^{2+} per metal binding group. Is the metal binding capacity greater or less than the mean soil Cu content?

(4) Table 1.4 in The Chemical Composition of Soils (Helmke, 2000) lists the geometric mean Cu content of US peat soils (Histosols): $193\,\text{mg}\,\text{g}^{-1}$. The organic matter content of peat soils is essentially 100%. Assume the metal binding capacity is identical with the Pahokee Peat Reference Humic Acid (IHSS sample 1R103H).

Compare the metal binding capacity of this peat with the Cu content of US peat soils, assuming one Cu^{2+} per metal binding group. Is the metal binding capacity greater or less than the mean soil Cu content?

(5) The International Humic Substances Society supplies reference samples to researchers (IHSS Products link at the website www.ihss.gatech.edu/). The Chemical Properties link leads you to data on the Elemental Composition and Acidic Functional Groups.

Using the oxygen content from the Elemental Composition page for the Suwannee River Standard Fulvic Acid (IHSS sample 1S101F), and assuming 10% of the total nitrogen is free amine (90% being amide), what percentage of the total Cu^{2+} binding capacity may be attributed to amine complexes.

(6) The Upper Poplar Creek, a mercury-polluted stream located on the US Department of Energy Oak Ridge reservation, has as dissolve organic carbon content of $DOC = 3\,\text{mg}\,\text{dm}^{-3}$. This aquatic organic matter contains $w(\text{C}) = 0.525\,\text{g}\,\text{g}^{-1}$ carbon and $w(\text{S}) = 6.5 \times 10^{-3}\,\text{g}\,\text{g}^{-1}$ sulfur.

Sulfur K-edge x-ray absorption spectra reveal 20% of the organic sulfur is thiol (R–SH) capable of forming stable complexes with Hg^{2+}. The dissolved Hg concentration in Upper Poplar Creek was measured at $c_{Hg} = 1.22 \times 10^{-6}\,\text{g}\,\text{dm}^{-3}$ in February 2009.

Given the atomic weight of sulfur and mercury, estimate the fraction of the mercury binding capacity of Poplar Creek—defined as the thiol content of the dissolved organic matter—saturated by Hg dissolved in the water column.

(7) Almendros et al. (1991) measured the nitrogen-15 nuclear magnetic resonance (*NMR*) spectrum of a compost material containing 8% nitrogen by weight. From the *NMR* spectrum, Almendros et al. (1991) estimates 8% of the total nitrogen is amine ($R{-}NH_2$), the form of nitrogen known to bind Cu^{2+}. The remaining nitrogen is amide (87%) and purine (5%).

Assuming each humic amine functional group binds one Cu^{2+} to form the complex $R{-}NH{-}Cu^{2+}$, estimate the maximum amount of Cu^{2+} this humic acid can bind.

(8) The International Humic Substances Society supplies reference samples to researchers (IHSS Products link at the website www.ihss.gatech.edu/). The Chemical Properties link leads you to data on the Elemental Composition. Using the oxygen content from the Elemental Composition page for the Waskish Peat Reference Humic Acid (IHSS sample 1R107H), estimate the mean oxidation state of carbon $\overline{OS}_C$. Assume the following nitrogen oxidation state: $OS_N = -3$.

(9) The International Humic Substances Society supplies reference samples to researchers (IHSS Products link at the website www.ihss.gatech.edu/). The Chemical Properties link leads you to data on the Elemental Composition. Using the oxygen content from the Elemental Composition page for the Elliot Soil Standard Humic Acid (1S102H) estimate the mean oxidation state of carbon $\overline{OS}_C$. Assume the following nitrogen oxidation state: $OS_N = -3$.

(10) Saliba et al. (2002) reports the following composition of corncob lignin: $C_9H_{9.84}O_{3.33}(OCH_3)_{0.38}$.

Determine the mean carbon oxidation state $\overline{OS}_C$ and carbon reduction $\overline{r}_C$ of corncob lignin.

(11) Fungal cell walls contain the polymer chitin (CAS Registry Number 1398-61-4) with chemical formula $(C_8H_{13}O_5N)_n$ (Fig. 2).

Verify the formal oxidation state of nitrogen $OS_N = -3$ and calculate the mean carbon oxidation state $\overline{OS}_C$ for chitin.

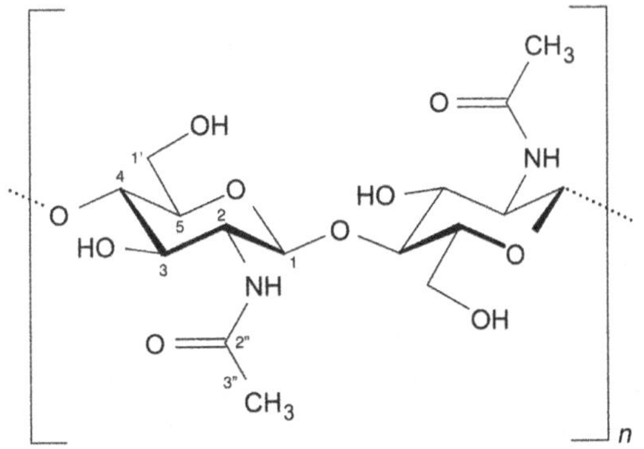

FIG. 2 Repeat unit of the chitin polymer (CAS Registry Number 1398-61-4), a characteristic component of fungal cell walls.

8 SURFACE CHEMISTRY AND ADSORPTION

(1) An experiment measuring adsorption of the triazine herbicide *Cynazine* by a Cheshire soil specimen yielded the results listed in Table 12 (Xing and Pignatello, 1996).

TABLE 12 The adsorption isotherm of adsorptive herbicide *Cyanazine* and adsorbent Cheshire soil

Dissolved $c_{A(aq)}$ $mg \cdot dm^{-3}$	Adsorbed n_A^{σ}/m $mg \cdot kg^{-1}$	Dissolved $c_{A(aq)}$ $mg \cdot dm^{-3}$	Adsorbed n_A^{σ}/m $mg \cdot kg^{-1}$
0.053	0.09	2.4	4.8
0.1	0.2	5.0	8.4
0.14	0.3	7.0	14.0
0.2	0.4	9.0	18.0
0.32	0.60	18.0	24.0
0.5	1.1	24.0	40.0
1.0	1.7	32.0	61.0
1.4	3.0		

Determine whether the adsorption isotherm is best represented by a partitioning isotherm model or a site-limited isotherm model. Estimate the appropriate adsorption coefficient ($K^{\ominus}_{s/w}$ or k_L) for *Cyanazine* adsorption by the Cheshire soil.

(2) An adsorption experiment of the chloroacetanilide herbicide *Alachlor* (CAS Registry Number 15972-60-8) by a shallow organic-rich aquifer near Piketon, Ohio yielded the results listed in Table 13. The organic carbon content of the aquifer is 0.02%.

(Source: Springer, A.E., 1994. Characterization and Modeling of Pesticide Fate and Transport in a Shallow Unconfined Aquifer. Ph.D. Diss. The Ohio State Univ., Columbus.)

TABLE 13 The adsorption isotherm of adsorptive chloroacetanilide herbicide *Alachlor* and adsorbent organic-rich aquifer

Dissolved	Adsorbed
$c_{A(aq)}$	n_A^{σ}/m
$mg \cdot dm^{-3}$	$mg \cdot kg^{-1}$
0.00	0.011
0.01	0.018
0.03	0.096
0.07	0.180
0.39	0.660
0.79	1.250

Determine the aquifer organic carbon-water partition coefficient $K^{\circ}_{oc/w}$ for *Alachlor* adsorption by the organic carbon fraction of the aquifer.

(3) An experiment measuring SO_4^{2-} adsorption by a by the aluminum oxide $\gamma - Al_2O_3(s)$ yielded the results listed in Table 14.

TABLE 14 The adsorption isotherm of adsorptive SO_4^{2-} and adsorbent $\gamma - Al_2O_3(s)$

Dissolved	Adsorbed
$c_{SO_4^{2-}}$	$\Gamma_{SO_4^{2-}}$
$mol \cdot dm^{-3}$	$mol \cdot m^{-2}$
1.40E−04	3.46E−07
3.00E−04	4.24E−07
7.05E−04	4.73E−07
1.12E−03	5.23E−07
2.13E−03	5.20E−07

Determine whether the adsorption isotherm is best represented by a partitioning isotherm model or a site-limited isotherm model. Estimate the appropriate adsorption coefficient ($K^{\circ}_{s/w}$ or k_L) for SO_4^{2-} adsorption by $\gamma - Al_2O_3(s)$.

(4) Table 15 (courtesy of T. Ranatunga and R.W. Taylor, Alabama A & M University) is from an experiment measuring Pb^{2+} adsorption at pH 4 by KAOLINITE (*Clay Minerals Society* specimen: KGa-1). Specimen KGa-1 has a specific surface area of $a_s = 9.6\,m^2\,g^{-1}$.

TABLE 15 The Adsorption Isotherm of Adsorptive Pb^{2+} and Adsorbent KAOLINITE $Al_2Si_2O_5(OH)_4(s)$

Dissolved $c_{Pb^{2+}}$ (mol dm^{-3})	Adsorbed $\Gamma_{Pb^{2+}}$ (mol m^{-2})	Dissolved $c_{Pb^{2+}}$ (mol dm^{-3})	Adsorbed $\Gamma_{Pb^{2+}}$ (mol m^{-2})
4.23E−07	1.30E−06	5.94E−05	2.22E−06
2.33E−06	1.77E−06	7.32E−05	2.28E−06
1.27E−05	1.91E−06	9.99E−05	2.37E−06
2.07E−05	2.09E−06	1.24E−04	2.52E−06

Determine whether the adsorption isotherm is best represented by a partitioning isotherm model or a site-limited isotherm model. Estimate the appropriate adsorption coefficient ($K^{\circ}_{s/w}$ or k_L) for $H_2PO_4^-$ adsorption by $Al_2Si_2O_5(OH)_4(s)$.

(5) Table 16 from Clark and McBride (1984) measures Cu^{2+} adsorption by IMOGOLITE $Al_2SiO_3(OH)_4(s)$ covering a pH range from about 4 to 7. The suspensions, which were prepared with 5 g dm^{-3} synthetic IMOGOLITE, also contained 50 mmol dm^{-3} $Ca(NO_3)_2(aq)$ and an initial concentration of 250 μmol dm^{-3} $Cu(NO_3)_2(aq)$.

TABLE 16 The Adsorption Edge for Adsorptive Cu^{2+} and Adsorbent IMOGOLITE $Al_2SiO_3(OH)_4(s)$

pH	% Cu^{2+} Adsorbed	pH	% Cu^{2+} Adsorbed
6.85	93.45	5.62	35.10
6.54	81.40	5.32	23.04
6.22	65.75	4.99	19.03
5.94	46.72	4.35	14.80

Determine adsorption edge parameters for Cu^{2+} adsorption by the least-squares fitting of the data to the nonlinear cation adsorption-edge model. You can determine the solution by "trial & error" to minimize the sum square error or use the *Solver* add-in in *Excel*.

(6) An *Alachlor* (Fig. 3) (CAS Registry Number 15972-60-8) adsorption experiment by a specimen collected from a shallow, organic-rich aquifer near Piketon, Ohio yielded the results listed in Table 17.

FIG. 3 *Alachlor* (CAS Registry Number 15972-60-8) is a wide-spectrum chloroacetanilide herbicide.

TABLE 17 The Adsorption Isotherm for Adsorptive *Alachlor* and Adsorbent Organic-Rich Aquifer Specimen

$c_{A(aq)}$ (mg dm^{-3})	n^{σ}_{A}/m(mg kg^{-1})	$c_{A(aq)}$ (mg dm^{-3})	n^{σ}_{A}/m(mg kg^{-1})
0.79	1.25	0.03	0.096
0.39	0.66	0.01	0.018
0.07	0.18	0.003	0.011

The organic carbon content of the aquifer is 0.02%. (Source: Springer, A.E., 1994. Characterization and Modeling of Pesticide Fate and Transport in a Shallow Unconfined Aquifer. Ph.D. Diss. The Ohio State University, Columbus.)

Estimate the Alachlor whole aquifer-water partition coefficient $K^{\ominus}_{s/w}$ and the organic carbon-water partition coefficient $K^{\ominus}_{oc/w}$ for the organic-carbon fraction based on a fit of the data.

(7) An *Alachlor* (CAS Registry Number 15972-60-8) adsorption experiment by a specimen collected from a shallow, organic-rich aquifer near Piketon, Ohio yielded the results listed in Table 18.

TABLE 18 The Adsorption Isotherm for Adsorptive *Alachlor* and Adsorbent Organic-Rich Aquifer Specimen

$c_{A(aq)}$ (mg dm^{-3})	n^{σ}_{A}/m(mg kg^{-1})	$c_{A(aq)}$ (mg dm^{-3})	n^{σ}_{A}/m(mg kg^{-1})
0.79	1.25	0.03	0.096
0.39	0.66	0.01	0.018
0.07	0.18	0.003	0.011

The organic carbon content of the aquifer is 0.02%. (Source: Springer, A.E., 1994. Characterization and Modeling of Pesticide Fate and Transport in a Shallow Unconfined Aquifer. Ph.D. Diss. The Ohio State University, Columbus.)

Estimate the retardation coefficient R_f for Alachlor transport through this organic-rich aquifer. Refer to Chapter 2 (*Chemical Hydrology*) for further details on the retardation coefficient model.

(8) The transport of 1,2-dibromoethane (Fig. 4) (CAS Registry Number 106-93-4) in an aquifer can be estimated using the organic carbon-water partition coefficient measured by any material containing natural organic matter. The adsorption of 1,2-dibromoethane by whole peat given in Table 19. The organic carbon content of the peat is $f_{oc} = 0.473$ kg kg^{-1}. (Source: Chiou, C.T., 2000, Environmental Science & Technology, 34L1254.)

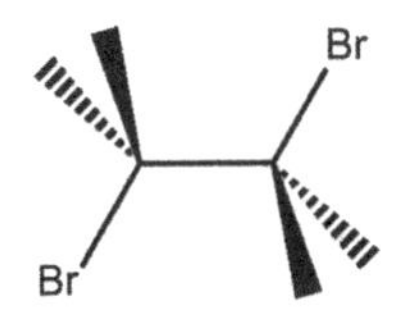

FIG. 4 1,2-Dibromoethane (CAS Registry Number 106-93-4).

TABLE 19 The adsorption isotherm for adsorptive 1,2-dibromoethane and adsorbent whole peat

$c_{A(aq)}$	n_A^σ/m	$c_{A(aq)}$	n_A^σ/m
$mg \cdot dm^{-3}$	$mg \cdot kg^{-1}$	$mg \cdot dm^{-3}$	$mg \cdot kg^{-1}$
21.1	369	83.9	1046
26.1	385	126.0	1377
30.4	415	181.7	2065
61.2	800		

The aquifer at Brookhaven National Laboratory (Upton, New York) is a compacted glacial till with a porosity of $\phi = 0.25$, specific discharge of $q_D = 1.22\,m\,day^{-1}$, and an organic carbon content of $f_{oc} = 0.0041\,kg\,kg^{-1}$.

Use the adsorption results in the table above to estimate the organic carbon-water partition coefficient $K^{\ominus}_{oc/w}$ and from the partition coefficient estimate the retardation coefficient R_f for 1,2-dibromoethane transport through the Long Island glacial till aquifer. Finally, using the specific discharge of the aquifer and the retardation factor, estimate the distance 1,2-dibromoethane will have traveled in the 8 years between 1999 and 2007.

9 REDUCTION-OXIDATION CHEMISTRY

(1) The two molecular structures below are flavin mononucleotide (FMN) and reduced flavin mononucleotide ($FMNH_2$). $FMNH_2$ serves as an electron donor in a variety of biochemical reduction-oxidation reactions (Fig. 5).

Identify which atoms undergo a change in oxidation state in FMN and $FMNH_2$ and calculate their formal oxidation state in each molecule.

(2) Balance the reduction half-reaction involving the anions $S_3O_6^{2-}$ and $S_2O_3^{2-}$.

(3) Balance the reduction half-reaction involving cystine $(SCH_2CH(NH_2)CO_2H)_2$ and cysteine $HSCH_2CH(NH_2)CO_2H$ (Fig. 6).

(4) Balance the reduction half-reaction involving the minerals HAUSMANITE $Mn_3O_4(s)$ and MANGANITE $MnOOH(s)$.

(5) The molecular structures of DDT (dichloro-diphenyl-trichloroethane, CAS Registry Number 50-29-3) to DDD (dichloro-diphenyldichloroethane, CAS Registry Number 72-54-8) are shown in Fig. 7.

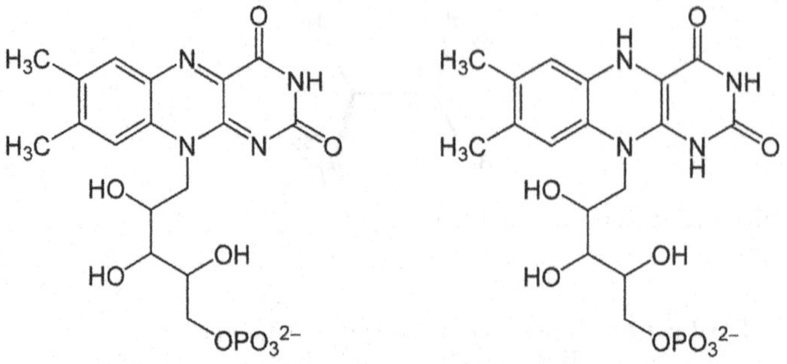

FIG. 5 Flavin mononucleotide (CAS Registry Number 146-17-8) and reduced flavin mononucleotide (CAS Registry Number 5666-16-0).

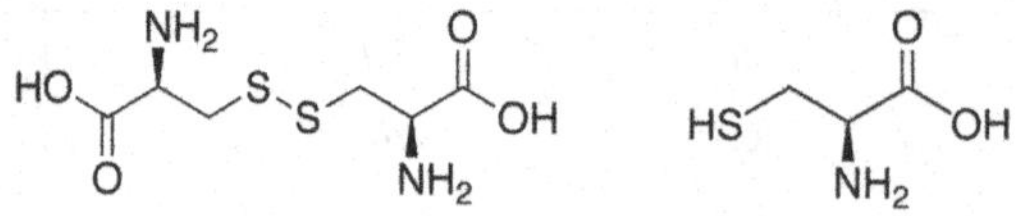

FIG. 6 Cystine (CAS Registry Number 56-89-3) and L-cysteine (CAS Registry Number 52-90-4).

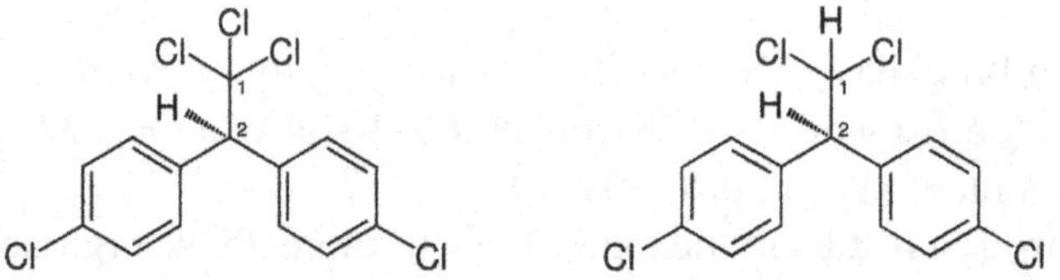

FIG. 7 DDT (dichloro-diphenyl-trichloroethane, CAS Registry Number 50-29-3) and DDD (dichloro-diphenyldichloroethane, CAS Registry Number 72-54-8).

Determine the formal oxidation state of the chlorinated methyl carbon in both compounds and write a balanced reduction half-reaction for reductive dechlorination.

(6) Given the relation between the Gibbs energy of reaction $\Delta_r G$ and the electrochemical potential E of a reaction, calculate the standard reduction potential E° for a balanced DDT dechlorination half-reaction (Table 20).

TABLE 20 Standard Gibbs Energy of Formation $\Delta_f G^{\circ}$ (298.15 K) for the Educts and Products in the DDT Dechlorination Half-Reaction

Specie	$\Delta_f G^{\circ}$ (298.15 K) (kJ mol^{-1})
DDT(aq)	+274.9
DDD(aq)	+238.7
H^+(aq)	0.0
Cl^-(aq)	−131.3

(7) Given the relation between the Gibbs energy of reaction $\Delta_r G$ and the electrochemical potential E of a reaction, calculate the standard reduction potential E° for a balanced reduction half-reaction involving the minerals HAUSMANITE $Mn_3O_4(s)$ and MANGANITE $MnOOH(s)$ (Table 21).

TABLE 21 Standard Gibbs Energy of Formation $\Delta_f G^{\circ}$ (298.15 K) for the Educts and Products in the Half-Reaction Involving HAUSMANITE and MANGANITE

Specie	$\Delta_f G^{\circ}$ (298.15 K) (kJ mol^{-1})
$Mn_3O_4(s)$	−1284.4
$MnOOH(s)$	−561.1
$H^+(aq)$	0.0
$H_2O(l)$	−237.3

(8) The (E, pH) Pourbaix electrochemical stability diagram (Fig. 8) shows the stability fields for several chromium solution and mineral species. This diagram was prepared using a total soluble chromium concentration of 50 ppb, the US EPA Maximum Contaminant Limit for chromium in drinking water.

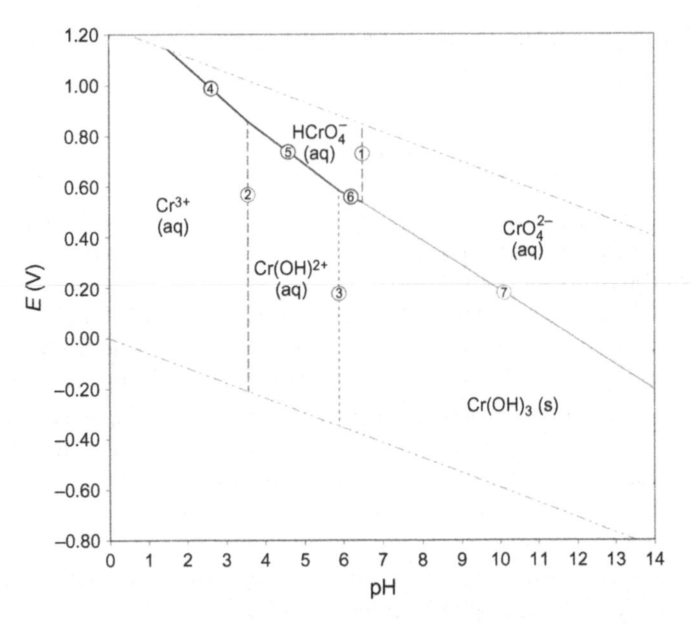

FIG. 8 Pourbaix electrochemical stability diagram for the components Cr–O–H and total dissolved chromium equal to 50 ppb (μg dm^{-3}).

Interpret the chemical significance of each numbered (E, pH) coordinate on the diagram.

(9) The (E, pH) Pourbaix electrochemical stability diagram (Fig. 9) shows the stability fields for several lead solution and mineral species. This diagram was prepared using a total soluble lead concentration of 15 ppb, the US EPA Maximum Contaminant Limit for drinking water.

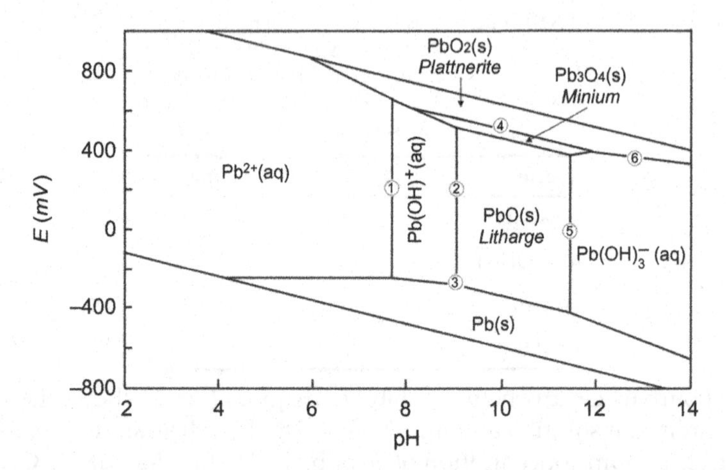

FIG. 9 Pourbaix electrochemical stability diagram for the components Pb–O–H and total dissolved lead equal to 15 ppb ($\mu g\,dm^{-3}$).

Interpret the chemical significance of each numbered (E, pH) coordinate on the diagram.

10 RISK ANALYSIS

(1) Fig. 10 plots the central estimate of the adverse (toxic) response from a series of doses. The dose (i.e., body burden) is plotted on a linear scale and the adverse response is a percentage of the exposed population.

Estimate the *benchmark dose (BMD)* corresponding to a 10% *benchmark response (BMR).*

(2) Fig. 11 plots the central estimate of the adverse (toxic) response from a series of doses. The dose (i.e., body burden) is plotted on a base-10 logarithmic scale (dose units: [$mg\,kg_{BM}^{-1}$]) and the adverse response is a percentage of the exposed population.

Toxicologists often plot dose on a logarithmic scale to improve the accuracy of threshold estimates. Estimate the *BMD* corresponding to a 10% *BMR*.

(3) For 6 years, a child has been drinking water containing 0.05 $mg\,dm^{-3}$ benzene (CAS Registry Number 71-43-2).

Estimate the *average daily dose (ADD)* for a child in this case, taking into account the nature of the toxicant (carcinogen or noncarcinogen).

(4) A construction worker, age 50 years, ingests soil containing 10 $mg\,kg^{-1}$ toluene (CAS Registry Number 108-88-3). The worker's oral soil ingestion rate is $IR = 330\,mg\,day^{-1}$. Work-related exposure duration of a typical construction worker in this case is 30 years.

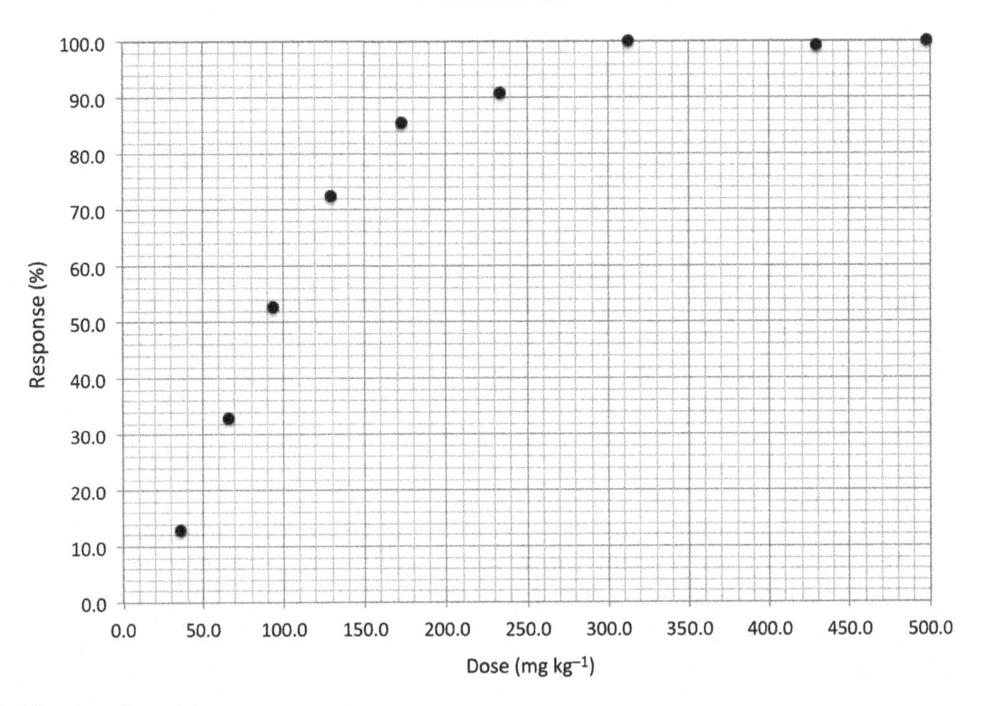

FIG. 10 Hypothetical dose-response curve.

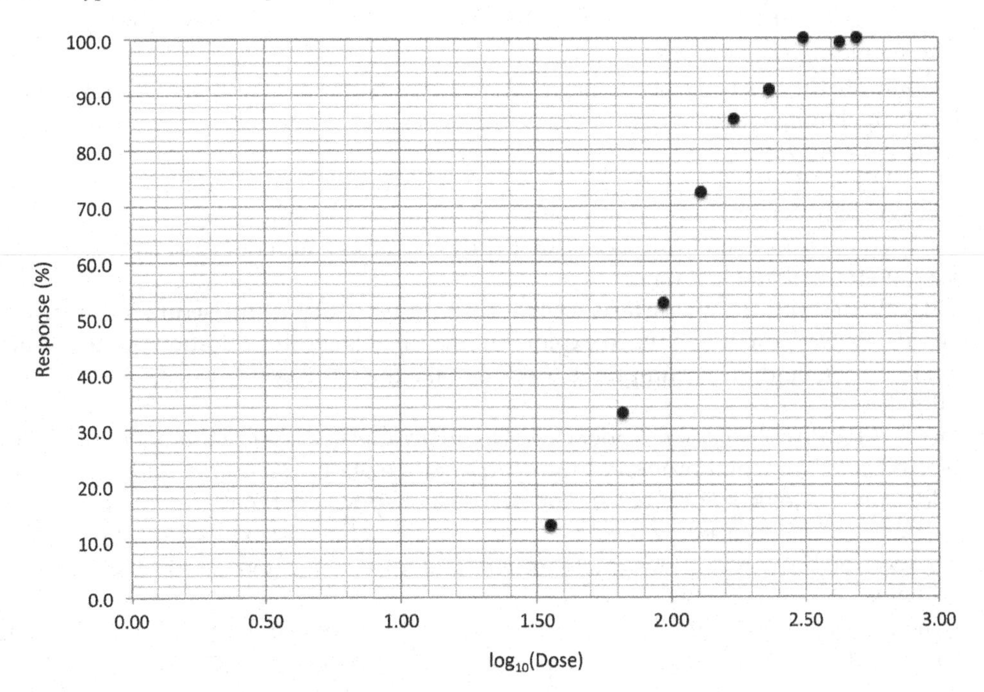

FIG. 11 Hypothetical dose-response curve.

Select (and justify) an appropriate work-related exposure frequency and estimate the *ADD* for this case, taking into account the nature of the toxicant (carcinogen or noncarcinogen).

(5) Estimate the *ADD* of methylmercury CH_3Hg^+ for an adult who eats 56 g of tuna daily containing 0.2 ppm CH_3Hg^+. The biological half-life of CH_3Hg^+ in humans is 70 days.

(6) An adult consuming 100 μg a day of a substance whose steady-state level in the body is later established to be $0.1\ mg\ kg_{BM}^{-1}$.

Estimate the biological half-life $t_{1/2}$ for elimination of this substance from the body.

(7) In 1996, the US EPA developed a new *reference dose* for methylmercury CH_3Hg^+: $RfD = 0.1\ mg\ kg_{BM}^{-1}\ day^{-1}$.

Estimate the amount of fish a 60 kg woman can safely eat each week if the average CH_3Hg^+ content in the fish consumed is 0.30 ppm.

(8) According to a 1992 survey, the drinking water in one-third of Chicago homes had lead levels of 10 ppb.

Identify the most sensitive receptor (resident child or resident adult) and calculate the lead *ADD* for a Chicago resident from exposure to drinking containing 10 ppb lead.

(9) Mackay (1982) reports a linear correlation between the *bio-concentration factor (BCF)* of organic compounds in fish and the octanol-water partition coefficients $K^{\circ}_{o/w}$

$$BCF = \frac{c_B}{c_W} \approx (0.048)\ K^{\circ}_{o/w}$$

The US EPA *PBT Profiler* (cf. Appendix 8.D, *Surface Chemistry and Adsorption*) compiles data, including $K^{\circ}_{o/w}$ values.

Compute the hexachlorobenzene (CAS Registry Number 118-74-1) *BCF* and calculate the hexachlorobenzene concentration of fish in waters containing 10 ppt dissolved hexachlorobenzene.

(10) The US EPA *PBT Profiler* (*cf.* Appendix 8.D, *Surface Chemistry and Adsorption*) tabulates selected environmental properties for a broad range of organic compounds. Among the tabulated environmental properties are the distribution and half-life of each compound in air, water, soil and sediment.

Briefly describe the characteristics that distinguish "soil" from "sediment" as used by the *PBT Profiler*. You will have to explore the site to answer this question but it will allow you to understand the assumptions used to make the tabulated estimates.

(11) The US EPA *PBT Profiler* (*cf.* Appendix 8.D, *Surface Chemistry and Adsorption*) tabulates selected environmental properties for a broad range of organic compounds.

Identify the toxicological end-point on a dose-response curve represented by the *chronic value ChV* as used by the *PBT Profiler*. You will have to explore the site to answer this question but it will allow you to understand the assumptions used to make the estimates.

(12) An adult Bengali has been drinking water from a tube wells in Lalpur, Nadia District, Bangladesh installed 35 years ago. The well water contains $c_W = 17\ mg\ dm^{-3}$ total dissolved arsenic.

Select an appropriate intake rate and exposure frequency to estimate the arsenic *ADD*. Based this *ADD* and a cancer slope factor of 1.5 (US EPA *Integrated Risk Information System*) estimate the *incremental excess lifetime cancer risk (ILECR)* for the 1860 inhabitants of Lalpur age 40 years or older resulting from 35 years of exposure.

(13) The lung cancer *ILECR* is estimated to be 10^{-4} resulting from chronic $^{222}Rn(g)$ exposure at a level of $10\,mBq\,dm^{-3}$. The *ILECR* due to $^{222}Rn(g)$ exposure in Dane County, Wisconsin can be estimated using the relation below

$$IELCR = (0.8 \cdot c_{(^{222}Rn)} + 1) \cdot (10^{-4})$$

Indoor air in Dane County, Wisconsin (US EPA Radon Zone 1) has $^{222}Rn(g)$ levels in excess of $0.148\,Bq\,dm^{-3}$. Estimate the *ILECR* and the number of ^{222}Rn lung cancer deaths for Dane County using the 2005 population: 458,106.

(14) The exposure factor $E_{i,j}$ for the shrew—the US EPA mammalian insectivore test specie—is given by the first expression below. The US EPA wildlife risk model assumes the mammalian insectivore food-source biota $B_{i,j}$ is exclusively *Eisenia fetida* (earthworm). The *soil-to-biota bioaccumuation* factor for cadmium uptake by earthworm is given by the second expression

$$E_{\text{worm, Cd}} = (S_{\text{Cd}} \cdot P_s + B_{\text{worm, Cd}}) \cdot FIR_{\text{worm}}$$

$$\ln B_{\text{worm, Cd}} = (0.795) \cdot \ln S_{\text{Cd}} + (2.114)$$

Table 22 lists exposure factor $E_{i,j}$ parameters for shrew: the biota ingestion rate FIR_{worm}, soil ingestion as a proportion of biota intake P_s and the cadmium *toxicity reference value* TRV_{Cd}.

TABLE 22 Exposure Factors and Cadmium Toxicity Reference Value TRV_{Cd} for Mammalian Insectivore: Shrew

Biota	FIR_{worm} ($kg_{dw}\,kg_{bw}^{-1}\,day^{-1}$)	P_s Fraction	TRV_{Cd} ($mg_{Cd}\,kg_{bw}^{-1}\,day^{-1}$)
Earthworm	0.209	0.03	0.770

By trial and error (best done using a spreadsheet), determine the risk-based *ecological soil screening level* S_{Cd}^{RB} for exposure of shrews to soil contaminated by cadmium. The risk-based cadmium ecological soil screening level S_{Cd}^{RB} must yield a cadmium *Hazard Quotient* HQ_{Cd} equal to unity

$$HQ_j = \frac{FIR_i \cdot E_{i,j}}{TRV_j} \equiv 1$$

References

Ahrens, L.H., 1954. The lognormal distribution of the elements. Geochim. Cosmochim. Acta 5, 49–73.

Almendros, G., Fruend, R., Gonzalez-Vila, F.J., Haider, K.M., Knicker, H., Luedemann, H.D., 1991. Analysis of carbon-13 and nitrogen-15 CPMAS NMR-spectra of soil organic matter and composts. FEBS Lett. 282 (1), 119–121.

Clark, C.J., McBride, M.B., 1984. Chemisorption of copper(II) and cobalt(II) on allophane and imogolite. Clays Clay Miner. 32 (4), 300–310.

Crandall, C.A., Katz, B.G., Hirten, J.J., 1999. Hydrochemical evidence for mixing of river water and groundwater during high-flow conditions, lower Suwannee River basin, Florida, USA. Hydrogeol. J. 7 (5), 454–467.

Dai, Z., Liu, Y., Wang, X., Zhao, D., 1998. Changes in pH, cation exchange capacity (CEC) and exchangeable acidity of some forest soils in southern China during the last 32-35 years. Water Air Soil Pollut. 108 (3–4), 377–390.

Dick, J.M., 2008. Calculation of the relative metastabilities of proteins using the CHNOSZ software package. Geochem. Trans. 9, 1–17.

Helmke, P.A., 2000. The chemical composition of soils. In: Sumner, M.E. (Ed.), Handbook of Soil Science. CRC Press, London, pp. B3–B24.

Mackay, D., 1982. Correlation of bioconcentration factors. Environ. Sci. Technol. 16 (5), 274–278.

Porębska, G., Ostrowska, A., Borzyszkowski, J., 2008. Changes in the soil sorption complex of forest soils in Poland over the past 27 years. Sci. Total Environ. 399 (1), 105–112.

Quirk, J.P., Schofield, R.K., 1955. The effect of electrolyte concentration on soil permeability. J. Soil Sci. 6, 163–178.

Saliba, E.O.S., Rodriguez, N.M., Pilo-Veloso, D., Morais, S.A.L., 2002. Chemical characterization of the lignins of corn and soybean agricultural residues. Arq. Bras. Med. Vet. Zootec. 54 (1), 42–51.

Xing, B., Pignatello, J.J., 1996. Time-dependent isotherm shape of organic compounds in soil organic matter: implications for sorption mechanism. Environ. Toxicol. Chem. 15 (8), 1282–1288.

Index

Note: Page number followed by *f* indicate figures *t* indicate tables, and *np* indicate footnotes.

A

Absorptive power, 148–149
Acetone, 179
Acid-base phenomena, 254–255
  chemistry fundamentals, 256–262
  exchangeable acidity, 254
  mineral solubility, 255
  nitrate and sulfate reduction, 321–323
  rock weathering, 255
  sodicity, 254–255
Acidity, water reference level, 260–262
Adenosine-5′-triphosphate (ATP), 352
Admicelles, 404–405
Adsorbent, 385
  environmental, 386–389
Adsorbent-specific surface area, 390
Adsorption complex, 386, 393
Adsorption edges, 424–426
Adsorption envelope experiment
  adsorption edges, 424–426
  interpreting, 426–428
  structure of, 429–430
  surface complexation models, 430–432
Adsorption isotherm
  area-based adsorption isotherms, 389–390
  chemical basis for, 409–411
  chemisorption, 392–394
  humic molecules, 404–405
  molecular and micellar solutions, 405–409
  physisorption, 392–394
Adsorptive, 385
Aliquots, 149
Alkali and sodic soils, 289–296
Alkalinity, 43–44, 255
  geochemical implications, 283–287
  and mineral acidity, 281–285
  quantifying, 282*b*
Aluminosilicate tetrahedra, 105
Aluminum, exchangeable, 313–317
  asymmetric exchange, 313–314
  neutralizing exchangeable soil acidity, 314–317
Aluminum, nonexchangeable, 317–318
American Petroleum Institute (API-23), 158
American Society for Testing and Materials (ASTM), 88
Amino acids, 345
Anaerobic respiration/anaerobiosis, 475
Anion adsorptives, 425
Antigo soil, 45, 69
Aqueous carbon dioxide, 257*b*
Aquicludes, 47
Aquifer, 47, 48*f*
Aquifuges, 47
Aquitards, 47
Area-based adsorption isotherms, 389–390
Arrhenius acid-base model, 258
Asymmetric exchange, 157–159, 169, 175
Asymmetric quotient expression, 159
Atmospheric deposition, 272–275
Atmospheric pressure, 73
Atterberg limits, 46
Atterberg plasticity limits, 115*b*
Average daily dose (ADD), 511
Avogadro constant, 125

B

Basicity, 255
  water reference level, 260–262
Belle Fourche montmorillonite, 122–124
Bernoulli equation, 70–71
Bicarbonate base hydrolysis, 287*t*
Big Bang, 6, 7*f*, 8–9
Binomial expansion, 81–82
Binomial theorem, 81
Bioaccumulation, 411
Bioconcentration factor (BCF), 411
Biomagnification factor (BMF), 411
Biosurfactants, 347–348
Body-burden, 498
Boltzmann-poisson, 193
Bond-valence model, 436
Bond-valence theory, 419
Bouwer, 65
Bragg's law, layer silicates, 138–140
Brønsted-Lowry acid-base model, 258–259
Bureau of Soils, 88

C

Calcium-saturated smectite clays, 136
Calvin cycle, 350
Capillary forces, 72–74, 129

Capillary fringe zone, 55
Carbon-13 and hydrogen-1 NMR spectroscopy, 375–376
Carbonate chemistry
  alkalinity and mineral acidity, 281–285
  aqueous carbon dioxide reference level, 281
  composite carbonic acid, 278*b*
  equilibrium reactions, 268–276
  pH, 280*f*
  RICE table, 267*t*
Carbonate rocks, 271
Carbon dioxide reference level, 283
Carbon fixation, 350–351
Carbon K-edge NEXAFS and STXM limitations, 377–378
Carbon K-edge X-ray absorption spectroscopy, 376
Carbon mineralization, 351–352
Carbon reduction, 342
Carbon uptake, 341
Carbon-use efficiency (CUE), 341
  for aerobic bacteria, 481
Carcinogenesis, 496*b*
Catabolism, 352
Cation adsorptives, 424
Cation exchange capacity (CEC), 159, 172*f*, 400
Cation substitution
  in silicate minerals, 109*b*
  smectite group minerals, 110, 111*t*
  in vermiculite, 110, 111*t*
ChemEQL, 257
Chemical hydrology
  capillary forces, 72–74
  cycle, 40–45
  day length calculation, 79–80
  empirical water retention, 76–79
  gradients, 70–72
  hydraulic heads, 70–72
  plate-theory model, 80–85
  predicting capillary rise, 74–76
  residence time, 40
  saturated zone, 48–55
  soil moisture balance, 60–64
  solute transport, 65–68
  unsaturated hydraulic conductivity functions, 76–79
  vadose zone, 55–58
  water resources, 40–45
Chemical rock weathering, 262–265
Chemisorption, 392–394
  at oxide mineral surfaces, 421–423
Chlorite group minerals, 111–114, 117*f*
Clapp-Hornberger functions, 76, 77*t*
Clay colloid chemistry
  clay mineral plasticity, 115–116
  clay plasticity and soil mechanical properties, 132–135
  smectite and vermiculite clay minerals
    crystalline swelling, 116–118
    electrolyte solutions and smectite gels, 121–129
    free swelling, 120–125
    swelling clay minerals, hydrated interlayer of, 118–120
Clay film thickness, 134, 135*f*
Clay gels, 130, 130*f*
Clay mineralogy and chemistry
  colloid chemistry
    clay mineral plasticity, 115–116
    clay plasticity and soil mechanical properties, 132–135
    smectite and vermiculite, hydration and swelling of, 116–125
  minerals
    chlorite group, 111–114
    mica group, 107–110
    neutral-layer, 106–107
    phyllosilicate, 105–106
    vermiculite and smectite group, 110–114
  mineral weathering
    Jackson weathering sequence, 91–93, 91*t*
    mineralogy, 90–91
  silicate minerals
    basicity and hydrolysis, 97–101
    coordination polyhedra, 94–96, 95*f*
    groups, 97
    iron(III) oxidation, 101
    pyribole and feldspar minerals, solid-state transformation of, 101–105
Clay mineral plasticity
  definition, 115
  halloysite clay, 116
  kaolinite clay, 116
  plastic deformation, 115
Clay plasticity and soil mechanical properties
  calcium-saturated smectite clays, 136
  clay-size fraction, 132
  granular-size fraction, 132
  mica-smectite, 132–133
  plasticity threshold, 133
  sandy clay loam, 134
  sandy loam, 134
  smectite clay swelling behavior, 136
  textural triangle, 133, 133*f*
Clay ribbon, 88, 89*f*
Clinoclore, 111–112
Closed-shell compounds, 353
Colloidal dispersion, 335–336
Colloidal properties, 376
Combustion, 272–275
Complete binary solid solution, 109
Contact angle, 74, 75*f*

Contaminant fate and transport
exposure pathway assessment, 506–507
factors affecting
by ground water, 526–527, 526*t*, 527*t*
by soils/sediments, 528, 528*t*
by surface water, 524–525, 524*t*, 525*t*
Continuous sequence, 349
Cosmological chemistry, 2–3
Coulomb forces, 386
Counter-ions, 386
Critical micelle concentration, 404
Crystalline swelling, 116–118
potassium-saturated montmorillonite clay, 118, 118*f*
sodium-saturated montmorillonite clay, 118, 118*f*
Crystallographic proton-adsorption model, 415–423
Cubic exchange parameter, 326
Cumulative distribution function (CDF), 494, 495*f*

D

Darcy's Law, 48–49, 48*b*, 65, 69
Day length calculation, 79–80
Degree of saturation, 47, 59, 76
Denitrification, 274, 321–322
Diffuse double-layer theory, 430
Dinitrogen oxide, 274*np*
Dioxygen oxidation, 352–353
Discharge ratio, 41
Dissociation, Arrhenius acid-base model, 258
Dissolved ions, 154
Dissolved iron pore-water concentration, 347
Dissolved organic matter (DOC), 344
Distributed-Reactivity model, 412–413
DNA, 348–349
Dose-response functions
gamma response functions, 496
logarithmic normal-distribution response functions, 496–497
normal-distribution response, 494
Weibull response functions, 494
Dry bulk density, 46
Dual-Domain model, 412–413
Dynamic equilibrium, 192

E

Earth's lithosphere
differentiation of, 14*f*
element abundance
biological processes, 20–21, 21*f*
concentration-frequency distributions, 21–22, 22*f*, 23*f*, 25*t*
geochemical processes, 20–21, 21*f*
law of proportionate effect, 25
sequential randomdilution model, 25
elemental composition
abundance, 4–6
atom mole fraction, 5
even-odd effect, 6*f*, 32–35
isotopes, 6–7
jigsaw puzzle pieces, 17–19
nuclear binding energy, 7–8, 7*f*
zigzag pattern, 32–35
logarithmic-normal distributions
canadian granite specimens, 26
histogram plot, 29
rock cycle, 17–19, 18*f*
soil formation, 19–20
Ecological risk assessment
ecological soil screening level, 522–523
wildlife risk model, 519–522
Electric potential energy, 193
Electrochemical potentials, 453
of aerobic electron transport chain, 486–487, 487*t*
for environmental and biological half-reactions, 486
fermentative anaerobic bacteria, 487–488
in soils and sediments
Pourbaix stability diagram, 459–470
stability diagrams, 457–459
Electrolyte solutions and smectite gels, 121–129
Electron magnetic moment, 353
Electron-pair donor, 386
Electron-pair receptor, 386
Electron transport chain, 352
Electrostatic forces, 118–120
Electrostatic valence bond, 419
Electrostatic valence principle, 418–419, 436–437
Empirical water retention, 76–79
Empirical water retention and unsaturated hydraulic conductivity functions
Clapp-Hornberger functions, 76
van Genuchten water retention function, 78–79
Equatorial coordinate system, 79*np*, 79
Equilibrium constant, 154
activity
Boltzmann-poisson, 193
concentrations, 193–194
empirical ion activity coefficient expressions, 194–195
ChemEQL input matrix
components, 235–238, 236*f*
species, 235–238
ChemEQL output matrix
data-file format, 238, 239*f*, 240*f*
chemical reactions, 190–191
complex
computer-based numerical models, 204, 204*t*
elementary validation, 204

Equilibrium constant *(Continued)*
fugacity, 234–235
Gibbs energy, 191–193, 192*f*
notation, 193
rice table method
of gases, 243–244
ionic compound, 240–241
simple
acetic acid hydrolysis, 199*t*
gas solubility, 202–203
perpetuating algebraic methods, 196
replicate solution chemistry, 195–197
weak monoprotic acid, 197–199
thermodynamic functions, 190–191
units, 193
water chemistry simulations, interpretion
logarithmic activity diagrams, 228–232, 229*f*, 230*f*, 231*f*
reaction quotients, 225–228
saturation indices, 225–228
water chemistry simulations, mineral solubility
aqueous solubility, 223–225
the Gibbs phase rule, 214–216
ion complexes effect, 214–216
pH effect, 211–214
water chemistry, validation
charge balance, 200, 205–206
ChemEQL components, 205*t*, 206*t*
computer-based numerical, applications, 204*t*
ion activity coefficients, 205–207
ion activity products, 207*t*, 208
ionic strength, 200, 205–207
Equilibrium quotient expressions, 153
Equivalent fraction-dependent selectivity coefficients, 181–185
Evaporite rocks, 271
Evapotranspiration, 60
Evapotranspiration models, 61–62
Exchangeable aluminum, 313–317
Exchangeable sodium and clay plasticity, 290–292
Exchange parameter, 158*np*, 158–160
Exchanger-bound ions, 154, 179
Exchange selectivity, 160–162, 323–324
Exocellular bioorganic compounds, 344–349
organic acids, 344–345
Exocellular enzyme, 345–346
Exothermic fusion, 8–9
Exposure pathway assessment
exposure elimination, 508–509, 508*b*
exposure points, 505–506, 506*t*
exposure routes, 504–505, 504*f*
fate and transport, 506–507
receptors, 503

F
Fayalite oxidation, 101
Fertilization, agricultural landscapes, 321–322
Field capacity, 56–57, 58*b*
Fixed-fugacity method, 208–211
Fluid dynamics, 142
Free swelling, 120–125
Na^+-saturated Belle Fourche montmorillonite pure water, 123–124, 124*f*
smectite clay minerals, 120
vermiculite, 120
Freezing-point depression, 152
Freundlich adsorption isotherm, 411–413
Fugacity, 234–235
Fulvic acids, 336
Fulvosäuren, 336*b*

G
Gaines-Thomas convention, 179–180
Gaines-Thomas equivalent-fraction convention, 153–155
Gaines-Thomas selectivity quotient expression, 156
Gapon convention, 153–155, 183–185
Garrels-Mackenzie bicarbonate-to-silica ratio, 267
Geodetic datum, 51–53
Geological chemistry, 2–3
Gibbs energy, 166, 167*f*, 167*t*, 168*f*, 179
Gibbs model, 268*b*
Gibbs phase rule, 218–221
Global carbon cycle, 350*f*
Global Ocean Survey, 349
Goethite surface, 436–439
Gradients, 70–72
Groundwater flow nets, 51–55

H
Halogens, 272
Hemimicelles, 348
Henry's Law, 177*f*, 178, 178*f*, 202–203, 203*t*
High-resolution transmission electron microscopy (HRTEM), 102
Hückel activity coefficient equation, 165
Human health effects
air inhalation equation, 532, 532*t*
chronic toxicities, 492, 493*t*
dose-response assessment
cancer mortality, 494
chronic lead exposure, neurological effects of, 494
dose-response functions, 494–497
low-dose extrapolation, 498–502
noncarcinogens, adverse physiological effects, 494
ecological risk assessment
ecological soil screening level, 522–523
wildlife risk model, 519–522

exposure factors
averaging time (AT), 510–511
exposure duration (ED), 510
inhalation rates, 509, 510*t*
soil ingestion, 509–510, 510*t*
exposure pathway assessment
exposure elimination, 508–509, 508*b*
exposure points, 505–506, 506*t*
exposure routes, 504–505, 504*f*
fate and transport, 506–507
receptors, 503
food ingestion equation, 531, 531*t*
intake estimates
average daily dose (ADD), 511
lifetime average daily dose (LADD), 512
mitigation, remedies for, 515–516, 516*t*
primary and secondary sources, 507
risk-based screening levels, 516–518, 517*t*
risk characterization
cumulative target risk (CTR), 513–514, 514*t*
hazard index (HI), 515
hazard quotient (HQ), 514–515
target cancer risk, 512–513
soil ingestion equation, 530, 530*t*
US federal risk assessment, 492, 492*f*
management and mitigation, 493
scientific research support, 493
water ingestion equation, 529, 529*t*
Humic acids, 336
Humic molecules, association behavior of, 404–405
Humic substances, 334–335
Humification process, 334
Humus, 334
Hydraulic conductivity, 49
Hydraulic heads, 70–72
Hydraulic pressure, 70
Hydrogen-bond donor-receptor pair, 386
Hydrogen bonds, 386, 419
Hydrogen ion transfer, Brønsted-Lowry acid-base model, 258–259
Hydrologic units, 47
Hydrophilic colloids, 434–436

I
Ideal and real exchange-complex mixtures, 179–181
Ideal exchange-complex mixtures, 180
Ideal gas law, 234
Ideal mixtures, 154, 175–176, 179–180
Igneous rocks, 97
Igneous silicate minerals, oxidation of iron(III) in, 101
Independent surface action principle, 410
Integral method, 170
Interlayer K^+ ions, 119
Interlayer swelling, 118
International Humic Substances Society (IHSS), 335
extraction protocol, 335*b*
International Union of Pure and Applied Chemistry (IUPAC), 2, 232, 319
Interpreting isotherm
asymmetric exchange, 157–159
effect of electrolyte concentration, 159–160
effect of ion selectivity, 160–162
influences on, 166–170
physical basis for, 164–166
symmetric exchange, 156–157
Interpreting isotherm, 155–157
Ion activity product (IAP), 225
Ion exchange
capacity, 171–172
characterization, 170–172
discovery of, 148–149
equivalent fraction-dependent selectivity coefficients, 181–185
experiments, 149–152
Gapon convention, 183–185
ideal and real exchange-complex mixtures, 179–181
interpreting isotherm, 155–157
non-linear least square fitting, 172–174
preparation, 170–172
selectivity constants, 152–155, 175–179
Ion exchange isotherm, 152–157, 166–170, 324–327
Ion-exchange isotherm model, 399–400
Ion exchange selectivity constant
ideal mixtures, 175–176
real mixtures, 176–179
Ion selectivity, 160–162
Iron-oxide mass fractions, 101
Iron-reducing bacteria, 478–480

J
Jackson weathering sequence
Atterberg system, 93
clay particle-size, 91
clay size fractions, 92
fine-silt-size fraction, 92
indicator minerals, 93
insoluble oxide minerals, 93
muscovite, 92
shrinkage limit (SL), 93
silicate minerals, crystal structure of, 93
silt and clay weathering stages, 91–92, 91*t*
Jigsaw puzzle pieces, 3–4

K

Knudson hypothesis, 496*b*

L

Langelier saturation index, 302–303
Langmuir conditions, 396, 399
Langmuir isotherm model, 394–398
Lewis acid-base theory, 422
Lignin precursor molecules, 344*f*
Limiting sodium adsorption ratio (LSAR), 304–309
Linear free energy principle, 410
Low-dose extrapolation
  carcinogenic agents
    benchmark dose method, 502
    incremental life-time cancer risk, 501
    slope factor (SF), 501
  noncarcinogenic agents
    average daily dose, 499
    benchmark–dose method, 499
    benchmark response, 499
    body-burden, 498
    lowest observable adverse effect level (LOAEL), 499
    no observable adverse effect level (NOAEL), 499
    reference dose (RfD), 500
Lowest observable adverse effect level, 498–499

M

Magic number, 34
Manure, 148
Marine environment, 322–323
Mass-based adsorption isotherms, 392
Maximum contaminant level (MCL), 3
Mean sea level (MSL), 51–54
Mechanical analysis, 88
Meniscus radius, 75
Metagenome, 349
Metamorphic rocks, 18, 97
Methanogenesis, 482
Mica group minerals
  cation substitution, 107–108
  granite, 107
  interlayer K^+ cations, 108–109
  talc and pyrophyllite, 107, 110*f*
Micellar solutions, 405–409
Microbial biomass growth, 341–343
Microbial electron transport chains
  of aerobic bacteria, 475–477
  of denitrifying bacteria, 477, 478*f*
  iron-reducing bacteria, 478–480, 479*f*
  by nitrate-reducing bacteria, 477
  sulfate-reducing bacteria, 480–481, 480*f*
Microbial growth factors, 345
Microbial respiration
  carbon use efficiency, 481–482
  electron transport chains, 475–481
  glucose catabolism, 473–475
  living organisms classification, 473, 473*t*
Mineral acidity, alkalinity and, 281–285
Mineral colloids, 386–389
Mineralization, 321
Mineral solubility
  acid-base, 255
  aqueous solubility of, 223–225
  Gibbs phase rule, 218–221
  ion complexes, 214–216
  of pH Effect, 211–214
Mineral weathering
  Jackson weathering sequence, 91–93, 91*t*
  mineralogy, 90–91, 90*t*
Mobile agent, 84–85
Mobile-stationary phase partitioning, 81*b*
Molar-to-molal conversion, 129
Molecular dynamics simulations, 119
  interlayer K^+ ions, 119, 119*f*
  interlayer Na^+ ions, 119, 120*f*
Molecular solution, adsorption by partitioning in, 405–409
Mole-fraction, 82
Montmorillonite, 150*np*, 150–151, 158

N

National Cooperative Soil Survey (NCSS), 78, 78*f*
National Oceanic and Atmospheric Administration (NOAA), 60, 257
National Soil Survey Center (NSSC), 133
  laboratory methods manual, 310*b*
Natric diagnostic horizon, 297*b*
Natural ion exchangers, 170–172
  measuring capacity, 171–172
  single cation, 171
Natural organic matter, 334
  chemical properties, 363–366
  stoichiometric composition, 337–341
Natural Resource Conservation Service (NRCS), 61
Negative hydraulic head, 74
Negative hydraulic pressure, 73
Nernst equation, 453–455
Neutral-layer minerals, 106–107, 107*f*, 108*f*
  kaolinite, 106, 108*f*
  layer stacking periodicity, 106
  pyrophyllite, 106–107
  serpentine group, 106, 108*f*
  talc $Mg_6(Si_4O_{10})_2(OH)_4$, 106
Newton-Rapson method, 327
Nicotinamide adenine dinucleotide phosphate (NADPH), 350–351

Nitrate fertilization, 322
Nitrate reduction, 322–323
  in denitrifying bacteria, 477
Nitrogen, 148
Nitrogen cycling, agricultural landscapes, 321–322
Nitrogen functional groups, 367–370
Nitrogen oxides, 274–275
Nonexchangeable aluminum, 317–318
Non-linear least square fitting, 172–174
  asymmetric exchange, 175
  symmetric exchange, 172–174
Nonpolar molecular surface area, 410
Nonpolar molecular volume, 410
No observable adverse effect level (NOAEL), 499
North Carolina Agricultural Experiment Station, 63
Nuclear binding energy, 7–8, 7*f*
Nuclear magic numbers, 34–35
Nucleosynthesis
  nuclear reactions
    alpha decay, 9
    beta-decay, 9
    electron capture, 9
    exothermic fusion reactions, 9
    neutron-capture, 9
    nuclear fusion, 9
    positron emission, 9
    S-process neutron capture, 9
  Uranium-235, 11

O

Oncogene, 497
Organic carbon, 410
Organic carbon turnover
  carbon fixation, 350–351
  carbon mineralization, 351–352
  organic-matter turnover models, 353–360
  oxidation, 352–353
  soil carbon pools, 362–363
Organic colloids, 386–389
Organic matter
  extraction, 334
  fractionation, 335–337
  as substrate, microbial growth, 341–343
Organic-matter turnover models, 353–360
Organic phase, 410
Osmotic coefficient ϕ, 143
Osmotic head, 152
Osmotic pressure, 121, 127
Oxidation, organic compounds by dioxygen, 352–353
Oxidation states, 140–141, 140*t*
  in goethite, 141, 141*t*
  in hedenbergite, 141, 141*t*
  in olivine, 141, 141*t*
Oxidative metabolism, 349
Oxide mineral surface, chemisorption at, 421–423
Oxygen functional groups
  acidometric titration, 365
  titratable weak acid groups, 363–365

P

Panther Creek montmorillonite, 112, 113*t*
Partition constant, 409
Partitioning, 84–85
Partitioning isotherm model, 401–403
Pascal's triangle, 81
Pauling's rules, 109
PBT profiler, 439
Periodic table
  elements organizes, 6
  isotopes table, 35*f*
  nuclear magic numbers, 34–35
  symbols, 6–7
Permanent wilting point, 56
pH-dependent surface charge
  crystallographic proton adsorption models, 415–423
  proton surface-charge experiment, 413, 414*f*
  weak acid-conjugate base proton adsorption model, 414–415
Phenol
  bond oxidation state assignments, 448, 449*t*
  carbon bonding sites, 448
  molecular structure, 448*f*
Phosmet, 69
Phosphorus functional groups, 370–372
pH scale, 319–320
Phyllosilicate layers, 105–106
Physisorption, 392–394
Piezometers, 49–51, 50*f*
Piezometric, 48, 70
Planetary accretion, geochemical behavior
  atmophiles, 12
  chalcophiles, 12
Planetary stratification
  chalcophilic elements, 14*f*, 15
  lithophilic elements, 15
  silicate-rich lithosphere, 13
Plant-available water content (AWC), 56, 58
Plastic density, 135
Plasticity index (PI), 115, 138
Plasticity threshold, 133
Plates and transfers, 83
Plate-theory models, 66*np*, 67, 80–85, 80*f*
  binomial expansion, 81–82
  mobile agent movement, 84–85
  partitioning, 84–85
  plates and transfers, 83

Plate-theory *(Continued)*
plate-theory models, 85
retardation-coefficient, 85
Porosity, 44–45
Position effect, 497*np*
Positive hydraulic pressure, 73
Potential evapotranspiration PET models, 61, 63
Pourbaix electrochemical stability diagrams
essential features, 459, 459*t*
precipitate-precipitate reduction boundary, 468
rules for interpreting, 468–470, 469*f*
solute-precipitate equilibrium boundary, 464–465
solute-precipitate reduction boundary, 465–467, 467*f*
solutesolute hydrolysis equilibrium boundary, 464
solute-solute reduction equilibrium boundary, 462–463
water stability limits, 460–462, 461*f*
Predicting capillary rise, 74–76
Prime Meridian, 79
Proton balance, 260
Proton motive force (PMF), 476
Proton surface-charge experiment, 413
Pyridine, 274, 274*f*
bond oxidation state assignments, 449, 450*t*
carbon bonding sites, 449
molecular structure, 450*f*
Pyroxene, 106
Pyrrole, 274, 274*f*

Q
Quadratic equation, 158, 175

R
Radiation-based models, 61
Radius ratio rule, 95–96, 96*f*
Random sequential-dilution
fluid-filled chamber, 37
Raoult's law, 175–176, 177*f*, 178*f*
Raoult's mole-fraction, 154
Real mixtures, 176–180
Reduction half reaction concept, 450
Reduction-oxidation chemistry
balancing reduction half-reactions
ascorbic and dehydroascorbic acid, 452, 452*f*
balancing steps, 451, 451*t*
factoring reaction, 451*b*
electrochemical potentials, 453
of aerobic electron transport chain, 486–487, 487*t*
for environmental and biological half-reactions, 486
fermentative anaerobic bacteria, 487–488
in soils and sediments, 457–470
formal oxidation states
assignment, 446–447
chemical rules, 447
methanogenic organisms, 482
microbial respiration
Calvin cycle balance sheet, 472, 472*t*
carbon use efficiency, 481–482
electron transport chains, 475–481
glucose catabolism, 473–475
living organisms classification, 473, 473*t*
Nernst equation, 453–455
platinum ORP electrode-potentials, 484–486, 485*t*
standard biological potentials, 455–456
without electron transfer, 483–484
Reference dose (RfD), 500, 500*b*
Residence time, 40–44, 43*t*
Residual sum-square (RSS), 173–174
Residuum, 4
Retardation-coefficient, 85
model, 85
transport model, 65–67
Rhizosphere, 344
Ribbon test, 88*b*
Rice table method, aqueous solubility
gases, 243–244
ionic compound, 240–241
minerals, 223–225
Rock cycle
chemical analysis of, 22
chemical cementation, 18
geologists classify, 17–18
metamorphic rock formations, 18
volcanic activity, 18
Rock weathering, acid-base phenomena, 255
Royal agricultural society, 148

S
Saturated zone, 44, 48–55
Saturation, 56
Saturation effect, 320
Saturation index (SI), 225
Schofield ratio law, 160
Sediments, 4
Selectivity constants, 152–155
Gaines-Thomas equivalent-fraction convention, 153–154
Gapon convention, 153–154
Vanselow mole-fraction convention, 153–154
Selectivity constants, 152–155, 175–179
Shear strength, quick clays, 291–292
Shotgun sequencing, 349
Siderophores, 346–347
Silicate minerals

basicity and hydrolysis of
chemical weathering processes, 100
inosilicate enstatite, 99
moles of charge, 97
orthosilicate anion, 99
pyrophyllite, 100
silicate anions, 97, 98*f*, 98*t*
silicate polyanions, 97, 98*t*
talc hydrolysis, 100
cation substitution in, 109*b*
coordination polyhedra
ball-and-stickmodels, 95, 95*f*
crystals diffract X-rays, 94
molecular orbital theory, 94
octahedron, 95
oxygen ions, 94
radius ratio rule, 95–96, 96*f*
tetrahedron, 95
X-ray diffractometers, 94
groups, 97
iron(III), oxidation of, 101
pyribole and feldspar minerals, solid-state transformation of
amphibole octahedral chains, 102, 103*f*
diffusion-limited dissolution mechanism, 102
HRTEM images, 102
I-beams, 102, 103*f*
palygorskite, 102–103
phyllosilicate minerals, 104, 105*t*
polysomatic defects, 102
protocrystalline, 104
pyroxene, 102, 103*f*
rate-limiting sites, 103–104
sepiolite, 102
sextuple-chain zippers, 104*f*
surface-site-limited adsorption models, 102
transmission electron microscopy (TEM), 101–102
Wadsley defects, 102
valence orbital hybridization, 93
valence shell electron-pair repulsion (VSPER), 93
Silicate rocks, 262–265
Single-adjustable parameter isotherm model
exchange, 169
symmetric exchange, 170
Single-adjustable-parameter isotherm model, 166–167
Single cation, 171
Single parameter isotherm models, 167
Single-point selectivity coefficient, 173
Site-limited adsorption isotherms, 390
Smectite and vermiculite clay minerals
crystalline swelling, 116–118
electrolyte solutions and smectite gels, 121–129
free swelling, 120–125
swelling clay minerals, hydrated interlayer of, 118–120
Smectite clay gels, 136*b*
Smectite clay swelling behavior, 136
Sodicity risk identification
exchange sodium percentage, 292–296
extreme alkalinity, 299–301
irrigation water, 301–304
limiting sodium adsorption ratio, 304–309
salinity, 298–299
USDA Natural Resource Conservation Service rates, 292
Sodium adsorption ratio (SAR), 155, 183, 293*b*
Soil acidity
active acidity, 309
cation-exchange capacity, 310*b*
neutralizing exchangeable, 314–317
total apparent acidity, 309
Soil adsorption coefficient, 440
Soil carbon pools, 362–363
Soil erodibility, 290–291
Soil fertility, 148
Soil mineral particles, 88
Soil moisture, 43
Soil moisture balance, 60–64
evapotranspiration models, 61–62
soil water balance, 63–64
Soil water balance, 63–64
Soil-water zone, 56
Solar system
accretion, 3
elemental abundance of, 5*f*, 6*f*
planetary formation
accretion, 12
stratification, 13–17
radioactive decay, 3
residuum, 4
sediments, 4
soil formation, 19–20
Sol-gel transition point, 137
Solid colloidal system, 137
Solid/gas (S/G) interfaces, 396
Solute transport
retardation-coefficient transport model, 65–67
saturated zone, 67–68
vadose zone, 68–69
Sorption by partitioning, 392
Spallation, 10
Specific surface excess ratios, 395

Spectator-ion, 261
charge balance, 261
Spectator oxides, 265
Spin, 353
Spin conservation and spin forbidden reactions, 378–380
Spin forbidden, 353
Spin-state, reactants, 352
Stability, 335–336
Stable colloidal dispersion, 434
Stable element, 2
Stable nuclides, 32–34
Standard biological potentials, 455–456
Static surface tension, 74
Stoke's law, 142–143
Stomata, 60
Streptomyces pilosus, 216–217
Submergence, 70
Submergence pressure, 73
Sulfate-reducing bacteria, 480–481
Sulfate reduction, 322–323
Sulfide minerals, 270
Sulfur functional groups, 372–375
Sulfur oxides, 272–274
Surface complexation models, 430–432
Surface excess amount, 390
Surface excess concentration, 390, 395
Swelling clay minerals, hydrated interlayer of, 118–120
Swelling limit, 46
Symmetric exchange, 156–157, 172–174
Symmetric exchange isotherm, 156–157
model, 169

T
Table of nuclides, 32–34
Tannins, 336*b*
Temperature-based PET models, 61
Terragenome Project, 349
Terrestrial carbon cycle. *See* Organic carbon turnover
Tetrahedral site geometry, 96
Tetrahedron, 95, 96*f*
Textural triangle, 133, 133*f*
Thermal fission, 10*t*, 11
Thermodynamic selectivity constant, 180
Thornthwaite potential evapotranspiration model, 62
Trioctahedral phyllosilicates, 106
Troposphere, 42*np*, 42–43
Tschermak substitution, 109, 111–112
Tumor-suppressor gene, 497

U
Unit cell formula, 112, 114
Unsaturated hydraulic conductivity functions, 76–79
Unstable elements, 2
Uranium-235, 11
US Department of Agriculture Natural Resource Conservation Service (NRCS), 133
US Environmental Protection Agency (USEPA), 3
US federal risk assessment paradigm, 492, 492*f*
management and mitigation, 493
scientific research support, 493
US Geologic Survey (USGS), 60
U-shaped osmotic cell, 126, 128*f*, 131–132, 132*f*

V
Vadose zone, 44, 55–58, 68–69
Valence-bond proton adsorption model, 417–420, 436–439
van Genuchten water retention function, 78–79, 78*f*
Vanselow mole-fraction convention, 153–155
van't Hoff factor *i*, 143
van't Hoff-Morse expression, 126–127
Vapor-pressure depression, 152
Vermiculite, 149–150, 150*np*
free swelling, 120
Vernal equinox, 79
Vulcanism, 272–275

W
Water budgets, 40–41, 42*t*
Water chemistry simulations
interpretion
logarithmic activity diagrams, 228–232, 229*f*, 230*f*, 231*f*
reaction quotients, 225–228
saturation index (SI), 225–228
mineral solubility
aqueous solubility, 223–225
fixed-fugacity scheme, 208
Gibbs phase rule, 214–216
ion complexes effect, 214–216
pH effect, 211–214
validation
charge balance, 200, 205–206, 247
database, 250
ion activity coefficients, 205–207, 248
ion activity products, 207*t*, 208, 248–250
ionic strength, 200, 205–207, 248
mass balance, 247
Water extractable organic matter (WEOM), 344*np*
Water reference level, 260–262, 260*b*
Water resources and hydrologic cycle

hydrologic units, 47
residence time, 41–44
volume and mass relations, 44–45
water budgets, 40–41
Water retention curve, 58–60
Water-to-clay mass ratio, 116, 118*f*, 121
Weak acid and base conjugates, 259–260
Weak acid-conjugate base proton adsorption model, 414–415
Weak monoprotic acid, 197–199
Wet bulk density, 46

X

X-ray absorption near-edge structure (XANES), 372–373
X-ray diffraction, 111–112
layer silicates, 138–140

Y

Yellow acids, 336

Z

Zylem, 60

Printed in the United States
By Bookmasters